"Genetics enjoy a high profile today in the minds of the intelligentsia and lay public alike. Fresh genetic results are often announced in newspapers with fanfare and draw immediate attention. No doubt they are "sexy," and, undeniably, we are in part products of our inheritance – our nature. However, only in part, and the ultimate story genetic reductionism tells always turns out to be more nuanced and complex than originally touted. Vigilant academics who follow the literature have come to understand that precious few heralded genetic findings survive follow-up scrutiny or replication. Where's the pushback? Well, here. In *The Heredity Hoax*, Richard Lerner and Gary Greenberg crystallize conceptual and methodological shortcomings of genetic reductionism in their own words as they expose problematic implications of genetic reductionist ideas for policy, programs, and social justice. Moreover, they skillfully marshal the words of proponents and critics to buttress their contentions. But, you ask, what perspective is left to supplant or, at least, supplement the genetic project? Lerner and Greenberg do not leave you in the lurch, but are persuasive that relational developmental systems – the coaction of nature and nurture – offer a more rational, productive, and optimistic framework for understanding human development. Want to be an intelligent consumer of the scientific literature in this grand debate of the 21st century? *The Heredity Hoax* is where to start."

—**Marc H. Bornstein**, *President Emeritus, Society for Research in Child Development; Editor,* Parenting: Science and Practice

"In this magnificent and timely volume Richard Lerner and Gary Greenberg have brought together a collection of chapters by the world's leading authorities that address the fatal misconceptions of classical genetic accounts of human development. Just as quantum theory has upended our classical view of the physical world, the various chapters in this volume systematically challenge and replace the determinist view of genetic heritability with a dynamic Relational Developmental Systems metatheoretical framework that revolutionizes our understanding of human development and its potential. This reframing of the dynamic nature of the relations between genetics and development has profound and potentially positive implications for our approaches to education, parenting and broader issues of social justice."

—**Larry Nucci**, *Professor, Graduate School of Education, University of California Berkeley*

"*The Heredity Hoax* is an important collection that battles genetic determinism on multiple fronts. Curated by two field leaders who have devoted their careers to this fight, this collection dismantles fallacies of simplistic reductionism in ethology, evolutionary biology, behavior genetics, and sociobiology. Not content with critique, Professors Lerner and Greenberg bring together perspectives to build a real and rigorous alternative framework that analyzes behavior as an emergent property of dynamic, relational, bio-social systems of organismal development. Connecting this new paradigm to policy frameworks, they show how bio-social science can take part in the optimistic cultivation of human flourishing and greater equality. This guide will be essential reading for newcomers to this field as well as experienced experts."

—**Aaron Panofsky**, *Professor, UCLA Institute for Society and Genetics, Public Policy, and Sociology; Director, Institute for Society and Genetics*

THE HEREDITY HOAX

This innovative and thought-provoking book integrates both new, authored material and reprints of existing literature that, together, provide a compelling narrative that reveals the fatally flawed science associated with genetic reductionist accounts of human behavior and development.

Through an interdisciplinary lens, it illuminates the dynamic nature of human development, empowering readers to question established notions, and embrace the complexity of our potential. Across the book, the work of top-tier scientists, from developmental, comparative, educational, and biological science illuminates theory and research converging on the conclusion that the multiple egregiously flawed work of genetic reductionists should be expunged from research pertinent to human development. The book challenges the prevailing reductionist narratives and their application to social policies, programs, and uses in media. Theoretically based and empirically rigorous, this multidisciplinary approach to human development will shine a light on the inequities in individuals or groups that suggest that specific genes do not enable them to succeed in life.

The Heredity Hoax invites graduate programs and advanced undergraduate courses on human development, human potential, epigenetics, and more to delve into the intricate interplay between genes, environment, and personal growth. This book will also serve as an unimpeachable source of evidence for researchers, educators, and social policymakers.

Richard M. Lerner is Bergstrom Chair in Applied Developmental Science and Director of the Institute for Applied Research in Youth Development in the Eliot Pearson Department of Child Study and Human Development at Tufts University.

Gary Greenberg is Professor Emeritus of Psychology at Wichita State University and Co-founder (with Ethel Tobach) of the International Society for Comparative Psychology; he has been a comparative psychologist for his entire career.

THE HEREDITY HOAX

CHALLENGING FLAWED GENETIC THEORIES OF HUMAN DEVELOPMENT

Edited by
RICHARD M. LERNER AND GARY GREENBERG

NEW YORK AND LONDON

Designed cover image: Shutter2U via Getty Images

First published 2025
by Routledge
605 Third Avenue, New York, NY 10158

and by Routledge
4 Park Square, Milton Park, Abingdon, Oxon, OX14 4RN

Routledge is an imprint of the Taylor & Francis Group, an informa business

© 2025 selection and editorial matter, Richard M. Lerner and Gary Greenberg; individual chapters, the contributors

The right of Richard M. Lerner and Gary Greenberg to be identified as the authors of the editorial material, and of the authors for their individual chapters, has been asserted in accordance with sections 77 and 78 of the Copyright, Designs and Patents Act 1988.

All rights reserved. No part of this book may be reprinted or reproduced or utilised in any form or by any electronic, mechanical, or other means, now known or hereafter invented, including photocopying and recording, or in any information storage or retrieval system, without permission in writing from the publishers.

Trademark notice: Product or corporate names may be trademarks or registered trademarks, and are used only for identification and explanation without intent to infringe.

ISBN: 9781032702933 (hbk)
ISBN: 9781032699578 (pbk)
ISBN: 9781032702988 (ebk)

DOI: 10.4324/9781032702988

Typeset in Minion Pro
by KnowledgeWorks Global Ltd.

CONTENTS

Preface   xi

SECTION I
FRAMING THE CHOICE: THE PSEUDO-EVIDENCE AND PESSIMISM OF GENETIC REDUCTIONS VS. THE EVIDENCE AND OPTIMISM DERIVED FROM RESEARCH FRAMED BY DYNAMIC, RELATIONAL DEVELOPMENT SYSTEMS-BASED CONCEPTS

    Editors' Introduction   3

1. Addressing the Heredity Hoax in Science and Society: A View of the Issues   13
   Richard M. Lerner and Gary Greenberg

2. The Fallacies and Failures of Genetic Reductionism: An Historical Review of Lessons Learned   21
   Gary Greenberg

SECTION II
METATHEORY AND THEORY ABOUT THE NATURE-NURTURE COACTION

    Editors' Introduction   35

3. Metatheory and the Primacy of Conceptual Analysis in Developmental Science   60
   David C. Witherington, Willis F. Overton, Robert Lickliter, Peter J. Marshall, and Darcia Narvaez

4. The Failure of Biogenetic Analysis in Psychology: Why Psychology Is Not a Biological Science   76
   Gary Greenberg

5. What Galton's Eugenics Has Wrought: Behavior Genetics
   and Heritability 91
   David S. Moore

SECTION III
THE CONCEPTS OF INSTINCT AND CRITICAL PERIODS

Editors' Introduction 105

6. Development Evolving: The Origins and Meanings of Instinct 119
   Mark S. Blumberg

7. Critical Period: A History of the Transition from Questions of When,
   to What, to How 127
   George F. Michel and Amber N. Tyler

8. Short Arms and Talking Eggs: Why We Should No Longer Abide the
   Nativist-Empiricist Debate 135
   *John P. Spencer, Mark S. Blumberg, Bob McMurray, Scott R. Robinson,
   Larissa K. Samuelson, and J. Bruce Tomblin*

SECTION IV
EVOLUTION

Editors' Introduction 151

9. Toward a New Developmental and Evolutionary Synthesis 170
   Gilbert Gottlieb

10. Précis of Evolution in Four Dimensions 187
    Eva Jablonka and Marion J. Lamb

11. Developmental Evolution 214
    Robert Lickliter

12. Evolving Evolutionary Psychology 221
    *Darcia Narvaez, David S. Moore, David C. Witherington, Timothy I. Vandiver,
    and Robert Lickliter*

13. Evolution beyond Neo-Darwinism: A new Conceptual Framework 242
    Denis Noble

Contents

SECTION V
BEHAVIOR GENETICS: HERITABILITY, TWIN STUDIES, ADOPTION STUDIES, AND IQ

    Editors' Introduction   257

14. From Gene to Organism: The Developing Individual as an Emergent, Interactional, Hierarchical System   279
    Gilbert Gottlieb

15. The Heritability Fallacy   290
    David S. Moore and David Shenk

16. The 1990 "Minnesota Study of Twins Reared-Apart" IQ Study: Ripe for Retraction?   299
    Jay Joseph

SECTION VI
SOCIOBIOLOGY

    Editors' Introduction   321

17. Sociobiology and the Theory of Natural Selection   351
    Stephen Jay Gould

18. Sociobiology and Human Development: Arguments and Evidence   359
    Richard M. Lerner and Alexander von Eye

SECTION VII
EPIGENETICS

    Editors' Introduction   383

19. Social Regulation of Human Gene Expression   391
    Steve W. Cole

20. Human Social Genomics   399
    Steven W. Cole

21. Behavioral Epigenetics   409
    David S. Moore

22. Dynamic Heredity   418
    Douglas Wahlsten

SECTION VIII
IMPLICATIONS FOR PROGRAMS AND POLICIES

    Editors' Introduction   425

23. *The Bell Curve* at 30: A Closer Look at the Within- and Between-Group IQ Genetic Evidence   433
    Jay Joseph and Ken Richardson

24. Using the Science of Learning and Development to Transform Educational Practice   477
    Linda Darling-Hammond, Lisa Flook, Channa Cook-Harvey, Brigid Barron, and David Osher

25. The Future of the Science of Learning and Development: Whole-Child Development, Learning, and Thriving in an Era of Collective Adversity, Disruptive Change, and Increasing Inequality   494
    Pamela Cantor and David Osher

26. Rejecting Genetic Reductionism and Embracing Relationism and the Complexity of Dynamic Systems   512
    Richard M. Lerner and Gary Greenberg

    Author Index   526
    Subject Index   546

PREFACE

At some point in the middle of the first decade of the 21st century, we began to talk with each other about the fact that genetic reductionist accounts of behavior and development continued to appear in prominent journals in several fields of study within psychology – for instance, cognitive, comparative, developmental, personality, and social – and within allied fields, such as economics, education, and sociology. Although there were several different instances of genetic reductionist ideas appearing in journals (behavioral genetics, sociobiology, evolutionary psychology, and evolutionary developmental psychology), all formulations "boil down to" (or "reduce to," if we wish to be ironic instead of colloquial) one basic claim: The genes inherited by a person at the moment of conception provide an unmodifiable *blueprint* for life, behavior, and development across the life span.

In addition to this counterfactual assertion, the ideas of genetic reductionists were problematically complicated by media translations of these claims. These translations were presented in newspaper and magazine articles, television and radio stories, and in the then newly burgeoning online news and social network platforms. The information associated with genetic reductionism pertained to human behaviors, skills, and experiences as varied as intellectual abilities, parenting, sexuality, childbearing, morality, character, religiosity, personality attributes, temperament, educational attainment, criminality, experiencing rape, male infidelity, having children out-of-wedlock, living in poverty, or even the proclivity to watch television.

In none of the genetic reductionist papers gaining peer-reviewed publication was there any specification of the specific gene or genes purported to cause these behaviors or experiences, and there was never any evidence of how the possession of these unidentified genes participated in a developmental process from conception to behavioral enactments or contextual experiences. In turn, in the public dissemination of this information it was a relative rarity to find reports of the conceptual and methodological criticism by scientists with a broad range of expertise pertinent to the claims of genetic reductionists. These critiques of genetic reductionist publications were presented by, in particular:

1. Cell biologists, geneticists, epigeneticists, evolutionary biologists, paleontologists, physical and evolutionary anthropologists, and mammologists (e.g., Lester Aronson, American Museum of Natural History; Sir Patrick Bateson, University of Cambridge and President of the Zoological Society of London; Steve Cole, University of California, Los Angeles; Marcus Feldman, Stanford University; Stephen Jay Gould, Harvard University; Mae-Wan Ho, Institute of Science and Society, London; Ruth Hubbard, Harvard University; Eva Jablonka, Massachusetts Institute of

Technology and Tel Aviv University; Donald Johanson, Arizona State University; Evelyn Fox Keller, Massachusetts Institute of Technology; Richard Lewontin, Harvard University; Owen Lovejoy, Kent State University; Benno Müller-Hill, University of Cologne; Denis Nobel, University of Oxford; T. C. Schneirla, American Museum of Natural History; and Ethel Tobach, American Museum of Natural History); and

2. Scholars from the physical sciences, mathematics, and economics who have provided incontrovertible logical, mathematical, and statistical proofs that the data generated from the research, and the interpretations derived from these data, by genetic reductionists were egregiously and irreconcilably flawed (e.g., P. W. Anderson, Bell Laboratories; Arthur Goldberger, University of Wisconsin; Nobel Laureate James Heckman, University of Chicago; David Layzer, Harvard University; Peter C. M. Molenaar, Penn State University; and Peter S. Saunders, King's College London).

Yet, the critiques continued to be largely ignored by genetic reductionists and the media continued to disseminate counterfactual information to the public. And we continued to be perplexed about how and why bad science continued to have a foothold in scientific journals and public dissemination.

By 2008, we concluded that the storylines told by genetic reductionists – whether labeling themselves as behavior geneticists, sociobiologists, evolutionary psychologists, or evolutionary developmental psychologists – and by segments of "the media" amounted to perpetrating a hoax in both science and society. We believed that the continuation of the hoax could eventuate in the weight of bad science compromising opportunities for engaging in good science and, as well, wreaking havoc on public policies and programs that sought a scientifically credible evidence base for a host of educational, physical and mental health, and social initiatives that could improve the lives of all youth, families, and communities.

We wondered what we could do to address the misinformation involved in the heredity hoax and thereby contribute something to preventing or at least diminishing the chances of bad science being used to create policies and programs. Being academics, it is certainly not surprising that we came up with the idea of producing a book, one that would be entitled "The Heredity Hoax."

Our next step was to decide what sort of book we would produce. Would it be a trade book, a co-authored scholarly book, an edited book that contained chapters written by some of the experts we have noted above, or a book of reprinted examples of the best science we knew of to counter the false claims and flawed methods of genetic reductionists? We started and then stopped several instances of all of these pathways forward because none of them seemed to be the right choice for us. Pauses between these attempted paths forward were occasionally short, but many were fairly long.

Time passed. One of us retired from his professorial position. We both grew old …

Then, merely 15 years after we had the initial ideas for *The Heredity Hoax*, we finally – in 2023 – hit upon a path that seemed right. It was a hybrid path, one that combined several of the paths we had considered: first, we could build on our past writings[1] about the significant and substantial conceptual and methodological problems with all of the above-noted instantiations of genetic reductionism. This task alone represented challenges because, in combination across the careers of both of us, our past writing extends more than 100 years! We nevertheless sought to extract from what we had written ideas that could be meshed with our present understanding of the shortcomings of all versions of genetic reductionism. Second, we could include in the book reprints of curated articles that reflect some of the best scientific work countering all of the instantiations of genetic reductionism that have been prominent across both the 20th century and now, the first quarter of the 21st century.

This two-pronged approach is what readers will encounter in *The Heredity Hoax: Promoting Social Justice Through Eliminating Genetic Reductionism from Science*. The presence of the two prongs of contributions to this book illustrates that the scholarship we present is substantially beyond the work of your two editors. In addition, this book exists because of the efforts of both students and colleagues. As we will illustrate, some of the individuals who started to work with us when they were in the first group became distinguished members of the second one by the time we undertook completing the book that readers of this preface are now encountering.

For instance, in 2008, when we began to first work on this project, we called upon three new doctoral students (working in the lab of RML) to help us generate all the articles we could find that we thought pertinent to our goals: Kristina Callina, Megan Mueller, and Christopher Napolitano. Because this effort occurred before Google Scholar or other major scholarly search engines existed, the work involved in this process was quite formidable. Today, Drs. Callina, Mueller, and Napolitano are successful scholars. At this writing they are, respectively, co-leading a research and evaluation consulting organization in the Greater Boston area, working as a professor in the Cummings School of Veterinary Medicine at Tufts University, and working as a professor in the Department of Educational Psychology of the College of Education at the University of Illinois, Urbana-Champaign. We are grateful for the great start they gave us in working on this book, and we hope that it will be helpful as they continue to pursue their scholarly career.

As we were completing the book, additional doctoral students (again in the lab of RML) provided significant assistance. Margaret Mackin and Kirsten Olander provided careful reference checking, making certain that anything cited in the text of a portion of the book appeared correctly in the accompanying References and, in turn, checking that any citation appearing in the Reference list was cited within the text. Natasha Keces reviewed every word on every line of the entire manuscript and made certain that there were no substantive or grammatical oddities present in the manuscript and that the publication style of the publisher was assiduously followed. If there remain any shortcomings in the book after these stellar students made these contributions to the manuscript, the fault lies with us and not with these early career scholars. We are in their debt.

We are in debt as well to many other people. Most significantly, we acknowledge our debt to the scholars who advised us about the content of this book, many of whom were also generous in allowing us to include their work in the sections of the book. We express our gratitude to these scientists, whose collective contributions to addressing the shortcomings of genetic reductionism operationalize the enactment of good science that advances scholarship and contributes to the use of science to optimize social justice and equity in the development of all people. These scholars include Mark S. Blumberg, Marc H. Bornstein, Pamela Cantor, Steve W. Cole, Mary Helen Immordino-Yang, Jay Joseph, Michael E. Lamb, Jacqueline V. Lerner, Robert Lickliter, George M. Michel, David S. Moore, David Osher, Willis F. Overton, Douglass Walsten, and David Witherington.

We also want to recognize the extraordinary contributions made by Jarrett M. Lerner, the Managing Editor at the Institute for Applied Research in Youth Development at Tufts University. His editorial acumen, knowledge, wisdom, and skills in addressing effectively and with excellence the challenges of undertaking and completing the sort of hybrid book we have produced are incomparable. We cannot imagine completing this work without his expert editorial assistance. We are deeply grateful to him. We are also grateful to our editor at Taylor and Francis, Molly Selby, and to her Editorial Assistant, Shivranjani Singh. Their support, guidance, and collaboration were essential contributions to this book, from its inception to its completion. We appreciate enormously all that they have done.

In recounting the long history but short past of this book, two people stood with us throughout the long projects, that is, our wives Jacquelinne V. Lerner and Patricia Greenberg, respectively. Fifteen years is a very long time to continue to hear that one's spouse continues to try to complete a project. Their unflagging support and generosity of spirit were essential to keep us moving forward. We owe them endless thanks.

A final note of thanks is due to the person whose intellectual abilities and leadership, integrity, and commitment to social justice provided a model of scholarly excellence for both of us while, at the same time, being a generous mentor, motivator, and friend. She was the person who brought us together and catalyzed a friendship and collaboration between us that has spanned, at this writing, almost half a century: Ethel Tobach, the late Curator *Emerita* of the American Museum of Natural History. With gratitude and love, we dedicate this book to her.

RML
Medford, MA

GG
Chicago, IL
February, 2024

Note

1. For instance, see Lerner (1976, 1986, 1992, 2002, 2015, 2018, 2021) and Greenberg (1981, 1982, 2004, 2006, 2011, 2013, 2015; Greenberg & Partridge, 2010).

References

Greenberg, G. (1981). Comment on U. Melotti. Towards a new theory of the origin of the family. *Current Anthropology*, 22, 631.

Greenberg, G. (1982). Ethology without instinct. Review of S. A. Barnett, *Modern ethology: The science of animal behaviour*. *Contemporary Psychology*, 27, 18–19.

Greenberg, G. (2004). R. I. P. Genetic determinism: Please. Review of G. Kaplan and L. J. Rogers (2003). *Gene worship*. New York: Other Press. *Developmental Psychobiology*, 46, 93–96.

Greenberg, G. (2006). The Emperor *still* has no clothes. [Review of the book, *Basic instinct: The genesis of behavior*, by M. Blumberg (2005).] *Developmental Psychobiology*, 48, 410–411.

Greenberg, G. (2011). The failure of biogenetic analysis in psychology: Why psychology is not a biological science. *Research in Human Development*, 8, 173–191.

Greenberg, G. (2013). A long way from genes to behavior. *International Journal of Developmental Science*, 7(2), 83–86.

Greenberg, G. (2015). The case against behavioral genetics. Review of A. Panofsky (2014). *Misbehaving science: Controversy and the development of behavior genetics*. Chicago: University of Chicago Press. *Developmental Psychobiology*, 57, 854–857.

Greenberg, G. & Partridge, T. (2010). Biology, evolution, and development. In W. F. Overton (Ed.), *Cognition, biology, and methods. Volume 1 of the Handbook of life-span development* (pp. 115–148). Editor-in-chief: R. M. Lerner. Wiley.

Lerner, R. M. (1976). *Concepts and theories of human development*. Addison Wesley Publishing Company.

Lerner, R. M. (1986). *Concepts and theories of human development* (2nd ed.). Random House.

Lerner, R. M. (1992). *Final solutions: Biology, prejudice, and genocide*. Penn State Press.

Lerner, R. M. (2002). *Concepts and theories of human development* (3rd ed.). Lawrence Erlbaum Associates.

Lerner, R. M. (2015). Eliminating genetic reductionism from developmental science. *Research in Human Development*, 12, 178–188.

Lerner, R. M. (2018). *Concepts and theories of human development* (4th ed.). Routledge.

Lerner, R. M. (2021). *Individuals as producers of their development: The dynamics of person ⇔ context coactions*. Routledge.

SECTION I

FRAMING THE CHOICE

The Pseudo-evidence and pessimism of genetic reductions vs. the evidence and optimism derived from research framed by dynamic, relational development systems-based concepts

SECTION

FRAMING THE CHOICE

EDITORS' INTRODUCTION

Developmental science seeks to *describe* (depict or represent), *explain* (account for), and *optimize* (maximize the probability of health and thriving) changes within an individual (*intraindividual change*) and of differences between people (*interindividual differences*) *across the life span* (Baltes et al., 1977; Lerner, 2018). However, what sorts of changes characterize an individual as he or she develops across the lifespan? Where do these changes come from? Do all of these changes pertain to the development of the person? Questions such as these are inevitably involved in all theoretical ideas about human development (Lerner, 2018; Raeff, 2016).

All scientific theories are derived from specific instances of philosophies. Philosophy has two general areas of concern: *Ontology*, or philosophical statements about what exists; and philosophical conceptions of how what exists can be known, or *epistemology*. *Paradigms* are philosophical ideas that combine ideas about what exists with ideas about how the things that exist can be known (Kuhn, 1962, 1970).

Paradigms are similar to what philosopher Stephen Pepper (1942) termed *world views*, that is conceptions that involve both ontological and epistemological statements and can be understood by considering what Pepper termed *root metaphors*. These metaphors are analogies that are used to illuminate the basic ideas that are involved in a specific worldview of paradigm.

An example that pertains to the major scientific paradigm across the past three centuries is the root metaphor of a *machine*. In what is typically referred to as the *mechanistic paradigm* (e.g., Overton, 2015; Overton & Reese, 1973; Reese & Overton, 1970), the ontological statement is that the natural world exists as a machine and a scientist can know how the world (the machine) works by understanding each of the parts of the machine and then additively combining this knowledge.

For instance, a car is a machine, and to understand the "nature" of a car, a researcher would need to discover how and when the parts of the car (the carburetor, the engine, the transmission, the battery, etc.) are known and then this knowledge is put together, the attributes of the car are revealed. Similarly, if a human can be ontologically regarded as machine-like, then the epistemological route to fully understanding what it means to be such a machine would be to reduce the whole – the human – to the parts comprising the human. The pursuit of such knowledge might start with understanding the different functions of the machine, that is, respiration, circulation, digestion, reproduction, sensory and motor functioning, etc.

However, what explains these functions? The epistemology ruling the mechanistic paradigm would lead a scientist to then *reduce* each function to organs, then to tissues, then to cells, and finally to the

DOI: 10.4324/9781032702988-2

chemicals comprising the cell and to the physical laws governing the combination of the chemicals. Ultimately, then, the mechanistic paradigm would see that at their root, each human is a machine that can be known by the laws of chemistry and physics. As well, all living creatures could be treated as machines that are knowable by reducing these structure and function to chemistry and physics and, as well, inanimate objects that exist in the natural world are also at their root composed of chemicals that function in accordance with the physical laws of the universe. As such, all science can be reduced to the same constituent phenomena, and these phenomena can be known by reducing the complexity of humans, or all flora and fauna, and in fact all of what exists in the universe to the laws of chemistry and physics.

One more philosophical tool is needed to move from the ideas within a paradigm to a scientifically useful (i.e., an empirically useful) set of ideas that can be productively used for research and discovery. This idea is termed a *metatheory*, which is a theory about theories (Lerner, 2018; Overton, 2015). A scientific metatheory is derived from a paradigm, and it is simply a set of ideas that define the components of a theory that would be consistent with a specific paradigm. A theory in science is a set of statements that integrates existing observations (empirically established "facts") and leads to the generation of additional facts.

In science, a theory is not something that is judged as either true or false. It is judged in regard to its empirical usefulness. Some criteria for determining usefulness are precision, scope, and deployability. Theories are more precise when they can integrate more specific findings and generate hypotheses that can predict more specific findings than other theories. Theories have greater scope when they can integrate more areas of research pertinent to a topic than is the case for other theories. Theories have great deployability when they can be used to explain observations with less use of additional ideas or assumptions than is the case with other theories (this criterion may be understood as a version of Occam's razor, which is the idea that explanations that are constructed with the smallest set of elements are better explanations; this idea is also known as the law of parsimony).

Many different theories can be derived from a paradigm. However, a metatheory of a paradigm indicates what ideas any theory associated with a specific paradigm must include or exclude. The metatheory for a mechanistic paradigm specifies that any theory derived from it should specify the machine-like nature of the to-be-explained entity that exists and then must specify the basic elements of explanation that need to be invoked to explain the entity. That is, what are the *essential* components of the machine, the foundational components that build all superordinate structures and functions of the parts of the machine?

In the example, we have noted that chemistry and physics are the essential elements to which the human machine could be reduced. Whereas mechanistically oriented scientists who study human behavior and development would not deny the ultimate importance of this level of reduction, less molecular referents have often been used in different theories. Examples are reduction to the elements of classically conditioned or operantly conditioned stimulus-response relations (e.g., Bijou & Baer, 1961; Skinner, 1938) or reduction to the genes received at conception (e.g., Buss, 2019; Dawkins, 1976; Plomin, 2018; Rushton, 2000).

Mechanistic Ideas in the Study of Human Development

Ideas associated with the mechanistic paradigm have been used to derive scientific theories and methods that have resulted in important advances in knowledge and useful translations of this knowledge into applications that have contributed positively to advances in human health and well-being. Of course, there are also well-known instances of scientific ideas and their applications being misused in regard to human welfare (e.g., Carmack et al., 2008; Clarke, 2021; Elliott, 2017; Rabuy, 2014).

Arguably, however, the major instance of the enactment of bad science and of its translation into applications to policies, social programs, and even discriminatory (or even worse) actions toward millions of human has occurred in regard to ideas, research, and actions taken on the basis of claims about the role of genes in human behavior and development (Gould, 1976; Greenberg, 2015; Hirsch, 1981; Lerner, 1992; Lifton, 1986; Moore, 2002, 2015b; Müller-Hill, 1988; Panofsky, 2014; Proctor, 1988).

Several different genetic reductionist conceptions of human behavior and development have been forwarded. Prominent examples that, at this writing, continue to be used to provide mechanistic accounts of human behavior and development include Behavior Genetics (e.g., Harden, 2021; Plomin, 2018), Sociobiology (e.g., Dawkins, 1976; Wilson, 1975; Wilson & Wilson, 2007), Evolutionary Psychology (Buss, 2019; Buss & von Hippel, 2018), and Evolutionary Developmental Psychology (e.g., Bjorklund, 2015, 2016; Bjorklund & Ellis, 2005).

The first of the two goals of this book is to critically discuss genetic reductionist models of the role of genes in humans' lives. All of these models have a common set of problems: (1) Overinterpretation of empirical findings, that is, making claims about the meaning or significance of their findings that go beyond what the data actually indicate; (2) Misinterpretation of empirical findings, that is, asserting that the data they present have one meaning (e.g., that provides evidence that differences between people on a score from a measure or analysis indicate that heredity has largely or wholly been a cause of the score or the results of an analysis when the sort of scoring they used or analysis they conducted cannot be used to determine causality; (3) Making claims about the role of genes that are counterfactual, that is, asserting that genes work in a specific way that, if true, would indicate that heredity plays the most significant role in one, some, or all of human behavior and development when, in fact, genes do not function as portrayed by the statements of genetic reductionists; and (4) Applying overinterpreted, misinterpreted, and counterfactual statements to policies and programs that affect education, training, health, reproductive rights, human rights, equity and social justice, and even the very lives of some people.

Together, these problems with genetic reductionist accounts of human behavior and development create egregious misconceptions about scientific facts about heredity and, specifically, of how genes play a part in our lives from conception through death (Moore, 2002, 2015a, 2015b). This misinformation has a pernicious effect on both science and society and, as such, is why we have entitled this book *The Heredity Hoax*. Our interest in countering the misinformation and mischief associated with the problems of genetic reductionism rationalizes the second goal of this book, which is challenging flawed genetic theories of human development (as reflected in the book's subtitle).

Science is a community endeavor. One instance of the role of the scientific community is represented by the standard of peer review. Scientific work cannot become archival knowledge until peers have evaluated the scholarly products of scientists. Such review makes science self-correcting and its knowledge cumulative across time and place. However, another instance of the community nature of science is to strive to supersede the faults and problems with a given approach to science with a better approach. This transcendence involves the second goal of this book. The scholarship that we present in this book to challenge models of genetic reductionism is associated with a metatheory derived from a different scientific paradigm. We discuss this paradigm and metatheory in the next section.

Concepts from Dynamic, Relational Developmental Systems (RDS) Metatheory

> There are more things in heaven and earth, Horatio,
> Than are dreamt of in your philosophy.
> (Shakespeare, 1604/1605, *Hamlet*, Act 1,
> Scene 5, Lines 167–168)

Over the course of the first quarter of the 21st century, the scholarly field that has been termed *developmental science* (e.g., Bornstein & Lamb, 2015) experienced a shift in the philosophy of science framing discourse (e.g., Overton, 2006, 2010, 2015). Increasingly, dynamic systems ideas have emerged as a focal lens for understanding human development across the life course (e.g., Cantor et al., 2021; Lerner, 2018; Mascolo & Bidell, 2020; Witherington & Lickliter, 2016). This emergence has been coupled with the broad rejection of ideas using reductionist notions in theoretical models of intraindividual change pathways across the life span.

Instead, theoretical models of human development have emerged to reflect the interest in the dynamic, that is, mutually influential, relations within an open, living system. These individual ⇔ context relations have been conceptualized as the basic process of human development. A dynamic metatheory, termed relational developmental systems (RDS), and the *process-relational* paradigm within which it is embedded have emerged within this scholarly context (Lerner, 2018, 2021; Lerner & Lerner, 2019; Overton, 2015).

Paradigm, Metatheory, and Epistemology in Human Development

We have explained that a metatheory is an ontological statement derived from a superordinate paradigm, that is, a metatheory specifies the concepts that must be included in any theory reflecting the paradigm of origin. By implication of what must exist within the ontological parameters of a specific metatheory (e.g., human behavior and development are composed of *dynamic* – that is, *mutually influential* – coactions across time and place among all variables for all levels of being, biological through sociocultural, existing within a system of relations constituting the ecology of human life), a metatheory also indicates what concepts cannot exist in a theory derived from the metatheory (e.g., a structurally or functionally meaningful construct that is not related to, for instance, or moderate in its action with, any other construct cannot be a part of a dynamic, relational system).

A concrete example of this point can be found in regard to the concept of a *gene*. Within a mechanistic metatheory, a biological construct such as a gene or a set of genes may be used to account for energizing and directing the function of the machine or, in the characterization used by Dawkins (1976) of a human, of the Lumbering Robot. In this account reduction to the specific gene or specific set of genes located within the robot could explain its behavior and, critically, the direction it takes to reproduce itself in future robots. In turn, within a metamodel that emphasized that dynamic coactions among constructs within the systems comprising an individual, a gene is no more or no less a part of the system than any other construct existing in the system. Within such a metamodel, a gene or a set of genes functions *only* through moderation by other facets of the system. As such, to understand the structure and function of the system, reductionism is eschewed as counterfactual, and dynamic relations within the coacting system must be studied.

This example provides a transition to a focus on the dynamic, relational development systems (RDS) metamodel and its place within a paradigm focused on dynamic systems (von Bertalanffy, 1933, 1968) and on the relations within such systems. As explained by Overton (2015), a burgeoning interest in dynamic, relational processes in human development has resulted in the emergence of a *dynamic, relational process paradigm*. This paradigm has superseded paradigms focusing either on mechanist and reductionist concepts or on organismic concepts (e.g., the fixed stages of development, Erikson, 1950, 1959; Freud, 1923, 1954; Kohlberg, 1978; Piaget, 1970); these latter concepts did not adequately treat the dynamics of coactions with time and place in creating intraindividual change or interindividual differences in intraindividual change, that is, changes that instantiate the potential specificity of each person's history of individual ⇔ context relations (Molenaar, 2004; Molenaar & Nesselroade, 2014, 2015; Rose, 2016). As discussed by von Bertalanffy (1933, 1968), a system is a complex, open organization of coacting variables that are also coacting with their contexts, which of course includes all other variables within the system. These coactions enable qualitatively new features of the system to emerge across time and place, and, therefore, the dynamics of the system create its continual evolution (e.g., Jablonka & Lamb, 2005) of agentic and autopoietic feedback processes.

Thus, the ideas in this paradigm focus on process (systematic changes in the developmental system), becoming (moving from potential to actuality; a developmental process with a past, present, and future; Whitehead, 1929/1978), holism (the meanings of entities and events derive from the temporally and ecologically textured context in which they are embedded; Bronfenbrenner & Morris, 2006; Cantor et al., 2021; Elder et al., 2015), and relational analysis (assessment of mutually influential relations in the developmental system). In sum, in this dynamic, relational process paradigm, the human is seen as inherently active, self-creating, self-organizing (autopoietic), self-regulating (agentic), nonlinear and complex, adaptive, and at least relatively plastic; that is, the person possesses the potential for systematic change through dynamic coactions with the context (Overton, 2015).

RDS metatheory includes ideas from this paradigm. Because of holism, RDS metatheory emphasizes the integration of different levels of organization, including biology, physiology, culture, physical ecology, and history, as involved in creating human life, behavior, and development over the life span (Lerner, 2018; Overton, 2015). In addition, because of the individual's agency, and in the service of adapting to the specific features of the specific settings within which the individual lives and develops, the individual acts on the world that is acting on the individual. Therefore, *each* person's life is structured by the experience of specific mutually influential coactions between levels of organization within the dynamic developmental system. These coactions vary across place (e.g., community, country, or culture) and across time (Bronfenbrenner & Morris, 2006; Elder, 1998; Elder et al., 2015). Because no two people, even monozygotic twins, experience the same system of individual ⇔ context coactions across ontogeny, the singularity, the specificity, of each human, is assured. As Rose (2016) puts it, each of us walks the road less traveled.

The "arrow of time," or temporality, is history, which is the broadest level within the ecology of human development. History imbues all other levels with change. Such change may be stochastic (e.g., nonnormative life or nonnormative historical events; Baltes et al., 2006) or systematic (e.g., history- or age-graded changes). The potential for systematic change constitutes a potential for *relative* plasticity (i.e., for malleability, the potential for systematic change in structure or function; Lerner, 1984, 2018) across the life course for individuals, families, and the broader ecology of human development. Such plasticity is regarded as a fundamental strength of humans.

Plasticity provides a basis for optimism that the course of development for all individuals may be enhanced (Lerner, 1984, 2018). If there is plasticity in every individual's developmental pathway, then policies and programs can be aimed at capitalizing on this plasticity to decrease social, educational, economic, and health disparities and to enhance the quality of life of all individuals. This implication of dynamic, RDS-based ideas enables developmental scientists to view with optimism the possibility of promoting individual ⇔ context relations that engender positive development, that is, a developmental pathway wherein an individual possessing positive attributes may strive to contribute in beneficial ways positively to the physical and social world that at the same time is supporting the well-being of the individual. Because there is plasticity in human development, there can also therefore be an intentionality around promoting social justice (Lerner & Overton, 2008).

However, in order to enhance the chances that this implication of RDS metatheory is instantiated through evidence derived from theoretically predicated research, it is important to be more specific about the concepts within RDS metatheory that should be used to frame theoretical models of human development. At the beginning of this section of the book, we noted that there are key questions that may be asked about human development, such as: What sorts of changes characterize an individual as he or she develops across the life span? Where do these changes come from? Do all of these changes pertain to the development of the person? Theories of development using ideas from RDS metatheory have specific answers to such questions.

Essentially, RDS metatheory-based ideas or, more simply, (for ease of exposition) RDS-based ideas, involve the view that human beings are active rather than passive and that humans are part of an integrated system of relations involving the individual and the ecology (the context) of human development; and it is this system of relations that is the source of the course of development across life (Raeff, 2016; Witherington & Lickliter, 2016). The integrated (relational) developmental system provides the necessary and sufficient conditions to account for the structure and function of development across the life course. As such, humans are self-constructors of their development; they are *autopoietic* (Lerner, 1982, 1984, 2012; Lerner & Busch-Rossnagel, 1981; Lerner & Callina, 2014; Lerner & Walls, 1999; Overton, 2015; Witherington & Lickliter, 2016).

RDS-based ideas therefore also emphasize that the world around the developing person – both the physical and the social ecology of human life – is active and changing. According to all RDS-based models of human development (see Overton & Molenaar, 2015, for examples), the basic process of development involves, then, the integration or fusion of actions within the relational development systems (Tobach & Greenberg, 1984). Specifically, then, in RDS-based models, there is an integration of variables across all levels of organization within the system; this integration involves (a) the actions of people in and on their world, and (b) the actions of the world on people. We use a bidirectional arrow – ⇔ – to represent these integrative and mutually influential relations. Integrated actions shape the quality and course of human behavioral, psychological, social, and cultural structure and function across the life span (Brandtstädter, 1998, 1999; Brandtstädter & Lerner, 1999; Lerner & Busch-Rossnagel, 1981; Mascolo & Fischer, 2015; Raeff, 2016; Witherington & Lickliter, 2016).

Sources of Action in Human Development

But, where do the actions that propel human development come from? Consistent with its conceptualization of the character of human life – of focusing on the integration of levels of organization (Aronson, 1984) – RDS-based ideas emphasize that the source of the actions involved in human development is derived from the mutually influential relations within the relational developmental system, for instance, between the individual and his or her context, represented as individual ⇔ context relations.

These dynamic (mutually influential) changes may involve both quantitative and qualitative changes in the processes of development. For instance, processes involved with a person's perceptual, motivational, or cognitive development undergo changes in kind or type (quality), and in amount, frequency, magnitude, or duration (quantity). This conception of change does not deny that there are some aspects of a person that remain the same throughout life; rather, it asserts that human development is a synthesis between processes that promote change and processes that promote constancy (Brim & Kagan, 1980; Lerner, 1985; Overton, 2015).

In other words, due to the integration of the organizational levels of human life, ranging from biology, through the individual and social relationships, to the community, institutional, cultural, and historical, RDS-based models indicate that the laws that govern the functioning of both constancy and change are relational ones; they pertain to processes of interrelated relations across all levels of organization within the developmental system. No one level, and no one variable within a level (e.g., genes), is privileged as the key to these integrated relations (Noble, 2015). *The process of development, therefore, cannot be reduced to one level or variable.* This rejection of reductionism, and more specifically, of genetic reductionism, is the key difference between models derived from RDS and genetic reductionist models associated with human development (e.g., behavioral genetics, sociobiology, or evolutionary developmental psychology).

An Overview of the Contributions to This Section

In the two selections included in this section of the book, the implications of this key difference between dynamic, RDS-based models of human development and genetic reductionist models are discussed in greater details. Lerner and Greenberg discuss implication for both the conduct of science and, as well, implications of the application of science to social policies and programs, both in formal and informal educational settings, and in regard to opportunities to live in a socially just and democratic world wherein there are equitable resources available for health, personal fulfillment, and family and community well-being on a habitable and safe planet.

In turn, Greenberg discusses the fallacies and failures of genetic reductionism. He provides an historical review of lessons learned, and this discussion illuminates why the heredity hoax has worked against the positive implications of RDS-based science discussed by Lerner and Greenberg. He offers ideas about dynamic RDS-based science which can counter the pernicious impact of counterfactual genetic-reductionist claims and, as such, demonstrates the wisdom in Kurt Lewin's famous statement, that there is nothing so practical as a good theory.

References

Aronson, L. R. (1984). Levels of integration and organization: A re-evaluation of the evolutionary scale. In G. Greenberg & E. Tobach (Eds.), *Evolution of behavior and integrative levels* (pp. 57–81). Erlbaum.

Baltes, P. B., Lindenberger, U., & Staudinger, U. M. (2006). Life span theory in developmental psychology. In R. M. Lerner (Ed.), *Handbook of child psychology: Theoretical models of human development* (6th ed., Vol. 1, pp. 569–664). John Wiley & Sons.

Baltes, P. B., Reese, H. W., & Nesselroade, J. R. (1977). *Life-span developmental psychology: Introduction to research methods*. Brooks/Cole.

Bijou, S. W., & Baer, D. M. (1961). *Child development: A systemic and empirical theory*, (Vol. 1). Appleton-Century-Crofts.

Bjorklund, D. F. (2015). Developing adaptations. *Developmental Review, 38*, 13–35.

Bjorklund, D. F. (2016). Prepared is not preformed: Commentary on Witherington and Lickliter. *Human Development, 59*(4), 235–241.

Bjorklund, D. F., & Ellis, B. J. (2005). Evolutionary psychology and child development: An emerging synthesis. In B. J. Ellis & D. F. Bjorklund (Eds.), *Origins of the social mind: Evolutionary psychology and child development* (pp. 3–18). Guilford.

Bornstein, M. H., & Lamb, M. E. (Eds.). (2015). *Developmental science: An advanced textbook* (7th ed.). Lawrence Erlbaum.

Brandtstädter, J. (1998). Action perspectives on human development. In W. Damon (Series Ed.) & R. M. Lerner (Vol. Ed.), *Handbook of child psychology: Theoretical models of human development* (5th ed., Vol. 1, pp. 807–863). Wiley.

Brandtstädter, J. (1999). The self in action and development: Cultural, biosocial, and ontogenetic bases of intentional self-development. In J. Brandtstädter & R.M. Lerner (Eds.), *Action and self-development: Theory and research through the life-span* (pp. 37–65). Sage.

Brandtstädter, J., & Lerner, R. M. (Eds.). (1999). *Action and self-development: Theory and research through the life-span*. Sage.

Brim, O. G., Jr., & Kagan, J. (Eds.). (1980). *Constancy and change in human development*. Harvard University Press.

Bronfenbrenner, U., & Morris, P. A. (2006). The bioecological model of human development. In W. Damon & R. M. Lerner (Eds.) & R. M. Lerner (Vol. Ed.), *Handbook of child psychology: Theoretical models of human development* (6th ed., Vol. 1, pp. 793–828). Wiley.

Buss, D. M. (2019). *Evolutionary psychology: The new science of the mind*. Taylor & Francis.

Buss, D. M., & von Hippel, W. (2018). Psychological barriers to evolutionary psychology: Ideological bias and coalitional adaptations. *Archives of Scientific Psychology, 6*(1), 148–158.

Cantor, P., Lerner, R. M., Pittman, K., Chase, P. A., & Gomperts, N. (2021). *Whole-child development, learning, and thriving: A dynamic systems approach*. Cambridge University Press.

Carmack, H. J., Bates, B. R., & Harter, L. M. (2008). Narrative constructions of health care issues and policies: The case of President Clinton's apology-by-proxy for the Tuskegee Syphilis Experiment. *Journal of Medical Humanities, 29*, 89–109. https://doi.org/10.1007/s10912-008-9053-5

Clarke, E. (2021). Indigenous women and the risk of reproductive healthcare: Forced sterilization, genocide, and contemporary population control. *Journal of Human Rights and Social Work, 6*, 144–147. https://doi.org/10.1007/s41134-020-00139-9

Dawkins, R. (1976). *The selfish gene*. Oxford University.

Elder, G. H. (1998). The life course and human development. In W. Damon (Series Ed.) & R. M. Lerner (Vol. Ed.), *Handbook of child psychology: Theoretical models of human development* (5th ed., Vol. 1, pp. 939–991). Wiley.

Elder, G. H., Shanahan, M. J., & Jennings, J. A. (2015). Human development in time and place. In M. H. Bornstein & T. Leventhal (Eds.), *Handbook of child psychology and developmental science: Ecological settings and processes in developmental systems* (7th ed., Vol. 4, pp. 6–54). Wiley.

Elliott, L. (2017). Victims of violence: The forced sterilisation of women and girls with disabilities in Australia. *Laws, 6*(8), 1–19. https://doi:10.3390/laws6030008

Erikson, E. H. (1950). *Childhood and society*. Norton.

Erikson, E. H. (1959). Identity and the life cycle. *Psychological Issues, 1*, 50–100.

Freud, S. (1923). *The ego and the id*. Hogarth Press.

Freud, S. (1954). *Collected works* (standard edition). Hogarth Press.

Gould, S. J. (1976). Grades and clades revisited. In R. B. Masterton, W. Hodos, & H. Jerison (Eds.), *Evolution, brain, and behavior: Persistent problems* (pp. 115–122). Lawrence Erlbaum Associates.

Greenberg, G. (2015). The case against behavioral genetics. Review of A. Panofsky (2014). Misbehaving science: Controversy and the development of behavior genetics. University of Chicago Press. *Developmental Psychobiology, 57*, 854–857.

Harden, K. P. (2021). *The genetic lottery: Why DNA matters for social equality*. Princeton University Press.

Hirsch, J. (1981). To "unfrock the charlatans." *Sage Race Relations Abstracts, 6*, 1–65.

Jablonka, E., & Lamb, M. (2005). *Evolution in four dimensions: Genetic, epigenetic, behavioral, and symbolic variation in the history of life*. MIT Press.

Kohlberg, L. (1978). Revisions in the theory and practice of moral development. *New directions for child development, 2*, 93–120.

Kuhn, T. S. (1962). *The structure of scientific revolutions*. University of Chicago Press.

Kuhn, T. S. (1970). *The structure of scientific revolutions* (2nd ed.). University of Chicago Press.

Lerner, R. M. (1982). Children and adolescents as producers of their own development. *Developmental Review, 2*, 342–370.
Lerner, R. M. (1984). *On the nature of human plasticity*. Cambridge University Press.
Lerner, R. M. (1985). Individual and context in developmental psychology: Conceptual and theoretical issues. In J. R. Nesselroade & A. von Eye (Eds.), *Individual development and social change: Explanatory analysis* (pp. 155–187). Academic Press.
Lerner, R. M. (1992). *Final solutions: Biology, prejudice, and genocide*. Penn State Press.
Lerner, R. M. (2012). Developmental science and the role of genes in development. *GeneWatch, 25*(1–2), 34–35. http://www.councilforresponsiblegenetics.org/genewatch/GeneWatchPage.aspx?pageId=413
Lerner, R. M. (2018). *Concepts and theories of human development* (4th ed.). Routledge.
Lerner, R. M. (2021). *Individuals as producers of their development: The dynamics of person ⇔ context coactions*. Routledge.
Lerner, R. M., & Busch-Rossnagel, N. A. (Eds.). (1981). *Individuals as producers of their development: A life-span perspective*. Academic Press.
Lerner, R. M., & Callina, K. S. (2014). The study of character development: Towards tests of a relational developmental systems model. *Human Development, 57*(6), 322–346.
Lerner, R. M., & Lerner, J. V. (2019). An idiographic approach to adolescent research: Theory, method, and application. In L. B. Hendry & M. Kloep (Eds.), *Reframing Adolescent Research* (pp. 25–38). Routledge.
Lerner, R. M., & Overton, W. F. (2008). Exemplifying the integrations of the relational developmental system: Synthesizing theory, research, and application to promote positive development and social justice. *Journal of Adolescent Research, 23*, 245–255.
Lerner, R. M., & Walls, T. (1999). Revisiting individuals as producers of their development: From dynamic interactionism to developmental systems. In J. Brandtstädter & R. M. Lerner (Eds.), *Action and self-development: Theory and research through the life-span* (pp. 3–36). Sage.
Lifton, R. J. (1986). *The Nazi doctors: Medical killing and the psychology of genocide*. Basic Books.
Mascolo, M. F., & Bidell, T. R. (Eds.). (2020). *Handbook of integrative developmental psychology: Festschrift for Kurt W. Fischer*. Routledge.
Mascolo, M. F., & Fischer, K. W. (2015). Dynamic development of thinking, feeling, and acting. In W. F. Overton, P. C. Molenaar, & R. M. Lerner (Eds.), *Handbook of child psychology and developmental science: Theory and method* (7th ed., Vol. 1, pp. 113–161). Wiley.
Molenaar, P. C. M. (2004). A manifesto on psychology as idiographic science: Bringing the person back into scientific psychology, this time forever. *Measurement, 2*(4), 201–218.
Molenaar, P. C. M., & Nesselroade, J. R. (2014). New trends in the inductive use of relation developmental systems theory: Ergodicity, nonstationarity, and heterogeneity. In P. C. Molenaar, R. M. Lerner, & K. M. Newell (Eds.), *Handbook of developmental systems and methodology* (pp. 442–462). Guilford Press.
Molenaar, P. C. M., & Nesselroade, J. R. (2015). Systems methods for developmental research. W. F. Overton, P. C. M. Molenaar, & R. M. Lerner (Eds.), *Handbook of child psychology and developmental science: Theory and method* (7th ed., Vol. 1, pp. 652–682). Wiley.
Moore, D. S. (2002). *The dependent gene: The fallacy of nature vs. nurture*. W. H. Freeman.
Moore, D. S. (2015a). The asymmetrical bridge. Book review of James Tabery's Beyond versus: The struggle to understand the interaction of nature and nurture. *Acta Biotheoretica, 63*(4), 413–427.
Moore, D. S. (2015b). *The developing genome: An introduction to behavioral epigenetics*. Oxford University Press.
Müller-Hill, B. (1988). *Murderous science: Elimination by scientific selection of Jews Gypsies, and others. Germany 1933–1945*. Oxford University.
Noble, D. (2015). Evolution beyond neo-Darwinism: A new conceptual framework. *The Journal of Experimental Biology, 218*, 7–13.
Overton, W. F. (2006). Developmental psychology: Philosophy, concepts, methodology. In W. Damon & R.M. Lerner (Series Eds.) & R.M. Lerner (Vol. Ed.), *Handbook of child psychology: Theoretical models of human development* (6th ed., Vol. 1, pp. 18–88). Wiley.

Overton, W. F. (2010). Life-span development: Concepts and issues. In W. F. Overton (Vol. Ed.), *Handbook of life-span development: Cognition, biology, and methods* (Vol. 1, pp. 1–29), Editor-in-chief: R. M. Lerner. Wiley and Sons.

Overton, W. F. (2015). Process and relational developmental systems. In W. F. Overton, P. C. M. Molenaar, & R. M. Lerner (Eds.), *Handbook of child psychology and developmental science: Theory and method* (7th ed., Vol. 1, pp. 9–62). Wiley.

Overton, W. F., & Molenaar, P. C. M. (2015). *Theory and method. Volume 1 of the handbook of child psychology and developmental science* (7th ed.), Editor-in-chief: Richard M. Lerner. Wiley.

Overton, W. F., & Reese, H. W. (1973). Models of development: Methodological implications. In J. R. Nesselroade & H. W. Reese (Eds.), *Life-span developmental psychology: Methodological issues* (pp. 65–86). Academic Press.

Panofsky, A. (2014). *Misbehaving science: Controversy and the development of behavior genetics*. University of Chicago Press.

Pepper, S. C. (1942). *World hypotheses: A study in evidence*. University of California Press.

Piaget, J. (1970). Piaget's theory. In P. H. Mussen (Ed.), *Carmichael's manual of child psychology* (3rd ed., Vol. 1, pp. 703–723). Wiley.

Plomin, R. (2018). *Blueprint: How DNA makes us who we are*. Allen Lane.

Proctor, R. N. (1988). *Racial hygiene: Medicine under the Nazis*. Harvard University.

Rabuy, B. (2014). Governor Brown, end sterilization abuse in California today. Prison Policy Initiative. https://www.prisonpolicy.org/blog/2014/09/12/sterilization/

Raeff, C. (2016). *Exploring the dynamics of human development: An integrative approach*. Oxford University Press.

Reese, H. W., & Overton, W. F. (1970). Models of development and theories of development. In L. R. Goulet & P. B. Baltes (Eds.), *Life-span developmental psychology: Research and theory* (pp. 115–145). Academic.

Rose, T. (2016). *The end of average: How we succeed in a world that values sameness*. HarperCollins Publishers.

Rushton, J. P. (2000). *Race, evolution, and behavior* (2nd Special Abridged ed.). Transaction Publishers.

Shakespeare, W. (1604/1605). *The Tragicall Historie of Hamlet Prince of Denmarke* (second quarto). Nicholas Ling, publisher, and James Roberts, printer.

Skinner, B. F. (1938). *The behavior of organisms*. Appleton.

Tobach, E., & Greenberg, G. (1984). The significance of T. C. Schneirla's contribution to the concept of levels of integration. In G. Greenberg & E. Tobach (Eds.), *Behavioral evolution and integrative levels*. Erlbaum.

von Bertalanffy, L. (1933). *Modern theories of development*. Oxford University Press.

von Bertalanffy, L. (1968). *General systems theory*. Braziller.

Whitehead, A. N. (1978). *Process and reality: Corrected edition*. The Free Press.

Wilson, D. S., & Wilson, E. O. (2007). Rethinking the theoretical foundation of sociobiology. *The Quarterly Review of Biology, 82*(4), 327–348.

Wilson, E. O. (1975). *Sociobiology: The new synthesis*. Harvard University Press.

Witherington, D. C., & Lickliter, R. (2016). Integrating development and evolution in psychological science: evolutionary developmental psychology, developmental systems, and explanatory pluralism. *Human Development, 59,* 200–234.

1.
ADDRESSING THE HEREDITY HOAX IN SCIENCE AND SOCIETY*
A View of the Issues
Richard M. Lerner† and Gary Greenberg‡

This book integrates both new, authored material and reprints of existing literature that, together, provide a compelling narrative that reveals the fatally-flawed science associated with genetic reductionist accounts of human behavior and development. Across the book the work of top-tier scientists, from developmental, comparative, educational, and biological science will illuminate theory and research converging on the conclusion that the multiple egregiously flawed work of genetic reductionists should be expunged from research pertinent to human development. In addition to being bad science, its application to social policies and programs and its use in popular media accounts, continues to serve as tools for the marginalization of and inequities to individuals or groups that are purported to possess genes that do not enable them to succeed in life.

The purpose of this book is actualized by including scientific papers that present readers with theoretically based and empirically rigorous information about a multidisciplinary approach to human development. Knowledge of human development derived from such work eschews reductionism and, instead, demonstrates the usefulness of dynamic relations among all levels of organization fused within the dynamic developmental system. These integrated levels of organization involve physical and physiological processes (e.g., genetic, neuronal, and hormonal processes); psychological and behavioral processes (e.g., cognitive, emotional, and behavioral processes that coact in the development of each individual's agency, purpose, identity, character, and moral development); and social relationships, societal institutions, and culture.

This book, in effect, provides a consensus statement by scientists, who, in their scholarship, both use and richly extend the conceptual and methodological facets of this dynamic, relational developmental systems approach to human development. The contributions to the literature that are used in this book have been chosen because, although providing technical information about biology, psychology, social relationships, and culture, also are written in a manner accessible to both academic and educated non-academic audiences.

The contributions present approaches to describing, explaining, and optimizing human development across the life span. These approaches stand in stark contrast to the core and pessimistic claim of genetic reductionism, that is: Each person's potential in life for thriving, health, success, achievement, fulfillment, and contributions to equity and social justice in their homes, communities, and nations is constrained by the specifics

---

* This chapter is published for the first time in this book.
† Tufts University, Medford, MA.
‡ Wichita State University, Wichita, KS.

of the genes inherited at birth. The approach to human development presented in the contributions in this book offer to science and to society reasons to be optimistic that developmental research may contribute to enhancing the health and positive development of all people *and* that science also may contribute to social justice and equity for all individuals and groups across the breadth of the life span.

The Importance of This Book

There are many actions that promote social justice and increase equity for all people. Recent events in our nation and world document that there are also many actions that decrease justice and equity. For more than 200 years the specious claims of genetic reductionists and their error-ridden research methodology have combined to claim that *all facets of human development and all differences between people in their development* can be explained by reducing the complexity of human life to the genes inherited at conception. But if everything a person does or becomes exists as an inherited blueprint residing in specific genes bestowed at conception, then there is no way that socially progressive policies and programs, equitable education, health and medical resources, or employment opportunities can successfully diminish or eliminate specific developmental outcomes of any individual or provide ways to improve human life for all people by changing inequitable social conditions.

Today, mainstream developmental science rejects such genetic reductionism. Most scientists studying human development use models and rigorous research methods to interrogate the mutually influential relations between individuals and their contexts – dynamic coactions usually symbolized as individual ⇔ context relations. Current science indicates that the contribution of genes to human development depends on the context in which they exist. Across the nation and world, both epigenetics and neuroscience research; developmental and comparative research; and educational science research document that human development is marked by enormous plasticity because of the specific dynamics of each person's individual ⇔ context relations. Using this knowledge, programs and policies can promote thriving in each individual. The contributions to this book champion good science by giving voice to how instances of such science coalesce to identify the logical, theoretical, and methodological flaws of different instantiations instances genetic reductionism. The contributions in this book also help create social understanding and opportunities for applications to programs and policies that help erase from science and society the counterfactual and dehumanizing pronouncements of genetic reductionists. In this way, this book contributes to helping create a more socially just world for all people across all of life.

The Timeliness of This Book: The Heredity Hoax Past and Present

At least since the time of the British polymath, Sir Francis Galton (1822–1911), heredity – the complement of genes received at conception – has been proffered as an explanation of a quite varied assortment of human characteristics, from genius or the lack thereof, academic abilities and educational attainment, physical, athletic, and artistic skills, personality, character, and temperament, morality or its absence, parenting, sexuality and sexual preferences, militarism and aggression, religiosity or spirituality, personal tastes (e.g., in art, music, or style of dress), and even preferences for leisure activities (e.g., the propensity to watch television). Whatever the hereditarian claim being made, the advocate for hereditary determination claims that the human characteristic in question can be explained by reducing the characteristic to one or more of the genes received by a person at conception. Simply, the hereditarian claim is that, in order to understand human life, whether for any one person, for a group, or for all people, all that must be done is to reduce human life to the specific genes, the specific DNA, of the individual, of the group, or of all humanity.

We believe that these hereditarian assertions are specious; they are based on misunderstandings of genetic activity and/or on counterfactual claims about heredity and, as well, they involve scientific methodology that is fatally flawed. As such, the idea that human behavior and development can be reduced to genes is useless in regard advancing scientific knowledge *and* to formulating social policies or programs. Such use, or even the recommendations that such use is appropriate, constitutes a hoax perpetrated on the public – the parents, educators, or policy makers, that want to use sound scientific knowledge to improve human life.

Across the professional lives of the scholars whose work is presented in this book, both the scientific literature and public discourse – in both mass media and in political rhetoric –includes claims that that the complement of genes received by a child at conception determined the inevitable course of the array of human characteristics we have noted already. These claims have been aimed at explaining the behavior and development of all humans and but, as well, these ideas have been used as the basis for accounting for differences between men and women, individuals of different racial groups, and people of various ethnic or cultural backgrounds.

Through the time period within which we prepared this book proposal, the pages of scholarly journals and of books authored by individuals with advanced scholarly training continue to propose ideas about the genetic determination of human attributes *and* of group differences in these attributes. These assertions have often been presented also in popular media reports reaching audiences of millions and, as well, the staffs of elected officials who use ideas of the genetic causation of specific human attributes as the evidence base for writing legislation that will purportedly protect the public from the pernicious effects of people possessing bad or inferior genes.

The infamous U. S. Supreme Court decision of 1927 that permitted the sterilization of a pregnant teenager is, sadly, only one instance of such sordid use of counterfactual genetic reductionist claims as a rationale for life-damaging treatment of marginalized people (Lerner, 2015). Supreme Court Justice Oliver Wendell Holmes wrote in support of a decision upholding a Virginia law that authorized sterilization of "mental defectives" without their consent. He agreed that a young woman, Carrie Buck, should be sterilized because she was unfit to reproduce. Raped, and now pregnant, her mental defect was evidenced by the fact that she was going to have a baby out of wedlock. Justice Holmes wrote:

We have seen more than once that the public welfare may call upon the best citizens for their lives. It would be strange if it could not call upon those who already sap the strength of the State for these lesser sacrifices often not felt to be such by those concerned, in order to prevent our being swamped with incompetence. It is better for all the world if, instead of waiting to execute degenerate offspring for crime or to let them starve for their imbecility, society can prevent those who are manifestly unfit from continuing their kind …. *Three generations of imbeciles are enough* (Buck v. Bell, 274 U.S.200, italics added).

And so Carrie Buck was sterilized. Although her pregnancy was not aborted and she even actually gave birth, she was kept from passing along her genes any further, so that her mental defectiveness could not be a further infliction on society. But Carrie Buck's experience was not unique. Doerr (2009) explained that "State laws permitting sterilization of individuals deemed unfit to reproduce—most commonly institutionalized persons with mental illness, or even conditions such as epilepsy—were common in the first half of the twentieth century" (p. 1). The Virginia law that resulted in Carrie Buck's forced sterilization was not repealed until 1974. However, before this law and the comparable laws in more than 30 other states in the U.S. were repealed, more than 65,000 people were forcibly sterilized in the United States—to protect society from them spreading their defective genes.

Moreover, doctors under contract with the California Department of Corrections and Rehabilitation forcibly sterilized nearly 150 female inmates from 2006 to 2010; the women targeted for sterilization were those deemed likely to return to prison in the future (Johnson, 2013). It was not until September 2014 that California Governor Jerry Brown signed a bill prohibiting forced sterilizations in prisons.

The reduction of the dynamic system of influences on human development to the role of the genes inherited by individuals or groups have a well-documented history of criticism by scientists from diverse areas of scholarship. However, as illustrated by the U.S. Supreme Court decision about Carrie Buck and by the forced sterilization of women that continued in the United States in the first two decades of the 21st century, there is at the same time a sorry history of the adoption of genetic reductionist ideas being translated into recommendations for applications to policies and programs that have negatively impacted millions of people across generations and around the world.

For instance, in the 1940s through the 1970s the Austrian physician and Nobel Laureate, Konrad Lorenz, presented flawed ideas about the "hard-wired" links between genes and behaviors to explain imprinting in precocial birds, fixed action patterns of fish, the militant enthusiasm of some people, and the moral inferiority of Jews (e.g., 1939, 1940a, 1940b, 1966, 1974). His ideas were then thoroughly countered and dismissed, for instance, by Lehrman (1953, 1970) and by Schneirla (1957, 1966). Nevertheless, the ideas resurfaced again in the heritability work of Jensen (1969, 1980) regarding racial differences in intelligence. Hernstein and Murray (1994) and Rushton (e.g., 2000) reiterated genetic reductionist arguments for the bases of racial differences in intelligence test scores.

In addition, genetic reductionist conceptions resurfaced in human sociobiological ideas about gender differences in sexuality and parenting (e.g., Dawkins, 1976; Freedman, 1979) and again in the postulation of Five Factor Theory, that there are five "big traits" (conscientiousness, agreeableness, neuroticism, openness to experience, and extraversion) that are fixed, stable, and biologically-set facets of personality. These purportedly fundamental facets of individual functioning are held to reflect "nature over nurture" and to involve attributes that "are more or less immune to environmental influences ... significant variations in life experiences have little or no effect on measured personality traits" (McCrae, Costa, et al., 2000, pp. 175–176). Moreover, these ideas exist in a field labeled evolutionary developmental psychology. For instance, in this field the idea of evolved probabilistic cognitive mechanisms (EPCMs) is used to identify purportedly innate propensities for development that reside in genes; these genes are said to have been shaped across eons by evolutionary processes to exert this control over cognition (e.g., Bjorklund, 2015; Bjorklund & Ellis, 2005; Del Giudice & Ellis, 2016). These EPCMs are claimed to control facets of higher levels of organization (e.g., cognition or social relationships). The claim is made that higher levels of organization exist only to manage the expression or release of the information contained in the essential, genetic level (see Witherington & Lickliter, 2016).

The articles included in this book document the bad science emblematic of these genetic reductionist claims. Given such documentation, it might be expected that a neophyte social or behavioral scientist – or a molecular geneticist happening on such egregiously flawed ideas – would infer that genetic reductionism could not be taken seriously by competent social and behavioral scientists. They might expect that such scientists would vociferously and visibly dismiss this thinking, in any form that it might occur. However, if these observers continued to pay attention to the literatures of these fields, they would learn that the presence of these ideas persists. As in the children's game, Whack-A-Mole, as soon as the failures of one instantiation of genetic reductionism are compellingly refuted, other instances of this problem-riddled conception, and another version of this idea, pops up.

Examples of this persistence abound. For instance, in 2014 journalist Nicholas Wade claimed that genes shape social behavior, manifested as behavioral traits; the expression of these traits are alleged to vary significantly among races. Wade argued that genetic inheritance accounts for racial differences in wealth and economic institutions. Wade also argued that, if racial groups are either poor or rich, the difference is because of the evolution of their genes, and not social discrimination, racism, lack of education, and so on. This book elicited considerable public attention despite the erudite and compelling criticism it has received, for instance, in a review of the book by Stanford University geneticist Marcus Feldman (2014).

Indeed, another instance of the popular dissemination of these genetic reductionist claims was written by developmental psychologist Jay Belsky. In what reads as a neo-eugenicist appeal, Belsky, writing an Op-Ed piece in the *New York Times Sunday Review* in November 2014, asserted that:

> What distinguishes children who prove more versus less susceptible—for better and for worse—to developmental experiences? There is no single factor, but genetics seems to play a role.... Should we seek to identify the most susceptible children and disproportionately target them when it comes to investing scarce intervention and service dollars? I believe the answer is yes.... One might even imagine a day when we could genotype all the children in an elementary school to ensure that those who could most benefit from help got the best teachers.

More recently, behavior geneticist Robert Plomin (2018) claimed in his book, *Blueprint*, that:

> Genetics is the most important factor shaping who we are (p. viii) and that the DNA revolution has made DNA personal by giving us the power to predict our psychological strengths and weaknesses from birth.
>
> (p. vii)

Plomin (2018) then asserts that, through genetic reductionism, scientists have:

> the ability to predict our psychological problems and promise from DNA.... Our future is DNA.
>
> (p. xiii)

In addition, in 2021 behavior geneticist Kathryn Paige Harden, in her book, *The genetic lottery: Why DNA matters for social equality*, asserted that:

> Yes, the genetic differences between any two people are tiny when compared to the long stretches of DNA coiled in every human cell. But these differences loom large when trying to understand why, for example, one child has autism and another doesn't; why one is deaf and another hearing; and ... why one child will struggle with school and another will not. Genetic differences between us matter for our lives. They cause differences in things we care about. Building a commitment to egalitarianism on our genetic uniformity is building a house on sand.
>
> (p. 19)

Moreover, lest anyone reading criticisms of genetic reductionist ideas be prone to accept the refutations of these ideas, evolutionary psychologists, David Buss and Willian von Hippel propose that genetically-based intellectual biases against genetic reductionism exist in the constitutional make-up of critics. Writing in a 2018 issue of

Archives of Scientific Psychology, a publication of the American Psychological Association, Buss and von Hippel (2018, p. 148) contend that:

> we argue that four interlocking barriers stand in the way of research scientists who seek to understand human social psychology. The first barrier is the political ideology of most social psychologists, which is typically on the left (or liberal) side of the spectrum. The second barrier is a view of human nature common among people on the political left, which is that we are born without any predilections to behave in a particular manner. According to this view, our mind is a blank slate at birth and is corrupted solely by the ills of bad environments or societies. The third barrier is a tendency to reject theories and findings that might contravene the "blank slate" view of human nature, particularly theories and findings that arise from evolutionary approaches to human behavior. *The fourth barrier is a collection of evolved tendencies that prevent investigators from being dispassionate seekers of scientific truth. These include our evolved tendency to be more focused on persuasion than truth-seeking, to be concerned with the maintenance of our prestige as scientists, and to form and maintain coalitions that compete with each other.* We provide initial evidence for some of these possibilities with data gathered from a survey of 335 established social psychologists. We conclude with the irony that our evolved psychology may interfere with the scientific understanding of our evolved psychology (italics added).

The Approach of This Book

Both the editors and the scientists whose work is reprinted in this book have been involved for several decades in directing their scholarship to counter the public and media attention paid to the popular audience books written by people such genetic reductionists as Robert Plomin and Kathryn Paige. Accordingly, the contributions of this book represent a broad consensus among scientists from different fields of biological, psychological, and other social and behavioral science fields about the role of genes in human development.

The editors believe that the combined impact of the contributions to the proposed book will be to provide its readers with both authoritative and accessible information that can be the means to erase the ongoing use of flawed and in many cases counterfactual statements about the role of genes in human development. As well, the expertise of the voices integrated in this book can provide a compelling counterpoint to the perniciously inequitable, undemocratic, and often racist policy ideas based on the bad science enacted by genetic reductionists.

To make this contribution to science and to society, the editors framed the book with two opening chapters that recount the nature of the heredity hoax perpetrated in science and in society by genetically-reductionist writing and recommendations and, as well, that provide a recounting of the ways in which genetic reductionism have affected the conduct of science, as seen through the lens of the career experiences of editor Gary Greenberg's career contributions to comparative psychology. The subsequent sections of the book are devoted to different instantiations of genetic reductionism in the history and current domains of science: Developmental theory, the concept of instinct, evolution, the concept of critical periods in development, the study of heritability and the methodology of behavior genetics, sociobiology, epigenetics, and the application of genetic reductionism to social policies and to human development programs.

Each section includes "classic" and current contributions by scientists providing the ontological, theoretical, methodological, and empirical scholarship that counters the claims made by genetic reductionists. Each section of the book is framed by an "Editors' Introduction" that in effect serves as a curation of the reprinted literature. The editors summarize the substance and implications of each of the contributions appearing in a section.

Within and across sections, readers will be informed about the theory-predicated and methodologically rigorous work that counters genetic reductionist ideas in specific areas of research. These contributions will provide up-to-date summaries of the knowledge that exists to reflect specific claims of genetic reductionists and, as well, to explain why the research methods employed by genetic reductions cannot be relied on as providing credible evidence about the use of genetic reductionism.

The final section of the book includes both reprinted articles and a final new chapter written by the editors. The focus of this section integrates the bases of the preceding contributions to the book in regard to constituting thorough refutation of the implications drawn by genetic reductionists for applications to policies and programs of their work. In the service of providing readers with the optimistic lens for understanding the potential for thriving and health among all people that we noted earlier in this prospectus, the material in this final section contrasts the constrained view of human development found in the examples of Belsky, Buss, Harden, Plomin noted earlier, with implications for enhancing the lives of all individuals that may derived from the policies and programs capitalizing on the relative plasticity of human development derived from systematic study of dynamic individual⇔context relations.

References

Belsky, J. (2014, November 30). The downside of resilience. *New York Times, Sunday Review*, p SR4.

Bjorklund, D. F. (2015). Developing adaptations. *Developmental Review, 38*, 13–35.

Bjorklund, D. F., & Ellis, B. J. (2005). Evolutionary psychology and child development: An emerging synthesis. In B. J. Ellis & D. F. Bjorklund (Eds.), *Origins of the social mind: Evolutionary psychology and child development* (pp. 3–18). Guilford.

Buck v. Bell, 274 U.S. 200 (1927). Retrieved from http://caselaw.lp.findlw.com/scripts/getcase.pl?court=US&vol=274&invol=200

Buss, D. M., & von Hippel, W. (2018). Psychological barriers to evolutionary psychology: Ideological bias and coalitional adaptations. *Archives of Scientific Psychology, 6*(1), 148–158.

Dawkins, R. (1976). *The selfish gene*. Oxford University.

Del Giudice, M., & Ellis, B. J. (2016). Evolutionary foundations of developmental psychopathology. In D. Cicchetti (Ed.), *Developmental psychopathology: Developmental neuroscience* (3rd. ed., Vol. 12, pp. 1–58). Wiley.

Doerr, A. (2009). Three generations of imbeciles are enough. Genomics Law Report. Retrieved from http://www.genomicslawreport.com/index.php/2009/06/25/three-generations-of-imbeciles-are-enough/

Feldman, M. (2014). Echoes of the past: Hereditarianism and a troublesome inheritance. *PLoS Genetics, 10*, e1004817.

Freedman, D. G. (1979). *Human sociobiology: A holistic approach*. Free Press.

Harden, K. P. (2021). *The genetic lottery: Why DNA matters for social equality*. Princeton University Press.

Hernstein, R. J., & Murray, C. (1994). *The bell curve: Intelligence and class structure in American life*. Free Press.

Jensen, A. R. (1969). How much can we boost IQ and scholastic achievement? *Harvard Educational Review, 39*, 1–123.

Jensen, A. R. (1980). *Bias in mental testing*. Free Press.

Johnson, C. G. (2013). Female inmates sterilized in California prisons without approval. Center for investigative reporting, 7.

Lehrman, D. S. (1953). A critique of Konrad Lorenz's theory of instinctive behavior. *Quarterly Review of Biology, 28*, 337–363.

Lehrman, D. S. (1970). Semantic and conceptual issues in the nature-nurture problem. In L. R. Aronson, E. Tobach, D. S. Lehrman, & J. S. Rosenblatt (Eds.), *Development and evolution of behavior: Essays in memory of T. C. Schneirla* (pp. 17–52). Freeman.

Lerner, R. M. (2015). Eliminating genetic reductionism from developmental science. *Research in Human Development, 12*, 178–188.

Lorenz, K. (1939). Über Ausfallserscheinungen im Instinctverhalten von Haustieren und ihre socialpsychologische Bedeutung. In O. Klemm (Ed.), *Charakter und Erziehung: 16. Kongress der Deutschen Gesellschaft für Psychologie in Bayreuth* (pp. 139–147). Leipzig: J. A. Barth.

Lorenz, K. (1940a). Durch Domestikation verursachte Störungen arteigenen Verhaltens. *Zeitschrift für angewandte Psychologie und Charakterkunde, 59,* 2–81.

Lorenz, K. (1940b). Systematik und Entwicklungsgedanke im Unterricht, *Der Biologe, 9,* 24–36.

Lorenz, K. (1966). *On aggression.* Harcourt, Brace & World.

Lorenz, K. (1974). Letter: Lorenz clarifies ideas. *Human behavior, September,* 6.

McCrae, R. R., Costa, P. T., Jr., Ostendorf, F., Angleitner, A., Hrebícková, M., Avia, M. D., Sanz, J., Sánchez-Bernardos, M. L., Kusdil, M. E., Woodfield, R., Saunders, P. R., & Smith, P. B. (2000). Nature over nurture: Temperament, personality, and life span development. *Journal of Personality and Social Psychology, 78,* 173–186.

Plomin, R. (2018). *Blueprint: How DNA makes us who we are.* Allen Lane.

Rushton, J. P. (2000). *Race, evolution, and behavior: A life history perspective* (2nd Special Abridged Edition). Charles Darwin Research Institute.

Schneirla, T. C. (1957). The concept of development in comparative psychology. In D. Harris (Ed.). *The concept of development* (pp. 78–108). University of Minnesota Press.

Schneirla, T. C. (1966). Instinct and aggression: Reviews of Konrad Lorenz, *Evolution and modification of behavior* (Chicago: The University of Chicago Press, 1965), and *On aggression* (New York: Harcourt, Brace & World, 1966). *Natural History, 75,* 16.

Wade, N. (2014) *A troublesome inheritance: Genes, race, and human.* Penguin Books.

Witherington, D. C., & Lickliter, R. (2016). Integrating development and evolution in psychological science: evolutionary developmental psychology, developmental systems, and explanatory pluralism. *Human Development, 59,* 200–234.

2.
THE FALLACIES AND FAILURES OF GENETIC REDUCTIONISM*
An Historical Review of Lessons Learned
Gary Greenberg†

The diverse group of authors and chapters in this book are linked by a set of common themes: Darwinian evolution (and epigenesis), the essential organizing principles of integrative levels and development, the significance of context and stimulus characteristics in the expression of behavior, the rejection that animals inherit unvarying behavior patterns (i.e., instincts), and crucially the repudiation of behavior genetics, that significant aspects of behavior in all animals (importantly in human beings) are determined (and programmed) by what genes those organisms inherit – in short, the contributors to this book dismiss the idea that some people are born smart and some are born stupid. I think it safe to say that our contributors understand biology (e.g., especially genetics) to be an important set of participating, but not determining, factors in the development of behavior.

This being the last professional thing I write I want to take advantage of being one of the editors of this book to trace my educational steps to here before getting to the substance of my contribution to this volume. I want to acknowledge those scholars whose ideas guided me here.

Still unsettled on a major at Brooklyn College in my junior year, the Strong Vocational Aptitude Test and several sessions with a few department advisors led me to enroll in Freshman Psychology. My friends, from whom I bought the textbooks, were sure I would find the non-human topics boring and instead promoted the human side of psychology, i.e., Freud and etc. Wrong! My Professor, David Raab, had us read three books, one of which was a reader co-edited by T. C. Schneirla (Crafts et al., 1950). Experimental Psychology at Brooklyn College was a hands-on research experience. Edward Girden's lectures and Woodworth and Schlosberg's (1955) textbook hooked me on a career as a research psychologist. I later enrolled in two courses from Professor Howard Moltz (Psychology of Learning and Comparative Psychology), who at the time I did not know was a Ph.D. student of T. C. Schneirla. It was only after completing my doctoral work at Kanas State University in 1971 that Schneirla came to play a critical role in my professional development.

Following my graduation from Brooklyn College in 1961, I was guided to graduate school at the University of Wichita by my professor for Advanced Experimental Psychology, Solomon Weinstock. He said I would find professors there who were Interbehaviorists, something I knew nothing about. It was only 46 years later, on reviewing a book about J. R. Kantor's Interbehaviorism "school" of psychology (Greenberg, 2008), that I finally discovered how he knew that about Wichita

---

* This chapter is published for the first time in this book.
† Wichita State University, Wichita, KS.

University – Weinstock was one of Kantor's doctoral students at Indiana University. Arriving in Wichita in 1962, I met two of my psychology professors who were also Kantor's doctoral students: David Herman and N. H. (Nicolas Henry) Pronko. Pronko was to become both my mentor and close friend in the years to come. While Weinstock, Herman, and Pronko are all long deceased, I believe they would be proud of the psychologist they helped me become.

After earning my MA degree at Wichita I found my way to the doctoral program at Kansas State University. My research advisor there was Fred Rohles. Interestingly, Fred had played a role in the earliest history of the American space program having trained two of our first space travelers, the chimpanzees Ham and Enos. His laboratory, the Institute for Environmental Research, was where I conducted my doctoral research on the effects of temperature and population density on aggression in two mouse strains (Greenberg, 1972a). (This research was prompted by the then recent outbreak of riots in major American cities, riots that took place in densely populated parts of those cities and in the hottest parts of summer.) The mice in my research were known to display different levels of aggression, high and low – a result it was believed of their genetics. My research, however, pointed to their aggression differences not to result from inherited aggression genetics, but rather from their different sensory thresholds. One strain was extremely sensitive to stimulation (high aggression mice), the other not so much (less aggressive mice).

Before beginning that research, I attempted to demonstrate a principle proposed by Schneirla that I learned about in my comparative psychology course from Howard Moltz who told us that in his imprinting experiments chicks never imprinted to objects that approached them, only to those that retreated from them. This is a reflection of Schneirla's (1939) Approach/Withdrawal hypothesis (A/W) which was to become an important way later I came to understand many aspects of behavior. [My attempts to imprint chicks were devastating failures (Greenberg, 1972b; Shapiro & Greenberg, 1974), something I discussed at a meeting with Eckhard Hess. Hess claimed the stimuli I was using to imprint chicks that were too small and not being perceived by them. I happened to have a film of the imprinting objects hitting the chicks as it approached and retreated from them. I remember saying something like "I think head bonking is clearly a readily perceptible event."]

One of my professors at Kansas State University was Harry Helson, who I feel comfortable identifying as being present at the founding of modern experimental psychology, his having taught at Cornell University along with Edward Titchener. Helson's (1964) important contribution to psychology is summarized in his book, *Adaptation Level Theory* (AL). An important feature of AL is that behavior is very much influenced by the context in which it occurs, importantly with respect to the intensity of stimuli influencing the behavior of interest. Relations between Schneirla's A/W hypothesis and Helson's AL theory seemed clear to me, especially as my thinking on these issues was developing and maturing. [Permit me this interesting aside about my rubbing shoulders with founders of modern experimental psychology. I was fortunate to tour Wolgang Kohler's lab at Dartmouth in 1965. Although he was not present I did observe his gold foil experimental set up. And my first attempts at imprinting were overseen by the biologist Al Guhl who first described the phenomenon of pecking orders in chickens in 1953 – e.g., Guhl & Ortman, 1953]

After completing my doctoral degree at Kansas State University in 1971, I fortuitously became associated with Ethel Tobach at the then influential Department of Animal Behavior at New York's American Museum of Natural History (Greenberg, Partridge, Weiss, & Pisula, 2004). Tobach was also a doctoral student of Schneirla's, and mainly because of her, I soon came under the influence of the approach to psychology he espoused.

At this stage in my education ideas which I had encountered and learned of were beginning to jell, especially connections between Kantor's and

Schneirla's formulations. Their ideas were summarized in an interesting paper by Lazar (1978) as follows:

> Schneirla's experimental work and Kantor's interbehavioral system complement one another. Their similar beliefs probably derive from commitments to naturalism, emphasis on comprehensive understanding in psychology, and the influence of Darwinian biology. Their naturalism explicitly emphasizes "integrative levels." Their "field" orientation eliminates biological reductionism and emphasizes comprehensive descriptions of empirical relationships among field factors. And, their developmental approach is integral to their explanation of behavior. (p. 177)

Ethel Tobach, with whom I was now working, was a giant in the field of comparative psychology (Greenberg, 2016). Accordingly, it is not a surprise that my association with her introduced me to scientists, important scientists, from many disciplines. But most importantly, it was she who immersed me in the psychology of Schneirla (Aronson et al., 1972; Schneirla, 1957). My association with Tobach resulted in our establishing the T. C. Schneirla Conference Series on the Evolution and Development of Behavior. This led to six conferences (and books – e.g., Greenberg & Tobach, 1984) which included contributions from a wide spectrum of scholars from multiple disciplines. The conference series was sponsored by a loosely formed group, the T. C. Schneirla Research Fund, which extended an invitation to the then Pennsylvania State University psychology professor Richard M. Lerner to become a member.

Meeting Lerner, and becoming familiar with his many contributions to developmental science (e.g. Lerner, 2018) closed the chain of ideas which brought me to edit this book with him. In this present book, The Heredity Hoax, the several links of the chain that brought me here include: Moltz, Pronko, Kantor, Helson, Tobach, Schneirla, and Lerner along with these accompanying ideas: context, evolution and epigenesis, integrative levels and emergence, approach/withdrawal, developmental systems theory (and others). Along the way, I encountered and was influenced by the works of Gilbert Gottlieb (e.g., Hood et al., 2010) and Zing Yang Kuo (1967), both of whom will appear frequently in this book (as will all of the crucial ideas I have already referenced). I mention them now to underscore their emphasis on the importance of how experiences enter importantly into behavioral ontogeny (e.g., Moltz, 1971). To be sure, many of those experiences are in Gottlieb's terms, nonobvious: "As for non obvious experiences, who could have dreamed that squirrel monkeys' innate fear of snakes derives from their earlier experience with live insects …? Or that chicks perceiving meal worms as edible morsels is dependent on their having seen their toes move…" (Gottlieb, 2001, p. 2)? And those influential developmental experiences begin long before hatching in chickens (Kuo, 1967) and similarly, prenatally in humans (e.g., Prechtl, 1988).

Now, at the end of my career, I am comfortable identifying myself as a developmental psychobiologist. As has previously been pointed out by others, the entire idea of development had been sidestepped and ignored and not really taken seriously by mainstream biologists and psychologists (Robert, 2004). Today, however, most understand psychology to be a developmental science, as the essays in this book make abundantly clear.

(Some) Principles of Psychology

I early came to understand psychology as a science, a natural science: "In other words, the teachings, talkings, walkings, 'thinkings,' 'imaginings,' etc. of people are just as *natural* as an object-in-gravitational-interaction-with-earth, one chemical substance-interacting-chemically-with-another, stomach-digesting-food, etc." (Pronko, 1947). Psychology is, of course, a higher science, and accordingly more complex in the hierarchy of the sciences than are physics and biology, which have been developing for far longer than the nascent science of psychology. Physics, for example, has been able to understand its phenomena with equations, biology almost so, though we still cannot write an equation for a peacock. But since psychology is still figuring out its variables, we cannot yet write equations. Nevertheless, we are far enough along in our development that we can specify principles which apply across the animal kingdom (Greenberg & Haraway, 2002). The themes imbedded in those principles are (1) the organizing principle of integrative levels, (2) that there is a tendency toward increased biological and behavioral complexity with recency of species appearance in evolutionary advance (e.g., Saunders & Ho, 1976), and (3) the contextual nature of determination of behavioral events. There are various descriptors of contemporary psychology, among them as a developmental science in which behavior is understood to be the result of the fusion of biological and psychosocial factors. Development is seen as a probabilistic outcome of a myriad of influences rather than as a product of preprogrammed genetic and biochemical causes. Nonlinear dynamic systems theory provides a theoretical language that is consistent with the description of behavioral development as I have come to understand (e.g., Partridge & Greenberg, 2010). Some of these principles are as follows:

Emergence

I began an earlier paper of mine (Greenberg, 2011, reprinted in Section 2 of this book) quoting Glinda, the Good Witch of the Wizard of Oz, who said, "It is always best to start at the beginning." Once again, I note that advice sometimes comes from unusual sources. In the present context, the beginning is the Big Bang, which is currently understood to be the beginning of *everything*!

I once pointed out that given enough time, hydrogen and helium, solely created at the beginning, become thinking beings – us. This has to do with the concept of emergence (Hodgson, 2000), an idea at the center of most, if not all, of the theoretical formulations of the various scholars who played a role in my path to this book. As Hodgson makes clear in his excellent discussion of the history and importance of the concept, at its simplest, emergence refers to the idea that properties within a complex system may not be reducible to constituent microelements. In other words, the whole is different than the sum of its parts. The astronomer Carl Sagan was noted for saying that, "All of the rocky and metallic material we stand on, the iron in our blood, the calcium in our teeth, the carbon in our genes were produced billions of years ago in the interior of a red giant star. *We are made of star stuff*" (Sagan, 1973, pp. 189–190, italics added).

Everyday examples of this process are found in water, composed of hydrogen and oxygen, neither of which have "wateriness" as a characteristic; and common table salt, made of sodium and chlorine, neither of which are themselves salty crystals (and one of which is a poison). Put together in different contexts, these substances become different – water and salt *emerge* from them. It is interesting how a physicist recently framed this idea: "…your genes alone aren't sufficient to define the person you are today. I mean all the necessary details that specify the way each part of your body, each single molecule, interacts with each other. That includes the countless little (and big) experiences that left marks in your brain, traces of the food you've eaten and air you've breathed, legacies of past illnesses, scars, and bruises. What makes you is this entire arrangement. Your you-ness, whatever exactly it is, emerges from the configuration of the particles you are constituted of" (Hossenfelder, 2022, p. 91).

Darwinian Evolution and Epigenesis

Evolution is an appropriate starting point for a discussion of the principles of psychology since in a sense everything in life can be said to begin here. Science speaks of cosmic evolution, the universe originating with the Big Bang, though, of course, the principles and mechanisms of cosmic evolution are different from those of biological evolution (Darwin, 1859). Biological evolution can be described as a change in the characteristics of populations of organisms over time. The concept applies to groups (species) and not to individuals; individuals develop over time but do not themselves evolve. Evolutionary changes are inherited – biologically and culturally – in the sense of being passed down across generations. Of course, the path to sentience is not inevitable, merely a result of emergent evolution.

The beauty of Darwin's understanding was captured by this statement of Ernst Mayr, one of the 20th century's leading evolutionary biologists:

> The most consequential change in man's view of the world, of living nature and of himself came with the introduction over a period of some 100 years beginning only in the 18th century, of the idea of change itself, of change over long periods of time: in a word, of evolution. Man's world view today is dominated by the knowledge that the universe, the stars, the earth and all living things have evolved through a long history that was not foreordained or programmed, a history of continual, gradual change shaped by more or less directional processes consistent with the laws of physics.
>
> (1979, p. 47)

Evolution, as understood today, is not the mechanical evolution of a generation ago but a creative evolution, i.e., growth, which is really such and which becomes more and more in the process. This growth is by its very nature epigenesis, suggesting that evolution is not a static process. This has recently been described as a developmental epigenesis (Lickliter & Witherington, 2017). There have been many definitions of epigenesis, including that of Kuo (1967), but the one that captures it best for this writer is that of probabilistic epigenesis by Moltz (1965):

> An epigenetic approach holds that all response systems are synthesized during ontogeny and that this synthesis involves the integrative influence of both intraorganic processes and extrinsic stimulative conditions. It considers gene effects to be contingent on environmental conditions and regards the genotype as capable of entering into different classes of relationships depending on the prevailing environmental context. In the epigeneticist's view, the environment is not benignly supportive, but actively implicated in determining the very structure and organization of each response system. (p. 44)

Anagenesis and Integrative Levels

Anagenesis is a concept that implies evolutionary progress (Aronson, 1984). As Gottlieb (1984, p. 454) put it: "The cardinal defining features of behavioral and psychological anagenesis [are] increases in ontogenetic plasticity and improvements in behavioral versatility, the latter through enhanced perceptual, cognitive, learning, social, and/or motor skills." The progress referred to here is evolutionary change, from simple to complex. This idea of hierarchy in evolution (i.e., an increase in complexity over geological time) was suggested by Saunders and Ho (1976) to be a second law of Darwinism. Indeed that the sciences themselves exist in such a hierarchy appears to have originated with Auguste Comte in the late 1800s (see Boorstein, 1998, p. 223) and was subsequently developed in the 20th century, conceptualized as the "concept of integrative levels." Aronson (1984) described this idea as a crucial organizing principle in science, which is "…a view of the universe as a family of hierarchies in which natural phenomena exist in levels of increasing organization and complexity" (p. 66). A contemporary physicist (Hossenfelder, 2022) has recognized this hierarchy in this way: "I will in the following refer to the areas of physics that study the fundamental laws as the *foundations of physics*. Everything else emerges from those fundamental laws, roughly in this order: atomic physics, chemistry, materials science, biology, psychology, sociology" (p. 87).

In their important book, Michel and Moore (1995) noted that T. C. Schneirla, who was among the preeminent comparative psychologists of the 20th century, applied this thinking to behavior invoking the idea of phyletic and behavioral levels, separated into two groups, one at which biological factors dominate behavior and one at which psychological and social principles become important. I have shown that these levels apply across the animal spectrum for all behaviors (Greenberg & Haraway, 2002).

The unity of the sciences is reflected in the increasing illustrations of the applications of physics to our understanding of complex behavior. "Unsurprisingly, ideas from physics strongly contribute to scientific explanations for how various animals find their orientation and navigate. The mechanisms are of a very diverse nature. ...In short, we can find that different disciplines of physics play a relevant role in animal navigation ...[including] mechanisms derived from quantum physics" (Sanjuan, 2023, p. 231). Given that I have always held that psychology is a natural science, it is relevant here that I include this long statement from Sanjuan (2023, pp. 231–232):

> Although the mechanisms of animal navigation are founded in principles and ideas from physics, I believe all this remains rather unknown to the physics community. After having decided on the subject, I was very excited to learn the recent news about the discovery of a long-lost letter from Albert Einstein addressed to Prof. Karl von Frisch.... on this subject .... In this unknown letter Einstein discusses the possibility of finding a link between physics and biology, writing, 'It is thinkable that the investigation of migratory birds and carrier pigeons may someday lead to the understanding of some physical process which is not yet known. (!)'

Psychology Is a Developmental Science

Whereas this is a book criticizing behavior genetics it is also a book about the concept of development, long ignored in psychology. Many of our authors were schooled about this by Schneirla's important chapter in Dale Harris' 1957 book, *The Concept of Development*. The full significance and impact of this principle is fully discussed and explored in the fourth edition of Richard M. Lerner's important book, *Concepts and Theories of Human Development* (2018).

I was surprised when I learned that an important concept in physics had a similarly important role to play in psychology – that of dynamic systems theory (Greenberg et al., 1999). Of course, that the systems approach is useful in biology as well is made clear in a recent book written from that perspective (Noble, 2006). Richard Lerner has made fruitful use of the systems approach in his ground breaking work on developmental issues, most significantly for Developmental Contextualism (e.g., Lerner, 1998) and, with Willis Overton, Relational Developmental Systems (Lerner & Overton, 2008; Overton & Lerner, 2014) Developmental Contextualism holds that organisms are fused with their environments, all features of which effect the developmental course of their behavior as well as their biology, making the idea of a genetic program unnecessary. The concept has its roots in the comparative developmental perspective of T. C Schneirla and stresses that bidirectional relations exist among the multiple levels of organization involved in human life (e.g., biology, psychology, social groups, and culture). The "rules" governing the developmental process are not locally encoded in some external control process but rather are derived from the mutual interactions of all the system variables as an organized whole. Thus, it is the process of development itself that drives the course that development takes. About this Lerner has held that individuals themselves produce and direct their own development (Lerner & Busch-Rossnagel, 1981).

Relational Developmental Systems is an organizing metatheory that recognizes the importance of *relations* between events. From this perspective, organisms are not simply collections of organs and other parts; rather they operate holistically, their parts are interdependent, regulating each other. The modus operandi of psychology is the study of dynamic relations between the multiple elements comprising psychological systems. The philosophical shift in psychological science away from a reductionistic and static orientation aimed at understanding the structural elements of behavior (i.e., psychological causes must be physically located in either a biological, psychological, or sociological entity) toward a dynamic, holistic, and relational orientation extends from meta-theory through the articulation of methodologies to the conceptualization of all psychological constructs. From this perspective, organisms are not separate from or independent of their environments but are *fused* with them. The environment, thus, is seen to be part of the organismic "whole."

Approach/Withdrawal (A/W)

First proposed by T. C. Schneirla in 1939 at a meeting of American Psychological Association the A/W concept is a way of understanding the origins of complex behavior in terms of simpler biphasic processes directly related to the quantitative characteristics of stimuli. Simply put, stimulus intensity directs behavior such that weak stimulation induces approach responses and intense stimulation induces withdrawal responses (Greenberg, 2017; Schneirla, 1939). The fundamental premise behind the approach/withdrawal concept is that these A/W response patterns underlie *all* complex adaptive responses and is a synthesis of several organizing concepts and principles, including stimulus intensity, levels of organization, plasticity and epigenesis, and maturation and experience. Each of these biphasic processes activated in an organism (from protozoa to higher vertebrates) depends on the nature of the stimulus and its intensity.

Why Am I Editing This Book?

I first became aware of the topic of this book from my earliest mentor, Henry Pronko (1980), who in an almost poetic analysis of "psycho-heredity" (1957) discussed, and analyzed, both sides of that term, which today we know as "behavior genetics," about which this book hopes to offer an important obituary. Some important themes addressed in this book are: it is not true that intelligence is inherited, swallows do not return to Capistrano, lemmings do not follow each other over cliffs down to the sea, bees do not dance in code, Monarch butterflies do not know their way to Mexico.

This is a book criticizing behavior genetics of which there are really two iterations (Greenberg, 2015; Panofsky, 2014). The first was originally envisioned by scientists in the 1950s who were attempting to establish a new field that would be devoted to the genetics of behavior. Because the eugenics of Nazi Germany was not far from their minds, they intended to avoid the controversial and political potential of the idea that behavior was a function of genetics: John Fuller and William Thompson, animal behavior psychologists, wrote the first text explicating the field in 1960, titled *Behavior Genetics*. Unfortunately, the original intent to avoid controversy did not last long leading to second iteration of behavior genetics, the one we are familiar with today which is the focus of the present book.

While many 20th-century psychologists played roles in behavior genetics' fall from grace, some original players came out of this history unscathed – Jerry Hirsch was one. While he was a founder of and early champion of behavior genetics, even serving as an editor of its namesake journal, *Behavior Genetics*, he was well known as an outspoken opponent of the political and racially charged controversies that dogged the discipline in the late 1900s. In fact, he had referred to behavior genetics as a science without scholarship (McGuire, 2008). [Forgive me for inserting a personal recognition of Hirsch, who I was honored to have introduced me for my Fellow's lecture at APA's Division 6 in 1996.]

Gilbert Gottlieb, an ardent critic of reductionism and genetic determinism, was among those comparative psychologists who understood psychology to be not a biological science, but, rather, a distinct psychological Science, as did Kantor and Schneirla. It is not a social science but a natural science, *a developmental science*, consistent and compatible with the principles of chemistry, biology, and physics. The legacy of Gottlieb is discussed and summarized in a 2010 book (Hood et al., 2010).

Among the reasons that behavior genetics is a flawed science is its blind adherence to the central dogma of molecular biology (Crick, 1970), that genetic information flows in one direction only – from inside to out, from the genotype to the phenotype. However, the central dogma is easy for teaching purposes but is not really the current understanding of what constitutes a gene. In his interesting, and very readable, polemic (his descriptor) against genetic reductionism, Noble (2006) points out that

> Strictly speaking ... to speak of a gene as 'the gene for x' is *always* incorrect. Many gene products, the proteins, must act together to generate biological functions at a high level. If we must use the expression 'gene for x' then we should at least add the plural and speak of 'genes for x'. Even this way of speaking is, however, seriously misleading. Not only do many genes co-operate in coding for the proteins that interact to produce any given biological function, each gene may also play a role in many different functions which makes it difficult to label genes with functions. (p. 9)

He further adds this: "Clearly, the simplistic view that genes 'dictate' the organism and its functions is just silly. It avoids the real challenge we face here, which is to understand the control processes that determine which proteins are produced ('expressed') and to what extent" (pp. 33–34). In addition, much of what we now know today about the differential individual genes has been understood as long 65 years ago. For example, Schneirla (1957) pointed out that in fruit flies, "...the same gene may influence the development of wing size and structure according to what temperature prevails during the development of the phenotype (p. 85)."

To be sure, arriving at this understanding is no easy task. In this context, it is worthwhile to heed the admonition of Lerner when he suggests that "We are at a point in the science of human development where we must move on to the more arduous task of understanding the integration of biological and contextual influences in terms of the developmental system of which they are a dynamic part (2004, p. 20)." Schneirla expressed a similar point in a different manner in his 1957 essay. In this context, I cite a book review about mental illness because of its parallels with the widespread belief in genetic reductionism: "We cling to reductive theories about brain chemistry because 'the reality – that mental illness is caused by an interplay between biological, genetic, psychological and environmental factors – is more difficult to conceptualize'" (Szalai, 2022, p. 9). Sloane said it more clearly in 1945 by referring to a "hypnotic spell that reductionism still casts over the scientific mind" (p. 218).

Indeed, the very meaning and identification of genes themselves is even now an open question. My earlier discussion of the role of genes in evolution and behavior (Greenberg & Partridge, 2010) was informed by E. F. Keller (2000), at the time perhaps the leading expert on what genes were. Interestingly, just what genes are has undergone several understandings since then (e.g., Gerstein et al., 2007). As they write, "To the early geneticists, a gene was an abstract entity..." (p. 670). Their discussion traces the meaning and understanding what a gene is through decades to conclude that, "The gene is a union of genomic sequences encoding a coherent set of potentially overlapping functional products...[there being] important implications of this definition" (p. 677).

I am confident that this is not yet the final word on this topic. It is, in fact, safe to say that the exact definition of the word gene has long been a source of scientific debate. The chapters in the current book all present careful criticisms of genetic determinism, a crucial tenet of which single genes are responsible for single traits, despite our still not fully understanding just what a gene is. But regarding this, "...each gene is a single player in a wonderfully intricate drama involving non-additive interactions of genes, proteins, hormones, food, and life experiences and leading to effects on a variety of cognitive and behavioral functions" (Berkowitz, 1996, p. 43).

We hope that this book will serve as an educational aid not only for students and the lay public but also for those scientists, psychologists, and biologists alike, who are simply unaware of recent developments in biology that render the standard accepted models as no longer valid as noted even by microbiologists such as Strohman (1997) and other psychologists (e.g., Gottlieb, 2004; Kaplan & Rogers, 2003; Lickliter & Honeycutt, 2003). This point is underscored by Mazzocchi (2008): "...the reductionist approach can no longer cope with both the enormous amount of information that comes from the so-called '-omics' sciences and technologies – genomics, proteomics, metabolomics and so on – and the astonishing complexity that they reveal" (p. 11).

I choose to ignore my own criticism here of genetic determinism, so well presented by the authors of the contents of this book, by myself and my co-editor, Richard Lerner, in our various publications. Instead, I conclude with this admonition from David Moore (2006), which sums up our thinking on the issues central to this book:

> Now, however, we understand that all of our characteristics – be they biological or psychological – are equally influenced by genetic and non-genetic factors. (p. 348)

References

Aronson, L. R. (1984). Levels of integration and organization: A re-evaluation of the evolutionary scale. In G. Greenberg and E. Tobach (Eds.), *Evolution of behavior and integrative levels* (pp. 57–81). Erlbaum.

Aronson, L. R., Tobach, E., Lehrman, D. S., & Rosenblatt, J. S. (Eds.). (1972). *Development and evolution of behavior: Essays in memory of T. C. Schneirla*. W. H. Freeman.

Berkowitz, A. (1996). Our genes, ourselves. *BioScience, 46* (1), 42–51.

Boorstein, D. J. (1998). *The seekers*. New York: Vintage.

Crafts, L. W., Schneirla, T. C., Robinson, E. E., & Gilbert, R. W. F. (1950). *Recent experiments in psychology* (2nd edition). McGraw Hill.

Crick, F. (1970). Central dogma of molecular biology. *Nature, 227*, 561–563.

Darwin, C. (1859). *The origin of species*. John Murray.

Fuller, J. L., & Thompson, W. R. (1960). *Behavior genetics*. Wiley.

Gerstein, M. B., Bruce, C., Rozowsky, J. S., Zheng, D., Du, J., Korbel, J. O., Emanuelsson, O., Zhang, Z. D., Weissman, S., & Snyder, M. (2007). What is a gene, post – ENCODE? History and updated definition. *Genome Research, 17*, 669–681.

Gottlieb, G. (1984). Evolutionary trends and evolutionary origins: Relevance to theory in comparative psychology. *Psychological Review, 91*, 448–456.

Gottlieb, G. (2001). The relevance of developmental-psychobiological metatheory to developmental neuropsychology. *Developmental Neuropsychology, 19*, 1–9

Gottlieb, G. (2004). Normally occurring environmental and behavioral influences on gene activity: From central dogma to probabilistic epigenesis. In C. G. Coll, E. L. Bearer & R. M. Lerner (Eds.), *Nature and nurture: The complex interplay of genetic and environmental influences on human behavior and development* (pp. 85–106). Erlbaum.

Greenberg, G. (1972a). The effects of ambient temperature and population density on aggression in two strains of mice, *Mus musculus*. *Behaviour, 42*, 119–131.

Greenberg, G. (1972b). Seven years of successful failures-to-replicate in an imprinting lab. Paper presented at the Animal Behavior Society, Terre Haute, Ind., October, 1972.

Greenberg, G. (2008). Psychology from the standpoint of an Interbehaviorist. A Review of B. D. Midgely and E. K. Morris (Eds.) (2006). *Modern Perspectives on J. R. Kantor and Interbehaviorism*, Reno, NV: Context Press. *The Psychological Record, 58*, 665–676.

Greenberg, G. (2011). The failure of biogenetic analysis in psychology: Why psychology is not a biological science. *Research in Human Development, 8*, 173–191.

Greenberg, G. (2015). The case against behavioral genetics. Review of A. Panofsky (2014). *Misbehaving science: Controversy and the development of behavior genetics*. Chicago: University of Chicago Press. *Developmental Psychobiology, 57*, 854–857.

Greenberg, G. (2016). In Memorium: Ethel Tobach 1921-2015. *American Psychologist, 71*, 75.

Greenberg, G. (2017). Approach/Withdrawal theory. In J. Vonk & T. K. Shackelford (Eds.), *Encyclopedia of animal cognition and behavior*. Springer International Publishing.

Greenberg, G., & Haraway, M. M. (2002). *Principles of comparative psychology*. Allyn and Bacon.

Greenberg, G., Partridge, T., Weiss, E., & Haraway, M. M. (1999). Integrative levels, the brain, and the emergence of complex behavior. *Review of General Psychology, 3*, 168–187.

Greenberg, G., Partridge, T., Weiss, E., & Pisula, W. (2004). Comparative Psychology, A New Perspective for the 21st Century: Up the Spiral Staircase. *Developmental Psychobiology, 44*(1), 1–15. https://doi.org/10.1002/dev.10153

Greenberg, G., & Partridge, T. (2010). Biology, evolution, and development. In W. F. Overton & R. M. Lerner (Eds.), *Handbook of life-span development: Cognition, biology, and methods* (Vol. 1, pp. 115–148). Wiley.

Greenberg, G., & Tobach, E. (Eds.). (1984). *Behavioral evolution and integrative levels*. Erlbaum.

Guhl, A., & Ortman, L. L. (1953). Visual patterns in the recognition of individuals among chickens. *The Condor, 55*(6), 287–298.

Helson, H. (1964). *Adaptation level theory: An experimental and systematic approach to behavior*. Harper.

Hodgson, G. M. (2000). The concept of emergence in social science: Its history and importance. *Emergence: A Journal of Complexity Issues in Organizations and Management, 2*(4), 65–77.

Hood, K. E., Halpern, C. T., Greenberg, G., & Lerner, R. M. (2010). Developmental systems, nature-nurture, and the role of genes in behavior and development: On the legacy of Gilbert Gottlieb. In E. K. Hood, C. T. Halpern, G. Greenberg, & R. M. Lerner (Eds.), *Handbook of developmental science, behavior, and genetics* (pp. 3–12). Wiley-Blackwell.

Hossenfelder, S. (2022). *Existential physics: A scientist's guide to life's biggest questions*. Viking.

Kaplan, G., & Rogers, L. J. (2003). *Gene worship: Moving beyond the nature/nurture debate over genes, brain, and gender*. Other Press.

Keller, E. F. (2000). *The century of the gene*. Harvard University Press

Kuo, Z. Y. (1967). *The dynamics of behavior development*. Random House.

Lazar, J. W. (1978). A comparison of some theoretical proposals of J. R. Kantor and T. C. Schneirla. *The Psychological Record, 24*, 177–190.

Lerner, R. M. (1998). Developmental contextualism. In G. Greenberg & M. M Haraway (Eds.), *Comparative psychology: A handbook* (pp. 88–97). Garland.

Lerner, R. M. (2004). Genes and the promotion of positive human development: Hereditarian versus developmental systems perspectives. In C. G. Coll, E. L. Bearer, & R. M. Lerner (Eds.), *Nature and nurture: The complex interplay of genetic and environmental influences on human behavior and development* (pp. 1–33). Lawrence Erlbaum Associates, Inc.

Lerner, R. M. (2018). *Concepts and theories of human development* (4th ed.). Routledge.

Lerner, R. M., & Busch-Rossnagel, N. A. (Eds.). (1981). *Individuals as producers of their development: A life-span perspective*. Academic Press.

Lerner, R. M., & Overton, W. F. (2008). Exemplifying the integrations of the relational developmental system: Synthesizing theory, research, and application to promote positive development and social justice. *Journal of Adolescent Research, 23*, 245–255.

Lickliter, R., & Honeycutt, H. (2003). Developmental dynamics: Toward a biologically plausible evolutionary psychology. *Psychological Bulletin, 129*, 819–835.

Lickliter, R., & Witherington, D. C. (2017). Towards a truly developmental epigenetics. *Human Development, 60*, 124–138.

Mayr, E. (1979). Evolution. *Scientific American, 239*(3), 46–55.

Mazzocchi, F. (2008). Complexity in biology. Exceeding the limits of reductionism and determinism using complexity theory. *EMBO Reports, 9*(1), 10–14.

McGuire, T. R. (2008). Jerry Hirsch (1922-2008): Obituary. *Genes, Brain & Behavior, 7*(8), 833–835.

Michel, G. F., & Moore, C. L. (1995). *Developmental psychobiology: An interdisciplinary science*. MIT Press.

Moltz, H. (1965). Contemporary instinct theory and the fixed action pattern. *Psychological Review, 72*(1), 27.

Moltz, H. (1971). *The ontogeny of behavior development.* Academic Press.

Moore, D. (2006). A very little bit of knowledge: Re-evaluating the meaning of the heritability of IQD. *Human development, 49,* 347–353.

Noble, D. (2006). *The music of life: Biology beyond the genes.* Oxford.

Overton, W. F., & Lerner, R. M. (2014). Fundamental concepts and methods in developmental science: A relational perspective. *Research in Human Development, 11*(1), 63–73.

Panofsky, A. (2014). *Misbehaving science: Controversy and the development of behavior genetics.* University of Chicago Press.

Partridge, T., & Greenberg, G. (2010). Contemporary ideas of physics and biology in Gilbert Gottlieb's epigenesis. In K. T. Hood, C. T. Halpern, G. Greenberg, & R. M. Lerner (Eds.), *Handbook of developmental science, behavior, and genetics* (pp. 166–202). Blackwell.

Prechtl, H. F. R. (1988). Developmental neurology of the fetus. *Bailliere's Clinical Obstetrics and Gynaecology, 2* (1), 21–36.

Pronko, N. H. (1947). A non-elementalistic approach to psychological data. *ETC: A Review of General Semantics, 4* (4), 285–289.

Pronko, N. H. (1957). "Heredity" and "environment" in biology and psychology. *The Psychological Record, 7,* 45–54.

Pronko, N. H. (1980). *Psychology from the standpoint of an interbehaviorist.* Brooks/Cole.

Robert, J. S. (2004). *Embryology, epigenesis, and evolution: Taking development seriously.* Cambridge University Press.

Sagan, C. (1973). *The Cosmic Connection: An Extraterrestrial Perspective.* Anchor Press/Doubleday.

Sanjuan, M. A. F. (2023). Physics of animal navigation. *The European Physical Journal: Special Topics, 232,* 231–235.

Saunders, P. T., & Ho, M. W. (1976). On the increase in complexity in evolution. *Journal of Theoretical Biology, 63,* 375–384.

Schneirla, T. C. (1939). A theoretical consideration of the basis for approach-withdrawal adjustments in behavior. *Psychological Bulletin, 37,* 501–502.

Schneirla, T. C. (1957). The concept of development in comparative psychology. In D. Harris (Ed.), *The concept of development* (pp. 78–108). University of Minnesota Press.

Shapiro, L. J., & Greenberg, G. (1974). Imprinting: Ornithological fact or psychological artifact? Paper presented at the American Ornithologist's Union, Norman, Ok., October, 1974.

Sloane, E. H. (1945). Reductionism. *Psychological Review, 52* (4), 214–223.

Strohman, R. C. (1997). The coming Kuhnian revolution in biology. *Nature Biotechnology, 15,* 194–200.

Szalai, J. (2022). State of mind: review of *Strangers to ourselves* by Rachel Aviv, *New York Times Book Review,* Sept 25, 2002. p. 9

Woodworth, R. S., & Schlosberg, H. (1955). *Experimental Psychology* (3rd ed.). Methuen.

SECTION II

METATHEORY AND THEORY ABOUT THE NATURE-NURTURE COACTION

EDITORS' INTRODUCTION

A multiplicity of theories, and a concern with the explanation of the processes of development, came to be predominant in developmental science by the beginning of the 1970s. Such concerns lead to the recognition that there is not just one way (one theory) to follow in attempting to put together the facts (the descriptions) of development. Rather, a pluralistic approach to such integration was seen as needed. When followed, such integrations may indicate that more descriptions are necessary. Thus, although observation (*empiricism*) is the basic feature of the *scientific method*, theoretical concerns guide descriptive endeavors. One gathers facts because one knows these facts will have a meaning within a particular theory. Moreover, since such theory-based research may proceed from any theoretical base, the data generated must be evaluated in terms of their use in advancing understanding of developmental change processes. These ideas burgeoned across the next three decades.

The 1970s, 1980s, and 1990s

The prominence of theory, the evaluation of theories by criteria of their usefulness in integrating the facts of development, and findings that developmental changes take many different forms at different points in time (and that such changes need to be understood from a diverse array of explanatory stances) led in the 1970s to an increasingly abstract concern with understanding the character of development. As a consequence, the decades of the 1970s, 1980s, and 1990s were characterized by the elaboration of numerous models of the association between the context of human life and the character of individual development.

At the same time, these models of person-context relations were being developed as frames for actual research about the linkages between individuals and their complex, multi-tiered settings. This research served as both a product and a producer of the enhancement of theories of person ⇔ context relations and of more nuanced understandings of the nature of the process through which human development was propelled by the associations individuals have with the ecology of human development.

Theory and Research about Infant Development

A major theoretical and empirical impetus for advances in the formulation of ideas pertinent to the growth, in the 1970s and 1980s, of a dynamic, relational developmental systems (RDS) metatheoretical orientation to the study of human development across the life span arose in the study of the first two years of life—infancy. The scholarship of two Michaels, Michael Lewis and Michael Lamb, were the chief architects of this advancement.

The work of Michael Lewis (e.g., 1972; Lewis & Feiring, 1978; Lewis & Lee-Painter, 1974; Lewis & Rosenblum, 1974; Pervin & Lewis, 1978) exemplifies the role that scholars of infant development played in devising models of person ⇔ context relations and of demonstrating their usefulness in research on human development. Building on the insights of Bell (1968) about the potential presence, in correlational data about socialization, of bidirectional influences between parents and children, Lewis and his colleagues launched a program of research that integrated model development with empirical research about infant–parent interaction.

For instance, in a book – *The Effect of the Infant on Its Caregiver* (Lewis & Rosenblum, 1974) – that represents a watershed event in the history of the study of human development through the use of person ⇔ context RDS-based models, Lewis argued that "Not only is the infant or child influenced by its social, political, economic and biological world, but in fact the child itself influences its world in turn" (Lewis & Rosenblum, 1974, p. xv) and maintained that "only through interaction can we study, without distortion, human behavior" (Lewis & Lee-Painter, 1974, p. 21). In his research with Lee-Painter, Lewis provided data supporting the use of a flow model of interaction in understanding, for instance, sequences of exchanges involving maternal and infant vocalizations as well as touch, looking, smiling, and play behaviors (e.g., Lewis & Lee-Painter, 1974, pp. 34–45).

Envisioning the relational, dynamic RDS-based models that would come to the fore in the study of human development by the end of the 20th century, Lewis and Lee-Painter (1974) foresaw that:

> What we need to develop are models dealing with interaction … or with the interaction independent of the elements …. This relational position not only requires that we deal with elements in interaction but also requires that we not consider the static quality of these interactions. Rather, it is necessary to study their flow with time … Exactly how this might be done is not at all clear. It may be necessary to consider a more metaphysical model, a circle in which there are neither elements nor beginnings/ends. (pp. 46–47)

One key instance of this influence arose in regard to the study of infant attachment. Here, the theory and research of Michael E. Lamb is a prime example of the use of person ⇔ context relational models in the study of infant attachment. Lamb and his colleagues (e.g., Lamb, 1977a, 1977b, 1977c, 1978a, 1978b; Lamb et al., 1985; Thompson & Lamb, 1986) approached the study of infant attachment within the context of the assumptions that:

1. Children have an influence on their "socializers" and are not simply the receptive foci for socializing forces.
2. Early sociopersonality development occurs in the context of a complex family system rather than in the context of the mother–infant dyad.
3. Social and psychological development is not confined to infancy and childhood but is a process that continues from birth to death.

(Lamb, 1978b, p. 137)

Within this conceptual framework, Lamb and his colleagues (e.g., Lamb et al., 1985) found that prior interpretations of infant attachment, which included "an emphasis on the formative significance of early experiences, a focus on unidirectional influences on the child, a tendency to view development within a narrow ecological context, and a search for universal processes of developmental change" (Thompson & Lamb, 1986, p. 1), were less powerful in accounting for the findings of attachment research than an interpretation associated with the sorts of person ⇔ context relational models burgeoning during the 1970s and 1980s. Accordingly, in a review of attachment research conducted through the mid-1980s, Lamb and his colleagues concluded that "reciprocal organism–environment influences, developmental plasticity, individual patterns of developmental change and broader contextual influences on development can better help to integrate and interpret the attachment literature, and may also provide new directions for study" (Thompson & Lamb, 1986, p. 1).

In essence, then, stimulated by scholars of infancy such as Michael Lewis and Michael Lamb, the study of human development during the 1970s and 1980s became increasingly focused on developing models, and conducting research, that enabled understanding of interactions, reciprocal influences, or bidirectional relations between individuals and the complex contexts within which they developed.

Damon (2006, pp. xiv) characterized these trends, as they were represented in the 1983, fourth edition of the *Handbook of Child Psychology* (Mussen, 1983):

> The grand old theories were breaking down. Piaget was still represented by his 1970 piece, but his influence was on the wane throughout the other chapters. Learning theory and psychoanalysis were scarcely mentioned. Yet the early theorizing had left its mark, in vestiges that were apparent in new approaches, and in the evident conceptual sophistication with which authors treated their material. No return to dust-bowl empiricism could be found anywhere in the set. Instead, a variety of classical and innovative ideas were coexisting: Ethology, neurobiology, information processing, attribution theory, cultural approaches, communications theory, behavioral genetics, sensory-perception models, psycholinguistics, sociolinguistics, discontinuous stage theories, and continuous memory theories all took their places, with none quite on center stage. Research topics now ranged from children's play to brain lateralization, from children's family life to the influences of school, day care, and disadvantageous risk factors. There also was coverage of the burgeoning attempts to use developmental theory as a basis for clinical and educational interventions. The interventions usually were described at the end of chapters that had discussed the research relevant to the particular intervention efforts, rather than in whole chapters dedicated specifically to issues of practice.

Accordingly, in order to understand the nature of changes in concepts and theories of human development that occurred from the 1970s through the 1990s, and how these changes resulted in a focus by the late 1990s and into the beginning of the 21st century on elaborating RDS-based theories of human development, it is important to focus on how the ideas involved in dynamic, RDS-based metatheory frames models of human development that were proven more useful than the ideas associated with mechanistic metatheory, especially in regard to genetic reductionist conceptions of human development.

The Role of Metatheory within Developmental Science

Reese and Overton (1970; Overton and Reese, 1973), among others (e.g., Lerner, 1976, 1978; Lerner & Kauffman, 1985, 1986; Riegel, 1975), pointed out that just as the facts and methods of science are to be understood as shaped by theory, scientific theories, in turn, are shaped by superordinate philosophies. Throughout the 1970s, repeated discussions occurred about how two major philosophical positions, the mechanistic and organismic models, shaped developmental theories (e.g., Lerner, 1976, 1978, 1979; Overton, 1973; Overton & Reese, 1973; Reese & Overton, 1970; Riegel, 1975, 1976a, 1976b; Sameroff, 1975). Each of these philosophical positions led to a different set, or "family," of theories.

For example, many mechanistic-type theories emphasized that even quite complex levels of human behavior can be reduced to rather simple elements: Basic stimulus-response (S-R) connections acquired through the "laws," or principles, of classical and operant conditioning (Baer, 1970, 1982; Bijou, 1976; Bijou & Baer, 1961; Skinner, 1938, 1950, 1971). Other mechanistic theories (e.g., Plomin, 1986; Rowe, 1994) sought to reduce social phenomena (e.g., parent-child relations, socialization) and psychological functioning (e.g., personality attributes, temperament style, or intelligence) to genetic inheritance (that is, to the complement of genes received at conception, the *genotype*).

In turn, many organismic-type theories emphasized that, as people develop, they pass through a universal and unchangeable sequence of qualitatively different phases, levels, or "stages," of development (e.g., Erikson, 1959, 1968; Freud, 1949, 1954; Piaget, 1950, 1970). Since each stage of development is different in kind from all others, organismic-oriented developmental scientists disagreed with mechanistically oriented ones about the appropriateness of reducing different levels (e.g., society, the family, and the individual) or different stages (e.g., the sensorimotor, preoperational, concrete operational, and formal operation stages posited by Piaget, 1970) to either one level (e.g., that of biology or, more specifically, genes) or to a common set of elements (e.g., stimulus-response connections formed through the "laws" of classical and operant conditioning), respectively.

The discussions prompted by the work of Reese and Overton (1970) and Overton and Reese (1973) involved, as well, consideration of the "family of theories" associated with each model. Although there are differences among family members (for example, Freud, in his organismic theory, emphasized emotional and personality development whereas Piaget, in his organismic theory, emphasized cognitive development), there is greater similarity among the theories within a family (e.g., the common stress on the qualitative, stage-like nature of development) than there is between theories associated with different families (e.g., mechanistically oriented Behavioristic theorists, such as Bijou and Baer, 1961, would deny the importance, indeed the reality, of qualitatively different stages in development).

Due to the philosophically based differences between families of theories derived from the organismic and mechanistic models, the 1970s, 1980s, and the early 1990s involved several discussions about the different stances held by members of one or another theoretical "family" regarding an array of key conceptual issues of development. Examples are the nature and nurture bases of development (Lehrman, 1970; Lerner, 1978; Overton, 1973); the quality, openness, and continuity of change (Brim & Kagan, 1980; Looft, 1973); appropriate methods for studying development (Baltes et al., 1977); and ultimately, the alternative truth criteria for establishing the "facts" of development (Dixon & Nesselroade, 1983; Reese & Overton, 1970).

This awareness of the philosophical bases of developmental theory, method, and data contributed to the consideration of additional models appropriate to the study of psychological development. In part, this consideration developed as a consequence of interest in integrating assumptions associated with theories derived from organismic and mechanistic models (Looft, 1973). For instance, Riegel (e.g., 1975, 1976a, 1976b) attempted to apply an historical model of development that seemed to include some features of organicism (e.g., the active organism) and some features of mechanism (e.g., the active environment). In turn, Riegel's interest in continual, reciprocal relations between an active organism and its active context (and not in either element per se), and the concern with these relations as they exist on all levels of analysis, formed a basis for his proposing a dialectical model of human development (Riegel, 1975, 1976a, 1976b).

Indeed, other developmental scientists, focusing too on the implications for theory of viewing distinct levels of analysis as reciprocally interactive, proposed related models, ones termed transactional (Sameroff, 1975, 1983), relational (Looft, 1973), or developmental contextual (Lerner, 1978, 1984, 1986). Laying the seeds of what would become, by the late 1990s, RDS-based models, this philosophically driven interest in bidirectional organism-context relations led several theorists to explore the application of a change-oriented contextual model to the collection and interpretation of developmental (and other psychological) data (see especially the volumes on contextualism edited by Hayes et al., 1993; Rosnow & Georgoudi, 1986).

The discussions about the influence of the organismic and mechanistic models led developmental psychologists to recognize that the stances scientists took in regard to key issues of human development—such as whether, because of the appropriateness of reducing all behavior to common elements, there is a sameness, or continuity, across life *or* whether, because of the existence of new stages, there is change, or discontinuity, across life—depended ultimately on philosophical positions. That is, developmental scientists recognized that a main (if not the ultimate) reason scientists had different positions regarding concepts and theories of development was that they were committed to different philosophies (e.g., see Kuhn, 1962, 1970; Overton, 1998, 2006, 2015). In other words, differences about these issues were underlain by non-empirical, philosophical differences and could not therefore be readily decided on the basis of data. Indeed, Reese and Overton (1970; Overton & Reese 1973) pointed out that developmental scientists working from different philosophical positions would have different truth criteria for establishing the "facts" of development, because what is a fact to one scientist may not be accepted as a legitimate or relevant fact by another. As a consequence, because of basic philosophical disagreements, disputes *across* philosophical positions could not be settled by facts.

In short, the interest that arose in the 1970s and that developed across the next two decades in the philosophical bases of theories of development also led many developmental scientists to explore the potential use of philosophies other than the organismic and mechanistic. The considerations of these ideas resulted in revised ways of thinking about the linkages between the developing individual and his or her changing context.

For instance, Klaus F. Riegel (1975, 1976a, 1976b) was in many ways both the intellectual leader of and catalyst for the exploration in the 1970s of the use of alternative models for the study of human development. This influence was the case, first, because he was a prolific and passionate writer—his book, *Psychology Mon Amour: A Countertext* (Riegel, 1978), being an excellent case in point—and, second, because he was editor of the journal *Human Development*, the prime outlet for theoretical scholarship in the field of human development.

Of the many important contributions of Riegel's scholarship, two are particularly pertinent to the present discussion. First, his dialectical model emphasized that the primary goal of a developmental analysis was the study of change, not stasis. Second, his model emphasized that any level of organization—from inner-biological, through individual-psychological and physical-environmental, to the sociocultural—influences and is influenced by all other levels. Thus, Riegel (1975, 1976a, 1976b) "developmentalized" and "contextualized" the study of the person by embedding the individual within an integrated and changing matrix of influences derived from multiple levels of organization.

Riegel (1973, 1975, 1976a, 1976b) proposed that dialectical philosophy could be used to devise a unique theory of development, one that did not focus on the organism (and, for instance, its genes or its maturation-guided progression through stages) or just on the environment (as, for instance, the source of the stimulation that provided the basis of S-R connections). Instead, Riegel (1975, 1976a, 1976b) hoped to forge a dialectical psychology that focused on the *relations* between developing organisms and their changing environments. Riegel emphasized that such relations involved continual conflicts among variables from several levels of "being" (or levels of organization of life phenomena). For example, he assumed that development involved constant changes among the multiple, reciprocally related inner-biological, individual-psychological, physical-environmental, and sociocultural levels of analysis.

Riegel's model of dialectic development was an important instance of the growing interest during this period in the interactive role of the changing physical and social context for human behavior and development. Riegel's ideas, as well as those of Sameroff (1975), Looft (1973), Lerner (1978, 1979), and others (e.g., Bronfenbrenner, 1977, 1979), were similar in their emphasis on change and context—and, to this extent, may be interpreted as being part of a common "family" of models. At this writing, these ideas are clearly linked to RDS metatheory. However, in the context of the 1970s and early 1980s what was clear was perhaps only that, as scholarship about this family of theories advanced, there were important distinctions among family members.

For instance, Riegel's (1975, 1976a, 1976b) ideas about context and change differed from those of other family members with respect to the format of change. The nature of dialectical change, which is always in the same direction, that of a synthesis between two "conflicting" opposites (termed thesis and antithesis), may be more compatible with the view of change found in organicism than that of philosophical position termed contextualism (Dixon et al., 1991; Pepper, 1942). Contextualism promotes a view of change that is dispersive, that is, that can occur in innumerable directions (Pepper, 1942). On the other hand, organismic change is always unidirectional; it is directed to a single end-point or goal (Pepper, 1942). Thus, when applied to the life span, Riegel's (1976b) dialectical view may have had more in common with organismic views (e.g., Alexander & Langer, 1990; Chapman, 1988a, 1988b; Piaget, 1970) than with contextual ones.

To counter this criticism, Riegel (1976b) tried to argue that dialecticism constituted a model of development distinct from organicism. In his view, the dialectical theory of cognitive development differed from the one of Piaget (1950, 1970). For example, whereas Piaget proposed that after the development of the last stage of development in his theory—a stage he termed "formal operations"—no new cognitive structure emerged, Riegel argued that the dialectic resulted in a fifth, open-ended stage of cognitive development. However, given that both the organismic model of Piaget and Riegel's dialectic model emphasized a single format and direction for developmental change, it was difficult for Riegel to maintain that at its core, in regard to the character of the main process of developmental change, the two positions were different.

Moreover, Riegel did not attend to the similarities and differences between his dialectical model and theories that emphasized the contextual philosophy or worldview (Pepper, 1942), although both sets of ideas emphasized change through individual-context relations. Given its problem with discriminating itself from organicism and the availability of a model for theory building—contextualism—which afforded a different, and more plastic view of change, the dialectical model of Riegel did not remain a conception of prime focus among developmental scholars beyond the 1970s and early 1980s. Nevertheless, attention to Riegel's ideas did facilitate the interest of the community of developmental scholars in considering other theoretical models of change through individual-context relations. Thus, at least in this respect, his dialectical model can be seen as compatible with the attention paid during these decades to contextualism (Hultsch & Hickey, 1978; Lerner & Kauffman, 1985, 1986; Lerner et al., 1980).

In contextualism, developmental changes occur as a consequence of reciprocal (bidirectional) relations between the active organism and active context. Just as the context changes the individual, the individual changes the context. As such, by acting to change a source of their own development—by being both products and producers of their context—individuals affect their own development (Bell, 1968; Bell & Harper, 1977; Lerner, 1982; Lerner & Busch-Rossnagel, 1981; Lewis & Rosenblum, 1974; Schneirla, 1957).

Contextualism found many adherents among developmental scientists across the 1970s to (at least the early) 1990s (Lerner et al., 1983), as well as many critics (e.g., see Capaldi & Proctor, 1999, for a review, and see Kendler, 1986, as an example). Nevertheless, because of the potential to provide ideas that could more usefully understand (e.g., account for more variance pertinent to) the dynamic (that is, the multi-level and bidirectional) relationships between the developing individual and variables associated with his or her biological, interpersonal, societal, cultural, and historical contexts, developmental scholars continued to explore the use of models of person-context relations associated with contextualism, if not specifically Riegel's dialecticism. Two major examples of such approaches were the bioecological model of human development (Bronfenbrenner, 1977, 1979) and the life-span developmental psychology perspective (e.g., Baltes et al., 1980). Both of these theoretical approaches to the relations between individuals and their contexts are instances of ideas that evolved from the contextualism of the 1970s and 1980s into the relationism associated by the late 1990s with concepts linked to RDS metatheory.

In addition to the bioecological and the life-span perspectives, other quite important instances of the influence of contextual thinking arose in the 1970s. Coming from a remarkably diverse array of intellectual traditions, these instances suggested that contextualism both offered a conceptual framework for asking ecologically meaningful questions and suggested methodological strategies for doing new and potentially more useful empirical research.

For example, in 1974 James J. Jenkins rejected the mechanistic model he had used to guide his associationist view of memory. He suggested that instead of this traditionally American approach to the study of memory, a contextual approach be adopted (Jenkins, 1974). He argued that "what memory is depends on context" (Jenkins, 1974, p. 789) and defended this view by presenting the results of several empirical studies that demonstrated that:

What is remembered in a given situation depends on the physical and the psychological context in which the event was experienced, the knowledge and skills that the subject brings to the context, the situation in which we ask for evidence for remembering, and the relation of what the subject remembers to what the experimenter demands.

(Jenkins, 1974, p. 793)

Jenkins (1974, p. 787) noted that, to deal adequately with all these sources of variation, means that "being a psychologist is going to be much more difficult than we used to think it to be." In part, this difficulty arises because there is no one mode of analysis, or methodological strategy, that suggests itself as always useful for assessment of all the levels of analysis involved at all historical moments in the memory process. Thus, not only is methodological pluralism promoted from this contextual perspective, but the criterion of usefulness must also be employed when deciding if a particular methodological strategy is appropriate. That is, reflective of the specificity principle (Bornstein, 2006, 2017), a scientist must decide: "What kind of an analysis of memory will be useful to you in the kinds of problems you are facing. What kinds of events concern you?" (Jenkins, 1974, p. 794). In other words, Jenkins (1974, p. 794) believed that:

The important thing is to pick the right kinds of events for your purposes. And it is true in this view that a whole theory of an experiment can be elaborated without contributing in an important way to the science because the situation is artificial and nonrepresentative in just the senses that determine its peculiar phenomena. In short, contextualism stresses relating one's laboratory problems to the ecologically valid problems of everyday life.

Thus Jenkins (1974) reaches a conclusion quite compatible with the one Bronfenbrenner (1977) reached. Clearly, the "spirit of the times" (the *zeitgeist*) in the 1970s set social and behavioral science on an intellectual course that prized the ecological validity of theory-predicated research.

In addition, Sarbin's (1977) dramaturgical model of psychological functioning had marked similarity to Riegel's (1975, 1976a, 1976b) dialectical model, as well as to features of the life-span perspective. This model is a technique that, through use of the notion of emplotment, attempts to capture the sequence of reciprocal events between individuals and their changing social contexts. Sarbin (1977) applied his contextualism model to the analysis of data sets pertinent to the genesis of schizophrenia, to the nature of hypnosis, and to the characteristics of imagination, in order to illustrate the integrative utility of contextually derived ideas. His work served to illustrate that contextual ideas can be useful in understanding an array of psychological processes, ranging from those associated with cognition and affect to those traditionally labeled as personality and social ones. Moreover, Sarbin emphasized that the interrelation among processes cannot only be integrated by contextual thinking but, in fact, needs to be appreciated if both adaptive and non-adaptive outcomes of person-context relations are to be understood. For example, Sarbin suggested that in the understanding of the bases of schizophrenia, the contextualist will, as compared to the mechanist, take:

> as his unit, not schizophrenia, not improper conduct, not the rules of society, but as much of the total context as he can assimilate. His minimal unit of study would be the man who acted as if he believed he could travel unaided through space *and* the person or persons who passed judgment on such claims.
>
> (Sarbin, 1977, p. 25)

Thus, as in Riegel's (1976a) model of crises being generated by conflicts among different developmental levels, Sarbin (1977) searched for the bases of adaptive and maladaptive functioning *not* within the realm of individual ("personological") functioning, but rather within the domain of the conflicts and crises created by the degrees of "goodness of fit" (Thomas & Chess, 1977) a person experiences in his or her relations with the social context. Sarbin also sees the relevance of his ideas to those put forth in other calls for contextualist thinking. In fact, Jenkins (1974), as well as Cronbach (1975) and Gergen (1973), made consonant appeals.

Indeed, these latter two papers are not the only instances of appeals for contextualism in the 1970s; other prominent examples may be cited. The *American Psychologist* is the journal of the American Psychological Association, designed to publish articles of current and broad interest to psychologists. The scholarship by Jenkins (1974), Riegel (1976a), and Bronfenbrenner (1977) were published in the *American Psychologist*, and in the last three years of the 1970s three additional papers appeared in the *American Psychologist* that, in different ways, made an appeal for contextualism.

Walter Mischel (1977), arguing for considering the role of context in understanding personality, suggested that, unless one considered the changing- and bidirectional-relations between people and their worlds, an adequate understanding of consistency and change in the person could not be attained. Petrinovich (1979) promoted "probabilistic functionalism"—an idea drawn from Egon Brunswik's (1955) notion of ecological validity—which called for an array of methodological strategies not dissimilar in intent to those suggested in calls for methodological pluralism put forth by contextual thinkers such as Bronfenbrenner (1977) and Jenkins (1974), among others (e.g., Lerner et al., 1980; and see also Lerner & Callina, 2014). Most interestingly, Albert Bandura (1978) reconceptualized his social-learning theory as involving causal processes that are based on reciprocal determinism. That is, consistent with key emphases in contextualism, Bandura asserted that, "from this perspective, psychological functioning involves a continuous reciprocal interaction between behavioral, cognitive, and environmental influences" (Bandura, 1978, p. 344).

As interest in contextualism grew in the 1970s and 1980s, the shortcomings of a completely contextual approach to human development became clearer as well (e.g., Overton, 1998). Based on Pepper's (1942) assertion that contextualism is a completely dispersive world hypothesis and, as such, provided no necessary systematicity or organization to successive changes across life (e.g., see Lerner & Kauffman, 1985), many developmental scientists began to seek ways to interrelate contextualism with organicism (e.g., Overton, 1984, 1991a, 1991b, 1994). These conceptual explorations resulted in an intellectual movement from contextualism to RDS-based ideas (Overton, 1998, 2006, 2010). In many cases, the substantive focus for this work was individual ⇔ context relations.

In sum, empirical findings emerging throughout the 1970s and 1980s indicated that organism-centered models of developmental change could not account for the multidirectionality of ontogenetic change. Instead, the context of human development needed to be incorporated into any adequate analysis of the diversity of developmental trajectories which was seen to characterize the life course. However, this context was not the simplistic, S-R environment of learning theorists (see White, 1970) or of those taking a reductionist and mechanist behavior-analytic approach to development (e.g., Bijou, 1976; Bijou & Baer, 1961).

Indeed, the multiple levels of the context, which seem linked to the individual level over the course of the life span, cannot be reduced to the molecular elements of any extant mechanistic-behavioristic theory (Lerner & Kauffman, 1985). Instead, organism and context may be seen as two distinct, yet inextricably linked, components of the *system of relationships comprising the ecology of human life* (e.g., Bronfenbrenner, 1979, 2005; Bronfenbrenner & Morris, 1998; Ford, 1987; Ford & Lerner, 1992).

Thus, and in support of the idea that research and theory during the 1970s, 1980s, and 1990s were mutually influential, the empirical findings about individual-context relations meshed quite well with the view of organism and context being forwarded in the dialectical, bioecological, and life-span views of human development. The view of human development that emerged from this empirical-theoretical synergy was one wherein theoretical reductionism was eschewed in favor of models that depicted changing, synthetic, and systematic relations among qualitatively distinct levels of analysis. By the mid-1990s neither organism nor context alone was regarded as sufficient to account for the course of individual development. Intellectual excitement about contextualism per se, as a possible metatheory or world hypothesis (Pepper, 1942) that alone could frame developmental science, had evolved (transformed) into intellectual excitement about ideas that integrated individual and context into a developmental system, one that focused on individual ⇔ relations. Thus, the combined influence of research and theory during these decades was to set the stage for the elaboration, in the mid-to-late 1990s and into the next century of theories that viewed individual and context as integrated systemically across life.

In sum, as the decade of the 1980s ended and 1990s progressed, Paul Mussen's (1970) view of developmental science at the beginning of the 1970s—that the field placed its emphasis on explanations of the process of development—was both validated and extended. Mussen alerted developmental scientists to the burgeoning interest not in either structure, function, or content per se but to change, to the processes through which change occurs, and thus on the means through which structures transform and functions evolve over the course of human life. His vision of and for the field presaged what emerged by the late 1990s to be at the cutting-edge of developmental theory: A focus on the process through which the individual's engagement with his or her context constitutes the basic process of human development.

Into the 21st Century: The Emergence of Relational Developmental Systems

The interest that had emerged by the end of the 1980s and first years of the 1990s in understanding the dynamic relation between individual and context was, during the mid-to-late 1990s and into the first decade of the 21st century, brought to a more abstract level, one concerned with understanding the character of the integration of the levels of organization comprising the context, or bioecology, of human development (Lerner, 1998a, 1998b, 2006a, 2006b). This concern was represented by reciprocal or dynamic conceptions of process, of how structures function and how functions are structured over time and, interestingly, by the elaboration of theoretical models that were not tied necessarily to a particular content domain but rather were focused on understanding the broader developmental system within which all dimensions of individual development emerged (e.g., Ford & Lerner, 1992; Gottlieb, 1992, 1997; Sameroff, 1983; Thelen & Smith, 1994, 1998, 2006). In other words, although particular empirical issues or substantive foci (e.g., biological development; perceptual and motor development; personality, affective, and social development; successful aging; wisdom; extraordinary cognitive achievements; intentional behavior and goal pursuit; language acquisition; the development of diverse children; psychological complexity; spiritual and religious development; or positive human development) lent themselves readily as exemplary sample cases of the processes depicted in a given theory (Lerner, 1998a, 2006a), the theoretical models that were forwarded within the mid-to-late 1990s and the early 2000s were superordinately concerned with elucidating the character of the individual ⇔ context (relational, integrative) developmental system (Lerner, 1998b, 2006b).

For example, as illustrated by most of the chapters in Volume 1 of the sixth edition of the *Handbook of Child Psychology* (Damon, 2006), a volume entitled "Theoretical Models of Human Development" (Lerner, 2006a), the theories forwarded by contributors illustrated that the interest and, arguably, the power of these instances of developmental theories lay in their ability to transcend a unidimensional portrayal of the developing person (e.g., the person seen from the vantage point of only cognitions, or emotions, or stimulus-response connections; for example, see Piaget, 1970; Freud, 1949; and Bijou & Baer, 1961, respectively). That is, in these theories the person was neither biologized, psychologized, nor sociologized. Rather, the individual was "systemized," that is, his or her development was conceptualized as embedded within an integrated matrix of variables derived from multiple levels of organization. Across these theories, development was conceptualized as deriving from the dynamic relations among the variables within this multi-tiered matrix.

Moreover, the theories represented in Volume 1 of the 2006 edition of the *Handbook* (Lerner, 2006a) did not use the polarities, or splits, that engaged developmental theory in the past, most notably nature/nurture. That is, the theories did not employ split depictions of developmental processes along what were argued to be conceptually implausible and empirically counterfactual lines (Gollin, 1981; Overton, 2006); the theories did not force counterproductive choices between false opposites. Rather, the theories were united by a common interest in gaining insight into the integrations that exist among the multiple levels of organization involved in human development (e.g., see Baltes et al., 2006; Benson et al., 2006; Brandtstädter, 2006; Bronfenbrenner & Morris, 2006; Cairns & Cairns, 2006; Elder & Shanahan, 2006; Fischer & Bidell, 2006; Gottlieb et al., 2006; Lerner, 2006b; Magnusson & Stattin, 2006; Oser et al., 2006; Overton, 2006; Rathunde & Csikszentmihalyi, 2006; Shweder et al., 2006; Spencer, 2006; Thelen & Smith, 2006; Valsiner, 2006).

As noted by Cairns and Cairns (2006, p. 155) in their historical review of developmental psychology within this volume of the *Handbook*:

> Today, the split conceptions of nature and nurture, and of the reductionist formulations associated with either a nature (e.g., sociobiology or behavior genetics) or a nurture (e.g., Behaviorism or functional analysis approaches) [perspective] have passed from the main stream of theoretical and scientific interest (e.g., see Gottlieb et al., 2006; Overton, 2006) and – through the lens of various

versions of developmental systems theories (e.g., see Fischer & Bidell, 2006; Magnusson & Stattin, 2006; Thelen & Smith, 2006) – scientific attention has focused on models and methods that now promise to begin to address the question of how "both causes work together" at the level of biology, interactions, and social networks.

Integrative, RDS-based theories had come to the fore of developmental science by 2006 and provided the field with models more complex than their organismic or mechanistic predecessors. These theoretical models were also more nuanced, more flexible, more balanced, and less susceptible to extravagant, or even absurd, claims (for instance, that "nature," split from "nurture" can shape the course of human development; that there is a gene for altruism, militarism, intelligence, and even television watching; or that, when the social context is demonstrated to affect development, the influence can be reduced to a genetic one; e.g., Lorenz, 1966; Plomin, 1986; Plomin et al., 1990; Rowe, 1994; Rushton, 1987, 2000). RDS-based theories had become, by the early years of the 21st century, clear indicators of the mainstream and distinctive features of the field. Indeed, the centrality of systemic and multidisciplinary thinking, spanning and integrating basic and applied scholarship, has been associated with a change in the very label of the field during this time period.

In addition to the seventh edition of the *Handbook of Child Psychology and Developmental Science* (Lerner, 2015c), more and more scholars of human development referred to their field as developmental science (e.g., see Cairns & Cairns, 2006; Magnusson & Stattin, 2006). Moreover, at least one leading graduate textbook in the field has changed its title from *Developmental psychology: An advanced textbook* (Bornstein & Lamb, 1999) to *Developmental science: An advanced textbook* (Bornstein & Lamb, 2005).

In sum, the change of name for the field studying the human life span reflects in large part key intellectual changes across about the past two decades or so at this writing: (a) the certain demise of split conceptions of the nature-nurture issue, and of reductionist approaches to either nature formulations (sociobiology, evolutionary developmental psychology [EDP], behavior genetics) or to nurture formulations (e.g., S-R [stimulus-response] models or functional analysis approaches) (Overton, 2006, 2010, 2015; Overton & Müller, 2013; Valsiner, 2006); (b) the ascendancy of a focus on RDS-based models, conceptions that seek to fuse systemically the levels of organization involved in the ecology of human development (from biology and physiology through culture and history; e.g., see Baltes et al., 2006; Elder et al., 2015; Gottlieb et al., 2006; Thelen & Smith, 2006; and the chapters across the four volumes of the 7th, 2015 edition of the *Handbook of Child Psychology and Developmental Science*; Lerner, 2015c); and (c) the emphasis on *relations* among levels, and not on the main effects of any level itself, as constituting the fundamental units of analysis of developmental analysis (e.g., see Brandtstädter, 2006; Bronfenbrenner & Morris, 2006; Fischer & Bidell, 2006; Magnusson & Stattin, 2006; Mascolo & Fischer, 2015; Noble, 2015; Rathunde & Csikszentmihalyi, 2006).

Implications of the RDS Metatheory for Developmental Science

The ascendancy of the process-relational paradigm and of RDS metatheory frame for the conduct of developmental science has been a product and a producer of a shift in the philosophy of science framing discourse within the field (Overton, 2003, 2006, 2010, 2015). RDS metatheory has served as a product and a producer of developmental-systems thinking, that has rejected the idea derived from the positivist and reductionist notion that the universe is uniform and permanent—that the study of human behavior should be aimed at identifying nomothetic laws that pertain to the generic human being. This idea has been replaced by an emphasis on the individual, on the importance of attempting to identify both differential and potentially idiographic laws as involved in the course of human life (e.g., Block, 1971; Magnusson, 1999a, 1999b; Molenaar & Nesselroade, 2014, 2015; Rose, 2016; Rose et al., 2013), and on regarding the individual as an active producer of his or her own development (Brandtstädter, 1999, 2006; Lerner, 1982; Lerner & Busch-Rossnagel, 1981; Lerner et al., 2015; Lerner et al., 2005; Mascolo & Fischer, 2015; Rathunde & Csikszentmihalyi, 2006). Similarly, the changed philosophical grounding of the field has altered developmental science from a field that enacted research as if time and place were irrelevant to the existence and operation of laws of behavioral development to a field that has sought to identify the role of contextual embeddedness and temporality in shaping the developmental trajectories of diverse individuals and groups (e.g., see Baltes et al., 2006; Bronfenbrenner & Morris, 2006; Elder et al., 1993; Elder & Shanahan, 2006; Elder et al., 2015).

Arguably, the most profound impact of the RDS metatheory on the practice of developmental science has occurred in the conceptualization of diversity and of interindividual differences in developmental trajectories (Bornstein, 2006, 2017; Spencer, 2006). From the perspective of the uniformity and permanence assumptions, individual differences—diversity—were seen, at best, through a lens of error variance, as prima facie proof of a lack of experimental control or of inadequate measurement. At worst, diversity across time or place, or in the individual differences among people, was regarded as an indication that a deficit was present. Either the person doing the research was remiss for using a research design or measurement model that was replete with error (with a lack of experimental control sufficient to eliminate interindividual differences), *or* the people who varied from the norms associated with the generic human being—the relations among variables that were generalizable across time and place—were in some way deficient (cf. Gould, 1981, 1996; Rose, 2016). They were, to at least some observers, less than normatively human (e.g., see Belsky, 2014). Things have changed, however.

From Deficit to Diversity in Developmental Science

For colleagues trained in developmental science within the 21st century, the prior philosophical grounding and associated philosophical assumptions about science may seem either unbelievably naive or simply quaint vestiges from an unenlightened past. In what, for the history of science, is a very short period (Cairns & Cairns, 2006), participants in the field of human development have seen a sea change that perhaps qualifies as a true paradigm shift in what is thought of as the nature of human nature and in the appreciation of time, place, and individual diversity for understanding the laws of human behavior and development (Bronfenbrenner & Morris, 2006; Elder & Shanahan, 2006; Overton, 2006, 2010, 2015; Shweder et al., 2006; Valsiner, 2006). With respect to the current book, the paradigm shift from materialism and the reductionist world view has brought us to the age of complexity science and

> The methods employed by complexity scientists [e.g., the editors and contributors to this book] … *achieve something that the reductionist approach is fundamentally incapable of*: They allow us to understand how nature's building blocks spontaneously self-assemble through a synergistic dance that creates wonderous emergent phenomena, like life, mind, and civilization.
>
> (Azarian, 2022, p. 4, italics added).

The publication in 1998 of the fifth edition of the *Handbook of Child Psychology*, edited by William Damon, may have been the first major reference work in developmental science that heralded that major contributions to the study of human development rejected the hegemony of positivism and reductionism. As evidenced by the chapters in all four volumes of the Damon (1998) *Handbook*, and arguably especially in Volume 1, *Theoretical Models of Human Development* (Lerner, 1998a), the majority of the scholarship then defining the cutting edge of the field of human development was associated with the sorts of RDS-based models of human development that fill the pages of the 2006 and 2015 editions of this volume of the *Handbook* and that, as projected by Cairns (1998), were at the threshold of their time of ascendancy within developmental science. As we indicated in our discussion of the subsequent chapter by Cairns and Cairns (2006), within less than a decade the prediction by Cairns (1998) had been instantiated.

The view of the world that emerged from the chapters in the fifth edition of Volume 1 of the *Handbook* (Lerner, 1998a), and that was confirmed across the corresponding chapters of the 2006 and 2015 editions of this work (including those chapters represented in earlier editions and those chapters new to an edition) is that the universe is dynamic and variegated. Time and place, therefore, are matters of substance, not error; and to understand human development, scholars must appreciate how variables associated with person, place, and time coalesce to shape the structure and function of behavior and its systematic and successive change.

Accordingly, diversity of person and context, and the idiographic and non-ergodic character of human development, have moved into the foreground of the analysis of human development (e.g., Lerner, 1991; Molenaar, 2004; Molenaar & Nesselroade, 2015; Rose, 2016). The dynamic, RDS-based perspective framing the study of human development at this writing does not reject the idea that there may be general laws of human development. Instead, there is an insistence on the presence of individual laws as well as a conviction that any generalizations about groups or humanity as a whole require empirical verification, not pre-empirical stipulation (Magnusson & Stattin, 2006; Molenaar & Nesselroade, 2015; Overton, 2006, 2015; Rose, 2016).

To paraphrase the insight of Kluckhohn and Murray (1948), made more than a half century ago, all people are like all other people, all people are like some other people, and each person is like no other person. Today, then, the science of human development recognizes that there are idiographic, differential, and nomothetic laws of human behavior and development (e.g., see Emmerich, 1968). Each person and each group possesses unique and shared characteristics that need to be the core targets of developmental analysis.

Differences, then, among people or groups are not necessarily indicators of deficits in one and strengths in the other (Spencer, 2006; Spencer et al., 2015). Certainly, it is not useful to frame the study of human development through a model that a priori sets one group as the standard for positive or normative development and regards another group, when different from the group set as the normative one, as therefore defined as being in deficit. If there is any remaining place in developmental science for a deficit model of humans, it is useful only for understanding the thinking of those individuals who continue to treat diversity as either by definition indicative of error variance or as necessarily reflective of a deficiency of human development.

Vestiges of Reductionist Models

Despite the contemporary emphasis on a RDS metatheory and on theories linked to it, the remnants of reductionism and deficit thinking still remain at the periphery of developmental science. These instances of genetic reductionism exist in behavior genetics (e.g., Harden, 2021; Rowe, 1994; Plomin, 2018, 2000; Plomin et al., 2016), in sociobiology (e.g., Rushton, 1999, 2000), and EDP (e.g., Bjorklund, 2015, 2016; Del Giudice & Ellis, 2016). These approaches constitute today's version of the biologizing errors of the past, such as eugenics and racial hygiene (Lerner, 1992, 2015a, 2015b; Proctor, 1988).

As explained by Cairns and Cairns (2006) and by Collins, et al. (2000), these ideas are no longer seen as part of the forefront of scientific theory. Nevertheless, their influence on scientific and public policy persists. Renowned biologists, working in the field of genetics and/or evolutionary biology, such as Bearer (2004), Edelman (1987, 1988), Feldman (e.g., 2014; Feldman & Laland, 1996), Ho (1984, 2010), Lewontin (2000), Müller-Hill (1988), and Venter (e.g., Venter et al., 2001); and eminent colleagues in comparative and biological psychology, such as Greenberg (e.g., Greenberg & Haraway, 2002; Greenberg & Tobach, 1984), Gottlieb (1997, 2004), Hirsch (1997, 2004), Lickliter (2016; Lickliter & Honeycutt, 2015); Michel (e.g., Michel & Moore, 1995), Moore (2015, 2016), and Tobach (1981, 1994; Tobach et al., 1974), alert us to the need for continued intellectual and social vigilance, lest such flawed ideas about genes and human development become the foci of public policies or social programs (Lerner, 2015a, 2015b).

Such applications of counterfactual ideas remain real possibilities, and in some cases unfortunate realities, due at least in part to what Horowitz (2000) described as the affinity of the "Person in the Street" to simplistic models of genetic effects on behavior. These simple and, I must emphasize, erroneous models are used by the Person in the Street to form opinions or to make decisions about human differences and potentials.

Genetic reductionism can, and has, led to views of diversity as a matter of the "haves" and the "have nots" (e.g., Belsky, 2014; Herrnstein & Murray, 1994; Rushton, 1999, 2000). There are, in this view, those people who manifest the normative characteristics of human behavior and development as the "haves," the people who possess (innately, it is presumed; e.g., Belsky, 2014) the attributes that make them healthy, adaptive, or resilient. Given the diversity-insensitive assumptions and research that characterized much of the history of scholarship in human development even into the late 1990s and the first decades of the 21st century, these normative features of human development were associated with middle-class, European American samples (Graham, 1992; McLoyd, 1998; Shweder et al., 2006; Spencer, 1990, 2006; Spencer et al., 2015). In turn, there are those people who manifest other characteristics, and these individuals were generally non-European American and non-middle-class. These individuals were regarded as the "have-nots." As such, if the former group is regarded as normative, then the characteristics of the latter groups are regarded as non-normative (Gould, 1996). When such an interpretation is forwarded, entry has thus been made down the slippery slope of moving from a description of between-group differences to an attribution of deficits in the latter groups (Lerner, 2015a, 2015b).

Such attribution is buttressed when seen through the lens of genetic reductionism because, in this conception, it must be genes that provide the final, material, and efficient cause of the characteristics of the latter groups (e.g., see Bjorklund, 2015, 2016; Plomin et al., 2016; Rowe, 1994; Rushton, 2000). Therefore, non-European American or non-middle-class groups are, in the fully tautological reasoning associated with genetic reductionism, behaviorally deficient because of the genes they possess, *and* because of the genes they possess, they have behavioral deficits (e.g., see Rushton, 2000). Simply, the ill-founded argument is that the genes that place one in a racial group are the genes that provide either deficits or assets in behavior, and one racial group possesses the genes that are assets, and the other group possesses the genes that are deficits (e.g., see Belsky, 2014, for such an argument).

Despite the theory and research that lends support to a dynamic conception of gene ⇔ experience coaction, some proponents of genetic reductionism maintain that concepts and methods regarding genes as separable from context are valid and overwhelmingly, or irrefutably, evident (e.g., Belsky, 2014; Bjorklund, 2015; Plomin et al., 2016). The media continue to tell this story and, perhaps more often than not, the Person in the Street is persuaded by it (Horowitz, 2000).

The challenge that such language use and public discourse represents is not merely one of meeting our scientific responsibility to amend incorrect dissemination of research evidence. Horowitz (2000) reminds us that an additional, and ethical, responsibility is to support social justice (see also Fisher et al., 2013; Fisher & Lerner, 2013; Lerner, 2015a, 2015b; Lerner & Overton, 2008). Horowitz emphasizes that such action is critical in the face of the simplistically seductive ideas and language of genetic reductionism, especially when coupled with the deficit model. Overton (2006) points also to the need to appreciate the subtlety of language to avoid loading our scientific language with phrases that, on a manifest level, may seem to reject the split thinking of genetic reductionism but, on a deeper, structural level, employ terms that legitimate the language of such thinking remaining part of scientific discourse. He notes:

> In its current split form no one actually asserts that matter, body, brain, genes or society, culture, and environment provide *the* cause of behavior or development: The background idea of one or the other being the privileged determinant remains the silent subtext that continues to shape discussions. The most frequently voiced claim is that behavior and development are the products of the *interactions* of nature and nurture. But interaction itself is generally conceptualized as two split-off pure entities that function *independently* in cooperative and/or competitive ways (e.g., Collins et al., 2000). As a consequence, the debate simply becomes displaced to another level of discourse. At this new level, the contestants agree that behavior and development are determined by *both* nature *and* nurture, but they remain embattled over the relative merits of each entity's essential contribution.
>
> (Overton, 2006, p. 33)

Similarly, he explains:

> Moving beyond behavior genetics to the broader issue of biology and culture, conclusions such as "contemporary evidence confirms that the expression of heritable traits depends, often strongly, on experience" (Collins et al., 2000, p. 228) are brought into question for the same reason. Within a relational metatheory, such conclusions fail because they begin from the premise that there are pure forms of genetic inheritance termed "heritable traits" and within relational metatheory such a premise is unacceptable.
>
> (Overton, 2006, p. 36)

Whereas contemporary developmental science rejects the philosophical, theoretical, and (in large part) methodological features of the split thinking associated with genetic reductionist approaches to human development, found in behavior genetics, EDP, sociobiology, subtle, and nuanced problems of language continue to suggest that these split approaches to human development remain legitimate. We have noted the potentially enormous negative consequences of such problematic language in our scientific discourse—especially if the Person in the Street believes that employing such terms means that the genetic reductionist ideas about social policy should be countenanced. As a consequence, developmental scientists must be assiduous and exact in the terms they use to explain why split conceptions in general, and genetic reductionist ones in particular, fail as useful frames for scientific discourse about human development.

Conclusions

By the end of the 20th century, and in the first ½ decades of the 21st century, the conceptually split, mechanistic, essentialist views, that had been involved in so much of the history of concepts and theories of human development, had been replaced by theoretical models that emphasized relationism and integration across all the distinct but fused levels of organization involved in human life. This dynamic synthesis of multiple levels of analysis is a perspective having its roots in systems theories of biological development (Cairns & Cairns, 2006; Gottlieb, 1992; Kuo, 1930, 1967, 1976; Novikoff, 1945a, 1945b; Schneirla, 1956, 1957; von Bertalanffy, 1933), and allows development to be understood as a property of systemic change in the multiple and integrated levels of organization (ranging from biology to culture and history) comprising human life and its ecology (Overton, 2015).

Moreover, as noted by Cairns and Cairns (2006), the interest in understanding person-context relations within an integrative, or systems, perspective has a rich history within developmental psychology as well as in developmental biology. For example, James Mark Baldwin (1897) expressed interest in studying development in context and thus in understanding integrated, multilevel, and hence interdisciplinary scholarship (Cairns & Cairns, 2006). These interests were shared as well by Lightner Witmer, the founder in 1896 of the first psychological clinic in the United States (Cairns & Cairns, 2006).

As well, Cairns and Cairns described the conception of developmental processes—as involving reciprocal interaction, bidirectionality, plasticity, and biobehavioral organization (all quite modern emphases)—as integral in the thinking of the founders of the field of human development. For instance, Wilhelm Stern (1914; see Kreppner, 1994) emphasized the holism that is associated with a developmental systems perspective about these features of developmental processes. In addition, other contributors to the foundations and early progress of the field of human development (e.g., John Dewey, 1916; Kurt Lewin, 1935, 1946; and even John B. Watson, 1928) emphasized the importance of linking child development research with application and child advocacy—a theme of contemporary relevance (Fisher et al., 2013; Lerner, 2012, Lerner et al., 1997, 2000a, 2000b; Lerner & Overton, 2008; Zigler, 1998). This orientation toward the application of developmental science is a contemporary view as well, derived from the emphasis on plasticity and temporal embeddedness within RDS-based theories.

In short, there has been a history of visionary scholars interested in exploring the use of ideas associated with RDS-based theories for understanding the basic process of human development and for applying this knowledge within the actual (ecologically valid) contexts of people to enhance their paths across life (Lerner & Callina, 2014). RDS-based theories have emerged from their historical roots to become, at this writing, the key conceptual frame associated with concepts and theories of human development.

An Overview of the Contributions to This Section

This section of the book includes three contributions that reflect the importance of the RDS-based frame for theory-predicted research in developmental science. Witherington et al. (2018) explain why metatheory is the prime conceptual tool of analysis and interpretation in developmental science. Through focusing on the study of epigenesis (including epigenetics), embodiment, and baselines for human nature and development, they discuss the main metatheoretical difference among theoretical models aiming to explain human development, the Cartesian-Split-Mechanistic paradigm and the Process-Relational paradigm. They explain how fundamental debates regarding these three areas of study derive from metatheoretical differences and not from theoretical differences.

In turn, Greenberg (2011) supports the ideas of Witherington et al. in regard to understanding the coactive role, and thus not a reductionist, primary role, of biology within holistic, embodied, and dynamic conceptions of the developmental process. In contrast to reductionist models such as behavior genetics or evolutionary psychology, he argues that development involves relational and probabilistically timed, and mutually influential coactions both within an organism and between organisms and their contexts. By reference to the ideas of emergence and self-organization from contemporary physics, he explains the importance of a dynamic, relational developmental systems approach to studying behavior and development.

Moore (2002) underscores the significance of understanding the role of metatheory in the analysis of human development by explaining how the faulty biologism of Sir Francis Galton, coupled with his foundational contributions to the field of statistics (e.g., regarding correlational analysis), combined to provide a basis for egregiously flawed and inhumane ideas about selective breeding of human beings, that is, eugenics. One seemingly biologically based statistical method used to support eugenic ideas is heritability analysis. Moore carefully explains why this method is not only conceptually and methodologically flawed but, as well, why it has nothing to do with the role of genes providing a basis of development within any person. Both this contribution from Moore and, as well, his contribution in Section 5 of this book provide the details about the shortcomings of heritability analyses.

References

Alexander, C. N., & Langer, E. J. (Eds.). (1990). *Higher stages of human development*. Oxford University Press.

Azarian, B. (2022). *The romance of reality: How the university organizes itself to create life, consciousness, and cosmic complexity*. BenBella Books.

Baer, D. (1970). An age-irrelevant concept of development. *Merrill-Palmer Quarterly of Behavior and Development, 16*, 238–245.

Baer, D. M. (1982). Behavior analysis and developmental psychology: Discussant comments. *Human Development, 25*, 357–361.

Baldwin, J. M. (1897). *Mental development in the child and the race*. Macmillan.

Baltes, P. B., Lindenberger, U., & Staudinger, U. M. (2006). Life span theory in developmental psychology. In R. M. Lerner (Ed.), *Handbook of child psychology: Theoretical models of human development* (6th ed., Vol. 1, pp. 569–664). John Wiley & Sons.

Baltes, P. B., Reese, H. W., & Lipsitt, L. P. (1980). Life-span developmental psychology. *Annual Review of Psychology, 31*, 65–110.

Baltes, P. B., Reese, H. W., & Nesselroade, J. R. (1977). *Life-span developmental psychology: Introduction to research methods*. Brooks/Cole.

Bandura, A. (1978). The self system in reciprocal determinism. *American Psychologist, 33*, 344–358.

Bearer, E. (2004). Behavior as influence and result of the genetic program: Non-kin rejection, ethnic conflict, and issues in global health care. In C. Garcia Coll, E. Bearer, & R. M. Lerner (Eds.), *Nature and nurture: The complex interplay of genetic and environmental influences on human behavior and development* (pp. 171–199). Erlbaum.

Bell, R. Q. (1968). A reinterpretation of the direction of effects in studies of socialization. *Psychological Review, 75*, 81–95.

Bell, R. Q., & Harper, L. V. (1977). *Child effects on adults*. Erlbaum.

Belsky, J. (2014, November 30). The downside of resilience. *New York Times, Sunday Review*, p SR4.

Benson, P. L., Scales, P. C., Hamilton, S. F., & Semsa, A., Jr. (2006). Positive youth development: Theory, research, and applications. In W. Damon & R. M. Lerner (Eds.), *Handbook of child psychology: Theoretical models of human development* (6th ed., Vol. 1, pp. 894–941). Wiley.

Bijou, S. W. (1976). *Child development: The basic stage of early childhood*. Prentice-Hall.

Bijou, S. W., & Baer, D. M. (1961). *Child development: A systemic and empirical theory* (Vol. 1). Appleton-Century-Crofts.

Bjorklund, D. F. (2015). Developing adaptations. *Developmental Review, 38*, 13–35.

Bjorklund, D. F. (2016). Prepared is not preformed: Commentary on Witherington and Lickliter. *Human Development, 59*, 235–241.

Block, J. (1971). *Lives through time*. Bancroft.

Bornstein, M. H. (2006). Parenting science and practice. In K. A. Renninger, I. E. Sigel, W. Damon, & R. M. Lerner (Eds.), *Handbook of child psychology: Child psychology in practice* (6th ed., Vol. 4, pp. 893–949). Wiley.

Bornstein, M. H. (2017). The specificity principle in acculturation science. *Perspectives in Psychological Science, 12*(1), 3–45.

Bornstein, M. H., & Lamb, M. E. (Eds.). (1999). *Developmental psychology: An advanced textbook* (4th ed.). Lawrence Erlbaum.

Bornstein, M. H., & Lamb, M. E. (Eds.). (2005). *Developmental science: An advanced textbook* (5th ed.). Lawrence Erlbaum.

Brandtstädter, J. (1999). The self in action and development: Cultural, biosocial, and ontogenetic bases of intentional self-development. In J. Brandtstädter & R.M. Lerner (Eds.), *Action and self-development: Theory and research through the life-span* (pp. 37–65). Sage.

Brandtstädter, J. (2006). Action perspectives on human development. In W. Damon & R. M. Lerner (Eds.), *Handbook of Child Psychology: Theoretical models of human development* (6th ed., Vol. 1, pp. 516–568). Wiley.

Brim, O. G., Jr., & Kagan, J. (Eds.). (1980). *Constancy and change in human development.* Harvard University Press.

Bronfenbrenner, U. (1977). Toward an experimental ecology of human development. *American Psychologist, 32,* 513–531.

Bronfenbrenner, U. (1979). *The ecology of human development: Experiments by nature and design.* Harvard University Press.

Bronfenbrenner, U. (2005). *Making human beings human: Bioecological perspectives on human development.* Sage.

Bronfenbrenner, U., & Morris, P. A. (1998). The ecology of developmental process. In W. Damon (Series Ed.) & R. M. Lerner (Vol. Ed.), *Handbook of child psychology: Theoretical models of human development* (5th ed., Vol. 1, pp. 993–1028). Wiley.

Bronfenbrenner, U., & Morris, P. A. (2006). The bioecological model of human development. In W. Damon & R. M. Lerner (Eds.) & R. M. Lerner (Vol. Ed.), *Handbook of child psychology: Theoretical models of human development* (6th ed., Vol. 1, pp. 793–828). Wiley.

Brunswik, E. (1955). Representative design and probabilistic theory in a functional psychology. *Psychological Review, 62,* 193–121.

Cairns, R. B. (1998). The making of developmental psychology. In W. Damon (Series Ed.) & R. M. Lerner (Vol. Ed.), *Handbook of child psychology: Theoretical models of human development* (5th ed., Vol. 1, 419–448). Wiley.

Cairns, R. B., & Cairns, B. D. (2006). The making of developmental psychology. In W. Damon & R. M. Lerner (Series Eds.) & R. M. Lerner (Vol. Eds.), *Handbook of child psychology: Theoretical models of human development* (6th ed., Vol. 1, pp. 89–165). Wiley.

Capaldi, E. J., & Proctor, R. W. (1999). *Contextualism in psychological research? A Critical Review.* Sage.

Chapman, M. (1988a). *Constructive evolution: Origins and development of Piaget's thought.* Cambridge University Press.

Chapman, M. (1988b). Contextuality and directionality of cognitive development. *Human Development, 31,* 92–106.

Collins, W. A., Maccoby, E. E., Steinberg, L., Hetherington, E. M., & Bornstein, M. H. (2000). Contemporary research on parenting: The case of nature and nurture. *American Psychologist, 55,* 218–232.

Cronbach, L. J. (1975). Beyond the two disciplines of scientific psychology. *American Psychologist, 30,* 116–27.

Damon, W. (Ed.). (1998). *Handbook of child psychology* (5th ed). Wiley.

Damon, W. (2006). Preface. In W. Damon & R. M. Lerner (Eds.), *Handbook of child psychology* (6th ed., pp. xi–xix). Wiley.

Del Giudice, M., & Ellis, B. J. (2016). Evolutionary foundations of developmental psychopathology. In D. Cicchetti (Ed.), *Developmental psychopathology: Developmental neuroscience* (3rd. ed., Vol. 12, pp. 1–58). Wiley.

Dewey, J. (1916). *Democracy and education: An introduction to the philosophy of education.* Macmillan.

Dixon, R. A., Lerner, R. M., & Hultsch, D. F. (1991). The concept of development in the study of individual and social change. In P. van Geert & L. P. Mos (Eds.), *Annals of theoretical psychology* (Vol. 7, pp. 279–323). Plenum.

Dixon, R. A., & Nesselroade, J. R. (1983). Pluralism and correlation analysis in developmental psychology: Historical commonalities. In R. M. Lerner (Ed.), *Developmental psychology: Historical and philosophical perspectives* (pp. 113–145). Erlbaum.

Edelman, G. M. (1987). *Neural Darwinism: The theory of neuronal group selection.* Basic Books.

Edelman, G. M. (1988). *Topobiology: An introduction to molecular biology.* Basic Books.

Elder, G. H., Jr., Modell, J., & Parke, R. D. (1993). Studying children in a changing world. In G. H. Elder, J. Modell, & R. D. Parke (Eds.), *Children in time and place: Developmental and historical insights* (pp. 3–21). Cambridge University Press.

Elder, G. H., Jr., & Shanahan, M. J. (2006). The life course and human development. In W. Damon & R. M. Lerner (Eds.). *Handbook of child psychology: Theoretical models of human development* (6th ed., Vol. 1, pp. 665–715). Wiley.

Elder, G. H., Jr., Shanahan, M. J., & Jennings, J. A. (2015). Human development in time and place. In M. H. Bornstein and T. Leventhal (Eds.), *Handbook of child psychology and developmental science: Ecological settings and processes in developmental systems* (7th ed., Vol. 4, pp. 6–54). Wiley.

Emmerich, W. (1968). Personality development and concepts of structure. *Child Development 39*, 671–690.

Erikson, E. H. (1959). Identity and the life cycle. *Psychological Issues*, *1*, 50–100.

Erikson, E. H. (1968). *Identity, youth, and crisis*. Norton.

Feldman, M. (2014). Echoes of the past: Hereditarianism and a troublesome inheritance. *PLoS Genetics*, *10*, e1004817.

Feldman, M. W., & Laland, K. N. (1996). Gene-culture coevolutionary theory. *Trends in Ecology and Evolution*, *11*, 453–457.

Fischer, K. W., & Bidell, T. R. (2006). Dynamic development of action and thought. In W. Damon & R. M. Lerner (Eds.), *Handbook of child psychology: Theoretical models of human development* (6th ed., Vol. 1, pp. 313–399). Wiley.

Fisher, C. B., Busch, N. A., Brown, J. L., & Jopp, D. S. (2013). Applied developmental science: Contributions and challenges for the 21st century. In R. M. Lerner, M. A. Easterbrooks, J. Mistry, & I. B. Weiner (Eds.), *Handbook of psychology: Developmental psychology* (2nd ed., Vol. 6, pp. 516–546). Wiley.

Fisher, C. B., & Lerner, R. M. (2013). Promoting positive development through social justice: An introduction to a new ongoing section of *Applied Developmental Science*. *Applied Developmental Science*, *17*(2), 57–59.

Ford, D. H. (1987). *Humans as self-constructing living systems*. Erlbaum.

Ford, D. H., & Lerner, R. M. (1992). *Developmental systems theory: An integrative approach*. Sage.

Freud, S. (1949). *Outline of psychoanalysis*. Norton.

Freud, S. (1954). *Collected works* (standard edition). Hogarth.

Gergen, K. J. (1973). Social psychology and history. *Journal of Personality and Social Psychology*, *26*, 309–20.

Gollin, E. S. (1981). Development and plasticity. In E. S. Gollin (Ed.), *Developmental plasticity: Behavioral and biological aspects of variations in development* (pp. 231–251). Academic Press.

Gottlieb, G. (1992). *Individual development and evolution: The genesis of novel behavior*. Oxford University Press.

Gottlieb, G. (1997). *Synthesizing nature-nurture: Prenatal roots of instinctive behavior*. Erlbaum.

Gottlieb, G. (2004). Normally occurring environmental and behavioral influences on gene activity: From central dogma to probabilistic epigenesis. In C. Garcia Coll, E. Bearer, & R. M. Lerner (Eds.), *Nature and nurture: The complex interplay of genetic and environmental influences on human behavior and development* (pp. 85–106). Erlbaum.

Gottlieb, G., Wahlsten, D., & Lickliter, R. (2006). The significance of biology for human development: A developmental psychobiological systems view. In W. Damon & R. M. Lerner (Eds.), *Handbook of child psychology: Theoretical models of human development* (6th ed., Vol. 1, pp. 210–257). Wiley.

Gould, S. J. (1981). *The mismeasure of man*. Norton.

Gould, S. J. (1996). *The mismeasure of man* (revised/expanded ed.). Norton.

Graham, S. (1992). "Most of the subjects were white and middle class": Trends in published research on African Americans in selected APA journals, 1970–1989. *American Psychologist*, *47*, 629–639.

Greenberg, G. (2011). The failure of biogenetic analysis in psychology: Why psychology is not a biological science. *Research in Human Development*, *8*, 173–191.

Greenberg, G., & Haraway, M. H. (2002). *Principles of comparative psychology*. Allyn & Bacon.

Greenberg, G., & Tobach, E. (Eds.). (1984). *Behavioral evolution and integrative levels*. Erlbaum.

Harden, K. P. (2021). *The genetic lottery: Why DNA matters for social equality*. Princeton University Press.

Hayes, S. C., Hayes, L. J., Reese, H. W., & Sarbin, T. R. (Eds.). (1993). *Varieties of scientific contextualism*. Context Press.

Herrnstein, R. J., & Murray, C. (1994). *The bell curve: Intelligence and class structure in American life.* Free Press.

Hirsch, J. (1997). Some history of heredity-vs-environment, genetic inferiority at Harvard (?), and The (incredible) Bell Curve. *Genetica, 99,* 207–224.

Hirsch, J. (2004). Uniqueness, diversity, similarity, repeatability, and heritability. In C. Garcia Coll, E. Bearer, & R.M. Lerner (Eds.), *Nature and nurture: The complex interplay of genetic and environmental influences on human behavior and development* (pp. 127–138). Erlbaum.

Ho, M. W. (1984). Environment and heredity in development and evolution. In M.-W. Ho & P. T. Saunders (Eds.), *Beyond neo-Darwinism: An introduction to the new evolutionary paradigm* (pp. 267–289). Academic Press.

Ho, M. W. (2010). Development and evolution revisited. In K. E. Hood, C. T. Halpern, G. Greenberg, & R. M. Lerner (Eds.), *Handbook of developmental systems, behavior and genetics.* (pp. 61–109). Wiley Blackwell.

Horowitz, F. D. (2000). Child development and the PITS: Simple questions, complex answers, and developmental theory. *Child Development, 71,* 1–10.

Hultsch, D. F., & Hickey, T. (1978). External validity in the study of human development: Theoretical and methodological issues. *Human Development, 21,* 76–91.

Jenkins, J. J. (1974). Remember that old theory of memory: Well forget it. *American Psychologist, 29,* 785–95.

Kendler, T. S. (1986). World views and the concept of development: A reply to Lerner and Kauffman. *Developmental Review, 6*(1), 80–95.

Kluckhohn, C., & Murray, H. (1948). Personality formation: The determinants. In C. Kluckhohn & H. Murray (Eds.), *Personality in nature, society, and culture* (pp. 53–69). Knopf.

Kreppner, K. (1994). William L. Stern: A neglected founder of developmental psychology. In R. D. Parke, P. A. Ornstein, J. J. Rieser, & C. Zahn-Waxler (Eds.), *A century of developmental psychology* (pp. 311–331). American Psychological Association.

Kuhn, T. S. (1962). *The structure of scientific revolutions.* University of Chicago Press.

Kuhn, T. S. (1970). *The structure of scientific revolutions* (2nd ed.). University of Chicago Press.

Kuo, Z.-Y. (1930). The genesis of the cat's response to the rat. *Journal of Comparative Psychology, 11,* 1–35.

Kuo, Z.-Y. (1967). *The dynamics of behavior development.* Random House.

Kuo, Z.-Y. (1976). *The dynamics of behavior development: An epigenetic view.* Plenum.

Lamb, M. E. (1977a). A reexamination of the infant social world. *Human Development, 20,* 65–85.

Lamb, M. E. (1977b). Father–infant and mother–infant interaction in the first year of life. *Child Development, 48,* 167–181.

Lamb, M. E. (1977c). The development of mother–infant and father–infant attachments in the second year of life. *Developmental Psychology, 13,* 637–648.

Lamb, M. E. (1978a). Qualitative aspects of mother– and father–infant attachments. *Infant Behavior and Development, 1,* 265–275.

Lamb, M. E. (1978b). Influence of the child on marital quality and family interaction during the prenatal, perinatal, and infancy periods. In R. M. Lerner & G. B. Spanier (Eds.), *Child influences on marital and family interaction: A life-span perspective* (pp. 137–163). Academic Press.

Lamb, M. E., Thompson, R. A., Gardner, W. P., & Charnov, E. L. (1985). *Infant–mother attachment.* Lawrence Erlbaum Associates.

Lehrman, D. S. (1970). Semantic and conceptual issues in the nature-nurture problem. In L. R. Aronson, E. Tobach, D. S. Lehrman, & J. S. Rosenblatt (Eds.), *Development and evolution of behavior: Essays in memory of T. C. Schneirla* (pp. 17–52). Freeman.

Lerner, R. M. (1976). *Concepts and theories of human development.* Addison-Wesley.

Lerner, R. M. (1978). Nature, nurture, and dynamic interactionism. *Human Development, 21,* 1–20.

Lerner, R. M. (1979). A dynamic interactional concept of individual and social relationship development. In R. L. Burgess & T. L. Huston (Eds.), *Social exchange in developing relationships* (pp. 271–305). Academic Press.

Lerner, R. M. (1982). Children and adolescents as producers of their own development. *Developmental Review, 2,* 342–370.

Lerner, R. M. (1984). *On the nature of human plasticity.* Cambridge University Press.

Lerner, R. M. (1986). *Concepts and theories of human development* (2nd ed.). Random House.

Lerner, R. M. (1991). Changing organism–context relations as the basic process of development: A developmental contextual perspective. *Developmental Psychology, 27*, 27–32.

Lerner, R. M. (1992). *Final solutions: Biology, prejudice, and genocide.* Penn State Press.

Lerner, R. M. (Eds.). (1998a). *Handbook of child psychology: Theoretical models of human development* (5th ed., Vol. 1). Wiley.

Lerner, R. M. (1998b). Theories of human development: Contemporary perspectives. In W. Damon (Series Ed.) & R. M. Lerner (Vol. Ed.), *Handbook of child psychology: Theoretical models of human development* (5th ed., Vol. 1, pp. 1–24). Wiley.

Lerner, R. M. (Ed.). (2006a). *Handbook of child psychology: Theoretical models of human development* (6th ed., Vol. 1), Editors-in-chief: W. Damon & R. M. Lerner. Wiley.

Lerner, R. M. (2006b). Developmental science, developmental systems, and contemporary theories of human development. In R. M. Lerner (Ed.), *Handbook of child psychology: Theoretical models of human development* (6th ed., Vol. 1, pp. 1–17). Editors-in-chief: W. Damon & R. M. Lerner. Wiley.

Lerner, R. M. (2012). Developmental science and the role of genes in development. *GeneWatch, 25*(1–2), 34–35.

Lerner, R. M. (2015a). Promoting social justice by rejecting genetic reductionism: A challenge for developmental science. *Human Development, 58*, 67–69.

Lerner, R. M. (2015b). Eliminating genetic reductionism from developmental science. *Research in Human Development, 12*, 178–188.

Lerner, R. M. (Ed.). (2015c). *Handbook of child psychology and developmental science* (7th ed.). Wiley.

Lerner, R. M., & Busch-Rossnagel, N. A. (1981). Individuals as producers of their development: Conceptual and empirical bases. In R. M. Lerner & N. A. Busch-Rossnagel (Eds.), *Individuals as producers of their development: A life-span perspective* (pp. 1–36). Academic Press.

Lerner, R. M., & Callina, K. S. (2014). Relational developmental systems theories and the ecological validity of experimental designs: Commentary on Freund and Isaacowitz. *Human Development, 56*, 372–380.

Lerner, R. M., Fisher, C. B., & Weinberg, R. A. (1997). Applied developmental science: Scholarship for our times. *Applied Developmental Science, 1*, 2–3.

Lerner, R. M., Fisher, C. B., & Weinberg, R. A. (2000a). Toward a science for and of the people: Promoting civil society through the application of developmental science. *Child Development, 71*, 11–20.

Lerner, R. M., Fisher, C. B., & Weinberg, R. A. (2000b). Applying developmental science in the twenty-first century: International scholarship for our times. *International Journal of Behavioral Development, 24*, 24–29.

Lerner, R. M., Hultsch, D. F., & Dixon, R. A. (1983). Contextualism and the character of developmental psychology in the 1970s. *Annals of the New York Academy of Sciences, 412*, 101–128.

Lerner, R. M., & Kauffman, M. B. (1985). The concept of development in contextualism. *Developmental Review, 5*, 309–333.

Lerner, R. M., & Kauffman, M. B. (1986). On the metatheoretical relativism of analyses of metatheoretical analyses: A critique of Kendler's comments. *Developmental Review, 6*, 96–106.

Lerner, R. M., Lerner, J. V., Bowers, E., & Geldhof, G. J. (2015). Positive youth development and relational developmental systems. In W. F. Overton & P. C. Molenaar (Eds.), *Handbook of child psychology and developmental science: Theory and method* (7th ed., Vol. 1, pp. 607–651), Editor-in-Chief: R. M. Lerner. Wiley.

Lerner, R. M., & Overton, W. F. (2008). Exemplifying the integrations of the relational developmental system: Synthesizing theory, research, and application to promote positive development and social justice. *Journal of Adolescent Research, 23*(3), 245–255.

Lerner, R. M., Skinner, E. A., & Sorell, G. T. (1980). Methodological implications of contextual/dialectic theories of development. *Human Development, 23*, 225–235.

Lerner, R. M., Theokas, C., & Jelicic, H. (2005). Youth as active agents in their own positive development: A developmental systems perspective. In W. Greve, K. Rothermund, & D. Wentura (Eds.), *The adaptive self: Personal continuity and intentional self-development* (pp. 31–47). Hogrefe & Huber Publishers.

Lewin, K. (1935). *A dynamic theory of personality.* McGraw-Hill.

Lewin, K. (1946). Behavior and development as a function of the total situation. In L. Carmichael (Ed.), *Manual of child psychology* (pp. 791–844). Wiley.

Lewis, M. (1972). State as an infant-environment interaction: An analysis of mother–infant behavior as a function of sex. *Merrill-Palmer Quarterly, 18*, 95–121.

Lewis, M., & Feiring, C. (1978). A child's social world. In R. M. Lerner & G. B. Spanier (Eds.), *Child influences on marital and family interaction: A life-span perspective* (pp. 47–66). Academic Press.

Lewis, M., & Lee-Painter, S. (1974). An interactional approach to the mother–infant dyad. In M. Lewis & L. A. Rosenblum (Eds.), *The effect of the infant on its caregivers* (pp. 21–48). Wiley.

Lewis, M., & Rosenblum, L. A. (Eds.). (1974). *The effect of the infant on its caregivers*. Wiley.

Lewontin, R. C. (2000). *The triple helix*. Harvard University Press.

Lickliter, R. (2016). Developmental evolution. *WIREs Cognitive Science, 8*(1–2), e1422. https://doi.org/10.1002/wcs.1422

Lickliter, R., & Honeycutt, H. (2015). Biology, development, and human systems. In W. F. Overton & P. C. M. Molenaar (Eds.), *Handbook of child psychology and developmental science: Theory and method* (7th ed., Vol. 1, pp. 162–207), Editor-in-chief: R. M. Lerner. Wiley.

Looft, W. R. (1973). Socialization and personality throughout the life-span: An examination of contemporary psychological approaches. In P. B. Baltes & K. W. Schaie (Eds.), *Life-span developmental psychology: Personality and socialization* (pp. 25–52). Academic Press.

Lorenz, K. (1966). *On aggression*. Harcourt, Brace & World.

Magnusson, D. (1999a). Holistic interactionism: A perspective for research on personality development. In L. A. Pervin & O. P. John (Eds.), *Handbook of personality: Theory and research* (2nd ed., pp. 219–247). The Guilford Press.

Magnusson, D. (1999b). On the individual: A person-oriented approach to developmental research. *European Psychologist, 4*, 205–218.

Magnusson, D., & Stattin, H. (2006). The person in context: A holistic-interactionist approach. In R. M. Lerner (Ed.), *Handbook of child psychology: Theoretical models of human development* (6th ed., Vol. 1, pp. 400–464). Editors-in-chief: W. Damon & R. M. Lerner. Wiley.

Mascolo, M. F., & Fischer, K. W. (2015). Dynamic development of thinking, feeling, and acting. In W. F. Overton & P. C. Molenaar (Eds.), *Handbook of child psychology and developmental science: Theory and method* (7th ed., Vol. 1, pp. 113–161), Editor-in-chief: R. M. Lerner. Wiley.

McLoyd, V. C. (1998). Children in poverty: Development, public policy, and practice. In W. Damon (Ed.), I. E. Sigel (Vol. Ed.) and K. A. Renninger (Vol. Ed.), *Handbook of psychology: Child psychology in practice* (Vol. 4). John Wiley & Sons Inc.

Michel, G. F., & Moore, C. L. (1995). *Developmental Psychobiology: An Interdisciplinary Science*. MIT Press.

Mischel, W. (1977). On the future of personality measurement. *American Psychologist, 32*, 246–54.

Molenaar, P. C. M. (2004). A manifesto on psychology as idiographic science: Bringing the person back into scientific psychology, this time forever. *Measurement, 2*(4), 201–218.

Molenaar, P. C. M., & Nesselroade, J. R. (2014). New trends in the inductive use of relation developmental systems theory: Ergodicity, nonstationarity, and heterogeneity. In P. C. Molenaar, R. M. Lerner, and K. M. Newell (Eds.), *Handbook of developmental systems and methodology* (pp. 442–462). Guilford Press.

Molenaar, P. C. M., & Nesselroade, J. R. (2015). Systems methods for developmental research. In W. F. Overton & P. C. M. Molenaar (Eds.), *Handbook of child psychology and developmental science: Theory and method* (7th ed., Vol. 1, pp. 652–682), Editor-in-chief: R.M. Lerner. Wiley.

Moore, D. S. (2002). *The dependent gene: The fallacy of nature vs. nurture*. W. H. Freeman.

Moore, D. S. (2015). *The developing genome: An introduction to behavioral epigenetics*. Oxford University Press.

Moore, D. S. (2016). Behavioral epigenetics. *WIREs Cognitive Science*. https://doi.org/10.1002/wcs.1333

Müller-Hill, B. (1988). *Murderous science: Elimination by scientific selection of Jews Gypsies, and others. Germany 1933-1945*. Oxford University.

Mussen, P. H. (Ed.). (1970). *Carmichael's manual of child psychology* (3rd ed.). Wiley.

Mussen, P. H. (Ed.). (1983). *Handbook of child psychology* (4th ed.). Wiley.
Noble, D. (2015). Evolution beyond neo-Darwinism: A new conceptual framework. *The Journal of Experimental Biology, 218*, 7–13.
Novikoff, A. B. (1945a). The concept of integrative levels and biology. *Science 101*, 209–215.
Novikoff, A. B. (1945b). Continuity and discontinuity in evolution. *Science 101*, 405–406.
Oser, F. K., Scarlett, W. G., & Bucher, A. (2006). Religious and spiritual development throughout the life span. In R. M. Lerner (Ed.), *Handbook of child psychology: Theoretical models of human development* (6th ed., Vol. 1, pp. 942–998). Editors-in-chief: W. Damon & R. M. Lerner. Wiley.
Overton, W. F. (1973). On the assumptive base of the nature-nurture controversy: Additive versus interactive conceptions. *Human Development, 16*, 74–89.
Overton, W. F. (1984). World views and their influence on psychological theory and research: Kuhn—Lakatos—Lauden. In H. W. Reese (Ed.), *Advances in child development and behavior* (Vol. 18, pp. 194–226). Academic Press.
Overton, W. F. (1991a). Historical and contemporary perspectives on developmental theory and research strategies. In R. Downs, L. Liben, & D. Palermo (Eds.), *Visions of aesthetics, the environment, and development: The legacy of Joachim Wohlwill* (pp. 263–311). Erlbaum.
Overton, W. F. (1991b). The structure of developmental theory. In H. W. Reese (Ed.), *Advances in child development and behavior* (Vol. 23, pp. 1–37). Academic Press.
Overton, W. F. (1994). The arrow of time and cycles of time: Concepts of change, cognition, and embodiment. *Psychological Inquiry, 5*, 215–237.
Overton, W. F. (1998). Developmental psychology: Philosophy, concepts, and methodology. In W. Damon (Series Ed.) & R. M. Lerner (Ed.), *Handbook of child psychology: Theoretical models of human development* (5th ed., Vol. 1, pp. 107–187). Wiley.
Overton, W. F. (2003). Development across the life span. In R. M. Lerner, M. A. Easterbrooks, & J. Mistry (Eds.), *Handbook of psychology: Developmental psychology* (Vol. 6, pp. 13–42), Editor-in-chief: I. B. Weiner. Wiley.
Overton, W. F. (2006). Developmental psychology: Philosophy, concepts, methodology. In W. Damon & R. M. Lerner (Series Eds.) & R. M. Lerner (Vol. Ed.), *Handbook of child psychology: Theoretical models of human development* (6th ed., Vol. 1, pp. 18–88). Wiley.
Overton, W. F. (2010). Life-span development: Concepts and issues. In W. F. Overton (Vol. Ed.), *Handbook of life-span development: Cognition, biology, and methods* (Vol. 1, pp. 1–29), Editor-in-chief: R. M. Lerner. Wiley and Sons.
Overton, W. F. (2015). Process and relational developmental systems. In W. F. Overton & P. C. M. Molenaar (Eds.), *Handbook of child psychology and developmental science: Theory and Method* (7th ed., Vol. 1, pp. 9–62), Editor-in-chief: R. M. Lerner. Wiley.
Overton, W. F., & Müller, U. (2013). Metatheories, theories, and concepts in the study of development. In R. M. Lerner, M.A. Easterbrooks, & J. Mistry (Eds.), *Handbook of psychology: Developmental psychology* (2nd ed., Vol. 6, pp. 19–58), Editor-in-chief: I. B. Weiner. Wiley.
Overton, W. F., & Reese, H. W. (1973). Models of development: Methodological implications. In J. R. Nesselroade and H. W. Reese (Eds.), *Life-span developmental psychology: Methodological issues* (pp. 65–86). Academic Press.
Pepper, S. C. (1942). *World hypotheses: A study in evidence*. University of California Press.
Pervin, L. A., & Lewis, M. (Eds.). (1978). *Perspectives in interactional psychology*. Plenum.
Petrinovich, L. (1979). Probabilistic functionalism: A conception of research method. *American Psychologist, 34*, 373–90.
Piaget, J. (1950). *The psychology of intelligence*. Harcourt Brace.
Piaget, J. (1970). Piaget's theory. In P. H. Mussen (Ed.), *Carmichael's manual of child psychology* (3rd ed., Vol. 1, pp. 703–723). Wiley.
Plomin, R. (1986). *Development, genetics, and psychology*. Erlbaum.
Plomin, R. (2000). Behavioural genetics in the 21st century. *International Journal of Behavioral Development, 24*, 30–34.
Plomin, R. (2018). *Blueprint: How DNA makes us who we are*. Allen Lane.

Plomin, R., Corley, R., DeFries, J. C., & Faulker, D. W. (1990). Individual differences in television viewing in early childhood: Nature as well as nurture. *Psychological Science, 1*, 371–377.

Plomin, R., Defries, J. C., Knopik, J. M., & Neiderhiser, J. M. (2016). Top 10 replicated findings from behavioral genetics. *Perspectives on Psychological Science, 11*(1), 3–23.

Proctor, R. N. (1988). *Racial hygiene: Medicine under the Nazis*. Harvard University.

Rathunde, K., & Csikszentmihalyi, M. (2006). The developing person: An experiential perspective. In R. M. Lerner (Ed.), *Theoretical models of human development. Volume 1 of handbook of child psychology* (6th ed., pp. 465–515). Editors-in-chief: W. Damon & R. M. Lerner. Wiley.

Reese, H. W., & Overton, W. F. (1970). Models of development and theories of development. In L. R. Goulet & P. B. Baltes (Eds.), *Life-span developmental psychology: Research and theory* (pp. 115–145). Academic.

Riegel, K. F. (1973). Dialectical operations: The final period of cognitive development. *Human Development, 16*, 346–70.

Riegel, K. F. (1975). Toward a dialectical theory of human development. *Human Development, 18*, 50–64.

Riegel, K. F. (1976a). The dialectics of human development. *American Psychologist, 31*, 689–700.

Riegel, K. F. (1976b). From traits and equilibrium toward developmental dialectics. In W. J. Arnold & J. K. Cole (Eds.), *Nebraska symposium on motivation* (pp. 348–408). University of Nebraska.

Riegel, K. F. (1978). *Psychology mon amour: A countertext*. Houghton Mifflin.

Rose, T. (2016). *The end of average: How we succeed in a world that values sameness*. HarperCollins Publishers.

Rose, L. T., Rouhani, P., & Fischer, K. W. (2013). The science of the individual. *Mind, Brain, and Education, 7*, 152–158.

Rosnow, R. L., & Georgoudi, M. (1986). *Contextualism and understanding in behavioral science: Implications for research and theory*. Praeger.

Rowe, D. C. (1994). *The limits of family influence: Genes, experience, and behavior*. Guilford Press.

Rushton, J. P. (1987). An evolutionary theory of health, longevity, and personality: Sociobiology, and r/K reproductive strategies. *Psychological Reports, 60*, 539–549.

Rushton, J. P. (1999). *Race, evolution, and behavior* (Special Abridged ed.). Transaction Publishers.

Rushton, J. P. (2000). *Race, evolution, and behavior* (2nd Special Abridged ed.). Transaction Publishers.

Sameroff, A. (1975). Transactional models in early social relations. *Human Development, 18*, 65–79.

Sameroff, A. J. (1983). Developmental systems: Contexts and evolution. In W. Kessen (Ed.), *Handbook of child psychology: History, theory, and methods* (Vol. 1, pp. 237–294). Wiley.

Sarbin, T. R. (1977). Contextualism: A world view for modern psychology. In J. K. Cole (Ed.), *Nebraska Symposium on motivation* (pp. 1–41). University of Nebraska Press.

Schneirla, R. C. (1956). Interrelationships of the innate and the acquired in instinctive behavior. In P. P. Grassé (Ed.), *L'instinct dans le comportement des animaux et de l'homme* (pp. 387–452). Mason et Cie.

Schneirla, T. C. (1957). The concept of development in comparative psychology. In D. B. Harris (Ed.), *The concept of development: An issue in the study of human behavior* (pp. 78–108). University of Minnesota Press.

Shweder, R. A., Goodnow, J. J., Hatano, G., LeVine, R. A., Markus, H. R., & Miller, P. J. (2006). The cultural psychology of development: One mind, many mentalities. In R. M. Lerner (Ed.). *Handbook of child psychology: Theoretical models of human development* (6th ed., Vol. 1, pp. 716–792). Editors-in-chief: W. Damon & R. M. Lerner. Wiley.

Skinner, B. F. (1938). *The behavior of organisms*. Appleton.

Skinner, B. F. (1950). Are theories of learning necessary? *Psychological Review, 57*, 211–220.

Skinner, B. F. (1971). *Beyond freedom and dignity*. Knopf.

Spencer, M. B. (1990). Development of minority children: An introduction. *Child Development, 61*, 267–269.

Spencer, M. B. (2006). Phenomenological variant of ecological systems theory (PVEST): A human development synthesis applicable to diverse individuals and groups. In W. Damon & R. M. Lerner (Eds.) & R. M. Lerner (Vol. Ed.), *Handbook of child psychology: Theoretical models of human development* (6th ed., Vol. 6, pp. 829–894). Wiley.

Spencer, M. B., Swanson, D. P., & Harpalani, V. (2015). Development of the self. In M. E. Lamb (Volume Ed.), *Handbook of child psychology and developmental science: Socioemotional processes* (7th ed., Vol. 3, pp. 750–793), Editor-in-chief: R. M. Lerner. Wiley.

Stern, W. (1914). *Psychologie der frühen Kindheit bis zum sechsten Lebensjahr*. Quelle & Meyer.
Thelen, E., & Smith, L. B. (1994). *A dynamic systems approach to the development of cognition and action*. MIT Press.
Thelen, E., & Smith, L. B. (1998). Dynamic systems theories. In W. Damon (Series Editor) & R.M. Lerner (Vol. Ed.), *Handbook of child psychology: Theoretical models of human development* (5th ed., Vol. 1, pp. 563–633). Wiley.
Thelen, E., & Smith, L. B. (2006). Dynamic systems theories. In R. M. Lerner & W. Damon (Eds.), *Handbook of child psychology: Theoretical models of human development* (6th ed., Vol. 1, pp. 258–312). Wiley.
Thomas, A., & Chess, S. (1977). *Temperament and development*. Brunner/Mazel.
Thompson, R. A., & Lamb, M. E. (1986). Infant–mother attachment: New directions for theory and research. In P. B. Baltes, D. L. Featherman, & R. M. Lerner (Eds.), *Life-span development and behavior* (Vol. 7, pp. 1–41). Erlbaum.
Tobach, E. (1981). Evolutionary aspects of the activity of the organism and its development. In R. M. Lerner & N. A. Busch-Rossnagel (Eds.), *Individuals as producers of their development: A life-span perspective* (pp. 37–68). Academic Press.
Tobach, E. (1994). …Personal is political is personal is political… *Journal of Social Issues, 50*, 221–224.
Tobach, E., Gianutsos, J., Topoff, H. R., & Gross, C. G. (1974). *The four horses: Racism, sexism, militarism, and social Darwinism*. Behavioral Publications.
Valsiner, J. (2006). Developmental epistemology and implications for methodology. In R. M. Lerner (Ed.). *Handbook of child psychology: Theoretical models of human development* (6th ed., Vol. 1, pp. 166–209). Editors-in-chief: W. Damon & R. M. Lerner. Wiley.
Venter, J. C., Adams, M. D., Myers, E. W., Li, P. W., Mural, R. J., [plus 270 others]. (2001). The sequence of the human genome. *Science, 291*, 1304–1351.
von Bertalanffy, L. (1933). *Modern theories of development*. Oxford University Press.
Watson, J. B. (1928). *Psychological care of infant and child*. Norton.
White, S. H. (1970). The learning theory approach. In P. H. Mussen (Ed.), *Carmichael's manual of child psychology* (3rd ed., Vol. 1., pp. 657–702). Wiley.
Witherington, D. C., Overton, W. F., Lickliter, R., Marshall, P. J., & Narvaez, D. (2018). Metatheory and the primacy of conceptual analysis in developmental science. *Human Development, 61*(3), 181–198.
Zigler, E. (1998). A place of value for applied and policy studies. *Child Development, 69*, 532–542.

3.
METATHEORY AND THE PRIMACY OF CONCEPTUAL ANALYSIS IN DEVELOPMENTAL SCIENCE

David C. Witherington*, Willis F. Overton†, Robert Lickliter‡, Peter J. Marshall†, and Darcia Narvaez§

Conceptual analysis is widely regarded as a necessary activity for conducting and advancing science [Kuhn, 1970; Machado & Silva, 2007; Overton, 2012; Wachtel, 1980; Watkins, 1975]. Unlike activities of empirical investigation (e.g., hypothesis generation, research design, implementation, and data analysis), conceptual analysis consists of *philosophical* investigation into "the well-foundedness of the conceptual structures (e.g., theories)" that serve to organize, make sense of, and explain our everyday observations and commonsense understandings of the world [Laudan, 1977, p. 48]. It involves examining and evaluating the cogency and coherence of theories, of the concepts that comprise them, of the hypotheses and conclusions drawn from them, and, most broadly, of conceptual argumentation [Laudan, 1977; Machado & Silva, 2007]. The philosophical feature of conceptual analysis entails, among other activities, what Strawson [1959] termed "descriptive metaphysics." This refers to the description of the most general features (ontological and epistemological) of our conceptual schemes concerning the nature of the world and the nature of knowing the world. Descriptive metaphysics is contrasted with "revisionary metaphysics," which attempts to revise our ways of thinking in an effort to establish an intellectually and morally perfect picture of the world.

Few scientists would seriously dismiss the value of conceptual analysis, especially given the postpositivist climate that pervades disciplines today. Yet, relative to empirical activity, the activity of conceptual analysis all too often assumes a marginalized status in science. This certainly holds true for the field of developmental science [Slife & Williams, 1995; Smedslund, 1991]. When push comes to shove, mainstream thought within our discipline still inclines toward Popper's [1959] instrumentalist tradition, wherein "hard data" serve as "the final and absolute privileged arbiter of truth" [Overton, 1998, p. 171]. Under this mindset, the role of conceptual analysis is to ensure that our theories are suitably packaged for rigorous empirical investigation. As such, conceptual analysis enters the realm of scientific activity as a necessary tool for evaluating the logical consistency, clarity, and testability (e.g., potential for falsification) of our theories and models; for identifying their lacunae; and for revealing implicit assumptions in our argumentation and conceptual presentation. But the "real" business of scientific decision-making does not arrive until the empirical testing of our theories commences. By virtue of developmental science's sustained reliance on instrumentalist doctrine, conceptual work in our discipline routinely subordinates to empirical activity and depends on "hard data" for

* University of New Mexico, Albuquerque, NM.
† Temple University, Philadelphia, PA.
‡ Florida International University, Miami, FL.
§ University of Notre Dame, Notre Dame, IN.

its very utility and scientific legitimacy [Lakatos, 1978; Overton, 2006, 2012].

Developmental science's second-class treatment of conceptual analysis is predicated on a mistaken conflation of concept and theory [Wakefield, 2007]. More to the point, it is predicated on an impoverished view of conceptualization in science, one that fails to discriminate between two distinct levels of conceptual inquiry: the *theoretical* and the *metatheoretical*. At the *theoretical* level, concepts involve a mode of understanding one step removed from science's most basic, observational level of commonsense understanding [Overton, 1998]. As such, *theoretical concepts*, and the theories that arise from them, serve as our scientific means of "organizing and reformulating observational understandings in a broader and more coherent fashion" [Overton, 2015, p. 14]. They are designed to be operationalized, expected to yield testable, observable predictions, and, as such, are subject to adjudication through empirical activity, in keeping with instrumentalist tradition.

However, a broader level of conceptualization necessarily frames both observational and theoretical levels of scientific discourse. It is the level of *metatheory*: a level of *pre-empirical* and *pre-theoretical* conceptual grounding within which both the empirical and theoretical activities of science operate. Metatheories involve a set of *background concepts* – various philosophical beliefs and assumptions that we, as humans and as scientists, hold concerning the nature of reality (ontology) and how we come to know that reality (epistemology) [Overton, 2015]. They establish what does and does not make sense to even consider or investigate in the observations that we make and the theories that we construct. All of our scientific work, therefore, necessarily presupposes, and is preconditioned by, the background concepts of metatheory [Overton, 2015]. Critically, this means that, unlike their theoretical counterparts, background concepts are *not* amenable to empirical investigation and adjudication; instead, they are, in Hacker's [2009] words, "grist for *philosophical* mills – not philosophical problems for experimental investigation" (p. 132).

In its conventional wisdom, mainstream developmental science still largely disregards the need to demarcate metatheoretical from theoretical levels of conceptualization. This disregard has obscured one of the most critical functions of conceptual analysis, namely the identification and explication of background concepts in science. And, as a result, one of conceptual analysis' most valuable lessons frequently goes unnoticed in our discipline. That lesson is simple but its implications profound: *many key theoretical debates within the sciences are actually metatheoretical debates and, as such, can only be resolved through conceptual analysis* [Hacker, 2009; Laudan, 1977; Wakefield, 2007].

Our purpose in writing this chapter is to instantiate this critical lesson. In the process, we hope to disabuse readers of the belief – deeply entrenched in our discipline – that all competing scientific theories can ultimately be adjudicated, completely and solely, through *empirical* activity. We begin by providing an overview of the metatheoretical divide that extensively, but all too often implicitly, frames theoretical discourse and debate within developmental science today. This is the divide between a Cartesian-Split-Mechanistic research paradigm [Lakatos, 1978], and a Process-Relational research paradigm – two fundamentally different ontological-epistemological frameworks of meaning through which we can view the theoretical concepts of our discipline [Overton, 2015; Overton & Lerner, 2014]. We then critically examine three influential theoretical concepts in developmental science – epigenesis, embodiment, and the notion of baselines for human nature and development – to reveal how seminal issues and debates within these domains of inquiry are driven by metatheoretical, not theoretical considerations, and therefore require conceptual, not empirical analysis, for resolution.

The Metatheoretical Divide in Developmental Science

An overview of the metatheoretical divide within developmental science today begins with the recognition that the metatheoretical level, in fact, consists of two levels of metatheory, arranged as a nested hierarchy (Fig. 3.1). The Cartesian-Split-Mechanistic and Process-Relational research paradigms represent the top level of the hierarchy and, as suggested, these paradigms are composed of ontological and epistemological concepts. Such concepts form the framing context for the next level of metatheory, as well as the framing context for the construction of theories and empirical methods. The second metatheoretical level, termed Mid-Range Metatheories or Metamodels, derives its concepts from the paradigm level. However, these midrange conceptual systems are less general than the paradigms and entail

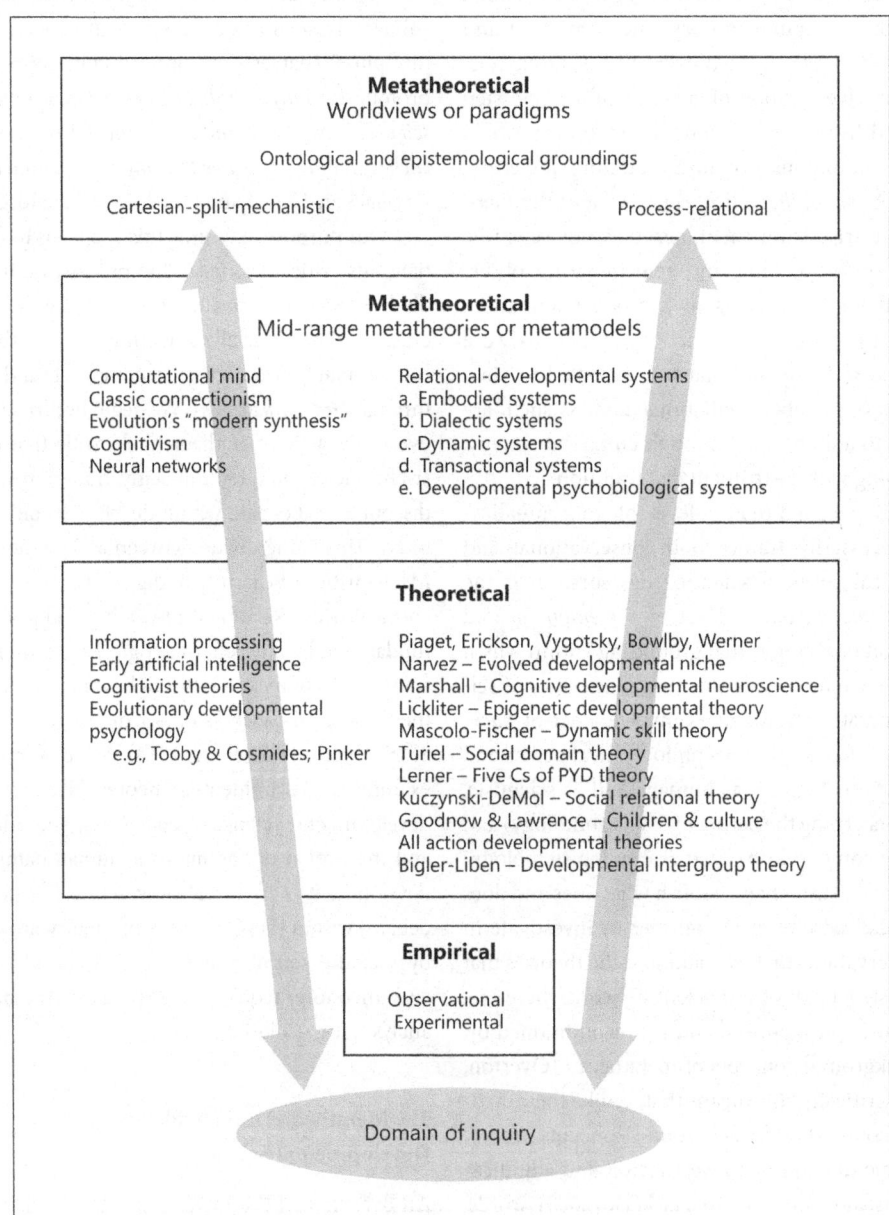

Fig. 3.1 Levels of empirical, theoretical, and metatheoretical scientific discourse.

principles that are identifiably more specific to the observational domains of interest. For example, nested within the Process-Relational research paradigm, the Relational Developmental Systems metamodel represents the human organism as a *dynamic holistic system,* while nested within the Cartesian-Split-Mechanistic research paradigm, the Computational Mind and Cognitivism metamodels represent the organism as an *input-output computational recording device.* Taken as a whole, paradigm and nested metamodels constitute the conceptual framework of any scientific research program, providing guiding principles for the construction of a variety of specific models and theories, as well as specific methods.

Until recently, the Cartesian-Split-Mechanistic research paradigm has been dominant in guiding theory construction and observational methods in developmental science. As the philosopher of science Imre Lakatos [1978] pointed out, this paradigm entails "Cartesian metaphysics, that is, the mechanistic theory of the universe – according to which the universe is a huge clockwork (and system of vortices) with push as the only cause of motion" (p. 47) and in which fundamental features of this world are split into dichotomous independent elements [Bernstein, 1983; Descartes, 1641/1996]. Conceiving of the universe in this fashion entails the ontological and epistemological concepts presented on the right side of Table 3.1. The Cartesian-Split-Mechanistic paradigm asserts that there is an ultimate bedrock reality (*atomism and realism*) that is *fixed* – in the sense of exhibiting no independent activity – and *material* in nature. This reality is also inherently *static and uniform.* Any observed change or organization is presumed to be explained by *extrinsic* (i.e., internal biological and external environmental) forces (i.e., causes). Arriving at this reality entails a *split understanding* of the world as independent elements: one constituting the fixed foundational reality and the other a derived appearance of reality. The movement from the apparent to the real is accomplished through the mechanism of *reductionism,* which entails objectively observing the phenomenological (i.e., apparently real) world and analyzing this world down to the reality of fixed, static, uniform, *monistic,* and *material* elements, as well as the forces operating on them.

In contrast to the Cartesian-Split-Mechanistic paradigm, the Process-Relational paradigm models the universe not after a clock, but rather after an *endogenously active and changing complex organic system.* Conceiving of the universe in this fashion entails the ontological and epistemological concepts presented on the left side of Table 3.1. The Process-Relational paradigm asserts that the universe is an *organic holistic system* in which every part (not element) implies every other part, and an alteration of any part would alter every other

Table 3.1 A Comparison of the Ontological and Epistemological Categories of the Process-Relational and Cartesian-Split-Mechanistic Metatheories

Paradigms

Process-Relational	Cartesian-Split-Mechanistic
Ontological categories	
Holism[a]	Atomism
Activity	Fixity
Nature as process	Nature as substance (matter)
Change-becoming Dialectic	Stasis-being
Necessary organization Structure-function relations	Uniformity/ contingent organization
Pluralistic universe	Monistic universe
Epistemological categories	
Holism[a]	Reductionism
Relational understanding	Split understanding
Multiple standpoints of analysis	Objectivism vs. subjectivism
Multiple forms of explanation	Efficient/material causal
Formal explanation Structure-function	
Final explanation	

Note: [a] Holism has both an ontological and an epistemological meaning.

part. Or to state this slightly differently, "holism" asserts that the identities (meanings) of entities and events derive from the context in which they are embedded. The whole is not an aggregate of discrete elements, but an organization of parts, each part being defined by its relations to other parts and to the whole. The holistic character of the system mandates a *relational* understanding where the focus is not on things, but on the relation among things. This contrasts with the split understanding of the Cartesian-Split-Mechanistic paradigm. Given this relational understanding, the paradigm also permits *multiple standpoints of analysis*, beyond the split objectivism vs. subjectivism, and *multiple forms of explanation*, beyond extrinsic causal explanation.

A key ontological feature of a holistic system is that it is *endogenously active.* There is no external force driving the system; rather, it is by its organic nature self-active. It is this activity of the system along with the system's *organization* that identifies the system as one that is constantly *changing* in a *dialectical* manner. System activity becomes *process* when it is holistically located in a temporal order of duration (i.e., the length of time the activity continues). As a consequence, in the Process-Relational paradigm, *nature is understood as process,* rather than the fixed matter of the Cartesian-Split-Mechanistic paradigm. Further, all actual objects – from subatomic particles to trees, houses, acts of persons – are understood as *acts of perception.* These objects, which are continually in the process of becoming and passing away, constitute the *real.* Actual objects cannot be reduced to something more real. Thus, the Process-Relational universe is pluralistic and not monistic, as in the Cartesian-Split-Mechanistic paradigm.

As stated earlier, when we move from the paradigm to the *metamodel* level of metatheory, concepts – while deriving from the paradigmatic level – are more specific to observational domains of interest. For developmental science, the observational level of interest is the living organism. Metamodels derived from the Cartesian-Split-Mechanistic paradigm (e.g., the computational mind, cognitivism) characterize living organisms as *inherently stable, fixed,* and *unchanging.* Both movement (behavior) and development are understood to be the result of *extrinsic* (i.e., internal biological and external environmental) *forces,* often termed *mechanisms, antecedent conditions,* or *independent variables.* The Cartesian organism is complicated, but not complex. It is complicated in the sense that it can be described in terms of independent pieces. Biology is a piece, culture – a piece, cognition, motivation, and affect are all pieces. Pieces combine – add together, interact – to form a whole that is no different than the sum of its pieces. There can be no real novelty, as apparent reality must, in principle, be reduced to the pieces. The Cartesian organism is *linear* both with respect to behavior and development; inputs are strictly proportional to outputs. Linearity means that the behavior and development of the organism are *deterministic* and, hence, in principle, completely *predictable.*

In contrast, metamodels derived from the Process-Relational paradigm all center on the concept of system, which is understood as an organization of parts, each part being defined by its relations to other parts and to the whole. *Relational-Developmental system* is the most inclusive of several such metamodels, including *developmental* systems [Ford & Lerner, 1992], *dynamic* systems [Witherington, 2015], *psychobiological* systems [Lickliter & Honeycutt, 2015], *dialectical* and *transactional* systems [Kuczynski & De Mol, 2015], and the metamodel of *enactivism* [Di Paolo, Buhrmann, & Barandiaran, 2017; Stewart, Gapenne, & Di Paolo, 2010].

The Relational-Developmental systems metamodel characterizes the living organism, which is itself a relational developmental system, as an *endogenously active and changing organization of part-whole relations.* The system requires no guiding forces to act, and its embodied acts operating coactively in a lived world of physical and sociocultural objects constitute the basic change process *(i.e., the sufficient condition of development).* Such capacity for change illustrates the *relative plasticity* of the system. The Relational-Developmental system is *self-creating (autopoietic, enactive)* in the sense that it operates according to its own processes. The system is also *self-organizing and self-regulating*. Self-organization refers to the fact that the system's order or organization (i.e., structure-function relations) proceeds from local embodied co-actions between smaller components of the system, and self-regulation connotes the fact that the embodied actions of the system regulate the environment, and the coacting environment regulates the system. The coaction of system and environment ensures that the system is completely contextualized and situated – i.e., time and place matter.

The Relational-Developmental system is complex in that the system is not decomposable into elements arranged in additive sequences of efficient-material causes and their effects. The system is, therefore, a *nonlinear* system. Any complex dynamic system is an *adaptive* system in the sense that the system (whole) coacts with its environment to grow (develop) from lesser to greater levels of complexity. It is this development of increasing complexity that introduces nonreducible novelty, or the emergence of systemic properties that are not characteristic of any of the parts of the system. Further, the search for efficient-material causes and mechanical mechanisms that is characteristic of the metamodels derived from the Cartesian-Split-Mechanistic paradigm is instead replaced in Relational-Developmental systems by the identification of dynamic action *patterns* – both in real-time as actual action events and across developmental time as organized, sequential, directional, relational developmental systems. This identification of action patterns and of the processes they entail is logically prior to the identification of necessary conditions and resources for development. It is, in fact, the specific relational developmental system under investigation that defines both conditions and resources.

With this brief, conceptual backdrop in place for understanding the metatheoretical divide in developmental science, we now turn to concrete instantiations of the divide from three important domains of investigation: the study of epigenesis in development; the study of embodiment and, specifically, embodied cognition; and the construction of baselines for what constitute typical human nature and development. For each domain, we will examine a central issue or point of theoretical debate – one that has inspired empirical activity designed to adjudicate its resolution – to reveal its foundations as a metatheoretical issue or debate, intractable to empirical scrutiny.

Epigenesis

Epigenesis refers to the increase in novelty and complexity of structure and function seen over the course of development. The roots of our present understanding of the concept of epigenesis can be traced back more than 2,500 years to the ancient Greeks. According to the Hippocratic school of ancient Greece, each fertilized egg was thought to contain all the organized structures of the adult organism, but in miniature form. Development simply involved growth of this preformed homunculus. From this perspective, development did not involve an increase in overall complexity over the course of the individual's lifetime, as all the parts and organs were present and in their proper form from the outset. This view was not without its critics. Based on his observations of animal embryonic development, Aristotle (384–322 BC) questioned this decidedly preformationist explanation of development, noting in his text *The Generation of Animals* that in their earliest stages eggs appeared formless and only gradually did embryonic structure take shape. He argued that adult parts were not present at the beginning of development, but rather appeared sequentially as development proceeds, a core tenet of what came to be known as epigenesis.

The refutation of a strict view of preformationism was given a substantial boost in the 18th century with the advent of improved microscopes. This new technology allowed careful observations of embryonic development by Caspar Friedrich Wolff (1733–1794), who documented that different organ systems differentiate and take form consecutively over the course of prenatal development. He emphasized that when organs first become observable, they do not appear in their final form. For example, the intestine of the chick embryo starts as a flat sheet and then becomes a tube. Wolff's findings were confirmed and extended several decades later by Karl Ernst von Baer (1792–1876), whose detailed descriptions of the embryological sequences of fish, birds, and mammals in his 1828 monograph, *On the Development of Animals*, provided an initial map of the process of differentiation, further documenting and providing support for the epigenesis view that development proceeds from the general to the more specific, from the simple to the complex. Von Baer's work supported Aristotle's original insights, providing compelling evidence that every step in development is possible only through the conditions preceding it. The more fine-grained observations allowed by better microscopes essentially resolved the preformationism/epigenesis debate, such that by the middle of the 19th century no scientist could reasonably argue for a strict form of preformationism, given the empirical evidence provided by the emerging science of embryology.

However, the processes or forces by which structural or functional transformation takes place over the course of development were still very open to debate and by the last decades of the 19th century, a new and more nuanced version of preformationism had emerged in biological thought. This view of development held that fertilized eggs contained an array of tiny substances that somehow specified and guided the development of adult form. Perhaps the most well-known version of this argument was promoted by the German biologist August Weismann (1834–1914). Weismann proposed that inherited "determinants," a preformed entity inherited in reproduction each new generation, regulated the course of development. Weismann rejected the notion of epigenesis in favor of his deterministic model of development in which properties of each cell were predetermined in the fertilized egg. In response to Weismann's model, the German zoologist Oscar Hertwig (1849–1922) argued that allocating the casual factors that regulate development to determinants that could not be directly observed was contrary to the scientific method. He argued that the way embryos respond plastically to changing environmental conditions and experimental manipulations posed substantial problems for Weismann's deterministic view. Nevertheless, Weismann's conceptual framework prevailed, thanks in part to the rediscovery of Mendel's work early in the 20th century and the rise of the gene-centric neo-Darwinian view of evolution that coalesced in the 1930s and 1940s. Indeed, Weismann's views provided the basic causal structure used to articulate ideas and distinctions between genotype and phenotype, heredity and development, and evolution and selection for most of the 20th century. A key assumption of these distinctions was that some phenotypic outcomes could be prespecified in the genes, independent of environmental factors and already determined at conception.

Nearly 50 years ago, Gilbert Gottlieb [1970] termed this prespecified conception of development *predetermined epigenesis*. From the predetermined epigenesis perspective, genetic activity gives rise to neural (and other) structures that begin to function when they become mature in the unidirectional sense of genetic activity → structure → function. This notion of a genetic program for development that is directly responsible for many of an individual's phenotypic characteristics was widely embraced across much of 20th-century biology and psychology. Concepts such as *instinctive* or *innate* or *hard-wired* behavior came from this framework and are still in play in some quarters of contemporary psychology, notably cognitive psychology and evolutionary psychology. In contrast to this predetermined epigenesis view of development, Gottlieb [1970] outlined a *probabilistic epigenesis* framework that emphasizes the holistic reciprocity of influences within and between levels of an organism's developmental manifold (genetic activity, neural activity, behavior, and the physical, social, and cultural influences of the external environment) and the ubiquity of gene-environment coaction in the realization of all phenotypes. In line with the evidence now available at all levels of analysis, probabilistic epigenesis holds that there are bidirectional influences within and between levels of analysis so that the appropriate formula for developmental analysis becomes genetic activity ↔ structure ↔ function. This insight is consistent with a Process-Relational research paradigm, with its emphasis on process, activity, change, emergence, and self-organization. Probabilistic epigenesis emphasizes that the roles played by any part process of a Relational-Developmental system – gene, cell, organ, organism, physical, social or cultural environment – is a function of all coacting parts and processes of the system. Gottlieb termed this perspective *relational causality* [Gottlieb & Halpern, 2004]. As such, probabilistic epigenesis provides a powerful antidote to enduring attempts to partition developmental cause or explanation into individual elements, such as nature or nurture, genes or environment, inherited or acquired (i.e., the split understanding of the Cartesian-Split-Mechanistic paradigm [Overton, 2015]).

Within the domains of perceptual and cognitive development, attempts at such partitioning or splitting have centered around the issue to what extent humans are innately prepared to interpret and act on the world and to what extent they rely on learning and experience. Such nativistic perspectives on perception and cognition typically rely (often implicitly) on the assumption of the poverty of the stimulus; that is, the developing individual simply displays too much knowledge or too much skill for experience or learning to be an adequate explanation of outcome. Thus, nativists have proposed that there is a core set of innate concepts – here the term *innate* meaning "biologically determined" rather than its strict definition "present at birth" – that provide the foundation for later learning [e.g., Carey & Markman, 1999; Landau, 2009; Spelke & Kinzler, 2007]. These core concepts are thought to be present in early infancy in the absence of obvious experience and are thus presumed to be biologically prespecified. For example, Spelke and Newport [1998] argued for differential roles of biology and experience, suggesting that a solution to the nature/nurture debate is the "thesis that human knowledge is rooted partly in biology and partly in experience and … that successful explanations of the development of knowledge will come from attempts to tease these influences apart" (p. 323).

The above quote highlights the assumption, still evident in cognitive and evolutionary psychology, as well as behavioral genetics, that "biology" and "experience" are somehow separate, independent elements that provide distinct additive contributions to human development. Importantly, this assumption is not an empirical issue to resolve, but rather is a conceptual one, centering on how developmental "interaction" is characterized. As Lerner and Overton [2017] point out, within a Cartesian-Split-Mechanistic paradigm, "interaction" is conceptualized as two or more entities that function independently in cooperative or competitive ways. For example, assumptions are made that some components (i.e., genes) serve as the informational source for construction of a developmental outcome, whereas other components (i.e., environment) merely serve to actualize (or not) that source information. In contrast, probabilistic notions of "interaction" – framed within a Process-Relational paradigm – emphasize the relational interpenetrations of hierarchical levels and processes, such that the components of a relational developmental system cannot be defined independently of the relations to other components and to the system as a whole. As such, the focus of probabilistic epigenesis notions of interaction is on the relations themselves, not what the components bring to these relations construed independently of the relations. From this framework, the influence that any component may have on other parts can extend to higher or lower levels, or remain at the same level.

Keller [2010] captured the tension between these contrasting conceptualizations of interaction, noting that:

> Not only is it a mistake to think of development in terms of separable causes, but it is also a mistake to think of development of traits as a product of causal elements interacting with one another. Indeed, the notion of interaction presupposes the existence of entities that are at least ideally separable – i.e., it presupposes an a priori space between component entities – and this is precisely what the character of developmental dynamics precludes. Everything we know about the processes of inheritance and development teaches us that the entanglement of developmental processes is not only immensely intricate, but it is there from the start. From its very beginning, development depends on the complex orchestration of multiple courses of action that involve coactions among many different kinds of ... components.
>
> (pp. 6–7)

Despite the fact that predetermined epigenesis and its unidirectional argument have been empirically refuted across multiple disciplines, conceptual vestiges of the preformationism/epigenesis debate remain in play in contemporary developmental psychology (more often than not, implicitly) because two fundamentally different conceptualizations of how to characterize the relations between biology and experience are routinely applied within our discipline. These alternate conceptualizations of gene-environment relations reviewed above are not limited to psychology, but also extend to biology and are particularly evident in the growing field of *epigenetics* [Witherington & Lickliter, 2017]. Despite the fact that empirical findings from epigenetics demonstrating the fallacy of partitioning genetic and environmental contributions to development could be viewed as an endorsement of Process-Relational metatheory, epigenetics research often emphasizes molecular levels of causality. Indeed, some quarters of epigenetics continue to regard genes and their environments as independent sources of information for the "shaping" of developing organisms [for examples, see Lickliter & Witherington, 2017]. This contemporary instantiation of the "interaction debates" [Tabery, 2014] demonstrates a key theme of our argument that such debates are metatheoretical in nature and as such require conceptual and not empirical analysis for successful resolution. Given that conceptualizations of development within contemporary epigenetics still routinely trade in a reductionist privileging of molecular over molar levels of explanation *and* in a continued reductionist focus on separate and distinct roles for genes and environment in any given developmental relation [Lickliter & Witherington, 2017], data alone will not resolve how to include a developmental point of view in contemporary epigenetics. To render epigenetics truly developmental, we need to dismiss the idea that any component of a system is a privileged informational source of new levels of phenotypic organization in development, relative to any other component of the system. This reconceptualization emphasizes that epigenetic processes are always emergent properties of historical and situated relations across multiple levels of biological organization.

Embodiment

The conceptual domain of embodiment concerns itself with the nature of the body's role in the functioning of the organism. Although the origins of research on embodiment go back much further, calls for a more embodied cognitive science became increasingly widespread in the early 1990s [e.g., Brooks, 1991; Hutchins, 1995; Varela, Thompson, & Rosch, 1991]. This paved the way for the field of "embodied cognition" and its growing prominence within psychological and developmental science today. Although there are various flavors of embodiment, proponents of an embodied approach to cognition are united in their dissatisfaction with the cognitivist metamodel of mind that arose during the "cognitive revolution" of the mid-to-late 20th century. In the cognitivist metamodel, the mind is an information processor that carries out computations over representations to which meaning has been preassigned. Such a view of the mind was encouraged by an emphasis on cognitive processing as a separate level of analysis [Marr, 1982] in combination with the functionalist idea that computational approaches to cognition could be pursued independently from considerations of the bodies and brains of living, acting individuals.

The central premise of embodiment presents a distinct challenge to the idea that cognition can be split off and studied at an isolated level of information processing. Instead, the processing of information can only be understood in the context of an active, agentive individual, which challenges the cognitivist notion of the individual as a passive receiver of information that is processed by a computational mind. In the late 1990s, one highly visible direction in embodied cognition attempted to extend the scope of computation to include objects in the local environment [Clark & Chalmers, 1998]. Instead of computations being carried out solely within the head, Clark and Chalmers proposed that they are carried out across a distributed system involving brain, body and trusted, local resources (e.g., a notebook or smartphone). However, on this view of the "extended mind," the body still only acts as a causal influence on the mind, and the account of cognition it provides is still fundamentally a computational one. Other approaches to the study of embodiment, in contrast, have rejected Cartesian grounded computational approaches and see the body as not merely a causal influence on cognition but as *constitutive of* cognition. These approaches generally fall under the umbrella of *enactivism*, a central premise of which is that the nature of the body an organism has – or more accurately, the nature of its embodiment – affords a range of possibilities for action, and it is this range of possibilities that gives rise to the particular world that is brought forth or "enacted" by the activity of the individual [Di Paolo et al., 2017; Stewart et al., 2010].

This question over whether bodies are constitutive of, or simply causally related to, cognitive processing remains a focal point of debate within scientific discourse on embodied cognition [Marshall, 2018; Menary, 2010; Rowlands, 2010; Wheeler, 2005]. However, the debate itself begs another, more pressing (but largely ignored) question: what level of conceptual inquiry – theoretical or metatheoretical – is appropriate for the debate's adjudication? Proponents on both sides routinely marshal empirical evidence to support their position and just as routinely frame the debate in terms of competing hypotheses [e.g., Aizawa, 2007; Block, 2005; Noe, 2004], Shapiro [2011], in an extensive review of the debate, has explicitly characterized the debate as theoretical in nature, situating it squarely within the realm of data interpretation and testability. Generally speaking, orthodox sentiment within embodied cognition circles seems inclined to regard the field's central debate as one of theoretical, rather than metatheoretical, concern. Such sentiment, however, is itself predicated on the philosophical-metatheoretical belief that psychological attributes, such as thinking, can be sensibly applied to parts of an organism – that it even makes sense to consider the possibility. And such a belief is rooted in the conceptual confusions of a Cartesian paradigm [Bennett & Hacker, 2003].

As noted by Di Paolo et al. [2017], considerations of embodiment are often "interpreted as amendments, improvements to the computational metaphor of the mind, but not as alternatives to it" (p. 13). However, viewing embodied cognition as an interesting add-on to mainstream theorizing fails to acknowledge that embodiment *as a concept* challenges the severing of mind and body that originated with Descartes and that forms the basis of the Cartesian-Split-Mechanistic research paradigm on which the metamodel of cognitivism is based. This incompatibility presents a distinct problem, one that can only be addressed by recognizing that the *metatheoretical* foundation of embodiment cannot be the "Cartesian organism" metamodel of cognitivism. A full consideration of embodiment, in other words, necessitates a radical, metatheoretical shift away from mainstream conceptualizations of mental life. Such a shift is evident in enactivist accounts. According to enactivism, there is no external world to be represented in the way that cognitivism assumes, and there is a rejection of the Cartesian notion that knowledge given by the external world is stored inside individual minds. In the enactivist account, the knower and the known world mutually specify each other. This idea has profound implications, since it suggests that the world experienced by an organism depends more on its own embodied activity than on internalizing the facts that are "given" by an independent external world. At this point, the complete incompatibility of embodiment with cognitivism and the metatheoretical assumptions on which it is based become very apparent, and a radically different account is required.

By emphasizing the codetermination of individual and environment, the enactivist perspective presents a radical solution to the metatheoretical problems of cognitivism. As such, adopting this perspective requires a change in metatheoretical frame, else it risks being untethered from scientific inquiry. One suggestion is that placing the study of embodiment in a relational metatheoretical frame provides a way forward [Marshall, 2018; Overton, 2008]. One specific way forward comes from the Process-Relational paradigm, in which embodiment is seen as a bridge construct relating three standpoints on the relation of the body to mental life. As discussed by Overton [2008], the body as form references the biological standpoint, the body as lived experience references a phenomenological or personal psychological standpoint, and the body as actively engaged with the world references a sociocultural or contextual standpoint. Seen in this way, embodiment bridges between a subpersonal level of ongoing physiological activity – the biological processes going on in the body at any given moment – and a personal level of acts that are intentional and goal-directed. It is through these acts that meaning is projected, and it is these acts that transform the objective world into the world the individual actually experiences. In contrast to the cognitivist perspective, a relational view of embodiment therefore allows consideration of how the world comes to have meaning for the developing individual [Marshall, 2016, 2018]. This rejection of the Cartesian split of cognitivism thus allows embodiment to play a key role in an integrative science of mental life that now enables considerations of identity, autonomy, and meaning.

Although much of the extant work on embodied cognition has not included a developmental aspect, it is clear from the above discussion that the construct of embodiment is closely intertwined with considerations of ontogeny. That said, there may be a bias in developmental science toward seeing notions of embodiment as most applicable to the study of infants. This bias perhaps stems from the notion that sensorimotor influences on cognition are more salient in the prelinguistic period, with later representational thought being more abstract and disconnected from the body (i.e., split off from embodiment). In addressing this bias, we should note that the construct of embodiment is indeed particularly relevant to the question of how intentionality, in terms of symbolic, reflective knowledge, feeling, and meanings, emerges from engaged and embodied actions in the first months and years of life [Overton, 2008]. However, embodiment should not be considered to be more salient to one particular stage of development. As a core construct in the Process-Relational paradigm, embodiment constitutes a necessary defining feature of *all* developmental change processes [Overton, 2015]. As such, it is the activity of the fully embodied individual that allows for the construction of meaning across the lifespan. Furthermore, the thread of embodiment runs not only throughout the individual lifespan, but across phylogenetic time [Marshall, 2016]. As we note below, achieving clarity in the understanding of evolutionary influences on human development can only come through closely considering embodied interrelations between brain, body, and culture. A wider acceptance of these deeper understandings about embodiment has the potential to inform and connect multidisciplinary research across all domains of human development.

Baselines for Human Nature and Development

A critical task for any responsible science is the construction of baselines, or starting points for measurement against which manipulated data or multiple samples can be compared. Developmental scientists routinely align the notion of baseline measurement with what constitutes normal or typical psychological functioning in some domain of interest and just as routinely take for granted that samples tested are representative of "normal" human beings who are, in turn, representative of their species. To gauge such baselines for normality – for what constitutes the "natural," species-typical human – they frequently employ statistical analyses of central tendencies, leaving the discipline with the hope that baselines can be established (and adjudicated) *empirically*, without further information. But such a hope flies in the face of a stark reality. Metatheoretical considerations guide the very question of what constitutes proper data for construction of a baseline measurement. After all, what researchers choose to empirically sample necessarily depends on a set of pre-empirical assumptions that guide their notions of "normality." How, for example, do oceanographers today avoid assuming that it is normal for oceans to be acidic or a soil biologist avoid the idea that farmland soils are typically nutrient poor? They take in the larger picture – a view across generations – and focus on the complexities of a well-functioning ecological system that is comprised of many entities. From this broad and deep perspective, they construct what a healthy ocean or healthy soil looks like. They construct the complexities of a healthy ecological system and the limited range in which it functions. They then come to regard the ecological systems that oceanography and soil science are working with today as fundamentally broken. So far, within psychology only those on the margins have noted the broken nature of humans studied by the discipline [e.g., Kidner, 2001]. Instead, the widespread psychopathologies documented throughout psychology are assumed to be "normal" for humanity by those who study them [e.g., Cicchetti & Toth, 2009].

What sort of philosophical-metatheoretical beliefs guide baseline construction in developmental science? Conventionally, psychologists have adopted a baseline for human nature grounded in notions of "civilization" and "progress." They have, in other words, appealed to assumptions shaped by Abrahamic religions as well as Greek and Enlightenment philosophies, assumptions that regard humanity's nature as discontinuous (i.e., split) from, and more intelligent than, that of other animals. In the realm of modern scientific inquiry, such a metatheoretical stance can be traced directly to Descartes and his over-intellectualized, rationalist assertions that only humans possess thought, true feeling, and conscious awareness [Malcolm, 1977]. It is a stance that regards the "natural" human as detached observer of both the world and her or his own body, a view that considers as baseline for everyday psychological functioning the Western ideal of the dispassionate, scientific or reflective attitude, considered the pinnacle of "civilized" functioning. Such a view is emblematic of the Cartesian-Split-Mechanistic paradigm and undoubtedly underlies assumptions – still all too common in developmental science – that date from WEIRD societies (Western, Educated, Industrialized, Rich, Democratic – 12% of the world's population [Henrich, Heine & Norenzayan, 2010]) are perfectly representative of baseline psychological functioning because they come from societies with the most "progress."

What is missing from this "civilization" and Cartesian-inspired perspective is a sense of humanity across time and *within* a natural ecology (e.g., a sense of humans as social mammals). For most of human genus history, an *indigenous* perspective has prevailed among human societies, one that is grounded in notions of "relation" and "embeddedness" [Narvaez, 2013; Four Arrows & Narvaez, 2016]. It is a perspective that considers humans as part of and partnered with Nature, not split off from or superior to it [Christen, Narvaez & Gutzwiller, 2017]. Nature preserves humans just as the ecologies of the landscape require careful attention and humble usage to preserve the well-being of the biocommunity. Such a focus of native science is one of partnership not dominance – of respectful relations [Cajete, 2000]. From its vantage point, "civilized" humans look uprooted and improperly reared, even crazy and immoral – a poor representative of humanity's potential [Sahlins, 2008]. In the realm of modern scientific inquiry, this relational alternative clearly maps onto the Process-Relational research paradigm.

Adopting a Process-Relational paradigmatic approach to the construction of species-typical baselines for human functioning and development necessitates a transdisciplinary, historically rich approach. From ethology, anthropology, and archeology, we learn that humanity spent 99% of its existence in small-band hunter-gatherer communities (SBHG). These societies have been studied all over the world by anthropologists in the past century or so, and they are described in first-contact diaries from explorers and others who invaded their lands. Interestingly, these societies display among themselves similar adult personalities. That is, SBHG are typically described as calm, content, generous, independent, and communal [Ingold, 2005; Narvaez, 2013]. One of the key influences for such uniformity in personality appears to be the similar ways young children are raised and develop. These societies provide humanity's developmental niche [Hewlett & Lamb, 2005].

Like other animals, humans evolved a developmental niche that matches up with typical developmental patterning of the young and that counts among the many things beyond genes that humans inherit from their ancestors [Gottlieb, 2002; Oyama, Griffiths & Gray, 2001; West-Eberhard, 2003]. Humans are social mammals, a lineage that emerged over 30 million years ago with an intensive niche that includes breastfeeding on request, caregiver responsiveness and affection, and self-directed social play. The human developmental niche adds to the mammalian nest multiple allomothers and positive social support for mother and baby, as well as soothing perinatal experiences. The human developmental nest is particularly intensive due to the move to bipedalism resulting in a narrowed pelvis, requiring children to be born highly immature – looking like fetuses of other animals until around 18 months of postnatal age [Trevathan, 2011]. Humans are highly malleable after birth, more so than chimpanzees, with multiple sensitive periods for development [Gómez-Robles, Hopkins, Schapiro & Sherwood, 2015]. At full-term birth, the baby emerges with only 25% of typical adult brain volume (typical, that is, of a healthy, young adult brain), and the baby's brain grows especially rapidly in the first year, with 90% of adult volume in place by age 5 [Trevathan, 2011]. Systems developed during these years include the stress response, vagus nerve, endocrine systems, neurotransmitters, immune system, and sociality. These are all influenced by the quality of the developmental nest [for reviews, see Narvaez, Panksepp, Schore & Gleason, 2013]. From current evidence, we can surmise that the evolved nest contributes significantly to the SBHG personality and even worldview.

Framing the "natural" human within a Relational-Developmental systems metamodel suggests that modern developmental science is studying and drawing conclusions from species-*atypical* individuals whose behavior represents adjustment to an *atypical* context from birth onwards – a context of domestication in the form of "civilization." According to Kidner [2001], modern industrial society encourages us both to deny that we are a particular kind of animal with an evolutionary history and to pretend that technology frees us from nature and its constraints. Developmental science has long played a role in maintaining this illusion of separation and detachment by implying that "the person studied as an isolated entity separate from culture or nature is either whole or healthy" and that "alternative forms of personhood are somehow necessarily deficient" [Kidner, 2001, p. 56]. In a time of planetary ruin by the dominant culture, it may be especially appropriate for psychologists to re-examine their metatheoretical framework for conceptualizing baselines.

Conclusion

For each of the three developmental domains reviewed in this chapter, conceptual analysis uncovers the metatheoretical underpinnings of a core issue or theoretical debate within the domain. We cannot overstate the critical importance of conceptual analysis as a necessary first step in any and all research activity. Only through conceptual analysis can questions in science that are legitimately subject to empirical analysis (the theoretical) be readily discriminated from those that require conceptual analysis for adjudication (the metatheoretical). And only when competing theories are established as metatheoretically compatible – are predicated, that is, on the same basic ontological and epistemological assumptions – can empirical investigation deliver theoretical resolution. It has become progressively clear that both the natural and social sciences are now awash in metatheoretical debate, given the prominence enjoyed in recent decades by the Process-Relational research paradigm as a viable metatheoretical alternative to the orthodoxy of a Cartesian-Split-Mechanistic paradigm [Kuhn, 1970; Latour, 2004; Lickliter & Honeycutt, 2015; Overton, 2015; Smolin, 2013]. However, longstanding adherence within developmental science to the instrumentalist tradition, coupled with the ever-increasing sway that state-of-the-art technology holds over disciplinary inquiry, make it easy to forget that some of our field's most critical problems can only be solved through conceptual analysis. Wittgenstein [1958] famously characterized psychology as a science predicated on "experimental methods and conceptual confusion" (p. xiv). Now more than ever, we need to take this characterization to heart – and repair it.

References

Aizawa, K. (2007). Understanding the embodiment of perception. *Journal of Philosophy*, *104*, 5–25.

Bennett, M.R., & Hacker, P.M.S. (2003). *Philosophical foundations of neuroscience*. Malden, MA: Blackwell.

Bernstein, R.J. (1983). *Beyond objectivism and relativism: Science, hermeneutics, and praxis*. Philadelphia, PA: University of Pennsylvania Press.

Block, N. (2005). Review of Alva Noe. *Journal of Philosophy*, *102*, 259–272.

Brooks, R.A. (1991). New approaches to robotics. *Science*, *253*, 1227–1232.

Cajete, G. (2000). *Native science: Natural laws of interdependence*. Santa Fe, NM: Clear Light.

Carey, S., & Markman, E.M. (1999). Cognitive development. In B.M. Bly & D.E. Rumelhart (Eds.), *Cognitive science* (pp. 201–254). New York, NY: Academic Press.

Christen, M., Narvaez, D., & Gutzwiller, E. (2017). Comparing and integrating biological and cultural moral progress. *Ethical Theory and Moral Practice*, *20*, 55.

Cicchetti, D., & Toth, S. (2009). The past achievements and future promises of developmental psychopathology: the coming of age of a discipline. *Journal of Child Psychology and Psychiatry*, *50*, 16–25.

Clark, A., & Chalmers, D. (1998). The extended mind. *Analysis*, *58*, 7–19.

Descartes, R. (1996). *Discourse on method and meditations on first philosophy* (D. Weissman, Ed.). New Haven, CT: Yale University Press.

Di Paolo, E., Buhrmann, T., & Barandiaran, X. (2017). *Sensorimotor life: An enactive proposal*. New York, NY: Oxford University Press.

Ford, D. H., & Lerner, R. M. (1992). *Developmental systems theory: An integrative approach*. Newbury Park, CA: Sage.

Four Arrows, & Narvaez, D. (2016). Reclaiming our indigenous worldview: A more authentic baseline for social/ecological justice work in education. In N. McCrary & W. Ross (Eds.), *Working for social justice inside and outside the classroom: A community of teachers, researchers, and activists* (pp. 93–112). In series, *Social justice across contexts in education* (S.J. Miller & L.D. Burns, Eds.). New York, NY: Peter Lang.

Gomez-Robles, A., Hopkins, W.D., Schapiro, S.J., & Sherwood, C.C. (2015). Relaxed genetic control of cortical organization in human brains compared with chimpanzees. *Proceedings of the National Academy of Sciences*, *12*, 14799–14804.

Gottlieb, G. (1970). Conceptions of prenatal development. In L.R. Aronson, D.S. Lehrman, E. Tobach, & J.S. Rosenblatt (Eds.), *Development and evolution* of behavior (pp. 111–137). San Francisco, CA: Freeman.

Gottlieb, G. (2002). On the epigenetic evolution of species-specific perception: The developmental manifold concept. *Cognitive Development, 17*, 1287–1300.

Gottlieb, G. & Halpern, C.T. (2004). A relational view of causality in normal and abnormal development. *Development and Psychopathology, 14*, 421–435.

Hacker, P.M.S. (2009). Philosophy: A contribution not to human knowledge but to human understanding. In A. O'Hear (Ed.), *The nature of philosophy, Royal Institute of Philosophy Supplement 65* (pp. 129–153). Cambridge, UK: Cambridge University Press.

Henrich, J., Heine, S.J., & Norenzayan, A. (2010). The weirdest people in the world? *Behavioral and Brain Sciences, 33*, 61–83.

Hewlett, B.S., & Lamb, M.E. (2005). *Hunter-gatherer childhoods: Evolutionary, developmental and cultural perspectives*. New Brunswick, NJ: Aldine.

Hutchins, E. (1995). How a cockpit remembers its speeds. *Cognitive Science, 19*, 265–288.

Ingold, T. (2005). On the social relations of the hunter-gatherer band. In R.B. Lee & R. Daly (Eds.), *The Cambridge encyclopedia of hunters and gatherers* (pp. 399–410). New York, NY: Cambridge University Press.

Keller, E.F. (2010). *The mirage of the space between nature and nurture*. Cambridge, MA: MIT Press.

Kidner, D.W. (2001). *Nature and psyche: Radical environmentalism and the politics of subjectivity*. Albany, NY: State University of New York.

Kuczynski, L., & De Mol, J. (2015). Dialectical models of socialization. In W.F. Overton & P.C. M. Molenaar (Vol. Eds.) & R.M. Lerner (Ed.-in-Chief), *Handbook of child psychology and developmental science. Vol. 1: Theory & Method* (7th ed., pp. 323–368). Hoboken, NJ: Wiley.

Kuhn, T.S. (1970). *The structure of scientific revolutions*, 2nd ed. Chicago, IL: University of Chicago Press.

Lakatos, I. (1978). *The methodology of scientific research programmes. Philosophical Papers, Vol. 1.* Cambridge, UK: Cambridge University Press.

Landau, B. (2009). The importance of the nativist-empiricism debate: Thinking about primitives without primitive thinking. *Child Development Perspectives, 3*, 88–90.

Latour, B. (2004). *Politics of nature*. Cambridge, MA: Harvard University Press.

Laudan, L. (1977). *Progress and its problems: Towards a theory of scientific growth*. Berkeley, CA: University of California Press.

Lerner, R.M., & Overton, W.F. (2017). Reduction to absurdity: Why epigenetics invalidates all Models involving genetic reduction. *Human Development, 60*, 107–123.

Lickliter, R., & Honeycutt, H. (2015). Biology, development, and human systems. In W.F. Overton, & P.C.M. Molenaar (Vol. Eds.) & R.M. Lerner (Ed.-in-Chief), *Handbook of child psychology and developmental science. Vol. 1: Theory & Method* (7th ed., pp. 162–207). Hoboken, NJ: Wiley.

Lickliter, R., & Withingerton, D.C. (2017). Towards a truly developmental epigenetics. *Human Development, 60*, 124–138.

Machado, A., & Silva, F.J. (2007). Toward a richer view of the scientific method: The role of conceptual analysis. *American Psychologist, 62*, 671–681.

Malcolm, N. (1977). *Thought and knowledge*. Ithaca, NY: Cornell University Press.

Marr, D. (1982). *Vision*. San Francisco, CA: W.H. Freeman.

Marshall, P.J. (2016). Embodiment and human development. *Child Development Perspectives, 10*, 245–250.

Marshall, P.J. (2018). Embodiment. In A.S. Dick, & U. Muller (Eds.), *Advancing Developmental Science: Philosophy, Theory, and Method* (pp. 29–40). New York, NY: Psychology Press.

Menary, R. (2010). *The extended mind*. Cambridge, MA: MIT Press.

Narvaez, D. (2013). The 99%-development and socialization within an evolutionary context: Growing up to become "A good and useful human being." In D. Fry (Ed.), *War, peace and human nature: The convergence of evolutionary and cultural views* (pp. 643–672). New York: Oxford University Press.

Narvaez, D., Panksepp, J., Schore, A., & Gleason, T. (Eds.) (2013). *Evolution, early experience and human development: From research to practice and policy*. New York, NY: Oxford University Press.

Noë, A. (2004). *Action in perception*. Cambridge, MA: MIT Press.

Overton, W.F. (1998). Developmental psychology: Philosophy, concepts, and methodology. In W. Damon (Series Ed.) & R.M. Lerner (Vol. Ed.), *Theoretical models of human development: Vol. 1, Handbook of child psychology* (5th ed., pp. 107–188). New York, NY: John Wiley & Sons.

Overton, W.F. (2006). Developmental psychology: Philosophy, concepts, methodology. In W. Damon, & R.M. Lerner (Series Ed.) & R.M. Lerner (Vol. Ed.), *Theoretical models of human development: Vol. 1, Handbook of child psychology* (6th ed., pp. 18–88). Hoboken, NJ: John Wiley & Sons.

Overton, W.F. (2008). Embodiment from a relational perspective. In W.F. Overton, U. Mueller, & J.L. Newman (Eds.), *Developmental perspectives on embodiment and consciousness*, (pp. 1–18). New York, NY: Erlbaum.

Overton, W.F. (2012). Evolving scientific paradigms: Retrospective and prospective. In L. L'Abate (Ed.), *Paradigms in theory construction* (pp. 31–66). New York, NY: Springer.

Overton, W.F. (2015). Process and relational developmental systems. In W.F. Overton & P.C. Molenaar (Eds.), *Handbook of child psychology and developmental science, Vol. 1: Theory and method* (pp. 9–62). Hoboken, NJ: Wiley.

Overton, W.F., & Lerner, R.M. (2014). Fundamental concepts and methods in developmental science: A relational perspective. *Research in Human Development*, 11, 63–73.

Oyama, S., Griffiths, P.E., & Gray, R.D. (2001). *Cycles of contingency: Developmental systems and evolution.* Cambridge, MA: MIT Press.

Popper, K. (1959). *The logic of scientific theory.* New York, NY: Basic Books.

Rowlands, M. (2010). *The new science of the mind: From extended mind to embodied phenomenology.* Cambridge, MA: MIT Press.

Sahlins, M. (2008). *The Western illusion of human nature.* Chicago, IL: Prickly Paradigm Press.

Shapiro, L. (2011). *Embodied cognition.* New York, NY: Routledge/Taylor & Francis.

Slife, B.D., & Williams, R.N. (1995). *What's behind the research? Discovering hidden assumptions in the behavioral sciences.* London, UK: Sage.

Smedslund, J. (1991). The pseudoempirical in psychology and the case for psychologic. *Psychological Inquiry*, 2, 325–338.

Smolin, L. (2013). *Time reborn: From the crisis in physics to the future of the universe.* New York, NY: Houghton Mifflin Harcourt.

Spelke, E., & Kinzler, K.D. (2007). Core knowledge. *Developmental Science*, 10, 89–96.

Spelke, E., & Newport, E. (1998). Nativism, empiricism, and the developmental of knowledge. In R.M. Lerner (Ed.), *Handbook of child psychology. Vol. 1. Theoretical models of human development* (5th ed., pp. 275–340). New York, NY: Wiley.

Stewart, J., Gapenne, O., & Di Paolo, E.A. (Eds.) (2010). *Enaction: Toward a new paradigm for cognitive science.* Cambridge, MA: MIT Press.

Strawson, P.F. (1959). *Individuals: An essay in descriptive metaphysics.* London, UK: Methuen & Co.

Tabery, J. (2014). *Beyond versus: The struggle to understand the interaction of nature and nurture.* Cambridge, MA: MIT Press.

Trevathan, W.R. (2011). *Human birth: An evolutionary perspective*, 2nd ed. New York, NY: Aldine de Gruyter.

Varela, F.J., Thompson, E., & Rosch, E. (1991). *The embodied mind: Cognitive science and human experience.* Cambridge, MA: MIT Press.

Wachtel, P.L. (1980). Investigation and its discontents: Some constraints on progress in psychological research. American Psychologist, 35, 399–408.

Watkins, J.W.N. (1975). Metaphysics and the advancement of science. *British Journal for the Philosophy of Science*, 26, 91–121.

Wakefield, J.C. (2007). Why psychology needs conceptual analysts: Wachtel's "discontents" revisited. *Applied & Preventive Psychology*, 12, 39–43.

West-Eberhard, M.J. (2003). *Developmental plasticity and evolution.* Oxford, UK: Oxford University Press.

Wheeler, M. (2005). *Reconstructing the cognitive world: The next step.* Cambridge, MA: MIT Press.

Witherington, D.C. (2015). Dynamic systems in developmental science. In W.F. Overton & P.C.M. Molenaar (Vol. Eds.) & R.M. Lerner (Ed.-in-Chief), *Handbook of child psychology and developmental science. Vol. 1: Theory & Method* (7th ed., pp. 63–112). Hoboken, NJ: Wiley.

Witherington, D.C., & Lickliter, R. (2017). Transcending the nature-nurture debates through epigenetics: Are we there yet? *Human Development*, 60, 65–68.

Wittgenstein, L. (1958). *Philosophical investigations* (G.E.M. Anscombe, transl.) (3rd ed.). Upper Saddle River, NJ: Prentice Hall.

4.
THE FAILURE OF BIOGENETIC ANALYSIS IN PSYCHOLOGY
Why Psychology Is Not a Biological Science
Gary Greenberg*

Many define psychology as a biological science and emphasize brains and genes as major determinants of behavior. Instead, it is argued here that psychology is a unique biopsychosocial science able to stand on its own. Biogenetic processes are indeed relevant but are simply participating, not causal, factors in behavioral origins. Long neglected by biologists and social scientists, the importance of developmental processes is emphasized. The author takes issue with behavior geneticists and argues that development is bidirectional—internal and environmental phenomena influence behavior—probabilistically. The author favors a relatively new model with roots in ideas from contemporary physics: emergence and self-organization—"relational developmental systems."

"It is always best to start at the beginning" said Glinda the *Wizard of Oz's* Good Witch. Advice sometimes comes from strange sources. I begin this chapter, then, with a definition of *psychology* as the biopsychosocial science of behavior. Doing so summarizes the essence of my contribution to this special issue of *Research in Human Development*. Although I will use this chapter as one more critique of behavior genetics, and therefore as a further attempt to "explode the gene myth" (e.g., Hubbard & Wald, 1993), I have criticized elsewhere the full extent of the biogenetic approach to psychology (Greenberg, 2007; Greenberg & Lambdin, 2007; Greenberg & Partridge, 2010). It is safe to say that most scientists, including most psychologists, believe psychology to be a biological science and that, as one prominent behaviorist put it, "Psychology ... can be completely explained in the language and data of neurophysiology—in principle if not in fact" (Uttal, 2005, p. 155). Of course, the general public is led to believe this view as well as this statement by a *New York Times* columnist makes clear:

> In the 1950s, the common view was that humans begin as nearly blank slates and that behavior is learned through stimulus and response. Over the ages, thinkers have argued that humans are divided between passion and reason, or between the angelic and the demonic. But now the prevailing view is that brain patterns were established during the millenniums when humans were hunters and gatherers, and we live with the consequences. Now, it is generally believed, our behavior is powerfully influenced by genes and hormones. Our temperaments are shaped by whether we happened to be born with the right mix of chemicals.
>
> (Brooks, 2006, p. 14)

* Wichita State University, Wichita, KS.

DOI: 10.4324/9781032702988-8

I was schooled in traditions that understood psychology differently—not as a biological science, but as a science that stands on its own, a unique psychological science, a natural science, consistent and compatible with the principles of other natural sciences. This perspective is nowhere made so clear as in a recent article by Overton and Müller (2013). This idea is, of course, not new being the position as well, of J. R. Kantor(1959), of B. F. Skinner (1953), and of T. C. Schneirla (Aronson, Tobach, Rosenblat, & Lehrman, 1972). I later became associated with Gottlieb's (2004) probabilistic epigenesis, Lerner's (1989, 1998) developmental contextualism, and Overton's (2010) relational, developmental systems view. From the perspective of these approaches to psychology, behavior is seen to be not a biological property of organisms, but a biopsychosocial property. Of course, we are biological organisms before we become psychosocial creatures; biology, however, is simply another participating factor in behavioral origins, not the causative factor.

The tendency to see wholes as the mere sums of their parts, that is, reductionism, still holds sway in many quarters of science. Today psychology and behavior are seen by many to be reducible to biogenetic substances—to brains and genes specifically. It is an unfortunate development that 21st-century psychology finds itself still hampered by reductionistic and counterfactual biological thinking: Of course, much of this reductionism is driven more by ideology than by pure science, as reflected in Lewontin's (1991) criticism of this controversy, which he titled *Biology as Ideology*. Part of the reasons for this biogenic focus was summed up by the executive director for Science of the American Psychological Association: "[Today's] newest age of reductionism is being fueled by the federal funding agencies, the Congress, and by the general public. Everyone seems to think that focusing on ever finer grains of sand will hasten cures for the worst of human afflictions and produce enormous leaps forward in our understanding of the human condition" (Breckler, 2006, p. 23).

This perspective was leant significant support by two major research efforts at the end of the 20th century: the Decade of the Brain (Tandon, 2000) and the Human Genome Project (2011). One important goal of these efforts was to elucidate the neural and genetic underpinnings of behavior. Thus, it may be understandable why biology, and more specifically why brains and genes, are seen to control behavior and why psychology is understood to be a biological science. Of course the biological foundations of behavior are indisputable; however, evolution, genetics, hormones, and neurophysiology are not, even together, the sole causative determinants of behavior. They are all necessary, although not sufficient, participating factors in the development of behavior.

The Decade of the Brain (Tandon, 2000) and the Human Genome Project (2011) purported to put to rest the search for the origins of behavior—to partition out the relative contributions to behavior of biology (nature) and environment (nurture), one assumption being that brains and genes are separable from and more important than environments and experiences. Although these two enormously expensive efforts yielded much significant and important information about the brain and the genome, their impact on our understanding of neural and genetic influences on behavior were minimal (Lewontin, 2000, 2011; Strohman, 1997). Although both efforts came down on the nature side of the nature/nurture equation,

> The Decade of the Brain has led to a realization that a comprehensive understanding of the brain cannot be achieved by a focus on neural mechanisms alone, and advances in molecular biology have made it clear that genetic expressions are not entirely encapsulated, that heritable does not mean predetermined.
>
> (Cacioppo, Bernston, Sheridan, & McClintock, 2000, p. 836)

In addition, both projects failed to "take development seriously" (Robert, 2004, p. xiii). Indeed, the idea of development has been sidestepped and neglected by biologists and psychologists.

There was, however, a small group of 20th-century psychologists who understood the proper role of developmental biological phenomena in behavioral origins and development (Gottlieb, 1992; Kantor, 1959; Kuo, 1967; Schneirla, 1957). Although there are important differences in the systems outlined by these important contributors to modern psychology, there are as well crucial shared ideas, including the significance of a multilevel approach, the importance of history and development, and the contextual nature of behavior. Gottlieb, being the most contemporary member of this group, had the advantage of being able to apply our more recent understanding of ideas in biology and the other sciences that we now know significantly affect behavior. Gottlieb stressed that development was bidirectional; internal cellular and external contextual factors all influence the development of behavior (as well as structures). These influences are not causal, but, rather, probabilistic.

Surprisingly, the perennial question of which contributes more to behavior, biology (nature), or experience (nurture) is not dead, a point made abundantly clear in Lewontin's (2011) recent review of Keller's 2010 book, *The Mirage of a Space Between Nature and Nurture*. It is worth quoting Lewontin:

> [We are in] an era when biological—and specifically, genetic—causation is taken as the preferred explanation for all human physical differences. Although the early and mid-twentieth century was a period of immense popularity for genetic explanations of class and race differences in mental ability and temperament, especially among social scientists, such theories have now virtually disappeared from public view, largely as a result of a considerable effort of biologists to explain the errors of those claims. The genes for IQ have never been found .... "DNA" has replaced "IQ" as the abbreviation of social import.
> (p. 26)

I don't usually find fault with Lewontin on such issues. He is, after all, one of the world's leading geneticists and has been consistently critical of biological determinism. But, though he went on to emphasize that genes are no longer seen to play the sole major role in biological disorders (diseases), I take issue with his assessment regarding the historical passing of ideas about genes and psychological characteristics. Diehard reductionists still exist, and they continue to try to find ways to split nature and nurture. Lerner (2004) agreed, pointing out that many behavioral scientists, psychologists especially, continue to believe that behavior genetics provides evidence for the inheritance of behaviors such as intelligence, parenting, morality, and even television viewing!

Research in molecular biology, in genetics, is neither easy nor inexpensive. It is, of course, necessary. Lerner (2004), for example, cautioned that

> We are at a point in the science of human development where we must move on to the more arduous task of understanding the integration of biological and contextual influences in terms of the developmental system of which they are a dynamic part.
> (p. 20)

Such arduous relational, developmental systems research in the last few decades has revealed much about the nature and functioning of genes (Hood, Halpern, Greenberg, & Lerner, 2010). This research has caused us to dispense with a number of ideas we once accepted as gospel. These ideas include the notion that single genes affect single traits: eye color, for example. Although some single gene/single traits are claimed to exist, this idea has been disputed and dismissed in literature now more than 80 years old (e.g., Jennings, 1924). In addition, the common mode is for genes to act in concert with others, that is, most characteristics are polygenetic. Genes exist in a cell that has many components, all of which function in a manner akin to chemicals in a test tube. Everything in the test tube affects everything else in that test tube; so too, everything in the cell affects genes.

Genes are, in essence, catalysts. Genes influence other genes, turning some on and some off. In addition, the chemistry of the cell is very much influenced by external factors—an obvious example being a person's diet. Campbell (1990) referred to this influence as "downward causation" an idea that reflects the bidirectional nature of processes from gene and environmental levels (e.g., Gottlieb, 1998; Weiss, 1973). Much of these external factors are essentially random, so that the developmental process is not predetermined, but, rather, probabilistic. Put another way,

> Since it has become evident that genes interact with their environment at all levels, including the molecular, there is virtually no interesting aspect of development that is strictly "genetic," at least in the sense that it is exclusively a product of information contained within the genes.
>
> (Elman et al., 1996, p. 21)

Of course, our genes are now known to be turned on and off throughout our lives, as a result of our varied experiences.

These recent findings in molecular biology challenge the central dogma of molecular biology (Crick, 1970), that genetic information flows in one direction only—from inside to out, from the genotype to the phenotype. Much contemporary research has shown this view to be false, but it is still not widely known, most significantly among psychologists. Although few psychologists are familiar with these findings, it is significant that many biologists are also unaware of recent developments in molecular biology that render the standard program of genetics as an unfolding of a set genetic code, no longer valid: "While this fact is not well known in the social and behavioral sciences, it is surprising to find that it is also not widely appreciated in biology proper … [!]" (Gottlieb, 2001, p. 47). Gottlieb was not alone in this assessment, as even a molecular biologist has noted (Strohman, 1997).

That this intellectual lacuna is true today, in 2011, is somewhat startling when we realize that some scientists were discussing genetics in these terms very early in the 20th century. H. S. Jennings, one of the pioneers of genetics, made the following statements in his 1924 article:

> But no single thing that the organism does depends alone on heredity or alone on environment; always both have to be taken into account.
>
> (p. 225)

> … it is not true that particular characteristics are in any sense represented or condensed or contained in particular unit genes. Neither eye color nor tallness nor feeble-mindedness, nor any other characteristic, is a unit character in any such sense.
>
> There is indeed no such thing as a "unit character," and it would be a step in advance if that expression should disappear.
>
> (228)

Surprisingly, that idea has not yet been vanquished.

> Open a book, read a newspaper, turn on the TV, read *Science* or *Nature* and you will find yourself bombarded with claims and counterclaims. Are there "genius" genes? If not those, then surely the "gay" ones? Is aggression the consequence of social and economic conditions, or is it a product of evolution? Are cognitive differences between men and women due to genetics or upbringing?
>
> (Oyama, Griffiths, & Gray, 2001, p. 1)

As Lewontin's review makes clear, the Human Genome Project taught us that genetics is far more complex that than we could have expected. Rather than furthering our understanding of the genetic basis of the biological and psychological characteristics of us, it has, as science often does, raised more questions than it answered. We have only recently learned that transcription of DNA to RNA protein producing factors is not one-to-one as we have long believed (Li et al., 2011). This finding, of course, deals another blow to traditional gene-centric thinking.

I have pointed out previously (Greenberg, 2005) that one of the more interesting things about behavior genetic analysis is the absence of any discussion as to how genetic influences might manifest themselves. No pathways are identified, though of course the pathway from genes to even structure is indirect and enormously complex. With respect to behavior, I have always found Skinner's (1966) views to reflect the true state of things: To the extent that we behave with structures we inherit, it may be possible to speak of the genetic or otherwise biological foundation of behavior. But although I have inherited two hands with a full complement of fingers I cannot play the piano. Slowly and gradually, out of a rich experience of the world, one builds a behavioral repertoire, including piano playing. As Moss (2003) pointed out, there is no explanation in attributing a trait, behavioral or structural, to genetics in light of what converging current research from several disciplines indicates.

The Ontological Structure of Psychology

If behavior is not under the direct control of biogenetic phenomena, what then accounts for its development? I think it is fair to say that all of those scholars I mentioned above (e.g., Gottlieb, Kantor, Kuo, Lerner, Overton, Skinner, & Schneirla) have understood behavior to result—that is, to develop—because after all, psychology is a developmental science, a "life-span developmental science" (Greenberg, Partridge, Mosack, & Lambdin, 2006; Lerner, 2002, 2011; Overton, 2006, 2010) from the dynamic fusion of several sets of factors (Greenberg & Haraway, 2002; Seay & Gottfried, 1978), including the phylogenetic, ontogenetic, experiential, cultural, and individual.

Phylogenetic Set

The organism's evolutionary status, what it is as a species. This set is embodied in Kuo's (1967) "principle of behavioral potentials" that suggests that each species has the potential to behave in species-typical ways. Of course, there is no guarantee that those potentials will be actualized. Thus, as Montagu (1952/1962) pointed out, "The wonderful thing about a baby ... is its promise" (p. 17)—we are born *Homo sapiens,* but we have to become human beings.

Ontogenetic Set

The development of an organism, from its embryonic state to its state as an adult and its eventual death. Again, the probabilistic nature of this ontogeny is underscored. Nothing in development—embryological or behavioral—is guaranteed by genes; nothing is preformed or preordained (Gottlieb, 1992; Nieuwkoop, Johnen, & Albers, 1985). It should be noted here that the developmental stage of an organism profoundly affects its behavior and the way in which it reacts to stimuli. For example, a baby in its crawling period can only get under the sink, but when she begins to walk care must be taken to fasten the kitchen drawers.

Experiential Set

"... [A]ll stimulative effects upon the organism through its life history" (Schenirla, 1957, p. 86). This concept refers as well to all actions initiated by the organism (Overton, 2006). Experience, then, is what happens to the organism and what it does. Kantor (1959) referred to this experiential history as the "reactional biography" (RB). The RB begins at conception and continues to be built up until the organism's death. Every stimulus and each act affects the organism and changes it, although some stimulation and some acts have much more profound and obvious effects than others. Learning, for example, is an important process in behavioral change, but it is nothing more than a special set of experiences.

Cultural Set

Organisms function in environments. The organism-environment relation forms a functional whole, and consequently environments are necessary features of the organism's biological and behavioral development. This relation is most obvious in humans, who have developed cultural systems (e.g., religion, dietary practices, social institutions) that affect behavioral development in multiple ways. But all living organisms, although perhaps at less complex levels, function within environments of their own making. Different species may inhabit different environments, eat different foods, and so on. This important point was stressed by the ethologist Jacob von Uexküll (1957) who termed the behavioral environment of an animal its *Umwelt*, its sensory-perceptual world (see Michel, 2010). Chimpanzees, for example, display different behavioral adaptations related to their unique environments (Matsuzawa, 1998). Individuals in two communities separated by only 10 km display markedly different behaviors. These differences include nest building, ant dipping, use of leaves for water drinking, food choices, and many others. These differences are less complex cultural traditions than are found in more complex species.

Individual Set

The uniqueness of each individual organism and how that uniqueness relates to its development. One animal may be more or less sensitive to sounds, or may have a developmental abnormality that limits its interactions with its world, or may be larger or smaller than its conspecifics, and so on. This set of factors recognizes the contribution of the individual's unique genotype and how the organism's biology, in dynamic interplay with contextual influences, may render it a different behaving creature than all others.

These five organizational sets provide the ontological structure of psychology. I am comforted by the use of a similar analysis by Overton (2006; Overton & Müller, 2013), one of the world's leading developmental psychologists, who used different labels but is substantially in agreement that several sets of factors, at different levels of analysis and influence, play a dynamic role in human development. We are especially in agreement with respect to the significance of the physical ideas of "fluid dynamic holism and associated concepts such as *self-organization, system*, and the synthesis of wholes" (Overton, 2006, p. 19) as they apply to understanding development.

The common theme that runs through these organizational sets is that temporal processes and relational constructs are the central conceptual features of each set. The challenge for the study of psychological development is to account for these dynamic relational processes that occur at multiple spatial and temporal streams, becoming manifest in the nexus of the individual organism. Although many scientists acknowledge the importance of multiple factors in behavior (although many still cling to the nature/nurture split) few recognize that these factors do not simply interact. Such a formulation would grant nature and nurture factors individual and independent significance in influencing behavior (Pronko, 1988). Rather, the dynamic interplay between these factors is a fusion (Tobach & Greenberg, 1984)—one cannot therefore say how much is determined by phytogeny, how much by ontogeny, how much by nature, how much by nurture, in much the same way we cannot determine how much of the area of a rectangle is a function of its width or its height. There is much to agree with Pronko's (1988) comment that "We must not neglect genetic and other biologic factors, but, instead of treating them as causal, we regard them as *aspects* of an integrated field event or events" (p. 78).

Development Is Probabilistic

The preceding critique of the biogenetic approach in psychology leads to a discussion of development, biological and psychological, not as something fixed or guaranteed, but as probabilistic. Nothing is guaranteed by the genes, something that contemporary molecular biology makes clear. That we have to fight to make this point universally accepted is remarkable given that it was understood by even the earliest pioneers of genetics. I again turn to H. S. Jennings (1924):

> The genes are simply chemicals that enter into a great number of complex reactions, the final upshot of which is to produce the completed body. The characters of the adult are no more present in the germ cells than is an automobile in the metallic ores out of which it is ultimately manufactured. [p. 230] ... What any cell shall become depends in fact on the conditions surrounding it: on its relation to the other cells. *Development, it turns out, is a continual process of adjustment to environment.*
>
> (p. 231, emphasis added)

We now understand that what a gene does is very much influenced by which other genes are being turned on or off at any particular time during development. In other words, genes do their work along with other genes, rather than individually. Genes, then, are not encapsulated and isolated from the environment, they are, rather, an integral part of that environment.

The Role of Development

Human development is complex, and our understanding of it invokes complex ideas, some of which come from the other sciences—biology of course, but also physics, especially complexity and dynamic systems theories (Partridge & Greenberg, 2010). Among the best discussions of development—its concepts, implications, and meanings—are those by Lerner (2002, 2011) and Overton (2006). I have in other places invoked the important idea of emergence (Greenberg, Partridge, & Ablah, 2006; Partridge & Greenberg, 2010), an idea from modern physics with parallels in Gestalt psychology (e.g., "The whole is different from the sum of the parts"). As Lerner (2004) points out, "The complexity of these [developmental] theories can be daunting [even] to scholars" (p. 1).

A common question today in theoretical papers regarding development, whether implicitly or explicitly posed, asks what is the role of biology, of brains and genes, in shaping development. Such questions take the form of hypotheses regarding the relationship between a given allele and trajectories of behavioral outcomes. In these situations, there is an underlying assumption that development is subsidiary to biological factors. In other words, it is assumed that ontogeny is a function primarily of phylogeny and that behavioral development is shaped by the organism's biology. Thus, biology (genes, neural circuits, hormones), is understood to be the guiding force that drives individual differences in the development of behavior. This view is a persistent problem with the behavior genetic approach to behavior it attempts to set values for the relative roles of the several sets of factors that influence behavior. This approach is, of course, interaction(e.g. Pronko, 1988), rather than fusion.

The Relational, Developmental Systems View of Psychology

However, a relational holistic position takes a dramatically different perspective on the relationship between biology and psychological development. From this perspective, development is an active system of processes superordinate to biology and evolution. Thus, it is not that genes and brains explain development, but that the developmental system explains the functioning of the gene, the brain, and even evolution at the level of individual ontogeny. The developmental system integrates biological functions into coordinated patterns which support behavior. It is, then, the process of development that shapes biological organization and provides a temporal context for biology-behavior-ecology interrelationships (Lerner & Bush-Rossnagel, 1981; Overton & Müller, 2013).

In endorsing this relational holistic position, I am proposing that the focus of study in developmental psychology should be on the pattern of interrelationships between biological structure, psychological states, and ecological contexts. A clear characterization of development is that organisms initially comprise relatively undifferentiated biological and behavioral features that over time become increasingly differentiated and reintegrated into a coherent biological and behavioral system (Overton, 2006, 2010). It is the probabilistic, epigenetic, and self-organizing principles of development (e.g., Gottlieb, 1992) within a dynamic ecological context that shape the processes of differentiation and integration that characterizes a given individual's genetic, neurological, and behavioral attributes, rather than the other way around.

Although many of these ideas have been discussed by earlier developmentalists, their treatment by Overton and Müller (2013) shows much more clearly the significance of how some of these concepts from physics bear directly on our contemporary understanding of psychological development. These concepts pertain to the ideas of system (that the parts of organisms function interdependently and that behavior is a process, and not a substance or thing, of the system), of hierarchy or directionality (as a fundamental principle of science that the universe exists in as a family of hierarchies in which natural phenomena exist in levels of increasing organization and complexity), of emergence and self-organization (that a corollary of the Big Bang theory is that given enough time hydrogen and helium become sentient beings), and of epigenesis (an holistic approach to understanding development, discussed in the next section). In Overton and Müller's language, developmental psychology as currently envisioned is in a post positivist era.

The Role of Epigenesis in Development

At least since the broad acceptance of the modern synthesis in biology the gene construct has served as the central biological organizing feature assumed to guide biological and behavioral development. However, the very notion of just what a gene is has changed since the end of the 20th century. It is no longer sufficient to speak of "the" gene; the term has come to mean different things to different people. The term *gene* is now understood to be shorthand for several different kinds of units. It may be that *gene* is not so much an identifiable thing as it a process involved in binding DNA to other factors which act together in polypeptide production. At its inception, and indeed, until only very recently, the gene, seemingly so concrete and definitive a structure, was nothing more than a hypothetical construct in a statistical equation (Burian, 1985, Keller, 2002). Behavior geneticists are, however, undaunted by this history of facts. They have continued to work under the false assumption that once the human genome was sequenced behavioral science would be able to incorporate genetic profiles into a general linear model calculus and be able to predict with a reasonable amount of statistical precision the general trajectory of behavioral development, especially those which had been demonstrated to be highly heritable and thus largely under genetic control.

However, it is now indisputable that "high heritability does not mean developmental fixity" (Lerner, 2002, p. 254) and that it does not equate to "largely under genetic control." These facts were recognized very early by Jennings (1924) and were emphasized by many contributors to Harris's (1957) important book, *The Concept of Development*. For example, in that volume, Schneirla (1957) pointed out that, "in experiments with the fruit fly . . . the same gene may influence the development of different wing size and structure according to what temperature prevails during the development of the phenotype" (p. 85). The assumption that genes control development was based on the premise that genes contained developmental information guiding biological development. Behavioral geneticists could then argue for genotypic control of the behavioral phenotype via the neurological endophenotype. However, as I have discussed above, the sequencing of the human genome (Venter, Adams, Myers, Li, Mural et al., 2001), has not yielded the scientific fruit for behavioral science that many leading behavioral geneticists envisioned. As a result there has been a growth of interest in epigenesis as a developmental process and of epigenetics as a mechanism through which genes and contexts transact through development.

Epigenesis has been described in a variety of ways, but none has been as well put as that by Moltz (1965):

> An epigenetic approach holds that all response systems are synthesized during ontogeny and that this synthesis involves the integrative influence of both intraorganic processes and extrinsic stimulative conditions. It considers gene effects to be contingent on environmental conditions and regards the genotype as capable of entering into different classes of relationships depending on the prevailing environmental context. In the epigeneticists' view, the environment is not benignly supportive, but actively implicated in determining the very structure and organization of each response system.
>
> (p. 44)

While the concept of epigenesis originated in biology, the usefulness of *probabilistic* epigenesis was recognized and promoted throughout the 20th century by psychologists such as Zing-Yang Kuo (1967), Gilbert Gottlieb (1992), and T. C. Schneirla (1957), although Schneirla never specifically employed the term *epigenesis* in his writing (Aronson et al. 1972). Probabilistic epigenesis has gained support from an exciting set of developments in contemporary science subsumed under the rubric of "dynamic systems theory and relational developmental systems theory," in which complex developmental processes are understood as composed of interrelations among many active system components of the whole developmental system, which I have discussed above. The implication of this position is that in a dynamic and changing environment, rather than genes specifying a particular developmental outcome, be it structural or behavioral, every outcome is an emergent result of the transaction between genes and their cellular, organismic, ecological, and temporal contexts. This view of epigenesis is epitomized by discoveries in biology that even identical genomes in extremely similar environments do not always follow the same developmental pathways. Ko and colleagues (Ko, Yomo, & Urabe, 1994), studying enzyme activity in bacteria, found that despite identical genomes and extremely uniform culture conditions, individual cells developed different levels of enzyme activity and grew into colonies of different size. Ko's studies showed that cell state in bacteria is determined not only by genotype and environment. Rather, "Changes of state can occur spontaneously, without any defined internal or external cause. By definition, these changes are epigenetic phenomena: dynamic processes that arise from the complex interplay of all the factors involved in cellular activities, including the genes" (Solé & Goodwin, 2000, p. 63).

Methodological Issues

As a result of the convergence of the ideas I discussed in this chapter, developmental psychologists now have at their disposal a conceptual architecture and an emerging methodology that is commensurate with their core theoretical principles (e.g., Molenaar, 2010). We find now that not only do empirical data (largely from experimental embryology and comparative psychology) indirectly support inferences about the role of integrated biopsychosocial systems in shaping phenotypic outcomes, but theoretical physicists and mathematicians have demonstrated that these same principles hold widely. Thus, there is now a set of methodological tools using systems methods that have the capability of testing many of the developmental systems postulates directly (Urban, Osgood, & Mabry, 2011). What is perhaps not especially surprising, unfortunately, is that though methodologies are routinely used in other systems disciplines (e.g., engineering, physics, biology, ecology) they "have been slower to diffuse in behavioral and social science" (Urban et al., 2011, p. 9).

The concepts of hierarchy, integrative levels and systems, self-organization, and emergence have been rather fully developed over the last quarter century and are being employed with considerable alacrity by scientists in many disciplines. The larger point here is that through experimental studies of developing organisms, it has become clear that the conceptualization of the gene as held by the central dogma was untenable; it would not explain empirical findings. Many sciences have a long history of physics envy, and as such a number of experimental disciplines drew from new ideas in physics emerging at the turn of the 20th century for an explanatory heuristic that was more consistent with the findings within their respective fields. Independently, but concurrently, theoretical physicists (e.g., Layzer, 1974, 2000) were continuing to develop a mathematical formalism with which to test hypotheses regarding the dynamics of hierarchically nested systems, complex systems with no centralized controls, and so on and also concluded that all systems, be they computational (i.e., bits of information), physical, biological, or social, displayed exactly the properties suggested by early theorists in both of these fields both sets of disciplines. Working from different approaches—one primarily inductive, the other primarily deductive—these respective sets of scholars found a convergent set of shared ideas which have profoundly greater explanatory capability and parsimony than those central to genocentric orientations like behavioral genetics and evolutionary psychology. The newly formed ideas are capable of empirical verification and are consistent with concepts from other sciences and may provide a more parsimonious model for lifespan development.

As I have discussed in another context (Partridge & Greenberg, 2010), one of the more important outcomes of this convergence is that we can now specify hypotheses directly corresponding to the key principles of this relational, developmental systems perspective (Overton & Müller, 2013) and test them using appropriate methodological tools. As molecular biologists are beginning to recognize "Key notions such as emergence, nonlinearity, and self-organization already offer conceptual tools that can contribute to transform and improve science" (Mazzocchi, 2008, p. 13).

Because "a major objective of developmental research is to study processes of change" (Nesselroade & Molenaar, 2010, p, 30), psychology as a developmental science requires more sophisticated methodologies to address its issues and ideas. Over the last decades important advances have been made in adapting state space and phase space portrait analyses to the level of data and measurement methodology common to psychology. Most psychophysiological data (e.g., electroencephalogram, or EEG) can be directly analyzed using such techniques. The developmentalist needs other techniques, some of which have been discussed in a recent special issue of this journal (Mabry & Urban, 2011). Other examples are those of Thelen and her colleagues (e.g., Thelen & Smith, 1998), who have used state and phase portrait analytic approaches to revolutionize the field of motor and perceptual development in infancy.

For the present, data at the behavioral level does not meet the requirements of these approaches directly. Yet important work adapting these analytic tools to be better suited to traditional psychological data has substantially bridged that gap (e.g., Granic & Hollenstein, 2003). For example, Lewis, Lamey, and Douglas (1999) characterized the developmental dynamics of two dimensions of early childhood socioemotional development: intensity of distress and attention to mother. Using a two-dimensional ordinal state space grid provided unparalleled insight into the dynamics of infant emotional development—insight that was not feasible with more traditional statistical analyses.

Sigmond Koch (1969) once asked whether a coherent science of psychology was even possible. Forty years later we know the answer to that question to be *yes*. I have maintained that psychology is, in fact, a natural science. Accordingly, the methods used, and in this context, especially by developmentalists, have become increasingly rigorous. Psychology is, of course, the science of individual behavior, and though we have a long history of statistical methods analyzing groups, it is now clear that as a developmental science, our focus is on the individual. An interesting analogy is proposed by Nesselroade and Molenaar (2010), in which the individual is likened to the Brownian movement of a single particle. As we follow a person across his or her life span, the relational developmental system model I discussed here allows us to discern the emergence of behaviors using newly developed methodologies.

A full description of these methods is beyond the scope of this chapter. Rather the aim here is to provide a brief description of a few of the analytic methods developed to study the properties of nonlinear dynamic systems, how these methods have been incorporated into related disciplines, and, most importantly, that the phenomena for which these methods were designed are conceptually identical to many of the central concepts. I have discussed linking these core concepts with analytic and methodological tools such as the use of cellular automata, Bayesian network analyses, state and phase portraits, state space grids, and nonlinear dynamic systems approaches to longitudinal covariance models are where the future of developmental science lies.

Concluding Comments

I began this chapter as a polemic against behavior genetics, and one could say I wandered somewhat from my topic and presented a discussion of psychology as a developmental science. My title is meant to reflect the uniqueness of psychology as a science and not that biological factors do not have a role to play in behavioral ontogeny. Indeed, as Overton (2006) made clear biological principles, from evolution to ecology (i.e., context), are important participating factors in behavior ontogeny. From the perspective I am promoting here, behavior is a process—not a thing or substance, a point discussed by Overton (2006) and Overton and Müller (2013).

The implication of this is that, though brains and genes are important, behavior is a phenomenon that emerges as the organism develops. Although many of the scholars I have discussed above understood the role of development, many contemporary scientists, especially behavioral scientists, have ignored development. A small, but growing group of influential scientists is leading the charge to reverse this trend, among them Lerner (2002, 2011) and Overton (2006, 2010). They are among the growing group of behavioral scientists spelling out the concepts and principles of developmental science.

When I was a graduate student at the University of Wichita, my mentor at the time, N. H. Pronko (see Pronko, 1973, 1980) expressed optimism that the point of view I espouse in this chapter would succeed. Although the going has been difficult, this optimism still prevails among an increasing number of behavioral scientists. At the 2003 meeting of the Society for the Study of Human Development, following a presentation by the late Gilbert Gottlieb, a member of the audience expressed concern that this approach was not widespread. Lerner responded by noting the several hundred members of the audience—professors and students—who agreed with Gottlieb's views. Lerner offered his prediction that those numbers can only increase. From where I sit today, the future looks rosy.

Acknowledgment

I thank Richard M. Lerner for his critical reading of an earlier draft of this chapter. My thinking on these issues has profited from years of intellectual discussions and collaborations with him.

References

Aronson, L. R., Tobach, E., Rosenblatt, J. R., & Lehrman, D. H. (Eds.). (1972). *Selected writings of T. C. Schneirla*. San Francisco, CA: Freeman.

Breckler, S. J. (2006). The newest age of reductionism. *Monitor on Psychology, 27*(8), 23.

Brooks, D. (2006, September 17). Is chemistry destiny? *New York Times*, p. WK14.

Burian, R. M. (1985). On conceptual change in biology: The case of the gene. In D. J. Depew & B. H. Weber (Eds.), *Evolution at a crossroads: The new biology and the new philosophy of science* (pp. 21–42). Cambridge, MA: MIT Press.

Cacioppo, J. T., Bernston, G. G., Sheridan, J. F., & McClintock, B. (2000). Multiple integrative analyses of human behavior: Social neuroscience and the nature of social and biological approaches. *Psychological Bulletin. 126*, 829–843.

Campbell, D. (1990). Level of organization, downward causation, and the selection-theory approach to evolutionary epistemology. In G. Greenberg & E. Tobach (Eds.), *Theories of the evolution of knowing: The T. C. Schneirla Conference Series, Volume 4* (pp. 1–18). Hillsdale, NJ: Erlbaum Associates.

Crick, F. (1970). Central dogma of molecular biology. *Nature, 227*, 561–563.

Elman, J. L., Bates, E. A., Johnson, M. H., Karmiloff-Smith, A., Parisi, D., & Plunkett, K. (1996). *Rethinking innateness: A connectionist perspective on development*. Cambridge, MA: MIT Press.

Gottlieb, G. (1992). *Individual development and evolution: The genesis of novel behavior*. New York, NY: Oxford University Press.

Gottlieb, G. (1998). Normally occurring environmental and behavioral influences on gene activity: From central dogma to probabilistic epigenesis. *Psychological Review, 105*, 792–802.

Gottlieb, G. (2001). A developmental psychobiological systems view: Early formulation and current status. In S. Oyama, P. E. Griffiths, & R. D. Gray (Eds.), *Cycles of contingency: Developmental systems and evolution* (pp. 41–54). Cambridge, MA: MIT Press.

Gottlieb, G. (2004). Normally occurring environmental and behavioral influences on gene activity: From central dogma to probabilistic epigenesis. In C. G. Coll, E. L. Bearer, & R. M. Lerner (Eds.), *Nature and nurture: The complex interplay of genetic and environmental influences on human behavior and development* (pp. 85–106). Mahwah, NJ: Erlbaum.

Granic, I., & Hollenstein, T. (2003). Dynamic systems methods for models of developmental psychopathology. *Development and Psychopathology, 15*, 641–669.

Greenberg, G. (2005). The limitations of behavior-genetic analyses: Comment on McGue, Elkins, Walden and Iocono (2005). *Developmental Psychology, 41*, 989–992.

Greenberg, G. (2007). Why psychology is not a biological science: Gilbert Gottlieb and probabilistic epigenesis. *European Journal of Developmental Science, 1*, 111–121.

Greenberg, G., & Haraway, M. M. (2002). *Principles of comparative psychology*. Boston, MA: Allyn & Bacon.

Greenberg, G., & Lambdin, C. (2007). Psychology is a behavioral science not a biological science: A discussion of the issue and a review of *Neural theories of mind; Why the mind brain problem may never be solved*, by William Uttal. *Psychological Record, 57*, 457–475.

Greenberg, G., & Partridge, T. (2010). Biology, evolution, and development. In W. F. Overton (Ed.), *Cognition, Biology, and Methods: Volume 1 of the Handbook of Life-span Development* (pp. 115–148). Hoboken, NJ: Wiley.

Greenberg, G., Partridge, T., & Ablah, E. (2006). The significance of the concept of emergence for comparative psychology. In D. Washburn (Ed.), *Primate perspectives on behavior and cognition* (pp. 81–97). Washington, DC: American Psychological Association

Greenberg, G., Partridge, T., Mosack, V., & Lambdin, C. (2006). Psychology is a developmental science. *International Journal of Comparative Psychology, 19*, 185–205.

Harris, D. B. (Ed.). (1957). *The concept of development: An issue in the study of human behavior*. Minneapolis, MN: University of Minnesota Press.

Hood, K. E., Halpern, C. T., Greenberg, G., &. Lerner, R. M. (Eds.). (2010). *Handbook of developmental science, behavior, and genetics*. Malden, MA: Wiley-Blackwell.

Hubbard, R. & Wald, E. (1993). *Exploding the gene myth: How genetic information is produced and manipulated by scientists, physicians, employers, insurance companies, and law enforcers.* Boston, MA: Beacon.

Human Genome Project. (2011). *ISCID encyclopedia of science and philosophy.* Retrieved October 20, 2011 from http://www.iscid.org/encyclopedia/Human-Genome-Project

Jennings, H. S. (1924). Heredity and environment. *Scientific Monthly, 19,* 225–238.

Kantor, J. R. (1959). *Interbehavioral psychology* (2nd ed.). Chicago, IL: Principia Press.

Keller, E. F. (2002). *Making sense of life: Explaining biological development with models, metaphors, and machines.* Cambridge, MA: Harvard University Press.

Ko, E. P., Yomo, T., & Urabe, I. (1994). Dynamic clustering of bacterial populations. *Physica D: Nonlinear Phenomena, 75,* 81–88.

Koch, S. (1969). Psychology cannot be a coherent science. *Psychology Today, 3*(4), 14, 64–68.

Kuo, Z. Y. (1967). *The dynamics of behavior development: An epigenetic view.* New York, NY: Random House.

Layzer, D. (1974). Heritability analyses of IQ scores: Science or numerology? *Science, 183,* 1259–1266.

Layzer, D. (2000). Comment on "Misconceptions of biometrical IQsts" by C. Capron et al. Unpublished manuscript. Cambridge MA: Harvard University Press.

Lerner, R. M. (1989). Developmental contextualism and the life-span view of person-context interaction. In M. Mornstein & J. S. Bruner (Eds.), *Interaction in human development* (pp. 217–239). Hillsdale, NJ: Erlbaum.

Lerner, R. M. (1998). Developmental contextualism. In G. Greenberg & M. M. Haraway (Eds.), *Comparative psychology: A handbook* (pp. 88–97). New York, NY: Garland.

Lerner, R. M. (2002). *Concepts and theories of human development* (3rd ed.). Mahwah, NJ: Erlbaum Associates.

Lerner, R. M. (2004). Genes and the promotion of positive human development: Hereditarian versus developmental systems perspectives. In C. G. Coll, E. L. Bearer, & R. M. Lerner (Eds.), *Nature and nurture: The complex interplay of genetic and environmental influences on human behavior and development* (pp. 1–33). Mahwah, NJ: Lawrence Erlbaum Associates.

Lerner, R. M. (2011). Structure and process in relational, developmental systems theories: A commentary on contemporary changes in the understanding of developmental change across the life span. *Human Development, 54,* 34–43.

Lerner, R. M., & Bush-Rossnagel, N. A. (1981). Individuals as producers of their development: Conceptual and empirical issues. In R. M. Lerner & N. A. Bush-Rossnagel (Eds.), *Individuals as producers of their development: A life-span perspective* (pp. 1–36). New York, NY: Academic Press.

Lewis, M. D., Lamey, A. V., & Douglas, L. (1999). A new dynamic systems method for the analysis of early socioemotional development. *Developmental Science, 2,* 457–475.

Lewontin, R. (1991). *Biology as ideology: The doctrine of DNA.* New York, NY: Harper Collins.

Lewontin, R. (2000). *It ain't necessarily so: The dream of the human genome and other illusions.* New York, NY: New York Review Books.

Lewontin, R. C. (2011, May 26). It's even less in your genes. *New York Review of Books, 23,* 26–27.

Li, M., Wang, I. X., Li, Y., Bruzel, A., Richards, A. L., Tuong, J. M., & Cheung, V. G. (2011). Widespread RNA and DNA sequence differences in the human transcriptome. *Science.* Retrieved from www.sciencexpress.org/19May2011/10.1126/science.1207018

Mabry, P. L., & Urban, J. B. (Eds.). (2011). Embracing Systems Science: New Methodologies for Developmental Science [Special Issue]. *Research in Human Development, 8*(1), 1–25.

Matsuzawa, T. (1998). Chimpanzee behavior. In G. Greenberg & M. M. Haraway (Eds.), *Comparative psychology: A handbook* (pp. 360–375). New York, NY: Garland.

Mazzocchi, F. (2008). Complexity in biology. Exceeding the limits of reductionism and determinism using the complexity theory. *EMBO Reports, 9*(1), 10–14.

Michel, G. F. (2010). The roles of environment, experience, and learning in behavioral development. In K. E. Hood, C. T. Halpern, G. Greenberg, & R. M. Lerner (Eds.), *Handbook of developmental science, behavior, and genetics* (pp. 123–165). Malden, MA: Wiley-Blackwell.

Molenaar, P. C. M. (2010). On the limits of standard quantitative genetic modeling of inter-individual variation: Extensions, ergotic conditions and a new genetic factor model of intra-individual variation. In K. E. Hood, C. T. Halpern, G. Greenberg, & R. M. Lerner (Eds.), *Handbook of developmental science, behavior, and genetics* (pp. 626–648). Malden, MA: Wiley-Blackwell.

Moltz, H. (1965). Contemporary instinct theory and the fixed action pattern. *Psychological Review, 72*, 27–47.

Montagu, A. (1962). Our changing conception of human nature. In *The humanization of man* (pp. 15–34). New York, NY: Grove Press.

Moss, L. (2003). *What genes can't do*. Cambridge, MA: MIT Press.

Nesselroade, J. R., & Molenaar, P. C. M. (2010). Emphasizing intraindividual variability in the study of development over the life span: Concepts and issues. In W. E. Overton (Ed.), *Cognition, biology, and methods. Volume 1 of the handbook of life-span development* (pp. 30–54). Hoboken, NJ: Wiley.

Nieuwkoop, P. D., Johnen, A. G., & Albers, B. (1985). *The epigenetic nature of early chordate development: Inductive interaction and competence*. Cambridge, UK: Cambridge University Press.

Overton, W. F. (2006). Developmental psychology: Philosophy, concepts, methodology. In W. Damon (Series Ed.) & R. M. Lerner (Vol. Ed.), *Theoretical models of human development. Handbook of child psychology*, 6th edition, *Vol 1* (pp. 18–88). New York, NY: Wiley.

Overton, W. F. (2010). Life-span development: Concepts and issues. In W. F. Overton (Ed.), *Cognition, biology, and methods. Volume 1 of the handbook of life-span development* (pp. 1–29). Hoboken, NJ: Wiley.

Overton, W. F. & Müller, U. (2013). Metatheories, theories, and concepts in the study of development. In R. M. Lerner, M. A. Easterbrooks, & J. Mistry (Eds.), *Comprehensive handbook of psychology: Developmental psychology* (Vol. 6). New York, NY: Wiley.

Oyama, S., Griffiths, P. E., & Gray, R. D. (Eds.). (2001). *Cycles of contingency: Developmental systems & evolution*. Cambridge, MA: MIT Press.

Partridge, T., & Greenberg, G. (2010). Contemporary ideas of physics and biology in Gilbert Gottlieb's epigenesis. In K. E. Hood, C T. Halpern, G. Greenberg, & R. M. Lerner (Eds.). *Handbook of developmental science, behavior, and genetics* (pp. 166–202). Malden, MA: Blackwell.

Pronko, N. H. (1973). *Panorama of psychology* (2nd ed.). Monterey, CA: Brooks/Cole.

Pronko, N. H. (1980). *Psychology from the standpoint of an interbehaviorist*. Belmont, CA: Wadsworth.

Pronko, N. H. (1988). Heredity versus environment. In N. H. Pronko (Ed.). *From AI to Zeitgeist: A philosophical guide for the skeptical psychologist* (pp. 75–80). New York, NY: Greenwood Press.

Robert, J. S. (2004). *Embryology, epigenesis, and evolution: Taking development seriously*. Cambridge, UK: Cambridge University Press.

Schneirla, T. C. (1957). The concept of development in comparative psychology. In D. B. Harris (Ed.), *The concept of development: An issue in the study of human behavior* (pp. 78–108). Minneapolis, MN: University of Minnesota Press.

Seay, B., & Gottfried, N. (1978). *The development of behavior: A synthesis of developmental and comparative psychology*. Boston, MA: Houghton Mifflin.

Skinner, B. F. (1953). *Science and human behavior*. New York, NY: Macmillan.

Skinner, B. F. (1966). The phylogeny and ontogeny of behavior. *Science, 153*, 1205–1213.

Solé, R., & Goodwin, B. (2000). *Signs of life: How complexity pervades biology*. New York, NY: Basic Books.

Strohman, R. C. (1997). The coming Kuhnian revolution in biology. *Nature Biotechnology, 15*, 194–200.

Tandon, P. N. (2000). The decade of the brain: A brief review. *Neurology India, 48*, 199–207.

Thelen, E., & Smith, L. B. (1998). Dynamic systems theories. In W. Damon & R. M. Lerner (Eds.), *Handbook of child psychology. Volume 1: Theoretical models of human development* (5th ed., pp. 563–634). Hoboken, NJ: John Wiley & Sons.

Tobach, E., & Greenberg, G. (1984). The significance of T. C. Schneirla's contribution to the concept of levels of integration. In G. Greenberg & E. Tobach (Eds.), *Behavioral evolution and integrative levels* (pp. 1–7). Hillsdale, NJ: Lawrence Erlbaum Associates.

Urban, J. B., Osgood, N. D., & Mabry, P. L. (2011). Developmental systems science: Exploring the application of systems science methods to developmental science questions. *Research in Human Development, 8*(1), 1–25.

Uttal, W. R. (2005). *Neural theories of mind: Why the mind-brain problem may never be solved.* Mahwah, NJ: Erlbaum.

Venter, J., Adams, M. D., Myers, E. W., Li, P. W., Mural, R. J., et al. (2001). The sequence of the human genome. *Science, 291,* 1304–1351.

von Uexküll, J. (1957). A stroll through the world of animals and men. In C. H. Schiller (Ed.), *Instinctive behavior* (pp. 5–80). New York, NY: International Universities Press.

Weiss, P. (1973). *The science of life.* New York, NY: Futura Publishing.

5.
WHAT GALTON'S EUGENICS HAS WROUGHT
Behavior Genetics and Heritability
David S. Moore[*]

> The famous Galtonian law of regression ... pretended to have established the laws of "ancestral influences" in mathematical terms. Now ... these laws of correlation have been put in their right place; such interesting products of mathematical genius may be social statistics *in optima forma,* but they have nothing at all to do with genetics or general biology! Their premises are inadequate for insight into the nature of heredity.
>
> —W. Johannsen (1911), p. 138

Two of my friends, who are sisters, tell the story of walking along the beach during sunset, eavesdropping on a conversation in which their children are avidly engaged. Although Sam is only six years old, he is already as tall as his eight-year-old cousin Jake. Commenting on this unusual state of affairs, Sam says, "I think I'm already as tall as you are!" Jake acknowledges this obvious truth, and says "Yeah. I think you're probably gonna be taller than me when we're grown-ups, because your parents are tall and my parents are short—these things are all in the gems."

Jake's lexical confusion notwithstanding, it is apparent that the idea that our genes determine our traits is firmly embedded in the fabric of our culture; how else could children with so few years of experience in the world already believe such things? This idea is such a part of our intellectual inheritance that it currently strikes many adults as intuitively obvious. But such ideas need to be carefully examined; in ages past, we believed the Earth was flat just because it looked that way. Is there similar (ultimately weak) "evidence" behind our intuitions about the origins of our traits? Certainly.

The ordinary lifelong observation that some traits seem to develop independently of our experiences is often taken as "evidence" that our genes must cause these traits. For example, traits that are present at birth, like five fingers on each of our two hands, seem to develop in the *absence* of experience; this contributes to the impression that they must be gentically determined." Similarly, some traits that are not present at birth—for instance, secondary sex characteristics like facial hair in men—nevertheless appear during adolescence even in the seeming absence of experience with specific environmental events. Some psychological characteristics, too, seem to develop independently of the conditions in which a person is reared. For example, orphaned human babies sometimes develop traits that are reminiscent of their deceased parents, even though the environments in which these babies develop provide no models for these characteristics. Such observations seem to imply that the appearance of certain traits is somehow predetermined—presumably by genes—since these traits do not seem to depend on specific experiences for their development.

In contrast, other traits are very obviously influenced by the ways in which children are

[*] Psychology Field Group, Pitzer College, Claremont California

nurtured. The specific language that you speak is probably the language that you heard spoken in the environment in which you developed. In many cases, your religious beliefs were shaped by the religious beliefs of your parents. Still other traits seem to be just a *little* impacted by the events of our lives; the extent to which you are shy might have changed somewhat as a result of your experiences, but if you are still relatively shy, you probably feel that nothing could ever turn you into the gregarious life of the party. We are left with a sense that there is a continuum of extents to which traits can be affected by experience. As a result, it seems reasonable to try to determine scientifically the extent to which specific traits can be influenced by environmental factors. Such a project might even involve trying to assign numbers to traits in order to represent their positions on the continuum: 100 percent could be defined to mean "completely open to environmental influence" and 0 percent could be defined to mean "not at all open to environmental influence."

The idea that it is possible to measure how *much* the environment contributes to our characteristics certainly appeals to common sense; when this idea first emerged in the 19th century, it appealed to the common sense of scientists as well. Consequently, scientists at the time began trying to develop statistical tools that could measure the extent to which particular traits can be affected by experience. Unfortunately, the work of these scientists reflected their 19th-century conceptualizations about the causes of traits, and these ideas have since turned out to be hopelessly simplistic. In the long run, the most important statistical tool to result from this work—the heritability statistic—wound up taking obsolete and completely untenable) conceptualizations about trait origins and firmly embedding them in our culture's belief systems. If you are under the impression that scientists have shown a certain characteristic to be caused by genetic factors, the ultimate source of this mistaken belief—whether or not you know it—was a study utilizing the logic underlying heritability statistics; these statistics undergird the widespread, but erroneous, belief that genes can determine traits. In fact, even though their name sounds like they should serve as a measure of a trait's "inheritability," heritability statistics do not even reflect the extent to which traits will be "passed down" from parents to their offspring. By and large, then, the public's current confusion about what genes can and cannot do can be traced back to the use—and misinterpretation—of these statistics.

Heritability statistics grew out of the work of one man: Francis Galton. While philosophers predating Galton contemplated the problem of trait origins as well, Galton alone is responsible for showing us the conceptual, methodological, and statistical path that has produced the data underlying the contemporary public's beliefs about trait origins. In fact, the basic conceptualization of the problem and the basic "scientific" method by which the problem is tackled has remained relatively unchanged since Galton first began working on it 130 years ago. And since the scientific underpinnings of our current misconceptions are rooted in Galton's ideas, laying them bare demands first revisiting Galton himself. But be forewarned: the confusion generated by a century of exposure to heritability statistics now runs so deep in people's minds that alleviating it will take us through some territory that might seem downright counterintuitive.

Galton

The fact that Galton was closely related to one of the most important thinkers of the 19th century likely played a role in drawing him to the problem of trait origins; Francis Galton and Charles Darwin were first cousins. Galton's mother, Violetta Darwin, was Charles's aunt. Galton was aware of his cousin's theory of evolution prior to most of the rest of the world; he later reported in an autobiography that Darwin's *On the Origin of Species* had stimulated his interest in how traits are inherited. In addition, the particular trait that Galton reported on in his first book—titled *Hereditary Genius* and published in 1869—was a trait that he called "eminence"; the fact that his cousin was famous no doubt contributed to his interest in the inheritability of this characteristic.

To study the inheritability of eminence, Galton searched biographical encyclopedias for information on the most distinguished people of 18th- and 19th-century England. His research revealed that a disproportionately large percentage of these "geniuses"—the country's most respected politicians, judges, lawyers, scientists, military commanders, artists, writers, and musicians— were related to one another by birth. As a result, Galton came to the novel-at-the-time conclusion that "heredity governed not only physical features but also talent and character." How, he then wondered, can we *explain* the observed fact that eminence runs in families?

In contemplating possible explanations, Galton considered the importance of only two broad factors: nature and nurture. Actually, he was the first scientist to use these particular terms, thereby defining the character of the debate that would ensue over the next century. Of particular significance, the distinction he made between nature and nurture was rather sharp. And since he felt "perfectly justified in attempting to appraise their relative importance" to the appearance of various traits, his subsequent work involved trying to devise scientific and statistical methods that would be suitable for doing just that. Such tools, he thought, would ultimately help him explain why certain traits run in families.

In 1883, with this goal in mind, Galton proposed the use—and outlined the details—of the twin study, a new approach designed to evaluate the relative contributions of nature and nurture to the appearance of traits. Galton believed that the study of twins would allow him to distinguish between "the effects of tendencies received at birth, and of those that were imposed by the special circumstances" of twins' unique lives. He thought that once he was able to distinguish between the effects of nature and nurture, he would then be able to determine the relative importance of each in the formation of the twins' traits.

By studying twins' (or their parents') responses to questionnaires, Galton determined that twins who were similar to one another as youths remained similar to one another as elderly adults, both in body and in mind, even when their later lives had taken them into different environmental circumstances. In addition, he observed that twins who were dissimilar to one another in youth—for instance, twins of different sexes— remained dissimilar in old age, even if they experienced an "identity of nurture." Galton concluded, "Nature is far stronger than Nurture within the limited range that I have been careful to assign the latter."

This result did not surprise Galton at all; in fact, eight years earlier, he had offered the rank speculation that "when nature and nurture compete for supremacy on equal terms ... the former proves the stronger." Six years before *that*, Galton had jumped to the same conclusion after studying his eminence data. Even though his eminence study—which was *not* a twin study—was not suitable for teasing apart nature and nurture, he concluded nonetheless that eminence runs in families because of nature, not nurture. In interpreting the results of this study, Galton ignored what is now obvious to us: in addition to sharing ancestors, Galton's eminent subjects were all raised in upper-class environments, with all of the educational, nutritional, and material advantages that such environments have always conferred. Clearly, regardless of the potential usefulness or validity of his new twin-study methodology, Galton had a preexisting bias that sometimes prevented him from seeing the contributions that environmental factors can make to the appearance of our traits.

The Genesis of Eugenics

Galton's conclusion that nature is more important than nurture was not his most notorious assertion. When he found that eminence—like physical traits such as height—ran in families, he quickly concluded that behavioral and personality traits could be artificially "selected," much as a rancher might produce an especially fast horse by selectively breeding a fast stallion with a fast mare. By 1869, Galton had already publicly presented his radical view that "it would be quite practicable to produce a highly gifted race of men by judicious marriages during several consecutive generations." Galton held this conviction even though he lacked any understanding at all of the *mechanism* by which behavioral and personality traits could be inherited.

Thus, after seeing in his 1883 twin data support for his preconceived notions about the primary importance of nature in trait formation, Galton coined the word "eugenics"—meaning "good in birth"—to refer to the "science" of improving humanity by selective breeding of the sort commonly employed on farm animals. Ultimately,

> [He] suggested that the state sponsor competitive examinations in hereditary merit, celebrate the blushing winners in public ceremony, foster wedded unions among them at Westminster Abbey, and encourage by postnatal grants the spawning of numerous eugenically golden offspring. (Some years later [Galton] would urge that the state rank people by ability and authorize more children to the higher- than to the lower-ranking unions.) The unworthy, Galton hoped, would be comfortably segregated in monasteries and convents, where they would be unable to propagate their kind.

A short 50 years later, this sort of thinking contributed to the brutal social policies of the Nazi regime and ultimately to the extermination of millions of innocent people.

But before the atrocities of the Nazis made plain the inherently evil nature of state-imposed eugenics programs, Galton's proposals didn't strike everyone as being particularly bad ideas. Scientists of the time understood that if Galton's hunch was correct—that behavioral and personality characteristics could be inherited like physical characteristics—this would be an important piece of information. After all, it is only if we understand how our traits are formed that we can conceive of intervening in their formation. And as remains the case today, many people in Galton's time believed that intervening to eradicate certain traits—symptoms of schizophrenia, perhaps, or mental retardation—was a good idea. The only questions are whether or not particular interventions are both effective and morally sound. And because Galton and his followers believed that "what Nature does blindly, slowly, and ruthlessly, man may do providently, quickly, and kindly," eugenics was seen as a reasonable approach to bringing some of society's problems under control. Thus, Galton and his followers set out to make eugenics a true science.

Before long, a new branch of biology called population genetics was born, built in part on Galton's ideas. But while it was a new branch of biology that needed Galton's ideas for its inception, it was Galton's extraordinary contributions to what was then a new branch of mathematics, called statistics, that fundamentally changed how scientists of *all* stripes could make sense of their observations of nature. And since Galton's statistical methods ultimately came to underlie the public's false confidence that some traits are "more genetic" than others, it is important to consider them in some detail.

Galton's Brainchild

At the 1884 International Health Exhibition in England, Galton began collecting data on people; before he was done, he had measured the weight, height, arm span, breathing power—basically anything that could be measured—of about nine thousand people, generating a huge data set. Poring over this data set, and paying particular attention to measurements he had obtained from adults *and* their parents, Galton stumbled onto a form of data analysis that he later called a "regression analysis." By 1888, he had figured out that a regression analysis is really just a particular form of a more general type of analysis that he called an analysis of "co-relations"; today, we know this as a correlational analysis.

Correlational analysis generates a statistic—a measurement—that reflects the strength of the relationship between two sets of numbers. For example, since weights and heights can both be represented as sets of numbers, Galton's correlational analysis can be used to evaluate how closely height and weight are related. Because weight ordinarily increases as height increases—taller people usually weigh more than shorter people—these two characteristics are said to be positively correlated. Similarly, because life span ordinarily decreases as cigarette consumption increases— smokers die, on average, at younger ages than do nonsmokers—these two characteristics are said to be negatively correlated. Of course, such an analysis can also reveal if two sets of numbers are completely *unrelated*; weight and vocabulary size among adults are unrelated to each other, since knowing how heavy an adult is does not even slightly improve your chances of guessing the size of that person's vocabulary.

Galton's initial uses of correlation were not in the service of understanding how genes per se contribute to traits. After all, when Galton was doing his work, the scientific community had not yet rediscovered Gregor Mendel's paper on "heritable factors," and the word "gene" would not be coined for several more years. As a result, Galton used correlational analyses only to address questions about the extent to which traits could be said to "breed true," that is, to be inherited *when environmental factors are held constant*. Farmers are, understandably, interested in the likelihood that breeding animals with particular traits will give rise to offspring possessing those traits; Galton's statistics provide this sort of information.

For example, to see if milk production levels breed true, Galton would measure the amount of milk produced by several cows and the amount of milk produced by their mature female offspring. If he then calculated that there is a high positive correlation between these two measurements, this would mean that cows who produced lots of milk were more likely than less productive cows to have offspring that, when mature, also produced lots of milk. Thus, detecting a correlation effectively allowed Galton to make reasonably accurate *predictions* about the traits likely to appear when the offspring of parents with known traits were allowed to develop *in environments similar to those in which their parents developed*. Before Galton, the statistical tools that would have allowed predictions in these sorts of situations did not exist. As might be obvious, correlational analyses like this one continue to be valuable today; they have useful applications in virtually every branch of contemporary science.

Before long, though, Gregor Mendel's paper on trait inheritance was rediscovered, the word "gene" was coined to refer to his "heritable factors," and Thomas Hunt Morgan demonstrated that genes on chromosomes are both influential in trait development and inheritable—actually passed physically from generation to generation. Thus, early in the 20th century, Galton's statistical techniques were modified slightly and ultimately applied to the task of trying to determine the extent to which genes contribute to the appearance of traits. Work on this task led to the development of the correlational statistic now known as "heritability." Today, this statistic remains the cornerstone of the branch of psychology called behavior genetics. And since the results of behavior genetics studies inform the decisions of those forging our public policies, it is very important to have an accurate understanding of what heritability means.

Seeing Double

A heritability estimate for a trait is a number that tells the extent to which differences in the appearance of the trait across several people can be "accounted for" by differences in their genes; computing these statistics is the aim of behavior geneticists. For example, the most widely publicized behavior genetics report of the 1990s was T. J. Bouchard and colleagues' finding that the heritability of IQ is .70. Given the definition of heritability, this means that 70 percent of the differences in IQ found among Bouchard's subjects could be "accounted for" by differences in their genes. But what exactly does that mean?

The best way to understand heritability estimates is to examine how they are produced in the first place. Heritability estimates are usually generated by analyzing data collected in studies of twins. Twins are important in the study of heritability because women can bear at least two different kinds—so-called identical twins and fraternal twins—and because the two types differ in the extent to which they have the same genes. The genetic differences between twin types result from events that occur around the time of their conception. Fraternal twins are conceived when two of a mother's unique eggs are fertilized at about the same time, each by a different, unique sperm. This gives rise to two unique zygotes, each of which ultimately develops into a unique person. Such twins are typically as different from each other as would be any two zygotes conceived by these parents; that is, they are as similar to—and as different from—each other as are regular siblings. Thus, the only difference between fraternal twins and an ordinary brother-sister pair is that fraternal twins begin their lives around the same time, whereas the lives of ordinary siblings usually begin a year or more apart. In contrast to fraternal twins, identical twins are produced when a single embryo splits in two, usually for unknown reasons. The resulting two embryos are made of cells that contain the exact same genetic material, and they normally develop into people that are unusually similar to each other in certain respects (for example, in their appearances).

The logic of twin studies rests on the fact that identical twins have identical genes, whereas fraternal twins do not. By producing two different types of twins, some with identical genes and some with genetic differences, nature has given us the raw materials for an "experiment"; if we find that both members of an identical twin pair almost always share a particular trait but that both members of a fraternal twin pair do not, then the pattern of variation in the trait can be "accounted for" by the pattern of variation in the genes. For example,

if—as appears to be the case—identical twins usually have similar IQs but fraternal twins sometimes do not, the differences among the fraternal twins must have developed because of *something* that was different for them but that was not different for the identical twins. And since fraternal twins have differences in their genes and identical twins do not, it is quite plausible that the "something" in question could be their genetic constitutions. In this way, behavior geneticists can sometimes trace patterns of variation in twins' traits to the patterns of genetic variation that characterize different twin types. Thus, if a behavior geneticist reports that IQ is highly heritable, what he means is that, by and large, individuals with the same genes have similar IQs and individuals with different genes have relatively different IQs.

The logic behind this sort of study is quite compelling; differences among fraternal twins that are *not* seen among identical twins *must* result, somehow, from differences present in the fraternal, but not the identical, twins' genes. So, what's the problem? The problems appear when we consider what heritability estimates do *not* mean. From this perspective, heritability statistics can be seen to be extraordinarily misleading.

Heritability: An Imposing Illusion

Reporting on an analysis of the heights of fraternal and identical twins, Robert Plomin—one of the leading behavior geneticists in America today—writes that the

> results indicate significant genetic effects. For these height data, heritability is estimated as 90 percent. This estimate ... indicates that, of the differences among individuals in height in the populations sampled, most of the differences are due to genetic rather than environmental differences among individuals.

Most people, upon encountering these words, would quite reasonably conclude that genetic factors must be more important than environmental factors in causing a person's height. After all, if genetic factors account for 90 percent of the differences in people's heights, it seems that environmental factors must not account for any more than 10 percent of this variation. And given this (mis)understanding, it would seem that it is probably difficult to influence highly heritable traits with manipulations of the environment. Thus, Plomin's statements, by seeming to suggest that people's heights are largely unaffected by their developmental environments, encourage us to conclude that people's heights are determined *prior to development,* by genes inherited from their parents.

Unfortunately, the fact of the matter is that these conclusions are most definitely not valid. Given the misleading language and concepts of behavior genetics, of course, one could be forgiven for coming to such erroneous conclusions. Nonetheless, the fact remains that heritability estimates are *not* measures of the importance of genes in the production of a person's traits, or of a trait's "openness" to environmental influence.

Heritability: Not about the Relative Importance of Genes

Heritability statistics do not reflect the relative importance of genes in causing traits, because—and this cannot be overemphasized—**heritability estimates tell us about what causes *variation* in traits; they tell us nothing at all about what causes traits themselves.** This difference seems incredibly subtle at first; if the difference between my height and your height can be traced to our differing genes, it certainly *seems* like this means that our respective genes determine our respective heights. I have discussed the difference between "accounting for variation" and "explaining causation" with many extraordinarily smart people who have never encountered it before, and when they first do encounter it, they are typically mystified; you're in good company if you feel the same way. But once you have given the issue plenty of thought, the difference seems about as subtle as a brick.

Consider a simple analogy drawn from nature: the formation of snowflakes. Snowflakes are formed only in the simultaneous presence of two factors, namely a temperature below 32 degrees Fahrenheit and a relative humidity high enough to allow for precipitation. Now, if on a given day, humidity is high at the North Pole but low at the South Pole, snow will fall only at the North Pole; *in this case, the variation in snowfall across the two locales can be accounted for completely by variation in relative humidity.* But such a circumstance certainly cannot be taken to mean that temperature is unimportant in causing snow. Thus, "accounting for variation" and "explaining causation" are profoundly different from each other. Humidity differences alone are enough to account for the differing snowfalls at the two poles in this example, but only because it is *always* cold enough for snow at both poles, not because coldness is unimportant in causing snow. Whenever a factor does not vary across situations (as temperature does not in the current example), it cannot "account" for variation in outcomes across those situations; still, this does not mean that the factor plays no role in causing the outcomes themselves.

This analogy illustrates how little an "account of variation" can tell us about causation. Snow is *caused* by two factors, even if all of the variation in its presence can be "accounted for" by variation in only one of those two factors. The same holds for our traits; it is quite possible, for instance, for genetic factors to "account" for 90 percent of the *differences* seen in people's heights, *without* genetic factors being any more important than environmental factors in *causing* people heights. *Regardless* of the heritability of height, a person's height is *caused* by both genetic and environmental factors. Twenty-five years ago, Richard Lewontin offered the following illustration of this point.

Lewontin asked us to imagine planting ordinary, genetically diverse seeds and then letting them grow to maturity in two different environments. Imagine that, within each environment, light, water, and nutrients are uniformly distributed but that one of the environments (environment A) provides its seeds with sufficient light, water, and nutrients, while the other environment (environment B) provides its seeds with just barely enough light, water, and nutrients to survive. At maturity, *all* of the height variation seen among the plants grown in environment A must be due to the genetic diversity that was originally present in the seeds; this *must* be the case, because all the seeds developed into mature plants in the *identical environment* (A). Thus, *none* of the variability in height can be accounted for by environmental variation (since there was none), and heritability of height in environment A must be 100 percent. Similarly, heritability of height in the deficient environment B is also 100 percent, since the height variation among the mature plants grown in this environment is also entirely attributable to the genetic diversity present in the seeds (again, because they all developed in the same environment). Thus, regardless of the environment in which the plants grew, the heritability of height is 100 percent in this example. If we didn't know any better, of course, a high heritability like this might have led us to think that plant height is 100 percent determined by genes. But examining the plants would tell a different story: even third graders know that plants grown in the deficient environment B can be counted on to be shorter, on average, than plants grown in the normal environment A. Thus, even if the heritability of a trait is very high—*in fact, even if it is 100 percent*—numerous *environmental* factors can nonetheless have overwhelmingly powerful effects on the trait's final appearance.

The converse of this story holds as well: a trait with a heritability of zero *cannot* be assumed to be *un*affected by genetic factors. Imagine taking one of the seeds from Lewontin's example and cloning it so as to produce a handful of genetically identical seeds. If we scattered these seeds in a variety of different soil types and let them develop, this time we would find *all* of the variation in the plants' heights to be accounted for by variation in the soils (since there is no genetic variation among the plants at all). In this case, then, the heritability of height would be calculated to be 0 percent.

But surely no one would argue that in this example genes have nothing to do with plants' heights! How could the height of a particular plant be *un*affected by the plant's genes, when genes participate in constructing the raw materials out of which plants are made? In fact, a trait *can* be importantly affected by genetic factors *even if it is not heritable at all*.

But if *perfectly* heritable traits are consequentially affected by both environmental and genetic factors, and if *non*heritable traits, likewise, are consequentially affected by both of these same factors, then what is the difference between heritable and nonheritable traits? In fact, *there is no difference at all*, at least, not in the extent to which heritable and nonheritable traits can be influenced by genetic and environmental factors.

To see this clearly, consider what it would mean if the heritability of hairstyles was .01, or 90 times less than the reported heritability of height. Is a heritable trait like height more influenced by genetic factors than is a nonheritable trait like hairstyle? No. In fact, *heritability estimates for height can be many times greater than heritability estimates for hairstyles, but this alone would* not *mean that a woman's height is more influenced by genetic factors than is her hairstyle*. Why not? Because heritabilities aside, genetic factors can profoundly influence the development of hair characteristics—curliness, for example—that can influence hairstyle decisions. By the same token, lower heritability estimates for hairstyles than for heights cannot be taken to mean that environmental factors impact hairstyles more than they impact heights; the height a woman attains in adulthood is affected by factors in her developmental environment—her diet, for example—just as the hairstyle she eventually dons is affected by environmental factors such as the advertising she is exposed to. Thus, although this might all seem extraordinarily counterintuitive, the fact is that heritable traits can be just as affected by environmental factors as can nonheritable traits, and nonheritable traits can be just as affected by genetic factors as can heritable traits. The disparity between what we assume heritability estimates tell us (given their name) and what they *really* tell us is *so* great that it is hard, at first, to make sense of it all. But this is true: since heritability estimates account for variation *and do not explain causation,* they tell us *nothing at all* about how genetic and environmental factors influence the development of our traits.

There are important practical consequences associated with the fact that heritability estimates do not tell us about what causes—or can affect—individuals' traits. For example, consider the following erroneous argument that makes use of Bouchard's high estimate of the heritability of IQ. Perhaps, some might argue, we should not bother devoting public resources to school programs designed to raise IQ scores, because the high heritability of IQ means that IQ is largely "genetic," and so not particularly open to environmental influence. The problem with this argument is that nothing could be farther from the truth. Since environmental manipulations can *profoundly* influence the development of traits that are even *more* heritable (even maximally heritable), heritability estimates cannot be appropriately used in this way.

Heritability: The Exquisite Specificity

At this point, you might find yourself saying, "Wait a minute! If the heritability of a trait tells us nothing about the extent to which genetic or environmental factors contribute to the trait's appearance, what use is knowledge of heritability?" A behavior geneticist might respond that heritability estimates do, at least, help us understand the source of *differences* among people. But even this claim is overstated, since it turns out that *heritability estimates cannot be generalized to situations different from the situation originally studied.* Instead, these statistics tell us only about the causes of differences among people "*in a particular population at a particular time*" *and in particular circumstances.*

Heritability estimates cannot be generalized because as soon as we start changing situations, the whole story changes, and the factors that once accounted for different outcomes might no longer do so. Consider again the formation of snowflakes. When we look at the causes of variation in snowfall at the poles, we find that the variation is accounted for by variation in relative humidity alone. But if we study variation in snowfall at several locations in an extremely humid Costa Rican rainforest, we will find that snowfall variation will be accounted for by variation in *temperature* alone. (Temperature varies in Costa Rica as a function of altitude; even in countries near the equator, snow falls on high mountaintops.) Thus, what "causes a difference" under one set of circumstances might not account for *any* of the variation detected in a different set of circumstances. As a result, accepting that a particular factor is a "cause" of a difference requires us to hold constant *every other factor* in a situation.

Heritability estimates must be understood in this same way. If *everybody* developed in environments just like those in which Bouchard's subjects developed, perhaps then we would be justified in thinking that a high heritability for IQ means that IQ differences are invariably accounted for by differences in genes. But we do *not* all develop in similar environments, and it remains likely that genetic differences that account for IQ variation in some environments do not account for IQ variation in others.

What is the value, then, of knowing what accounts for *differences* among people *in a unique situation,* if it is within our power to easily change situations? Understood this way, it is not clear that it is at all useful to know what accounts for differences. Knowing how a full complement of influential factors collectively *causes* individuals' traits is the more important goal, by far. For this reason, I focus in this book exclusively on questions about the *causes* of traits, and not on questions about the sources of differences among individuals.

Heritability: Unrelated to "Inheritability"

The upshot of this situation is that even though "heritability" sounds like it would reflect how "inheritable" a trait is, in a world of changeable environments, heritability estimates do not, in fact, do any such thing. The truth is that the statistical concept of heritability bears almost no resemblance whatsoever to our intuitive concept of "inheritability."

Consider first a trait that seems like it is part of our "biological inheritance," namely the presence of five fingers on each of our hands. Given our confidence that we inherit this trait, most of us would probably expect this trait to be highly heritable. But as the philosopher Ned Block has explained, most of us would be mistaken:

> The *heritability* of number of fingers and toes in humans is almost certainly very *low*. What's going on? If you look at cases of unusual numbers of fingers and toes, you find that most of the variation is environmentally caused, often by problems in fetal development. For example, when pregnant women took thalidomide [a drug later implicated in the production of birth defects] ... many [of their] babies had fewer than five fingers and toes. And if we look at numbers of fingers and toes in adults, we find many missing digits as a result of accidents.

But if most of the variation in people's digit numbers can be traced to variation in environmental factors, then little of the variation in digit number is "accounted for" by genetic variations; in fact, human genes only rarely—if ever—contribute to the appearance of just four fingers on a hand. Thus, heritability of this trait would be calculated to be rather low. For similar reasons, many common traits—possessing opposable thumbs, four limbs, or teeth located inside your mouth—are not particularly heritable, even though it is inconceivable to most of us that these traits are not "inherited." Is it just a minor inconvenience that "heritability" and "inheritability" sound so much alike when in fact they mean very different things? I think not; this terminology leaves most of us quite confused.

Consider the opposite side of the same coin: What do heritability estimates tell us about traits that we intuitively sense are *not* inheritable? Block addressed this situation as well, writing "Some years ago when only women wore earrings [in America], the heritability of having an earring was high." It seems as if this can't be right; we all know that earring wearing is *very* affected by cultural factors, and so is unlikely to be "inherited" according to any traditional definition of this word. Nonetheless, earring-wearing behavior in 1950s America was, in fact, highly heritable. What's going on here?

This counterintuitive finding reflects the fact that in mid-century America, there actually were certain genes that were found in most earring wearers and that were hardly ever found in non-earring wearers. Think about that for a second. Does this mean that there actually are genes "for" earring wearing? Not at all, *because correlation tells us nothing at all about causation;* genes whose presence is correlated with earring wearing need not play *any direct role* at all in causing that behavior. As it happens, the genetic factors consistently found in earring wearers in the 1950s were two X chromosomes, which characterize almost all women and very few men. But X chromosomes do not *cause* earring wearing any more than do ovaries (the presence of which were also highly correlated with earring wearing in the 1950s). Instead, since only women wore earrings at that time—*for cultural reasons*—there was a high correlation between the presence of two X chromosomes and the presence of jewelry dangling from the ears.

Amazingly, the discovery that specific genes are highly correlated with the appearance of a trait is enough for behavior geneticists to conclude that the trait is heritable, *even if the appearance of the trait obviously depends on cultural factors.* Why is that? Because for behavior geneticists, the goal is to predict the appearance of traits by looking at genes, and if there is a high correlation between the presence of some genes—any genes!—and the presence of a trait, then looking at the genes allows them to make predictions about the trait's appearance. For behavior geneticists, the fact that X chromosomes do not directly *cause* earring wearing or that cultural factors *do* have a direct causal role in producing this behavior are simply not of interest. As far as they are concerned, if there is a *correlation* between earring wearing and the presence of certain genes—however *in*directly related to earring wearing those genes are—then earring wearing is heritable, even if environmental factors play essential roles in causing the trait. Knowing what effects nongenetic factors have on a trait's appearance is simply not of interest to behavior geneticists, who consequently just ignore these factors.

Take note: this is not just an abstract argument without implications. In fact, it would be entirely possible using behavior geneticists' correlational methods to find that genetic variations among American blacks and whites "account for" most of the variation in their IQ scores, *but this would not mean that the IQ differences are caused by the genetic differences*. Instead, as Block's earring example shows, a finding that IQ is heritable could simply mean that people with genes that contribute to different skin colors develop different IQs because of how society treats people with different skin colors. In the language of behavior genetics, genes that contribute to skin color differences could fully "account for" IQ differences, even if these genes influence IQ *only* via racist attitudes and behaviors present in our society.

Galton's Legacy

So where does all of this leave us? First, heritability estimates tell us nothing at all about what causes an individual's traits. Second, heritability estimates do not reflect the extent to which traits are impervious to environmental influences. Third, heritability estimates apply only to people in a particular population, at a particular time, and in particular circumstances; they can *never* be appropriately generalized to other populations. Finally, despite its evocative name, heritability estimates do *not* tell us how likely it is that parental traits will be *inherited* by their offspring. Instead, heritability estimates are capable of indicating—counterintuitively—that characteristics like finger number are not heritable and that characteristics like earring-wearing behavior are.

Ultimately, then, heritability estimates are not at all what they appear to be. If, in the future, you hear that a specific trait—let's take alcoholism, for example—is highly heritable, try to resist the temptation to immediately think several things that you might otherwise think. First, even if alcoholism were *perfectly* heritable, this would not mean that it will necessarily afflict the children of alcoholics; even *perfectly* heritable traits are affected by the environment, so raising the children of alcoholics in an environment different from the one in which their alcoholic parents were raised could make all the difference. Second, if alcoholism is heritable, this does not mean that it is caused by genetic factors; it needn't even necessarily be the case that genetic *differences* between alcoholic and nonalcoholic individuals *directly* contribute to the differences in their drinking patterns. And, finally, if alcoholism is heritable among Iowans, it need not be the case that it is heritable among Ohioans. (I am not simply being recalcitrant here: heritability estimates calculated for one population *do not apply* to another population.) These are truths that are widely misunderstood by a public that nonetheless avidly consumes heritability data.

Given these problems, is there some *other* measure that better captures our intuitive notion of "inheritability?" Unfortunately, there is not. The heritability estimate, with all of its inescapable problems, is the statistic that comes closer than any other to capturing this intuitive notion. In fact, *the scientific data that underlie the public's misperception that some traits are more "in the genes" than others are those that come from heritability studies; there are scant, if any, other scientifically collected data underlying this conviction*. Individuals certainly have anecdotal "evidence" behind their intuitions that not all traits are equally inheritable, but in the past 100 years, the main scientific approach to studying inheritance has been the one pioneered by Galton. Unfortunately, the foregoing analysis of heritability requires us to view the endeavors and claims of Galton's followers with skepticism.

Heritability estimates are misleading at best, and the interpretations they encourage are simply wrongheaded at worst; they do not illuminate the origins of our traits as we might otherwise have assumed. P. R. Billings and colleagues agreed with this assessment when their analysis of the twin study literature led them to conclude that "identical twin studies offer no convincing evidence of the genetic basis of human behavior."

Actually, the situation is even worse than it appears. I have been describing how the logic behind *an idealized, perfect* twin study would still limit how we could properly interpret the study's results. Lamentably, there are other logical problems that bedevil *actual* twin studies conducted in the real world.

SECTION III

THE CONCEPTS OF INSTINCT AND CRITICAL PERIODS

EDITORS' INTRODUCTION

Readers of this book may be familiar with the concept of instinct, and this being a book against the genetic determinism of behavior it is fitting that we have included a section devoted to the criticism of this particular approach to understanding behavioral origins. "It is almost *de rigeur* to begin any discussion of instinct by deploring the word's ambiguity and restricting its use, if it be tolerated at all, to some carefully defined set of behaviors" (Diamond, 1974, p. 238). Before offering what psychologists of our persuasion understand what is meant by instinct we begin our discussion with these compelling statements by Frank Beach (1955):

> The basic ideas underlying a concept of instinct probably are older than recorded history …. They have been controversial ideas and they remain so today. Nevertheless, the instinct concept has survived in almost complete absence of empirical validation. (p. 401) … No bit of behavior can ever be fully understood until its ontogenesis has been described. (p. 407)

Of course, these statements, made in 1955 and 1974, reflect the basic themes championed by all contributors to this book.

The concept of instinct is perplexing. The concept may seem to both describe and explain the appearance of complex behaviors in different species but, when a scientist considers the concept with more than passing scrutiny, the explanatory use of the term evaporates and, as well, even the descriptive use of the term becomes problematic. In his very readable argument against nativism and genetic accounts of behavior Blumberg (2005) provides examples of painstaking (e.g. "arduous") research examples showing that developmental (i.e., experiential and epigenetic) factors can *always* be found to underlie what has mistakenly been attributed to instincts. Yes, there are many behaviors that appear to be innate and instinctive, but there are so many animal species and so many behaviors and only so many scientists to study them, So, for example, herding behaviors by dogs has never been the subject of serious empirical investigation without which the easy answer is "it must be instinctive."

Blumberg was a student of Howard Moltz, himself a student of T. C. Schneirla, and Moltz's (1965) contribution to this discussion is worth citing here. Accordingly, Blumberg and Moltz discuss and provide empirical evidence for *alternative* routes to the stereotyped characteristics of so-called instinctive (in Moltz's terminology *fixed action patterns*) behavior. As Moltz pointed out, "Schneirla has repeatedly emphasized … stereotypy can arise from causes which operate directly and are immediately apparent

DOI: 10.4324/9781032702988-11

or from causes which are considerably more subtle" (p. 38) or in Gottlieb's terms which are non-obvious precursors of behavior (Gottlieb, 2001). In his contribution to the present book Blumberg refers to Lorenz's comment regarding the stereotypy of head scratching by birds: "I do not see how to explain this clumsy behavior unless we admit that is inborn." In another context Blumberg refers to Lorenz's inability to understand other possible origins of these behaviors as "more than a touch of intellectual laziness" (Blumberg, 2005, p. 203). There being so many meanings routinely ascribed to "instinct," it is easy to see some invoke more than one meaning at any given time.

Accordingly, it should not surprise us to see how murky the instinct concept has become. How did the murkiness of the concept of instinct arise? What can be done to bring light (clarity) into the discussion of this concept? The notion of instinct is perhaps most often associated with the work of Konrad Lorenz, reviewed historically, and somewhat critically, by Richards (1974). Beginning in the 1930s, Lorenz, an Austrian-born zoologist and physician, studied specific types of behavior he termed *instinctive behavior*. By this term, Lorenz seemed to mean behavior that is preformed in the genotype. Lorenz contended that humans inherit a genotype and built into this genotype is a "limited range of possible forms in which an identical genetic blueprint can find its expression in phylogeny" (cited by Blest, 1966).

In essence, then, Lorenz contended that there is a fixed and invariant relation between specific genetic inheritance and specific behaviors, and this correspondence is what he meant by instinctive behavior. Specific behaviors are preformed, or at least predetermined, and, thus, they are innate; they are built into the organism through genetic inheritance (the genotype) and, thus, these behaviors are simply unavailable to any environmental influence.

More specifically, Lorenz saw specific inherited properties of nervous system structures as innate. Specific groups of neurons, he claimed, had built into them specific, distinctive properties (Lehrman, 1970). They obtained these properties directly from the genotype, with experience having no influence. For example, as Lehrman (1970, p. 24) pointed out, one such innate property of a given neural structure is "its ability to select, from the range of available possible stimuli, the one which specifically elicits its activity, and thus the response seen by the observer." That is, in the view of Lorenz, specific nervous system structures come with the innate ability to select out specific stimuli from the environment; these are the stimuli that elicit (bring forth) the built-in (predetermined) functional component of the structure, that is, the response (Lorenz, 1965).

Because, as Lorenz (1965) contended, experience plays no role in the presence of this instinctive behavior, a scientist does not have to bother with the issue of how the relation between the stimuli and responses comes to be established. All one has to say is that the behavior is there because it is innate. Then, one simply "explains" that innate behavior comes this way. Thus, to Lorenz, no further analysis was needed. In advancing this argument, Lorenz "solved" the problems of behavioral development by simply avoiding them—by defining them away.

In essence, then, Lorenz (1940a, 1940b, 1965) argued that genetic inheritance represented a "blueprint" for the development and final level and form of behavior—a flawed concept resurrected by behavior geneticist Robert Plomin in his 2018 book, *Blueprint*, discussed in the first section of the present book. In both the earlier and later versions of this concept of blueprint, the term is used to represent a set of directives that were unalterable by environment, experience, learning, socialization, and so on (cf. Lehrman, 1953, 1970). This genetic inheritance was believed to be able to circumscribe behavior so severely because it led directly to the formation of an instinct—a predetermined, innate, and unmodifiable pattern of behavior specific to the species within which it exists. The behaviors associated with this instinct are then not capable of environmental, experiential modification.

Thus, to Lorenz, and to Plomin (2018), behavior is constrained by instincts; variation in behavior beyond the limits imposed by the genetically fixed instinct is not possible. Such a conception of genetic influence precludes, then, a process analysis of the ways in which genetic and environmental variables contribute coactively to behavioral development. In other words, Lorenz's (1940a, 1940b, 1965) conception of instinct precludes a consideration of how organismic and/or contextual processes may contribute to the development and organization of behavior. His conception eliminates any use for studying how behavior may be altered or enhanced.

From the perspective of comparative psychologist T. C. Schneirla (1957, 1966), however, there are several problems inherent in Lorenz's ideas about instincts. By making a distinction between what is innate and what comes about through the environment and by implying that there exists a genetic blueprint that imposes fixed constraints on development, Lorenz opted for the "which one?" (nature or nurture) question, which Anastasi (1958) rejected as inadequate because there would be no place to see the effects of genes if there was no environment and there would be no organism in an environment if there were no genes. Simply, the two are completely integrated in all biological life. Thus, given Schneirla's dynamic, relational developmental systems (RDS) related conception of gene ⇔ context relations, the notion of innate, or instinctive, behavior as formulated by Lorenz (1965) is not scientifically useful for the following reasons:

1. Nature and nurture are inextricably bound; it is inappropriate to assert that genes can directly produce human behavior (e.g., Gottlieb, 1996, 1997, 2004; Lickliter, 2016; Moore, 2015a, 2016). Nature variables need the supportive, facilitative influence of experiential factors in order to contribute to behavior. In turn, of course, experience needs nature variables with which to coact; as such, behavior development occurs through gene ⇔ context relations within the integrated, relational developmental system (Overton, 2015).

2. Because of this interdependency, it is inappropriate to speak of "innate" as meaning developmentally fixed—that is, to speak of specific behavior as being unavailable to environmental influence or to say that an organism must develop certain behaviors because it inherited a certain genotype (Lehrman, 1970, p. 23). The nature ⇔ nurture relation is more complex. Genes play a role in human development only through these coactions. The role of genes in human development will be different under different environmental (experiential) conditions. Therefore, it is incorrect to speak of a genetic blueprint. Simply, there is no isomorphism between genotype and eventual behavior.

Conclusions

Lorenz used the terms *innate* or *instinctive* to refer to behavior that is genetically fixed and, therefore, unavailable to environmental influence. However, from the perspective of adherents to dynamic, relational developmental systems-based models of human development, the notion of instinct can be rejected as being overly simplistic, as being based on faulty logic and, most important, as ignoring the problems and issues of behavioral development. To study the problems of behavioral development, developmental scientists must avoid terms such as *innate* (as employed by Lorenz, 1940a, 1940b, 1965, 1966, or Plomin, 2018). Such terms end scientific investigation by simply saying that a behavior develops in a specific way because the organism is built that way. Thus, the use of the terms *innate* or *instinctive* avoids assessing the processes by which behavior develops and, hence, is of little, if any, scientific use.

Perhaps the most succinct summary of the criticisms that can be leveled against the use of the term instinct was made by another one of T. C. Schneirla's former students, Daniel Lehrman. In a classic paper, published in 1953, Lehrman noted:

> The "instinct" is obviously not present in the zygote. Just as obviously it is present in the behavior of the animal after the appropriate age. The problem for the investigator who wishes to make a closer analysis of behavior is: How did the behavior come about? The use of "explanatory" categories such as "innate" and "genetically fixed" obscures the necessity of investigating developmental *processes* in order to gain insight into the actual mechanisms of behavior and their interrelations. The problem of development is the problem of the development of new structures and activity patterns from the resolution of the interaction of *existing* structures and patterns, within the organism and its internal environment, and between the organism and its outer environment. At any stage of development, the new features emerge from the interactions within the *current* stage and between the *current* stage and the environment. The interaction out of which the organism develops is *not* one, as is often said, between heredity and environment. It is between *organism* and environment! And the organism is different at each stage of its development. (p. 345)

Although the theoretical position of Lorenz, Plomin, and other writers using genetic reductionist ideas (e.g., Harden, 2021) is egregiously flawed conceptually and is, as well, empirically counterfactual, these ideas were given and are still given great attention and, often significant approval. For instance, Lorenz was awarded the Nobel Prize in Medicine or Physiology in 1973, and Harden's (2021) book, *The Genetic Lottery*, received widespread public media attention despite strong scientific objections to her ideas (e.g., Feldman & Riskin, 2022). Accordingly, given the continuing attention given to the flawed ideas of genetic reductionists, it is important to discuss another key instance of the concepts reductionists use to try to bring credibility to their version of the role of genes in human life and development.

However, before turning a discussion of the critical-periods hypothesis, we believe it is important to close this discussion of instinct with the introduction of Schneirla's (1939) approach/withdrawal (A/W) hypothesis. This hypothesis constitutes a set of organizing principles that provides an account of behavioral origins in terms of biphasic processes which are based on the characteristics and effects of stimuli. Briefly, the A/W concept is based on the premise that approach and withdrawal are two basic response patterns underlying all complex adaptive responses and, as such, involves a synthesis of several organizing principles and concepts (Raines & Greenberg, 1998). The concept proposes that behavior, in the full array of species, is directed toward weak sources of stimuli (approach) and away from (withdrawal) strong sources of stimulation. The concept has been shown to apply to the full spectrum of animals and to a full array of behaviors (Greenberg & Haraway, 2002).

Lorenz (1937; see too Richards, 1974) famously studied imprinting—the following response of newly hatched ducks and chickens to follow their mothers—*and* the responses of various young birds to cardboard dummies of raptors and other flying birds, termed the hawk-goose phenomenon (Schleidt et al., 2011). These behaviors were understood by Lorenz to be instinctive because they were present at hatching and without any learning or other obvious experiential precursors.

However, both of these behaviors can be accounted for by using Schneirla's A/W idea without relying on instinctive explanations. Schneirla (1965) showed how the A/W idea could account for the apparent instinctive behavior of goslings running from hawk silhouettes. When approached from the goose configuration, the long thin neck enters the visual field gradually, while from the hawk configuration, the short neck and extended wings enter the visual field abruptly, stimulating a larger retinal area all at once, where the larger area of retinal stimulation equals a more intense visual stimulus invoking withdrawal by the goslings (on this point, see also Burghardt, 1973).

Finally, with respect to understanding the effects of stimulus intensity (e.g., A/W) on imprinting Howard Moltz (1963) reported the following:

> When initial approach movements are exhibited ... it is significant to note that they are invariably directed toward the [imprinting] object as the object retreats but not as it moves toward the animal ... During the past 5 years we have worked with over 600 birds and I cannot think of a single instance in which approach was initiated as the object moved toward the bird; indeed, when travelling in that direction, the object will occasionally evoke withdrawal responses. (pp. 125–126)

So much for the innate and instinctive accounts of imprinting and the hawk/goose phenomenon. We think it important to close our discussion of *instinct* with this admonishment and warning by Lehrman (1953, p. 359) which is crucial to main themes of the present book:

> Any instinct theory which regards "instinct as immanent, preformed, inherited, or based on specific neural structures is bound to divert the investigation of behavior development from fundamental analysis and the study of developmental problems. Any such theory of "instinct" inevitably tends to short-circuit the scientist's investigation of intraorganic and organism-environment developmental relationships with underlie the development of "instinctive" behavior.

The Critical-Periods Hypothesis

The notion of critical periods in development was formulated in embryology. Within that area of science, the idea was advanced that the various parts of the whole organism (e.g., various organs or organ systems) emerge in a fixed sequence; more importantly, it was held that the parts that develop in a fixed sequence do so with just a certain amount of time allowed for each part to develop. It was believed that there was an overall timetable of development, and each part of the whole organism had its own fixed time of emergence, set by maturation which, in turn, was a process fixed by genes. Each part of the whole, then, had a critical period in which to develop (e.g., see Scott, 1962).

This concept specifies that a part of the organism that is in its critical period can easily be stimulated. Such a part is highly responsive to both facilitating and disruptive influences. Thus, if the part does not develop normally or appropriately during its critical period, it will never have a second chance. Because the time limits of development are invariably fixed by maturation, even if the part does not develop, the focus of development will change. It will shift to another organ system, in accordance with the predetermined timetable of development, and that different organ system will then be in its critical period of development. Hence, any part that does not develop appropriately during its own critical period will not have another chance in life.

Similarly, in human development such a critical period idea refers to a time in the life span of a person during which it is crucial for a particular feature of development to emerge. The period is crucial because specific maturational processes then occurring would allegedly place time limits on the development (Rosenblatt et al., 1961; Schneirla & Rosenblatt, 1961, 1963). For example, within his psychosocial theory of development, Erik Erikson (1950, 1959, 1968) divided the human life span into eight stages, each of which may be interpreted as consistent with the definition of a maturation-based critical period. The eight stages in Erikson's theory may be regarded as critical periods because each emerges in accordance with a maturational "ground plan," a developmental scheme that is built into the person (Erikson, 1959). Thus, Erikson maintains that, in the first year of life, the infant must develop a certain degree of a "sense of trust." If the infant does not develop this feeling at the time when it is supposed to develop, not only will there never be another chance but also the rest of that person's development will be unfavorably altered.

Clearly, the critical-periods hypothesis places primary dependence for healthy development on an intrinsic, maturation-determined timetable. What this formulation clearly indicates, then, is that maturation in and of itself sets critical time limits for development; there are periods in an organism's development that are circumscribed by maturation, and the time limits of these periods are somehow not related to experiential factors. However, from Schneirla's (1957) and others' RDS-based probabilistic-epigenetic position (Gottlieb, 1970, 1983, 1991, 1992, 1997, 2004; Gottlieb, 1997; Tobach, 1981; Tobach & Greenberg, 1984), such a conception of critical periods is untenable. Rather than emphasizing the independent contribution of maturation, Schneirla would opt to investigate the process by which maturation and experience coact to enable a specific development to take place at a specific time in ontogeny. For instance, in rejecting the idea of critical periods, Schneirla and Rosenblatt (1963) noted that, in their theory of behavior development,

> We conclude that factors of maturation may differ significantly in their influence upon ontogeny, both in the nature and in the timing of their effects, according to what relations to the effects of experience are possible under the existing conditions. (p. 1113)

Schneirla did not say that specific developments were not critical for some later developments. He would agree to some extent with other researchers concerned with the critical-periods notion (e.g., Scott, 1962) that there are critical phases of life, for instance for the development of learning. He would agree that what is learned at a certain time in an organism's ontogeny may indeed be important or even essential for whatever follows (Schneirla & Rosenblatt, 1963; Scott, 1962). That is, Schneirla and Rosenblatt (1963) noted that:

> In discussing his concept of critical periods, Scott [1962] reports us as having "suggested that there are critical stages of learning – that what has been learned at a particular time in development may be critical for whatever follows.
>
> Although we are not disposed to dispute this broad statement, it is not ours. In our view, any such sentence should have a more comprehensive context, to the effect that what the young animal may attain in behavior at any phase of ontogeny depends upon the outcome of earlier development in its every aspect. (p. 1110)

Accordingly, Schneirla's critique of the critical periods hypothesis is based on his contention that the use of the term "critical" means only that what happens at "Time 1" in an organism's life may be very important—in fact, foundational or even essential—for what can or will happen at "Time 2." Such an assertion merely describes a relation between events that occur at two different times in life; it makes no statement about whether the first event was determined by maturation alone or by coaction between maturation and experience (Bateson, 1983).

It is the source of the "criticalness" in development about which Schneirla argued. Simply, time limits for development that are fixed by maturation, arising without the contribution of experience, are inconsistent with his probabilistic–epigenetic position. Rather, Schneirla proposed a theory that placed "emphasis upon the *fusion* of maturation (growth-contributed) and experience (stimulation-contributed) processes at different stages in behavior ontogeny, together with the … contribution both of maturation and experience …, as well as the interrelations of these contributions" (Schneirla & Rosenblatt, 1963, p. 288). Indeed, Howard Moltz (1973), a leading student of Schneirla's found experimentally that the time limits of specific purportedly critical periods (e.g., involving the immediately after hatching "following" behaviors of some species of birds) *could* be altered through specific manipulations of the birds' early visual experiences (e.g., Moltz & Stettner, 1961). Similarly, Salzen (1998) noted that "imprinting could be obtained by exposure to the imprinting stimulus at ages past that of the normal end of the sensitive or critical period, provided that the sensory experience of the neonate had been severely restricted" (p. 568).

Weak and Strong Versions of the Hypothesis

Of course, Schneirla's view is not the only one that exists in regard to the meaning and bases of the concept of critical periods in development. Indeed, over the course of numerous reviews of the concept (e.g., Bateson, 1979, 1983; Colombo, 1982; Connolly, 1972; Hess, 1973; Hinde, 1962; Nash, 1978; Scott, 1962; Thorpe, 1961), several definitions of critical period were forwarded. These definitions may be divided in several ways. For instance, Krashen (1975) distinguished between strong and weak versions of the critical-periods hypothesis. Consistent with McGraw (1943), Colombo (1982) states the hypothesis in its weak form:

> A critical period is a time during the life span of an organism in which the organism may be affected by some exogenous influence to an extent beyond that observed at other times. Simply, the organization is more sensitive to environmental stimulation during a critical period than at other times in its life. (p. 261)

Similarly, in Krashen's (1975) view, the weak version of the hypothesis states that there are periods in life when the development of a system can best be furthered by particular stimulation but that the system's development can, nevertheless, still occur after such a period.

In essence, then, in the weak form of the hypothesis, the critical period is really only a *sensitive period*, one wherein particular experiences may most readily or efficiently promote development of a system (e.g., cognition, vision, or language); nevertheless, similar or perhaps distinct experiences can foster the system's development after such a period, albeit perhaps with the requirement that the experience (e.g., the stimulus, an intervention program) be more intense or of greater duration in order to result in comparable development (Greenberg & Haraway, 2002; MacDonald, 1985).

Thus, this form of the hypothesis indicates that critical periods are not so critical after all, and that they are little more than labels applied to the well-known and hardly controversial observations that:

1. When a system is developing it needs stimulation to allow it to adequately do so (e.g., if humans were totally deprived of light stimulation their visual system would not develop; Hebb, 1949), or simply that, as Schneirla (1957) explained, the human development system needs to be active to function adequately; and
2. It is easier to influence a system—for better or worse—when it is in a state of development than after it has been fully organized (MacDonald, 1985).

In sum, in the weak form of the critical-periods hypothesis, particular experiences play a non-contingent (Moltz, 1973) role in development; although they are not absolutely necessary for adequate or healthy development to occur, particular experiences can enhance development due to the greater efficiency of their influence at a particular time.

With such not-quite-critical critical periods, developmental deficits produced by the lack of a particular experience (e.g., language deficits due to the absence of an adequate language model) may be overcome by experiences in later life. If recovery of function can occur, this recovery means that whereas a given period may be *optimal* (Moltz, 1973) for the development of a particular function, it is not a critical time for this development. As Bateson (1983, p. 8) puts it, "Once the mechanisms protecting behavior from change are stripped away by suitable treatment, change resulting from renewed plasticity is once again possible" (see too Bateson, 2015, 2016).

In turn, in the strong form of the critical-periods hypothesis, particular stimulation is needed at a particular time in order for normal development to proceed; in other words, if the appropriate stimulation does not occur when it is supposed to in life, then what will occur is "an irrevocable result not modifiable in subsequent development" (Scott, 1962). Thus, in such a formulation the organism *needs* certain stimulation for its continued normal development, and given inappropriate experience, it is *vulnerable to,* or *at risk for,* abnormal development during such a period (Colombo, 1982). Simply, for such a period, no recovery of function by later experience is possible (Krashen, 1975) and, as such, experience during this period has a "contingent" role (Moltz, 1973); that is, it is absolutely necessary for normal development.

The original instance of the strong version of the critical-periods hypothesis derives from the work of Konrad Lorenz (1937). Lorenz introduced the concept of "imprinting" to describe what he believed to be an irrevocable social bond, or attachment, formed by newly hatched precocial birds (e.g., birds such as ducks or geese, that can move sufficiently to follow other animals immediately after birth). These birds followed the first moving object they saw (usually their mothers) during the first hours after birth. Although Moltz and Stettner (1961) were able to manipulate (i.e., extend) the time period for imprinting through altering the visual experience of such birds (by placing a latex hood over their eyes that, although allowing light to come through, did not enable them to see any patterns or shapes), Lorenz (1937, 1965) claimed, nevertheless, that these first few hours were the critical period for imprinting to occur.

On a personal note, one of us (GG) was a student in a comparative psychology course taught by Howard Moltz at the time his 1961 study was published. I recall to this day the discussion he generated by asking us where we thought he was able to find suitable latex hoods for his newly hatched ducklings. The surprising answer was that he used store bought condoms for this purpose, attesting to the inventiveness and creativeness often required of this type of research.

The evidence that exists to support the reality of such strong critical periods may be questioned, however. Colombo (1982) summarized data pertinent to the existence of strongly defined critical periods in regard to four areas of research: imprinting in birds, social development in rhesus monkeys, language acquisition in humans, and binocular vision development in mammals. Colombo (1982) noted:

> Nearly every demonstration of a critical period in behavioral development during the past 50 years has been followed by a demonstration of some behavioral recovery from the effects of critical period exposure or deprivation. The first example was with avian imprinting, in which Lorenz's (1937) claims of a tightly bounded period during which a permanent parent–offspring relationship was formed were rigorously tested. Subsequent evidence suggested that the critical period was not as temporally distinct (Brown, 1974) nor were the effects of stimulation within it as irreversible (e.g., Ratner & Hoffman, 1974; Salzen & Meyer, 1968) as Lorenz had originally thought.

(Bateson, 1966)

After observing the results of social isolation during the first year of life, Harlow (1959, 1965) suggested the existence of a critical period for the development of social behavior in the rhesus monkey lasting (in one version) from birth to 250 days. The critical stimulus was apparently what he called "contact comfort," the absence of which during this early period resulted in permanent social/psychological maladjustment. Later, however, a series of experiments (Mason & Kenney, 1974; Novak & Harlow, 1975; Suomi & Harlow, 1972) demonstrated that with special interventions and patience, the adverse effects of deprivation during this period could be overcome.

Language acquisition was another major developmental process to which critical period theory was applied, only to have that application subsequently questioned. Elaborating on a suggestion by Penfield and Roberts (1959) and through the use of data on early and late unilateral brain damage (e.g., Basser, 1962) and the development of language (Lenneberg et al., 1964), Lenneberg (1967, 1969) hypothesized a period of receptiveness to language lasting from ages 2 to 12. Language could be most easily acquired during this period; after this period, acquisition of a first language would be extremely difficult, if not impossible. The absolute irreversibility of the period's effects has been somewhat disconfirmed by subsequent investigation of acquisition after linguistic deprivation (Curtiss, 1977; Curtiss et al., 1975; Fromkin et al., 1974) and of second language learning (McLaughlin, 1977).

In initial studies of the critical period for the development of binocular vision, during which monocular deprivation resulted in anatomical degeneration of the deprived eye's pathways, complete domination of cortical physiology by the deprived eye, and apparent blindness of the deprived eye (e.g., Hubel & Wiesel, 1970) no recovery of function was reported (Blakemore & Van Sluyters, 1974; Hubel & Wiesel, 1970; Wiesel & Hubel, 1965). Subsequent studies, however, demonstrated that recovery in at least the behavioral aspects of visual function could be obtained after the end of the period (e.g., Baxter, 1966; Chow & Stewart, 1972; Cynader et al., 1976; Mitchell et al., 1977; Timney et al., 1978; Mitchell, 1978). It is worth noting, however, that this recovery has yet to be demonstrated in primates (Crawford et al., 1975; von Noorden et al., 1970). (pp. 268–269)

Moreover, Colombo (1982) reviewed additional evidence that both the presumed onsets and terminations of critical periods—that is, the times in life when these periods are believed to begin and end—are influenced by variables both endogenous and exogenous to the organism. Thus, the time limits of these periods are not as fixed and sudden as Lorenz (1937, 1965) maintained or as impervious to contextual influences as Lorenz also believed. Colombo (1982) indicated that rather than a sudden and dramatic onset of sensitivity to a specific stimulus, sensitivity rises gradually to a peak and then gradually declines. These changes can be manipulated; for example, manipulation in regard to binocular visual development can occur by altering the amount of light in the rearing environment. We have noted that similar perceptual stimulation manipulations can alter the imprinting period in birds (e.g., Moltz & Stettner, 1961; Salzen, 1998). In addition, pharmacological manipulations can extend the imprinting period of birds or even prevent it from occurring at all (Colombo, 1982).

Conclusions

There is no good evidence to support the strong version of the critical-periods hypothesis, as for instance advanced by Lorenz (1937, 1965). Nature variables (e.g., an instinct, DNA) *do not* prescribe fixed time limits within the life span, wherein specific stimulation must occur for normal development to proceed. Rather, when a specific portion of the relational developmental system is in a period of marked development, it is especially responsive to influences by variables "outside" the specific portion system but of course endogenous to the overall, holistic, integrated system. Indeed, Colombo's (1982) conclusions regarding the character of critical periods in development are comparable to those that reflect Schneirla's (1957) perspective. Colombo noted that "the emergence of a critical period ... is based on, and may be predicted by ..., the interaction of dynamic, developing systems, and as much effort should be directed toward identifying those systems and their interactions as toward identifying the period itself" (Colombo, 1982, p. 270).

Subsequent discussions of the critical-periods hypothesis support Colombo's (1982) views (e.g., Michel, 2010; Michel & Tyler, 2005). For instance, Michel and Tyler (2005) note that the history of research on critical periods indicates that the questions posed by researchers have changed. They have altered from the question of whether there a critical period to ones about what controls the developmental processes that shape criticality for specific facets of development (e.g., language, vision, neural development, etc.). Michel and Tyler (2005) emphasize that this focus on process has enabled researchers to move to a more sophisticated approach to research, one that seeks to understand specific ways in which past and current experiences shape ensuing experiences. This developmental focus moves the study of criticality in development to an assessment of the trajectory of experiences in the lives of organisms, and how these pathways frame the importance of prior features of the trajectory for subsequent development.

An Overview of the Contributions to This Section

Blumberg (2016) discusses the origins and meanings of instinct. He explains that this term was applied to phenomena in animals that seemed difficult to explain by reference to then contemporary concepts of learning (e.g., migration or reproduction patterns of turtles or birds). The intention of a person labeling such phenomena as instinctual was to explain (or explain away) the behaviors. Blumberg discusses why labeling complex behavior as instinctual, or genetically determined, fails to explain its development. He explains that these behaviors are outcomes of species-typical experiences occurring within reliable ecological contexts, that is, of complex organism-context coactions.

Michel and Tyler (2005) extend Blumberg's (2016) use of complex individual-context coactions across time and place to the analysis of critical periods in development. Rejecting simplistic reference to time (or age, which is simply time since birth) as a causal basis of phenomena labeled as critical periods, Michel and Tyler focus on a process involving how an individual's history of past and current experiences shape development. Such a change in focus affords researchers the means to interrogate how to modify experiences in the service of enhancing development.

The ideas forwarded by both Blumberg (2016) and Michel and Tyler (2005) reflect dynamic, relational developmental systems-based alternatives to genetic reductionist conceptions of instinct or critical periods in behavior and development. Spencer et al. (2009) emphasize that any continuation of the nature-nurture debate actually serves to provide reductionists with a claim that their notions are still viable scientific alternatives to dynamic systems conceptions of human development. Spencer et al. explain that such debating serves to distract attention from the reality of developmental systems, from embracing the concept of epigenesis, and from an ecologically valid understanding of the nature of contextual influences and of the subtlety of individual context coactions, that is, the view that development emerges via cascades of interactions across multiple levels of causation, from genes to environments. Using examples from studies of imprinting, spatial cognition, and language development, they call for a focus in developmental science on the dynamics of the process involved in individual-context coactions across time and place.

References

Anastasi, A. (1958). Heredity, environment, and the question "how?" *Psychological Review, 65*, 197–208.

Basser, L. S. (1962). Hemiplegia of early onset and the faculty of speech with special reference to the effects of hemispherectomy. *Brain, 85*, 427–60.

Bateson, P. (2015). Ethology and human development. In W. F. Overton & P. C. Molenaar (Eds.), *Theory and method. Volume 1* of the *handbook of child psychology and developmental science* (7th ed., pp. 208–243), Editor-in-Chief: R. M. Lerner. Wiley.

Bateson, P. (2016). Robustness and plasticity. *WIREs Cognitive Science, 8*(1–2), e1386. https://doi.org/10.1002/wcs.1386

Bateson, P. P. G. (1966). The characteristics and context of imprinting. *Biological Reviews, 41*, 177–220.

Bateson, P. P. G. (1979). How do sensitive periods arise and what are they for? *Animal Behavior, 27*, 470–86.

Bateson, P. P. G. (1983). The interpretation of sensitive periods. In A. Oliverio & M. Zapella (Eds.), *The behavior of human infants* (pp. 57–70). Plenum.

Baxter, B. (1966). Effect on visual deprivation during postnatal maturation of the electroencephalogram of the cat. *Experimental Neurology, 14*, 224–237.

Beach, F. (1955). The descent of instinct. *The Psychological Review, 62*(6), 401–410.

Blakemore, C., & Van Sluyters, R. C. (1974). Reversal of the physiological effects of monocular deprivation in kittens: Further evidence for a sensitive period. *Journal of Physiology, 248*, 663–716.

Blest, A. D. (1966). Learning, instinct and evolution. *Nature, 212*, 564.

Blumberg, M. (2005). *Basic instinct: The genesis of behavior.* Thunder's Mouth Press.

Blumberg, M. S. (2016). Development evolving: The origins and meanings of instinct. *WIREs Cognitive Science, 8*(1–2), e1371. https://doi.org/10.1002/wcs.1371

Brown, R. T. (1974). Following and visual imprinting in ducklings across a wide age range. *Developmental Psychobiology, 8*, 27–33.

Burghardt, G. (1973). Instinct and innate behavior: Toward an ethological psychology. In J. A. Nevin (Ed.), *The study of behavior: Learning, motivation, emotion, and instinct* (pp. 323–400). Scott Foresman.

Chow, K. L., & Stewart, D. L. (1972). Reversal of structural and functional effects of long-term visual deprivation in the cat. *Experimental Neurology, 34*, 409–433.

Colombo, J. (1982). The critical period concept: Research, methodology and theoretical issues. *Psychological Bulletin, 91*, 260–275.

Connolly, K. (1972). Learning and the concept of critical periods in infancy. *Developmental Medicine and Child Neurology, 14*, 705–714.

Crawford, M. L., Blake, R., Cool, S. J., & von Noorden, G. K. (1975). Physiological consequences of unilateral and bilateral eye closure in macaque monkeys: Some further observations. *Brain Research, 84*, 150–154.

Curtiss, S. (1977). *Genie: A psycholinguistic study of a modern day "wild child."* Academic Press.

Curtiss, S., Fromkin, V., Rigler, M., Rigler, D., & Krashen, S. (1975). An update on the linguistic development of Genie. In D. Dato (Ed.), *Developmental psycholinguistics: Theory and applications* (pp. 145–157). Georgetown University Press.

Cynader, M., Berman, N., & Hein, A. (1976). Recovery of function in cat visual cortex following prolonged deprivation. *Experimental Brain Research, 25*, 139–156.

Diamond, S. (1974). Four hundred years of instinct controversy. *Behavior Genetics, 4*(3), 237–252.

Erikson, E. H. (1950). *Childhood and society*. Norton.

Erikson, E. H. (1959). Identity and the life cycle. *Psychological Issues, 1*, 50–100.

Erikson, E. H. (1968). *Identity, youth, and crisis*. Norton.

Feldman, M. W., & Riskin, J. (2022). Why biology is not destiny. *The New York Review*, https://www.nybooks.com/online/2022/04/02/an-evolving-view-of-inheritance/

Fromkin, V., Krashen, S., Curtiss, S., Rigler, D., & Rigler, M. (1974). The development of language in Genie: A case of linguistic isolation beyond the "critical period." *Brain and Language, 1*, 81–107.

Gottlieb, G. (1970). Conceptions of prenatal behavior. In L. R. Aronson, E. Tobach, D. S. Lehrman, & J. S. Rosenblatt (Eds.), *Development and evolution of behavior: Essays in memory of T. C. Schneirla* (pp. 111–137). Freeman.

Gottlieb, G. (1983). The psychobiological approach to developmental issues. In M. M. Haith & J. Campos (Eds.), *Handbook of child psychology: Infancy and biological bases* (Vol. 2, pp. 1–26). Wiley.

Gottlieb, G. (1991). The experiential canalization of behavioral development: Theory. *Developmental Psychology, 27*, 4–13.

Gottlieb, G. (1992). *Individual development and evolution: The genesis of novel behavior*. Oxford University Press.

Gottlieb, G. (1997). *Synthesizing nature-nurture: Prenatal roots of instinctive behavior*. Erlbaum.

Gottlieb, G. (2001). The relevance of developmental-psychobiological metatheory to developmental neuropsychology. *Developmental Neuropsychology, 19*, 1–9.

Gottlieb, G. (2004). Normally occurring environmental and behavioral influences on gene activity: From central dogma to probabilistic epigenesis. In C. G. Coll, E. L. Bearer, & R. M. Lerner (Eds.). *Nature* and *nurture: The complex interplay of genetic and environmental influences on human behavior and development* (pp. 85–106). Erlbaum.

Greenberg, G., & Haraway, M. M. (2002). *Principles of Comparative Psychology*. Allyn & Bacon.

Harden, K. P. (2021). *The genetic lottery: Why DNA matters for social equality*. Princeton University Press.

Harlow, H. F. (1959). Love in infant monkeys. *Scientific American, 200*, 68–74.

Harlow, H. F. (1965). Total isolation: Effects on macaque monkey behavior. *Science, 148*, 666.

Hebb, D. O. (1949). *The organization of behavior*. Wiley.

Hess, E. H. (1973). *Imprinting*. Van Nostrand Reinhold.

Hinde, R. A. (1962). Sensitive periods and the development of behavior. *Little Club Clinic in Developmental Medicine, 7*, 25–36.

Hubel, D. H., & Wiesel, T. N. (1970). The period of susceptibility to the physiological effects of unilateral eye closure in kittens. *Journal of Physiology, 206*, 419–436.

Krashen, S. D. (1975). The critical period for language and its possible bases. *Annals of the New York Academy of Sciences, 263*, 211–224.

Lehrman, D. S. (1953). A critique of Konrad Lorenz's theory of instinctive behavior. *Quarterly Review of Biology, 28*, 337–363.

Lehrman, D. S. (1970). Semantic and conceptual issues in the nature-nurture problem. In L. R. Aronson, E. Tobach, D. S. Lehrman, & J. S. Rosenblatt (Eds.), *Development and evolution of behavior: Essays in memory of T. C. Schneirla* (pp. 17–52). Freeman.

Lenneberg, E. (1967). *The biological basis of language*. Academic Press.

Lenneberg, E. (1969). On explaining language. *Science, 165*, 635–643.

Lenneberg, E., Nichols, I., & Rosenberger, E. F. (1964). Primitive stages of language development in mongolism. In D. Mc Rioch & A. Weinstein (Eds.), *Disorders of communications. Research publications of the Association for Research in Nervous and Mental Diseases* (Vol. 42, pp. 119–137). Williams & Wilkins.

Lickliter, R. (2016). Developmental evolution. *WIREs Cognitive Science, 8*(1–2), e1422. https://doi.org/10.1002/wcs.1422

Lorenz, K. (1937). Imprinting. *Auk, 54*, 245–273.

Lorenz, K. (1940a). Durch Domestikation verursachte Störungen arteigenen Verhaltens. *Zeitschrift für angewandte Psychologie und Charakterkunde, 59*, 2–81.

Lorenz, K. (1940b). Systematik und Entwicklungsgedanke im Unterricht, *Der Biologe, 9*, 24–36.

Lorenz, K. (1965). *Evolution and modification of behavior*. University of Chicago Press.

Lorenz, K. (1966). *On aggression*. Harcourt, Brace & World.

MacDonald, K. (1985). Early experience, relative plasticity, and social development. *Developmental Review, 5*, 99–121.

Mason, W. A., & Kenney, M. (1974). Redirection of filial attachments in rhesus monkeys: Dogs as surrogate mothers. *Science, 183*, 1209–1211.

McGraw, M. B. (1943). *The neuromuscular maturation of the human infant*. Columbia University Press.

McLaughlin, B. (1977). Second language learning in children. *Psychological Bulletin, 84*, 438–59.

Michel, G. F. (2010). The roles of environment, experience, and learning in behavioral development. In K. E. Hood, C. T. Halpern, G. Greenberg, & R. M. Lerner (Eds.). *Handbook of developmental systems, behavior and genetics* (pp. 123–165). Wiley Blackwell.

Michel, G. F., & Tyler, A. N. (2005). Critical period: A history of the transition from questions of when, to what, to how. *Developmental Psychobiology, 46*, 156–162.

Mitchell, D. E. (1978). *Recovery of vision in monocularly and binocularly deprived kittens.* Paper presented at the Eleventh Symposium of the Center of the Visual Science, June, NY, Rochester.

Mitchell, D. E., Cynader, M., & Movshon, J. A. (1977). Recovery from the effects of monocular deprivation in kittens. *Journal of Comparative Neurology, 176*, 53–64.

Moltz, H. (1963). Imprinting: An epigenetic approach. *Psychological Review, 70*(2), 125–138.

Moltz, H. (1965). Contemporary instinct theory and the fixed action pattern. *Psychological Review, 72*(1), 27–47.

Moltz, H. (1973). Some implications of the critical period hypothesis. *Annals of the New York Academy of Sciences, 223*, 144–146.

Moltz, H., & Stettner, L. J. (1961). The influence of patterned-light deprivation on the critical period for imprinting. *Journal of Comparative and Physiological Psychology, 54*, 279–283.

Moore, D. S. (2015a). *The developing genome: An introduction to behavioral epigenetics.* Oxford University Press.

Moore, D. S. Moore, D.S. (2016). Behavioral epigenetics. *WIREs Systems Biology and Medicine, 9*, e1333. doi:10.1002/wsbm.1333

Nash, J. (1978). *Developmental psychology: A psychobiological approach.* Prentice-Hall.

Novak, M. A., & Harlow, H. F. (1975). Social recovery of monkeys isolated for the first year of life: I. Rehabilitation and therapy. *Developmental Psychology, 11*, 453–65.

Overton, W. F. (2015). Process and relational developmental systems. In W. F. Overton & P. C. M. Molenaar (Eds.), *Handbook of child psychology and developmental science*, volume 1: *theory and method* (7th ed., pp. 9–62), Editor-in-chief: R. M. Lerner. Wiley.

Penfield, W., & Roberts, L. (1959). *Speech and brain mechanisms.* Princeton University Press.

Plomin, R. (2018). *Blueprint: How DNA makes us who we are.* Allen Lane.

Raines, S., & Greenberg, G. (1998). Approach/withdrawal theory. In G. Greenberg & M. M. Haraway, eds. *Comparative psychology: A handbook.* (pp. 74–81). Garland Publishing.

Ratner, A. M., & Hoffman, H. S. (1974). Evidence for a critical period for imprinting in khaki campbell ducklings (*Anas platyrhynchos domesticus*). *Animal Behavior, 22*, 249–255.

Richards, R. J. (1974). The innate and the learned: The evolution of Konrad Lorenz's theory of instinct. *Philosophy of Social Science, 4*, 111–133.

Rosenblatt, J. S., Turkewitz, G., & Schneirla, T. S. (1961). Early socialization in the domestic cat as based on feeding and other relationships, between female and the young, In B. M. Foss (Ed.), *Determinants of infant behaviour* (pp. 51–74). Methuen.

Salzen, E., & Meyer, C. C. (1968). Reversibility of imprinting. *Journal of Comparative and Physiological Psychology, 66*, 269–275.

Salzen, E. A. (1998). Imprinting. In G. Greenberg & M. M. Haraway (Eds.), *Comparative psychology: A handbook* (pp. 562–575). Garland Publishing.

Schleidt, W., Shalter, M. D., & Moura-Neto, H. (2011). The hawk/goose story: The classical ethological experiments of Lorenz and Tinbergen, revisited. *Journal of Comparative Psychology, 125*(2), 121–133. https://doi.org/10.1037/a0022068

Schneirla, T. C. (1939). A theoretical consideration of the basis for approach-withdrawal adjustments in behavior. *Psychological Bulletin, 37*, 501–502.

Schneirla, T. C. (1957). The concept of development in comparative psychology. In D. Harris (Ed.), *The concept of development* (pp. 78–108). University of Minnesota Press.

Schneirla, T. C. (1965). Aspects of stimulation and organization in approach/withdrawal processes underlying vertebrate behavioral development. *Advances in the Study of Behavior, 1*, 1–74. Reprinted in L. R. Aronson, E. Tobach, D. D. Lehrman, & J. S. Rosenblatt (Eds.) (1971). *Selected writings of T. C. Schneirla* (pp. 344–412). W. H. Freeman.

Schneirla, T. C. (1966). Instinct and aggression: Reviews of Konrad Lorenz, Evolution and modification of behavior (Chicago: The University of Chicago Press, 1965), and On aggression (New York: Harcourt, Brace & World, 1966). *Natural History, 75*, 16.

Schneirla, T. C., & Rosenblatt, J. S. (1961). Behavioral organization and genesis of the social bond in insects and mammals. *American Journal of Orthopsychiatry, 3*, 223–253.

Schneirla, T. C., & Rosenblatt, J. S. (1963). Critical periods in behavioral development. *Science, 139*, 1110–1114.

Scott, J. P. (1962). Critical periods in behavioral development. *Science, 138*, 949–958.

Spencer, J. P., Blumberg, M. S., McMurray, B., Robinson, S. S., Samuelson, L. K., & Tomlin, J. B. (2009). Short arms and talking eggs: Why we should no longer abide the nativist–empiricist debate. *Child Development Perspectives, 3*(2), 79–87.

Suomi, S. J., & Harlow, H. F. (1972). Social rehabilitation of isolate-reared monkeys. *Developmental Psychology, 6*, 487–496.

Thorpe, W. (1961). Sensitive periods in the learning of animals and man: A study of imprinting. In W. Thorpe & C. Zangwill (Eds.), *Current problems in animal behavior* (pp. 194–224). Cambridge University Press.

Timney, B., Mitchell, D. E., & Griffin, F. (1978). The development of vision in cats after extended periods of dark-rearing. *Experimental Brain Research, 31*, 547–560.

Tobach, E. (1981). Evolutionary aspects of the activity of the organism and its development. In R. M. Lerner & N. A. Busch-Rossnagel (Eds.), *Individuals as producers of their development: A life-span perspective* (pp. 37–68). Academic Press.

Tobach, E., & Greenberg, G. (1984). The significance of T. C. Schneirla's contribution to the concept of levels of integration. In G. Greenberg & E. Tobach (Eds.), *Behavioral evolution and integrative levels*. Erlbaum.

von Noorden, G. K., Dowling, J. E., & Ferguson, D. C. (1970). Experimental amblyopia in monkeys: I. Behavioral studies of stimulus deprivation in amblyopia. *Archives of Ophthalmology, 84*, 206–214.

Wiesel, T. N., & Hubel, D. H. (1965). Extent of the recovery from the effects of visual deprivation in kittens. *Journal of Neurophysiology, 28*, 1060–1072.

6.
DEVELOPMENT EVOLVING
The Origins and Meanings of Instinct
Mark S. Blumberg[*,†]

Introduction

Every complex behavior challenges us to identify its origins. How do birds know to migrate south for the winter? How do border collies know to herd sheep? How do sea turtles find their way back home to the beach on which they hatched? As a shorthand—as an aid to communication—we might talk about a migratory instinct, a herding instinct, or a homing instinct. Such labels may seem gratifying, but it is an illusory gratification. Scratch the surface of any complex, adaptive behavior and one is confronted with a seemingly endless array of hard questions spanning evolutionary and developmental time, the intricacies of ecological and social experience, and the machinations of the nervous system with its billions of neurons. The more we dive into these matters, the harder it is to settle on any clear notion of what an instinct actually is. As Patrick Bateson[1] has pointed out, this conceptual confusion about *instinct* is reflected in the many meanings that are routinely ascribed to it, including:

- present at birth,
- not learned

[*] Departments of Psychological & Brain Sciences, The DeLTA Center, The University of Iowa, Iowa City, IA.
[†] Department of Biology, The DeLTA Center, The University of Iowa, Iowa City, IA.

- developed before it is used
- unchanged once developed
- shared by all members of a species
- adapted during evolution
- served by a distinct module in the brain
- attributable to genes

Scientists often unknowingly invoke more than one of these meanings at any given time, and may even unwittingly switch between meanings in a single article. This isn't just a matter of lazy thinking. The murkiness of the term reflects actual confusion about the subject. No one doubts the existence of species-typical behaviors, and we can all agree that any science of behavior must endeavor to make sense of them. But there is an unsettling gulf between widely accepted assumptions surrounding *instinct* and the actual science available to explain it.

The Ethological Approach to Instinct

The modern study of instinct began in the 1930s with the emergence of ethology. Ethology is a subdiscipline of zoology devoted to understanding behavior in its natural context. One of the founders of ethology, Konrad Lorenz, popularized this new discipline for the general public with his many famous images of "imprinted" ducklings walking behind the bearded Austrian as if he were their mother. In 1973, the young science of ethology received a significant vote of approval when three of its founders—Lorenz, Niko Tinbergen, and Karl von Frisch—received the Nobel Prize in Physiology or Medicine.

Lorenz aimed to do for behavior what Charles Darwin's evolutionary insights did for bones. Writing in *Scientific American* in 1958[2], Lorenz begins with a familiar discussion of the evolution of forelimbs: "A whale's flipper, a bat's wing and a man's arm are as different from one another in outward appearance as they are in the functions they serve. But the bones of these structures reveal an essential similarity of design. The zoologist concludes that whale, bat and man evolved from a common ancestor" (p. 119).

Lorenz then makes his critical transition from bones to behavior: "[I]s it not possible that beneath all the variations of individual behavior there lies an inner structure of inherited behavior which characterizes all the members of a given species, genus or larger taxonomic group—just as the skeleton of a primordial ancestor characterizes the form and structure of all mammals today" (p. 119)?

As his first example, he cites head-scratching in birds, which he observes to be perfectly consistent from bird to bird: produced by crossing a hindlimb *over the wing* so as to reach the head (Fig. 6.1). Lorenz exclaims that most birds scratch

Fig. 6.1 Head-scratching in a dog and a European bullfinch. Konrad Lorenz used scratching in these two very different species to argue for the notion that behavior is shaped by evolution. With respect to head-scratching, he stated unequivocally that it "is part of their genetic heritage and is not shaped by training" (p. 119). From Lorenz, 1958[2].

using "precisely the same motion" (p. 120)! He then turns to other vertebrates, including mammals, and notes that they also scratch in the same way. For Lorenz, only one conclusion could be drawn from similar behaviors expressed by such different animals: "I do not see how to explain this clumsy action unless we admit that it is inborn. Before the bird can scratch, it must reconstruct the old spatial relationship of the limbs of the four-legged common ancestor which it shares with mammals" (p. 120). In other words, scratching in dogs, birds, and other animals is the ultimate instinct: ancient, pre-programmed, and immutable.

The Response to Lorenz

Decades of subsequent research have since taught us to be skeptical of Lorenz's broad assertions about the origins of behavior. For one thing, head-scratching turns out to be more flexibly produced than Lorenz assumed. Burtt and Hailman[3], for example, reported that small, young birds typically scratch their heads by moving a leg *under* a wing. Moreover, some adults will use the overwing method when perching and will *switch* to the underwing method in flight. Based on these and other observations, they suggested that a bird's method of scratching depends not on pre-programmed instructions but on the bird's posture, balance, and center of gravity *at any given moment*[4]. Terms such as *hardwired* and *innate* gloss over the fact that scratching depends on context—on multiple factors acting in real time. By changing context, we reveal how flexible a behavior can be.

Writing in *Scientific American*, in an article cunningly titled "How an instinct is learned," Hailman[5] challenged Lorenz's fundamental notion of instinct: "The term 'instinct,' as it is often applied to animal and human behavior, refers to a fairly complex, stereotyped pattern of activity that is common to the species and is inherited and unlearned. Yet, braking an automobile and swinging a baseball bat are complex, stereotyped behavioral patterns that can be observed in many members of the human species, and these patterns certainly cannot be acquired without experience. Perhaps stereotyped behavior patterns of animals also require subtle forms of experience for development" (p. 241). Hailman meticulously demonstrated the influence of such subtle forms of experience through his investigations of pecking in newly hatched sea gulls.

Hailman's perspective is a forerunner to today's *developmental systems* approach to the origins of abilities, traits, and behaviors[6]. The striking observation that guides the developmental systems approach is that *processes*—sometimes obvious, sometimes subtle—give rise to the emergent properties of each individual's behavior. DNA plays a critical role in these processes, but does not by itself create traits. Accordingly, instincts are not preprogrammed, hardwired, or genetically determined; rather, they emerge each generation through a complex cascade of physical and biological influences[7-9]. (This process-oriented developmental perspective is has long been referred to as *epigenesis*. This term should not be confused with *epigenetics*, which refers specifically to the study of how non-genetic factors influence gene expression (see David Moore's discussion of behavioral epigenetics in Chapter 21 of this book).

Lorenz's instinct concept did not adequately consider the roles that development and experience play in the emergence of species-typical behaviors and in the transmission of behavior across generations. Even Lorenz's explanation for the phenomenon that is most closely associated with him—visual imprinting in ducklings—has undergone significant modification over the years. Whereas Lorenz believed that hatchlings come into the world equipped with a single learning program that simply needs to be activated by an appropriate stimulus, subsequent research shows that imprinting comprises two independent processes[10].

The first process entails a predisposition for chicks to orient toward stimuli that resemble the head and neck region of a generic mother hen; under natural conditions, this predisposition typically results in the chick orienting toward its own mother. The second process entails the acquisition of detailed information about the stimulus; again, under natural conditions, this process typically results in the chick learning about its mother. Interestingly, this two-process model has been applied to the problem of how human infants develop their ability to recognize faces (for a recent review, see Johnson et al.[11])

Gilbert Gottlieb spent much of his career investigating another form of imprinting—*auditory* imprinting—in which newly hatched chicks and ducklings are attracted to the mother's call[8]. Because the behavior of hatchlings seemed to be expressed without any obvious experience with the mother or her call, this adaptive behavior was thought to be an instinct. However, Gottlieb pursued this question in a way that no one else had before him by asking whether embryos obtain critical experiences *while still in the egg*. Amazingly, he found that they do: Embryos vocalize from within the egg, and these vocalizations shape the development of the auditory system in a way that is critical for their post-hatching attraction to the mother's call. Gottlieb also found that he could make a hatchling of one species prefer the maternal call of another species by manipulating its earlier embryonic experiences. Thus, even *prenatal* experiences shape the development of species-typical behavior, often in subtle and non-obvious ways.

Gravity as an Inheritance

Inheritance was once strictly defined as the passing on, upon one's death, of money, property, debts, and other earthly possessions. In contrast, within the biological sciences, inheritance has become synonymous with the transmission of DNA from one generation to the next. A developmental systems perspective, however, encourages a broader definition of inheritance to include all of the biological and environmental factors that influence individual development, especially those that are reliably transmitted. By this view, DNA is certainly part of our inheritance, but so are all the species-specific cytoplasmic factors in the egg that are passed from mother to daughter. And so are the numerous environmental factors in which every biological system develops, including (but not limited to) temperature, oxygen, carbon dioxide, atmospheric pressure, and gravity.

Consider gravity, which exerts its effects everywhere and continually. It shapes and orders life on our planet: A tree's trunk is rooted in the ground and its leaves point skyward, where birds fly with their bellies directed back toward the ground. Behavioral responses to gravity are universally expressed, being found in unicellular organisms and mammals. For example, as many pet owners can attest, a cat falling upside-down will gracefully flip itself over and land on its paws. This *righting response* is made possible by the vestibular system, which includes an apparatus in our inner ear that detects changes in linear and angular acceleration. As a cat falls, the system detects the changes in acceleration and activates muscles throughout the body to flip the cat right-side up before it hits the ground.

Rat pups exhibit the righting response at birth. In a variant of the cat-falling-to-the-ground test, experimenters release a pup upside-down in a tank of warm water. The typical behavior of a pup in this water immersion test is to flip over immediately and land right-side up, a demonstration of an already-functioning vestibular system. But is this system a hardwired and ancient instinctive response to a perennially reliable feature of life on our planet? For researchers, this was a particularly difficult question to answer, as it is not possible to simply turn gravity on and off at will.

To circumvent this problem, April Ronca, Jeffrey Alberts, and their colleagues flew pregnant rats on the NASA Space Shuttle during a period of gestation when the vestibular system is developing[12]. These pregnant rats returned to Earth two days before delivering their offspring, which were then compared to "ground controls" that were gestated normally on Earth. These researchers observed a variety of behavioral and neuroanatomical changes to the vestibular system resulting from gestation in microgravity. For example, whereas the ground-control pups exhibited normal responses in the water immersion test, the pups gestated in space often failed to even attempt to flip over, falling to the bottom of the tank on their backs (see Video 1).

Interestingly, after a week of experience in Earth's gravity the pups' righting responses were no longer impaired, which raises the question of whether isolation from Earth's gravity across the entire period of vestibular system development would lead to more lasting effects. Regardless, the lesson from this research is clear: As with Gottlieb's mallard ducklings, the presence of complex and adaptive behavior at birth tells us very little about the developmental importance of environmental factors to that behavior. Clearly, even the "simplest" instincts develop, and do so in response to numerous factors that we inherit from our parents, including the gravitational environment of our parents' home planet.

Anomalous Individuals and Developmental Plasticity

Ethology generally emphasizes species-typical behavior in natural settings. But focusing on the behavior of typically formed animals can also engender the illusion that behavioral development is a highly scripted and predetermined process. In contrast, the study of anomalous creatures—whether they arise through physical or genetic manipulation or alteration of the developmental environment—can provide key insights otherwise unavailable[13]. Critically, anomalous creatures also help us to better understand the processes the guide typical development.

For example, Johnny Eck was a performer best known for his role in the 1932 cult classic movie, *Freaks*. Born with a condition known as *amelia*, his legs were exceedingly short and functionless.

Like other individuals with this condition, Eck learned to walk using his hands. As he demonstrates repeatedly in *Freaks*, Eck's locomotion was fluid and graceful. Eck could walk down steps and climb ladders (https://www.youtube.com/watch?v=z4aET2RGG5Q). He used his hands the way most humans use their feet.

Similarly, Faith is a dog that was born in Oklahoma City with short, functionless forelimbs (https://www.youtube.com/watch?v=oSB9aBMayxU). As has occasionally been documented in animals with this condition, Faith learned to walk upright on her hind limbs. But this is not merely a circus trick, as Faith's body grew in such a way to make upright walking possible, including a curved spine that shifted forward her center of mass. Thus, incredibly, Faith accomplished in one brief lifetime what has long been considered the crowning achievement of human evolution. Perhaps even more striking is Duncan, a boxer with malformed hind legs that walks and runs on his fore legs (https://www.youtube.com/watch?v=xaM-xXgl4Bs).

Johnny Eck, Faith, and Duncan force us to reconsider our standard ideas about normal and abnormal, typical and atypical, well formed and deformed. These individuals grew into their bodies and learned to use them in highly functional ways. In fact, the *process* by which they learned to move their bodies is no different from the process by which all animals do.

To see this, let's now return to the realm of typical development and consider the diverse patterns of locomotion in mammals: From quadrupedal walking, trotting, and galloping to bipedal walking and hopping. Across all rodent species, all of these locomotor patterns are observed and there is a clear relationship between the shape of an animal's body—its *morphology*—and the pattern of locomotion that it displays. In fact, at each stage of development, as an animal's morphology changes, its locomotor pattern changes as well.

For example, jerboas are desert rodents that, as adults, have very long hind legs and exhibit bipedal walking and hopping gaits[14]. One might think that jerboas instinctively exhibit these gaits, but studies of the development of locomotion in this species tell a different story: As newborns, jerboas have similarly proportioned limbs as other rodents and they exhibit locomotor patterns that are identical to other newborn rodents with similar shapes (Fig. 6.2). But as jerboas grow and their hind legs lengthen disproportionately, their locomotor patterns change accordingly. Specifically, as the hind legs grow longer than the fore legs, jerboas pass through an awkward stage where they struggle to accommodate their overly long legs. Later in development as their hind legs gain strength, they are able to lift themselves up and walk and hop about.

In other rodent species, such as rats and gerbils, we see similar patterns relating the shape and size of a body to the locomotor patterns expressed[15]. All rodent species examined thus far pass through a series of locomotor patterns that reflect their specific morphologies at each age. As bodies change and species-typical morphologies emerge, locomotor patterns diverge. As with head-scratching in birds, posture, balance, and center of mass—all intimately linked with morphology—determine how we move.

Demonstrating close correspondences between body morphology and behavior does not necessarily mean that behavior flows from morphology. A skeptic might respond by saying that evolution ensured that behavior and morphology develop in a synchronous way without actually influencing one another. But let's not forget Johnny Eck, Faith, and Duncan: the locomotor patterns in these individuals cannot be due to any preprogramming of behavior because their behavior reflects unique solutions to unique, species-*atypical* bodies. In other words, individual behaviors emerge from individual development. Whether typically or atypically formed, we all must learn through individual experience to use the bodies that we have—not the bodies that we were "supposed" to have.

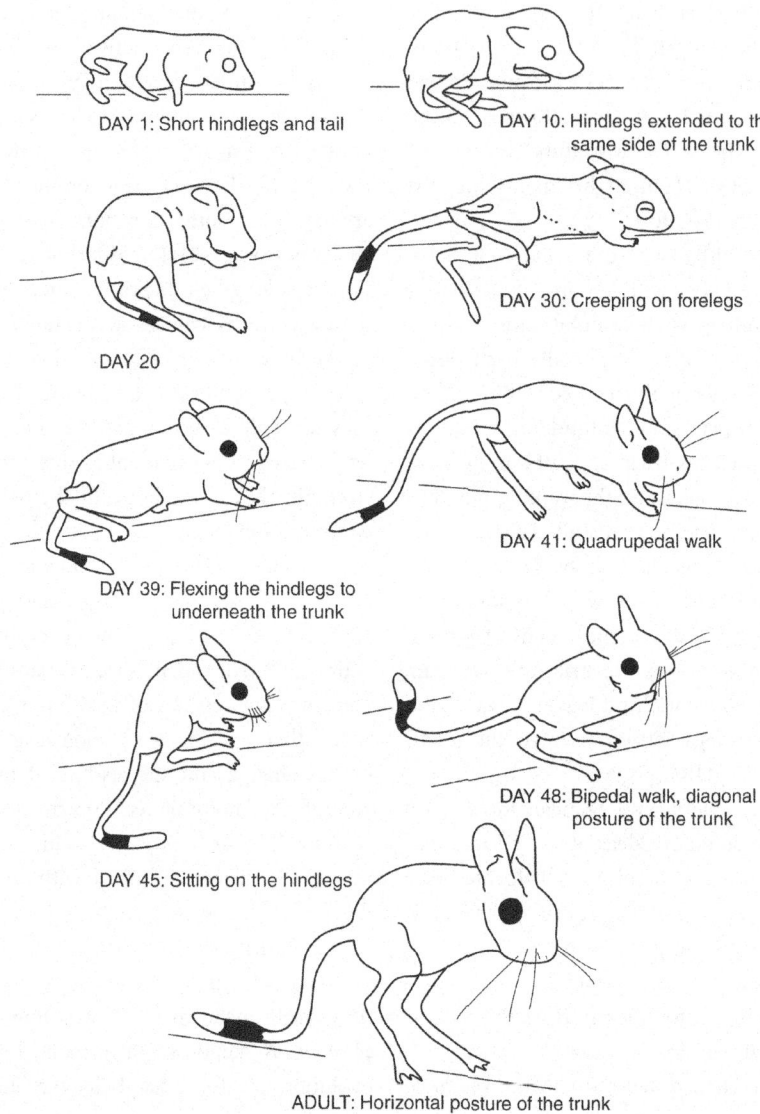

Fig. 6.2 The limbs of jerboas change dramatically across early development and their locomotor patterns change in lock-step. As newborns, these desert rodents look much like other rodents and they move around similarly as well. As their hind legs elongate, they crawl around very awkwardly. Finally, with gaining strength, they can walk and hop upright. From Eilam & Shefer[14].

Conclusions

History teaches us that we always learn important, critical details about a behavior by asking about its development. When Gottlieb saw that hatchlings are attracted to the maternal call, he could have stopped his investigation there and simply labeled the behavior an instinct. Instead, he asked the next question, revealed the developmental process that gives rise to the behavior, and ultimately taught us something general and profound about the nature of development and its often non-obvious causes.

Species-typical behaviors can begin as subtle predispositions in cognitive processing or behavior. They also develop under the guidance of species-typical experiences occurring within reliable ecological contexts. Those experiences and ecological contexts, together comprising what has been called an *ontogenetic niche*, are inherited along with parental genes[16]. Stated more succinctly, environments are inherited—a notion that shakes the nature-nurture dichotomy to its core. That core is shaken still further by studies demonstrating how even our most ancient and basic appetites, such as that for water, are learned[17]. Our natures are acquired.

None of this should be taken to mean that all behaviors are equally malleable. On the contrary, behaviors lie along a continuum from highly malleable or plastic to highly rigid or robust[18] (See Patrick Bateson, 2002). Our challenge, then, is to move beyond the age-old practice of applying dichotomous labels to behaviors[19]. Instead, we should focus more on understanding the developmental contexts and conditions in which a behavior is more or less malleable.

So the next time you see a marvelous and complex behavior—such as a border collie herding sheep or birds flying south for the winter—try to resist the temptation to label it as *instinctive, hardwired, genetic,* or *innate*. By foregoing a label and digging deeper, you will open yourself to consideration of the myriad of factors that shape who we are and why we behave the way we do.

Supplementary Material

Refer to Web version on PubMed Central for supplementary material.

Acknowledgments

Preparation of this chapter was made possible in part by grants from the National Institutes of Health (R37-HD081168; R01-MH050701).

References

1. Bateson P. The corpse of a wearisome debate. Science. 2002; 297:2212.
2. Lorenz KZ. The evolution of behavior. Sci Am. 1958; 199:67–74. passim.
3. Burtt EH, Hailman JP. Head-Scratching among North-American Wood-Warblers (Parulidae). Ibis. 1978; 120:153–170.
4. Burtt EH, Bitterbaum EJ, Hailman JP. Head-Scratching Method in Swallows Depends on Behavioral Context. Wilson Bulletin. 1988; 100:679–682.
5. Hailman JP. How an instinct is learned. Scientific American. 1969; 221:98–106.
6. Oyama S, Griffiths PE, Gray RD. Cycles of Contingency: Developmental Systems and Evolution. MIT Press; Cambridge: 2003.
7. Blumberg M. Basic Instinct: The Genesis of Behavior. Thunders Mouth Press; New York, New York: 2006.
8. Gottlieb G. Synthesizing Nature-Nurture: Prenatal Roots of Instinctive Behavior. Lawrence Erlbaum; Mahwah, NJ: 1997.
9. Johnston TD, Edwards L. Genes, interactions, and the development of behavior. Psychol Rev. 2002; 109:26–34.
10. Bolhuis JJ, Honey RC. Imprinting, learning and development: from behaviour to brain and back. Trends Neurosci. 1998; 21:306–311.
11. Johnson MH, Senju A, Tomalski P. The two-process theory of face processing: modifications based on two decades of data from infants and adults. Neurosci Biobehav Rev. 2015; 50:169–179.
12. Ronca AE, Fritzsch B, Bruce LL, Alberts JR. Orbital spaceflight during pregnancy shapes function of mammalian vestibular system. Behav Neurosci. 2008; 122:224–232.
13. Blumberg, M. Freaks of Nature: What Anomalies Tell Us About Development and Evolution. Oxford University Press; New York: 2009.
14. Eilam D, Shefer G. The developmental order of bipedal locomotion in the jerboa (Jaculus orientalis): pivoting, creeping, quadrupedalism, and bipedalism. Dev Psychobiol. 1997; 31:137–142.
15. Eilam D. Postnatal development of body architecture and gait in several rodent species. J Exp Biol. 1997; 200:1339–1350.

16. West MJ, King AP. Settling Nature and Nurture into an Ontogenic Niche. Developmental Psychobiology. 1987; 20:549–562.
17. Hall WG, Arnold HM, Myers KP. The acquisition of an appetite. Psychological Science. 2000; 11:101–105.
18. Bateson P, Gluckman P. Plasticity, Robustness, Development and Evolution. Cambridge University Press; Cambridge: 2011.
19. Johnston TD. The Persistence of Dichotomies in the Study of Behavioral-Development. Developmental Review. 1987; 7:149–182.

Further Reading

20. Blumberg MS, Freeman JH, Robinson SR, editors. Oxford handbook of developmental behavioral neuroscience. Oxford University Press; New York: 2010.
21. Lehrman DS. A critique of Konrad Lorenz's theory of instinctive behavior. Q Rev Biol. 1953; 4:337–363.
22. Spencer J, Blumberg MS, McMurray R, Robinson S, Samuelson L, Tomblin J. Short arms and talking eggs: Why we should no longer abide the nativist-empiricist debate. Child Dev Perspect. 2009; 3:79–87.
23. West-Eberhard MJ. Developmental plasticity and evolution. Oxford University Press; New York: 2003.

7.
CRITICAL PERIOD
A History of the Transition from Questions of When, to What, to How
George F. Michel* and Amber N. Tyler*

Development is an historical phenomenon in which previous events affect the manifestation of both current and subsequent events and current events become the previous events that affect subsequent events. Hence, there is a serial order to developmental phenomena that has a cumulative aspect. Consequently, development must be defined by the illumination of the factors creating and governing the serial order and the processes of change and stability of that order over time (Michel & Moore, 1995).

The serial order of development reflects both logical and empirical characteristics. For example, given that cell division is a doubling process, a two-cell stage must logically precede a four-cell stage. However, a 16-cell stage can precede a subsequent 8-cell stage only if there is empirical evidence of cell death between the 16- and 8-cell stage events. Thus, development is a serially ordered process that is identifiable across time, but it is not defined by time. This distinction is often overlooked in research. Too often, especially in research on psychological characteristics, comparison across time (as represented by age) serves as the only indication that the research is developmental. A brief historical examination of the concept of critical period may help illustrate why time should not be a defining characteristic of development.

Morphological and behavioral characteristics of an individual emerge over time. Development typically exhibits both regularity in the serial order for the appearance of specific characteristics (stages) and regularity in the time from conception for when the "stages" in the serial order occur. The regularities of order and time are identified by their similarity across individuals and the patterns of behavioral and morphological development are identified in part by both the sequence and time course of the appearance of new characteristics. For example, "limb buds" appear before "limbs" and "stepping" appears before "walking" and the time from conception to the appearance for each of these characteristics is fairly similar across members of the same species. Since the time of appearance of each morphological structure is closely associated with the "age" (time since fertilization of the organism), it is not surprising that the timing of development (the when) became a major focus of research. Thus, cross-individual and within-species regularity of the appearance of morphological structures and the within species regularity of the time (or age) of appearance of such structures helped create the field of embryology and developmental biology. The regularity in age of appearance within a species rapidly became the marker of normal development and any variations in age of appearance marked abnormal development.

Early in the history of embryology, investigators wanted to determine whether the embryo

---

* Department of Psychology, University of North Carolina Greensboro, Greensboro, NC.

was an "unfolding" of some preformed being or a constructed entity. Manipulations shortly after the initial divisions of the zygote, demonstrated that dividing the zygote resulted in the development of two half-organisms. This seemed to support the notion of an unfolding of some preformed entity. Subsequent studies demonstrated that such division resulted in the development of two separate whole individuals. This seemed inconsistent with the notion of a preformed entity. Debate about the contradictions of these studies was ended when it was discovered that the timing of the division (or age and stage of the organism at the time of division) was associated with the difference in outcome. Thus, there seemed to be a period of time that was critical to whether or not dividing the embryo would result in the development of two half-organisms or two whole-organisms.

Spemann (1938/67) demonstrated that the developmental outcome (what type of tissue characteristics a cell begins to manifest) of embryonic cells is determined by their location. Signals from the local environment induce cells to adopt a particular developmental "fate". That is, a normal organism (e.g., a frog) will develop even when its tissue has been rearranged (e.g., skin and brain cells exchanged during the gastrula stage). However, the timing of the exchange became critical when it was noted that if the exchange occurred when the cells were in the late gastrula stage, the organism developed with inappropriately placed patches of tissue. The cells seemed committed to a particular fate. Thus, questions about "when" (i.e., the age of the embryo when manipulations did or did not affect the outcome of development) became the main focus of experimental embryology.

Ultimately, this led to the establishment of the field of teratology—the investigation of factors that would disrupt "normal" (the regularities of) development. Developmental questions focused on whether or not there was a critical period for exposure to certain events which would affect the course of development and, if there was a critical period, what were its time/age boundaries. Embryologists identified critical periods for the exposure to many atypical environmental (particularly chemical) conditions that had profound effects on the morphological development of the organism. They discovered, also, that certain manipulations had negative consequences despite the age of the embryo at the time of the manipulation. Thus, the latter did not exhibit a critical period for their impact.

As embryology grew as an experimental/manipulation science, the definition of "abnormal" development became directly related to any variation in the typically regular age of developmental events. "Normal" development was defined as the typical sequence of events occurring at the typical age/time. Failure of the normal sequence, as a result of exposure to conditions at particularly sensitive periods or as a result of not being exposed to appropriate conditions at their optimal times, was abnormal development. Hence, a maturational time-table became a guide to the distinction of normal and abnormal development. Physicians adopted developmental milestones to represent normal development against which abnormal development could be identified. Anything that delayed or advanced aspects of the typical pattern of development was defined as producing abnormal development.

In contrast to considering such variability in the timing of developmental processes as simply abnormal, de Beer (1958) and Gould (1977) argued that such variability was the foundation for the evolution of species variability. Comparisons of the patterns of development among closely related species revealed that the origin of morphological differences, with important adaptive consequences, derived from differences in the timing of typical sequences of developmental events. Since these related species were not necessarily at a disadvantage in living, both de Beer and Gould considered the variability in the course and timing of developmental events as a "natural" aspect of living systems.

Of course, specific developmental outcomes can be more or less advantageous for the individual. If some outcome is disadvantageous for the individual, it should prompt attempts at rehabilitation (for those with the disadvantageous outcome) and prevention (for those who may be at risk for such an outcome). Placing too much emphasis on the timing of development could result in a more pessimistic approach to the building rehabilitation and prevention programs (Bateson, 1979; Bateson & Hinde, 1987). The metaphor of development as a train moving on a track according to a time-table with a limited number of switching stations can lead to the notion that redirection of the outcome (rehabilitation) is unlikely because a switching station has been passed. Knowledge of those factors responsible for the sequential order of developmental events and those that influence when they occur can generate a more optimistic approach.

As biologists began to investigate the "natural" behavior of animals (sometimes referred to as instinctive behavior), some investigators chose to pose the issue of behavioral development within the framework of developmental biology. For example, Konrad Lorenz (1937/57) subjected to more systematic investigation the common barnyard phenomenon that ducklings, goslings, and chicks often mistakenly follow a human (usually a child) instead of following their parent shortly after hatching. To find out why, he chose the developmental biological technique of manipulating the age at which newly hatched ducklings and goslings were exposed to various potential parental substitutes. These studies led him to conclude that there was a critical period for the development of the pattern.

During a particular age period (within hours after hatching), the young bird had to be exposed to a moving object and thereafter it would follow that same object. Subsequently in adulthood, objects with characteristics similar to those of the object followed, were courted for mating. This process of forming a perceptual pattern for selecting mates from exposure shortly after hatching was translated in to English as "imprinting". Lorenz's definition of imprinting included a rigidly defined critical period. Lorenz proposed that the onset and offset of this period was determined by intrinsic processes of development under control of the organism's genes. Therefore, the age of occurrence of the critical period could vary across the species but vary little within a species. Lorenz proposed that critical periods might exist for the development of other species-typical behavior patterns.

Indeed, research on bird song by W. H. Thorpe (e.g., 1961) and Peter Marler (e.g., 1970) revealed that males of many species of birds acquire their species-typical song pattern by hearing the song of their father when nestlings. For some species, there seemed to be a critical age period for exposure to the species-typical song. Isolation from that song during that age and exposure to the song subsequently did not result in the development of the species-typical song.

Not long after Lorenz's work on the critical periods for the establishment of social behaviors in birds, John Paul Scott (1962) reported a critical period for the development of socialization in dogs and cats. Unless encouraged to interact with humans at a particular early age, certain breeds of dogs and cats would not be able to be socialized to human interaction. For other breeds the critical period appeared to be much longer or non-existent. Since this pattern seemed to mimic the pattern of filial imprinting reported by Lorenz, Scott proposed that critical periods for socialization may underlie the social development of many species of bird and mammal, including humans. Harry Harlow (c.f., Harlow & Harlow, 1965) also reported a series of studies on the development of social abilities in Rhesus monkeys that indicated that social deprivation during the infant monkey's first three months had rather catastrophic long term developmental consequences. Similar deprivation after six months of age resulted in relatively mild developmental disturbances.

In the late 1950s, Austin Riesen (c.f., 1975) reported a series of studies with monkeys and cats that showed that the absence of light or patterned visual stimulation during infancy led to blindness in adults. He went on to describe much of the anatomy and physiology that was disrupted by the patterned light deprivation which was responsible for the blindness. The functional blindness was created only when kittens were reared without patterned light during their first four months. A similar period of four months without patterned light after the kittens were older than 6 months did not affect their sight. Hence there seemed to be a critical period for the development of functional sight. Similar results of rearing without patterned light were obtained with monkeys.

Drawing on the results of imprinting, birdsong development, development of visual ability, and socialization, combined with the evidence on language development in humans and the effects of various kinds of brain damage on such development, Eric Lenneberg (1967) proposed a critical period for language development. Indeed, his proposal included a critical period for the development of both language and the specialization of the left hemisphere for language processing. Interestingly, Leneberg proposed a 10–12 year window (from birth to puberty) for the critical period of language acquisition and the development of hemispheric specialization for language skills.

Similarly drawing on the literature of ethology and animal behavior, Bowlby (1969) proposed a critical period for the formation of an attachment relationship between the mother and child which if disrupted resulted in the development of adult psychopathology. From birth to approximately three months, the infants can recognize their caregivers but they do not seem to be socially attached specifically to them. From 3 to 6 months, infants exhibit caregiver preference. However, the period from 6 months to 3 years seemed to be critical for the formation of an attachment relationship that would become the basis for all future social partnerships and the capacity for the individual to form emotional bonds with others or to exhibit either sympathy or empathy.

Not long after Leneberg's and Bowlby's publications, critical periods began to be proposed for the development of a host of human characteristics including sensory and perceptual abilities, social skills, motor skills, language and second language acquisition, critical reasoning skills, etc.

(Bornstein, 1989). Most often the evidence for the critical period consisted of some relatively minimal demonstration that a weakness of skill was associated with an unusual event or a particular kind of experiential deprivation that occurred early in the individual's life. Educators had long proposed that educational experiences should be restricted to certain age periods because it is believed that these periods represent the time when children are "ready" for such experiences to have their developmental impact. The empirical investigations of the concept of critical period provided support for such proposals.

While critical periods were flourishing in the study of behavior, embryologists had come to focus less on the timing or age of the exposure and more on processes that were associated with time or age. That is, time and the status of the developmental process were intimately related for those aspects of development that exhibited the regularity that attracted investigation by developmental biologists. For example, a morphological structure such as the corpus callosum in the mammalian brain is composed of hundreds of millions of the axons of pyramidal cells of the cortex. Pyramidal cells on the right side of the cortex project their axons to the left side of the cortex and pyramidal cells on the left side project their axons to the right side of the cortex. The criss-crossing of these axons over the third ventricle create the corpus callosum. The timing of the growth of the projections that create the corpus callosum is well specified and several events can disrupt the formation of this structure. Thus, studies focused on the "when" of development revealed a critical period during which several factors (including some known teratogens) can disrupt the formation of the corpus callosum.

However, as developmental biologists examined this critical period it was discovered that a major factor that created the criticality of the timing was the formation of certain forms of glial cells called "bridge" cells at the border of the third ventricle. When the axons approached the area of the third ventricle, if the bridge cells were present, the axons would extend across the ventricle to the opposite side. If the bridge cells were not present, then the axons continued to project to other areas within their own hemisphere (c.f., Silver, Lorenz, Wahlsten, & Coughlin, 1982).

The bridge cells themselves had a relatively delimited time of existence and, as with all developmental phenomena, this time was a consequence of previous events and the current conditions. Bridge cells are formed at one time from certain glial cells and then after some time (and for reasons not yet completely known) they die. As this process was examined, it became clear that the formation of the corpus callosum could be disrupted by any combination of factors that sped-up or slowed-down the growth of the axons, sped-up the formation or the death of the bridge cells, or caused bridge cells to form in the wrong place in the brain. Hence, developmental biologists began to examine the criticality of the sequence of events for the formation of some morphological characteristic. They asked questions about what characteristics composed the sequence of events of development and how these were disrupted by the manipulations that seemed to mark a critical period. Questions of "when" were replaced by questions of "what".

In the study of imprinting and bird song, questions of "when" also were being replaced by questions of "what". Many studies showed that the onset and ending of the critical period for imprinting could be altered by various sorts of environmental manipulations. For example, dark rearing would delay the offset of the critical period. The onset of the "critical period" seemed to be determined by the sensory and motor abilities of the individuals and these had developmental courses that were influenced by the manipulation of certain ubiquitous experiences and conditions which then altered the onset of the critical period. Moreover, the end of the critical period for song learning and imprinting seemed to be a self-terminating process in some species (Bateson, 1987; Eales, 1985; ten Cate, 1989). That is, the acquisition of the percept of the object or the song during the process of imprinting or song acquisition prevented any further acquisition of other percepts.

Patrick Bateson (1966, 1987) showed that the termination of the critical period could be created by the experimental conditions used to test the critical period. Chicks would imprint on the perceptual characteristics of their cages if their deprivation from exposure to "imprintable objects" did not occur in darkness. Thus, having been exposed to a stimulus, a percept would be acquired for the discrimination of familiar from non-familiar. Familiar stimuli would be approached and attended to whereas the exposed chick would withdraw from unfamiliar stimuli and thereby fail to become familiar with its characteristics. With this knowledge, other investigators showed that deprivation during the critical period could be overcome by specific patterns of exposure at later ages. Critics claimed that such manipulations only affected "taming" and not imprinting. Nevertheless, the filial and mating preferences of certain species could be altered by manipulations that occurred outside of the supposed critical period for imprinting.

Investigations of the critical period for bird song acquisition also revealed that some of the criticality depended on the experimental conditions of investigation. Species who were exposed only to the sound of a song exhibited a critical period for acquisition that did not occur if they were exposed to the presence of a singer. In some species, the singer needed to be a familiar companion whereas in other species the singer could be a stranger (c.f., Petrinovich, 1988). Again, the investigation of bird song acquisition was replacing the "when" of exposure with investigations of "what": What was it about the conditions of the individual that made timing relevant?

Interestingly, at the time of Scott's 1962 report of critical periods for socialization, Schneirla and Rosenblatt (1963) reported that the development of species-typical social behavior in cats exhibited a cascading set of events in which the experiences associated with each social event was critical for the cascade. That is, the kitten begins to adjust to the mother at birth using the behavioral systems available, in part through the prenatal fetus-mother relationship. As new behavioral systems emerge through changes in both the kitten's and mother's systems, new social adjustments are made. By separating kittens from their mother and litter mates and keeping them on an artificial brooder, Schneirla and Rosenblatt showed that the kittens had difficulty adjusting to the mother at reunion. The mother's system had changed during the separation, in part by her adjustment to the remaining kittens in the litter, and the kitten's system had been altered by the separation and adjustment to the brooder. At reunion, the kitten behaves toward the mother in ways more appropriate to an earlier phase of their interaction, thereby producing discordant interactions between them.

All separated kittens manifested difficulties in their social adjustments, no matter when during development their separation occurred. No single period during development appeared to be more crucial for social adjustment than any other. However, the particular difficulties manifested at reunion depended on the level of social ability achieved by the kitten before separation and the degree of changes in the mother induced by her experiences during the separation. The degree of difficulty at reunion depends on the degree of discrepancy between the level of the kitten's social ability and that of its mother. Thus, the timing of the events seemed to emerge from the logic of the sequence rather than from some intrinsic clock. Hence, Schneirla and Rosenblatt argued against the notion of critical periods in development and more for critical sequences for the emergence of specific social skills and abilities. Although this research was a serious challenge to research focused on questions of "when" in the development of behavior, it had little impact at the time.

One reason for the lack of impact was that critics argued that the Schneirla and Rosenblatt study was not the appropriate type of experiment for evaluating whether there is a sensitive period in cat socialization (i.e., whether the impact of a controlled stimulus varies with the age of the individual). Schneirla and Rosenblatt did not examine whether age modified the response to a specific type of stimulus. Instead, they deconstructed the process of social development to identify why separations for different lengths of time and at different stages in the relationship between the mother cat and her offspring led to different developmental outcomes in the social relationship between the mother and offspring.

Scott's (1962) study examined the responsiveness to humans of puppies who were at different ages when they were first exposed to a human. Hence, it met the critics' criterion for the identification of a critical period. Although Scott proposed that there was a critical period for socialization in dogs, only the socialization of puppies to humans was examined. More importantly, the age of the puppy at exposure to humans seemed to be the explanation for why "socialization" did or did not fail.

Obviously, social behavior involves social interactions between individuals. That is, since the mother behaves differently at different points in the relationship (which does occur across time), the reunion at different "times" presents a different stimulus to the kitten. However, it is only as a result of the Schneirla and Rosenblatt (1963) study that the discrepancies between mother and offspring can be identified as the reason for the failure of the reestablishment of the relationship. Once those discrepancies are identified, research can be begin to seek ways of providing compensatory manipulations that promote a developmental pathway that results in the desired outcome. This is important to the rehabilitation aspect of developmental research and it was put to excellent use by Mason in his research on the types of social and non-social experiences that rehabilitate the social skills of Rhesus monkeys who were raised on a cloth "mother" (Mason & Capitanio, 1988). Nevertheless, the Schneirla and Rosenblatt study was dismissed as not appropriately investigating the concept of critical period

Of course, it was (and is) still clear that at some periods (stages) of development, exposure to certain conditions more easily affected development than exposure to those same conditions at other periods (stages). However, the notion of some process controlled by the timing or age of the individual did not seem appropriate for comprehending the historical contingency of developmental phenomena. Hence, the notion of critical period was replaced by the notion of a "sensitive" period. Sensitive periods are not clock-like, built-in or predetermined periods in development but are themselves the product of development. Thus, we should expect the variability in onset/offset (timing), specificity, etc. evident in their study. Replacement of critical periods by sensitive periods should operate as a "promissory note" that future research will be designed to reveal exactly why the development of some characteristic was sensitive to a particular pattern of experience at a particular time in the individual's life. Replacing "critical" with "sensitive" marked the recognition that once the "what" of development was discovered, timing alone would not be critical for manipulating the developmental outcome.

Nevertheless, the use of the concept of sensitive period permitted many investigators to retain the notion that there are biological processes that normally unfold at a certain age and that they dictate the neural response to experience. Therefore, even if the timing of these biological processes was sensitive to experience, this did not imply that age/time was not important. These investigators argue that such factors just make it much more difficult to define the temporal limits of sensitive periods. By adding the notion that any sensitive period may be a reflection of multiple sequential sensitive periods that interact with one another, these investigators believe that they have accounted for why different manipulations and dependent measures often yield different results about the temporal limits of the sensitive period. That is, different stimuli engage specific neural processes to a greater or lesser degree than others, and different dependent measures are more or less affected by the various interacting processes that impact responsiveness to a particular experience. Thus, sensitivity may appear greater or lesser depending upon the salience of the stimuli delivered during the experimental manipulation and the pattern of overlap among interacting sensitive periods.

However, in a somewhat Ptolemaic manner, it may be that it is only the attempt to retain the importance of time as a defining aspect of development that prompts such a convoluted notion of sensitive period. All phenotypic traits (including sensitive periods of experiential vulnerability) are generated during individual ontogeny because particular aspects of the temporal and spatial arrangements of individuals and their contexts reliably occur at times when the organism is in particular developmental states, having had a particular developmental past (Michel & Moore, 1978). Therefore, when accepted as simply a promissory note for further investigation of exactly why some events affect development at certain stages more than at other stages rather than as an explanation of development, sensitive periods need not become burdened with unnecessary complexity.

In the past two decades, developmental biology has shifted from questions of "when" and "what" creates the regularities of development to questions of "how" regularity is achieved. Research is focused on how processes of intra- and intercellular communication produce the various "pathways" of development that result in both individual differences and inter-individual similarities of characteristics. Revealing these mechanisms will permit both the identification of developmental pathways with potentially unacceptable outcomes and identification of places in the pathway in which either simple or complex interventions can be undertaken to establish more acceptable pathways. In other words, investigations of the "how" of morphological development is providing us with more sophisticated control techniques and opportunities. Thus, age/time is no longer the defining aspect of developmental phenomena (c.f., Gilbert, 2003).

There are too few current investigations of the "how" of behavioral development. Most investigations of the "how" of development have relied upon the presumed indirect behavioral consequences of developmental biological investigations of the development of gross brain structures and their general physiology. That is, the "how" of development is thought to require reduction to a physiological or molecular level of biological investigation. However, as we become more sophisticated in the understanding of organism-environment interaction in behavioral development, we will begin to achieve greater understanding of the pathways responsible for individual differences and inter-individual similarities of psychological characteristics (c.f., Gilbert, 2001). Such understanding will provide us with greater control of our offspring's destiny and greater responsibility for our exercise of such control.

References

Bateson, P. P. G. (1966). The characteristics and context of imprinting. Biological Reviews, 41, 177–220.

Bateson, P. P. G. (1979). How do sensitive periods arise and what are they for? Animal Behaviour, 27, 470–486.

Bateson, P. P. G. (1987). Imprinting as a process of competitive exclusion. In J. P. Rauschecker & P. Marler (Eds.), Imprinting and cortical plasticity: Comparative aspects of sensitive periods (pp. 151–168). New York: Wiley.

Bateson, P. P. G., & Hinde, R. A. (1987). Developmental changes in sensitivity to experience. In M. H. Bornstein (Ed.), Sensitive periods in development (pp. 19–34). Hillsdale, NJ: Erlbaum.

Bornstein, M. H. (1989). Sensitive periods in development: Structural characteristics and causal interpretations. Psychological Bulletin, 105, 179–197.

Bowlby, J. (1969). Attachment and loss (Vol. 1). Attachment. New York: Basic Books.

de Beer, G. R. (1958). Embryos and ancestors (3rd Ed.). London: Oxford University Press.

Eales, L. A. (1985). Song learning in zebra finches: Some effects of song model availability on what is learned and when. Animal Behaviour, 33, 1293–1300.

Gilbert, S. F. (2001). Ecological developmental biology: Developmental biology meets the real world. Developmental Biology, 233(1), 1–12.

Gilbert, S. F. (2003). Developmental biology (7th Ed.). Sunderland, MA: Sinauer Associates, Inc.

Harlow, H. F., & Harlow, M. K. (1965). The affectional systems. In A. M. Schrier, H. F. Harlow, & F. Stollnitz (Eds.), Behavior of nonhuman primates (Vol. 2, pp. 287–334). New York and London: Academic Press.

Lenneberg, E. H. (1967). The biological foundations of language. New York: Wiley.

Lorenz, K. (1937/57). The conception of instinctive behavior. In C. H. Schiller (Ed. and Trans.), Instinctive behavior (pp. 129–175). New York: International Universities Press.

Marler, P. (1970). A comparative approach to vocal learning: Song development in white-crowned sparrows. Journal of Comparative and Physiological Psychology, 71, 1–25.

Mason, W. A., & Capitanio, J. P. (1988). Formation and expression of filial attachment in rhesus monkeys raised with living and inanimate mother substitutes. Developmental Psychobiology, 21, 401–430.

Michel, G. F., & Moore, C. L. (1978). Biological perspectives in developmental psychology. Monterey, CA: Brooks/Cole.

Michel, G. F., & Moore, C. L. (1995). Developmental psychobiology: An interdisciplinary science. Cambridge, MA: The MIT Press.

Petrinovich, L. (1988). The role of social factors in white-crowed sparrow song development. In T. R. Zentall & B. G. Galef (Eds.), Social learning (pp. 255–278). Hillsdale, NJ: Erlbaum.

Riesen, A. (1975). The developmental neuropsychology of sensory deprivation. New York: Academic Press.

Schneirla, T. C., & Rosenblatt, J. S. (1963). "Critical periods" in the development of behavior. Science, 139, 1110–1115.

Scott, J. P. (1962). Critical periods in behavioral development. Science, 138, 949–958.

Silver, J., Lorenz, S. E., Wahlsten, D., & Coughlin, J. (1982). Axonal guidance during development of the great cerebral commissures: Descriptive and experimental studies, in vivo, on the role of preformed glial pathways. Journal of Comparative Neurology, 210, 10–29.

Spemann, H. (1938/67). Embryonic development and induction. New York: Hafner.

ten Cate, C. (1989). Behavioral development: Toward understanding processes. In P. P. G. Bateson & P. Klopfer (Eds.), Perspectives in ethology (Vol. 8, pp. 243–269). New York: Plenum Press.

Thorpe, W. H. (1961). Bird song. Cambridge: Cambridge University Press.

8.
SHORT ARMS AND TALKING EGGS
Why We Should No Longer Abide the Nativist-Empiricist Debate
John P. Spencer, Mark S. Blumberg*, Bob McMurray*,
Scott R. Robinson*, Larissa K. Samuelson*, and J. Bruce Tomblin†*

Introduction

Spelke and Kinzler (2007) recently described developmental science as a struggle between two dichotomous groups. On one end sit the "blank slaters," who view the brain as an unconstrained general learning device; on the other sit the evolutionary psychologists, who view the brain as an amalgam of special-purpose learning devices (see also Pinker, 2002). Although Spelke's nativist views typically align with those of evolutionary psychologists (e.g., see Spelke & Newport, 1998), Spelke and Kinzler propose a middle ground according to which "humans are endowed neither with a single, general-purpose learning system nor with myriad special-purpose systems and predispositions" (p. 89). Instead, they suggest "that humans are endowed with a small number of separable systems of core knowledge. New, flexible skills and belief systems build on these core foundations" (p. 89).

We applaud Spelke and Kinzler's move to a middle ground, but we cannot meet them in this particular place. As we argue here, developmental scientists should no longer embrace "endowments," "primitives," "core knowledge," "essences" (Gelman, 2003), or other static concepts that

* Department of Psychology, University of Iowa, Iowa City, IA.
† Department of Communication Sciences and Disorders, University of Iowa, Iowa City, IA.

devalue developmental process. After all, "endowments" are bestowed, not developed. Similarly, "primitives" are *not developed or derived from anything else*. These nativist concepts originated in the rationalist tradition of Plato, Descartes, and Kant (Spelke & Newport, 1998), and thus nativists assume that relationships between developmental antecedents and consequents are rational and transparent (Blumberg, 2005; Johnston, 1987). This leads, in turn, to nativists' overly narrow conception of experience (Lehrman, 1953).

We wish to locate developmental science in new territory, where we invoke only grounded processes to explain the remarkable transformations of development. To move in this direction, we must accept some inconvenient truths—inconvenient in the sense that they make our task as scientists considerably more difficult. First, development is often a nonobvious process that does not easily conform to our intuitions or rational expectations (Gottlieb, 1997). For example, what rationalist analysis would have predicted that the quantity of stimulation provided by a mother rat to her pups would affect gene expression (Weaver et al., 2004), brain development (Liu et al., 2000), and adult sexual behavior (Moore, 1995)? Such examples should broaden our conception of what qualifies as relevant experience.

Second, we cannot sidestep the complexity of development by invoking evolutionary causation. The nativist appeal to endowments and primitives is an attempt to move beyond the here-and-now to an evolutionary past that prescribes adaptive outcomes. We, too, embrace evolutionary theory, but the fact that organisms evolved does not remove the need to explain developmental process, because brain and behavior are shaped through development, not programmed before development (Blumberg, in press; West-Eberhard, 2003). In Gottlieb's words, evolution involves "selection for the entire developmental manifold" (1997, p. 76).

Third, although the nativist-empiricist debate has been rich and scholarly at times, too often nativist stories persist even when their supporting studies are demonstrably flawed (see Blumberg, 2005). And, critically, the lessons that nativists should learn when these interpretations turn out to be insufficient fail to temper the next round of nativist claims. Rather, nativists routinely extrapolate well beyond the data, making bold claims about time points not directly under investigation. For instance, Marcus (2001) described a habituation study with 4-month-olds, concluding "it seems likely that at least some of the machinery that infants use in this task is innate" (p. 370), but he presents no evidence to support this claim. Indeed, he goes further: a "reason for believing that something is innate is that there may be no other satisfying account for how a given piece of knowledge could arise" (p. 371). We contend that we can find more satisfying accounts through rigorous developmental analyses that embrace process, complexity, and evolutionary history (see Lehrman, 1953; Oyama, Griffiths, & Gray, 2001). We take up this charge in the present chapter.

The sections below focus on three areas of research—imprinting, spatial cognition, and language development—that justify our negative appraisal of nativism and the nativist-empiricist debate and illustrate the value of a developmental systems perspective. Our examples span low and high-level cognition with humans and non-human animals, and they represent domains that have been central to nativist accounts. We could have chosen additional examples or reviewed important critiques of nativist claims (e.g., Clearfield & Mix, 1999; Haith, 1998; Jones, 1996), but such choices would not have altered our central theme: development is an epigenetic process that entails cascades of interactions across multiple levels of causation, from genes to environments (Johnston & Edwards, 2002). Many factors routinely shape development, from the ordinary—such as the length of a child's arm—to the extraordinary—such as the vocalizations an embryonic duck produces within its egg. Our hope—and our challenge to young scientists reading this chapter—is that one day we will achieve a science that is firmly grounded in developmental process.

Imprinting

Imprinting is widely viewed as an iconic example of an innately specified behavior (e.g., Spelke & Newport, 1998). As Lorenz (1935) and other classical ethologists described it, filial imprinting is a rapid form of learning that involves the establishment of perceptual and social preferences after a brief exposure to visual cues during early development. For example, when a duckling is exposed to its mother immediately after hatching, the duckling approaches and follows her. Similarly, if the hatchling is exposed to a red wooden box on wheels, it directs its approach and following responses toward the box. Beyond its significance for the young bird, imprinting has a much broader impact on species recognition and social preferences (ten Cate, 1994).

Imprinting clearly involves learning during early development, but is it necessary to invoke a special, innate learning mechanism? Although early ethologists emphasized the uniqueness of imprinting and went to great pains to distinguish it from other forms of learning, subsequent research has softened this stance. Today, researchers no longer view it as fundamentally different from other forms of perceptual learning (Bateson, 1966; Bolhuis, 1991; Klopfer, 1973). For example, we now know that the once-rigid critical period during which imprinting must be established varies in duration and depends on contextual factors both within and outside of the learning environment (Bolhuis & Honey, 1998; Horn, 2004). Most investigators now refer to a *sensitive* period in which the quantity and quality of sensory experience has a strong influence on the strength and reversibility of the imprinted preference (Bolhuis & Trooster, 1988).

But the modern understanding of imprinting goes well beyond issues of timing toward a deep understanding of developmental process. Analysis of neural mechanisms suggests that imprinting entails two distinct processes, one involving generalized learning of stimulus features of an imprinted object, and the second involving a predisposition to approach novel objects of the general form of members of the same species (Horn & McCabe, 1984; Bolhuis & Honey, 1998). The notion of a predisposition might seem to fit neatly within the nativist ethos. Indeed, it is true that a chick reared in total darkness is still predisposed to approach a stuffed hen, thereby ruling out a necessary role for visual experience. However, chicks reared in total darkness develop the predisposition only if they receive one of several nonspecific experiences, such as running in a wheel or exposure to the hen's maternal assembly call (Bolhuis, Johnson, & Horn, 1985). In other words, *nonspecific, nonvisual* factors—factors outside the realm of what is typically considered relevant postnatal experience—can promote development of a *visual* predisposition, even though there is no obvious relationship between the predisposition and the nature of the experience. This example highlights that a "predisposition," like any other characteristic of an animal, must *develop*, and it is important to study the process through which this occurs.

Perhaps the strongest nativist claims about imprinting have been based on the preference expressed by naïve hatchlings for the maternal call of their species. As Gottlieb (1997) relates, ducklings hatched from eggs incubated in isolation show a species-appropriate preference toward the maternal call of their species, and this auditory bias facilitates imprinting to associated visual cues. Lorenz was quick to attribute this preference to innate, species-specific auditory recognition governed by genes. Gottlieb, however, experimentally demonstrated that the preference was not expressed by hatchlings that were incubated in isolation *and* devocalized, and therefore deprived of all prenatal auditory experience (that is, maternal and sibling vocalizations as well as their own vocalizations). Indeed, self-stimulation from embryonic vocalizations tunes the auditory system and establishes a bias that shapes the later preference for the maternal call (Gottlieb, 1997). In this way, embryos—so-called talking eggs—help create their own species-typical environment.

Although the hatchling's auditory preference depends on prior experience, a nativist might argue that the embryonic vocal behavior that shapes auditory development is innate. This, of course, poses a problem of infinite regress for any explanation of developmental process, and it does not represent a logically valid source of evidence for innateness. Rather, as Gottlieb's work beautifully illustrates, it always remains for further empirical work to resolve the factors—genetic, neural, organismal, environmental—that contribute to the ontogeny of each attribute of the organism at each point in developmental time.

The study of imprinting has revealed the nonobvious nature of development: behaviors are constructed through a cascade of developmental interactions, including influences of the environment that are both inherited and constructed (Goldstein, King, & West, 2003; Jablonka & Lamb, 2005; West, King, & Arberg, 1988). In light of our accumulated knowledge about imprinting and the broader view of experience that Lehrman (1953) and Gottlieb handed down to us that goes far beyond the notion of *relevance* (see also Oyama et al., 2001), the nativist focus on abilities that "are observed in the absence of any visual experience in newborn humans, infants, or newly hatched chicks" (Spelke & Kinzler, 2007, p. 89) is out of line with the empirical record and is uninformed by the lessons of the past (see Blumberg, 2005).

Spatial Cognition

Our examination of imprinting highlights the benefits of a developmental systems view. Here, we demonstrate how nativist claims within the domain of spatial cognition extend beyond the data and fail to appreciate the subtlety of developmental process.

Dead-Reckoning

Dead-reckoning is a navigational process that establishes one's current location based on past locations and movement history. According to Spelke and Newport (1998), dead reckoning is a core, innate ability whose developmental appearance does not rely on postnatal experience. Although a few studies appear to support this claim, a careful review reveals otherwise.

One study that proponents of nativism cite (even though the study's author did not advance a nativist interpretation) concerns young alpine geese navigating homeward from a distance of 40–100 meters (von Saint Paul, 1982). Because these goslings had never left the home area, Spelke and Newport (1998) claimed that they "do not learn to dead reckon by trial and error" (p. 312) and that their navigational ability must, therefore, be innate. But does this mean the geese had been deprived of all relevant experience that might support learning? These birds were 35–40 days old when tested, they took daily walks before testing within a 30 by 500 meter home range, and they were trained to return to the nest across distances of several meters. Such experience seems relevant to us.

A second example limited the role of postnatal experience by testing two-day-old chicks (Regolin, Vallortigara, & Zanforlin, 1995). Researchers placed chicks in the central corridor of an apparatus facing an object on which they had imprinted (visible through a window) and were free to walk around. When chicks left the corridor, they showed a strong preference to walk into rooms that were closer to the imprinted object even though the object was not visible. Although chicks learn a lot in the first two days after hatching (see, e.g., Hailman, 1967, 1969), it is not clear what accounts for their performance. Regolin and colleagues (1995) suggested that chicks might use "inborn" knowledge, but they also suggested that chicks might adopt a simple perceptual-motor strategy: "if you turned right (left) before the goal disappeared, then turn right (left) to find it again" (p. 198). Spencer and Dineva (2009) tested this possibility using a computer simulation of a random walk process with one constraint: as the simulated chick exited the virtual corridor—but while the object was still in view—it got a small "push" toward the goal. The best-fitting run of this simulation reproduced Regolin et al.'s results, showing that a detailed consideration of simple processes can obviate the need for inborn knowledge.

What about research with humans? Multiple studies have shown that infants' dead-reckoning abilities emerge gradually after the onset of independent locomotion (e.g., Cornell & Heth, 1979; Lepecq & Lafaite, 1989; Rider & Rieser, 1988). Indeed, this motor milestone has a profound influence on infants' navigation through space to find hidden objects (Bai & Bertenthal, 1992; Clearfield, 2004; Horobin & Acredolo, 1986), how they represent objects (Kermoian & Campos, 1988), and even how they represent socioemotional experiences (Bertenthal, Campos, & Kermoian, 1994; Campos et al., 2000). And experiments giving prelocomotor infants early locomotor experience in infant "walkers" have shown that experience contributes *directly* to these changes (e.g., Bertenthal, Campos, & Barrett, 1984).

Nevertheless, nativists often cite one study (McKenzie, Day, & Ihsen, 1984) as evidence of innate dead-reckoning abilities before the onset of independent locomotion (e.g., Spelke & Newport, 1998), even though this study did not examine dead-reckoning as it's normally defined because infants were not moved from one location to another. Instead, infants were trained only to *turn* toward a particular target marked by a distinctive cue (which six-month-old infants did at above-chance levels). Learning to track rotational movements is one component of dead-reckoning, but dead-reckoning requires more. That said, what enabled infants to track rotational movements in this study? One likely factor is their experience sitting with support and sitting independently, motor skills that develop between 1–5 and 4–7 months, respectively (Bayley, 2006). In this context, the results of this study do not support claims that dead-reckoning (or even a component of dead-reckoning) is innate.

Spatial Reorientation

Dead-reckoning works well provided that one can track and update a representation of movement through the environment. What happens, however, if one is disoriented? It is now accepted that humans and nonhuman animals reorient using the geometry of a space (Cheng & Newcombe, 2005). According to Spelke and colleagues (Hermer & Spelke, 1994, 1996; Hermer-Vazquez, Spelke, & Katsnelson, 1999; Spelke & Kinzler, 2007), animals have this ability because they possess an innate, encapsulated geometric module.

Brown, Spetch, and Hurd (2007) examined whether use of geometry for reorientation is innate by rearing fish in a circular space, but testing them in a rectangular space. When there were no distinctive featural cues, fish used the geometry of the rectangular space to search for an exit in the corners diagonally opposite one another. This means fish do not need exposure to a rectangular space *during rearing* to orient using geometry. But is it the case that no experience is needed? Brown and colleagues (2007) cannot answer this question because the fish had 8–12 days of training *in the rectangular space* before testing. Chiandetti and Vallortigara (2008) investigated whether three-day-old chicks reared in either rectangular or nonrectangular spaces use geometry when first placed in a rectangular space. Chicks did not show statistically robust geometric biases until after 50 training trials (Chiandetti, personal communication), suggesting that *some* experience is necessary.

What about the second aspect of the geometric module claim: is reliance on geometry "encapsulated"? This claim stems from evidence that animals, children, and adults fail to use unambiguous nongeometric information (such as the color of a wall) to reorient under some conditions (Hermer & Spelke, 1994, 1996). Studies show, however, that fish and birds use nongeometric information to reorient (Brown et al., 2007; Cheng & Newcombe, 2005), and young children and adults do as well, provided the room in which the task is conducted is large and the nongeometric cues provide stable landmark information (Hupbach & Nadel, 2005; Hupbach, Hardt, Nadel, & Bohbot, 2007; Learmonth, Nadel, & Newcombe, 2002; Ratliff & Newcombe, 2008).

Recently, Lee, Shusterman, and Spelke (2006) argued that an associative strategy could explain many of these findings. They provided several tests of this claim; however, all tests used small-scale geometric cues and small, moveable hiding containers. When researchers repeat a variant of the Lee et al. experiment in a large space with stable landmarks, people once again reorient using nongeometric cues (Newcombe, Ratliff, Shallcross, & Twyman, in press).

In summary, there is no compelling evidence to support nativist accounts of spatial cognition. Rather, this domain offers numerous examples of emergence and developmental change (see Plumert & Spencer, 2007). Indeed, research showing the direct influence of locomotor experience on infants' spatial understanding provides some of the strongest evidence that perception, action, cognition, and emotion co-develop in infancy (for discussion, see Bertenthal & Campos, 1990).

Language Development

As they have with imprinting, researchers have emphasized the uniqueness of language and the need for special capacities and constraints to guide the learner to correct linguistic structure (Chomsky, 1959). Such constraints are often described as innate, fixed factors, external to the learning system. By treating constraints as innate and fixed, however, researchers oversimplify their developmental origins, and by treating them as external, they ignore the interactivity of learning systems. In addition, approaches based on innate constraints rely too heavily on a rationalist analysis of language that overlooks the cascade of mutually dependent processes that affect learning and development (see Elman et al., 1996; Christiansen, Dale, Ellefson, & Conway, 2002). Although a systems view does not posit that learning is unconstrained, "constraints" in this framework are not fixed initial conditions. Rather, they arise out of the complex systems that co-develop with language.

Phonology, Grammar, and Domain-General Learning

Work on language learning was stymied for decades by claims that general-purpose learning mechanisms were insufficient for language development. The landscape shifted, however, when Saffran, Aslin, and Newport (1996) demonstrated that, after only brief exposure, nine-month-olds implicitly acquire surface statistics that are useful for segmenting words. We now know that infants and adults can learn a range of statistics that underlie phonetic categories (Maye, Weiss, & Aslin, in press; Maye, Werker & Gerken, 2002), phonology (Newport & Aslin, 2004; Saffran & Thiessen 2003), and grammar (Gomez, 2002; Mintz, 2002, 2003; Saffran, 2003; Thompson & Newport, 2007). Such computations are not limited to language, suggesting a domain-general mechanism (Creel, Newport, & Aslin, 2004; Fiser & Aslin, 2002a, 2002b).

Statistical learning provides a clear alternative to nativist views, yet nativist ideas continue to permeate debates about this form of learning. Nativist arguments stem from a rationalist analysis of statistical learning that assumes that learners count statistics independently and accurately (e.g., Remez, 2005). This results in a huge set of possible statistics and many units over which they could be computed. Thus, statistical learning must be *constrained* to consider appropriate statistics (Marcus & Berent, 2003; Newport & Aslin, 2004; Saffran, 2003; Yang, 2004). Some assume that such constraints are fixed, endowed, and language-specific (Spelke & Newport, 1998; Yang, 2004).

Such views do not accurately characterize realistic learning systems, however. Connectionist networks can capture statistics of sequences and contextual dependencies (e.g., Elman, 1990) and are capable of computing multiple statistics simultaneously. Subtle statistical relationships work together to permit learning of abstract notions (such as verb class; Christiansen & Monaghan, 2006). Additionally, learning can happen in fits and starts, showing dramatic nonlinearities over development (Abbs, Gupta, Tomblin, & Lipinski, 2007) that are exquisitely sensitive to the developmental history of the system (Altmann, 2002; MacDonald & Christiansen, 2002).

Nativist analyses of statistical learning also oversimplify the content being learned, assuming a one-to-one mapping between statistics and linguistic structure. But statistics can show intricate dependencies, such as when variation in one class of statistics points learners to a second class of statistics. For example, relationships in grammars jump over embedded elements, such as number marking on verbs. In English, verb tense must agree with the preceding noun (e.g., "She walks" but "they walk Ø"), but the tense marker always appears *after* the verb—the relationship between nouns and tense marker must skip the adjacent verb. Classical analyses suggest that this poses difficulty for association learning, because the adjacent statistics (for example, noun-verb, verb-marker) are not useful—the learner must disregard adjacent relationships and discover the appropriate nonadjacent relationship. How can the system choose the correct class of statistics without prior knowledge? Gomez (2002) used artificial grammars to show that adults and infants can identify the correct nonadjacent statistics if adjacent transitional probabilities are variable, and therefore undependable, cues (for related results, see Rost & McMurray, 2009; Yu & Smith, 2007). Thus, rather than noise, variability is critical to learning statistics in context.

Other factors, such as social context, also play a key role in focusing the learner on particular statistics. For example, infant-directed speech changes the statistics of word segmentation (Kempe, Brooks, & Gillis, 2005) and vowel categories (Kuhl, Andruski, Chistovich, & Chistovich, 1997), a natural consequence of speaking clearly and simplifying the vocabulary (Cutler & Butterfield, 1990; Krause & Braida, 2003). These examples highlight that the learner does not require an innate push toward specific statistics—the rich social milieu can provide a scaffold for language development (Deacon, 1997; Goldstein et al, 2003).

Nativists also claim that statistics alone cannot account for the patterned representations of language; algebraic rules are also necessary. Algebraic rules operate over symbols, rather than specific perceptual or linguistic items (over which statistics are computed). Marcus, Vijayan, Bandi Rao, and Vishton (1999) attempted to demonstrate algebraic rule-learning in human infants when surface statistics were unavailable. Because infants learned the rule, Marcus and colleagues claimed that rule-learning must be innate and linked specifically to language (see Marcus, Fernandez, & Johnson, 2007). On the basis of unsuccessful modeling, they further argued that rule-learning cannot emerge from a statistical learning device. But their pessimism was premature: statistical learning models like simple recurrent networks can do the trick (Altmann & Diennes, 1999; Seidenberg & Elman, 1999). Altmann (2002), for instance, found that a model given prior experience with language (similar to infants' experiences in the home) showed rule-like behavior.

The only other support for innate rule-learning comes from Peña, Bonatti, Nespor, and Mehler (2002), who demonstrated that adults extract rule-like nonadjacent statistics under some conditions. However, Perruchet, Tyler, Galland, and Peereman (2005) analyzed in detail the stimuli Peña and colleagues used and found that *adjacent* statistics supporting the apparent rule-learning are available. This redundancy of statistics at surface and deeper levels is a feature of real language (e.g., Monaghan, Chater, & Christiansen, 2005) and represents a nonobvious source of developmental change.

The examples above reveal the step-by-step, dynamic nature of statistical learning. In this context, we contend that innate constraints are unnecessary and fixed constraints of *any* kind are unlikely because the kinds of things that modulate learning develop along with language. For example, some think that perceptual systems provide fixed constraints for statistical learning, allowing the learning of certain statistics and the prevention of others (Creel et al., 2004; Fiser, Scholl, & Aslin, 2007). For this to work, however, perceptual systems must be stable during language development. They are not: in cases where perceptual processes affect statistical learning, the hypothetical perceptual "constraints" themselves develop (Johnson, Amso, & Slemmer, 2004; Sussman, Wong, Horvath, Winkler, & Wang, 2007).

Word Learning

Another classic rationalist argument for specialized language mechanisms originated with Quine (1960), who proposed that a child presented with a visual scene and a novel word faces an infinite number of possible interpretations. Thus, children must be innately constrained to consider only some of the possible meanings of a novel word.

Recent work suggests, however, that general cognitive processes and a cascade of developmental processes move children step by step from slow and deliberate to fast and efficient word learning. For example, Yu, Smith, Christensen, and Pereira (2007) examined Quine's problem, not from the perspective of an outside adult observer, but from the child's own perspective using head-mounted cameras. The result: the child's view is much more focused than previously thought, with only one object in view at any given moment. This narrow focus occurs because the child's smaller body and shorter arms keep objects close to the eyes. This fresh look at Quine's problem suggests that language-specific constraints are unnecessary: the problem is greatly simplified through the physical constraint of short arms!

Short arms can get children to the correct referent, but they cannot build a lexicon with categories that span individual instances. For that, nativists argue that children need constraints and innate knowledge to help them carve up the world (Markman, 1991; Soja, Carey, & Spelke, 1991). But do they? Work on the development of one well-studied word-learning bias—the shape bias—shows that becoming an effective word learner is an emergent product of basic attentional learning.

The "shape bias" refers to children's (and adults') tendency to generalize a novel name for a novel solid object to other solid objects on the basis of similarity in shape (Landau, Smith, & Jones, 1988). Smith and colleagues (2002) have proposed that the shape bias develops out of statistical regularities among the words and categories in the early noun vocabulary via general processes of attentional learning. Languages present regularities among linguistic devices, object properties, and perceptual category organization. For example, English distinguishes between objects that are countable and those that are not via the use of different determiners: countable things are preceded by "a," "another," and number words, whereas uncountable things are preceded by "some" and "more." Moreover, there are consistent associations between classes of nouns and object properties: count nouns generally refer to rigid objects that have solid surfaces, straight edges, and sharp corners and are organized into categories based on object shape, whereas mass nouns usually refer to nonsolid substances with irregular shapes that are organized into categories based on material similarity (Samuelson & Smith, 1999). Thus, as children begin learning a vocabulary of individual words, they are also regularly exposed to, and learn, a rich set of statistical regularities among words, object properties, and category organizations (Samuelson & Smith, 1999; Smith, Colunga, & Yoshida, 2003).

These regularities are the basis for learned associations between naming contexts and object properties that come to mechanistically shift children's attention to the correct features of novel referents (Smith, 1999; Smith & Samuelson, 2006).

The shape bias, then, simplifies the word learning situation and thereby aids vocabulary development (Samuelson, 2002; Smith et al., 2002), but it is not innate. Rather, it is the emergent product of a step-by-step cascade in which children move from individual name-referent pairings, to generalizations within categories, to generalizations that span similarities across categories. And, importantly, this cascade is grounded in general processes—detection of statistical regularities and learned associations.

In summary, nativist accounts of language development rely heavily on rationalist arguments for specialized learning mechanisms. Contemporary theory illustrates, however, how these arguments underestimate the computational power of even simple mechanisms (see McMurray, 2007), particularly when they are embedded within a developmental history.

Conclusions

As developmental scientists who work in a variety of domains and have been trained in diverse traditions, we share a profound dismay that our field has been consumed for so long by the nativist-empiricist debate. We hope to spur our colleagues and the next generation of scholars to seek new ground—not by compromising on the quantity and quality of "core knowledge systems" and "primitives" (Spelke & Kinzler, 2007), but by demanding an end to ungrounded claims about origins. This requires that we jettison the false dichotomies of the past (Johnston, 1987) and embrace a truly modern view of developmental process and developmental systems (Elman et al., 1996; Gottlieb, 1997; Oyama et al., 2001; Thelen & Smith, 1994).

We re-emphasize that a developmental systems view is not the classical counterpoint to the nativist program—we are not arguing for a return to empiricism and notions of a "blank slate." After all, the notion of a "blank slate" is just as poorly grounded as claims about "primitives" and "essences." Rather, developmental science should acknowledge that development does not begin at birth; that it is a complex, historical process; and that the relationship between cause and consequence is often nonobvious. There is no easy way around such inconvenient truths. Viewing the topic through this lens, researchers can have legitimate interests in characterizing the abilities of newborns—they are certainly fascinating!—but such characterizations do not provide privileged insight into origins. Human infants have attained certain abilities at birth, just as they have attained other abilities one day prior to birth and one day after. To lose sight of this fact is to lose sight of development itself.

What is the way forward? First, we should hold each other to a higher standard when evaluating claims about origins without direct evidence. Some of us may examine fetuses, others newborns, still others toddlers, adolescents, adults, and the aged. We can justify any choice. But what we cannot justify is studying one time point and then making unsubstantiated—or worse, unsubstantiable—claims about prior points in time. One may think such claims are justified when there appear to be no relevant prior experiences that can account for the observed behavior. But the overly narrow conception of experience that nativists offer withers away in light of evidence that nonobvious experiences critically shape the development of behavior (Gottlieb, 1997).

Second, we must invest in the future. Today's young developmental scientists have at their disposal an incredible array of sophisticated tools that form the backbone of cutting-edge, interdisciplinary research. However, these tools cannot replace the need for equally sophisticated training in contemporary developmental theory. For the sake of this generation, it is time to retire the nativist-empiricist dialogue and encourage a new dialogue that is forward looking and grounded in a modern view of developmental process.

Acknowledgments

We would like to thank Karla McGregor, Amanda Owen, the members of the Delta Center, and five anonymous reviewers for helpful comments on earlier versions of this chapter. This work was supported by NSF HSD 0527698 awarded to John P. Spencer, an Independent Scientist Award (MH66424) from the National Institute of Mental Health awarded to Mark S. Blumberg, and NICHD 045713 awarded to Larissa K. Samuelson. To join the discussion of this chapter and associated commentaries, visit the Delta Center's Facebook page: www.facebook.com/home.php#/group.php?gid=20773264989

References

Abbs B, Gupta P, Tomblin JB, Lipinski J. A behavioral and computational integration of phonological, short-term memory and vocabulary acquisition processes in nonword repetition. In: MacNamara, D.; Trafton, G., editors. Proceedings of the twenty-ninth annual conference of the Cognitive Science Society; Austin, TX: Cognitive Science Society; 2007. p. 59–64.

Altmann G. Learning and development in neural networks—the importance of prior experience. Cognition. 2002; 85:B43–B50.

Altmann G, Diennes Z. Rule learning by seven-month-old infants and neural networks. Science. 1999; 284:875.

Bai DL, Bertenthal BI. Locomotor status and the development of spatial search skills. Child Development. 1992; 63:215–226.

Bateson PPG. The characteristics and context of imprinting. Biological Review. 1966; 41:177–220.

Bayley N. Bayley scales of infant and toddler development. 3rd ed. Harcourt Assessment; San Antonio, TX: 2006.

Bertenthal B, Campos JJ. A systems approach to the organizing effects of self-produced locomotion during infancy. In: Rovee-Collier, C.; Lipsitt, L., editors. Advances in infancy research. Ablex; Norwood, NJ: 1990. p. 1–60.

Bertenthal BI, Campos JJ, Barrett K. Self-produced locomotion: An organizer of emotional, cognitive, and social developments in infancy. In: Emde, R.; Harmon, R., editors. Continuities and discontinuities in development. Plenum; New York: 1984. p. 175–210.

Bertenthal BI, Campos JJ, Kermoian R. An epigenetic perspective on the development of self-produced locomotion and its consequences. Current Directions in Psychological Science. 1994; 5:140–145.

Blumberg MS. Basic instinct: The genesis of behavior. Thunder's Mouth Press; New York: 2005.

Blumberg MS. Freaks of nature: What anomalies tell us about development and evolution. Oxford University Press; Oxford: 2009.

Bolhuis JJ. Mechanisms of avian imprinting: A review. Biological Reviews of the Cambridge Philosophical Society. 1991; 66:303–345.

Bolhuis JJ, Honey RC. Imprinting, learning and development: from behaviour to brain and back. Trends in Neuroscience. 1998; 21:306–311.

Bolhuis JJ, Johnson MH, Horn G. Effects of early experience on the development of filial preferences in the domestic chick. Developmental Psychobiology. 1985; 18:299–308.

Bolhuis JJ, Trooster WJ. Reversibility revisited: Stimulus-dependent stability of filial preference in the chick. Animal Behaviour. 1988; 36:668–674.

Brown AA, Spetch ML, Hurd PL. Growing in circles: Rearing environment alters spatial navigation in fish. Psychological Science. 2007; 18:569–573.

Campos JJ, Anderson DI, Barbu-Roth MA, Hubbard EM, Hertenstein MJ, Witherington D. Travel broadens the mind. Infancy. 2000; 1:149–220.

Cheng K, Newcombe NS. Is there a geometric module for spatial orientation? Squaring theory and evidence. Psychonomic Bulletin & Review. 2005; 12(1):1–23.

Chiandetti C, Vallortigara G. Is there an innate geometric module? Effects of experience with angular geometric cues on spatial re-orientation based on the shape of the environment. Animal Cognition. 2008; 11:139–146.

Chomsky N. A review of B. F. Skinner's verbal behavior. Language. 1959; 35(1):26–58.

Christiansen MH, Dale RAC, Ellefson MR, Conway CM. The role of sequential learning in language evolution: computational and experimental studies. In: Cangelosi, A.; Parisi, D., editors. Simulating the evolution of language. Springer; New York: 2002. p. 165–188.

Christiansen M, Monaghan P. Discovering verbs through multiple cue integration. In: Hirsh-Pasek, K.; Golinkoff, RM., editors. Action meets word: How children learn verbs. Oxford University Press; New York: 2006. p. 88–110.

Clearfield MW. The role of crawling and walking experience in infant spatial memory. Journal of Experimental Child Psychology. 2004; 89:214–241.

Clearfield MW, Mix KS. Number versus contour length in infants' discrimination of small visual sets. Psychological Science. 1999; 10:408–411.

Cornell EH, Heth CD. Response versus place learning by human infants. Journal of Experimental Psychology: Human Learning and Memory. 1979; 5:188–196.

Creel SC, Newport EL, Aslin RN. Distant melodies: Statistical learning of non-adjacent dependencies in tone sequences. Journal of Experimental Psychology: Learning, Memory, and Cognition. 2004; 30:1119–1130.

Cutler A, Butterfield S. Durational cues to word boundaries in clear speech. Speech Communication. 1990; 9:485–495.

Deacon T. The symbolic species: The co-evolution of language and the brain. Norton; New York: 1997.

Elman J. Finding structure in time. Cognitive Science. 1990; 14:179–211.

Elman J, Bates E, Johnson M, Karmiloff-Smith A, Parisi D, Plunkett K. Rethinking innateness: A connectionist perspective on development. MIT Press; Cambridge, MA: 1996.

Fiser J, Aslin RN. Statistical learning of higher-order temporal structure from visual shape-sequences. Journal of Experimental Psychology: Learning, Memory, and Cognition. 2002a; 28:458–467.

Fiser J, Aslin RN. Statistical learning of new visual feature combinations by infants. Proceedings of the National Academy of Sciences. 2002b; 99:15822–15826.

Fiser J, Scholl BJ, Aslin RN. Perceived object trajectories during occlusion constrain visual statistical learning. Psychological Bulletin and Review. 2007; 14:173–178.

Gelman SA. The essential child: Origins of essentialism in everyday thought. Oxford University Press; Oxford: 2003.

Goldstein MH, King AP, West MJ. Social interaction shapes babbling: Testing parallels between birdsong and speech. Proceedings of the National Academy of Sciences. 2003; 100(13):8030–8035.

Gómez RL. Variability and detection of invariant structure. Psychological Science. 2002; 13:431–436.

Gottlieb G. Synthesizing nature-nurture: Prenatal roots of instinctive behavior. Lawrence Erlbaum Associates; Mahwah, NJ: 1997.

Hailman JP. Behaviour. E.J. Brill; Leiden: 1967. The ontogeny of an instinct: The pecking response in chicks of the laughing gull (*Larus atricilla* L.) and related species.

Hailman JP. How an instinct is learned. Scientific American. 1969; 221:98–106.

Haith MM. Who put the cog in infant cognition? Is rich interpretation too costly. Infant Behavior and Development. 1998; 21:167–179.

Hermer L, Spelke E. A geometric process for spatial reorientation in young children. Nature. 1994; 370:57–59.

Hermer L, Spelke E. Modularity and development: The case of spatial reorientation. Cognition. 1996; 61:195–232.

Hermer-Vazquez L, Spelke E, Katsnelson A. Sources of flexibility in human cognition: Dual-task studies of space and language. Cognitive Psychology. 1999; 39:3–36.

Horn G. Pathways of the past: the imprint of memory. Nature Reviews Neuroscience. 2004; 5:108–113.

Horn G, McCabe BJ. Predispositions and preferences: Effects on imprinting of lesions to the chick brain. Animal Behaviour. 1984; 32:288–292.

Horobin K, Acredolo LP. The role of attentiveness, mobility history, and separation of hiding sites on stage IV search behavior. Journal of Experimental Child Psychology. 1986; 41:114–127.

Hupbach A, Hardt O, Nadel L, Bohbot V. Spatial reorientation: Effects of verbal and spatial shadowing. Spatial Cognition & Computation. 2007; 7:213–226.

Hupbach A, Nadel L. Reorientation in a rhombic environment: No evidence for an encapsulated geometric module. Cognitive Development. 2005; 20:279–302.

Jablonka E, Lamb MJ. Evolution in four dimensions: Genetic, epigenetic, behavioral, and symbolic variation in the history of life. MIT Press; Cambridge: 2005.

Johnson S, Amso D, Slemmer J. Development of object concepts in infancy: Evidence for early learning in an eye tracking paradigm. Proceedings of the National Academy of Sciences. 2004; 100(18):10568–10573.

Johnston TD. The persistence of dichotomies in the study of behavioral development. Developmental Review. 1987; 7:149–182.

Johnston TD, Edwards L. Genes, interactions, and the development of behavior. Psychological Review. 2002; 109:26–34.

Jones SS. Imitation or exploration? Young infants' matching of adults' oral gestures. Child Development. 1996; 67:1952–1969.

Kempe V, Brooks P, Gillis S. Diminutives in child-directed speech supplement metric with distributional word segmentation cues. Psychonomic Bulletin & Review. 2005; 12:145–151.

Kermoian R, Campos JJ. Locomotor experience: A facilitator of spatial cognitive development. Child Development. 1988; 59:908–917.

Klopfer PH. On behavior: Instinct is a Cheshire cat. J. B. Lippincott; New York: 1973.

Krause J, Braida L. Acoustic properties of naturally produced clear speech at normal speaking rates. Journal of the Acoustical Society of America. 2003; 115(1):362–378.

Kuhl PK, Andruski JE, Chistovich I, Chistovich L. Cross-language analysis of phonetic units in language addressed to infants. Science. 1997; 277:684–686.

Landau B, Smith LB, Jones SS. The importance of shape in early lexical learning. Cognitive Development. 1988; 3:299–321.

Learmonth AE, Nadel L, Newcombe NS. Children's use of landmarks: Implications for modularity theory. Psychological Science. 2002; 13:337–341.

Lee SH, Shusterman A, Spelke ES. Reorientation and landmark-guided search by young children. Psychological Science. 2006; 17(7):577–582.

Lehrman DS. A critique of Konrad Lorenz's theory of instinctive behavior. The Quarterly Review of Biology. 1953; 4:337–363.

Lepecq JC, Lafaite M. The early development of position constancy in a no-landmark environment. British Journal of Developmental Psychology. 1989; 7:289–306.

Liu D, Diorio J, Day JC, Francis DD, Meaney MJ. Maternal care, hippocampal synaptogenesis and cognitive development in rats. Nature Neuroscience. 2000; 3(8):799–806.

Lorenz K. Der kumpan in der umwelt des vogels. Journal für Ornithologie. 1935; 83:137–213. 289–413.

MacDonald M, Christiansen MH. Reassessing working memory: Comment on Just and Carpenter (1992) and Waters and Caplan (1996). Psychological Review. 2002; 109:35–54.

Marcus GF. Plasticity and nativism: Towards a resolution of an apparent paradox. In: Wermter, S.; Austin, J.; Willshaw, D., editors. Emergent neural computational architectures based on neuroscience. Springer; Heidelberg: 2001. p. 368–382.

Marcus GF, Berent I. Are there limits to statistical learning? Science. 2003; 300:53–54.

Marcus G, Fernandez K, Johnson S. Infant rule learning facilitated by speech. Psychological Science. 2007; 18(5):387–391.

Marcus GF, Vijayan S, Bandi Rao S, Vishton PM. Rule learning by seven-month-old infants. Science. 1999; 283:77–80.

Markman E. The whole object, taxonomic and mutual exclusivity assumptions as initial constraints in word meaning. In: Byrnes, JP.; Gelman, SA., editors. Perspectives on language and thought: Interrelations in development. MIT Press; Cambridge: 1991. p. 72–106.

Maye J, Weiss DJ, Aslin RN. Statistical phonetic learning in Infants: Facilitation and feature generalization. Developmental Science. 2008; 11:122–134.

Maye J, Werker JF, Gerken L. Infant sensitivity to distributional information can affect phonetic discrimination. Cognition. 2002; 82:101–111.

McKenzie BE, Day RH, Ihsen E. Localization of events in space: young infants are not always egocentric. British Journal of Developmental Psychology. 1984; 2:1–9.

McMurray B. Defusing the childhood vocabulary explosion. Science. 2007; 317:631.

Mintz TH. Category induction from distributional cues in an artificial language. Memory & Cognition. 2002; 30:678–686.

Mintz TH. Frequent frames as a cue for grammatical categories in child directed speech. Cognition. 2003; 90:91–117.

Monaghan P, Chater N, Christiansen MH. The differential role of phonological and distributional cues in grammatical categorization. Cognition. 2005; 96:143–182.

Moore CL. Maternal contributions to mammalian reproductive development and the divergence of males and females. Advances in the Study of Behavior. 1995; 24:47–118.

Newcombe NS, Ratliff KR, Shallcross W, Twyman AD. Developmental Science. Young children's use of features to reorient is more than just associative: Further evidence against a modular view of spatial processing. in press.

Newport EL, Aslin RN. Learning at a distance: I. Statistical learning of non-adjacent dependencies. Cognitive Psychology. 2004; 48:127–162.

Oyama S, Griffiths PE, Gray RD, editors. Cycles of contingency: Developmental systems and evolution. MIT Press; Cambridge: 2001.

Peña M, Bonatti L, Nespor M, Mehler J. Signal-driven computations in speech processing. Science. 2002; 298:604–607.

Perruchet P, Tyler M, Galland N, Peereman R. Learning nonadjacent dependencies: No need for algebraic computations. Journal of Experimental Psychology: General. 2005; 133(4):573–583.

Pinker S. The blank slate: The modern denial of human nature. Viking; New York: 2002.

Plumert JM, Spencer JP. The emerging spatial mind. Oxford University Press; Oxford, UK: 2007.

Quine WV. Word and object. MIT Press; Cambridge: 1960.

Ratliff KR, Newcombe N. Reorienting when cues conflict: Using geometry and features following landmark displacement. Psychological Science. 2008; 19:1301–1307.

Regolin L, Vallortigara G, Zanforlin M. Object and spatial representations in detour problems by chicks. Animal Behavior. 1995; 49:195–199.

Remez R. Perceptual organization of speech. In: Pisoni, D.; Remez, R., editors. Handbook of Speech Perception. Blackwell; Oxford, UK: 2005. p. 28–50.

Rider EA, Rieser JJ. Pointing at objects in other rooms: Young children's sensitivity to perspective after walking with and without vision. Child Development. 1988; 59:480–494.

Rost G, McMurray B. Speaker variability augments phonological processing in early word learning. Developmental Science. 2009; 12:339–349.

Saffran J. Statistical language learning: Mechanisms and constraints. Current Directions in Psychological Science. 2003; 12:110–114.

Saffran JR, Aslin RN, Newport E. Statistical learning by 8-month-old infants. Science. 1996; 274:1926–1928.

Saffran JR, Thiessen ED. Pattern induction by infant language learners. Developmental Psychology. 2003; 39:484–494.

Samuelson LK. Statistical regularities in vocabulary guide language acquisition in connectionist models and 15-20-month-olds. Developmental Psychology. 2002; 38:1016–1037.

Samuelson LK, Smith LB. Early noun vocabularies: Do ontology, category organization and syntax correspond? Cognition. 1999; 73:1–33.

Seidenberg M, Elman JL. Do infants learn grammar with algebra or statistics? Science. 1999; 284:434–435.

Smith LB. Children's noun learning: How general learning processes make specialized learning mechanisms.

In: MacWhinney, B., editor. The emergence of language. Laurence Erlbaum Associates; Mahwah, NJ: 1999. p. 277–303.

Smith LB, Colunga E, Yoshida H. Making an ontology: Cross-linguistic evidence. Cognitie Creier Comportament. 2003; 7(1):61–90.

Smith LB, Jones SS, Landau B, Gershkoff-Stowe L, Samuelson LK. Object name learning provides on-the-job training for attention. Psychological Science. 2002; 13:13–19.

Smith LB, Samuelson LK. An attentional learning account of the shape bias: Reply to Cimpian & Markman (2005) and Booth, Waxman & Huang (2005). Developmental Psychology. 2006; 42:1339–1343.

Soja NN, Carey S, Spelke ES. Ontological categories guide young children's inductions of word meaning: Object terms and substance terms. Cognition. 1991; 38:179–211.

Spelke ES, Newport EL. Nativism, empiricism, and the development of knowledge. In: Damon, W.; Lerner, RM., editors. Handbook of child psychology. Volume 1: Theoretical models of human development. John Wiley & Sons; New York: 1998. p. 275–340.

Spelke ES, Kinzler KD. Core knowledge. Developmental Science. 2007; 10:89–96.

Spencer JP, Dineva E. Taking a random walk through nativist claims about spatial cognitive development. 2009. Manuscript in preparation

Sussman E, Wong R, Horvath J, Winkler I, Wang W. The development of the perceptual organization of sound by frequency separation in 5–11 year old children. Hearing Research. 2007; 225(1–2): 117–127.

ten Cate C. Perceptual mechanisms in imprinting and song learning. In: Hogan, JA.; Bolhuis, JJ., editors. Causal mechanisms of behavioural development. Cambridge University Press; Cambridge: 1994. p. 116–146.

Thelen E, Smith LB. A dynamic systems approach to the development of cognition and action. MIT Press; Cambridge: 1994.

Thompson SP, Newport EL. Statistical learning of syntax: The role of transitional probability. Language Learning and Development. 2007; 3:1–42.

von Saint Paul, U. Do geese use path integration for walking home? In: Papi, F.; Wallraff, HG., editors. Avian navigation. Springer; New York: 1982. p. 298–307.

Weaver IC, Cervoni N, Champagne FA, D'Alessio AC, Sharma S, Seckl JR, et al. Epigenetic programming by maternal behavior. Nature Neuroscience. 2004; 7:847–854.

West MJ, King AP, Arberg AA. The inheritance of niches: The role of ecological legacies in ontogeny. In: Blass, EM., editor. Handbook of behavioral neurobiology. Vol. 9. Plenum; New York: 1988. p. 41–62.

West-Eberhard MJ. Developmental plasticity and evolution. Oxford University Press; Oxford: 2003.

Yang C. Universal grammar, statistics or both? Trends in Cognitive Sciences. 2004; 8:451–456.

Yu C, Smith L. Rapid word learning under uncertainty via cross-situational statistics. Psychological Science. 2007; 18:414–420.

Yu C, Smith LB, Christensen M, Pereira A. Proceedings of the 29th Annual Meeting of the Cognitive Science Society. Lawrence Erlbaum Associates; Mahwah, NJ: 2007. Two views of the world: Active vision in real-world interaction.

Bertenthal B, Campos JJ. A systems approach to the organizing effects of self-produced locomotion during infancy. In: Rovee-Collier, C.; Lipsitt, L., editors. Advances in infancy research. Ablex; Norwood, NJ: 1990. p. 1–60.

Creel SC, Newport EL, Aslin RN. Distant melodies: Statistical learning of non-adjacent dependencies in tone sequences. Journal of Experimental Psychology: Learning, Memory, and Cognition. 2004; 30:1119–1130.

Marcus GF, Vijayan S, Bandi Rao S, Vishton PM. Rule learning by seven-month-old infants. Science. 1999; 283:77–80.

Newport EL, Aslin RN. Learning at a distance: I. Statistical learning of non-adjacent dependencies. Cognitive Psychology. 2004; 48:127–162.

SECTION IV

EVOLUTION

EDITORS' INTRODUCTION

Within the relational developmental system, the organism has a distinct influence on the multilevel context that is influencing it. That is, the organism is an active contributor to its own development (e.g., Lerner, 1982, 2018; Lerner et al., 2005; Schneirla, 1957). Clearly, then, at least among mammals (in contrast to insects), and certainly among humans (Tobach & Schneirla, 1968), development is both biological and social. In fact, no form of life as we know it comes into existence independent of other life. No animal lives in total isolation from others of its species across its entire life span (Tobach, 1981; Tobach & Schneirla, 1968). In other words, the change process in human development involves integrated coactions across the biological-through social (and cultural and historical; Elder et al., 2015; Raeff, 2016) levels of organization; these levels comprise the ecology of human development (e.g., Bronfenbrenner, 2005).

Moreover, Gottlieb (1970, 1997, 2004) explained that the process of ontogenetic development is related to the emergence in phylogenetic history of this individual ⇔ context process. It is important, then, to consider this link between evolution (phylogeny) and ontogeny—both to explicate the character of developmental change and to illustrate the difference between how dynamic, relational developmental systems (RDS)-based theories envision the link between ontogeny and phylogeny (e.g., Lickliter & Witherington, 2017; Witherington & Lickliter, 2016) and how this link is approached by proponents of genetic reductionism, for instance, by proponents of evolutionary developmental psychology (EDP) (e.g., Bjorklund, 2015, 2016; Bjorklund & Ellis, 2005).

Probabilistic Epigenesis and Human Evolution

Early humans were relatively defenseless, having neither sharp teeth nor claws. Coupled with the dangers of living in the open African savanna, where much of early human evolution occurred, group living was essential for survival (Masters, 1978; Washburn, 1961). Therefore, human beings were more likely to survive if they acted in concert with the group than if they acted in isolation. Human characteristics that support social relations (e.g., attachment and empathy) may have helped human survival over the course of human evolution (Hoffman, 1978; Hogan et al., 1978; Sahlins, 1978).

Biological survival requires meeting the demands of the environment or attaining a goodness of fit (Chess & Thomas, 1984, 1999; Lerner & Lerner, 1983, 1989; Thomas & Chess, 1977) with the context. Because this environment is populated by other members of one's species, adjustment to (or fit with) these other organisms is a requirement of survival (Tobach & Schneirla, 1968).

DOI: 10.4324/9781032702988-16

Given this biological fusion with the social ecology of human development, it is not surprising to learn that several scholars having ideas associated with dynamic, RDS-based theories believed that human evolution has promoted the link between biological and social functioning (e.g., Featherman & Lerner, 1985; Gould, 1977). In other words, the ontogenetic integration of human biological and social levels of organization has been shaped by the evolutionary history of humans.

The scholarship of Stephen J. Gould (1977) has provided singular contributions to the understanding of this linkage between ontogeny and phylogeny. A discussion of his ideas advances understanding of the relevance of human evolution to the individual ⇔ context relations that propel individual development across the life span.

Gould's Views of Ontogeny and Phylogeny: Evolutionary Bases of Individual ⇔ Context Relations

As evident from the title of his book, *Ontogeny and Phylogeny* (1977), Stephen J. Gould had an abiding interest in detailing the relation between ontogeny and phylogeny. He contended "That some relationship exists .... Evolutionary changes must be expressed in ontogeny, and phyletic information must, therefore, reside in the development of individuals" (Gould, 1977, p. 2). However, this point in itself is obvious and unenlightening for Gould. What makes the study of the relation between ontogeny and phylogeny interesting and important is that there are "changes in developmental timing that produce parallels between the stages of ontogeny and phylogeny" (Gould, 1977, p. 2).

Discussing the relation between ontogeny and phylogeny may raise the hackles (read: "Haeckels") of many scientists trained in human development. The recapitulation ideas of Haeckel (1868), especially as they were adopted by G. Stanley Hall (1904), have long been in disfavor. "It is, of course, now known that Haeckel erred in formulating this so-called law. Indeed, some hold that he deliberately falsified his reports (Greenberg & Mosack, 2004, p. 909)." However, while not true in the form Haeckel originally proposed, this biogenetic "law" still has general acceptance in biology and more specifically in embryology (Richardson & Keuck, 2002). Recapitulation has been used, most recently by evolutionary psychologists, to support racism, criminal anthropology, and early conceptions of child development (e.g., Greenberg & Mosack, 2004).

Haeckel's theory of recapitulation involves the idea that the process of evolution occurred through a change in the timing of developmental events; this change in timing created a universal acceleration of development that pushed ancestral, adult forms into the juvenile stages of descendants. For example, Haeckel (1868) interpreted the gill slits of human embryos as characteristics of ancestral adult fish that had been compressed into the early stages of human ontogeny through this universal process of acceleration of developmental rates in evolving lines.

It is unfortunate for the scientific study of links between ontogeny and phylogeny that scientists came to regard Haeckel's concept of recapitulation and, even more, the entire topic of the connection between phylogeny and ontogeny as ideas that should not be addressed within evolutionary biology. As Gould (1977, p. 2) explained, "Haeckel's biogenetic law was so extreme, and its collapse so spectacular, that the entire subject became taboo." The absence of scientific attention was problematic because Gould (1977) noted that alternative formulations of the relation between ontogeny and phylogeny could avoid the shortcomings of Haeckel's (1868) formulation. Indeed, Gould (1977) offered such an alternative, one that provided a different conception of the evolutionary basis of individual ⇔ context relations. According to Gould (1977), this alternative is the key to human evolution and to human plasticity. In order to understand this alternative, it is important to introduce three interrelated terms: heterochrony, neoteny, and pedomorphosis.

Evolution: Editors' Introduction

According to Gould (1977), evolution occurs when ontogeny is altered in one of two ways. First, evolution occurs when new characteristics are introduced, within any period of development, which then have varying influences on later developmental stages. The second way in which evolution occurs is when characteristics that are already present undergo changes in developmental timing. This second means by which phyletic change occurs is termed heterochrony. Specifically, heterochrony is changes in the relative time of appearance and rate of development of characteristics already present in ancestors.

In human evolution, a specific type of heterochrony has been predominant; as a consequence, the changes that were associated with human plasticity occurred. The type of heterochrony that has characterized human evolution is neoteny, which is a slowing down, a retardation, of development of selected somatic organs and parts. Heterochronic changes are regulatory effects; that is, they constitute "a change in rate for features already present" (Gould, 1977, p. 8). Gould (1977) maintained that neoteny has been a—and probably the—major determinant of human evolution.

For example, delayed growth has been found to be important in the evolution of complex and flexible social behavior and, interrelatedly, it has led to an increase in cerebralization by prolonging into later human life the rapid brain-growth characteristics of higher vertebrate fetuses. As such, this general evolutionary retardation of human development (slowing down the rate of development in comparison to ancestral species) has resulted in adaptive features of ancestral juveniles being retained. That is, a key characteristic of human evolution is paedomorphosis, or phylogenetic change involving retention of ancestral juvenile characteristics by the adult. In other words, Gould (1977) noted:

> Our paedomorphic features are a set of adaptations coordinated by their common efficient cause of retarded development. We are not neotenous only because we possess an impressive set of paedomorphic characters; we are neotenous because these characters develop within a matrix of retarded development that coordinates their common appearance in human adults ... [and these] temporal delays themselves are the most significant feature of human heterochrony. (pp. 397, 399)

But what are some of the paedomorphic–neotenous characteristics? How do they provide an evolutionary basis of human plasticity and individual ⇔ context relations? Gould (1977) himself answered these questions, and, in so doing, indicated that humans' evolving plasticity both enabled and resulted from their embeddedness in a social and cultural context. Gould, (1977) noted:

> In asserting the importance of delayed development ... I assume that major human adaptations acted synergistically throughout their gradual development ... The interacting system of delayed development—upright posture—large brain is such a complex: delayed development has produced a large brain by prolonging fetal growth rates and has supplied a set of cranial proportions adapted to upright posture. Upright posture freed the hand for tool use and set selection pressures for an expanded brain. A large brain may, itself, entail a longer life span. (p. 399, italics added)

and

> Human evolution has emphasized one feature of ... common primate heritage—delayed development, particularly as expressed in late instruction and extended childhood. This retardation has reacted synergistically with other hallmarks of hominization—with intelligence (by enlarging the brain through prolongation of fetal growth tendencies and by providing a longer period of childhood

learning) and with socialization (by cementing family units through increased parental care of slowly developing offspring). It is hard to imagine how the distinctive suite of human characters could have emerged outside the context of delayed development. (p. 400)

Thus, in linking neoteny with reciprocal relations between brain development and sociocultural functioning, Gould (1977) made an argument of extreme importance for comparative–developmental and sociocultural–intergenerational analyses of human development. The role of the former type of analysis is raised in regard to species differences (heterochrony) in the ontogeny of brain organization and their import for levels of plasticity finally attained across life. In other portions of the evolutionary biology literature and in the anthropology literature, there is support for the link suggested by Gould (1977) between plastic brain development and human sociocultural functioning.

Individual ⇔ Context Relations in Evolution: Paleoanthropological Perspectives

Several ideas in anthropology suggest that humans have evolved to manifest social dependency (e.g., Tobach, 1981; Tobach & Schneirla, 1968). The course and context of evolution was such that it was more adaptive to act in concert with the group than in isolation. For example, Masters (1978) noted that early hominids were hunters. These ancestors evolved from herbivorous primates under the pressure of climatic changes that caused the African forest to be replaced with savanna. Masters speculated that the large brains of humans (1978, p. 98), may be the result of cooperation among early hominids and, hence, in an evolutionary sense, the human brain is a social organ. Indeed, he believed that, with such evolution, the "central problem" in anthropological analysis—that of the origin of society—may be solved. Washburn (1961) appeared to agree. He noted that the relative defenselessness of early humans (lack of fighting teeth, nails, or horns), coupled with the dangers of living on the open African savanna, made group living and cooperation essential for survival (Hogan, et al., 1978; Washburn, 1961).

There is some dispute in anthropological theory as to whether material culture or specific features of social relations, such as intensified parenting, monogamous pair bonding, nuclear family formation, and, thus, specialized sexual-reproductive behavior, were superordinate in these brain-behavior evolutionary relations. For example, some paleoanthropologists have maintained the idea that there are five characteristics that separate human beings from other hominids: large neocortex, bipedality, reduced anterior dentation with molar dominance, material culture, and unique sexual and reproductive behavior (e.g., of all primates only the human female's sexual behavior is not confined to the middle of her monthly menstrual cycle; Fisher, 1982a). Some paleoanthropologists believe that early human evolution was a direct consequence of brain expansion and material culture. However, Lovejoy (1981), among others (e.g., Johanson & Edey, 1981), believes that:

Both advanced material culture and the Pleistocene acceleration in brain development are sequelae to an already established hominid character system, which included intensified parenting and social relationships, monogamous pair bonding, specialized sexual-reproductive behavior, and bipedality. (p. 348)

Other debates also exist. For instance, the roles that continual sexual receptivity and loss of estrus played in the evolution of human pair bonding are controversial and complex (e.g., Belsky et al., 1991; Bjorklund & Shackelford, 1999; Ellis et al., 2012; Fisher, 1982b; Harley, 1982; Isaac, 1982; Swartz, 1982; Washburn, 1982). Such debate, however, exists in the midst of the general consensus indicated earlier:

that the social functioning of hominids (be it interpreted as dyadic, familial, or cultural) was reciprocally related to the evolution of the human brain. Many evolutionary biologists appear to reach a similar conclusion.

For example, summarizing a review of literature pertaining to the character of the environment to which organisms adapt, Lewontin and Levins (1978) stressed that reciprocal processes between organism and environment were involved in human evolution; as such, this leads to a view that human functioning is one source of its own evolutionary development. Lewontin and Levins (1978) stated that:

> The activity of the organism sets the stage for its own evolution ... The labor process by which the human ancestors modified natural objects to make them suitable for human use was itself the unique feature of the way of life that directed selection on the hand, larynx, and brain in a positive feedback that transformed the species, its environment, and its mode of interaction with nature. (p. 78)

Moreover, not only did Lovejoy (1981) and Fisher (1982a) give graphic accounts of the history of the role of hominid social behavior in human evolution, but—in specific support of Gould's (1977) views—they also showed how the complex social and physical facets of this evolution led to human neoteny. Interestingly, whereas Fisher and (especially) Lovejoy viewed the ecological presses that led to the evolution of social behaviors as eventuating in bipedalism and then rapid brain development, they nevertheless both saw these links in more of a circular than a linear framework.

For instance, Lovejoy (1981) noted that it was not just that ecological changes led to social relationships, which in turn led to bipedalism, and, in turn, to brain evolution. Instead, social relationships that led to brain evolution were then themselves altered when larger-brained and more plastic organisms were involved in them; in turn, new social patterns may have extended humans' adaptive presses and opportunities into other arenas, ones fostering further changes in the brain, in social embeddedness, and so forth. Indeed, as Johanson and Edey (1981) described Lovejoy's (1981) position, it is one that requires the examination of the process

> of a complex feedback loop in which several elements interact for mutual reinforcement ... If parental care is a good thing, it will be selected for by the likelihood that the better mothers will be more apt to bring up children, and thus intensify any genetic tendency that exists in the population toward being better mothers. But increased parental care requires other things along with it. It requires a greater IQ on the part of the mother; she cannot increase parental care if she is not intellectually up to it. That means brain development—not only for the mother, but for the infant daughter, too, for someday she will become a mother.
>
> In the case of primate evolution, the feedback is not just a simple A–B stimulus forward and backward between two poles. It is multipole and circular, with many features to it instead of only two—all of them mutually reinforcing. For example, if an infant is to have a large brain, it must be given time to learn to use that brain before it has to face the world on its own. That means a long childhood. The best way to learn during childhood is to play. That means playmates, which, in turn, means a group social system that provides them. But if one is to function in such a group, one must learn acceptable social behavior. One can learn that properly only if one is intelligent. Therefore, social behavior ends up being linked with IQ (a loopback), with extended childhood (another loop), and finally with the energy investment and the parental care system which provide a brain capable of that IQ, and the entire feedback loop is complete.

All parts of the feedback system are cross-connected. For example: if one is living in a group, the time spent finding food, being aware of predators and finding a mate can all be reduced by the very fact that one is in a group. As a consequence, more time can be spent on parental care (one loop), on play (another) and on social activity (another), all of which enhance intelligence (another) and result ultimately in fewer offspring (still another). The complete loop shows all poles connected to all others.

(pp. 325–326)

Conclusions about Paleoanthropological Perspectives of Individual ⇔ Context Relations

Our discussion of the links between Gould's (1977) ideas pertinent to the role of neotenous heterochrony in the evolution of human plasticity has involved, as well, a discussion of the role of reciprocal relations between organisms and their contexts in human evolution. In other words, neoteny provides adaptive advantages for members of both older and younger generations. Considering children first, the neoteny of the human results in the newborn child being perhaps the most dependent organism found among placental mammalian infants (Gould, 1977). Moreover, their neoteny means that this dependency is extraordinarily prolonged, and this elongation requires intense parental care for the child for several years.

The plasticity of childhood processes, which persists among humans for more than a decade, thus entails a history of necessarily close contact with adults and places an "adaptive premium … on learning (as opposed to innate response) … unmatched among organisms" (Gould, 1977, p. 401). Gould agrees with de Beer (1959), who stated that for the human:

Delay in development enabled him to develop a larger and more complex brain, and the prolongation of childhood under conditions of parental care and instruction consequent upon memory-stored and speech-communicated experience, allowed him to benefit from a more efficient apprenticeship for his conditions of life. (p. 930)

The neoteny of humans, their prolonged childhood dependency on others, and their embeddedness in a social context composed of members of the older generation who both protect them and afford them the opportunity to actualize their potential plasticity allow members of a new birth cohort to adapt to the conditions and presses particular to their historical epoch.

Such development in a new cohort also has evolutionary significance for members of the older cohort. Gould (1977) pointed out that neoteny and the protracted period of dependent childhood may have led to the evolution of features of adult human behavior (e.g., parental behavior). The presence of young and dependent children requires adults to be organized in their adult ⇔ adult and adult ⇔ child relations in order to support and guide the children effectively. Furthermore, since the period of childhood dependency is so long, it is likely that human history tended to involve the appearance of later-born children before earlier-born children achieved full independence (Gould, 1977). Gould (1977, p. 403) saw such an occurrence as facilitating the emergence of pair bonding, and further saw "in delayed development a primary impetus for the origin of the human family."

Several lines of evidence—from human development, evolutionary biology, sociology, and anthropology—converge to suggest the use of a dynamic, RDS-based approach to the dynamic links between human development across the life span (ontogeny) and to the developmental changes across generational and historical time that constitute human evolution (Jablonka & Lamb, 2005; Lewontin & Levins, 1978). However, there exist alternative views of the sources of human ontogeny and phylogeny that rely on genetic reductionist conceptions of ontogeny and phylogeny. We consider such formulations next.

Evolutionary Psychology and Evolutionary Developmental Psychology

The evidence about embodiment, plasticity, and epigenetics that accounts for the character of evolutionary and developmental change understandably elicits skepticism about and, even more, the rejection of the "extreme nature" claims of genetic reductionists (Rose & Rose, 2000). For instance, evolutionary psychology (EP) claims that "everything from children's alleged dislike of spinach to our supposed universal preferences for scenery featuring grassland and water derives from [the] mythic human origin in the African savannah" (Rose & Rose, 2000, p. 2). These claims are predicated on the basis of the assertion that one can explain:

> all aspects of human behaviours, and thence culture and society, on the basis of universal features of human nature that found their final evolutionary form during the infancy of our species some 100–600,000 years ago. Thus for EP, what its protagonists describe as the 'architecture of the human mind' which evolved during the Pleistocene is fixed, and insufficient time has elapsed for any significant subsequent change. In this architecture there have been no major repairs, no extensions, no refurbishments, indeed nothing to suggest that micro or macro contextual changes since prehistory have been accompanied by evolutionary adaption.
>
> (Rose & Rose, 2000, p. 1)

Clearly such assertions within EP (e.g., Buss, 2019a, 2019b; Buss & von Hippel, 2018) are inconsistent with the now quite voluminous evidence in support of the epigenetic character of evolution and ontogeny (e.g., Lickliter, 2016; Lickliter & Honeycutt, 2015; Noble, 2015; Richardson, 2007), of the multiple, integrated dimensions of evolution (Jablonka & Lamb, 2005), and of the role of the organism's own agency and of culture in creating change within and across generations (Gottlieb, 1997, 1998, 2002/2014, 2004).

Simply, proponents of EP (get the nature of evolution quite wrong (pun intended). They fail to appreciate the autopoietic character of the holistic, dynamic, and integrated relational developmental system, and therefore they adhere to an atavistic and incorrect view of the role of genes in this self-constructing system. As Noble (2015) explains, the appropriate understanding of genes within this system involves:

> two fundamental concepts. The first one is the distinction between active and passive causes. Genes are passive causes; they are templates used when the dynamic cell networks activate them. The second concept is that there is no privileged level of causation. In networks, that is necessarily true, and it is the central feature of what I have called the theory of biological relativity, which is formulated in a mathematical context.
>
> (Noble, 2015)

I will illustrate the second point in a more familiar nonmathematical way. Take some knitting needles and some wool. Knit a rectangle. If you don't knit, just imagine the rectangle. Or use an old knitted scarf. Now pull on one corner of the rectangle while keeping the opposite corner fixed. What happens? The whole network of knitted knots moves. Now reverse the corners and pull on the other corner. Again, the whole network moves, though in a different way. This is a property of networks. Everything ultimately connects to everything else. Any part of the network can be the prime mover, and be the cause of the rest of the network moving and adjusting to the tension. Actually, it would be better still to drop the idea of any specific element as prime mover. It is networks that are dynamically functional.

Now knit a three-dimensional network. Again, imagine it. You probably don't actually know how to knit such a thing. Pulling on any part of the three-dimensional structure will cause all other parts to move.... It doesn't matter whether you pull on the bottom, the top or the sides. All can be regarded as equivalent. There is no privileged location within the network. (p. 11)

Noble (2015) adds that this conception of the role of genes within the dynamic, integrated relational developmental system is consonant with the ideas of Ho and Saunders (1979; see also Ho, 2010, 2013). He notes their view that:

The intrinsic dynamical structure of the epigenetic system itself, in its interaction with the environment, is the source of non-random variations which direct evolutionary change, and that a proper study of evolution consists in the working out of the dynamics of the epigenetic system and its response to environmental stimuli as well as the mechanisms whereby novel developmental responses are canalized.

(Ho & Saunders, 1979, p. 573)

Nevertheless, despite biologists presenting concepts and data enumerating the errors about evolution and genetics involved in EP (e.g., Richardson, 2007), examples of misguided scholarship about evolution (phylogeny) and its relation to ontogeny continue to appear in the literature. An example of this flawed scholarship is a variant of EP, EDP.

Evolutionary Developmental Psychology

Bjorklund and Bering (2002, p. 347) state that EDP "focuses on the adaptive nature of psychological mechanisms built into the brains of juveniles, some of which may serve immediate demands at different stages of development, and some of which serve preparatory roles for maturity." There are several examples of the flawed claims of proponents of EDP (e.g., Bjorklund, 2015, 2016; Bjorklund & Ellis, 2005; Del Giudice & Ellis, 2016), and one representative example is what is termed "paternal investment theory" (Belsky, 2012; Belsky et al., 1991; Draper & Harpending, 1982, 1988). For instance, Ellis et al. (2012) claim that:

paternal investment theory links low male parental investment to more aggressive and hypermasculine behavior in sons and more precocious and risky sexual behavior in daughters (Draper & Harpending, 1982, 1988). The assumption is that natural selection has designed boys' and girls' brains to detect and encode information about their fathers' social behavior and role in the family as the basis for calibrating socio-sexual development in gender-specific ways. (p. 32)

The purported mechanism for what Ellis et al. (2012) term this evolutionary-developmental phenomenon is that there is:

a unique role for fathers in regulating daughters' sexual behavior. The theoretical basis for emphasizing father-effects is (a) that the quality and quantity of paternal investment is—and presumably always has been—widely variable across and within human societies; (b) this variation recurrently and uniquely influenced the survival and fitness of children during our evolutionary history...; and (c) variability in paternal investment, much more than maternal investment, was diagnostic of the local mating system (degree of monogamy vs. polygyny) and associated levels of male-male competition...The mating system is important because more polygynous cultures and subcultures are characterized by heightened male intrasexual competition, dominance-striving, and violence, with concomitant diminution of paternal involvement and investment (Draper & Harpending, 1982, 1988). In turn, female reproductive strategies in this context are biased toward earlier sexual debut, reduced reticence in selecting mates, and devaluation of potential long-term relationships with high-investing males, all of which translate into more RSB [risky sexual behavior]. (p. 32)

However, such embodiment of the individual and of his or her plastic developmental biological, psychological, and behavioral processes within the relational developmental system provides a basis for epigenetic changes across generations (e.g., see Moore, 2015a, 2016), that is, for changes in gene ⇔ context relations within one generation being transmitted to succeeding generations. As such, the "Just So" stories approach to purported behavioral outcomes of evolution (Gould, 1981) used by sociobiology is shared by proponents of EDP; such an approach is conceptually flawed in that it ignores contemporary scholarship about evolutionary processes and their impact on ontogeny (e.g., Gissis & Jablonka, 2011; Ho, 2010, 2013; Ho & Saunders, 1979; Lickliter, 2016; Lickliter & Honeycutt, 2015; Meaney, 2010, 2014), and are therefore empirically counterfactual.

Embodiment provides the basis for epigenetic change within the life span of an individual; such change involves qualitative discontinuity across ontogeny in relations among biological, psychological, behavioral, and social variables. Evidence for the plasticity of human development within the integrated levels of the ecology of human development makes biologically reductionist accounts of parenting, offspring development, or sexuality implausible, at best, and entirely fanciful, at worst.

For instance, Bateson (2016) notes that, "The robust mechanisms [of evolution] that make *species* different from each other also impact processes that make *individuals* distinct from one another. Children both influence their environment and are influenced by it" (p. 1). Bateson (2016) goes on to explain that:

Recent discoveries in genetics and epigenetics have given us profound new insight into the development of the individual—an understanding marked by the dynamic interplay of *robustness* and *plasticity*. Robustness is profound and real: All humans develop certain predictable traits, and nobody will ever confuse an adult human with an adult howler monkey. At the same time, humans have a remarkable capacity for specialization and change that emerges very early in development in response to individual experiences, educational opportunities, and culture. Importantly, robustness and plasticity cannot be cleanly separated; certainly, one should not think of them in the same way as the discredited dichotomy of innate versus acquired. This is because plasticity in its many forms depends on underlying robust processes—a point illustrated by the history of behavioral biology (p. 1)… the key point is that the genotype of an individual can be expressed very differently depending upon the developmental environment.

(see Lickliter, 2016, p. 3)

However, as explained by Witherington and Lickliter (2016), the arguments of EDP proponents constitutes, in essence, an essentialist approach that sees genes as the provider of the key information determining the substance of robustness of human development; the role of the non-essentialist level—the developmental environment in the terms of Bateson (2016) is, to proponents of EDP, only the control of the emergence (or not) of genetically based phenomena. That is, the role in human development of the levels of organization higher than the genetic one is only to manage the expression or release of the information contained in the essential, genetic level. In short, in the essentialist approach of EDP there is a Cartesian-like split between the ultimate cause of development—pre-organism existing information, shaped by evolution (phylogeny), and inserted into the organism at conception through the content of a gene—and the instantiation of the information, which depends on the vicissitudes of everyday life, the ebb and flow of relations between the organism and its context across ontogeny (Witherington & Lickliter, 2016).

In turn, RDS-based models of human development embrace complexity (without reducing it to an essentialist entity), and the autopoietic process of development itself is the source of structure and function of the organism (Noble, 2015). There is no pre-existing information split off from the developmental process, and no essential level of organization to which complex higher levels are to be reduced (Witherington & Lickliter, 2016).

Witherington and Lickliter (2016) emphasize that the concept of emergence is of fundamental importance in understanding RDS-based approaches and how they differ from essentialist approaches, such as EDP (e.g., Bjorklund, 2015; Bjorklund & Ellis, 2005; Del Giudice & Ellis, 2016). They explain that a fundamental idea in the EDP approach is that there are entities, evolved probabilistic cognitive mechanisms (EPCMs), that pre-exist the organism and frame its development, with developmental environments across ontogeny—for example, Bronfenbrenner's conception of nested systems comprising the ecology of human development (e.g., Bronfenbrenner & Morris, 2006)—just determining what is placed within the frame—much like a building contractor frames one's house but the owners of the house fill in the frame by acting to select paint colors, appliances, floor coverings, etc. In contrast, and as also emphasized by Witherington and Lickliter (2016; see too Bateson, 2015; Lickliter, 2016; Lickliter & Honeycutt, 2015; Mascolo, 2013; Overton, 2015; van Geert & Fischer, 2009; Witherington, 2011, 2015), Raeff (2016, pp. 12–13) in the RDS-based view:

> behavior emerges out of interrelations among "ongoing processes intrinsic to the system" (Lewis, 2000, p. 38). Claiming that human functioning emerges through interrelations among intrinsic constituent processes means that one does not have to involve external, Antecedent, or independent factors to explain what people do. In addition, the concept of emergence stands in explicit contrast to any conceptualization of behavior and development as predesigned or predestined by, for example, genetics or how the brain is "hardwired." Rather, what a person does emerges, or is always coming into being, through the ongoing dynamics of constituent processes.

In short, at this writing developmental science includes two very different approaches to the complexity of the integrated, multilevel, interrelated changes that everyone within the field agrees characterizes human ontogeny. One approach is an essentialist, genetic reductionist model, and the other approach is the RDS-based model. What are we to make of these two approaches to human development and its evolutionary bases? Are both useful frames for the study of human development? If so, then how should research proceed? If not, why? And again, how should research about human development proceed?

Two Approaches to Developmental Complexity

Given the features of the essentialist and the RDS approaches that Witherington and Lickliter (2016) explain in careful detail, a key question must be addressed in evaluating their respective usefulness: Are the characteristics of an individual (a) features deriving from the constituent processes of the developmental system or are they (b) an outcome of the developmental system acting on something that preexists and that merely awaits expression, should the organism happen to grow up in an environment "typical" for its species? From an EDP perspective, Del Giudice and Ellis (2016) contend that "while [sic] evolved mechanisms prepare an organism for life in a species-typical environment, they are not preformed or specified in advance by a rigid genetic program" (p. 7).

But where do these "mechanisms" exist and in what form? From an essentialist perspective, they must exist prior to the existence of the organism that houses them during its ontogeny. Presumably these mechanisms *must* be located in the gametes of parents. But how did the information or process constituted by these "mechanisms" come to reside in the gametes?

This information or material—or whatever it is—must have come through the germ line of the parents' parents (so we are now going to the grandparental generation for an answer to the question of the origin of the EPCMs in the development of a given, "target" individual). But the same question continues to be needed to be asked of this grandparent generation, of the one prior to it, and so on through an infinite regress that keeps the question being pushed further back in history without any definitive empirical verification. Because of this infinite regress, EDP sets up an argument that cannot be falsified by any developmental data pertinent to a target individual's life span, because there always has to be an appeal made to a former generation as the source of the "whatever."

Moreover, the idea of this whatever—for the purposes of illustration, we will label it a homunculus—can only pertain to something that could actually exist *if* genes and context are split entities and, as such, if genes were then conceived of as entities that contained the homunculus, *and*, as well, if modern work in biology pertaining to epigenetics was irrelevant (e.g., Lester et al., 2016; Moore, 2015a, 2016; Meaney, 2010). Of course, in such a formulation, the homunculus could only be released if the gene was turned on sufficiently, and here proponents of EDP claim that, for such an occurrence, for the homunculus to be instantiated, the "correct" context, a species typical one, needs to be present for at least some (unspecified) portion of ontogeny (e.g., again, see Del Giudice & Ellis, 2016, p. 7).

But here lie problems of circularity of reasoning becoming coupled with an argument already fatally flawed by the use of the unfalsifiable postulation of an infinite regress: If one sees the homunculus, then it *must* be the case that there was a species-typical environment because there would be no other way for the homunculus to appear. In turn, if one does not see the homunculus, then it *must* be that it is absent because there was not a species-typical environment within which the organism developed. Ironically, the postulation of the existence of this prior-to-being homunculus cannot be falsified by any direct empirical evidence pertinent to the purported evolutionary (phylogenetic) history that created it. Phylogeny is not studied, and instead, reference is made to an unassessed ontogenetic history that is inferred to have existed because of the presence or absence of some behaviors that are claimed to reflect the also-never-assessed evolutionary history!

Indeed, it is ironic that the only recourse proponents of EDP have to prove their phylogenetic case is to appeal to an ontogenetic developmental process that is regarded by them to have no ultimate causal efficacy, but only the capacity to facilitate expression of an entity caused by a phylogenetic process. The morass of logical problems and appeals to impossible-to-document histories make the cornerstone idea of EDP—EPCMs—as useful a scientific concept as the homunculus label we have applied to it.

Importantly, developmental science has been subjected to these problematic formulations before the advent of EDP. That is, the logical and empirical shortcomings of the EDP concept of EPCMs are comparable to the fatal flaws associated with the other formulations of essentialist thinking in developmental science. All of these formulations become counterfactual because of not being able to marshal the empirical evidence that is needed to support their claims about scientific usefulness.

For instance, for the nurture-reductionism of Skinner (1971) to work as a comprehensive explanation of the behavior of organisms (Skinner, 1938), there must be an S for every R. However, as pointed out by Bowers (1973), one of the key reasons that Skinner's approach fails is the problem of the missing S. Simply, research has failed to identify an S (i.e., a discriminative stimulus, which has the status of a secondary reinforcing stimulus) for every R that exists. Yet, such an S is stipulated by Skinner to be needed to elicit operant behavior. If such stimuli are *the* causes of operant behaviors in any given situation, then how can empiricists hold that the S-R formulation (S^D-R-S^R) is useful when there are so many Rs for which there are no Ss to be seen? They cannot. As such, the radical behaviorism of Skinner (1938, 1971) is reduced to a view that must be accepted on the basis of faith (that there must have been an S somewhere) and not on empirical evidence.

The concept of the fixed action pattern (FAP) formulated by Lorenz (1937a, 1937b, 1965) is an example of a nature-essentialist formulation, one that is similar to the EPCMs postulated by proponents of EDP (e.g., Bjorklund, 2015; Bjorklund & Ellis, 2005; Del Giudice & Ellis, 2016). Oddly, these EDP proponents are either unaware of this similarity or have elected to not note it because of the several logical and empirical problems with Lorenz's concept or, perhaps as well, because of Lorenz's own history of problems with his use of nativist ideas (e.g., see Lerner, 1992). In any case, Lorenz (1937a, 1937b, 1965) used the FAP to illustrate his concept of instinct, which he claimed was a behavior that could be observed when the individual experienced a specific "releasing" stimulus—that is, when the organism encountered a certain stimulus that "triggered" a given instinct. Lorenz posited the existence of an "innate releasing mechanism" (IRM), a hypothetical mechanism believed to involve a set of receptor cells that released the instinctual behavior pattern when activated by a specific environmental stimulus.

Blumberg (2016) provided an important critique of Lorenz's concept of instinct. He notes that, instincts "are not satisfactorily described as inborn, pre-programmed, hardwired, or genetically determined. Rather, research in this area teaches us that species-typical behaviors *develop*—and they do so in every individual under the guidance of species-typical experiences occurring within reliable ecological contexts" (Blumberg, 2016, p. 1). Blumberg also illustrates the bases for this view by pointing to an article by Hailman (1969), and explains that:

> Writing in *Scientific American*, in an article cunningly titled 'How an instinct is learned,' Hailman challenged Lorenz's fundamental notion of instinct: 'The term "instinct," as it is often applied to animal and human behavior, refers to a fairly complex, stereotyped pattern of activity that is common to the species and is inherited and unlearned. Yet, braking an automobile and swinging a baseball bat are complex, stereotyped behavioral patterns that can be observed in many members of the human species, and these patterns certainly cannot be acquired without experience. Perhaps stereotyped behavior patterns of animals also require subtle forms of experience for development' (p. 241). Hailman meticulously demonstrated the influence of such subtle forms of experience through his investigations of pecking in newly hatched sea gulls.

> Hailman's perspective is a forerunner to today's *developmental systems* approach to the origins of abilities, traits, and behaviors. The striking observation that guides the developmental systems approach is that *processes*—sometimes obvious, sometimes subtle—give rise to the emergent properties of each individual's behavior. DNA plays a critical role in these processes, but does not by itself create traits. Accordingly, instincts are not preprogrammed, hardwired, or genetically determined; rather, they emerge each generation through a complex cascade of physical and biological influences. (p. 2)

Lorenz (1937a, 1965), however, does not agree with the RDS-based recasting of the concept of instinct; instead, Lorenz argues that experience over the course of an organism's life (its ontogeny) had no role in the shaping the development of a presumed neural structure that enabled the IRM to occur. Instead, the key, innate (instinctual) feature of such a neural structure was "its ability to select, from the range of available possible stimuli, the one which specifically elicits its activity, and thus the response seen by the observer" (Lehrman, 1970, p. 24). The response to the innate structure was an FAP.

The classic example of a FAP involves the male three-spined stickleback fish (Lorenz, 1965). When this fish encounters another male three-spined stickleback with a red belly, the fish displays a set of behaviors indicative of threat. In contrast, when the fish encounters a female with a swollen (but non-red) belly, the male displays the behavior pattern indicative of mating.

However, a problem with the foundational argument and definition of the FAP exists. Lorenz admitted that if the appropriate releasing stimulus was not encountered for some period of time, then the FAP could occur spontaneously. That is, it "might go off in vacuo, as if dammed energy burst through containing valves" (Richards, 1987, p. 531).

It seemed obvious to Lorenz (1965) that the FAP with the three-spined sticklebacks was a behavior clearly shaped by evolution, given what he saw as the importance of the threat or mating displays by the male fish for, respectively, warding off competitors for female fish and for engagement with a possible mate if a competitor for the mate swam away in the face of the FAP. However, the problematic facet of this and other examples of FAPs (Lehrman, 1970; Richards, 1987), which in effect might be termed an evolved probabilistic *behavioral* mechanism (EPBM), is the spontaneous enactment of the behavior. This spontaneous behavior would occur with no evolutionary-relevant stimulus (a male or female conspecific) in view to engage the purported innate neural structure housing the IRM. Thus, the purported phylogenetic antecedent that explained the EPBP only explained it in some cases, at some times, in some contexts.

Similarly, proponents of EDP can only say that if the EPCM occurs normatively, then the context was species typical. If the EPCM is not expressed normatively then this manifestation of behavior is taken as proof for the existence of a context (and typically a *never measured* context) that was atypical. Like Skinner (1971), in regard to the postulation that a relevant (releasing) S must have been present if one saw an R (Bowers, 1973), the outcome in the formulation of EDP, the appearance (or lack thereof) of the EPCM, explains the antecedent that purportedly explains the outcome!

Thus, the ideas of the proponents of EDP (e.g., Bjorklund, 2015, 2016; Bjorklund & Ellis, 2005; Giudice & Ellis, 2016) converge with those involved in other essentialist formulations. Whether developmental scientists are discussing EPCMs or EPBMs, they will confront the same problem: Neither concept is linked to a developmental process that identifies the essentialist "mechanisms" independent of its ontogenetic emergence or their subsequent display in ontogeny after its emergence. Developmental scientists cannot say that these homunculi, EPCMs, EPBMs, FAPs, or whatever, are always there, independent of context, because the context instantiates them (or does not instantiate them as the case may be). However, the instantiation by the context will be different under different environmental conditions (e.g., species-typical vs. atypical, or red underbelly present vs. red underbelly absent), and therefore the whatever (e.g., the homunculus or EPCM) cannot be known to exist in any form without the context.

Conclusions

There is no way of knowing the purported evolutionary-based "whatever" independent of the ontogenetic context! Simply, then, and at best, the foundational concept of EDP (the EPCM), the essentialist approach to the complexity of human development, is entirely non-empirical and gratuitous. At worst, the concept of EPCM is so fraught with logical, conceptual, and empirical problems that using it as a basis for research in human development is a scholarly dead end.

The objections that proponents of EDP have to RDS-based approaches to the complexity of human development (e.g., Bjorklund, 2016) reflects at best a lack of understanding of the dynamics of the relational developmental system (e.g., see Raeff, 2016; Witherington & Lickliter, 2016). Whatever the basis of the objections of proponents of EDP, however, their views have resulted in their invention of the equivalent of a homunculus to explain—or, perhaps better, to try to explain away—the holistic and autopoietic features of the relational developmental system.

An Overview of the Contributions to This Section

The contributions presented in this section begin with an excerpt written by the renowned comparative psychologist, Gilbert Gottlieb; it is the final chapter of his book, *Synthesizing nature-nurture: Prenatal roots of instinctive behavior* (Gottlieb, 1997). In this piece, Gottlieb explains that a solely genetic interpretation of an individual's structure (morphological features) or function (an individual's physiological, psychological, behavioral, and social attributes) is too narrow. Both phylogeny and ontogeny involve the integration (the fusion across time and place) of both bottom-up processes (genetic and cell-to-cell coactions) and top-down processes (psychological, behavioral, and contextual influences (see too Witherington, 2011, 2014, 2015; Witherington & Heying, 2013; Witherington & Lickliter, 2016). Gottlieb's concluding point is that:

> To fill the enormous gap between molecular neurobiology and psychological function, it will also be necessary to have mutual respect for top-down as well as bottom-up approaches—it is not enough to understand how the brain develops from a sheerly internalist perspective.
>
> (Gottlieb, 1997, p. 163)

Jablonka and Lamb (2007) note that, although Darwin recognized that life circumstances contribute to any variation among individuals in the hereditary influences, the evolution of theory about evolution became centered on causation due to random gene mutations. Jablonka and Lamb explain advances in evolutionary biology indicate that the complete reliance on random variation of genes to account for evolution is inadequate. The context or individuals moderates evolutionary change. Failure to recognize this contextual influence in theories of evolution obviates the usefulness of such conceptions because they omit consideration of data from disciplines from molecular biology to cultural studies, showing that genes are influenced by environmental variation. They also note that the transmission of variation across generations is not completely the outcome of genetic variation. Their book, *Evolution in Four Dimensions* (2005), presents evidence for four dimensions of evolution: Genetic, epigenetic, behavioral, and symbol-based. All dimensions coact in the production and generation of cross-generation transmission, and for humans, the fourth dimension—which involves the symbol systems and institutions of culture—is uniquely and particularly important.

Lickliter (2016) reflects the ideas discussed by Jablonka and Lamb and underscores that points that scholarship in evolutionary biology reflects a burgeoning understanding of the complex dynamics of genetic functioning and expression and of the probabilistic nature of development across time and place. Lickliter emphasizes that a dynamic, relational developmental systems approach to evolution is at the cutting edge of contemporary biological and social/behavioral science.

Building on the changing evidence base in the study of evolutionary biology that is discussed by Gottlieb (1997), Jablonka and Lamb (2005), and Lickliter (2016), Narvaez et al. (2022) discuss how evolutionary theory and the study of psychological development might have a mutually informative association. They contrast what they describe as "Narrow Evolutionary Psychology" (NEP), which proponents claim constitutes a scientific revolution akin to a Copernican or Darwinian paradigm shift involving a neo-Darwinian adaptationist framework within evolutionary biology, and the computationalist "mind-as-computer" framework within cognitive science, with a more inclusive, developmental evolutionary psychology theory (DEPTH). The latter approach builds on the role that epigenesis plays in the course of human development and derives ideas and evidence for dynamic coactions from the fields of developmental neuroscience, anthropology, and cognitive archeology. Narvaez explains that the DEPTH approach has the added value framing applied scholarship that can address and contribute to remedying the challenges facing human life by recognizing the fundamental significance of the relative plasticity of human behavior and development and the significance of mutually influential individual-context coactions across time and place.

Noble (2015) underscores the fundamental importance of revising or replacing the Modern Synthesis (neo-Darwinist) theory of evolution that has been in place for about 100 years at this writing. As emphasized in all contributions in this section, Noble explains that this revision in evolutional biology must occur to align the field with contemporary findings from several scientific fields. In addition, however, concepts that obscure the process of evolution and its implications for behavior and development must be dropped or substantially revised. He focuses on such Neo-Darwinist concepts as gene, selfish, code, program, blueprint, book of life, replicator, and vehicle, explaining that such terms confuse conceptual and empirical issues. For instance, he points out that the concept of "gene" has changed from denoting a necessary cause (operationalized in regard to an inheritable phenotype) to becoming only an empirically testable hypothesis (in regard to causation by DNA sequences).

Noble proposes an alternative conceptual framework that eliminates the problematic concepts such as those he discussed and offers an integrated, that is, a dynamic systems view of evolution. As such, then, Noble's conceptions align with the ideas of Gottlieb (1997), Jablonka and Lamb (2007), Lickliter (2016), and Narvaez et al. (2022) presented in this section.

References

Bateson, P. (2015). Ethology and human development. In W. F. Overton & P. C. Molenaar (Eds.), *Theory and method. Volume 1 of the handbook of child psychology and developmental science* (7th ed., pp. 208–243), Editor-in-chief: R. M. Lerner. Wiley.

Bateson, P. (2016). Robustness and plasticity. *WIREs Cognitive Science*. https://doi.org/10.1002/wcs.1386

Belsky, J. (2012). The development of human reproductive strategies: Progress and prospects. *Current Directions in Psychological Science, 21*(5), 310–316.

Belsky, J., Steinberg, L., & Draper, P. (1991). Childhood experience, interpersonal development, and reproductive strategy: An evolutionary theory of socialization. *Child Development, 62*, 647–670.

Bjorklund, D. F. (2015). Developing adaptations. *Developmental Review, 38*, 13–35.

Bjorklund, D. F. (2016). Prepared is not preformed: Commentary on Witherington and Lickliter. *Human Development, 59*(4), 235–241.

Bjorklund, D. F., & Bering, J. M. (2002). The evolved child: Applying evolutionary developmental psychology to modern schooling. *Learning and Individual Differences, 12*(4), 347–373.

Bjorklund, D. F., & Ellis, B. J. (2005). Evolutionary psychology and child development: An emerging synthesis. In B. J. Ellis & D. F. Bjorklund (Eds.), *Origins of the social mind: Evolutionary psychology and child development* (pp. 3–18). Guilford.

Bjorklund, D. F., & Shackelford, T. K. (1999). Differences in parental investment contribute to important differences between men and women. *Current Directions in Psychological Science, 8*(3), 86–89.

Blumberg, M. S. (2016). Development evolving: The origins and meanings of instinct. *WIREs Cognitive Science*. https://doi.org/10.1002/wcs.1371

Bowers, K. S. (1973). Situationism in psychology: An analysis and a critique. *Psychological Review, 80*, 307–336.

Bronfenbrenner, U. (2005). *Making human beings human: Bioecological perspectives on human development*. Sage.

Bronfenbrenner, U., & Morris, P. A. (2006). The bioecological model of human development. In W. Damon, & R. M. Lerner (Eds.), & R. M. Lerner (Vol. Ed.), *Handbook of child psychology: Vol. 1. Theoretical models of human development* (6th ed., pp. 793–828). Wiley.

Buss, D. M. (2019a). *Evolutionary psychology: The new science of the mind*. Taylor & Francis.

Buss, D. (2019b). *Evolutionary psychology: The new science of the mind*. Taylor & Francis.

Buss, D. M., & von Hippel, W. (2018). Psychological barriers to evolutionary psychology: Ideological bias and coalitional adaptations. *Archives of Scientific Psychology, 6*(1), 148–158.

Chess, S., & Thomas, A. (1984). *The origins and evolution of behavior disorders: Infancy to early adult life*. Brunner/Mazel.

Chess, S., & Thomas, A. (1999). *Goodness of fit: Clinical applications from infancy through adult life*. Brunner/Mazel.

de Beer, G. R. (1959). Paedomorphosis. *Proceedings of the XV International Congress of Zoology, 15*, 927–930.

Del Giudice, M., & Ellis, B. J. (2016). Evolutionary foundations of developmental psychopathology. In D. Cicchetti (Ed.), *Developmental psychopathology*, Vol. 12: *Developmental neuroscience* (3rd. ed., pp. 1–58). Wiley.

Draper, P., & Harpending, H. (1982). Father absence and reproductive strategy: an evolutionary perspective. *Journal of Anthropological Research, 38*, 255–273.

Draper, P., & Harpending, H. (1988). A sociobiological perspective on the development of human reproductive strategies. In K. B. MacDonald (Ed.), *Sociobiological perspectives on human development* (pp. 340–372). Springer-Verlag.

Elder, G. H., Jr., Shanahan, M. J., & Jennings, J. A. (2015). Human development in time and place. In M. H. Bornstein & T. Leventhal (Eds.), *Handbook of child psychology and developmental science, vol. 4: Ecological settings and processes in developmental systems* (7th ed., pp. 6–54), Editor-in-chief: Richard M. Lerner. Wiley.

Ellis, B. J., Schlomer, G. L., Tilley, E. H., & Butler, E. A. (2012). Impact of fathers on risky sexual behavior in daughters: A genetically and environmentally controlled sibling study. *Development and Psychopathology, 24*, 317–332.

Featherman, D. L., & Lerner, R. M. (1985). Ontogenesis and sociogenesis: Problematics for theory about development across the lifespans. *American Sociological Review, 50*, 659–676.

Fisher, H. E. (1982a). Of human bonding. *The Sciences, 22*, 18–23, 31.

Fisher, H. E. (1982b). Is it sex? Helen E. Fisher replies. *The Sciences, 22*, 2–3.

Gissis, S. B., & Jablonka, E. (Eds.). (2011). *Transformations of Lamarckism: From subtle fluids to molecular biology*. MIT Press.

Gottlieb, G. (1970). Conceptions of prenatal behavior. In L. R. Aronson, E. Tobach, D. S. Lehrman, & J. S. Rosenblatt (Eds.), *Development and evolution of behavior: Essays in memory of T. C. Schneirla* (pp. 111–137). Freeman.

Gottlieb, G. (1997). *Synthesizing nature-nurture: Prenatal roots of instinctive behavior*. Psychology Press.

Gottlieb, G. (1998). Normally occurring environmental and behavioral influences on gene activity: From central dogma to probabilistic epigenesis. *Psychological Review, 105*, 792–802.

Gottlieb, G. (2002/2014). *Individual development and evolution: The genesis of novel behavior*. Psychology Press.

Gottlieb, G. (2004). Normally occurring environmental and behavioral influences on gene activity: From central dogma to probabilistic epigenesis. In C. Garcia Coll, E. Bearer, & R. M. Lerner (Eds.), *Nature and nurture: The complex interplay of genetic and environmental influences on human behavior and development* (pp. 85–106). Erlbaum.

Gould, S. J. (1977). *Ontogeny and phylogeny*. Belknap Press of Harvard University Press.

Gould, S. J. (1981). *The mismeasure of man*. Norton.

Greenberg, G., & Mosack, V. (2004). Recapitulation. In Fisher, C. B. & Lerner, R. M. (Eds). *Encyclopedia of Applied Developmental Science*, Sage Publications (pp. 909–911).

Haeckel, E. (1868). *Naturliche Schopfungsgeschichte*. Georg Reimer.

Hailman, J. P. (1969). How an instinct is Learned. *Scientific American, 221*, 98–106.

Hall, G. S. (1904). *Adolescence: Its psychology and its relations to psychology. anthropology, sociology, sex, crime, religion, and education*. Appleton.

Harley, D. (1982). Models of human evolution. *Science, 217*, 296.

Ho, M. W. (2010). Development and evolution revisited. In K. E. Hood, C. T. Halpern, G. Greenberg, & R. M. Lerner (Eds.), *Handbook of developmental systems, behavior and genetics* (pp. 61–109). Wiley Blackwell.

Ho, M. W. (2013). No genes for intelligence in the fluid genome. In R. M. Lerner & J. B. Benson, (Eds.), *Advances in Child Development and Behavior: Embodiment and epigenesis: Theoretical and methodological issues in understanding the role of biology within the relational developmental system. Part B. Ontogenetic dimensions* (pp. 67–92). Elsevier.

Ho, M. W., & Saunders, P. T. (Eds.). (1979). Beyond neo-Darwinism – An epigenetic approach to evolution. *Journal of Theoretical Biology, 78*, 573–591.

Hoffman, R. F. (1978). Developmental changes in human infant visual-evoked potentials to patterned stimuli recorded at different scalp locations. *Child Development, 49*(1), 110–118.

Hogan, R., Johnson, J. A., & Emler, N. P. (1978). A socioanalytical theory of moral development. *New Directions for Child Development, 2*, 1–18.

Isaac, G. L. (1982). Models of human evolution. *Science, 217*, 295.

Jablonka, E., & Lamb, M. (2005). *Evolution in four dimensions: Genetic, epigenetic, behavioral, and symbolic variation in the history of life*. MIT Press.

Jablonka, E., & Lamb, M. (2007). Précis of evolution in four dimensions. *Behavioral and Brain Sciences, 30*(4), 353–365. https://doi.org/10.1017/S0140525X07002221

Johanson, D. C., & Edey, M. A. (1981). *Lucy: The beginnings of humankind*. Simon & Schuster.

Lehrman, D. S. (1970). Semantic and conceptual issues in the nature-nurture problem. In L. R. Aronson, E. Tobach, D. S. Lehrman, & J. S. Rosenblatt (Eds.), *Development and evolution of behavior: Essays in memory of T. C. Schneirla* (pp. 17–52). Freeman.

Lerner, J. V., & Lerner, R. M. (1983). Temperament and adaptation across life: Theoretical and empirical issues. In P. B. Baltes & O. G. Brim, Jr. (Eds.), *Life-span development and behavior* (Vol. 5, pp. 197–231). Academic Press.

Lerner, J. V., & Lerner, R. M. (1989). Introduction: Longitudinal analyses of biological, psychological, and social interactions across the transitions of early adolescence. *Journal of Early Adolescence, 9*, 175–180.

Lerner, R. M. (1982). Children and adolescents as producers of their own development. *Developmental Review, 2*, 342–370.

Lerner, R. M. (1992). *Final solutions: Biology, prejudice, and genocide*. Penn State Press.

Lerner, R. M. (2018). *Concepts and theories of human development* (4th ed.). Routledge.

Lerner, R. M., Theokas, C., & Jelicic, H. (2005). Youth as active agents in their own positive development: A developmental systems perspective. In W. Greve, K. Rothermund, & D. Wentura (Eds.), *The adaptive self: Personal continuity and intentional self-development* (pp. 31–47). Hogrefe & Huber Publishers.

Lester, B. M., Conradt, E., & Marsit, C. (2016). Introduction to the special section on epigenetics. *Child Development, 87*, 29–37.

Lewis, M. D. (2000). The promise of dynamic systems approaches for an integrated account of human development. *Child Development, 71*, 36–43.

Lewontin, R. C., & Levins, R. (1978). Evolution. *Encyclopedia Einaudi* (Vol. 5). Einaudi.

Lickliter, R. (2016). Developmental evolution. *WIREs Cognitive Science, 8*(1–2), e1422. https://doi.org/10.1002/wcs.1422

Lickliter, R., & Honeycutt, H. (2015). Biology, development, and human systems. In W. F. Overton & P. C. M. Molenaar (Eds.), *Handbook of child psychology and developmental science, volume 1: Theory and method* (7th ed., pp. 162–207), Editor-in-chief: R. M. Lerner. Wiley.

Lickliter, R., & Witherington, D. C. (2017). Towards a truly developmental epigenetics. *Human Development, 60*, 124–138.

Lorenz, K. (1937a). The companion in the bird's world. *Auk, 54*, 245–273.

Lorenz, K. (1937b). Über den Begriff der Instinkthandlung, *Folia Biotheoretica, 2*, 17–50.

Lorenz, K. (1965). *Evolution and modification of behavior*. University of Chicago Press.

Lovejoy, C. O. (1981). The origin of man. *Science, 211*, 341–350.

Mascolo, M. (2013). Developing through relationships: An embodied coactive systems framework. In R. M. Lerner & J. B. Benson (Eds.), *Advances in child development and behavior: Embodiment and epigenesis: Theoretical and methodological issues in understanding the role of biology within the relational developmental system. Part B. Ontogenetic dimensions* (pp. 185–225). Elsevier.

Masters, R. D. (1978). Jean-Jacques is alive and well: Rousseau and contemporary sociobiology. *Daedalus, 107*, 93–105.

Meaney, M. (2010). Epigenetics and the biological definition of gene x environment interactions. *Child Development, 81*, 41–79.

Meaney, M. (2014). Epigenetics offer hope for disadvantaged children. *Child & Family Blog*. https://childandfamily-blog.com/epigenetics-offer-hope-disadvantaged-children/

Moore, D. S. (2015a). *The developing genome: An introduction to behavioral epigenetics*. Oxford University Press.

Moore, D. S. (2015b). The asymmetrical bridge. Book review of James Tabery's Beyond versus: The struggle to understand the interaction of nature and nurture. *Acta Biotheoretica, 63*(4), 413–427.

Moore, D. S. (2016). Behavioral epigenetics. *WIREs Systems Biology and Medicine, 9*, e1333. https://doi.org/10.1002/wsbm.1333

Narvaez, D., Moore, D. S., Witherington, D. C., Vandiver, T. I., & Lickliter, R. (2022). Evolving evolutionary psychology. American Psychologist, 77(3), 424–438.

Noble, D. (2015). Evolution beyond neo-Darwinism: A new conceptual framework. *Journal of Experimental Biology, 218*, 7–13.

Overton, W. F. (2015). Process and relational developmental systems. In W. F. Overton & P. C. M. Molenaar (Eds.), *Handbook of child psychology and developmental science, Volume 1: Theory and method* (7th ed., pp. 9–62), Editor-in-chief: R. M. Lerner. Wiley.

Raeff, C. (2016). *Exploring the dynamics of human development: An integrative approach*. Oxford University Press.

Richards, R. (1987). *Darwin and the emergence of evolutionary theories of mind and behavior*. University of Chicago Press.

Richardson, M. K., & Keuck, G. (2002). Haeckel's ABC of evolution and development. *Biological Review, 77*, 495–528.

Richardson, R. C. (2007). *Evolutionary psychology as mas maladapted psychology*. MIT Press.

Rose, H., & Rose, S. (2000). Introduction. In H. Rose & S. Rose (Eds.), *Alas Poor Darwin: Arguments against evolutionary psychology* (pp. 1–13). Vintage.

Sahlins, M. D. (1978). The use and abuse of biology. In A. L. Caplan (Ed.), *The sociobiology debate*. Harper & Row.

Schneirla, T. C. (1957). The concept of development in comparative psychology. In D. B. Harris (Ed.), *The concept of development: An issue in the study of human behavior* (pp. 78–108). University of Minnesota Press.

Skinner, B. F. (1938). *The behavior of organisms*. Appleton.

Skinner, B. F. (1971). *Beyond freedom and dignity*. Knopf.

Swartz, D. (1982). Is it sex? *The Sciences, 22*, 2.

Thomas, A., & Chess, S. (1977). *Temperament and development*. Brunner/Mazel.

Tobach, E. (1981). Evolutionary aspects of the activity of the organism and its development. In R. M. Lerner & N. A. Busch-Rossnagel (Eds.), *Individuals as producers of their development: A life-span perspective* (pp. 37–68). Academic Press.

Tobach, E., & Schneirla, T. C. (1968). The biopsychology of social behavior of animals. In R. E. Cooke & S. Levin (Eds.), *Biologic basis of pediatric practice* (pp. 68–82). McGraw-Hill.

van Geert, P., & Fischer, K. W. (2009). Dynamic systems and the quest for individual-based models of change and development. In J. P. Spencer, M. S. C. Thomas, & J. L. McClelland (Eds.), *Toward a new grand theory of development: Connectionism and dynamic systems theory re-considered: Connectionism and dynamic systems theory reconsidered* (pp. 313–336). Oxford University Press.

Washburn, S. L. (1961). *Social life of early man*. Wenner-Gren Foundation for Anthropological Research.

Washburn, S. L. (1982). Is it sex? *The Sciences, 22*, 2.

Witherington, D. C. (2011). Taking emergence seriously: The centrality of circular causality for dynamic systems approaches to development. *Human Development, 54*, 66–92.

Witherington, D. C. (2014). Self-organization and explanatory pluralism: Avoiding the snares of reductionism in developmental science. *Research in Human Development, 11*(1), 22–36.

Witherington, D. C. (2015). Dynamic systems in developmental science. In W.F. Overton & P.C. Molenaar (Eds.), *Handbook of child psychology and developmental science. Volume 1: Theory and method* (7th ed., pp. 63–112), Editor-in-chief: R. M. Lerner. Wiley.

Witherington, D. C., & Heying, S. (2013). Embodiment and agency: Toward a holistic synthesis for developmental science. In R. M. Lerner & J. B. Benson, (Eds.), *Advances in Child Development and Behavior: Embodiment and epigenesis: Theoretical and methodological issues in understanding the role of biology within the relational developmental system. Part A. Philosophical, theoretical, and biological dimensions* (pp. 161–192). Elsevier.

Witherington, D. C., & Lickliter, R. (2016). Integrating development and evolution in psychological science: evolutionary developmental psychology, developmental systems, and explanatory pluralism. *Human Development, 59,* 200–234.

9.
TOWARD A NEW DEVELOPMENTAL AND EVOLUTIONARY SYNTHESIS*
Gilbert Gottlieb[†]

Just as I was wrapping up this work I became aware of an odd convergence—from cognitive psychology and from developmental biology—that stimulated me to think about what the future might hold. Strange as it may seem, prominent authors from the fields of cognitive psychology and developmental neurobiology were homing in independently on the same conclusion; namely, that the idea of a sheerly genetic determination of form and function is too narrow and should be replaced by the broader concept of an embryological "morphogenetic field," which includes genes but goes beyond them to include cell-cell coactions as well. A sheerly genetic determination would thus be replaced by a cellular neurobiological determination. This is an improvement, of course, but it ignores the feed downward influences from the environment and behavior. It is still a strictly bottom-up approach to neurobiological and psychological development.

I first describe and comment on the specific proposals from cognitive psychologists and then do the same for developmental biologists.

Developmental Cognitive Psychology

The cognitive psychologists want to redefine *innate* in light of their appreciation of the relevance of biology, particularly neurobiology, to developmental

---

* Gottlieb, G. (1997). Toward a new developmental and evolutionary synthesis. In G. Gottlieb. *Synthesizing nature-nurture: Prenatal roots of instinctive behavior* (pp. 143–163). Psychology Press.
† University of North Carolina at Chapel Hill

psychology. To my mind, they have misappropriated the concept of innateness in the service of their discovery of neurobiology. They don't have a theory of development unless they can use the innateness concept, which, in their hands, represents a too-narrow appreciation of the contribution of neurobiology to behavior/psychology, in the sense of biologically inspired predispositions or biases (Spelke & Newport, in press) or biological constraints (Elman et al., 1996). The latter define *innate* "to refer to putative aspects of brain structure, cognition or behavior that are the product of interactions internal to the organism … this usage of the term does not correspond to genetic or coded in the genes." Their Table 1.2 specifies that innate outcomes are the consequence of molecular and cellular interactions.

Because Spelke and Newport (in press) use the usual narrow definition of experience (obvious contributory organism-environment encounters) and not the broader definition of functional activity advocated here, they actually end up with the old notion of innate (= not only independent of frank learning but independent of experience). For Spelke and Newport, the resolution of the age-old innate (nature)-acquired (nurture) dichotomy is to accept it as valid, not in the sense that innate = genetically determined but, rather, innate = biologically predisposed, a higher order, morphogenetic field concept. The resolution I have opted for in this volume is to accept that certain developmental outcomes are species-typical or species-specific, adaptive, and responsive to a narrow class of stimulation in the absence of prior exposure to these configurations (independent of frank learning but not independent of experience, broadly defined). Who would have dreamed that squirrel monkeys' innate fear of snakes could derive from their experience with live insects (Masataka, 1994)? Or that chicks' perception of mealworms as food items depends on their having seen their own toes move (Wallman, 1979)?

Therefore, in the place of frank associationistic (S-R, S-S) learning, I hold open the likelihood of nonobvious experiential background factors as indispensable contributors to development. In this scheme, so-called biological predispositions or constraints are seen in a broader coactional context.

To my mind, it is only one step up from genetic determination to cite cellular-biological predispositions as sufficient causes of psychological outcomes. There remain nonobvious organism-environment experiences and two-way traffic in the developmental-psychobiological system whether one cares to accept them or not. That is the appropriate resolution of the nature-nurture dichotomy, to my way of thinking. Given the empirical demonstrations of the nonobvious prenatal experiential canalization of species-specific behavior described earlier, and the sensory stimulation of gene expression reviewed earlier, biologically encapsulated definitions of innate or instinctive predispositions or constraints (Elman et al., 1996) are no longer tenable. This is all the more true for human cognitive development, given that all of the sensory systems are capable of function prior to birth in humans (Gottlieb, 1971b). Gene expression is being activated by sensory stimulation, as well as other factors, in the fetal period. Behavioral embryological approaches to human development may also reveal organism-environment coactions that are relevant to the psychological adaptations of the infant (e.g., prenatal auditory experience leading to the newborn's selective auditory response to its mother's voice: DeCasper & Spence, 1986). Because Elman et al. are interested in language development, it is a pity that they do not recognize the prenatal auditory experience of speech in the human fetus as a developmental-psychobiological constraint. This issue is renewed more broadly by Locke (1993) and by Turkewitz (1988).

Developmental and Evolutionary Biology

A genuine paradigm shift seems to be gradually taking hold in the study of developmental biology, and it has implications for the study of evolution. Although I recognize the shift as momentous, I feel that it does not go quite far enough, most likely because of limitations in training, self-learning, and expertise that we all share.

For many years the evolutionary changes in physiology, morphology, behavior, and psychology have been ascribed to changes in gene frequencies in populations of interbreeding individuals, meaning that a change in the genetic composition of a population was believed to be required for an evolutionary change in a phenotype. This view of genetic change—phenotypic change jumps right over development. A few brave souls, who were significantly out of step with the times, speculated that changes in genes brought out opportunities for novel coactions during embryonic development, and it was the consequence of these new developmental coactions that brought about changes in morphology that were recognized as evolutionary changes (e.g., de Beer, 1930, 1958; Garstang, 1922; Goldschmidt, 1933, 1952). That view has now been elaborated, with considerably more empirical support, in the hands of a small number of biologists who see the embryological "morphogenetic field" as the appropriate focus of both developmental and evolutionary understanding (e.g., Edelman, 1992; Gilbert et al., 1996; Holliday, 1990).

This is a substantial change in the focal level of analysis, moving from a primary focus on genes to cell-cell coactions, which of course includes the genes but goes several levels beyond (above) them. Although the coactions are usually seen as vertically and horizontally bidirectional, the biologists, as do the cognitive psychologists, stop short of including behavior and the external environment as part of their developmental systems analysis. An important consideration in understanding this shortcoming is that no one is yet competent to move comfortably (i.e., knowledgeably) between levels. This is why broader interdisciplinary and multidisciplinary collaborations will be necessary in the future if behavior and psychological functioning are to become a genuine part of the new developmental and evolutionary synthesis.

Given the present context (the newness of the emancipation from a gene-centered focus to the cellular level in developmental cognitive psychology, and in evolutionary and developmental biology), it is perhaps understandable that the present new synthesis must remain for the time being a strictly bottom-up approach as far as a neural, behavioral, and psychological understanding are concerned. When George Miklos (1993, p. 851) wrote, "Control of the genome means control of development, and control of development means control of behaviors," he declared a strategy for an experimental program of research that, in my opinion, will one day have to be joined by a program of study moving in the opposite direction if we are to achieve an adequate understanding of the development and evolution of genes, nervous systems, and behaviors. It is now known that light induces changes in the circadian rhythms of behavioral activity in fruit flies by affecting gene expression of the proteins that are implicated in setting the circadian clock (Lee, Parikh, Itsukaichi, Bae, & Edery, 1996; Myers, Wager-Smith, Rothenfluh-Hilfiker, & Young, 1996). That should alert even the most conservative bottom-up person to the reality of top-down influences during the normal course of development.

To try to lend a hand in what one day will surely be a magnificent synthesis, I offer the following brief recap of a macro developmental, behaviorally driven scheme of evolutionary change, a top-down approach meant to mesh with the bottom-up approach (Gottlieb, 1992).

The Relationship of Development to Evolution

The insightful concept that changes in individual development are the basis for evolution was raised originally by St. George Mivart (1871) in his book *On the Genesis of Species*. William Bateson (1894) favored the idea that phenotypic variation was developmentally inspired, but little was done to further work out the details until Walter Garstang (1922) and Gavin de Beer (1930, 1958) delivered their respective coups de grâce to Ernst Haeckel's recapitulation doctrine, and, from another side entirely, Richard Goldschmidt (1933, 1952) hypothesized that changes in early embryonic development would be necessary for evolution to occur. Although Garstang and de Beer were interested in showing the importance of various kinds of ontogenetic changes to evolution generally, Goldschmidt, having become convinced of the impossibility of neo-Darwinian microevolution producing a new species, had come to view developmental macromutation as essential to the production of the large differences necessary for speciation.

The foregoing scientists were the principal ones to establish the developmental basis of evolutionary change. The view that I wish to propose in this chapter builds on their pioneering insight, but it is different in a very important way. To be specific, Garstang, de Beer, and Goldschmidt, in agreement with the proponents of the modern synthesis, believed that a genetic change or mutation is necessary to bring about the developmental changes that lead to evolution. The point I wish to advocate is that there is so much untapped potential in the existing developmental system (including the genes) that evolution can occur without changing the genetic constitution of a population. Such changes may eventually lead to a change in genes (or gene frequencies), but evolution will have already occurred at the phenotypic level before the genetic change occurs. According to the present viewpoint, genetic change is a secondary or tertiary consequence of enduring behavioral changes brought about initially by supragenetic alterations of normal or species-typical development.

For those readers that were taught (or learned on their own) that changes in gene frequencies drive evolutionary changes in morphology, it may come as a surprise that there is no relationship

Toward a New Developmental and Evolutionary Synthesis

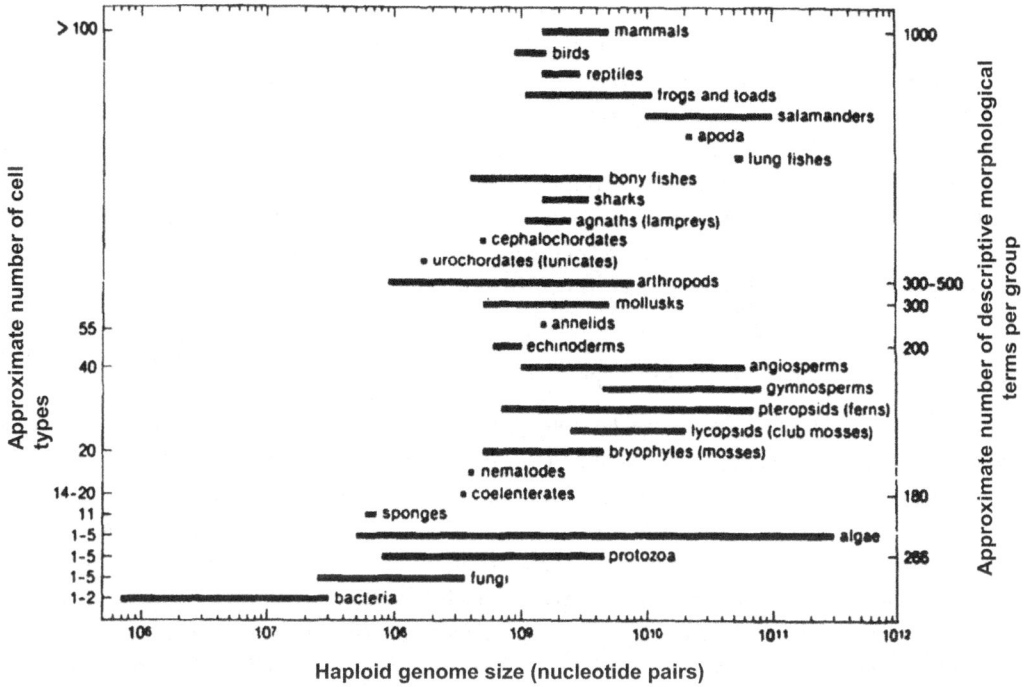

Fig. 9.1 The C-value paradox: absence of a relationship between genome size and morphological complexity. The bars show the ranges of genome sizes for various categories of organisms. The somewhat subjective ordering of categories is from morphologically most simple at the bottom to most complex at top. Two estimates of complexity are given: Approximate numbers of cell types in the body of some groups are indicated on the left vertical axis, and approximate numbers of morphological descriptive terms for certain groups are indicated on the right vertical axis. C-value data from Sparrow, price, and Underbrink (1972). Figure from Raff and Kaufman (1983), Reprinted by permission.

between morphological complexity and genome size. If evolution is entirely gene based, one would have thought it would take more genes to build more complex bodies, but that is not the case, as can be seen in Fig. 9.1. If there were a relationship between genome size and morphological complexity, you would see an increase in genome size moving from the lower left of the diagram to the upper right of the diagram in a more or less straight line.

For our present developmental psychobiological purposes, it is even more relevant to know that there is no relationship between the number of genes coding for protein and the number of neurons in the nervous system. As shown in Table 9.1, although mice and humans each have approximately 70,000 coding genes, mice have

Table 9.1 Approximate Number of Genes and Neurons in Different Lineages

	Genes	Neurons
Chordates		
Mus musculus	70,000	40 million
Homo sapiens	70,000	85 billion
Nematodes		
Caenorhabditis Elegans	14,000	302
Arthropods		
Drosophila Melanogaster	12,000	250,000

Source: From Miklos and Edelman (1996). Reprinted by permission. The actual number of neurons in *C. elegans* is known to be 302. Otherwise, the other figures in this table are approximations.

approximately 40 million neurons and humans have about 85 *billion* neurons. At the invertebrate level, while the round worm (*Caenorhabdhitis elegans*) and the fruit fly (*Drosophila melanogaster*) have around 12,000–14,000 coding genes, the roundworm has 302 neurons and the fruit fly 250,000 neurons. Thus, there is good reason to seek the answer to evolution above the level of the genes, in the total developmental system.

The Induction of Behavioral Neophenotypes

In a book closing out his underappreciated but otherwise illustrious research career as a broadly based developmental scientist, Zing-Yang Kuo (1976) coined the term *behavioral neophenotype* to refer to momentous behavioral changes or deviations from normality that could be brought into existence by altering the usual conditions of an animal's early development or experience. Kuo's purpose for advocating the creation of behavioral novelties was to show that species-typical behavior was rather more highly modifiable than anyone believed and not rigidly or narrowly fixed by genetic constraints. To make his point, Kuo did such things as "create" a male dog that had no reproductive interests in female dogs in heat and, further, actively prevented other males from engaging in sexual behavior with such females. Kuo chose to tamper with reproductive behavior to make his point all the more telling: The usual or normal behavioral predilection for male dogs to copulate with females in heat (and thereby perpetuate the species) is a result of their having been exposed to usual or normal developmental conditions, not to instincts dictated solely by their genetic endowment. Kuo's argument that the establishment of behavioral neophenotypes by altering developmental circumstances should be one of the major aims of experimental animal psychology has not caught on because it seemed to many to be nonbiological. I hope that by supplying a broader rationale the significance of behavioral neophenotypes will be better appreciated.

If my presentation of the evidence to this point has been persuasive, the reader should be convinced that internal and external coactions during individual development create the resulting phenotype. This will, of course, be just as true for behavior as for anatomy or physiology. According to the viewpoint being developed in this chapter, the ease or difficulty of creating behavioral neophenotypes, and the directions of most ready behavioral change, would allow us to assess the immediately present evolutionary potential of a species. Naturally, we will expect to find that some species possess much greater immediate behavioral malleability or plasticity than other species (e.g., wood ducklings vs. mallards), and that in itself will be informative about the pace and range of immediate evolutionary potential in those species.

But I am getting a little bit ahead of my story. I am certainly not being original in suggesting that behavioral innovations lead the way to evolutionary change. A number of biologists of various persuasions have resuscitated Lamarck's notion of the centrality of behavioral change to evolution (e.g., Bonner, 1983; Hardy, 1965; Larson, Prager, & Wilson, 1984; Leonovicova & Novak, 1987; Mayr, 1982; Piaget, 1978; Plotkin, 1988; Reid, 1985; Sewertzoff, 1929; Wyles, Kunkel, & Wilson, 1983). What is new and not yet widely appreciated is the supragenetic means (neophenogenetic pathway described next) by which normal or usual development can be altered so as to produce a behavioral neophenotype that is likely to lead to evolutionary change. (In an excellent chapter on this topic, P. Bateson [1988] has come to much the same conclusion.)

In essence, as is widely recognized, what needs to happen to bring about evolution is the production of animals that live differently from their forebears. Living differently, especially living in a different place, will subject the animals to new stresses, strains, and adaptations that will eventually alter their anatomy and physiology (without necessarily altering the genetic constitution of the changing population). The new situation will call forth previously untapped resources for anatomical and physiological change that are part of each species' already existing developmental

Table 9.2 Three Possible Stages in Evolutionary Pathway Initiated by Behavioral Neophenotype

I: Change in Behavior	II: Change in Morphology	III: Change in Genes
First stage in evolutionary pathway: Change in ontogenetic development results in novel behavioral shift (behavioral neopheotype), which encourages new environmental relationships.	Second stage of evolutiory change: New environmental relationships bring out latent possibilities for morphological-physiological change. Somatic mutation or change in genetic regulation may also occur, but a change in structural genes need not occur at this stage.	Third stage of evolutionary change: Resulting from long-term geographic or behavioral isolation (separate breeding populations). It is important to observe that evolution has already occurred phenotypically before stage III is reached. Modern neo-Darwinism, however, does not consider evolution to have occurred unless there is a change in genes or gene frequencies.

Source: From Gottlieb (1992). Reprinted by permission.

adaptability. At some time further down the road it is possible the genetic makeup of the evolving population may change, but by the time that happens (if it does) the new behavioral, anatomical, and physiological changes will already be in place. The neophenogenetic pathway for evolutionary change is thus seen as (a) an alteration of development leading to a significant change in behavior, followed by (b) a change in morphology, and, eventually, possibly (c) a change in genetic composition of the population. Consistent with their view of the strictly genetic determination of the phenotype, adherents of the modern synthesis would consider that evolution occurred only if and when step c was achieved. From the present point of view, enduring transgenerational changes in behavior and morphology (i.e., phenotypic evolution) have occurred by step b, without the necessity of adding to, subtracting from, or otherwise changing the original genetic composition of the population. The present view holds that genes are part of a very flexible and highly adaptable developmental system, but that genes do not determine the features of the mature organism. Consequently, from this point of view, evolution involves changes in the developmental system (of which the genes are an essential part), but not necessarily changes in the genes themselves. For instance, it is entirely consistent with the present proposal that alterations in development may cause genes to become active in the developmental process that were heretofore quiescent. It is well accepted among developmental geneticists that only a very small portion of the genome is expressed during individual development, so there is always present a large untapped genetic resource that can be brought to surface under abnormal (non-species-typical) developmental circumstances, whether internal or external to the organism. The behavioral neophenogenetic pathway of evolutionary change is depicted in Table 9.2.

Determinants of Behavioral Plasticity

The present theory lays great store in the malleability or adaptability of organisms, especially the higher vertebrates (birds and mammals—more on this latter point later). The creation of behavioral neophenotypes is necessarily dependent on the existence of some degree of behavioral plasticity or adaptability. Thus, the determinants of behavioral plasticity are an important consideration. One key limiting component of plasticity is the nervous system, particularly the brain, and the other is the developing organism's early experiences. These two components are in lockstep: Larger brained species can make more of their early experiences, and early experiences affect the maturation and size of the brain. Thus, the most conspicuous developmental route to increasing behavioral plasticity and creating behavioral neophenotypes is through early experiential alterations (including nutrition) that have positive effects on enhancing the maturation of the brain.

Beginning in the 1950s, developmental psychobiologists began in earnest to study the influence of early rearing experiences on enhancing the nervous system and later exploratory behavior and problem-solving ability, the latter two interrelated forms of behavior and psychological functioning being of most relevance in engendering the sort of evolutionary progression described previously and depicted in Table 9.2. For the present purposes, we are most interested in the developmental conditions that produce the sorts of behavioral plasticity that would enhance the likelihood of an individual (a) being able to survive by behavioral means in a drastically changed environment or (b) whose behavior would be likely to bring it into a new environment, thus precipitating the anatomical and physiological changes in stage II in Table 9.2.

Led by the pioneering experiments of Seymour Levine (1956) and Victor Denenberg (1969), a large number of studies showed that the unusual, perhaps stressful, experience of subjecting young rodents to handling by human beings during early development resulted in producing relatively stress-resistant animals, ones that would be capable of exploration (instead of freezing) and adaptive learning when faced with a completely strange and unfamiliar environment in adulthood. The research of Levine (1962) and his collaborators showed that the axis between the adrenal and pituitary glands was enhanced by the handling experience, and this anatomical-physiological change was correlated with the handled animal being able to tolerate greater stress in adulthood. As shown by Denenberg (1964) and his colleagues, the handling experience had to occur early in development if it was to be effective. Animals subjected to the same experience at older ages did not benefit from the experience, as indicated by later tests of resistance to stress and of exploratory behavior. A particularly important feature of these experiments is that the effect wrought in one generation persists into the next generation, even though the genes have not been altered (Denenberg & Rosenberg, 1967).

In a series of experiments with a rather different purpose, Hymovitch (1952) showed in a definitive manner how variations in early experience are crucial to later *problem solving* in adulthood. He reared young rats under four conditions and then later tested them in the Hebb-Williams maze. The animals were housed individually in:

1. A stovepipe cage (which permitted little motor or visual experience).
2. An enclosed running or activity wheel (which permitted a lot of motor activity but little variation in visual experience).
3. A mesh cage that restricted motor activity but allowed considerable variation in visual experience as it was moved daily to different locations in the laboratory.
4. The fourth group of animals contained 20 animals that were reared socially in a so-called free environment box that was very large (6 × 4 ft) compared to the other conditions, and was fitted with a number of blind alleys, inclined runways, small enclosed areas, apertures, and so on, that offered the rats a wide variety of opportunities for motor and visual exploration and learning in a complex physical environment.

The animals lived in these four environments from about 27 days of age to 100 days of age, at which time testing in the Hebb-Williams maze was completed.

Although rearing in the stovepipe and the enclosed running wheel led to the same level of poor performance, rearing in the mesh cage and the free environment led to the same level of good performance over 21 days of testing in the Hebb-Williams maze. All of the groups also showed the same level of improvement over the 3 weeks of testing, so the animals reared in the mesh cages and free environment began functioning at a superior level early in testing.

Next, in order to determine whether it was the early experience in each environment that made for the differences between the groups, Hymovitch repeated the experiment with four groups of animals that differed in *when* they had the free-environment or stovepipe experience: One group had the free-environment experience from 30 to 75 days of age and then were placed in the stovepipe for 45 days; a second group had the stovepipe experience from 30 to 75 days and then had the free-environment experience for 45 days; a third group remained in the free environment throughout the experiment; and a fourth group remained in their normal laboratory cages throughout the experiment (these would be the most thoroughly or consistently deprived from the standpoint of motor and visual experience).

The animals that experienced the free environment early and the stovepipe later in life performed just as well as the animals that remained in the free environment throughout the experiment. The crucial finding is that the animals who experienced the stovepipe environment early and the free environment later in life performed as poorly as the animals that remained in their normal cages throughout the experiment (the most deprived group). It is important to note that these differences in problem-solving ability were not in evidence when Hymovitch challenged the rats with a simpler, alley maze, more like the ones that were in wide use in most animal learning laboratories at the time. It was only when they were challenged by the much more difficult Hebb-Williams series of problems that the differences in problem-solving ability were in evidence.

Forgays and Forgays (1952) undertook to replicate Hymovitch's important findings and also to determine (a) whether the "playthings" in the free environment were crucial and (b) why the mesh-cage-reared animals did so well without direct experience of interacting with the multifarious objects in the free environment. They found indeed that the "playthings" (inclined planes, blind alleys, etc.) were essential to the superior performance of the free-environment animals and that the mesh-cage-reared animals only do as well when their cages are moved about frequently so that they visually encounter a considerable degree of varied environmental input, including the opportunity to watch the animals in the free environment with the playthings.

It was not long before these early experience studies were extended to other animals, including nonhuman primates, where social isolation and otherwise highly restricted, deprived rearing conditions were employed. Indeed, even in primates with relatively large brains, the normal or usual variety of experiences early in life was critical for the appearance of normal exploratory and learning abilities later in life. Deprived infants showed severe deficiencies in their later behavior (Harlow, Dodsworth, & Harlow, 1965). Just having a large brain is insufficient for the development and manifestation of the superior problem-solving skills characteristic of primates (Mason, 1968; Sackett, 1968). Thus, behavioral plasticity that is essential to behavioral neophenogenesis is dependent on variations in early experience as well as possessing a large brain.

The conditions that favor the appearance of a behavioral neophenotype are severe or species-atypical alterations in environmental contingencies early in life. These changed contingencies can arise in two ways in animals living in nature:

1. Some sort of physical or geographical change happens *to* (is forced on) the animal (a disruption of habitat, climatic change, and so on).
2. Probably more frequently, the migration of the animal into a somewhat different habitat based on normal exploratory behavior.

The large-brained animals that are more likely to withstand (1) and commit (2) are ones that have had not only traditional but nontraditional variations in their early experience. To put it the other way around, exposure to conservative or narrow social and physical environmental contingencies early in development will make animals less likely to withstand (1) and unlikely to perpetrate (2).

These predictions on evolutionary readiness, as it were, follow from the results of the early experience studies reviewed previously. There is a developmental dynamic that causes animals to prefer the familiar and thus to strive to reinstate earlier life situations or repeat versions of their early life experiences in adulthood. Consequently, animals that have had considerable variation in their early social and physical experiences will tend to seek out such variation in adulthood—just what is needed to heighten exploratory behavior and encourage novelty seeking! Although actual developmental experiments have not yet been done to show that animals (including humans) that have had considerable variation in their early experience will tend to seek out novel experiences as adults, there are two studies of adult mammals and birds that show that novelty is a psychological dimension of experience that can be abstracted such that animals so trained will consistently prefer to interact with novel rather than familiar objects or situations when given a choice (Honey, 1990; Macphail & Reilly, 1989). From the present theoretical standpoint, it would be most valuable to validate the developmental induction of novelty-seeking behavior in later life through the experience of considerable variation early in life.

Another way, albeit indirect, to test the behavioral neophenogenetic hypothesis about evolutionary readiness is to examine the exploratory behavior and rate of evolutionary change in large-brained versus smaller brained species. The cerebral component of behavioral neophenogenesis predicts a higher degree of exploratory behavior in large-brained species versus small-brained species and a consequent faster rate of evolutionary change in larger versus smaller brained species. The modes of behavioral neophenogenesis reviewed thus far are summarized in Table 9.3.

Table 9.3 Modes of Behavioral Neophenogenesis

Unusual (e.g., "handling") and enriched early experiences lead to:

1. Increased resistance to stress
2. Increased brain size
3. Enhanced exploratory behavior
4. Enhanced problem solving (learning ability) In adulthood

These would aid adaptation should (a) the organism's usual environment change drastically and (b) would also support the seeking out of new habitats. In the absence of environmental change: (1) and (2) are often invoked as the initial stages of evolution.

Source: From Gottlieb (1992). Reprinted by permission.

Exploratory Behavior and Rate of Evolutionary Change in Large- versus Small-Brained Species

After J. Huxley (1957), Bernhard Rensch (1959), and other evolutionary biologists agreed on the pinnacle status of birds and mammals based on considerations of ontogenetic and behavioral plasticity, a comparative psychologist named Harry Jerison (1973) produced a monumental tome, *Evolution of the Brain and Intelligence,* in which he was able to show that birds and mammals are in a class by themselves as far as the evolution of brain:body ratio is concerned. As can be seen in Fig. 9.2, at any given body weight, birds and mammals have a higher brain weight than all species of lower vertebrates at the same body weight. Consequently, according to the ideas already developed in this chapter, we should expect to see, by and large, greater behavioral plasticity in birds and mammals than in lower vertebrates. As it happens, that prediction does accord rather well with learning ability, conceived of as a species' ability to show forms of learning above the level of conditioning: Sensory preconditioning and learned stimulus configuring are possible only in birds and mammals (Razran, 1971, p. 221).

From the present standpoint, one of the most interesting findings in an ambitious experimental study of exploratory behavior in a very large variety of vertebrates is the clear superiority of all

Toward a New Developmental and Evolutionary Synthesis

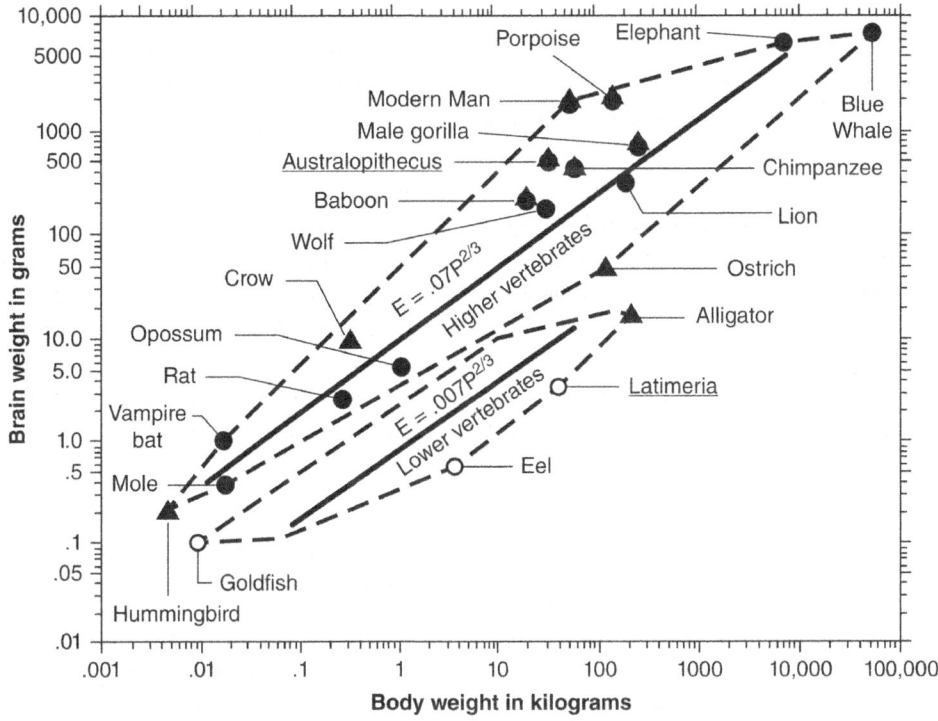

Fig. 9.2 Brain: body ratios of birds and mammals ("higher vertebrates") vs. fish, amphibians, reptiles ("lower vertebrates"). From Jerison (1969). Reprinted by permission from the University of Chicago Press.

mammalian forms tested (Fig. 9.3). In their study, Glickman and Sroges (1966) studied over 300 animals in over 100 different species, certainly the most extensive survey of exploratory behavior ever undertaken, so these results are most impressive from the standpoint of the consistency of mammalian superiority over the other vertebrate species.

It has been our contention that exploratory behavior—when a species is sufficiently plastic to initiate it—places the individual in a different niche facing different selective (adaptive) demands and thereby brings out latent morphological changes that then allow a genetically based evolutionary change to follow in its wake, as summarized in Table 9.2.

This sort of scheme allows evolution to proceed at a much more rapid rate than is the case when a species must await a severe environmental change or catastrophe such as oceans drying up or months of darkness, to prune individuals

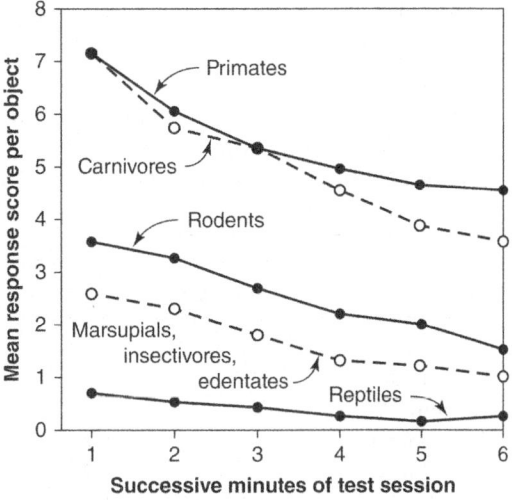

Fig. 9.3 The number of exploratory approaches made to novel objects by a large number of mammalian and reptilian species. From Glickman and Sroges (1966). Reprinted by permission from E. J. Brill.

that are capable of adapting to the change from those that are not. Such extreme environmental changes are rare and widely spaced in time, compared to the frequency and tempo of evolution that could be instigated by behavioral neophenotypes. Consequently, given the relationship between large brains and behavioral plasticity, behavioral neophenogenesis predicts that species with large brains should show evidence of a faster evolutionary pace than species with smaller brains. That prediction accords rather well with the finding of Wyles et al. (1983) of an almost perfect correlation between relative brain size and rate of anatomical evolution for a large number of vertebrate species.

As shown in Table 9.4, humans, the group with the largest relative brain size (Fig. 9.2), show the fastest rate of anatomical evolution, with the larger and older hominoid groups ranking second in brain size and rate of evolution. What is perhaps most interesting is that the relatively recently evolved songbirds rank just below hominoids and well above other mammals and other birds on both relative brain size and rate of anatomical evolution. Finally, the classifications of "Other mammals" and "Other birds" show a relatively larger brain size than the lower vertebrates (lizards, frogs, salamanders) and a corresponding faster tempo of evolutionary change. Consonant with the present theory, but without the developmental component, Wyles and colleagues invoked behavioral innovation, as well as a large gene pool, as the major driving force for the observed differences in rates of anatomical evolution:

> Behavioral innovation refers to the nongenetic (or genetic) origin of a new skill in a particular individual, leading it to exploit the environment in a new way. . . . [The] nongenetic propagation of new skills and mobility in large populations will accelerate anatomical evolution by increasing the rate at which anatomical mutants of potentially high fitness are exposed to selection in new contexts.
>
> (Wyles et al., 1983, p. 4396)

Stage II: Change in Morphology without Change in Genes

The present viewpoint takes advantage of the well-accepted fact that only a very small proportion of an individual's genotype participates in the developmental process. Thus, behavioral and morphological phenotypic changes can be immediately instigated by a change in an individual's developmental conditions. In our view, a change in developmental conditions activates heretofore quiescent genes, thus changing the usual developmental process and resulting in an altered behavioral or morphological phenotype. Consequently, stage II in the evolutionary pathway (Table 9.2) holds that the new environmental relationships bring out latent possibilities for morphological-physiological change in advance of the usual criterion of evolution: a change in structural genes or gene frequencies in the population (stage III). It is very exciting that this course of events corresponds to what is known about the correlation of morphological and genetic change in the evolutionary record. According to the best measurement techniques currently available, it would appear that morphological change antedates structural genetic and chromosomal change among the major vertebrate groups (Larson et al., 1984; Sarich, 1980, and references therein). Such a state of affairs is precisely compatible with the notion that behavioral neophenogenesis initiates a process

Table 9.4 Brain Size in Relation to Rate of Anatomical Evolution

Taxonomic Group	Relative Brain Size	Anatomical Rate
Homo	114	>10
Hominoids	26	2.5
Songbirds	23	1.6
Other mammals	12	0.7
Other birds	4.3	0.7
Lizards	1.2	0.25
Frogs	0.9	0.23
Salamanders	0.8	0.26

From Wyles, Kunkel, and Wilson (1983). Reprinted by permission.

of anatomical change that culminates only later in genic and chromosomal change. Because bringing extragenetic or supragenetic considerations into the evolutionary process, even in its introductory phases, is unorthodox, I would like to point out that this view can be integrated with population-genetic thinking and the modern synthesis (see next section), if the reality or plausibility of the early supragenetic stages I and II can be granted. Because those working on the problem of the apparent lack of synchrony in morphological and chromosomal evolution have themselves looked to behavior (specifically, social behavior) as the means whereby morphological change may be accelerated over chromosomal change in the course of evolution (Larson et al., 1984), the present author's contribution may be seen in the addition of the developmental dimension and its influence on the larger category of behavioral plasticity (exploratory behavior, problem solving) that is so essential to the early supragenetic phases of evolution. It is now acknowledged in many different quarters, both within and without the modern synthesis, that the time has come to include the role of individual development in evolution (e.g., P. Bateson, 1988; Futuyma, 1988; Gilbert et al., 1996; Goodwin, 1984; Ho & Saunders, 1982; Johnston & Gottlieb, 1990; Løvtrup, 1987; Oyama, 1985; Rosen & Buth, 1980; Thomas, 1971).

Integration of Individual Development into the Population-Genetic Model

Finally, to more explicitly integrate the present theory into the modern synthesis, the evolutionary pathway described here (Table 9.2) is consistent with the idea championed by Parsons (1981), among others, that a behaviorally mediated ecological independence precedes reproductive isolation when populations within a species first begin to split off (speciate). That is, these authors see behaviorally mediated changes in habitat selection (stage II in my scheme), especially microhabitat preferences, as the first step in the pathway to eventual speciation (reproductive isolation).

According to this view, when reproductive isolation eventually occurs, it is based on a later developing genetic incompatibility (stage III here) between originally homogeneous gene pools. According to the present scenario, the cause of the original behavioral divergence (e.g., preferences for different temperatures, humidities, light intensities, diets, oviposition sites, and mating sites in fruit flies) would be found in differences in the developmental histories (stage I) of the individuals showing the divergent behaviors. The stage I or developmental contribution to behavioral differences mediating speciation events has not yet been widely appreciated in the literature of evolutionary biology, even where novel shifts in behavior are seen as the key to speciation. That is to say, the causes of the novel shifts in behavior have received little attention as yet (Parsons, 1981, p. 230). The present scenario, specifically the induction of a behavioral neophenotype through a change in developmental conditions, offers an explanation for the novel behavior.

In summary, although the architects of the modern synthesis all agree that the mechanisms of evolution are mutation or genetic recombination, selection, migration, and eventual reproductive isolation, the present work describes how migration (invasion of new habitats or niches) may occur without mutation (or recombination) and selection first initiating a change in genes or gene frequencies. Both laboratory and field research indicate that reproductive isolation (i.e., incipient speciation) can occur without major genetic alterations (reviews in Bush, 1973; Singh, 1989, pp. 445–446). The usual scenario—theoretically speaking—is that a major genetic alteration is the necessary *first* step in speciation (Mayr, 1954). Bush's review indicates that behavior can lead the way to speciation with genetic changes coming into play only later on, which is consonant with the evolutionary model presented here. In this respect, it is noteworthy that Mayr (1988, Essay 28, pp. 541–542) now advocates a pluralistic view of the factors that instigate evolutionary change, one that might not entirely reject the present approach.

Bolder Speculation

A bolder and much more radical proposal would hold out for the possibility of morphological change initiated by behavioral change alone (i.e., without a shift in the physical environment). Tradition is recognized as an important component of animal as well as human behavior (Bonner, 1980), so what would be required under this more radical evolutionary scenario would be a change in some behavioral tradition, especially one affecting the rearing experience of young animals in the process of growing up, so that the subsequent behavior of the developing animals would be altered without necessarily causing them to leave their usual physical environment or niche. So far, all of the behaviorally mediated evolutionary scenarios have assumed an ecological change. The more radical notion of a strictly behaviorally mediated morphological change leading to speciation without an ecological or environmental shift has not been put forward before, at least to my knowledge.

Lamarck's (1809/1984) behaviorally mediated morphological evolution was said by him to be stimulated by a physical environmental change or stressor. The controversial Lamarckian experiments by Steele (1979) involved a morphological response to an environmental change. The possibility of strictly behaviorally inspired morphological change in the absence of stage II (Table 9.2) has not been previously entertained. Based on the evidence and other considerations discussed in this monograph, I think it remains a theoretical possibility. For example, a change in behavior almost inevitably brings about a change in social interactions, and some authors theorize that changes in social interaction prompted the evolution of human intelligence (reviewed in Byrne & Whiten, 1988). But these scenarios are couched within the terms of the modern synthesis (genetic change brought about by natural selection leads to evolutionary change in intelligence). In the framework developed here, an evolutionary change in brain structure and function might well occur without a genetic change in the population. There is such a tremendous amount of currently unexpressed developmental potential that behavioral and anatomical evolution would seem possible without the necessity of new genetic variations produced by mutation and genetic recombination.

How Can Changes Arising in One Generation Persist across Generations?

It is appropriate to ask how the new phenotypic changes can be preserved from one generation to the next if there has not been a mutation or new genetic recombination. The answer is that the transgenerational stability of new behavioral and morphological phenotypes is preserved by the repetition of the developmental conditions that gave rise to them in the first place. Because genes are a part of the developmental system and cannot make traits by themselves, this same requirement (repetition of developmental conditions) holds for the transgenerational perpetration of new phenotypes stemming from mutation and genetic recombination, even though that requirement often goes unrecognized or unspecified in the modern synthesis account of evolutionary change.

Although I am unable to propose a specific molecular mechanism whereby the organism's new experiences activate previously inactivated DNA (to get the expression of previously inactive genes), the present proposal obviously assumes such a mechanism. For example, something akin to the "homeobox" would fit the bill (Edelman & Jones, 1993; Gehring, 1987; Ingham, 1988). Some such mechanism is required for the Bolder Speculation to work. The activation of previously inactive genes must be occurring when, for example, avian oral epithelial cells grow a "mammalian" tooth under altered developmental conditions (Kollar & Fisher, 1980).

It seems clear to me that the Bolder Speculation applies to our own species: Earlier, we had suggested that *Homo erectus* could have possibly evolved into *Homo sapiens* through a dramatic change in rearing practices (Gottlieb, Johnston, & Scoville, 1982).

It remains to be seen whether the behavioral activation of new anatomical features from a previously unexpressed developmental potential is more widely applicable, or whether the more usual scenario is a behaviorally initiated ecological shift bringing out latent morphological change (stage II in Table 9.2). In either event, a developmentally wrought behavioral change would play an important role in evolution.

Conclusion

The present theory of behavioral neophenogenesis is manifestly a theory of vertebrate evolution, particularly of the higher vertebrates (birds and mammals), where the role of early experience in enhancing brain size, learning ability, exploratory behavior, and resistance to stress has been experimentally demonstrated in a number of species. The impact of early experience on later behavior is not without import in invertebrates (e.g., Jaisson, 1975; McDonald & Topoff, 1985) and lower vertebrates, but such developmental studies are few in number, so the question remains more open on the role of behavioral neophenogenesis in the evolution of invertebrate and lower vertebrate forms (fish, amphibians, reptiles). However, our general concept of neophenogenesis (Johnston & Gottlieb, 1990), which is a more global developmental theory of phenotypic evolution, would seem to have broad application in the evolutionary arena.

Finally, it is gratifying to see the concept of probabilistic epigenesis being used so constructively in some quarters of infant development (Bertenthal, Campos, & Kermoian, 1994), cognitive psychology (Bidell & Fischer, 1997), song learning in birds (Logan, 1992; West, King, & Freeberg, 1994), and medical genetics (review by Strohman, 1993), as well as in penetrating critiques of theories of development in biology, psychology, and sociology (Ford & Lerner, 1992; Gariepy, 1995; Lerner & Kauffman, 1985; Logan, 1985; Michel & Moore, 1995; Shanahan et al., 1997; van der Weele, 1995).

With a view toward what the future might hold with respect to theory and experiment in development and evolution, as I noted in the introduction to this chapter, there has been a recent movement in some quarters of cognitive development and in developmental and evolutionary biology to shift the focus of analysis from the genetic level to cell-cell interactions (the morphogenetic field). However, this is still a strictly bottom-up, internalist view of development (van der Weele, 1995) that ignores the contribution of embryonic/fetal behavior and the prenatal environment in constructing the organism. Thus, in order to truly move ahead in our theories and experiments, it behooves us to include the behavioral and environmental levels in a constructive, bidirectional view of epigenesis and its bearing on evolution, and to pursue such behaviorally and environmentally inspired inquiries into the prenatal stages of gene expression and neural development. Multidisciplinary training and interdisciplinary research will be necessary if the discipline of developmental psychobiology is to meet its promise of unifying biology and behavioral science (Michel & Moore, 1995).

To fill the enormous gap between molecular neurobiology and psychological function, it will also be necessary to have mutual respect for top-down as well as bottom-up approaches—it is not enough to understand how the brain develops from a sheerly internalist perspective.

References

Bateson, P. (1988), The active role of behaviour in evolution. In M.-W. Ho & S. W. Fox (Eds.), *Evolutionary processes and metaphors* (pp. 191–208). London: Wiley.

Bateson, W. (1894). *Materials for the study of variation, treated with especial regard to discontinuity in the origin of species*. New York: Macmillian.

Bertenthal, B. J., Campos, J. J., & Barrett, K. C. (1984). Self-produced locomotion. In R. N. Emde & R. J. Harmon (Eds.), *Continuities and discontinuities in development* (pp. 175–210). New York: Plenum.

Bidell, T. R., & Fischer, K. W. (1997). Between nature and nurture: The role of human agency in the epigenesis of intelligence. In R. Sternberg & E. Grigorenko (Eds.), *Intelligence: Heredity and environment* (pp. 193–242). New York: Cambridge University Press.

Bonner, J. T. (1980). *The evolution of culture in animals.* Princeton, N. J.: Princeton University Press.

Bonner, J. T. (1983). How behavior came to affect the evolution of body shape. *Scienta, 118,* 175–183.

Bush, G. L. (1973). The mechanism of sympathetic host race formation in the true fruit flies (*Tephritidae*). In M. J. D. White (Ed.), *Genetic mechanisms of speciation in insects* (pp. 3–23). Dordrecht, Holland: D. Reidel.

Byrne, R. W., & Whiten, A. (Eds.) (1988). *Machiavellian intelligence: Social experience and the evolution of intellect in monkeys, apes, and humans.* Oxford: Oxford University Press.

DeCasper, A. J., &Spence, M. J. (1986). Prenatal maternal speech influences newborns' perception of speech sounds. *Infant behavior and Development, 9,* 133–150.

Denenberg, V. H. (1964). Critical periods, stimulus input, and emotional reactivity: A theory of infantile stimulation. *Psychological Review, 71,* 335–351.

Denenberg, V. H. (1969). The effects of early experience. In E. S. E. Hafez (Ed.), *The behaviour of domestic animals* (2nd ed., pp. 95–130). Baltimore: Williams & Wilkens.

Denenberg, V. H., & Rosenberg, K. M. (1967). Nongenetic transmission of information. *Nature, 216,* 549–550.

Edelman, G. M. (1992). *Bright air, brilliant fire: On the matter of the mind.* New York: Basic Books.

Edelman, G. M., & Jones, F. S. (1993). Outside and downstream of the homeobox. *Journal of Biological Chemistry, 268,* 20683–20686.

Elman, J. L., Bates, E. A., Johnson, M. H., Karmiloff-Smith, A., Parisi, D., & Plunkett, K. (1996). *Rethinking innateness: A connectionist perspective on development.* Cambridge, MA: MIT Press.

Ford, D. H., & Lerner, R. M. (1992). *Developmental systems theory: An integrative approach.* Newbury Park, CA: Sage.

Forgays, D. G., & Forgays, J. W. (1952). The nature of the effect of free-environmental experience in the rat. *Journal of Comparative and Physiological Psychology, 45,* 322–328.

Futuyma, D. J. (1988). Sturm und Drang and the evolutionary synthesis. *Evolution, 42,* 217–226.

Gariépy, J.-L. (1995). The evolution of a developmental science: Early determinism, modern interactionism, and a new systemic approach. *Annals of Child Development, 11,* 167–224.

Garstang, W. (1922). The theory of recapitulation: A critical re-statement of the biogenetic law. *Journal of the Linnean Society of London, Zoology, 35,* 81–101.

Gehring, W. J. (1987). Homeoboxes in the study of development. *Science, 236,* 1245–1252.

Gilbert, S. F., Optiz, J. M., & Raff, R. A. (1996). Resynthesizing evolutionary and developmental biology. *Developmental Biology, 173,* 357–372.

Ginty, D. D., Bading, H., & Greenberg, M. E. (1992). Trans-synaptic regulation of gene expression. *Current Opinion in Neurobiology, 2,* 312–316.

Glickman, S. E., & Sroges, R. W. (1966). Curiosity in zoo animals. *Behaviour, 26,* 151–188.

Golden-Meadow, S. (1997). The resilience of language in humans. In C. T. Snowden & M. Hausberger (Eds.), *Social influences on vocal development* (pp. 293–311). New York: Cambridge University Press.

Goldschmidt, R. (1933). Some aspects of evolution. *Science, 78,* 539–547.

Goldschmidt, R. (1952). Evolution, as viewed by one geneticist. *American Scientist, 40,* 84–98, 135.

Goodwin, B. C. (1984), A relational or field theory of reproduction and its evolutionary implications. In M.-W. Ho & P. T. Saunders (Eds.), *Beyond neo-Darwinism: An introduction to the new evolutionary paradigm* (pp. 219–242). London: Academic Press.

Gottlieb, G. (1971). Ontogenesis of sensory function in birds and mammals. In E. Tobach, L. R. Aronson, & E. Shaw (Eds.), *The biopsychology of development* (pp. 67–128). New York: Academic Press.

Gottlieb, G. (1992). *Individual development and evolution: The genesis of novel behavior.* New York: Oxford University Press.

Gottlieb G, Johnston, T. D., Scoville R. P. (1982). Conceptions of development and the evolution of behavior. *Behavioral and Brain Sciences, 5*(2), 284–284.

Hardy, A. C. (1965). *The living stream.* London: Collins.

Harlow, H. F. (1958). The nature of love. *American Psychologist, 13*(12), 673–685.

Harlow, H. F., Dodsworth, R. O., & Harlow, M. K. (1965). Total social isolation in monkeys. *Proceedings of the National Academy of Sciences USA, 54,* 90–96.

Ho, M.-W., & Saunders, P. T. (1982). The epigenetic approach to the evolution of organisms – With notes on its relevance to social and cultural evolution. In H. C. Plotkin (Edd.), *Learning, development, and culture: Essays in evolutionary epistemology* (pp. 343–361). London: Wiley.

Holliday, R. (1990). Mechanisms for the control of gene activity during development. *Biological Reviews, 65,* 431–471.

Honey, R. C. (1990). Stimulus generalization as a function of stimulus novelty and familiarity in rats. *Journal of Experimental Psychology: Animal Behavior Processes, 16*(2), 178–184.

Huxley, J. S. (1957). The three types of evolutionary progress. *Nature, 180,* 454–455.

Hymovitch, B. (1952). The effects of experimental variations on problems solving in the rat. *Journal of Comparative and Physiological Psychology, 45,* 313–321.

Ingram, P. W. (1988). The molecular genetics of embryonic pattern formation in *Drosophlia*. *Nature, 335,* 25–34.

Jaisson, P. (1975). L'impregnation dans l'ontogenese des comportements de soins aux cocons chez la jeune formi rousse (*Formica polyctena* Forst). *Behaviour, 52,* 1–37.

Jerison, H. J. (1969). Brain evolution and dinosaur brains. *American Naturalist, 103,* 575–588.

Jerison, H. J. (1973). *Evolution of the brain and intelligence*. New York: Academic Press.

Johnston, T. D., & Gottlieb, G. (1990). Neophenogenesis: A developmental theory of phenotypic evolution. *Journal of Theoretical Biology, 147,* 473–495.

Kollar, E. J., & Fisher, C. (1980). Tooth induction in chick epithelium: Expression of quiescent genes for enamel synthesis. *Science, 207,* 993–995

Kuo, Z.-Y. (1976). *The dynamics of behavior development* (enlarged ed.). New York: Plenum.

Lamarck, J. B. (1984). *Zoological philosophy: An exposition with regard to the natural history of animals*. Chicago: University of Chicago Press. (Original work published in French, 1809)

Larson, A., Prager, E. M., & Wilson, A. C. (1984). Chromosomal evolution, speciation and morphological change in vertebrates: The role of social behavior. *Chromosomes Today, 8,* 215–228.

Lee, C., Parikh, V., Itsukaichi, T., Bae, K., & Edery, I. (1996). Resetting the Drosophila clock by photic regulation of PER and a PER-TIM complex. *Science, 271,* 740–744.

Leonovicová, V., & Novák, V. J. A. (Eds.). (1987). *Behavior as one of the main factors of evolution*. Pvaha, Czechoslovakia: Czechoslovak Academy of Sciences.

Lerner, R. M., & Kauffman, M. B. (1985). The concept of development in contextualism. *Developmental Review, 5,* 309–333.

Levine, S. (1962). The effects of infantile experience on adult behavior. In A. J. Bachrach (Ed.), *Experimental foundations of clinical psychology* (pp. 139–169). New York: Basic Books.

Locke, J. L. (1993). *The child's path to spoken language*. Cambridge, MA: Harvard University Press.

Logan, C. A. (1985). Development as explanation. *Applied Developmental Psychology, 2,* 1–32.

Logan, C. A. (1992). Developmental analysis in behavioral systems: The case of bird song. *Annals of the New York Academy of Sciences, 662,* 102–117.

Løvtrup, S. (1987). *Darwinism: Refutation of a myth*. Beckenham, England: Croom Helm.

Macphail, E. M., & Reilly, S. (1989). Rapid acquisition of a novelty versus familiarity concept by pigeons (*Columba livid*). *Journal of Experimental Psychology: Animal Behavior Processes, 15,* 242–252.

Masataka, N. (1994). Effects of experience with live insects on the development of fear of snakes in squirrel monkeys, *Saimiri sciurens*. *Animal Behaviour, 46,* 741–746.

Mason, W. A. (1968). Early social deprivation in the nonhuman primates: Implications for human behavior. In D. Glass (Ed.), *Biology and behavior: Environmental influences* (pp. 70–101). New York: Rockefeller University Press.

Mayr, E. (1954). Change of genetic environment and evolution. In J. Huxley, A. C. Hardy, & E. B. Ford (Eds.), *Evolution as a process* (pp. 157–180). London: Allen and Unwin.

Mayr, E. (1982). *The growth of biological thought*. Cambridge, MA: Harvard University Press.

McDonald, P., & Topoff, H. (1985). Social regulation of behavioral development in the ant, *Novomessor albisetosus* (Mayr). *Journal of Comparative Psychology, 99,* 3–14.

Michel, G. F., & Moore, C. L. (1995). *Developmental psychobiology: An interdisciplinary science*. Cambridge, MA: MIT Press.

Miklos, G. L. G. (1993). Molecules and cognition: The latterday lessons of levels, language, and lac. Evolutionary overview of brain structure and function in some vertebrates and invertebrates. *Journal of Neurobiology, 24*, 842–890.

Miklos, G. L. G., & Edelman, G. M. (1996). *Complexity in gene function and morphogenesis: Major challenges of the post-sequence era.* Unpublished manuscript.

Mivart, St. G. (1871). *On the genesis of species.* London: Macmillan.

Myers, M. P., Wager-Smith, K., Rothenfluh-Hilfiker, A., & Young, M. W. (1996). Light induced degradation of TIMELESS and entrainment of the *Drosophila* circadian clock. *Science, 271,* 736–740.

Oyama, S. (1985). *The ontogeny of information.* Cambridge, England: Cambridge University Press.

Parsons, P. A. (1981). Habitat selection and speciation in Drosophila. In W. R. Atchley & D. S. Woodruff (Eds.), *Evolution and speciation: Essays in honor of M. J. D. White* (pp. 219–240). Cambridge, England: Cambridge University Press.

Piaget, J. (1978). *Behavior and evolution.* New York: Pantheon Books.

Plotkin, H. C. (Ed.). (1988). *The role of behavior in evolution.* Cambridge, MA: MIT Press.

Raff, R. A., & Kaufman, T. C. (1983). *Embryos, genes, and evolution.* New York: Macmillan.

Razran, G. (1971). *Mind in evolution.* New York: Houghton Mifflin.

Reid, R. G. B. (1985). *Evolutionary theory: The unfinished synthesis.* Ithaca, NY: Cornell University Press.

Rensch, B. (1959). *Evolution above the species level.* New York: Columbia University Press.

Rosen, D. E., & Buth, D. G. (1980). Empirical evolutionary research versus neo-Darwinian speculation. *Systematic Zoology, 29,* 300–308.

Sackett, G. P. (1968). Abnormal behavior in laboratory-reared rhesus monkeys. In M. W. Fox (Ed.), *Abnormal behavior in animals* (pp. 293–331). Philadelphia: W. B. Saunders.

Sarich, V. (1980). A macromolecular perspective on the material basis of evolution. In L. K. Piternick (Ed.), *Richard Goldschmidt: Controversial geneticist and creative biologist* (pp. 27–31). Boston: Birkhauser.

Sewertzoff, A. N. (1929). Directions of evolution. *Acta Zoologica* (Stockholm), *10,* 59–141.

Shanahan, M. J., Valsiner, J., & Gottlieb, G. (1997). Developmental concepts across disciplines. In J. Trudge, M. J. Shanahan, & J. Valsiner (Eds.), *Comparisons in human development* (pp. 34–71). New York: Cambridge University Press.

Singh, R. S. (1989). Population genetics and the evolution of species related to Drosophila melanogaster. *Annual Review of Genetics, 23,* 425–453.

Sparrow, A. H., Price, H. J., & Underbrink, A. G. (1972). A survey of DNA content per cell and per chromosome of prokaryotic and eukaryotic organisms: Some evolutionary considerations. *Brookhaven Symposium on Biology, 23,* 451–494.

Spelke, E. S., & Newport, E. L. (1998). Nativism, empiricism, and the development of knowledge. In R. M. Lerner (Ed.), *Handbook of child psychology: Theoretical models of human development* (Vol. 1, 5th ed., pp. 275–340). New York: Wiley.

Strohman, R. C. (1993). Book reviews. *Integrative Physiological and Behavioral Science, 28,* 99–110.

Thomas, C. A. (1971). The genetic organization of chromosomes. *Annual Review of Genetics, 5,* 237–256.

Turkewitz, G. (1988). A prenatal source for the development of hemispheric specialization. In D. L. Molfese & S. J. Segalowitz (Eds.), *Brain lateralization in children* (pp. 73–81). New York: Guilford Press.

Wallman, J. (1979). A minimal visual restriction experiment: Preventing chicks from seeing their feet affects later responses to mealworms. *Developmental Psychobiology, 12,* 391–397.

van der Weele, C. (1995). *Images of development: Environmental causes in ontogeny.* Unpublished doctoral dissertation, Vrije University, Amsterdam.

West, M. J., King, A. P., & Freeberg, T. M. (1994). The nature and nurture of neophenotypes: A case history. In L. A. Peal (Ed.), *Behavioral mechanisms in evolutionary ecology* (pp. 238–257). Chicago: University of Chicago Press.

Wyles, J. S., Kunkel, J. G., & Wilson, A. C. (1983). Birds, behavior, and anatomical evolution. *Proceedings of the National Academy of Sciences USA, 80,* 4394–4397.

10.
PRÉCIS OF *EVOLUTION IN FOUR DIMENSIONS**
Eva Jablonka[†] and Marion J. Lamb[‡]

1. Introduction

Since its beginning in the early 19th century, the history of evolutionary theory has been a stormy one, marked by passionate and often acrimonious scientific arguments. It began with Lamarck, who 200 years ago presented the first systematic theory of evolution, but it was largely through the influence of Darwin's *On the Origin of Species* (Darwin 1859; henceforth *Origin* in this chapter) that evolution took center stage as the foremost integrating theory in biology. In the late 19th and early 20th centuries, the theory went through neo-Darwinian, neo-Lamarckian, and saltational upheavals, but eventually it achieved a 60-year period of relative stability through what is commonly known as the Modern Synthesis. The Modern Synthesis, which began to take shape in the late 1930s and has been updated ever since, was a theoretical framework in which Darwin's idea of natural selection was fused with Mendelian genetics. The stability it gave to Darwinian theory was the result of the elasticity biologists allowed it. By giving up some initial assumptions about strict gradualism, by tolerating selective neutrality, by accepting that selection can occur at several levels of biological organization, and by other adjustments, the Modern Synthesis was made to accommodate much of the avalanche of molecular and other data that appeared in the second half of the 20th century.

One thing that most mid- and late-20th century evolutionists were unwilling to incorporate into their theory was the possibility that the generation of new variations might be influenced by environmental conditions, and, hence, that not all inherited variation is "random" in origin. During the first 50 years of the Modern Synthesis's reign, Lamarckian processes, through which influences on development could lead to new heritable variation, were assumed to be non-existent. When induced variations eventually began to be recognized, they were downplayed. Developmental processes in general were not a part of the Modern Synthesis, and until recently developmental biology had little influence on evolutionary theory. This is now changing, and as knowledge of developmental mechanisms and the developmental aspects of heredity are incorporated, a profound, radical, and fascinating transformation of evolutionary theory is taking place.

In *Evolution in Four Dimensions* (Jablonka & Lamb 2005; henceforth mentioned as *E4D* in this précis), we followed the traditional 20th-century heredity-centered approach to evolutionary theory and looked at how new knowledge and ideas about heredity are influencing it. We described four different types of heritable variation

---

* Jablonka, E., & Lamb, M. (2007). Précis of Evolution in Four Dimensions. Behavioral and Brain Sciences, 30(4), 353-365. doi:10.1017/S0140525X07002221
† Cohn Institute for the History and Philosophy of Science and Ideas, Tel Aviv University, Tel Aviv, Israel.
‡ 11 Fernwood, Clarence Road, London, United Kingdom.

(genetic, epigenetic, behavioral, and symbolic), some of which are influenced by the developmental history of the organism and therefore give a Lamarckian flavor to evolution. By systematically analyzing and discussing the processes involved, we examined the role and prevalence of induced variations, arguing that they are important and versatile and that the theory of evolution and studies based on it will remain deficient unless they are fully incorporated. Since the book was completed in 2004, a lot of new material has been published, and we refer to some of it in this précis.

We had several aims in writing *E4D*. One was to provide an antidote to the popular DNA-centered view of evolution. Many people have been convinced by eminent popularizers that the evolution of every trait – whether cellular, physiological, morphological, or behavioral – can be and should be explained in terms of natural selection acting on small variations in DNA sequences. In *E4D*, we tried to explore a different and, we believe, better type of explanation, which is based on behavioral ecology, experimental psychology, and cultural studies, as well as modern molecular biology. Because we wanted to catch the attention of lay people who are interested in evolution, we tried to reduce the amount of jargon used and made use of unconventional illustrations and thought experiments to explain our views. We also used the old philosophical device of a dialogue with a "devil's advocate," whom we called Ifcha Mistabra ("the opposite conjecture" in Aramaic), to explore the premises and difficulties of the approach we described. Obviously, in this précis for professional scientists we do not try to reproduce these stylistic features of the book.

2. The Transformations of Darwinism

We started *E4D* with a historical introduction in which we described some of the shifts in ideas that we think are important for understanding how and why biologists arrived at the gene- and DNA-centered view of heredity and evolution that prevails today. We began with Darwin, who gave his "laws" of biology in the closing paragraph of the *Origin*:

> These laws, taken in the largest sense, being Growth with Reproduction; Inheritance which is almost implied by reproduction; Variability from the indirect and direct action of the external conditions of life, and from use and disuse; a Ratio of Increase so high as to lead to a Struggle for Life, and as a consequence to Natural Selection, entailing Divergence of Character and the Extinction of less-improved forms.
>
> (Darwin 1859, pp. 489–90)

Darwin's laws were very general. How reproduction, growth, and inheritance are realized in different biological systems, how variability is generated, and what types of competitive interactions are important, all had to be qualified. Evolutionary biology since Darwin can be seen as the history of the qualification of these processes. As the quotation from the *Origin* makes clear, Darwin included "use and disuse" as a cause of variability: he accepted that there are Lamarckian processes in evolution.

August Weismann's version of Darwinism, disapprovingly dubbed "neo-Darwinism" by Romanes, is an important part of the history of evolutionary thinking, and its influence can still be seen in contemporary views of heredity and evolution. Unlike Darwin, Weismann gave natural selection an exclusive role in evolution, ruling out change through the inherited effects of use and disuse or any other form of the inheritance of acquired somatic (bodily) characters. His reasons for doing so were partly the lack of evidence, but also the difficulty of envisaging any mechanism through which the inheritance of acquired characters could occur. Certainly, Weismann's own elaborate theory of heredity and development did not allow it.

Weismann believed that there is a sharp distinction between cells of the soma, which are responsible for individual life, and germline cells, which are responsible for producing sperm and eggs. Only germline cells have all the hereditary determinants necessary for producing the next generation. As Weismann saw it, there was no way in which information from body cells could be transferred to germline cells: He assumed (incorrectly) that development and differentiation involve quantitative and qualitative changes in the cells' nuclear contents, and that, as far as heredity is concerned, the soma is a dead end.

One of Weismann's great achievements was to recognize the source of some of the heritable variation that Darwin's theory of natural selection required. He saw how meiosis and the sexual processes could bring together different combinations of the parents' hereditary determinants, thereby producing differences among their offspring. However, that still left the problem of the origin of new variants. It surprises many people to discover that Weismann, the great opponent of Lamarckism, thought that the source of all new variation was accidental or environmentally induced alterations in the germline determinants.

Weismann's ideas and those of his supporters and rivals were debated vigorously during the late 19th century. His elaborate theory of heredity and development was never popular – and turned out to be largely wrong – but elements of it were influential during the foundation of genetics at the beginning of the 20th century and consequently became embedded in the Modern Synthesis. The distinction Weismann made between soma and germline, his claim that somatic changes could not influence the germline, and his belief that heredity involved germline-to-germline continuity, helped to provide the rationale for studying heredity in isolation from development. One part of Weismann's thinking that was soon forgotten, however, was the idea that new germline variation originates through environmental induction.

In the early 20th century, most of the pioneers of the young science of genetics consciously ignored development and focused on the transmission and organization of genes. The Danish geneticist Johannsen provided the conceptual basis for modern genetics by distinguishing between the genotype and phenotype. The genotype is the organism's inherited potential (the ability to develop various characters), while the phenotype is the actualization of this potential in a particular environment. Hence, the phenotype is by definition the consequence of the interaction between the genotype and the environment. Johannsen's unit of heredity, the gene, was not a representative of the phenotype or a trait, but rather, a unit of information about a particular potential phenotype. Genes were generally assumed to be very stable, although through occasional accidents they changed (mutated) to new alleles. At the time, what a gene was materially was unknown, and how the phenotype was realized was a complete mystery. But for Johannsen and his fellow geneticists, the abstract concept of the gene meant that "Heredity may then be defined as *the presence of identical genes in ancestors and descendants*" (Johannsen 1911, p. 159, his emphasis).

This view of heredity became part of the "Modern Synthesis" of the late 1930s, in which ideas and information from paleontology, systematics, studies of natural and laboratory populations, and especially, from genetics, were integrated into the neo-Darwinian framework. Some of the assumptions on which the Synthesis was based were: (1) Heredity takes place through the transmission of germline genes, which are discrete and stable units located on nuclear chromosomes. They carry information about characters. (2) Variation is the consequence of the many random combinations of alleles generated by sexual processes; usually, each allele has only a small phenotypic effect. (3) New alleles arise only through accidental mutations; genes are unaffected by the developmental history of the organism, and changes in them are not specifically induced by the environment, although the overall rate of change might be affected. (4) Natural selection occurs between individuals (although selection between groups was not explicitly ruled out). Theoretical models of the behavior of genes in populations played a key role in the Synthesis, and Theodosius Dobzhansky, one of its leading figures, proclaimed that "evolution is a change in the genetic composition of populations" (Dobzhansky 1937, p. 11). This view was not shared by everyone, but the voices of embryologists and others who believed that heredity involves more than genes, were seldom heard and were generally ignored by evolutionists.

The advent of molecular biology in the 1950s meant that the Modern Synthesis version of Darwinism was soon being updated to incorporate the new discoveries. At first, these discoveries seemed only to reinforce the basic tenets of the Modern Synthesis. The gene, the unit of heredity, was seen as a sequence of nucleotides in DNA which coded for a protein product that determined some aspect of the phenotype (or sometimes for an RNA molecule with a functional role in information processing). The seemingly simple mechanism of DNA replication explained the fidelity of inheritance. Information encoded in a gene's DNA sequence was first transcribed into RNA, and then translated into the amino acid sequence of a protein. According to Francis Crick's central dogma, information can never flow from a protein back to RNA or DNA sequences, so developmental alterations in proteins cannot be inherited. This, of course, was soon being seen as a validation of the neo-Darwinian view that "acquired characters" could have no role in evolution. Changes in DNA sequence – mutations – arise only from rare mistakes in replication or from chemical and physical insults to DNA. Although specific mutagens might increase the overall mutation rate, all changes were assumed to be blind to function. As molecular biology developed, DNA began to be seen as more than coded information for making proteins. Because its sequences carry regulatory and processing information that determines which protein is made where and when, DNA assumed a more directive role – it was seen as a plan for development, a program.

Some modifications of the original Modern Synthesis had to be made. It transpired that many variations in the amino acid sequences of proteins (and many more variations in DNA sequences) make no phenotypic difference: some genetic variations seem to be selectively neutral. Moreover, there are genes located in the cytoplasm, which do not obey Mendel's laws. It was also recognized that there are internal processes, such as the movements of "jumping genes" (transposons), that generate mutations. However, the Modern Synthesis version of neo-Darwinism was elastic enough to accommodate these findings.

Modern Synthesis neo-Darwinism took an interesting twist in the 1970s as a result of the attention biologists had been giving to the long-standing problem of the evolution and persistence of "altruistic" traits, which decrease the fitness of the individuals displaying them. For our purposes here, the solutions that were reached are less important than the broader effect the debate had, which was to lead to an even greater focus on the gene not only as a unit of heritable variation, but also as a unit of selection. Richard Dawkins developed and popularized this gene-centered view of evolution in *The Selfish Gene* (Dawkins 1976), and subsequently, it was adopted by most biologists. The gene was depicted as the unit of heredity, selection, and evolution. According to Dawkins, individual bodies live and die, but for evolutionary purposes they should be seen simply as vehicles, as carriers of genes. A gene is a replicator, an entity that is copied in a way that is independent of any changes in the vehicle that carries it, and adaptive evolution occurs only through the selection of germline replicators. Cultural evolution takes place through the spread of cultural replicators, which Dawkins called memes (see sect. 6).

The sketch we have just given shows that the historical route to the present gene-centered view of Darwinism has been evolutionary, in the sense that modifications that happened early on became the basis for what happened later. At an early stage, developmental aspects became vestigial and the significance of the germline grew disproportionately large; this form of the theory eventually became adapted to an environment dominated first by genetics and then by molecular biology, so that first the gene, and then DNA, was seen as the source of all hereditary information. We believe that recent data and ideas may mean that the gene- and DNA-centered form of Darwinism is heading for extinction, and in *Evolution in Four Dimensions* we have suggested the sort of Darwinian theory that may replace it. It is a theory that sees DNA as a crucial heritable developmental resource, but recognizes that DNA is not the only resource that contributes to heredity. New discoveries in cell and developmental biology and in the behavioral and cognitive sciences mean that it no longer makes sense to think of inheritance in terms of almost invariant genes carrying information about traits encoded in DNA sequences. First, the genome has turned out to be far more flexible and responsive than was previously supposed, and the developmental processes that result in phenotypic traits are enormously complicated. Second, some transmissible cellular variations, including variations that are transmitted through the reproductive cells, are the result of spontaneous or induced epigenetic changes, rather than differences in DNA. Third, for animals, behaviorally transmitted information plays a significant role in evolution. Fourth, as is already well-recognized, symbolic culture has powerful evolutionary effects in humans. All types of heritable variations and their interactions with each other and the environment have to be incorporated into evolutionary theorizing. This is particularly important for scientists trying to understand the evolutionary basis of human behavior, who throughout the history of evolutionary ideas have been active and passionate participants in the major debates.

3. From Genes to Development to Evolution: A Complex Relationship

In the early days of genetics, the characters chosen for analysis were largely those that could be interpreted in terms of genes that behaved according to Mendel's laws of segregation. It soon became clear, however, that the relationship between genes and characters is complex: It is not a one-to-one relationship but, rather, a many-to-many relationship. An allelic difference in a single gene can lead to many character differences, and what is seen depends on the external environment, the internal cellular environment, the other alleles present in the genome, and the level at which the analysis is made. Furthermore, several different alleles, often located in different parts of the genome, may, as a combination, collectively affect a character. Often a variation in a single gene makes no difference to the phenotype.

Although these facts became obvious quite early on in the 20th century, the temptation to see a simple causal relation between genes and characters was not resisted. As we are well aware, the idea of simple genetic causality has been politically misused – most horribly by German eugenicists in the 1930s and 1940s, but in other places and at other times too. The attraction of simple linear causation is still present: It is not uncommon to read reports in the popular press about the discovery of a "gene for" obesity, criminality, religiosity, and so on. Many non-geneticists believe that knowledge of a person's complete DNA sequence will enable all their characteristics to be known and their problems predicted. This widespread belief in "genetic astrology" leads to many unrealistic hopes and fears – fears about cloning and stem cells, for example, and hopes that genetically engineered cures for all individual ills and social evils are just around the corner.

As molecular biology developed, it did at first seem that the relationships between genes and biochemical characters might be simple. A small change in a gene's DNA sequence was seen to lead to a corresponding change in a protein's amino acid sequence, which eventually caused a change in one or more characteristics of the organism. In some of the so-called "monogenic" diseases, for example, a simple DNA change makes a qualitative difference in a protein, which leads to the malfunction of the system of which it is a part. However, it turned out that even in these cases the effects of the DNA change are often context dependent. Sickle cell disease is a paradigmatic example of a small DNA change (a single nucleotide substitution) that leads to an amino acid change in a protein (a subunit of hemoglobin), which results in a large phenotypic change (very severe anemia). Many of the details of how this substitution caused these changes were worked out in the early days of molecular biology. More recent studies have shown, however, that the severity of the disease depends markedly on other factors, including which alleles of other genes are present (Bunn 1997). Some Bedouin Arabs, for example, show only relatively mild symptoms, because they carry an unusual allele of a different gene that counteracts the effects of sickle alleles.

Even at the molecular level, the relationship between DNA, RNA, and proteins has turned out to be vastly more complex than originally imagined. First, most DNA does not code for proteins at all. Only about 2% of human DNA codes for proteins, and the current estimate of the number of protein-coding genes is around 25,000 (about the same as for the mouse – and not many more than for the nematode worm). Second, the RNA products of DNA transcription come in a variety of lengths and organizations. Because of various processes that occur during and after transcription, an RNA transcript often corresponds to several different proteins – sometimes hundreds. Third, much of the DNA that does not code for proteins is nevertheless transcribed into RNA. We know the functions of some of this RNA – it has many, including enzymatic and regulatory ones – but for much of it we are still very much in the dark about what, if anything, it does. Fourth, there are DNA sequences that are not transcribed at all (or so it is believed): Some act as binding sites for regulators, some act as structural elements, and others have no known function and may be genomic parasites. Fifth, DNA can be changed during development. It can be cut up, sewn together, and moved around. Sequences in some cells undergo amplification, or bits are deleted, or they are rearranged, as happens, for example, in the immune system. These are developmental changes, executed by the cell's own genetic-engineering kit. Sixth, not only does RNA have messenger, enzymatic, and regulatory functions, but it can also act as hereditary material which is replicated and passed on from mother cells to daughter cells, including germ cells.

Evelyn Fox Keller (2000) has described how the meaning of the term "gene" changed during the 20th century, arguing that it had lost much of its clarity. What has happened in molecular biology in the first few years of the 21st century emphasizes this even more (Pearson 2006). It seems that "gene," the "very applicable little word" coined by Johannsen, can no longer be used without qualification.

What the new knowledge about the relation between DNA and characters shows is that thinking about the development of traits and trait variations in terms of single genes and single-gene variations is inappropriate. It is cellular and intercellular networks (which include genetic networks) that have to be considered. If the effects of small changes in DNA base sequence (classical gene alleles) are highly context dependent and often, when considered in isolation, have, on average, no phenotypic consequences, then the unit underlying phenotypic variation cannot be the classical gene. A shift in outlook is needed. The concept of information in biology, which was inspired by and based on the notion of genes that carry information in their DNA sequences, needs to be changed and cast in more developmental and functional terms (Jablonka 2002). Because it is phenotypes, the products of development, that are selected, and heritably varying phenotypes are the units of evolution, the evolutionary implications of all the developmental resources that contribute to heritable phenotypes have to be considered. Moreover, since it is recognized that regulated DNA changes occur within a generation, the possibility that the mechanisms underlying such developmental modifications may also generate variations that are transmitted between generations cannot be ignored (Shapiro 1999).

Research on the origins of DNA variation challenges the idea that all variations in DNA (mutations) are blind or "random" (see Chapter 3 of *E4D*). The term "random mutation" is a problematic one that is used in several somewhat different ways. It is used to mean that mutations (1) are not highly targeted, that is, that identical (or very similar) changes in DNA do not occur in many different individuals within a population (although there are some "hot spots" in the genome where mutations are more likely than elsewhere); (2) are not developmentally or environmentally induced, that is, that identical changes in conditions do not result in identical mutations; and (3) are not adaptive, that is, that they do not increase the chances that the individuals carrying them will survive and reproduce. Each of these three senses in which mutation has been assumed to be random has been questioned. Mainly as a result of work in microorganisms, fungi, and plants, it is now recognized that some mutations may be targeted, induced, and adaptive.

The flavor of the data coming from this research can be appreciated from a few examples showing that DNA sequence variations can be both highly targeted and condition dependent. Under conditions of genomic stress, such as when two genomes from different sources meet (e.g., when plant hybridization occurs), there can be repeatable and wide-ranging, yet specific, genomic and chromosomal changes (for an eye-opening example, see Levy & Feldman 2004). Because hybridization is thought to be of major importance in plant evolution, the global modifications that hybridization induces are of great interest and importance. Nutritional or heat stress in plants can also lead to specific, repeatable changes in particular DNA sequences. Certain microorganisms have what look like adaptive stress responses: Data from studies of the mutation rates in bacteria indicate that both the overall mutation rate and the mutation rate of specific genes may be increased in stressful conditions, and that these increases improve their chances of survival. The idea that these are evolved adaptive mechanisms is being actively explored (Caporale 2003). The mechanisms proposed do not make adaptive changes a certainty, but they do increase the chances that a DNA variation generated by the evolved systems that respond to stress will lead to a better-functioning phenotype.

How extensive and significant evolved mutational mechanisms are in animals is not yet clear, mainly because little relevant research has been done. Induced mutational processes are certainly part of the mammalian immune response, and there are hints that stress reactions similar to those found in plants may occur in mammalian germ cells (Belyaev & Borodin 1982), but little is known. Nevertheless, induced mutation is potentially enormously important for humans. If, as seems likely, bacterial pathogens exposed to pharmacological stresses have sophisticated mutation-generating mechanisms that enable them to adapt and survive, then a detailed understanding of these mechanisms is essential if we are to have a chance of combating the growing problem of drug resistance.

4. Epigenetic Inheritance

In the first part of *E4D* (Chapters 2 and 3), we showed that the genetic inheritance system, based on DNA, is not as simple as is commonly assumed. Not only is the relationship between variations in DNA sequences and variations in biochemical and higher-level traits more complex, but the idea that all DNA changes arise through random mistakes is wrong. Heredity involves more than DNA, however, and in the second part of *E4D* we looked at heritable variations that have little to do with DNA sequence differences. These variations are described as "epigenetic," and the systems underlying them are known as epigenetic inheritance systems. Like almost everything else in the biological world, these systems depend on DNA, but, by definition, epigenetic variations do not depend on DNA variations.

The term "epigenetic inheritance" is used in two overlapping ways. First, epigenetic inheritance in the broad sense is the inheritance of phenotypic variations that do not stem from differences in DNA sequence. This includes cellular inheritance (see the second usage), and body-to-body information transfer that is based on interactions between groups of cells, between systems, and between individuals, rather than on germline transmission. Body-to-body transmission takes place through developmental interactions between mother and embryo, through social learning, and through symbolic communication.

Second, *cellular* epigenetic inheritance is the transmission from mother cell to daughter cell of variations that are not the result of DNA differences. It occurs during mitotic cell division in the soma, and sometimes also during the meiotic divisions in the germline that give rise to sperm or eggs. Therefore, offspring sometimes inherit epigenetic variations. In both soma and germline, transmission is through chromatin marks (non-DNA parts of chromosomes and DNA modifications that do not affect the sequence or code), various RNAs, self-templating three-dimensional structures, and self-sustaining metabolic loops (Jablonka & Lamb 1995).

In *E4D* we treated cellular epigenetic inheritance separately (Chapter 4) from body-to-body information transmission, and divided the latter into transmission through social learning (Chapter 5) and transmission through symbolic systems (Chapter 6). As happens so often in biology, some phenomena did not fit neatly into any of these three categories. In particular, it was difficult to know where to put information inherited through routes such as the placenta or milk, and where to put the ecological legacies that offspring receive from their parents and neighbors. In this précis, we describe these important routes of information transfer in Section 4.2.

4.1. Cellular Epigenetic Inheritance

It is easiest to explain what epigenetic inheritance is about by using its most important and obvious manifestation – the maintenance of determined and differentiated states in the cell lineages of multicellular organisms. Most of the cells in an individual have identical DNA, yet liver cells, kidney cells, skin cells, and so on are very different from each other both structurally and functionally. Furthermore, many cell types breed true: mother skin cells give rise to daughter skin cells, kidney cells to kidney cells, and so on. Since they have exactly the same DNA, and since the developmental triggers that made them different in the first place are usually no longer present, there must be mechanisms that actively maintain their differing gene expression patterns, structural organization, and complex metabolic states and enable them to be transmitted to daughter cells. These mechanisms are known as epigenetic inheritance systems (EISs). Their study is a fast-moving area of research, because not only is epigenetic inheritance a central aspect of normal development, it is also increasingly being recognized as being of great importance in cancer and other human diseases. In addition, it is responsible for the transmission of some normal and pathological variations between generations.

Cellular epigenetic inheritance is ubiquitous. All living organisms have one or more mechanism of cellular epigenetic inheritance, although not all mechanisms are shared by all organisms. In non-dividing cells such as nerve cells, there is no epigenetic inheritance, but there is epigenetic cell memory: Certain functional states and structures persist dynamically for a very long time. This cell memory seems to involve the same epigenetic mechanisms as those that underlie epigenetic inheritance (Levenson & Sweatt 2005).

There are at least four types of EIS:

1. Self-sustaining feedback loops. When gene products act as regulators that directly or indirectly maintain their own transcriptional activity, the transmission of these products during cell division results in the same states of gene activity being reconstructed in daughter cells.
2. Structural inheritance. Pre-existing cellular structures act as templates for the production of similar structures, which become components of daughter cells.
3. Chromatin marking. Chromatin marks are the proteins and small chemical groups

(such as methyls) attached to DNA, which influence gene activity. They segregate with the DNA strands during replication and nucleate the reconstruction of similar marks in daughter cells.
4. RNA-mediated inheritance. For example, silent transcriptional states are actively maintained through repressive interactions between small, transmissible, replicating RNA molecules and the mRNAs to which they are partially complementary.

These four types of EIS are interrelated and interact in various ways. For example, RNA-mediated gene silencing seems to be closely associated with DNA methylation, a chromatin marking EIS, and some chromatin marks may be generated through structural templating processes. The categories are therefore crude, and there are probably other types of non-DNA cellular inheritance as well.

The epigenetic information that a cell receives depends on the conditions that ancestral cells have experienced – on which genes have been induced to be active, which proteins are present, and how they are organized. Passing on induced changes in epigenetic states is crucial for normal development. Unfortunately, transmitting cellular epigenetic changes can also have pathological effects, as it does with some cancers and during aging.

Heritable epigenetic modification sometimes affects whole chromosomes. This is the case in female mammals, where all (or almost all) of one of the two X chromosomes in each cell is inactivated during early embryogenesis, and this state is then stably inherited by all daughter cells in the lineage. Inactivation is brought about by chromatin remodeling and RNA-mediated epigenetic mechanisms. During mitotic cell division, the epigenetic state of the active and inactive X is very stable. However, during gametogenesis the inactive X is reactivated, so the different epigenetic states are not transmitted through meiosis to the next generation.

Sometimes epigenetic states that are mitotically inherited are reset, rather than abolished, during meiosis. A well-known example is genomic imprinting, in which the epigenetic state of a gene, chromosomal domain, or whole chromosome depends on the sex of the transmitting parent (and thus on whether the germ cells undergo oogenesis or spermatogenesis). The chromatin marks on genes inherited from the father are different from those on maternally derived genes, and consequently whether or not a particular gene is expressed may depend on the sex of the parent from which it was inherited. This has had interesting evolutionary consequences (Section 7), the outcome of which is that when the imprinting system goes wrong in humans, the resulting disorders mainly affect growth and behavioral development (Constância et al. 2004).

With imprinting, the epigenetic state is reset when the chromosome goes through the opposite sex, but there is increasing evidence that some epigenetic variations are neither abolished nor reset during meiosis. They are transmitted and affect offspring, just like DNA variations. Indeed, often they were at first assumed to be conventional gene mutations. The number and variety of examples of these transgenerationally transmitted epigenetic variations is increasing rapidly. One case that we described in *E4D* was that of mice with an epigenetically inherited phenotype that includes yellow coat color, obesity, and a propensity for cancer. The degree of expression of this phenotype is inherited, and is correlated with the chromatin mark (extent of methylation) associated with a particular DNA sequence. What is interesting about this case is that the phenotypes of offspring (and the underlying marks) can be changed by altering the mother's nutrition during gestation (Dolinoy et al. 2006). Other, comparable cases of induced effects are being investigated. A recent series of experiments with rats has shown how some industrial compounds that are endocrine disruptors can cause epigenetic changes in germline cells that are associated with testis disease states; the changes are inherited for at least four generations (Anway et al. 2005). In humans, Marcus Pembrey and his colleagues (2006) are studying the transgenerational effects of smoking and food supply in the male line, and have concluded from their analysis of body mass and mortality that some mechanism for transmitting epigenetic information must exist.

We could catalogue many more examples of transgenerational epigenetic inheritance in animals, but most of the best examples are found in plants. The scope and evolutionary importance of this type of inheritance in plants is well recognized and is receiving a lot of attention from botanists (e.g., Rapp & Wendel 2005). There may be good evolutionary reasons why plants show so much epigenetic inheritance. In contrast to most animals, where the germline is segregated off quite early in embryogenesis, the germline of plants is repeatedly derived from somatic cells (which is why we can propagate flowering plants by taking cuttings). Consequently, epigenetic states established during the development of the plant soma may sometimes persist and be transmitted to the next generation. This may be of adaptive significance. Animals can adjust to new circumstances behaviorally, whereas plants do not have this option and use non-behavioral strategies. We argued in E4D that induced epigenetic changes and their inheritance may do for plants what learnt behaviors and their transmission do for animals.

Although we think that EISs are particularly important in plants, we believe that epigenetic variation is significant in the evolution of all groups, including vertebrates. Unlike most genetic variations, commonly epigenetic variations are induced, are repeatable, are reversible, and often occur at a higher rate than gene mutations. These properties make their effects on evolution very different from those of genetic variations: Evolutionary change can be more rapid and have more directionality than gene-based models predict.

4.2. Developmental Endowments and Ecological Legacies

It is not clear how much information in addition to that transmitted through DNA sequences is passed to offspring by the germline cell-to-cell route. It used to be assumed that the size of sperm means they can carry little information other than that in DNA, but it is now acknowledged that fathers transmit a lot through the cellular epigenetic routes we have just described. Mothers have additional routes of information transfer through materials in the egg and, in mammals, through the womb and milk. Both parents can also transfer information through feces, saliva, and smells. The transmission of epigenetic information by body-to-body routes has been recognized in many different species of animals, and also in plants (Mousseau & Fox 1998). In all body-to-body inheritance of this type, variations are not transmitted through the germline. Rather, offspring receive materials from their parents that lead them to reconstruct the conditions that caused the parents to produce and transfer the material to them, and thus they pass on the same phenotype to their own descendants.

The long-term effects of prenatal conditions and early parental care on human physiology are attracting increasing attention. A mother's nutrition during pregnancy, for example, is known to have profound effects on the health of her offspring when they are adults (Bateson et al. 2004; Gluckman & Hanson 2005). Sometimes the effects are surprising: for example, malnutrition during pregnancy increases the likelihood of obesity and related problems in adult offspring. There are interesting evolutionary theories about why this occurs (Gluckman & Hanson 2005). However, we are more interested in cases in which a phenotype that was induced during early development is later transmitted (or has the potential to be transmitted) to the individual's own offspring and subsequent generations, since it is then justifiable to speak about the "inheritance" of the induced trait. Examples of this type of heredity were recognized in animals many years ago (Campbell & Perkins 1988), and there is now some evidence that it occurs in humans (Gluckman & Hanson 2005). Most cases involve body-to-body transmission through the uterine environment. In E4D, we used the example of lines of Mongolian gerbils in which a male-biased sex ratio and aggressive female behavior is perpetuated, probably because the mother's phenotype reconstructs a testosterone-rich uterine environment that induces the same hormonal and behavioral state in her daughters.

Animals continue to receive information from their mother (and sometimes father) after birth. In *E4D* we used the results of experiments with European rabbits to illustrate the variety of routes through which youngsters acquire information about their mother's food preferences. These experiments showed that information is transmitted during gestation (presumably through the placenta or uterine environment), while suckling (either through milk or the mother's smell), and by eating the mother's feces. The substances transferred enable the young to reconstruct their mother's food preferences. When they leave the burrow, knowing what is good and safe to eat is an obvious advantage.

Even when an animal becomes independent of the direct influences of its parents, it may inherit information from past generations because it occupies an ecological niche that they created. By affecting the development and behavior of animals as they grow up, the nature of the niche created in one generation may lead to the reconstruction of the same type of niche in the next. Odling Smee et al. (2003) have described many examples of niche-construction activities in groups ranging from bacteria to mammals, and Turner (2000) has given some dramatic examples of sophisticated ecological engineering by animals. The paradigmatic example of niche-construction is the dam built by beavers, and the inheritance and maintenance of the dam and the environment it creates by subsequent generations. Ecological inheritance of this type is the result of developmental processes that are reconstructed in every generation. From the niche-constructing organism's point of view, the ancestrally constructed environment provides it with a developmental resource, and through its activity, the organism, in turn, bequeaths a similar resource to its offspring.

5. Animal Traditions: Transmission through Socially Mediated Learning

It is very difficult to erect boundaries between epigenetic and behavioral inheritance. In *E4D*, we classified information transmission through the transfer of substances – a category of inheritance that Sterelny (2004) has called "sample-based inheritance" – with behavioral inheritance, because commonly, body-to-body substance transmission is the outcome of how parents behave. In this précis, we have grouped body-to-body information transfer with germline cell-to-cell epigenetic inheritance, because in both cases, information transfer is through material substances. Both ways of classifying inheritance seem to have legitimacy, although neither is entirely satisfactory.

In Chapter 5 of *E4D*, as well as considering transmission involving the transfer of materials, we looked at the transfer of visual or auditory information through socially mediated learning. No one doubts that socially mediated learning can have long-term, transgenerational effects that can sometimes lead to traditions, but for many years the amount and scope of this type of information transfer in nonhuman animals have been underplayed, and its evolutionary implications neglected. Only recently have animal traditions been given a more central role. There are now a number of new studies (e.g., Hunt & Gray 2003; Rendell & Whitehead 2001; Whiten et al. 2005) and several books about it (e.g. Avital & Jablonka 2000; Fragaszy & Perry 2003; Reader & Laland 2003).

In *E4D*, we distinguished between two types of socially mediated learning – non-imitative and imitation-based social learning – and used some well-known examples to illustrate them. For non-imitative social learning leading to an animal tradition, we used the ability of tits to open milk bottles. In parts of England and elsewhere, this behavior spread rapidly because naïve tits learnt, when in the presence of experienced individuals, that milk bottles are a source of food. A less familiar case is the tradition of opening pine cones and eating the inner kernels that developed in black rats living in Jerusalem-pine forests in Israel. In this case, maternal behavior provides conditions that enable the young to acquire this new and rather complex practice. Another time-honored example is that of the Koshima macaques, who learnt to wash sweet potatoes from an innovative young female. In all of these three cases, imitation was probably not involved – naïve animals learnt what to do from experienced individuals by being exposed to their behavior and its effects, but they did not learn how they did it. They seem to have learnt how to do these things through their own trials and errors, with the social environment providing selective cues and opportunities for learning.

With imitative learning, animals learn both what to do and how to do it by observing the way experienced individuals behave. Humans are great vocal and motor imitators, of course, but vocal imitation is also well developed in songbirds and cetaceans, and these vocal traditions have received a lot of attention. Motor imitation, on the other hand, seems to be much less common, although it is not clear that there is not some degree of motor imitation in social mammals.

Information transmission by the body-to-body route, whether through substances or through behavior, has very different properties from transmission by the genetic and epigenetic cell-to-cell route. First, with the exception of information transmitted in the egg and, in mammals, in utero (which, with today's technology, need not be an exception), body-to-body transmission is not always from parents to offspring. Information can be inherited from foster parents and, with imitative and non-imitative social learning, from related or unrelated members of the group or even from other species. Second, with behavioral transmission, in order for a habit, skill, preference, or other type of knowledge to be transmitted, it has to be displayed. There is no latent information that can skip generations as there is with the genetic system. Third, unlike most new information transmitted by the cellular route, new behaviorally transmitted information is not random or blind. What an innovating individual transmits depends on its ability to learn something by trial-and-error or by other methods and to reconstruct, adjust, and generalize it. The potential receiver of information is not a passive vessel, either: Whether or not information is transferred depends on the nature of the information and the experiences of the receiving animal.

In some cases, socially mediated learning may involve a combination of different transmission routes. These can cooperatively and synergistically combine to reinforce and stabilize the behavior pattern. Following Avital and Jablonka (2000), we argued in *E4D* that traditions – behavior patterns that are characteristic of an animal group and are transmitted from one generation to the next through socially mediated learning – are very common. They can affect many aspects of an animal's life, from habitat choice, to food preferences and food handling, predation and defense, and all aspects of mating, parenting, and social interactions with other group members. Social learning, especially early learning, has very strong, long-term effects, and some traditions are very stable. They can evolve through cumulative additions and alterations, with one behavior being the foundation on which another is built. Different behaviors may reinforce each other, creating a stable complex of behaviors – in other words, a lifestyle. We suggested that such cultural evolution might be partly responsible for complex behaviors, such as bower-building by bowerbirds, which are usually regarded as exclusively a result of the stability of genetic resources.

Social learning that does not involve symbolic communication is as common in humans as in other mammals. Aspects of our food preferences, our choices of habitat and mate, our parenting style, and pair bonding are based on learning mechanisms that we share with other animals. However, in humans, every aspect of life is also associated with symbol-based thinking and communication, particularly through language. Because the symbolic system enables an expansion of information transmission that is so great and so different, we have treated it as a dimension of heredity in its own right.

6. Symbol-Based Information Transmission

Similar to other inheritance systems, the symbolic system enables humans to transmit information to others, but in this special case it also enables humans to communicate with themselves: the symbolic mode of communication is a mode of thought. It permeates everything that humans do, from the most mundane activities to the most sublime.

In *E4D* (Chapter 6), we stressed the special properties of symbolic communication, using the linguistic system as our main example. We defined a symbolic system as a rule-bound system in which signs refer to objects, processes, and relations in the world, but also evoke and refer to other symbols within the same system. Symbolic communication extends the quality, quantity, and range of the information transmitted, and, as symbols are units of meaning (words, sentences, images, vocal units, etc.), they are amenable to combinatorial organization, which can be recursive and theoretically unlimited in scope. However, combinatorial potential is not sufficient for a developed symbolic system: The rules that underlie and organize symbols into a system must ensure that most combinations will not be nonsensical, must allow rapid evaluation (at all levels – truth value, emotional value, action directive), and thus must have functional consequences. The symbolic system of communication enables reference not only to the here and now, but to past, future, and imaginary realities. It profoundly affects behavior by enabling reference to the not-here and not-now. This qualitatively extends the range of possibilities of symbolic communication. Because reference to past and future allows direct references to the relations between causes (past) and effects (present or future), as well as reference to abstract (i.e., logical) relations, symbolic systems enormously extend the potential for transmitting information. They also lead to a requirement for learning, because their own elements and structure undergo updating as the system becomes more sophisticated and is applied to new domains of life and thought.

Language is an excellent example of a symbolic system of communication, but so too are mathematics, music, and the visual arts. The various symbolic systems are, however, different – the type of modularity in each system, the "mobility" of the "units," and the types of principles binding the system together are not the same and apply to different levels of individual and social organization. Symbolic information, like all information transmitted behaviorally, can be passed to unrelated individuals, but unlike the type of information discussed in the last section, it can also remain latent and unused for generations (most obviously with written words). In the latter respect, as well as in the wealth of variations that are possible, it is like the DNA-based system.

The work of anthropologists and social scientists has shown that cultural evolution rivals DNA-based evolution in its range and complexity. However, the two popular theories that dominate discourse on the evolution of culture – memetics and evolutionary psychology – provide what many see as unsatisfactory explanations of culture and the way it changes. We believe that this is because both are based on neo-Darwinian models of evolution that do not incorporate the developmental aspect of cultural innovation and transmission. Other approaches, such as that taken by Richerson and Boyd (2005), make development much more central and acknowledge the direct effects of developmental learning mechanisms on cultural evolution. Memetics is a theory of culture which was developed in analogy with, and as an extension of, the selfish gene view of Richard Dawkins. It is based on the idea that cultural units of information (memes) reside in the brain, are embodied as localized or distributed neural circuits, have phenotypic effects in the form of behaviors or cultural products, and move from brain to brain through imitation (Dawkins 1982). Memes are "replicators" and are comparable to genes. From our perspective, there is one basic problem with the meme concept, and this is that it ignores development as a cause of cultural variation. The assumption that the meme can be seen as a replicator, rather than as a trait that is the result of development, is false. How can a circuit in the brain, which is developmentally constructed during learning, be seen as anything other than a phenotypic trait? If we accept, as we must, that the brain circuit underlying a facet of culture is a developmentally reconstructed trait, then we have to accept that it is sensitive to environmental influences and that acquired (learned) modifications in it (and its many physiological correlates) are transmitted to others. The distinction between cultural "replicators" and cultural "phenotypes" is simply untenable.

Even focusing on "symbolic" memes, which can be communicated without concomitant actions (humans can pass on a command but not implement it), does not solve the problem, because development still cannot be ignored. Symbols and symbolic-system rules must be learnt, and learning is an aspect of development. Most imitation and the use of symbols is not machine-like – it is not blind to function, but is governed by understanding and by perceived goals. It is impossible to ignore the instructional aspects of the generation of new memes, which are central to the symbolic system. We therefore think that although memetics rightly stresses the autonomy of cultural evolution and the complexity of interrelations between memes, it is inadequate as an evolutionary theory of culture because of the false dichotomy that it has created between cultural memes and cultural phenotypes.

We are also critical of most versions of evolutionary psychology. Evolutionary psychologists stress the universal aspects of human-specific propensities and behavior, including cultural behavior. They focus on the genetically evolved basis of the human cultural ability. This, of course, is important. However, it leads to assumptions and inferences about the evolved structure of the mind and the evolved genetic basis of psychological strategies, which we think are very problematical. The main problem is the downplaying of the autonomy of cultural evolution and the conjecture that the diverse behavioral strategies are underlain by specifically selected genetic networks. In *E4D*, we illustrated the problem with a thought experiment that shows how purely cultural evolution could lead to a universal and stable cultural product (literacy) that has all the properties that would indicate to some that it has a specifically selected genetic basis, which it certainly does not.

We conclude that genetic and cultural selective processes are important in human evolution, but they cannot be considered independently from the social construction processes at the individual and group levels that have been recognized and emphasized by the social sciences. Development, learning, and historical construction are central to the generation of cultural entities, to their transmissibility, and to their selective retention or elimination.

7. Putting Humpty Dumpty Together Again: Interactions between Genetic, Epigenetic, Behavioral, and Symbolic Variations

In the first two parts of *E4D* we described the genetic, epigenetic, behavioral, and symbolic systems of information transfer, stressing the relative autonomy of each. When looking at evolution, an analysis that focuses on a single system of transmission is appropriate for some traits, but not for all. Every living organism depends on both genetic and epigenetic inheritance, many animals transmit information behaviorally, and humans have an additional route of information transfer through symbol-mediated communication. These four ways of transmitting information, with their very different properties, mechanisms, and dynamics, are not independent, and their interactions have been important in evolution. The third part of *E4D* was an attempt to "put Humpty Dumpty together again" by looking at the interrelationships and evolutionary interactions between the different inheritance systems. At present, only a few have been worked out, and even those are only partially understood. However, the cases that have been studied show that there is a surprising richness in the multidisciplinary approach.

We started (Chapter 7) with a discussion of the direct and indirect interactions between the genetic and epigenetic systems. It is obvious that changes in DNA sequences must affect chromatin marks. A mutation changing a cytosine to thymine, for example, may abolish a potential cytosine methylation site. Similarly, changes in control sequences may affect the binding affinity of protein and RNA regulatory elements, and thus directly influence the epigenetic inheritance of states of gene activity. Even greater effects are seen when cells suffer a genomic shock, such as the DNA damage that follows irradiation: For several generations, the descendants of irradiated parents have elevated somatic and germline mutation rates, an effect that has been attributed to induced heritable changes in epigenetic marks on the genes involved in maintaining DNA integrity (Dubrova 2003). In plants, hybridization, another type of genomic shock, causes targeted epigenetic (and genetic) changes at particular chromosomal sites and in certain families of sequences. These sites and sequences are altered in a specific and predictable way, and the modifications are transmitted across generations (e.g., see Levy & Feldman 2004).

Not only do genetic changes affect epigenetic variations, but epigenetic variations affect DNA sequences. Changes in chromatin marks affect the mobility of transposable elements and the rate of recombination, so they affect the generation of genetic variation. Ecological factors such as nutritional stresses or temperature shocks can lead to targeted changes in both chromatin and DNA, and often the epigenetic changes are primary; they probably act as signals that recruit the DNA-modifying machinery (Jorgensen 2004). Direct interactions between the genetic and epigenetic systems seem to be of importance in plant adaptation and speciation (Rapp & Wendel 2005), but ecological and genomic stresses may also have direct effects on the evolution of animals (Badyaev 2005; Fontdevila 2005). The burst of interacting genetic and epigenetic variations that is induced by stress suggests that the rate of evolutionary change may be far greater than is assumed in most models of evolution.

As well as their direct influences on the generation of genetic variation, EISs have enormous indirect effects on evolution through genetic change. Without efficient epigenetic systems that enable lineages to maintain and pass on their characteristics, the evolution of complex development would have been impossible. However, efficient epigenetic inheritance is a potential problem for multicellular organisms, because each new generation usually starts from a single cell – the fertilized egg – and that cell has to have the capacity to generate all other cell types. We believe that past selection of genetic and epigenetic variations that improve the capacity of potential germline cells to adopt or retain a totipotent state may help to explain the evolution of features in development, such as (1) the relatively early segregation and quiescent state of the germline in many animal species; (2) the difficulty of reversing the differentiated state of their somatic cells; and (3) the mechanisms that erase chromatin marks during gametogenesis and early embryogenesis. The evolution of cellular memory necessitated the evolution of timely forgetting!

Not everything is forgotten, however. As we have already indicated, the new embryo does have epigenetic legacies from its parents, including those known as genomic imprints. We think that originally these may have been a by-product of the different ways that DNA is packaged in the sperm and egg, which resulted in the two parental chromosomes in the zygote having different chromatin structures. Some of these differences were transmitted during cell division and affected gene expression, so when and where a gene was expressed depended on whether it was transmitted through the mother or the father. When this was disadvantageous, selection would have favored genes in the parents and offspring that eliminated the differences, but occasionally the difference was exploited. Haig and his colleagues have suggested how the conflicting influences of parents in polygamous mammals may have led to the evolution of imprints and imprinting mechanisms that have effects on embryonic growth and development (Haig 2002). Epigenetic inheritance may also have had a key role in the evolution of mammalian sex chromosomes and some of their peculiarities, such as the relatively large number of X-linked genes associated with human brain development and the overrepresentation of spermatogenesis genes on the X (Jablonka 2004c).

There is a general sense in which the nongenetic inheritance systems can affect genetic evolution. In new environmental conditions, all organisms can make developmental adjustments through cellular epigenetic changes; animals can also make behavioral modifications, and humans can solve problems using their symbolic systems. If conditions persist, natural selection will favor the most well-adjusted phenotypes and the genes underlying them – the genes whose effects lead to a more reliable, faster, developmental adjustment, or the ones with fewer undesirable side-effects. Waddington (1975), whose work we discussed in some detail in E4D, coined the term genetic assimilation to describe the process through which natural selection of existing genetic variation leads to a transition from an environmentally induced character to one whose development becomes increasingly independent of the inducing conditions. A more inclusive concept, genetic accommodation, has been suggested by West-Eberhard (2003). Genetic accommodation includes not only cases in which developmental responses become, through selection, more canalized (less affected by changes in the environment and the genome), but also cases in which they become dependent on different or additional features of the environment, which leads to altered or increased developmental plasticity. We think this concept is valuable, but at the time we wrote E4D we had not fully accommodated to it, so we used it more sparingly than we would do now, and framed most of our discussion in terms of genetic assimilation.

Genetic assimilation can occur only when the developmental response is called for repeatedly over many generations, which happens either (i) because the environmental change persists (e.g., a long-lasting climatic change), or (ii) because the organism's activities lead to increased ecological stability (e.g., through a constructed niche such as the beavers' dam), or (iii) through intergenerational epigenetic inheritance. In the last case, the transmitted cellular epigenetic state, behavior, or culture provides the transgenerational continuity necessary to effect significant genetic change.

In Chapter 7 we described several experiments showing how induced cellular epigenetic changes in organisms ranging from yeasts to mammals can reveal previously hidden genetic variation whose selection can lead to evolutionary change. The molecular bases of some of these examples of genetic assimilation have been worked out. In one particularly interesting case in the fruit fly *Drosophila*, the selectable variations that the inducing agent revealed were not previously cryptic genetic variations but new epigenetic variations.

Genetic assimilation can occur not only with environmentally induced changes in form, but also with persistent changes in behavior. In E4D (Chapter 8), we described Spalding's old (1873; reproduced in Haldane 1954) but entertaining scenario of a learned response (talking in parrots) that through selection for improvements in learning become an instinct. We went on to show how when previously learned behaviors are genetically assimilated and hence become more "automatic," this may enable the animal to learn an additional pattern of behavior because the former learning effort is no longer necessary. Avital and Jablonka (2000) called this process the assimilate– stretch principle and suggested that it could explain how lengthy and complex sequences of "innate" behaviors have evolved.

As with other learned behaviors, human culture has affected genetic evolution. A well-known example is the way in which the domestication of cattle led to changes in the frequency of the gene that enables adult humans to absorb the milk sugar lactose. As cattle were domesticated, milk became a potential source of energy, but adult humans, like most mammals, cannot break down lactose, so unprocessed milk causes indigestion and diarrhea. Nevertheless, drinking fresh milk has definite advantages in certain populations – most notably those in northern countries, where sunlight is in short supply and vitamin D is therefore scarce. Lactose, like vitamin D, enables calcium (which is plentiful in milk) to be absorbed from the intestine, and hence prevents rickets and osteomalacia. Consequently, in northern countries, people who carried the uncommon allele that enabled them to break down lactose when adults were healthier, and through natural selection this allele became the most common one. The beneficial effects of milk drinking in northern populations are reflected in their myths, which presumably have an educational value and further encourage the dairying culture and milk-drinking habit.

A good example (which we did not use in *E4D*) of a cultural change that has guided genetic change is the effect of the cultural spread of sign language among congenitally deaf people (Nance & Kearsey 2004). Until the invention and use of sign language, deaf people were cognitively, socially, and economically handicapped, and rarely had children, but once sign language began to be used and they became cognitively adept, many of their social disadvantages disappeared. Naturally, they tended to marry other people with whom they could communicate. As a result of deaf-by-deaf marriage and the improved chances of their surviving and having children, in the United States the frequency of people with the most common type of deafness, connexin deafness, has doubled over the last 200 years. Nance and Kearsey suggest that the evolution of speech in the hominid lineage may have been promoted by a comparable process, in which those with effective oral communication chose others who were similarly endowed and in this way speeded up the fixation of genes affecting speech and speech-dependent characteristics.

Cultural practices probably affected not only the spread of genes underlying oral communication, but also the cumulative evolution of the language capacity itself. In *E4D* we argued that neither the Chomskians nor the functionalists provided a satisfactory explanation of this. The explanation we offered took as its starting point the suggestion that linguistic communication involves the grammatical marking of a constrained set of core categories that describe who did what to whom, when, and how. Following Dor and Jablonka (2000), we argued that the ability to rapidly learn to recognize and mark these categories evolved through partial genetic assimilation. There was a continual interplay between the cultural and genetic systems in which the invention and transmission of linguistic rules that were useful (e.g., the distinction between the categories of one/more-than-one) was at first cultural. Because individuals who had a genetic constitution that made learning the rule more reliable, rapid, and effective had an advantage, partial genetic assimilation occurred. Further linguistic innovation and spread led to more genetic assimilation. Thus, as they accumulated, the basic rules of language became very easy to learn. We believe that this type of process, in which cultural innovation and spread comes first and genetic change follows, has been important not only in the evolution of the language capacity, but also in the evolution of other aspects of human cognitive capacities.

8. The Evolution of Information-Transmission Systems

We argued in *E4D* that there are four types of heredity system that can produce variations that are important for evolution through natural selection. Some of the variations they transmit seem to be goal-directed: they arise in response to the conditions of life and are targeted to particular functions. In the penultimate chapter (Chapter 9), we looked at the evolutionary origins of these systems that enable "the educated guess" – systems that limit the search space and increase the likelihood that some of the variations generated will be useful. There is no great mystery about their evolution: they arose through natural selection as a side-effect or modification of functions that evolved for other purposes. For example, stress-induced mutation probably evolved as a modification of mechanisms that were originally selected to repair DNA, and targeted mutation arose through the selection of DNA sequences that are prone to repair and replication errors.

The evolution of epigenetic inheritance systems, which are found in all organisms, must have begun in simple unicellular organisms. Some types of EIS, such as transmission of self-sustaining feedback loops and certain structural elements, would be automatic by-products of selection for the maintenance of cellular structures and functions. With others, such as chromatin marks and RNA-based inheritance, their evolution may have been tied up with the selection of mechanisms for the packaging and protection of DNA, and for defending the cell against foreign or rogue DNA sequences. Once adaptive epigenetic systems were in place in the cell, in certain conditions the ability to pass on their adapted state was an advantage. The environments that would favor transmitting existing epigenetic states to daughter cells are those that fluctuate, but not too often. When environmental conditions change very frequently, cells adapt physiologically; when the change occurs over a very long time-span (hundreds of generations), cells can adapt genetically; but when the changes occur on the intermediate time scale (every 2 – 100 generations), passing on the existing epigenetic state (having cell memory) is beneficial. Daughter cells get "free" information from their parents and do not have to spend time and energy finding appropriate responses themselves.

Behavioral transmission also results in progeny getting selected useful information from their parents, and is also of advantage in environments that fluctuate. The body-to-body transmission of various substances through the egg, uterus, milk, feces, and so on is probably inevitable, but when it is advantageous to the young, selection would favor genetic changes that made the transmission and the response to it more reliable. Similarly, socially mediated learning is inevitable when youngsters learn in social conditions, but it became a major route of information transfer through selection for paying attention to and learning from those from whom the young can acquire information about what is good to eat, how to find it, how to avoid predators, and so on. In the hominid lineage, the social system resulted in communication traditions that led to selection for genetically better communicators and better ways of communicating. Ultimately, partial genetic assimilation of the ability to learn useful vocal and gestural signs and rules produced the relatively easy-to-learn symbolic systems of human societies.

The origins of all the non-genetic inheritance systems, which sometimes transmit induced and targeted information to daughter cells or organisms, are unexceptional. However, the effects they had were dramatic. We argued in *E4D* that some of the greatest evolutionary transitions were built on new ways of transmitting information, which opened up new ways of adapting to the conditions of life. The transition from unicells to multicellular organisms with several types of cell would be impossible without quite sophisticated EISs; behavioral information transmission was crucial for the formation of complex social groups; and in the primate lineage, the emergence of symbolic communication led to the explosive cultural changes we see in human societies.

9. Conclusions

At the beginning of this précis we suggested that evolutionary theory is undergoing a profound change. Instead of the DNA-centered version of Modern Synthesis Darwinism that dominated the latter part of the 20th century, a new version of evolutionary theory is emerging, in which:

> i. heredity is seen as the outcome of developmental reconstruction processes that link ancestors and descendants and lead to similarity between them. It includes both function-blind replication processes (such as DNA replication) and reconstruction processes that depend on and are determined by function. As Oyama (1985) and Griffiths and Gray (1994) have argued,

DNA is a crucial, but not exclusive, heritable developmental resource.

ii. units of heritable variation are genes (alleles), cellular epigenetic variations (including epialleles), developmental legacies transmitted by the mother during embryogenesis, behavioral legacies transmitted through social learning, symbolic information, and ecological legacies constructed by ancestral generations. All can be thought of as "units" of heredity, although commonly they are not very discrete.

iii. new heritable variation can be purely fortuitous in origin and blind to function (like most classical mutations), but some is directed, produced as a developmentally constructed response to the environment.

iv. units of selection or targets of selection are what James Griesemer (2002) terms reproducers. These are entities that display differential reproduction – mainly individuals, but also groups and species, and, in the precellular world, replicating molecules and molecular complexes.

v. units of evolution are heritably varying types (mainly types of traits), the frequency of which changes over evolutionary time.

vi. evolution occurs through the set of processes that lead to changes in the nature and frequency of heritable types in a population.

One of the main things we wanted to establish in *E4D* is that there is a wealth of data showing the richness and variety of heredity processes. Epigenetic inheritance is present in all organisms: It is not an unusual and bizarre exception to the rules of heredity, but an important, mainstream, hereditary process. Behavioral inheritance is an uncontroversial mode of information transmission in social animals, and symbols are central to human life and hominid evolution. All these modes of transmission lead to transgenerational phenomena and processes that are of huge practical importance for medicine, for agriculture, for ecology, and for conservation issues. It is clearly not possible to reduce heredity and evolution to genes, not just because the interrelationships are very complicated (which they are), but because of the partial autonomy of different systems of inheritance. Although the view we suggest is in some ways more complex than the gene-based view, it leads to more realistic and often simpler alternative interpretations of developmental and evolutionary events and processes.

As biologists recognize that the concepts of heredity and evolution have to go beyond DNA and "selfish genes," and acknowledge that behaviorally and culturally transmitted variations have been significant in the evolution of animals and man, some of their antagonism towards the social sciences may disappear. Incorporating a broader concept of heredity into evolutionary thinking may also help to remove some of the social scientists' prejudices about biological interpretations of human behaviors and societies. In future, a biologist will need to be more of a social scientist, and a social scientist will need to be more of a biologist.

We predict that in 20 years time, the late 1990s and the first decade of the 21st century will be seen as revolutionary years for evolutionary theory. The effects of the synthesis that is now emerging, and which incorporates development, will be comparable, we believe, to the revolutionary change that followed the introduction of Mendelian genetics into evolutionary thinking during the Modern Synthesis of the late 1930s. Like the former synthesis, the emerging "post-Modern" synthesis is the result of a collective effort. It brings together the mass of information coming from the many branches of molecular biology, developmental biology, medicine, ecology, hybridization studies, experimental studies of behavior, developmental and social psychology, the cognitive sciences, anthropology, and sociology. The new version of evolutionary theory can no longer be called neo-Darwinian, because it includes, in addition to the neo-Darwinian process of selection of randomly generated small variations, significant Lamarckian and saltational processes. Whatever it is called, a new transformed Darwinian theory is upon us.

References

[The letters "a" and "r" before author's initials stand for target article and response references, respectively.]

Aboitiz, F. (1988) Epigenesis and the evolution of the human brain. *Medical Hypotheses* 25(1):55–9. [CMH]

Aboitiz, F. (1992) Mechanisms of adaptive evolution. Darwinism and Lamarckism restated. *Medical Hypotheses* 38(3):194–202. [CMH]

Aboitiz, F. (1995) Working memory networks and the origin of language areas in the human brain. *Medical Hypotheses* 44(6):504–506. [CMH]

Aboitiz, F., Garcia, R. R., Bosman, C. & Brunetti, E. (2006) Cortical memory mechanisms and language origins. *Brain and Language* 98(1):40–56. [CMH]

Anway, M. D., Cupp, A. S., Uzumcu, M. & Skinner, M. K. (2005) Epigenetic transgenerational actions of endocrine disruptors and male fertility. *Science* 308:1466–69. [aEJ]

Aoki, K., Wakano, J. Y. & Feldman, M. W. (2005) The emergence of social learning in a temporally changing environment: A theoretical model. *Current Anthropology* 46:334–40. [AM]

Arbib, M. A. (2005) From monkey-like action recognition to human language: An evolutionary framework for neurolinguistics. *Behavioral and Brain Sciences* 28(2):105–24. Available at: http://www.bbsonline.org/Preprints/Arbib-05012002 [AS]

Ashikawa, I. (2001) Surveying CpG methylation at 5′-CCGG in the genomes of rice cultivars. *Plant Molecular Biology* 45:31–9. [rEJ]

Avital, E. & Jablonka, E. (2000) *Animal traditions: Behavioural inheritance in evolution.* Cambridge University Press. [arEJ]

Badyaev, A. V. (2005) Stress-induced variation in evolution: From behavioural plasticity to genetic assimilation. *Proceedings of the Royal Society of London B: Biological Sciences* 272:877–86. [aEJ]

Baglioni, C. (1967) Molecular evolution in man. In: *Proceedings of the Third International Congress of Human Genetics,* Chicago, IL, September 5–10, 1966, ed. J. F. Crow & J. V. Neel. The Johns Hopkins University Press. [CLB]

Baptista, L. F. & Trail, P. W. (1992) The role of song in the evolution of passerine diversity. *Systematic Biology* 41(2):242–47. [ZF]

Bates, E., Benigni, L., Bretherton, I., Camaioni, L. & Volterra, V. (1979) *The emergence of symbols: Cognition and communication in infancy.* Academic Press. [ASW]

Bateson, P., Barker, D., Clutton-Brock, T., Deb, D., D'Udine, B., Foley, R. A., Gluckman, P., Godfrey, K., Kirkwood, T., Lahr, M. M., McNamara, J., Metcalfe, N. B., Monaghan, P., Spencer, H. G. & Sultan, S. E. (2004) Developmental plasticity and human health. *Nature* 430:419–21. [aEJ]

Baylin, S. B. & Ohm, J. E. (2006) Epigenetic gene silencing in cancer—a mechanism for early oncogenic pathway addiction? *Nature Reviews Cancer* 6:107–16. [rEJ]

Belyaev, D. K. (1979) Destabilizing selection as a factor in domestication. *Journal of Heredity* 70:301–308. [rEJ]

Belyaev, D. K. & Borodin, P. M. (1982) The influence of stress on variation and its role in evolution. *Biologisches Zentralblatt* 100:705–14. [aEJ]

Belyaev, D. K., Ruvinsky, A. O. & Borodin, P. M. (1981a) Inheritance of alternative states of the fused gene in mice. *Journal of Heredity* 72(2):107–12. [ZF]

Belyaev, D. K., Ruvinsky, A. O. & Trut, L. N. (1981b) Inherited activation-inactivation of the star gene in foxes: Its bearing on the problem of domestication. *Journal of Heredity* 72(4):267–74. [ZF]

Bergstrom, C. T. & Lachmann, M. (2004) Shannon information and biological fitness. *Proceedings of the IEEE Workshop on Information Theory,* San Antonio, TX, October 24–29, 2004, pp. 50–54. [rEJ]

Blackmore, S. J. (1999) *The meme machine.* Oxford University Press. [SB]

Blackmore, S. J. (2001) Evolution and memes: The human brain as a selective imitation device. *Cybernetics and Systems* 32:225–55. [SB]

Boughman, J. W. (1998) Vocal learning by greater spear-nosed bats. *Proceedings of the Royal Society B: Biological Sciences* 265(1392):227–33. [ZF]

Boyd, R. & Richerson, P. J. (1985) *Culture and the evolutionary process.* University of Chicago Press. [AM, LS]

Boyd, R. & Richerson, P. J. (1995) Why does culture increase human adaptability? *Ethology and Sociobiology* 16:125–43. [AM]

Boyd, R. & Richerson, P. J. (2005) *The origin and evolution of cultures.* Oxford University Press. [AM]

Breland, K. & Breland, M. (1961) The misbehavior of organisms. *American Psychologist* 16:681–84. [ZF]

Bridgeman, B. (2005) Action planning supplements mirror systems in language evolution. *Behavioral and Brain Sciences* 28(2):129–30. [AS]

Brisson, D. (2003) The directed mutation controversy in an evolutionary context. *Critical Reviews in Microbiology* 29(1):25–35. [TED]

Britten, R. J. (1986) Rates of DNA sequence evolution differ between taxonomic groups. *Science* 231: 393–98. [CLB]

Bryant, J. M. (2004) An evolutionary social science? A skeptic's brief, theoretical and substantive. *Philosophy of the Social Sciences* 34:451–92. [AM]

Bunn, H. F. (1997) Pathogenesis and treatment of sickle cell disease. *New England Journal of Medicine* 337: 762–69. [aEJ]

Burdge, G. C., Slater-Jefferies, J. L., Torrens, C., Phillips, E. S., Hanson, M. A. & Lillycrop, K. A. (2007) Dietary protein restriction of pregnant rats in the F_0 generation induces altered methylation of hepatic gene promoters in the adult male offspring in the F_1 and F_2 generations. *British Journal of Nutrition* 97:435–39. [rEJ]

Bybee, J. (2006) From usage to grammar: The mind's response to repetition. *Language* 82:711–33. [ASW]

Campbell, J. H. & Perkins, P. (1988) Transgenerational effects of drug and hormonal treatments in animals: A review of observations and ideas. *Progress in Brain Research* 73:535–53. [aEJ]

Caporale, L. (2003) Natural selection and the emergence of a mutation phenotype: An update of the evolutionary synthesis considering mechanisms that effect genome variation. *Annual Review of Microbiology* 57:467–85. [aEJ]

Caramelli, D., Lalueza-Fox, C., Vernesi, C., Lari, M., Casoli, A., Mallegni, F., Chiarelli, B., Dupanloup, I., Bertranpetit, J., Barbujani, G. & Bertorelle, G. (2003) Evidence for a genetic discontinuity between Neandertals and 24,000-year-old anatomically modern Europeans. *Proceedings of the National Academy of Sciences USA* 100:6583–97. [SL]

Carlson, W. B. (2000) Invention and evolution: The case of Edison's sketches of the telephone. In: *Technological innovation as an evolutionary process*, ed. J. Ziman, pp. 137–58. Cambridge University Press. [AM]

Casadesús, J. & Low, D. (2006) Epigenetic gene regulation in the bacterial world. *Microbiology and Molecular Biology Reviews* 70:830–56. [rEJ]

Cavalli-Sforza, L. L. & Feldman, M. W. (1981) *Cultural transmission and evolution*. Princeton University Press. [AM]

Cervera, M.-T., Ruiz-García, L. & Martínez-Zapater, J. (2002) Analysis of DNA methylation in *Arabidopsis thaliana* based on methylation-sensitive AFLP markers. *Molecular Genetics and Genomics* 268:543–52. [rEJ]

Chan, T. L., Yuen, S. T., Kong, C. K., Chan, Y. W., Chan, A. S. Y., Ng, W. F., Tsui, W. Y., Lo, M. W. S., Tam, W. Y., Li, V. S. W. & Leung, S. Y. (2006) Heritable germline epimutation of MSH2 in a family with hereditary nonpolyposis colorectal cancer. *Nature Genetics* 38:1178–83. [rEJ]

Chappell, J. M. & Sloman A. (2007) Natural and artificial meta-configured altricial information-processing systems. *International Journal of Unconventional Computing.* 3(3):211–39. Available at: http://www.cs.bham.ac.uk/research/projects/cosy/papers/#tr0609. [AS]

Chong, S. & Whitelaw, E. (2004) Epigenetic germline inheritance. *Current Opinion in Genetics and Development* 14:692–96. [rEJ]

Constância, M., Kelsey, G. & Reik, W. (2004) Resourceful imprinting. *Nature* 432:53–7. [aEJ]

Croft, W. (2000) *Explaining language change: An evolutionary approach*. Pearson Education. [LS]

Crow, J. F. (2001) The beanbag lives on. *Nature* 409:771. [AM]

Cullis, C. A. (2005) Mechanisms and control of rapid genomic changes in flax. *Annals of Botany* 95:201–206. [rEJ]

Darlington, C. D. (1939) *The evolution of genetic systems.* Cambridge University Press. [rEJ]

Darwin, C. (1859) *On the origin of species by means of natural selection, or the preservation of favoured races in the struggle for life.* John Murray. [aEJ, AM]

Dautenhahn, K. & Nehaniv, C. L. (2002) *Imitation in animals and artifacts.* MIT Press. [LS]

Dawkins, R. (1976) *The selfish gene.* Oxford University Press. [SB, aEJ, ZF, LS]

Dawkins, R. (1982) *The extended phenotype.* Freeman/Oxford University Press. [SB, arEJ]

Dawkins, R. (2006) *The god delusion.* Bantam. [rEJ]

de Boysson-Bardies, B. & Vihman, M. M. (1991) Adaptation to language: Evidence from babbling and first words in four languages. *Language* 67:297–319. [ASW]

Deacon, T. (1997) *The symbolic species: The co-evolution of language and the human brain.* Penguin. [SB]

DeCasper, A. J. & Fifer, W. P. (1980) Of human bonding: Newborns prefer their mothers' voices. *Science* 208:1174–76. [ASW]

DeCasper, A. J. & Prescott, P. A. (1984) Human newborns' perception of male voices: Preference, discrimination, and reinforcing value. *Developmental Psychobiology* 17:481–91. [ASW]

Deeckea, V. B., Ford, J. K. B. & Spong, P. (2000) Dialect change in resident killer whales: Implications for vocal learning and cultural transmission. *Animal Behaviour* 60(5):629–38. [ZF]

Delage, J. & Goldsmith, M. (1912) *The theories of evolution*, trans. A. Tridon (from the 1909 French edition). Palmer. [rEJ]

Dennett, D. C. (1978) *Brainstorms: Philosophical essays on mind and psychology*. MIT Press. [AS]

Dennett, D. C. (1995) *Darwin's dangerous idea*. Penguin. [SB]

Dobzhansky, T. (1937) *Genetics and the origin of species*. Columbia University Press. [aEJ]

Dolinoy, D. C., Weidman, J. R., Waterland, R. A. & Jirtle, R. L. (2006) Maternal genistein alters coat color and protects A^{vy} mouse offspring from obesity by modifying the fetal epigenome. *Environmental Health Perspectives* 114:567–72. [aEJ]

Donald, M. (1991) *Origins of the modern mind*. Harvard University Press. [LG]

Donald, M. (1993) Précis of *Origins of the modern mind* (with commentary). *Behavioral and Brain Sciences* 16(4):737–91. [LG]

Dor, D. & Jablonka, E. (2000) From cultural selection to genetic selection: A framework for the evolution of language. *Selection* 1:33–55. [arEJ]

Dubrova, Y. E. (2003) Radiation-induced transgenerational instability. *Oncogene* 22:7087–93. [aEJ]

Feldman, M. W. & Cavalli-Sforza, L. L. (1989) On the theory of evolution under genetic and cultural transmission with application to the lactose absorption problem. In: *Mathematical evolutionary theory*, ed. M. W. Feldman. Princeton University Press. [AM]

Feldman, M. W. & Laland, K. N. (1996) Gene-culture coevolutionary theory. *Trends in Ecology and Evolution* 11:453–57. [AM]

Fontdevila, A. (2005) Hybrid genome evolution by transposition. *Cytogenetic and Genome Research* 110:49–55. [aEJ]

Foss, J. (1992) Introduction to the epistemology of the brain: Indeterminacy, micro specificity, chaos, and openness. *Topoi* 11:45–58. [JF]

Foss, J. (2000) *Science and the riddle of consciousness: A solution*. Kluwer. [JF]

Fragaszy, D. M. & Perry, S. (2003) *The biology of traditions*. Cambridge University Press. [aEJ]

Gabora, L. (2003) Contextual focus: A cognitive explanation for the cultural transition of the Middle/Upper Paleolithic. In: *Proceedings of the 25th Annual Meeting of the Cognitive Science Society*, Boston, MA, July 31–August 2, 2003, ed. R. Alterman & D. Hirsch. Erlbaum. [LG]

Gabora, L. (2004) Ideas are not replicators but minds are. *Biology and Philosophy* 19:127–43. [LG, rEJ]

Gabora, L. (2006) Self-other organization: Why early life did not evolve through natural selection. *Journal of Theoretical Biology* 241(3):443–50. [LG]

Gallistel, C. R. (1999) The replacement of general-purpose learning models with adaptively specialized learning modules. In: *The cognitive neurosciences*, ed. M. Gazzaniga, pp. 1179–91. MIT Press. [TED]

Gibson, J. J. (1979) *The ecological approach to visual perception*. Erlbaum. [AS]

Gilmore, J. H., Lin, W., Prastawa, M. W., Looney, C. B., Sampath, Y., Vetsa, K., Knickmeyer, R. C., Evans, D. D., Smith, J. K., Hamer, R. M., Lieberman, J. A. & Gerig, G. (2007) Regional gray matter growth, sexual dimorphism, and cerebral asymmetry in the neonatal brain. *The Journal of Neuroscience* 27:1255–60. DOI:10.1523/JNEUROSCI.3339-06.2007 [AS]

Ginsburg, S. & Jablonka, E. (in press) The transition to experiencing. I. Limited learning and limited experiencing. II. The evolution of associative learning based on feelings. *Biological Theory*. [rEJ]

Gluckman, P. & Hanson, M. (2005) *The fetal matrix: Evolution, development and disease*. Cambridge University Press. [arEJ]

Gluckman, P. D., Hanson, M. A. & Beedle, A. S. (2007) Non-genomic but trans-generational inheritance of disease risk. *BioEssays* 29:145–54. [rEJ]

Goldstein, M. H., King, A. P. & West, M. J. (2003) Social interaction shapes babbling: Testing parallels between birdsong and speech. *Proceedings of the National Academy of Sciences USA* 100:8030–35. [ASW]

Goldstein, M. H. & West, M. J. (1999) Consistent responses of human mothers to prelinguistic infants: The effect of prelinguistic repertoire size. *Journal of Comparative Psychology* 113:52–58. [ASW]

Grant-Downton, R. T. & Dickinson, H. G. (2005) Epigenetics and its implications for plant biology. 1. The epigenetic network in plants. *Annals of Botany* 96:1143–64. [rEJ]

Grant-Downton, R. T. & Dickinson, H. G. (2006) Epigenetics and its implications for plant biology. 2. The "epigenetic epiphany": Epigenetics, evolution and beyond. *Annals of Botany* 97:11–27. [rEJ]

Griesemer, J. (2000) The units of evolutionary transition. *Selection* 1:67–80. [rEJ]

Griesemer, J. (2002) What is "epi-" about epigenetics? *Annals of the New York Academy of Sciences* 981:97–110. [aEJ]

Griffiths, P. E. & Gray, R. D. (1994) Developmental systems and evolutionary explanation. *Journal of Philosophy* 91:277–304. [aEJ]

Gulevich, R. G., Oskina, I. N., Shikhevich, S. G., Fedorova, E. V. & Trut, L. N. (2004) Effect of selection for behavior on pituitary-adrenal axis and proopio-melanocortin gene expression in silver foxes (*Vulpes vulpes*). *Physiology and Behavior* 82:513–18. [ZF]

Hager, J. B. & Wolf, R. (2006) A maternal-offspring coadaptation theory for the evolution of genomic imprinting. *PLoS Biology* 4(12):e380. [rEJ]

Haig, D. (2002) *Genomic imprinting and kinship*. Rutgers University Press. [aEJ]

Haldane, J. B. S. (1954) Introducing Douglas Spalding. *British Journal of Animal Behaviour* 2:1–11. [aEJ]

Hallpike, C. R. (1986) *The principles of social evolution*. Clarendon. [AM]

Healy, S. D., de Kort, S. R. & Clayton, N. S. (2005) The hippocampus, spatial memory and food hoarding: A puzzle revisited. *Trends in Ecology and Evolution* 20(1):17–22. [ZF]

Hirsch, H. V. B. & Spinelli, D. N. (1970) Visual experience modifies distribution of horizontally and vertically oriented receptive fields in cats. *Science* 168:869–71. [BB]

Hunt, G. R. & Gray, R. D. (2003) Diversification and cumulative evolution in New Caledonian crow tool manufacture. *Proceedings of the Royal Society of London B: Biological Sciences* 270:867–74. [aEJ]

Ingold, T. (2000) The poverty of selectionism. *Anthropology Today* 16:1–2. [AM]

Ingold, T. (2007) The trouble with "evolutionary biology." *Anthropology Today* 23:3–7. [AM]

Jablonka, E. (2002) Information: Its interpretation, its inheritance, and its sharing. *Philosophy of Science* 69:578–605. [arEJ]

Jablonka, E. (2004a) Epigenetic epidemiology. *International Journal of Epidemiology* 33:929–35. [rEJ]

Jablonka, E. (2004b) From replicators to heritably varying phenotypic traits: The extended phenotype revisited. *Biology and Philosophy* 19:353–75. [rEJ]

Jablonka, E. (2004c) The evolution of the peculiarities of mammalian sex chromosomes: An epigenetic view. *BioEssays* 26:1327–32. [arEJ]

Jablonka, E. & Lamb, M. J. (1990) Lamarckism and ageing. *Gerontology* 36:323–32. [rEJ]

Jablonka, E. & Lamb, M. J. (1995) *Epigenetic inheritance and evolution: The Lamarckian dimension*. Oxford University Press. [arEJ, AM]

Jablonka, E. & Lamb, M. J. (2005) *Evolution in four dimensions: Genetic, epigenetic, behavioral, and symbolic variation in the history of life*. MIT Press. [SB, CLB, BB, TED, ZF, JF, LG, CMH, arEJ, SL, AM, AS, LS, ASW]

Jablonka, E. & Lamb, M. J. (2006a) Evolutionary epigenetics. In: *Evolutionary genetics*, ed. C. W. Fox & J. B. Wolf, pp. 252–64. Oxford University Press. [rEJ]

Jablonka, E. & Lamb, M. J. (2006b) The evolution of information in the major transitions. *Journal of Theoretical Biology* 239:236–46. [rEJ]

Jablonka, E. & Lamb, M. J. (2007) The expanded evolutionary synthesis – a response to Godfrey-Smith, Haig, and West-Eberhard. *Biology and Philosophy* 22:453–72. [rEJ]

Jablonka, E., Oborny, B., Molnár, E., Kisdi, E., Hofbauer, J. & Czárán, T. (1995) The adaptive advantage of phenotypic memory in changing environments. *Philosophical Transactions of the Royal Society of London* B350:133–41. [rEJ]

Jaenisch, R., Hochedlinger, K., Blelloch, R., Yamada, Y., Baldwin, K. & Eggan, K. (2004) Nuclear cloning, epigenetic reprogramming, and cellular differentiation. *Cold Spring Harbor Symposia on Quantitative Biology* 69:19–27. [rEJ]

Jirtle, R. L. & Skinner, M. K. (2007) Environmental epigenomics and disease susceptibility. *Nature Reviews Genetics* 8:253–62. [rEJ]

Johannsen, W. (1911) The genotype conception of heredity. *American Naturalist* 45:129–59. [aEJ]

Jorgensen, R. A. (2004) Restructuring the genome in response to adaptive challenge: McClintock's bold conjecture revisited. *Cold Spring Harbor Symposia on Quantitative Biology* 69:349–54. [aEJ]

Kacelnik, A., Chappell, J., Weir, A. A. S. & Kenward, B. (2006) Cognitive adaptations for tool-related behaviour in New Caledonian crows. In: *Comparative cognition: Experimental explorations of animal intelligence*, ed. E. A. Wasserman & T. R. Zentall, pp. 515–28. Oxford University Press. Available at: http://www.cogsci.msu.edu/DSS/2004-2005/Kacelnik/Kacelnik_etal_Crows.pdf [AS]

Kaplan, C. A. & Simon, H. A. (1990) In search of insight. *Cognitive Psychology* 22:374–419. [AM]

Karmiloff-Smith, A. (1994) Précis of *Beyond modularity: A developmental perspective on cognitive science*. *Behavioral and Brain Sciences* 17(4):693–745. Available at: http://www.bbsonline.org/documents/a/00/00/05/33/index.html [AS]

Keller, E. F. (2000) *The century of the gene*. Harvard University Press. [arEJ]

Kisilevsky, B. S., Hains, S. M. J., Lee, K., Xie, X., Huang, H., Zhang, K. & Wang, Z. (2003) Effects of experience on fetal voice recognition. *Psychological Science* 14:220–24. [ASW]

Knox, M. R. & Ellis, T. H. (2001) Stability and inheritance of methylation states at PstI sites in *Pisum*. *Molecular Genetics and Genomics* 265:497–507. [rEJ]

Kukekova, A. V., Trut, L. N., Oskina, I. N., Johnson, J. L., Temnykh, S. V., Kharlamova, A. V., Shepeleva, D. V., Gulievich, R. G., Shikhevich, S. G., Graphodatsky, A. S., Aguirre, G. D. & Acland, G. M. (2007) A meiotic linkage map of the silver fox, aligned and compared to the canine genome. *Genome Research* 17(3):387–99. [ZF]

Kukekova, A. V., Trut, L. N., Oskina, I. N., Kharlamova, A. V., Shikhevich, S. G., Kirkness, E. F., Aguirre, G. D. & Acland, G. M. (2004) A marker set for construction of a genetic map of the silver fox (*Vulpes vulpes*). *Journal of Heredity* 95(3):185–94. [ZF]

Kussell, E., Kishony, R., Balaban, N. Q. & Leibler, S. (2005) Bacterial persistence: A model of survival in changing environments. *Genetics* 169:1807–14. [rEJ]

Kussell, E. & Leibler, S. (2005) Phenotypic diversity, population growth, and information in fluctuating environments. *Science* 309:2075–78. [rEJ]

Lachmann, M. & Jablonka, E. (1996) The inheritance of phenotypes: An adaptation to fluctuating environments. *Journal of Theoretical Biology* 181:1–9. [rEJ]

Lachmann, M., Sella, G. & Jablonka, E. (2000) On the advantages of information sharing. *Proceedings of the Royal Society of London B: Biological Sciences* 267:1287–93. [rEJ]

Laland, K. N., Kumm, J. & Feldman, M. W. (1995) Gene-culture coevolutionary theory – a test-case. *Current Anthropology* 36:131–56. [AM]

Lamb, M. J. (1994) Epigenetic inheritance and aging. *Reviews in Clinical Gerontology* 4:97–105. [rEJ]

Leakey, R. (1984) *The origins of humankind*. Science Masters/Basic Books. [LG]

Levenson, J. M. & Sweatt, J. D. (2005) Epigenetic mechanisms in memory formation. *Nature Reviews Neuroscience* 6:108–18. [aEJ]

Levy, A. A. & Feldman, M. (2004) Genetic and epigenetic reprogramming of the wheat genome upon allopolyploidization. *Biological Journal of the Linnean Society* 82:607–13. [aEJ]

Lewis, K. (2007) Persister cells, dormancy and infectious diseases. *Nature Reviews Microbiology* 5:48–56. [rEJ]

Lindberg, J., Björnerfeldt, S., Saetre, P., Svartberg, K., Seehuus, B., Bakken, M., Vilà, C. & Jazin, E. (2005) Selection for tameness has changed brain gene expression in silver foxes. *Current Biology* 15(22):915–26. [ZF]

Lipo, C. P., O'Brien, M. J., Collard, M. & Shennan, S., eds. (2005) *Mapping our ancestors: Phylogenetic approaches in anthropology and prehistory*. Aldine. [AM]

Liu, B. & Wendel, J. (2003) Epigenetic phenomena and the evolution of plant allopolyploids. *Molecular Phylogenetics and Evolution* 29:365–79. [rEJ]

Maturana, H. & Varela, F. J. (1973) *De Maquinas y Seres Vivos*: Editorial Universitaria. [CMH]

Maynard Smith, J. (2000) The concept of information in biology. *Philosophy of Science* 67:177–94. [rEJ]

Maynard Smith, J. & Szathmáry, E. (1995) *The major transitions in evolution*. Freeman/Oxford University Press. [CLB, rEJ]

Mayr, E. (1982) *The growth of biological thought*. Belknap Press. [rEJ]

McClintock, B. (1984) The significance of responses of the genome to challenge. *Science* 226:792–801. [rEJ]

Meaney, M. J. (2001) Maternal care, gene expression, and the transmission of individual differences in stress reactivity across generations. *Annual Review of Neuroscience* 24:1161–92. [rEJ]

Meltzoff, A. N. (1988) Imitation, objects, tools, and the rudiments of language in human ontogeny. *Human Evolution* 3:45–64. [ASW]

Meltzoff, A. N. & Moore, M. K. (1977) Imitation of facial and manual gestures by human neonates. *Science* 198:75–78. [ASW]

Mesoudi, A. (2007) Using the methods of social psychology to study cultural evolution. *Journal of Social,*

Evolutionary and Cultural Psychology 1: 35–58. [AM]

Mesoudi, A., Whiten, A. & Laland, K. N. (2004) Is human cultural evolution Darwinian? Evidence reviewed from the perspective of *The Origin of Species*. *Evolution* 58:1–11. [AM]

Mesoudi, A., Whiten, A. & Laland, K. N. (2006) Towards a unified science of cultural evolution. *Behavioral and Brain Sciences* 29:329–83. [AM]

Moore, J. H. (1994) Putting anthropology back together again: The ethnogenetic critique of cladistic theory. *American Anthropologist* 96:925–48. [AM]

Morange, M. (2002) The relations between genetics and epigenetics: An historical point of view. *Annals of the New York Academy of Sciences* 981(1):50–60. [ZF, rEJ]

Mousseau, T. A. & Fox, C. W. (1998) *Maternal effects as adaptations*. Oxford University Press. [aEJ]

Nance, W. E. & Kearsey, M. J. (2004) Relevance of connexin deafness (DFNB1) to human evolution. *American Journal of Human Genetics* 74:1081–87. [aEJ]

Neisser, U. (1976) *Cognition and reality*. Freeman. [AS]

Odling Smee, F. J., Laland, K. N. & Feldman, M. W. (2003) *Niche construction: The neglected process in evolution*. Princeton University Press. [AM, aEJ]

Oyama, S. (1985) *The ontogeny of information: Developmental systems and evolution*. Cambridge University Press. [arEJ]

Oyama, S., Griffiths, P. E. & Gray, R. D., eds. (2001) *Cycles of contingency: Developmental systems and evolution*. MIT Press. [AM]

Pál, C. (1998) Plasticity, memory and the adaptive landscape of the genotype. *Proceedings of the Royal Society of London B: Biological Sciences* 265:1319–23. [rEJ]

Pearson, H. (2006) What is a gene? *Nature* 441:398–401. [aEJ]

Pembrey, M. E., Bygren, L. O., Kaati, G., Edvinsson, S., Northstone, K., Sjöström, M., Golding, J. & the ALSPAC study team (2006) Sex-specific, male-line transgenerational responses in humans. *European Journal of Human Genetics* 14:159–66. [aEJ]

Pepperberg, I. M. (2004) Pepperberg, I. M. (2001, September). Lessons from cognitive ethology: Animal models for ethological computing. In *Proceedings of the First International Workshop on Epigenetic Robotics, Modeling Cognitive Development in Robotic Systems* (pp. 5–12).

Pogribny, P., Tryndyak, V. P., Muskhelishvili, L., Rusyn, I. & Ross, S. A. (2007) Methyl deficiency, alterations in global histone modifications, and carcinogenesis. *Journal of Nutrition* (Suppl.) 137:216–22. [rEJ]

Poole, J. H., Tyack, P. L., Stoeger-Horwath, A. S. & Watwood, S. (2005) Animal behaviour: Elephants are capable of vocal learning. *Nature* 434(7032):455–56. [ZF]

Popova, N. K. (2006) From genes to aggressive behavior: The role of serotonergic system. *BioEssays* 28:495–503. [rEJ]

Rando, O. J. & Verstrepen, K. J. (2007) Timescales of genetic and epigenetic inheritance. *Cell* 128:655–68. [rEJ]

Rapp, R. A. & Wendel, J. F. (2005) Epigenetics and plant evolution. *New Phytologist* 168:81–91. [arEJ]

Rassoulzadegan, M., Grandjean, V., Gounon, P., Vincent, S., Gillot, I. & Cuzin, F. (2006) RNA-mediated non-mendelian inheritance of an epigenetic change in the mouse. *Nature* 441:469–74. [rEJ]

Reader, S. M. & Laland, K. N. (2003) *Animal innovation*. Oxford University Press. [aEJ]

Rendell, L. & Whitehead, H. (2001) Culture in whales and dolphins. *Behavioral and Brain Sciences* 24:309–82. [aEJ]

Richards, E. J. (2006) Inherited epigenetic variation – revisiting soft inheritance. *Nature Reviews Genetics* 7:395–401. [rEJ]

Richerson, P. J. & Boyd, R. (2005) *Not by genes alone: How culture transforms human evolution*. University of Chicago Press. [aEJ]

Riddle, N. C. & Richards, E. J. (2002) The control of natural variation in cytosine 5 methylation in *Arabidopsis*. *Genetics* 162:355–63. [rEJ]

Rivera, M. C. & Lake, J. A. (2004) The ring of life provides evidence for a genome fusion origin of eukaryotes. *Nature* 431:152–55. [AM]

Rodenhiser, D. & Mann, M. (2006) Epigenetics and human disease: Translating basic biology into clinical applications. *Canadian Medical Association Journal* 174:341–48. [rEJ]

Rodin, S. N., Parkhomchuk, D. V. & Riggs A. D. (2005) Epigenetic changes and repositioning determine the evolutionary fate of duplicated genes. *Biochemistry (Moscow)* 70:559–67. [rEJ]

Rosenberg, S. M. (2001) Evolving responsively: Adaptive mutation. *Nature Reviews Genetics* 2:504–15. [AM]

Runciman, W. G. (1998) The selectionist paradigm and its implications for sociology. *Sociology* 32:163–88. [SB]

Sangster, T. A., Lindquist, S. & Queitsch, C. (2004) Under cover: Causes, effects and implications of Hsp90-mediated genetic capacitance. *BioEssays* 26:348–62. [rEJ]

Serre, D., Langaney, A., Chech, M., Maria Teschler-Nicola, M., Paunovic, M., Mennecier, P., Hofreiter, M., Possnert, G. & Pääbo, S. (2004) No evidence of Neandertal mtDNA contribution to early modern humans. *PLoS Biology* 2:e57. DOI:10.1371/journal.pbio.0020057. [SL]

Shapiro, J. A. (1999) Genome system architecture and natural genetic engineering in evolution. *Annals of the New York Academy of Sciences* 870:23–35. [aEJ]

Sherry, D. F. & Galef, B. G. (1984) Cultural transmission without imitation: Milk bottle opening by birds. *Animal Behavior* 32:937–38. [SB]

Shi, R., Werker, J. F. & Morgan, J. L. (1999) Newborn infants' sensitivity to perceptual cues to lexical and grammatical words. *Cognition* 72:11–21. [ASW]

Siegal, M. L. & Bergman, A. (2006) Canalization. In: *Evolutionary genetics*, ed. C. W. Fox & J. B. Wolf, pp. 235–51. Oxford University Press. [rEJ]

Sloman, A. (1979) The primacy of non-communicative language. In: *The analysis of meaning: Informatics 5 Proceedings ASLIB/BCS Conference, Oxford, March 1979*, pp. 1–15, ed. M. MacCafferty & K. Gray. Available at: http://www.cs.bham.ac.uk/research/projects/cogaff/81-95.html [AS]

Sokal, R. R., Oden, N. L., Walker, J. & Waddle, D. M. (1997) Using distance matrices to choose between competing theories and an application to the origin of modern humans. *Journal of Human Evolution* 32:501–22. [SL]

Sperber, D. (1996) *Explaining culture: A naturalistic approach*. Blackwell. [rEJ]

Spinelli, D. N., Pribram, K. H. & Bridgeman, B. (1970) Visual RF organization of single units in the visual cortex of monkey. *International Journal of Neurosciences* 1:67–74. [BB]

Steels, L. (2004) Analogies between genome and language evolution. In: *Artificial life IX*, ed. J. Pollack, pp. 200–206. MIT Press. [LS]

Sterelny, K. (2004) Symbiosis, evolvability and modularity. In: *Modularity in development and evolution*, ed. G. Schlosser & G. Wagner, pp. 490–516. University of Chicago Press. [aEJ]

Suddendorf, T. & Corballis, M. C. (2007) The evolution of foresight: What is mental time travel and is it unique to humans? *Behavioral and Brain Sciences* 30(3):299–351. [AM]

Szathmáry, E. (2006) The origins of replicators and reproducers. *Philosophical Transactions of the Royal Society B* 361:1761–76. [LS]

Takeda, S. & Paszkowski, J. (2006) DNA methylation and epigenetic inheritance during plant gametogenesis. *Chromosoma* 115:27–35. [rEJ]

Tomasello, M. (2003) *Constructing a language: A usage-based theory of language acquisition*. Harvard University Press. [ASW]

Tomasello, M. & Barton, M. (1994) Learning words in nonostensive contexts. *Developmental Psychology* 30:639–50. [ASW]

Trehub, A. (1991) *The cognitive brain*. MIT Press. [AS]

Trut, L. N. (1999) Early canid domestication: The farm-fox experiment. *American Scientist* 87(2):160–69. [ZF]

Trut, L. N., Plyusnina, I. Z. & Oskina, I. N. (2004) An experiment on fox domestication and debatable issues of evolution of the dog. *Russian Journal of Genetics* 40:644–55. [rEJ]

Tuite, M. F. & Cox, B. S. (2006) The [PSIþ] prion of yeast: A problem of inheritance. *Methods* 39:9–22. [rEJ]

Turner, J. S. (2000) *The extended organism: The physiology of animal-built structures*. Harvard University Press. [aEJ]

Van Speybroeck, L., Van de Vijver, G. & De Waale, D., eds. (2002) *From epigenesis to epigenetics: The genome in context*. Annals of the New York Academy of Sciences, Vol. 981. [Whole volume.] [rEJ]

Varela, F. J. (1979) *Principles of biological autonomy*. Elsevier/North Holland. [CMH]

Vastenhouw, N. L., Brunschwig, K., Okihara, K. L., Müller, F., Tijsterman, M. & Plasterk, R. H. A. (2006) Long-term gene silencing by RNAi. *Nature* 442:882. [rEJ]

Vetsigian, K., Woese, C. & Goldenfeld, N. (2006) Collective evolution and the genetic code. *Proceedings of the National Academy of Sciences USA* 103:10696–701. [LG]

Waddington, C. H. (1975) *The evolution of an evolutionist*. Edinburgh University Press. [aEJ]

Wade, N. (2002) Comparing mouse genes to man's and finding a world of similarity. *The New York Times*, December 5, pp. A1, A34. [CLB]

Walczak, A. M., Onuchic, J. N. & Wolynes, P. G. (2005) Absolute rate theories of epigenetic stability. *Proceedings of the National Academy of Sciences USA* 102:18926–31. [rEJ]

Wallman, J., Turkel, J. & Trachtman, J. (1978) Extreme myopia produced by modest change in early visual experience. *Science* 201:1249–51. [BB]

Wang, Y., Lin, X., Dong, B., Wang, Y. & Liu, B. (2004) DNA methylation polymorphism in a set of elite rice cultivars and its possible contribution to intercultivar differential gene expression. *Cellular and Molecular Biology Letters* 9:543–56. [rEJ]

Watson, J. S. (1966) The development and generalization of contingency awareness in early infancy: Some hypotheses. *Merrill-Palmer Quarterly* 12:123–35. [ASW]

Watson, J. S. (1985) Contingency perception in early social development. In: *Social perception in infants*, ed. T. M. Field & N. A. Fox, pp. 157–76. Ablex. [ASW]

Weaver, I. C. G., Cervoni, N., Champagne, F. A., D'Alessio, A. C., Sharma, S., Seckl, J. R., Dymov, S., Szyf, M. & Meaney, M. J. (2004) Epigenetic programming by maternal behaviour. *Nature Neuroscience* 7:847–54. [rEJ]

Wells, M. J. & Wells, J. (1957) The function of the brain of *Octopus* in tactile discrimination. *Journal of Experimental Biology* 34(1):131–42. [ZF]

West-Eberhard, M. J. (2003) *Developmental plasticity and evolution*. Oxford University Press. [arEJ, AM]

Whiten, A., Horner, V. & de Waal, F. B. M. (2005) Conformity to cultural norms of tool use in chimpanzees. *Nature* 437:737–40. [aEJ]

Whiten, A. & van Schaik, C. P. (2007) The evolution of animal "cultures" and social intelligence. *Philosophical Transactions of the Royal Society B: Biological Sciences* 362(1480):603–20. [ZF, rEJ]

Wickner, R. B., Edskes, H. K., Ross, E. D., Pierce, M. M., Baxa, U., Brachmann, A. & Shewmaker, F. (2004) Prion genetics: New rules for a new kind of gene. *Annual Review of Genetics* 38:681–707. [rEJ]

Wilson, E. O. (1975) *Sociobiology: The New Synthesis*. Wiley. [ZF]

Woit, P. (2002) Is string theory even wrong? *American Scientist* 90(2):110–12. [ZF]

Yoo, C. B. & Jones, P. A. (2006) Epigenetic therapy of cancer: Past, present and future. *Nature Reviews Drug Discovery* 5:37–50. [rEJ]

Zilberman, D. & Henikoff, S. (2005) Epigenetic inheritance in *Arabidopsis*: Selective silence. *Current Opinion in Genetics and Development* 15:557–62. [rEJ]

Zordan, R. E., Galgoczy, D. J. & Johnson, A. D. (2006) Epigenetic properties of white– opaque switching in *Candida albicans* are based on a self-sustaining transcriptional feedback loop. *Proceedings of the National Academy of Sciences USA* 103:12807–12. [rEJ]

Zuckerkandl, E. & Cavalli, G. (2007) Combinatorial epigenetics, "junk DNA", and the evolution of complex organisms. *Gene* 390:232–42. [rEJ]

11.
DEVELOPMENTAL EVOLUTION*
Robert Lickliter[†]

Biologists and psychologists are re-thinking the long-standing premise of genes as the primary cause of development, a view widely embraced in 20th-century biology. This shift in thinking is based in large part on: (1) the growing appreciation of the complex, distributed regulatory dynamics of gene expression and (2) the growing appreciation of the probabilistic, contingent, and situated nature of development. We now appreciate that what actually unfolds during individual development represents only one of many possibilities. This expanded focus on the developmental process, often referred to as a *developmental systems* approach, has far-reaching implications for developmental and evolutionary theory, including new ways of thinking about the consequences of activity and experience, the emergence of novel properties or traits, the nature and extent of heredity, and the Origins of phenotypic variability. © 2016 Wiley Periodicals, Inc.

Introduction

Does development influence evolution? Over the last 150 years, the answer to this question has taken many forms. It might seem obvious that knowledge of development would be necessary to understand evolution. Indeed, many biologists held this view during much of the 19th century. In fact, Darwin viewed characters or traits as resulting from changes in the process of individual growth and reproduction,[1] insisting that all inheritance must be a product of both the transmission and the development of traits. However, little was known about the process of development during Darwin's time and for many years thereafter. Ultimately, the dominant school of evolutionary thought in the 20th century—known today as the Modern or Neo-Darwinian Synthesis—abandoned any interest in development as a contributor to evolution.[2]

Although the many mysteries and complexities involved in the processes of development are far from solved, we now know that these processes involve widely distributed interactions across many levels of the individual and its environment. Scientists working with species as divergent as fruit flies, angler fish, macaques, and humans[3-5] have provided compelling evidence over the last several decades that the development of any physical or behavioral trait is the result of a complex web of co-actions among the organism's genes, molecular interactions within and across cells, and the nature and sequence of the physical, biological, and social environments in which the individual develops.

This growing appreciation of the interactive, distributed, and contingent nature of development has inspired many biologists and psychologists to reconsider the long-standing premise of genes as

---

* Lickliter, R. (2016). Developmental evolution. *WIREs Cognitive Science 2016*. doi: 10.1002/wcs.1422.
† Department of Psychology, Florida International University, Miami, FL.

DOI: 10.4324/9781032702988-19

the primary cause of development, a core tenet of 20th-century biology. This realization has, in turn, fostered the growth of research programs focused on identifying how the relations among genetic and nongenetic factors both guide and constrain the course of development.[6,7] The implications of this broader-based approach to our understanding of development—referred to as a *developmental systems* approach—are considerable and far-reaching, including novel ways of thinking about the roles of activity and experience in development, the emergence of novel properties and traits, the nature and extent of heredity, and the origins of phenotypic variability. Given the breadth of these implications, I focus in this essay on one particular dimension—the links between developmental and evolutionary change.

The Roles of Development in Evolution

Darwin's theory of evolution, described in the *Origin of Species,* was one of descent with modification. He did not, however, explain the origins of such modifications, a fact pointed out by St. George Mivart in 1871.[8] Mivart recognized that natural selection can account for the preservation and increase of phenotypic traits within a population, as Darwin proposed, *but not for their origin.* This is where development enters the picture. Although Mivart did not identify development as a source of novel traits or characters, growing evidence from molecular, developmental, and evolutionary biology have indicated that this can be the case.

Broadly speaking, development contributes to evolution in two important ways: (1) It generates the reliable reproduction of phenotypes across generations and (2) it introduces phenotypic variations and novelties of potential evolutionary significance.[9] In the first case, the process of development constrains phenotypic variation such that the traits and characters sifted through the filter of natural selection are not random or arbitrary. This is the *regulatory* function of development in evolution. It involves the physical properties of biological materials (including genes and chromosomes), and the temporal and spatial limitations on the coactions of internal, external, and ecological factors. These constraints collectively serve to restrict the 'range of the possible' in phenotypic form and function. For example, the limited number of body plans observed across animal species highlights the regulatory or constraining role of development. Most animals have a body form that is bilaterally symmetrical, meaning that it can be divided into matching halves along a central axis. This bilateral arrangement, with one of each sensory channel (eye, ear, nostril) and limb pair on either side, is seen in nearly all animals, including humans.

Importantly, developmental influences can also vary across individuals, and the complex interactions of these influences across all levels, from biochemical to cultural, can result in modified phenotypic outcomes. For example, the developmental environment (e.g., temperature) can determine the sexual phenotype in some species of reptiles and fish, induce morphological changes that allow individuals to better escape predation in several amphibian species, and determine caste affiliation in some social insect species (see Gilbert[10] for discussion and additional examples). This production of phenotypic change, called the *generative* function of development, has significant implications for understanding the mechanisms of evolutionary change.[11] In particular, the generative function of development provides a source of phenotypic variation upon which natural selection can act.[12] Simply put, evolutionary novelties largely originate in the process of development.

This idea of the importance of development to evolution has a long, complex history in both biology and psychology. Following the thread of an idea proposed by French naturalist Etienne Geoffroy Saint-Hilaire in the early 1800s, the embryologist Gavin de Beer[13] proposed that evolutionary change can only come about by changes in development. However, for de Beer et al. working on the relation between development and evolution during the first half of the 20th century, modifications in development that could initiate evolutionary changes were thought to be the result of random genetic mutation, genetic drift, or genetic recombination. Environmental factors were not typically considered in discussions of evolutionary change (even by those focusing on the importance of development) because it was thought that nongenetic factors could not be reliably replicated across generations and therefore could not provide a basis for the heritable variation upon which natural selection could act. However, individuals of most animal species are typically raised by parents of the same species, in an environment or developmental niche that has been occupied by that species for many generations. This continuity of early experience across generations serves to surround the developing organism in a physical, biological, and social environment that is as characteristic of its species as is its genotype. Gottlieb's[14] decades of research on the development of species identification in ducklings provides an elegant example of how nonobvious this continuity of early experience can be. His program of research in behavioral embryology documented how the features and patterns of transgenerationally recurring prenatal sensory experience, including self-stimulation, both guide and constrain young ducklings' attention, perception, and learning during the prenatal and postnatal periods of development. This species-typical experience effectively ensures successful species identification after hatching, without appeals to 'instinctive,' 'innate,' or 'genetically determined' behavior (see Blumberg, Development evolving: the origins and meanings of instinct, *WIREs Cogn Sci,* also in the collection How We Develop).

Moving on from Narrow Views of Evolutionary Change

Missing the importance of development to evolution in biology and psychology during most of the 20th century came largely from a narrow definition of evolution as *a change in the genetic composition of populations*.[15] In this framework, development was seen as the process by which genotypic specification was translated into the phenotypic traits of individuals, including their anatomy, physiology, and behavior. This narrow perspective on both development and evolution was promoted by several generations of biologists, and was based on three widely held underlying assumptions regarding development and heredity:

1. Instructions for building organisms reside in genes.
2. Genes are the exclusive means by which these instructions are faithfully transmitted from one generation to the next.
3. There is no meaningful feedback from the environment or the experience of the organism to its genes.

These three assumptions fit neatly within the conceptual framework of population genetics, which is concerned with how genetic mutation, genetic recombination, and natural selection could lead to changes in gene frequencies in a population. In the mid-20th century, the architects of the 'Modern Synthesis' of evolutionary biology (including Theodosius Dobzhansky, Julian Huxley, Ernst Mayr, and George Gaylord Simpson) promoted these three assumptions and saw no need to integrate development into their collective attempts to synthesize the tenets of Darwinism and Mendelism.[16] The gene-centered framework of the Modern Synthesis of evolutionary biology effectively sidestepped the issue of development while also downplaying any role of the environment in the evolutionary process. If genes contain all the necessary information for phenotypes and if events and experiences during individual development could not directly influence the phenotypic traits of offspring, then internal factors (genes) clearly had to have priority over external factors when attempting to explain both development and evolution.

In recent decades, these assumptions about development and heredity have all been called into question. Specifically, growing evidence from molecular biology, neuroscience, and behavioral ecology, among others, indicates that genes are *not* the only source of inheritance across generations. In addition to genetic inheritance, epigenetic inheritance, behavioral inheritance, environmental inheritance, and cultural inheritance are increasingly recognized as contributing to transgenerational phenotypic stability.[12,17]

Reassessing the Links between Development and Evolution

Expanding the definition of inheritance to include developmental resources beyond the genes is reshaping theory and research across the evolutionary and developmental biology. For example, evolutionary developmental biology (often referred to as *evo-devo*) is a growing field of research that involves a partnership among evolutionary, developmental, and molecular biologists to better integrate our understanding of developmental processes and evolutionary change. Unlike the gene-centered emphasis of the Modern Synthesis, evo-devo views evolution as changes in developmental processes rather than simply changes in gene frequencies. Evo-devo thus deals with how developmental processes can affect and effect evolutionary change. This perspective motivates a wide range of research questions, including how modifications in developmental processes can lead to novel phenotypes, the role of developmental plasticity in evolution, and how ecological factors influence developmental and evolutionary change.[18] For example, the discovery of the homeotic Hox gene family in vertebrates in the 1980s motivated researchers to empirically assess the roles of gene duplication and gene regulation in the evolution of morphological diversity. One of the most surprising findings from this early work in evo-devo was the realization that the genes involved in the morphological form of animals as different as a fruit fly, a mouse, and a human being are fundamentally the same.[19] The differences in form across these different species turn out to be due less to differences in genes and more to how the genes are regulated during embryonic development.

Like evo-devo, the rapidly growing field of epigenetics is also leading to the reassessment of links between development and evolution (see Moore, Behavioral epigenetics, *WIREs Syst Biol Med,* also in the collection How We Develop). Epigenetics is typically defined as changes in gene expression and function that are not due to changes in the DNA sequence. More broadly, it is the study of how the environment can affect the genome of the individual during its development, as well as the development of its descendants, without change in the coding sequence of the DNA itself. Evidence from epigenetics research is calling into question longstanding assumptions regarding the fundamental and privileged role of genes in development, heredity, and evolution. In particular, growing evidence from epigenetic research contradicts all three widely held underlying assumptions regarding development and heredity, namely: (1) instructions for building organisms resides in the genes, (2) genes are the exclusive means by which these instructions are transmitted from one generation to the next, and (3) there is no significant feedback from the environment or the experience of the organism to its genes. These assumptions have been called into question through numerous demonstrations of the environmental regulation of gene expression and cellular function, as well as the varied effects of sensory stimulation and social interaction on neural and hormonal responsiveness.[20]

The epigenetic approach to evolutionary issues is broadening our understanding of phenotypic plasticity, which may be defined broadly as the ability of an organism to modify its phenotype in response to its environment (see Bateson, Robustness and plasticity in development, *WIREs Cogn Set,* also in the collection How We Develop). The capacity for phenotypic plasticity was long considered by most biologists to be genetically determined. However, the rich interplay between genes and their environments demonstrated by contemporary epigenetic research has revealed a range of mechanisms whereby developing individuals can modify their morphology, physiology, or behavior in response to the features of their developmental context, suggesting additional pathways to evolutionary change.

For example, timing of hormone production and the sensitivity of organs and tissues to the presence of hormones can be readily altered by features of the environment and both can result in significant changes in morphology and behavior.[21,22] Many organisms' nervous systems monitor their environment and can rapidly change or adjust the hormonal milieu within the organism. The presence and levels of hormones in turn alter gene expression patterns, which in turn contribute to the maintenance or modification of phenotypes.

Research on desert locusts provides a striking example of the links between context, gene expression, and phenotypic plasticity.[23] The desert locust (*Schislocerca gregaria*) is usually cryptic in color (green) and solitary. In this form it avoids other locusts and flies alone at nighttime. However, under certain shifts in climatic conditions that result in an increase in desert vegetation, the number of locusts can explode, triggering a rapid increase in population density that leads to transformation of their color (now black and bright yellow) and a dramatic change in their social behavior (Fig. 11.1). With the increase in vegetation, normally solitary locusts form bands of hoppers that eventually form swarms consisting of billions of locusts; these swarms can cause catastrophic damage to agricultural crops. This transformation in color and social behavior involves morphological, physiological, and behavioral changes that have been traced to the action of numerous chemical messengers and more than 500 genes. Anstey et al.[23] have shown that a key agent in this remarkable transformation is the neurotransmitter serotonin, which is synthesized in the locust's thoracic nervous system in response to multiple sensory cues (touch, smell, or sight) provided by social contact with other locusts when population density increases rapidly. Within as little as 2 h of increased proximity to other locusts, elevated serotonin levels switch behavior from mutual aversion to mutual attraction, allowing the formation of enormous locust swarms within just a matter of days.

Fig. 11.1 Solitary and gregarious morphs of the desert locust. This dramatic and rapidly produced difference in both appearance and behavior is triggered by increased social stimulation.

Such demonstrations of phenotypic plasticity in response the environmental change have led some developmental and evolutionary biologists to propose the idea that an accumulation of hidden genetic variation and developmental potential can present itself when developing organisms are challenged by unusual developmental conditions.[11] A striking example of this hidden variation is the demonstration that chickens can be induced to grow teeth.[24] Under typically occurring prenatal conditions, when the chick embryo's oral epidermis and oral mesenchyme cells interact, the embryo grows the usual, species-typical chick beak (Fig. 11.2, left). However, when the chick embryo's oral epidermis is experimentally placed in contact with mouse mesenchyme during embryogenesis, the embryo produces a mammalian tooth rather than a chick beak (Fig. 11.2, right). This startling turn of events hints at the enormity of hidden developmental-genetic potential, which can provide developing organisms a key pathway for modulating phenotypic change in response to changing environmental

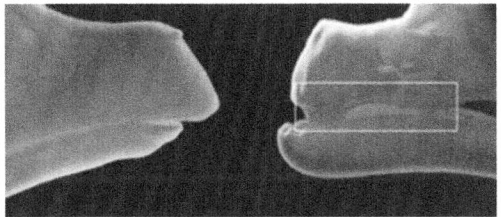

Fig. 11.2 Normal and experimentally modified beaks of chicken embryos. The presence of teeth is highlighted in the embryo on the right, an example of hidden developmental-genetic potential.

conditions. Genes interact with other constituents of the cell, which interacts with other cells in the organism, which interacts with other organisms. It is out of this dynamic, multileveled process that phenotypes emerge. In other words, phenotypes are the outcomes of the whole developmental system, comprising the organism embedded within its particular genetic and cellular make-up, and in its specific physical, biological, and social environments.

Conclusion

Because of the variability of developmental resources available across different environments, and because only a portion of the genome is actually expressed, what is realized during the course of individual development is only one of many possible outcomes. This is a core tenet of *probabilistic epigenesis,* the view that neither physical nor behavioral development can have a predetermined outcome. Consistent with this probabilistic view of development, there is now considerable evidence that parents transfer to offspring a variety of non-genetic factors in reproduction that can directly influence phenotypic outcomes, including DNA methylation patterns, chromatin marking systems, RNA-interference, cytoplasmic chemical gradients, and a range of sensory stimulation necessary for normal development.

The realization that control for the course of development does not reside in any one factor or component, but rather in the nature and dynamics of the relations among factors *internal* and *external* to the organism, shifts conceptualizing development away from the gene-centered, prespecified framework assumed for most of the last century and toward an appreciation of development as a situated and historical process that is dependent on developmental resources distributed across the organism-environment system. Further, it is the process of development itself that produces the phenotypic variation that is screened by natural selection. This insight provides a rationale for the necessity of bridging developmental and evolutionary accounts of phenotypic change.

Further Reading

Blumberg MS. *Freaks of Nature: What Anomalies Tell Us about Development and Evolution.* New York: Oxford University Press; 2010.

Lickliter R. Developmental evolution and the origins of phenotypic variation. *BioMol Concepts* 2014, 5: 343–352.

Moore DS. *The Developing Genome: An Introduction to Behavioral Epigenetics.* New York: Oxford University Press; 2015.

Oyama S. *The Ontogeny of Information.* 2nd ed. Durham, NC: Duke University Press; 2000.

References

1. Bowler PJ. *Evolution: The History of an Idea.* Berkeley, CA: University of California Press; 1989.
2. Mayr E, Provine W. *The Evolutionary Synthesis.* Cambridge, MA: Harvard University Press; 1980.
3. Champagne FA. Epigenetic mechanisms and the transgenerational effects of maternal care. *Front Neuroendocrinol* 2008, 29:386–397.
4. Hirsch HV, Tompkins L. The flexible fly: experience-dependent development of complex behaviors in Drosophila melanogaster. *J Exp Biol* 1994, 195:1–18.
5. Piferrer P, Martines P, Ribas L, Vinas A, Diaz N. Functional genomic analysis of sex determination in teleost fish. In: Saroglia M, Liu Z, eds. *Functional Genomics in Aquaculture.* Ames, IA: Oxford University Press; 2012.
6. Bateson P, Gluckman P. *Plasticity, Robustness, Development, and Evolution.* Cambridge: Cambridge University Press; 2011.

7. Minelli A, Pradeu T. *Towards a Theory of Development*. New York: Oxford University Press; 2014.
8. Mivart SG. *On the Genesis of Species*. Macmillan: London; 1871.
9. Alberch P. The generative and regulatory roles of development in evolution. In: Mosakowski D, Roth G, eds. *Environmental Adaptation and Evolution*. Stuttgart: Fischer-Verlag; 1982, 19–36.
10. Gilbert SF. Ecological developmental biology: developmental biology meets the real world. *Dev Biol* 2001, 233:1–12.
11. Gottlieb G. Normally occurring environmental and behavioral influences on gene activity: from central dogma to probabilistic epigenesis. *Psychol Rev* 1998, 105:792–802.
12. West-Eberhard M. *Developmental Plasticity and Evolution*. New York: Oxford University Press; 2003.
13. de Beer G. *Embryology and Evolution*. Oxford: Oxford University Press; 1930.
14. Gottlieb G. *Synthesizing Nature-Nurture: Prenatal Roots of Instinctive Behavior*. Mahwah, NJ: Erlbaum; 1997.
15. Dobzhansky T. *Genetics and Origins of Species*. New York: Columbia University Press; 1937.
16. Huxley J. *Evolution: The Modern Synthesis*. London: George Allen and Unwin; 1942.
17. Jablonka E, Lamb MJ. *Evolution in Four Dimensions: Genetic, Epigenetic, Behavioral, and Symbolic Variation in the History of Life*. Cambridge, MA: MIT Press; 2005.
18. Hall BK, Olson W. *Keywords and Concepts in Evolutionary Developmental Biology*. Cambridge, MA: Harvard University Press; 2003.
19. Carroll S. *Endless Forms Most Beautiful: The New Science of Evo Devo*. New York: W.W. Norton & Co; 2005.
20. Gilbert SF, Epel D. *Ecological Developmental Biology*. Sinauer: Sunderland, MA; 2009.
21. Crews D. Epigenetics and its implications for behavioral neuroendocrinology. *Front Neuroendocrinol* 2008, 29:344–357.
22. Gilbert S, Epel D. *Ecological Developmental Biology: Integrating Epigenetics, Medicine, and Evolution*, 2nd ed. Sunderland Associates: Sunderland, MA; 2015.
23. Anstey ML, Rogers SM, Ott SR, Burrows M, Simpson SJ. Serotonin mediates behavioral gregarization underlying swarm formation in desert locusts. *Science* 2009, 323:627–630.
24. Kollar EJ, Fisher C. Tooth induction in chick epithelium: expression of quiescent genes for enamel synthesis. *Science* 1980, 207:993–995.

12.
EVOLVING EVOLUTIONARY PSYCHOLOGY

Darcia Narvaez*, David S. Moore†, David C. Witherington‡, Timothy I. Vandiver‡, and Robert Lickliter§

Which evolutionary theory can best benefit psychological theory, research, and application? The most well-known school of evolutionary psychology has a narrow conceptual perspective (a.k.a., "Narrow Evolutionary Psychology" or NEP). Proponents of NEP have long argued that their brand of evolutionary psychology represents a full-fledged scientific revolution, with Buss (2020) recently likening NEP's scientific impact to that of a Copernican or Darwinian paradigm shift. However, NEP stands on two traditions that are now the subjects of serious debate and revision: the neo-Darwinian adaptationist framework within evolutionary biology, and the computationalist "mind-as-computer" framework within cognitive science. Although NEP calls itself revolutionary, the significant revolutions taking place today in both evolutionary biology and cognitive science reveal NEP to be rooted in the orthodoxies of the past. We propose a more inclusive, developmental evolutionary psychology theory (DEPTH) better suited for our field in multiple ways, from acknowledging epigenesis to incorporating developmental science. To discern appropriate baselines for human nature and for human becoming, one must integrate developmental neuroscience, anthropology, and cognitive archeology. To be of value in addressing and remedying the challenges facing humanity, psychological theories must recognize the central importance of our plasticity, evolved developmental niche, and deep history.

* Department of Psychology, University of Notre Dame, Notre Dame, IN.
† Psychology Field Group, Pitzer College, Claremont, CA.
‡ Department of Psychology, University of New Mexico, Albuquerque, NM.
§ Department of Psychology, Florida International University, Miami, FL.

Public Significance Statement

Understanding humans and our behavior requires unpacking our evolutionary heritages and developmental pathways. We inherit much more than genes. Biology, environment, and culture are seamlessly intertwined, in both individual development and evolution. Every individual constructs their personhood through real-time engagement with the world, so it matters what kind of relational experiences the individual has. In contrast to the most prominent evolutionary psychology theory that emphasizes the stranglehold of humanity's evolutionary past, a developmental evolutionary psychology theory orients to dynamic development in the present, taking epigenetics and plasticity seriously. This approach is better able to guide research and practice, and free people from the disempowering belief in biological determinism.

Few psychologists would seriously question the central role of an evolutionary framework in explaining psychological functioning. In fact, evolutionary theory has long informed efforts in psychology to understand mind and behavior, from the recapitulationist-inspired doctrines of G. Stanley Hall and Sigmund Freud to the ethological frameworks of Konrad Lorenz and John Bowlby (Boring, 1950). Proponents of modern evolutionary psychology celebrate the field's empirical yield, its theoretically driven hypothesis generation, the novelty of its predictions, and the heuristic value of its interpretative framework; these contributions are lauded as clear demonstrations of evolutionary psychology's invaluable role within psychological science (e.g., Buss, 1999; Confer et al., 2010; Lewis et al., 2017; Tooby & Cosmides, 2015). Many proponents even proclaim that evolutionary psychology is uniquely positioned to conceptually unify psychology (Buss, 2020; Confer et al., 2010).

Maintaining an evolutionarily informed psychology is critically important, but is the current field of evolutionary psychology up to the task? The broad discipline of evolutionary psychology is a blend of several diverse applications of evolutionary theory to human mind and behavior, including Human Behavioral Ecology, Cultural Evolution (including gene-culture coevolution theory), Social Constructivist approaches, Evolutionary Developmental Psychology, and Developmental Systems approaches (e.g., Bjorklund & Pellegrini, 2002; Gottlieb, 1992; Griffiths & Gray, 1994; Gurven, 2018; Moore, 2006; Scher & Rauscher, 2003; Sear et al., 2007). Ploeger et al. (2008a) aptly described evolutionary psychology, in this most general sense, as "more a collection of point of views [sic], which are not necessarily consistent with one another, than it is a coherent theory" (p. 41). Indeed, this collection of viewpoints reveals some deep-seated conceptual differences in perspective over ideas as foundational as the role that natural selection plays in evolution (Heyes, 2000; Scher & Rauscher, 2003). Too much conceptual heterogeneity and even metatheoretical division are evident in this broadly construed evolutionary psychology to adequately unify psychological science.

In contrast, one branch of evolutionary psychology has proffered a conceptually unified platform that its adherents tout as "revolutionary"—a grand metatheory not just for all of evolutionary psychology but for all of psychological science (Buss, 2015, 2020; Tooby & Cosmides, 2015). This more narrowly defined evolutionary psychology entertains its fair share of theoretical debates (e.g., debates over the extent of modularity and over domain-specificity in human psychological mechanisms), but these debates unfold within a shared conceptual framework of ideas concerning the nature of the evolutionary process and psychological functioning (Confer et al., 2010). Nonetheless, this shared conceptual framework has also inspired decades-long "vigorous opposition" (Buss, 2020, p. 5) from critics both within and outside of psychological science, including from those who consider themselves members of the broad evolutionary psychology community (e.g., Barrett et al., 2014; Scher & Rauscher, 2003; Stotz, 2014).

As the most prominent public face of the field, this narrower brand of evolutionary psychology—dubbed "narrow evolutionary psychology" (NEP) by Scher and Rauscher (2003)—continues to represent what most psychologists consider to be "mainstream" evolutionary psychology (Confer et al., 2010; Lewis et al., 2017; Ploeger et al., 2008b). Proponents of NEP tout the conceptual pedigree of their approach by tracing its origins to particular 20th century developments in the disciplines of evolutionary biology and cognitive science (Tooby & Cosmides, 2015). From evolutionary biology, NEP has embraced adaptationism and inclusive fitness theory, privileging natural selection as a major cause of current human functioning, and privileging genes as principal units of selection. From cognitive science, NEP has embraced computationalism and its information-processing conceptualization of the mind/brain. As foundational principles, adaptationism and computationalism establish the shared conceptual framework within which NEP operates, a framework that

NEP proponents espouse as a "cogent metatheory for understanding the complexities of the human mind" (Buss, 2020, p. 1).

A close look at NEP's foundational principles, however, shows them to be inconsistent with contemporary thinking in both evolutionary biology (e.g., Bolhuis et al., 2011) and cognitive science (e.g., Newen et al., 2018). NEP's core assumptions concerning the nature of the evolutionary process rely on adaptationist ideas, while its core assumptions concerning the architecture of the human mind rely on computationalist ideas. However, adaptationism and computationalism—the conceptual "pillars" that support NEP—are now the subject of serious, mainstream debate and fundamental revision within the very scientific disciplines to which NEP owes its conceptual allegiance. This raises critical questions about the adequacy of NEP's conceptual framework and the extent to which it can do justice to the complexities of the human mind.

A useful evolutionary psychology needs to be grounded in conceptual advances, not just empirical output. At the level of sheer empirical generativity, few can seriously challenge NEP's prowess. But data, findings, and the theories that frame them are of inherently limited value when they emanate from flawed conceptual presuppositions. As Hogan (2001) has succinctly noted, "all the empiricism in the world cannot salvage a bad idea" (p. 27). As we argue below, within both evolutionary biology and cognitive science, the constructs of gene-centered adaptationism and computationalism are looking more and more like bad ideas.

Fortunately, as we have indicated, alternative approaches to evolutionary psychology exist, ones that can advance the field beyond NEP's increasingly antiquated conceptual roots (e.g., Barrett et al., 2014; Buller, 2005; Lickliter & Honeycutt, 2013; Scher & Rauscher, 2003). But to adequately inform psychological science, evolutionary psychology needs a new, unifying set of "first principles" abstracted from some of the conceptual diversity endemic to evolutionary psychology, broadly construed. In what follows, we first examine NEP, the most prominent version of evolutionary psychology, and describe how its conceptual foundations are currently mired in controversy. Then we address alternative, up-to-date perspectives that should inform the shape of the bedrock foundations of any emerging evolutionary psychology. Finally, we discuss what kinds of alternative principles better meet the needs of the day theoretically and practically, identifying more appropriate baselines for human normality, with attention to critical developmental processes. We conclude with a discussion of how a developmental evolutionary psychology better addresses contemporary and urgent psychological questions.

Pillar I: Neo-Darwinism, Adaptationism, and Contemporary Evolutionary Biology

Neo-Darwinism, or the "Modern Synthesis" (Huxley, 1942), was the theory that defined evolutionary biology through much of the 20th century (e.g., Laland et al., 2015; Pigliucci, 2007). This understanding of evolution blended ideas drawn from Darwinian views of evolution and Men- delian views of genetics. It represented a population-level approach to evolutionary change, emphasizing the principles of natural selection and adaptation, differences in survival and reproduction (fitness), and heredity. Collectively, these features of the neo-Darwinian approach supported a narrative in which environments "pose" well-defined problems for organisms to solve, and the individuals best able to survive and reproduce in those environments are the individuals whose traits represent the best "adaptation" to the problems posed. Changes from one generation to the next in a population's genetics occur as "genes for adaptive traits" spread by natural selection through succeeding generations. Unabashedly gene-centric (see Mayr, 1961, 1982), the Modern Synthesis promoted a strong form of selectionism—the belief that natural selection, acting on gene frequencies in a population, is the primary cause of both evolutionary change and the stability of species-typical traits observed within populations.

Neo-Darwinism remains the primary theoretical foundation underlying contemporary biology (Mayr, 2001); it has proven to be a remarkably flexible theory able to generate and explain an enormous amount of empirical data (Coyne, 2009; Ellegren & Sheldon, 2008). However, scientists and philosophers concerned with *theoretical* biology have been expressing concerns about neo-Darwinism since at least the late 1950s (Gould, 1980, 2002; Jablonka & Lamb, 1995; Moore, 2002; Noble, 2015; Pigliucci, 2007; Tanghe et al., 2018; Waddington, 1957). These concerns have often focused on a consequence of the fact that the 20th century architects of the Modern Synthesis—knowing that they were not yet able to explain how developmental processes give rise to phenotypes—chose to finesse their predicament by defining evolution as a process that affects gene frequencies across populations (Moore, 2002). This move allowed them to construct neo-Darwinism as a theory strictly about the roles of genes in evolution (Jablonka & Lamb, 1995). As a result, neo-Darwinism ignores the role of development in evolution, despite the acknowledged fact that developmental processes should play a central role in any comprehensive theory of biology (Moore, 2008a). To many theorists, this situation seems untenable, suggesting that traditional neo-Darwinism will not be able to stand the test of time (e.g., Blumberg, 2009; Laland et al., 2015; Noble, 2015; Stotz, 2014; Tanghe et al., 2018). Indeed, *Nature* recently published a debate titled "Does Evolutionary Theory Need a Rethink," which raised serious concerns about the neo-Darwinian perspective (Laland et al., 2014; Wray et al., 2014). Nonetheless, NEP's evolutionary concepts remain rooted in the latter, contested theoretical framework. Consistent with the neo-Darwinian tradition, proponents of NEP continue to privilege genes as the primary targets of natural selection.

NEP and the Conceptual Agenda of Neo-Darwinism

According to neo-Darwinism, natural selection is what "delivers" adaptations across generations by means of gene selection; this is the central idea underlying adaptationist thinking. In other words, natural selection is characterized as a mechanical, antecedent force, capable of shaping individual development by inherited mechanisms. Applying this to psychology, NEP theorists assume that psychological adaptations can be explained solely with reference to natural selection of genetic variations inherited from prior generations (Confer et al., 2010). For example, innate rules of perception and cognition are presumed to be prespecified in the genes as a result of selection pressures in our ancestral past (Tooby & Cosmides, 2015). This prespecification assumption central to NEP allows its proponents to claim that each human comes into the world with "innate ideas or a priori concepts" (Tooby & Cosmides, 2015, p. 7). However, this assumption that prespecifications for "evolved" human traits can exist in the genome—in advance of the real-time developmental processes that elicit said traits—is inconsistent with biologists' current understanding that all physical and psychological traits must be constructed during individual development, whether those traits have an acknowledged evolutionary history or not (Gottlieb, 1992; Lickliter, 2008; Lickliter & Berry, 1990; Oyama, 1985). Specifically, all phenotypes come into being via developmental processes that involve the coaction of deeply entangled biological and nonbiological factors (e.g., DNA, epigenetic marks, nutritional factors, and social environments, to name just a few) that mutually influence one another as development takes place. Simply put, human capacities are neither genetically nor environmentally specified, but emerge within processes of development (Ingold, 2006). Contemporary evolutionary biology is well aligned with this fact, particularly evolutionary developmental biology (see Hall, 2012). NEP, however, is not.

In contrast to their sociobiological predecessors (Wilson, 1975), NEP proponents claim to not be gene-centric and, therefore, to not promote genetic determinism. Rather, they endorse a form of weak interactionism in which phenotypic development is thought to reflect the additive operation of two separate sources of information, one that is internal, formative, and relatively fixed (genes) and one that is external, supportive, and relatively variable (environment). This conceptual separation of causal factors that arise from genes and those that arise from the environment is, however, indefensible in light of contemporary biological theories and data (Moore, 2002). It is now widely understood across the life sciences that gene activity is regulated by systemic factors in and above the levels of molecular and cellular activity, and that many of these factors originate outside of the organism. As a result, genetic and environmental factors cannot be meaningfully partitioned (Lickliter, 2009; Moore, 2015).

Most biologists are no longer preoccupied with, and are even skeptical of, adaptationist thinking, and the closely associated idea that genes hold privileged status in the developmental construction of traits. Any sharp focus on adaptationism has been criticized by biologists and philosophers of science since the late 1970s (e.g., Buller, 2005; Gould & Lewontin, 1979; Richardson, 2007; Shapiro & Epstein, 1998) and much more nuanced understandings have since permeated biological theory (Godfrey-Smith, 2001; Millstein, 2007; Orzack & Forber, 2017). Biologists now hold that many phenotypes are better explained with reference to nonadaptive forces. In this "pluralism" perspective, adaptation is considered only one cause of a trait's evolution; other factors include developmental constraints and historical contingency, which are nonadaptive forces. In other words, natural selection, central to adaptation, is now recognized by biologists as only one of several mechanisms of evolutionary change. Nevertheless, for proponents of NEP, "identifying the [evolved] adaptive functions of psychological mechanisms" has continued to be an "indispensable" goal since the 1980s (Buss, 2020, p. 2). Consequently, NEP and its underlying conceptual tenets have been the subject of a great deal of criticism (e.g., Bolhuis & Wynne, 2009; Lloyd & Feldman, 2002; Rose & Rose, 2000), with a range of arguments challenging this approach. (Trenchant critiques leveled at NEP are available in Buller, 2005; Fodor, 2000; Gould, 2002; Lewontin, 2000; Lynch, 2007; and Richardson, 2007.) We now consider how, even beyond the limitations of adaptationism, the neo-Darwinian framework within evolutionary biology on which the pioneers of NEP built their field is now being disputed among evolutionary biologists themselves.

The Evolution of Evolutionary Biology and Its Challenge to Neo-Darwinism

The conceptual foundations of neo-Darwinism have been a source of debate within evolutionary biology since well before the emergence of NEP in the 1980s. In particular, contemporary biologists and philosophers of science are now focused on several phenomena and ideas that have collectively undermined the older—and in some cases, discredited—assumptions that are foundational for NEP theorists. Three important examples of these phenomena and ideas are: the blurred distinction between proximate and ultimate causes of phenotypes (Laland et al., 2013); the role of niche construction in evolution (Bolhuis et al., 2011; Laland et al., 2015; Lloyd & Feldman, 2002); and the importance of developmental plasticity to evolutionary change (e.g., Laland et al., 2014).

Proximate and Ultimate Causes of Phenotypes

A conceptual cornerstone of NEP has been the dichotomization of causal explanation for phenotypic traits into phylogenetic (evolutionary) or ontogenetic (developmental) factors, operating on ultimate and proximate levels, respectively. Phylogenetic factors are assumed to have operated on the individual's ancestors and to have delivered to the current generation gene specifications for traits that are merely "read out" during development. In contrast, ontogenetic factors are understood to operate during the individual's development and to trigger or interfere with the expression of the phylogenetically delivered gene-based specifications. Proponents of NEP claim they are concerned only with the ultimate causation of human behavior and its function or adaptive value (Buss, 1999; Gaulin & McBruney, 2004), allowing the field to effectively sidestep the role of development in understanding the phenotype.

However, converging evidence from the last 30 years of biological science, particularly from epigenetics and evolutionary developmental biology, indicates that the decoupling of proximate and ultimate levels of explanation is not tenable, because genetic and environmental contributions to development cannot be viewed independently (Lickliter & Berry, 1990). Instead, "proximate causes are themselves often also evolutionary causes" (Laland et al., 2015, p. 6), since "development is a direct cause of why and how adaptation and speciation occur" (Laland et al., 2014, p. 164); this is the case because developmental processes affected by niche construction and developmental constraints have been implicated in changes to both the rate and direction of evolution. NEP's continued application of the proximate/ultimate distinction, and its neglect of development, reify the misleading assumption that natural selection, acting on previous generations, delivers a set of genetic specifications (ultimate causes) to the current generation, and that contexts and experiences (proximate causes) independently trigger the unfolding of prespecified traits. In fact, genetic and experiential contributors to phenotypes are interrelated at all stages of development, and neglecting either type of factor leads to a profound distortion of how traits are built (Lickliter & Witherington, 2017; Moore, 2015). The architects of the Modern Synthesis—like the NEP theorists that followed them—assigned no role for developmental processes in evolution, but the discovery that developmental processes always affect the emergence of adaptive traits means that a new conceptualization of evolutionary explanation is needed.

Niche Construction

The neo-Darwinian idea that environments "pose" problems for organisms to solve has been undermined dramatically by the recognition that organisms are not exposed to environments at random and need not respond to their environments in a passive way. Instead, organisms effectively inherit many aspects of their environments from their parents (Griffiths & Gray, 1994; Lickliter, 2008; West & King, 1987). This is how manatees and minnows come to inhabit an aquatic environment while elephants and human beings come to inhabit a terrestrial environment. In addition, organisms play important roles in creating their environments. Some animals actively alter their habitats; for example, birds build nests, beavers build dams, humans build houses, and ants create gardens in which they grow their fungal food. All organisms also effectively "construct" their own environments in a more passive way, as Lewontin and Levin (1997) noted: "every terrestrial organism is surrounded by a shell of warm moist air produced by its own metabolism, a shell that constitutes its most immediate 'environment'" (p. 97). Niche-construction theory has helped biologists acknowledge that evolution is not primarily about solving preexisting problems that have been "posed" to organisms. The challenges faced by individual organisms are often not independent of those same organisms, and even commonly self-imposed. Accordingly, biologists now recognize that organisms have some influence over their own evolution, rendering questionable the idea that phenotypes reflect "solutions" to problems posed by environments that preexist and are independent of the organisms themselves. For humans, genes have coevolved with niche construction and culture. For example, the introduction of livestock and pastoralism enabled lactose toleration past childhood (Beja-Pereira et al., 2003) and shifted societies from matrilineal to patrilineal inheritance (Holden & Mace, 2003). Findings like this have led some theorists to surmise that cultural factors may play "an active, leading role in the evolution of genes" (Richerson et al., 2010, p. 8985). Because neo-Darwinists explicitly reject the idea that organisms can influence their own evolution, the phenomenon of niche construction again suggests that a different conceptualization of evolutionary explanation is needed.

Developmental Plasticity

In part because human brains are far more malleable than previously realized, NEP's commitment to the idea of a genetically inherited universal human nature has not fared much better than the idea that we inhabit organism-independent, problem-posing environments. Of course, most human brains share a remarkable number of similarities because 99.9% of the human genome is common to all of us (National Human Genome Research Institute, 2018) and all people develop in environments that share many similarities (e.g., all normal human developmental environments are characterized by the presence of caregivers, shelter, community support, the use of language, etc.). However, it is now clear that brains are remarkably plastic in the face of idiosyncratic experiences. For example, experience playing a musical instrument has significant effects on the structure and function of the brain (Poldrack, 2018), and growing up with only a right cerebral hemisphere (i.e., with only half of a normal-sized brain) can nonetheless lead to relatively normal neurological, linguistic, and cognitive outcomes by adolescence (Asaridou et al., 2020). It is now apparent that our experiences affect our brains via the regulation of gene expression (Moore, 2017) and synaptic connectivity (Bolhuis et al., 2011), rendering the idea of a universal genetically inherited human nature increasingly unlikely. As noted below, humanity's particular immaturity and plasticity in early life means that early experiences actually contribute to shaping the brain and its functions for the long term (e.g., Schore, 2019; Shonkoff & Phillips, 2000).

Importantly, developmental plasticity (along with developmental constraints) is now recognized as an essential contributor to evolutionary processes (Burman, 2019; Carroll, 2005; West-Eberhard, 2003). The discovery that cultural practices played a crucial role in the evolution of human DNA that contributes to adults' ability to digest milk (Durham, 1991; Tishkoff et al., 2007; see also Moore, 2008b) strongly suggests that developmental plasticity permits the retention of juvenile phenotypes into adulthood when those phenotypes prove to be adaptive later in life. This sort of finding has led some theorists to argue that developmental plasticity might even "play a central directing role in evolution" (Wilson & Laland, 2016), an observation that has significant implications for evolutionary science, implications that encourage a rethinking of how we conceptualize evolutionary explanation.

The Extended Evolutionary Synthesis

Many of the aforementioned challenges to the tenets of neo-Darwinism are formally represented in the "Extended Evolutionary Synthesis" that has emerged in the last 15 years (Laland et al., 2015; Pigliucci, 2007). A few examples here will suffice. First, the Extended Evolutionary Synthesis (EES) challenges the traditional idea, still prevalent in the NEP literature, that natural selection is of paramount importance in explaining adaptation, and that selection pressures in the Pleistocene epoch are responsible for traits that characterize contemporary humans, even though today's people emerged in the context of swiftly changing environments. In the clever words Bolhuis (2005) used to reject this outmoded idea, "we're not Fred or Wilma" (p. 706). Second, in contrast to the evolutionary model underlying NEP, proponents of the EES see organisms as actively involved in constructing their own environments, giving them some influence over their own evolution (Bolhuis et al., 2011; Laland et al., 2014, 2015). Third, the EES elevates the role of development in evolution, recognizing that developmental processes are responsible for the novel phenotypes on which natural selection can operate (Laland et al., 2014). Most of the many theorists who have called for revisions to the Modern Synthesis (e.g., Jablonka & Lamb, 2005; Moore, 2002, 2008b; Noble, 2006, 2015; Odling-Smee et al., 2003; Oyama et al., 2001; Pigliucci, 2007; Stotz, 2014; Witherington & Lickliter, 2016) agree that the adaptationism and the gene/environment dichotomy at the heart of NEP's evolutionary thinking is problematic, in that it distorts how evolutionary processes actually contribute to observed behavioral characteristics. Nevertheless, NEP theorists have mostly ignored niche construction and the role of development in evolution, and they have given short shrift to developmental plasticity. Most biologists would agree that the narrow focus on natural selection and adaptation that characterizes NEP is simply not reflective of contemporary evolutionary biology.

Pillar II: Computationalism and Contemporary Cognitive Science

The first pillar of NEP's approach—its neo-Darwinian, adaptationist focus—has been on questionable conceptual ground for some time, given widespread challenges that have arisen from within evolutionary biology itself. The second pillar of NEP's approach—computationalism, or the view of cognition as computation—has enjoyed much steadier support over the years, having informed and directed orthodox thought in cognitive science since the field's origins in the "cognitive revolution" of the 1950s (Bruner, 1990; van Gelder, 1995). NEP's signature appeal to inherited cognitive mechanisms that respond to ancestral rather than present environmental conditions is firmly rooted in this type of computationalism. Nonetheless, the computationalist pillar of NEP's approach is now showing signs of significant conceptual instability in the wake of a decade's worth of increasingly mainstream challenges from cognitive scientists themselves.

Computationalism's basic refrain has long assumed axiomatic status within many if not most psychological circles: all acts of cognition, even in their most rudimentary form, involve information processing functionally akin to what digital computers do. According to computationalist doctrine, cognitive activity intercedes between an organism's sensory inputs and behavioral, motoric outputs. Furthermore, cognitive activity is understood to consist of brain-based, subpersonal (e.g., outside of conscious awareness) mechanisms that operate, in rule-governed, algorithmic fashion, on some form of representational content (e.g., internally stored information that "stands for" the world itself) to yield mandates for an organism's behavior. In effect, cognition is viewed as internalized problem-solving, a centralized, in-the-head, decision-making activity for controlling human behavior.

Computationalist models conceptualize cognitive functioning as "wholly realized by systems and mechanisms inside the brain," as an intracranial activity isolated and detached from the continuous, real-time, fully interdependent perceptual-motor engagement of organisms with their environments (Kiverstein, 2018, p. 19; van Gelder, 1995; Wheeler, 2005). Brains, in other words, are viewed as housing cognitive functioning, but since brains have no direct access to the world that an organism inhabits, cognition is conceptualized as brains' computations on representational stand-ins for that world (Wheeler, 2005). Thus, for computationalists, an organism's behaviors do not constitute cognition; they simply reflect the products of an interiorized competence, of brain-based transformations of given sensory input values. This means that the intelligent behavior of an organism is necessarily preceded by, and distinct from, mechanisms of cognition themselves (Hutto & Myin, 2013). In computationalism, cognition acts on representational content, not on the world itself.

NEP and the Conceptual Agenda of Computationalism

Proponents of NEP wholeheartedly embrace the computationalism of orthodox cognitive science, so much so as to contend that "the brain is not just like a computer. It is a computer" (Tooby & Cosmides, 2015, p. 19). Notwithstanding the contrary view of neuroscientists (e.g., Panksepp & Panksepp, 2000), they argue that human brains consist of numerous, functionally specialized mechanisms, or information processing modules termed "evolved psychological mechanisms," each of which evolved via natural selection to maximize inclusive fitness by solving recurrent problems characteristic of the environments early humans inhabited during the Miocene and Pleistocene geological epochs (Bjorklund & Pellegrini, 2002; Tooby & Cosmides, 2015). Such mechanisms, recently described by Buss (2020) as "procedures inside the head" that process environmental input and generate behavioral output (p. 3), thoroughly exemplify the interiorized problem-solving status of the computational mind.

It is important to note that NEP extends the bedrock principles of computationalism in two key ways. First, proponents take the detached, brain-based cognizer of orthodox cognitive science and further remove that "central executive" from interdependent organism-environment transactions that occur in real time. Proponents of NEP insist that human behavior is, in fact, frequently not a response to the present environment but to conditions that existed in our prehistoric past, creating a somewhat irrational disconnect between present conditions and behavior. They suggest that information is processed using programs that act not on content corresponding to the current environment but on content related to ancestral conditions. By virtue of their contention that we are born with representational content corresponding to these ancestral conditions, proponents of NEP explicitly resurrect "the necessity of 'innate ideas'" (Tooby & Cosmides, 2015, p. 56).

Second, by adopting the "massive" modularity hypothesis that is a core aspect of their theories, NEP proponents take the notion of modularity in cognitive functioning and magnify it considerably, positioning these "domain-specific, content-rich programs specialized for solving ancestral problems" as starting points for guiding and constraining information processing in human development (Tooby & Cosmides, 2015, p. 47). In this way, they carve out a distinctly nativist stance for the inherent modularity of computationalism (van Gelder, 1995). Modules are viewed as species-typical mechanisms that arise from genetic information (under "normal" environmental circumstances) and that serve as necessary preconditions for the experience of individual development. Each evolved module is treated as a preformed entity that appears independently of development, already laden with specific representational "knowledge" corresponding to those domains that presented distinct adaptive challenges for our ancestors (Tooby & Cosmides, 2015).

Computationalism's traditionally dominant and largely unquestioned status in cognitive science has long provided NEP's proponents with conceptual security—a security borne of the orthodox wisdom of the cognitive revolution. In recent years, however, computationalism itself—the very idea that cognition is computation—has become the subject of thoughtful challenges that are increasingly being given serious consideration within mainstream cognitive science.

The Rise of 4E Cognition—Embodied, Embedded, Extended and Enactive—and Its Challenge to Computationalism

Cognitive science in the 21st century has borne witness to a heightened period of self-critique (Newen et al., 2018). The assumption that cognition is information processing is not only open to active debate within cognitive science but is also at risk of being overthrown entirely.

Since the earliest days of computationalism's ascendency, repudiations of its various principles have arisen from the ranks of ecological psychology, dynamic systems theory, robotics, and phenomenologically inspired treatments such as enactivism (e.g., Brooks, 1991; Bruner, 1990; Gibson, 1979; van Gelder, 1995; Varela et al., 1991). Such longstanding repudiations, however, remained largely at the fringes of cognitive science. Not until the last decade or so have these alternatives to computationalism become "a staple and familiar feature of the cognitive science landscape" (Hutto & Myin, 2018, p. 95), relegating computationalism to simply one alternative among many. Though evolutionary psychology, in either its broad or narrow formulations, need not embrace any of these alternatives in order to maintain internal consistency, it is simply no longer the case that its foundational cognitive theories must be aligned with computationalism.

Cognitive science today is taking seriously *4E cognition,* a family of perspectives that conceptually ground cognition in the embodiment of agents and their embeddedness in the world. The *Oxford Handbook of 4E Cognition* (Kiverstein, 2018; Newen et al., 2018) reveals three distinct, robust, and competing conceptualizations of mind: embedded, extended and enactive approaches. The "embedded" sector of today's conceptual landscape remains committed to the view of cognition as brain-constituted and thoroughly computational in nature. Proponents of this territory, however, take seriously the complex mutuality of causal relations that obtain among brains (as seats of cognition), the bodies that house them, and the worlds in which those bodies are embedded (Kiverstein, 2018).

In contrast to embedded conceptualizations of mind, the "extended" sector of today's 4E landscape rejects the "neurocentric internalism" of orthodox computationalism and, in the process, extends the boundaries of what counts as cognitive activity beyond the brain (Wheeler, 2014, p. 378). For proponents of this territory, facets of body and world (e.g., eye movements, hand gestures, tool use, writings, cultural artifacts) do not simply affect cognitive activity in a causal sense but can, in fact, be partially constitutive of cognitive activity itself (Newen et al., 2018). Extended theorists, however, still maintain a basic allegiance to the fundamentally computational nature of cognition; they merely distribute such computations (and their representational content) across brain, body, and world (Kiverstein, 2018; Wheeler, 2005, 2014).

Like their extended counterparts, the third, "enactive" sector of today's 4E landscape rejects interiorized, brain-centric notions of mind. But enactive proponents go one step further by also rejecting the foundational status of computation and representation in the functioning of minds. Proponents of enactivism repudiate the purely intellectualized approach to cognition that computationalism entails (Hutto & Myin, 2013, 2017). In orthodox computationalism, activities of computation resemble logical and mathematical operations, modeled after the deliberative, calculation-based decision making that constitutes developmentally sophisticated, consciously reflective, analytic modes of thought (Tallis, 2004). For enactive proponents, behavior is constitutive of cognition, and the dynamics of an organism acting in real time, inextricably engaged with and coupled to its real-world context, should serve as conceptual grounding for our understanding of what cognition is (Di Paolo, 2009; Thompson, 2007). Cognition at its most basic is something that organisms do in practice, through their embeddedness in the world; it is not an internal, behind-the-scenes computational processing of representations or subsequent driver of bodily movements (Hutto & Myin, 2018; Newen et al., 2018). For these proponents, cognition is practical, procedural knowing: "skillful know-how in situated and embodied action" (Engel et al. 2013, p. 202).

In contrast to orthodox computationalism, both extended and enactive conceptualizations of mind feature prominently in what Engel and colleagues (2013) have termed the "pragmatic turn" in cognitive science—a robust and revolutionary trend toward an "action-oriented paradigm," bolstered by considerable empirical evidence (p. 202). This represents a true revolution in psychological science today that is being waged not through conservation of old ideas but through ongoing enactivist challenges to the fundamental tenets of computationalism. Even less radical strains of 4E cognition, such as extended conceptualizations of mind, reject the exaggerated, Pleistocene-program version of computationalism to which proponents of NEP continue to adhere (Clark, 2003).

Toward an Integrative Developmental Evolutionary Psychology Theory

Within evolutionary psychology, the privileged, standard-bearer status that has allowed NEP proponents to articulate the field's conceptual underpinnings no longer seems tenable. As challenges to neo-Darwinist adaptationism in evolutionary biology and to computationalism in cognitive science continue to mount, well-established alternative approaches to evolutionary psychology, within psychology's own broad ranks, warrant renewed consideration. In fact, a number of evolutionary psychologists have long decried NEP's adaptationist and computationalist grounding, promoting instead a more complex, dynamic, enactivist grounding for the discipline (e.g., Barrett et al., 2014; Bolhuis et al., 2011; Burman, 2019; Lyon, 2006; Scher & Rauscher, 2003). These voices are in tune with current trends in both evolutionary biology and cognitive science. As such, they should be foregrounded in future conceptual discourse within evolutionary psychology. The field's scientific currency demands it. In what follows, we offer some initial considerations on how to move forward with a developmental evolutionary psychology theory—DEPTH—that is responsibly grounded in modern evolutionary biology and cognitive science and that takes seriously what it means to be a malleable social mammal with a lengthy childhood—what it means, in other words, to be human.

The goal of integrating the evolutionary sciences into psychology should be to understand the unique evolutionary pathway that has brought humanity to its present moment (Henley et al., 2019; Small, 2008). A truly evolutionarily informed psychology needs to include an understanding of the human species, its deep history as a social mammal situated in a broader, interrelated web of life, its multiple inheritances, its developmental processes, its basic needs, and the multiple systems involved in meeting those needs during development (Ingold, 2004; Narvaez, 2014). It would involve an understanding of people as dynamic, complex systems who self-organize in coordination with their experiences in the world (e.g., Thelen & Smith, 1994). It would integrate knowledge of epigenetics and plasticity in shaping human beings. It would synthesize understandings of an individual's unique functional adaptation to their life circumstances and of the species' evolutionary adaptations. It would have a sense of human potential and what optimal neurobiological functioning looks like in a wide variety of different contexts. NEP provides none of this. Central to this transdisciplinary endeavor is, at the very least, the examination of the role of niche provision in human development, the establishment of baselines for the range of species-typical human behaviors evident in the world today, and the integration of developmental plasticity and epigenetics. All play a role in the foundations for human functioning and could provide substantial dividends when systematically studied.

Examine Niche Provision

NEP's focus on the inheritance of evolved psychological mechanisms distracts from accruing empirical evidence about how early relational experience (Organism × Environment) constructs the person. Although niche *construction* has been acknowledged as part of our extragenetic inheritances (Odling-Smee, 1988), niche *provision* is an inheritance as well (Stotz & Narvaez, 2018). Every animal develops in a niche that contributes to the form of its physical, behavioral, and psychological characteristics (Gottlieb, 1998; West & King, 1987). Humanity's evolved developmental niche (EDN) provides resources required for the construction of, for example, a healthy body (that resists infection), coordinated intelligence (critical for learning to find local food sources), and sociality (a critical feature of human adaptation), (Narvaez, 2014; Narvaez et al., 2013, 2014, 2016). Humans are social mammals, a line that emerged 20–40 million years ago with intensive early parenting practices (Konner, 2005). Human neonates are particularly immature, looking much like fetuses of other primates until 18 months of age (Trevathan, 2011); this observation led anthropologist Ashley Montagu to suggest that an external womb (exterogestation) was needed during that time (Montagu, 1986). The human EDN for young children provides the resources needed for a healthy, well-functioning psychosocial neurobiology, including soothing perinatal experiences; extended on-request breastfeeding; maternal and allomaternal responsiveness to infant needs, including affectionate touch (keeping baby calm during rapid growth); positive climate of support for mother and child; and self-directed play in the natural world with multi-aged playmates throughout childhood (Hewlett & Lamb, 2005; Hrdy, 2009). In the absence of EDN support, various systems may not develop properly, undermining the development of later-developing systems (Knudsen, 2004). For example, early life stress appears to impair the stress response for the long term (Lupien et al., 2009). Stress reactivity, a signal of dysregulation rooted in early life stress or trauma (van der Kolk, 2014; Shonkoff et al., 2012), makes social relations difficult and prosocial behavior even more so (Porges & Carter, 2010), in part because the stress response tends to focus energy on survival and self-protection (Sapolsky, 2004).

Establish Species-Typical Baselines

To embrace an evolutionary perspective, one must take a deep history view: "Phylogenetic history must be added [to explanations of the human psyche]; otherwise one fails to explain the peculiar potency that ontogenetic and cultural factors have in the shaping of the human mind as opposed to that of other animals" (Henley et al., 2019, p. 527). Important for human self-understanding is our emergence from social mammals whose offspring-rearing practices, such as breastfeeding and touch, led to increasing numbers of successfully reproducing progeny across generations, and so were retained for tens of millions of years (Konner, 2005). Civilization has been around no more than 20,000 years, or 1% of the existence of the genus *Homo*. Many people in industrialized societies, which represents a still smaller fraction of that 1%, raise children very differently than do people in traditional societies around the world, including a variety of small-band hunter-gatherer societies, which represent the type of society in which humans lived for over 1.9 million years (Lee & Daly, 2005). These traditional societies all share certain identifiable commonalities, in particular the early experiences they provide their offspring (Hewlett & Lamb, 2005). Data are accruing to demonstrate how important early experience, particularly components of the EDN, is for fostering well-functioning neurobiology, which grounds psychological functioning (Narvaez, 2014; van der Kolk, 2014; Witherington et al., 2018). For example, childhoods that are more EDN-consistent are associated with better physiological regulation (Tarsha et al., 2020) and mental health (Narvaez, Woodbury, et al., 2019; Narvaez et al., 2016).

It is important to be clear on baselines for drawing conclusions about human development. It may be time to rethink using individuals with an EDN-inconsistent childhood as prototypical specimens for gathering information on the human species generally. Lacking a perspective of deep history, many scholars, including NEP theorists, take as normal human nature the characteristics of individuals from industrialized and westernized societies, including dysregulation, selfishness, and aggression (e.g., Dawkins, 1976; Thornhill & Gangestad, 1996), even though their early experiences can be viewed as EDN-inconsistent (Narvaez et al., 2013). Anthropologists have noted a different set of characteristics among individuals raised within EDN-consistent contexts such as intuitive cooperation and generosity, high autonomy with high communalism, and with no coercion and little competition (e.g., Ingold, 2005; Sorenson, 1998). Many characteristics that NEP proponents study (because they are assumed to represent human nature) such as concern for fatherhood and mate selection (e.g., see Buss, 1994), are not apparent among nomadic foragers, who tend to be promiscuous and matrifocal rather than patriarchal (Hrdy, 2009; Marlowe, 2004). Moreover, small-band hunter-gatherer groups often consist of kin and nonkin, with membership changing by the day, with interdependence across groups as the norm (Hill et al., 2011).

The EDN has been around for 20–40 million years as part of the social mammalian line, with slight revisions by hominids (e.g., multiple responsive caregivers, variation on breastfeeding length). Consequently, the EDN could be used to develop an exceptionally good heuristic for determining a baseline for species-typical development, rather than selecting a baseline out of thin air or assuming that research participant behavior is adaptive (Narvaez & Witherington, 2018). Generalizing from contemporary human behavior in industrialized societies where EDN-inconsistent childhoods are widespread—as psychological research, including work in NEP, often does (Henrich et al., 2010)—may misinform psychologists about human potential.

Integrate Developmental Plasticity and Epigenetics

In addition to using misleading baselines, NEP misleads on the causes of contemporary human behavior by looking for these causes in genetic inheritance while ignoring developmental factors (Ingold, 2006; Lewontin et al., 1984; Moore, 2002; Noble, 2006; Oyama, 1985). Evolutionary theory should be used to advance our understanding of human beings as malleable social mammals who undergo a lengthy period of development (2 to 3 decades). The naïve view of phenotype causation advanced by NEP is simply inadequate for the task. For example, genetic action is a much more complex story than acknowledged by NEP proponents. Gene regulation involves not only the activation of a gene, but the creation of mRNA via splicing processes and the insertion, deletion, or exchange of single nucleotides of the RNA, all of which can be characterized as a type of molecular epigenesis involving more than the genome alone (Stotz & Griffiths, 2018). Molecular epigenesis involves "recruitment, activation and transportation of transcription, splicing, and editing factors [which] renders them functional and allows the environment to have specific effects on gene expression" (Stotz & Griffiths, 2018, p. 110). Thus, gene regulation involves not only genetics, but also, via epigenetics, factors much further afield, including hormones, diets, parenting, and influences from the broader social environment. Beyond genes and gene-related effects, humans inherit cell and body plans (Margulis, 1998), culture and ecology (Jablonka & Lamb, 2005), as well as developmental systems for raising the young (Gottlieb, 1998), all of which are deeply integrated and together escalate the influence of experience on development while increasing the importance of developmental plasticity to evolutionary change. Psychological research and theory have begun to attend to the impact of developmental system differences in light of basic needs fulfillment (e.g., for warm responsive care in infancy) on shaping psychosocial neurobiological functioning (e.g., attachment; Schore, 2019). Initial data suggest that humans might be much more epigenetically shaped and more plastic than other animals, especially early in life (Gómez-Robles et al., 2015). The complexity of inheritances and the dynamic nature of development and human plasticity are not addressed by NEP, but need to be by a more biologically credible theory of evolutionary psychology.

These three realms of study overlap and require systematic investigation but hold promise for reshaping evolutionary psychology in helpful ways. First, organisms self-organize around experience, so it matters what that experience entails. For example, human infants organize their psychosocial neurobiology around the care received, for better or for worse. Adverse childhood experiences (ACES) are on the rise in the United States, leading to illness and early death (Felitti & Anda, 2005). Denial of the evolved developmental niche in early life may provoke toxic stress, which is known to impair multiple neurobiological systems with potentially long-term effects on physiological and psychological health (e.g., Lanius et al., 2010). NEP does not contribute to our understanding of these processes, but DEPTH very well could. Second, as noted by others, psychology (along with some other disciplines) has often used samples from WEIRD (Western, European, Industrialized, Rich, and Democratic) countries when seeking baselines for species typicality. This practice undergirds discrimination toward non-WEIRD populations, despite the fact that non-WEIRD populations maintain more cultural and biological diversity, a hallmark of evolutionary processes (Díaz et al., 2019). NEP seems to draw their baselines from similarly skewed samples. Third, if we understand the nuances of the neurobiological plasticity of the human animal, unlike NEP we can avoid the fatalism of a limited developmental focus and better attend to how individuals self-organize through epigenetic processes. When we understand ontogenetic mechanisms, we can establish what the relations between optimal neurobiological development and psychological functioning look like, and what is necessary for flexible resilience. We can focus on prevention as well as intervention to enhance human potential.

In response to a deeper evolutionary perspective on the self-organizing of human beings, psychologists may also need to address the sociopolitical systems that undermine or condone ill being, especially among the less privileged. By taking advantage of the broader developmental-evolutionary approach we advocate, psychologists can employ a deep-history evolutionary perspective to help determine how to promote the well-being of humans and nonhumans and to foster a sustainable future (Kidner, 2001).

Conclusion

The goal of integrating evolutionary sciences into psychology should be to understand the unique evolutionary and developmental paths of human beings. A developmental evolutionary psychology theory (DEPTH) can do just that. How do human beings come to be shaped by processes that are influenced by our evolutionary history, but also by more proximal contingencies in our developmental histories, and the current state of our bodies and contexts? Although ideas and phenomena considered by narrow evolutionary psychology (NEP) theorists are worth considering, we have indicated NEP's restricted focus and deficient conceptual foundations. So much is left out. It is quite problematic that many psychologists and nonpsychologists have come to think of NEP as the only way in which to bring evolutionary ideas to bear on psychological questions. Although NEP bills itself as a revolutionary theory applicable to all of psychological science, it is in fact built on an outdated framework and, as such, should not be considered the agreed-upon perspective of all theorists interested in seeing evolutionary thought represented in psychology. Instead, it is time to recognize the heterogeneity of ideas available about how evolution should be brought to bear on psychological questions (Scher & Rauscher, 2003). Evolutionary processes are vital for explaining human behavior and integrating across biology, culture, and development. But an evolutionary psychology too out of step with current developments in evolutionary biology, cognitive science, genomics, and other disciplines cannot adequately advance such important interdisciplinary integrations. NEP's reliance on neo-Darwinian adaptationism, staunch computationalism, and neglect of appropriate baselines for human normality and development all render NEP inadequate for handling the all-important application of evolutionary theory to the study of human psychological functioning and well-being.

Narrow evolutionary psychology's focus on survival and reproduction leaves out *thriving*, which is required for natural selection—a family line must thrive to out-compete rivals over generations. What is important for thriving offspring? The Evolved Developmental Niche may establish a "cultural commons" for shaping human nature toward cooperation and openness, crucial features of our adaptive past (Hrdy, 2009). Societies that provide the EDN are characterized by self-regulation, cooperation, social and emotional intelligence, and humble self-confidence (Fry, 2013; Ingold, 2005; Lee, 2018; Sorenson, 1998). One proposal is to reset the baseline for normal human nature away from the "nasty and brutish" perceptions advanced by writers in the last millennium (Hobbes, 1651; Spencer, 1850). Once we shift our focus to prosocial human characteristics and to developmental processes—to DEPTH—we will be motivated to study how such traits are shaped initially by early life experience (when human beings are particularly developmentally plastic via epigenetic processes), and subsequently maintained by ongoing needed support throughout life.

The questions that a developmental evolutionary psychology would ask are almost the flip side of what is typically studied by narrow evolutionary psychologists. Rather than puzzling about altruism, the question would be what sorts of evolutionary and developmental processes explain why there is little or no concern about "altruism" in deeply cooperative societies that do not perceive "other" to any large degree, but experience unity with a sentient world full of persons, even rivers, mountains, and winds as a part of the commonself or Ecological Self (Harvey, 2017; Narvaez, Four Arrows, et al., 2019). What evolutionary and developmental processes led some humans (the dominant ones today) to become so uncooperative with the natural world, so much so that they/we have broken previously resilient ecologies all over the planet? Why did they begin to think of themselves as separate from and superior to the natural world, unlike most prior humans and societies of the world? For a species whose sociality across species and with kin and nonkin has been adaptive, how do people become accustomed to disconnection, distrust, and antisociality? These are questions that could be considered and addressed by a more integrated, broad-based evolutionary psychology. Narrow evolutionary psychology is not up to the task, but a developmental evolutionary psychology theory is.

References

Asaridou, S. S., Demir-Lira, Ö. E., Goldin-Meadow, S., Levine, S. C., & Small, S. L. (2020). Language development and brain reorganization in a child born without the left hemisphere. *Cortex*, *127*, 290–312. https://doi.org/10.1016/j.cortex.2020.02.006

Barrett, L., Pollet, T. V., & Stulp, G. (2014). From computers to cultivation: Reconceptualizing evolutionary psychology. *Frontiers in Psychology*, *5*, 867. https://doi.org/10.3389/fpsyg.2014.00867

Beja-Pereira, A., Luikart, G., England, P. R., Bradley, D. G., Jann, O. C., Bertorelle, G., Chamberlain, A. T., Nunes, T. P., Metodiev, S., Ferrand, N., & Erhardt, G. (2003). Gene-culture coevolution between cattle milk protein genes and human lactase genes. *Nature Genetics*, *35*(4), 311–313. https://doi.org/10.1038/ng1263

Bjorklund, D. F., & Pellegrini, A. D. (2002). *The origins of human nature: Evolutionary developmental psychology*. American Psychological Association. https://doi.org/10.1037/10425-000

Blumberg, M. S. (2009). *Freaks of nature: What anomalies tell us about development and evolution*. Oxford University Press.

Bolhuis, J. J. (2005). We're not Fred or Wilma. *Science*, *309*(5735), 706–706. https://doi.org/10.1126/science.1115209

Bolhuis, J. J., Brown, G. R., Richardson, R. C., & Laland, K. N. (2011). Darwin in mind: New opportunities for evolutionary psychology. *PLoS Biology*, *9*(7), e1001109. https://doi.org/10.1371/journal.pbio.1001109

Bolhuis, J. J., & Wynne, C. D. L. (2009). Can evolution explain how minds work? *Nature*, *458*(7240), 832–833. https://doi.org/10.1038/458832a

Boring, E. G. (1950). *A history of experimental psychology*. Appleton-Century-Crofts.

Brooks, R. A. (1991). Intelligence without representation. *Artificial Intelligence*, *47*(1–3), 139–159. https://doi.org/10.1016/0004-3702(91)90053-M

Bruner, J. (1990). *Acts of meaning*. Harvard University Press.

Buller, D. J. (2005). *Adapting minds: Evolutionary psychology and the persistent quest for human nature*. MIT Press.

Burman, J. T. (2019). Development. In R. J. Sternberg & W. Pickren (Eds.), *The Cambridge handbook of the intellectual history of psychology* (pp. 287–317). Cambridge University Press. https://doi.org/10.1017/9781108290876.012

Buss, D. M. (1994). *The evolution of desire: Strategies of human mating*. Basic Books.

Buss, D. M. (1999). *Evolutionary psychology: The new science of mind*. Pearson.

Buss, D. M. (2015). *Evolutionary psychology: The new science of mind* (5th ed.). Routledge. https://doi.org/10.4324/9781315663319

Buss, D. M. (2020). Evolutionary psychology is a scientific revolution. *Evolutionary Behavioral Sciences, 14*(4), 316–323. https://doi.org/10.1037/ebs0000210

Carroll, S. B. (2005). *Endless forms most beautiful: The new science of evo devo*. Norton.

Clark, A. (2003). *Natural born cyborgs: Minds, technologies, and the future of human intelligence*. Oxford University Press.

Confer, J. C., Easton, J. A., Fleischman, D. S., Goetz, C. D., Lewis, D. M., Perilloux, C., & Buss, D. M. (2010). Evolutionary psychology. Controversies, questions, prospects, and limitations. *American Psychologist, 65*(2), 110–126. https://doi.org/10.1037/a0018413

Coyne, J. A. (2009). Evolution's challenge to genetics. *Nature, 457*(7228), 382–383. https://doi.org/10.1038/457382a

Dawkins, R. (1976). *The selfish gene*. Oxford University Press.

Di Paolo, E. (2009). Extended life. *Topoi, 28*(1), 9–21.

Díaz, S., Settele, J., Brondizio, E., Ngo, H. T., Guèze, M., Agard, J., Arneth, A., Balvanera, P., Brauman, K. A., Butchart, S. H. M., Chan, K. M. A., Garibaldi, L. A., Ichii, K., Liu, J., Subramanian, S. M., Midgley, G. F., Miloslavich, P., Molnar, Z., Obura, D., … Zayas, C. (Eds.). (2019). *IPBES summary for policymakers of the global assessment report on biodiversity and ecosystem services*. The IPBES Secretariat.

Durham, W. H. (1991). *Coevolution: Genes, culture, and human diversity*. Stanford University Press.

Ellegren, H., & Sheldon, B. C. (2008). Genetic basis of fitness differences in natural populations. *Nature, 452*(7184), 169–175. https://doi.org/10.1038/nature06737

Engel, A. K., Maye, A., Kurthen, M., & König, P. (2013). Where's the action? The pragmatic turn in cognitive science. *Trends in Cognitive Sciences, 17*(5), 202–209. https://doi.org/10.1016/j.tics.2013.03.006

Felitti, V. J., & Anda, R. F. (2005). *The adverse childhood experiences (ACE) study*. Centers for Disease Control and Kaiser Permanente.

Fodor, J. A. (2000). *The mind doesn't work that way*. MIT Press. https://doi.org/10.7551/mitpress/4627.001.0001

Fry, D. (Ed.). (2013). *War, peace and human nature*. Oxford University Press. https://doi.org/10.1093/acprof:oso/9780199858996.001.0001

Gaulin, S. J., & McBruney, D. H. (2004). *Evolutionary psychology* (2nd ed.). Prentice Hall.

Gibson, J. J. (1979). *The ecological approach to visual perception*. Houghton Mifflin.

Godfrey-Smith, P. (2001). Three kinds of adaptationism. In S. H. Orzack, & E. Sober (Eds.), *Adaptationism and optimality* (pp. 335–357). Cambridge University Press.

Gómez-Robles, A., Hopkins, W. D., Schapiro, S. J., & Sherwood, C. C. (2015). Relaxed genetic control of cortical organization in human brains compared with chimpanzees. *Proceedings of the National Academy of Sciences of the United States of America, 122*(48), 14799–14804. https://doi.org/10.1073/pnas.1512646112

Gottlieb, G. (1992). *Individual development and evolution: The genesis of novel behavior*. Oxford University Press.

Gottlieb, G. (1998). Normally occurring environmental and behavioral influences on gene activity: From central dogma to probabilistic epigenesis. *Psychological Review, 105*(4), 792–802. https://doi.org/10.1037/0033-295x.105.4.792-802

Gould, S. J. (1980). Is a new and general theory of evolution emerging? *Paleobiology, 6*(1), 119–130. https://doi.org/10.1017/S0094837300012549

Gould, S. J. (2002). *The structure of evolutionary theory*. Harvard University Press. https://doi.org/10.2307/j.ctvjsf433

Gould, S. J., & Lewontin, R. C. (1979). The spandrels of San Marco and the Panglossian paradigm: A critique of the adaptationist programme. *Proceedings of the Royal Society of London, Series B: Biological Sciences, 205*(1161), 581–598. https://doi.org/10.1098/rspb.1979.0086

Griffiths, P. E., & Gray, R. D. (1994). Developmental systems and evolutionary explanation. *The Journal of Philosophy, 91*, 277–304. https://doi.org/10.2307/2940982

Gurven, M. D. (2018). Broadening horizons: Sample diversity and socioecological theory are essential to the future of psychological science. *Proceedings of the National Academy of Sciences of the United States of America, 115*(45), 11420–11427. https://doi.org/10.1073/pnas.1720433115

Hall, B. K. (2012). *Evolutionary developmental biology* (2nd ed.). Springer.

Harvey, G. (2017). *Animism: Respecting the living world* (2nd ed.). C. Hurst & Co.

Henley, T., Rossano, M., & Kardas, E. (Eds.). (2019). *Handbook of cognitive archaeology: A psychological framework*. Routledge. https://doi.org/10.4324/9780429488818

Henrich, J., Heine, S. J., & Norenzayan, A. (2010). The weirdest people in the world? *Behavioral and Brain Sciences*, *33*(2–3), 61–83. https://doi.org/10.1017/S0140525X0999152X

Hewlett, B. S., & Lamb, M. E. (2005). *Hunter-gatherer childhoods: Evolutionary, developmental and cultural perspectives*. Aldine.

Heyes, C. (2000). Evolutionary psychology in the round. In C. Heyes & L. Huber (Eds.), *The evolution of cognition* (pp. 3–22). MIT Press.

Hill, K. R., Walker, R. S., Bozicevic, M., Eder, J., Headland, T., Hewlett, B., Hurtado, A. M., Marlowe, F., Wiessner, P., & Wood, B. (2011). Coresidence patterns in hunter-gatherer societies show unique human social structure. *Science*, *331*(6022), 1286–1289. https://doi.org/10.1126/science.1199071

Hobbes, T. (1651). *Leviathan*. Clarendon Press.

Hogan, R. (2001). Wittgenstein was right. *Psychological Inquiry*, *12*, 27.

Holden, C. J., & Mace, R. (2003). Spread of cattle led to the loss of matrilineal descent in Africa: A coevolutionary analysis. *Proceedings of the Royal Society of London Series B: Biological Sciences*, *270*(1532), 2425–2433. https://doi.org/10.1098/rspb.2003.2535

Hrdy, S. (2009). *Mothers and others: The evolutionary origins of mutual understanding*. Belknap Press.

Hutto, D. D., & Myin, E. (2013). *Radicalizing enactivism: Basic minds without content*. MIT Press.

Hutto, D. D., & Myin, E. (2017). *Evolving enactivism: Basic minds meet content*. MIT Press. https://doi.org/10.7551/mitpress/9780262036115.001.0001

Hutto, D. D., & Myin, E. (2018). Going radical. In A. Newen, L. De Bruin, & S. Gallagher (Eds.), *The Oxford handbook of 4E cognition* (pp. 95–115). Oxford University Press.

Huxley, J. (1942). *Evolution: The modern synthesis*. Allen & Unwin.

Ingold, T. (2004). Beyond biology and culture: The meaning of evolution in a relational world. *Social Anthropology*, *12*(2), 209–221.

Ingold, T. (2005). On the social relations of the hunter-gatherer band. In R. B. Lee & R. Daly (Eds.), *The Cambridge encyclopedia of hunters and gatherers* (pp. 399–410). Cambridge University Press.

Ingold, T. (2006). Against human nature. In N. Gonteir, J. P. van Bendegen, & D. Aerts (Eds.), *Evolutionary epistemology, language, and culture* (pp. 259–281). Springer.

Jablonka, E., & Lamb, M. J. (1995). *Epigenetic inheritance and evolution: The Lamarckian dimension*. Oxford University Press.

Jablonka, E., & Lamb, M. J. (2005). *Evolution in four dimensions: Genetic, epigenetic, behavioral, and symbolic variation in the history of life*. MIT Press.

Kidner, D. W. (2001). *Nature and psyche: Radical environmentalism and the politics of subjectivity*. SUNY Press.

Kiverstein, J. (2018). Extended cognition. In A. Newen, L. De Bruin, & S. Gallagher (Eds.), *The Oxford handbook of 4E cognition* (pp. 19–40). Oxford University Press.

Knudsen, E. I. (2004). Sensitive periods in the development of the brain and behavior. *Journal of Cognitive Neuroscience*, *16*(8), 1412–1425. https://doi.org/10.1162/0898929042304796

Konner, M. (2005). Hunter-gatherer infancy and childhood: The! Kung and others. In B. Hewlett & M. Lamb (Eds.), *Hunter-gatherer childhoods: Evolutionary, developmental and cultural perspectives* (pp. 19–64). Aldine Transaction.

Laland, K. N., Odling-Smee, J., Hoppitt, W., & Uller, T. (2013). More on how and why: Cause and effect in biology revisited. *Biology & Philosophy*. *28*(5), 719–745. https://doi.org/10.1007/s10539-012-9335-1

Laland, K. N., Uller, T., Feldman, M. W., Sterelny, K., Müller, G. B., Moczek, A., Jablonka, E., & Odling-Smee, J. (2015). The extended evolutionary synthesis: Its structure, assumptions and predictions. *Proceedings of the Royal Society of London Series B: Biological Sciences*, *282*(1813), 20151019. https://doi.org/10.1098/rspb.2015.1019

Laland, K. N., Uller, T., Feldman, M. W., Sterelny, K., Müller, G. B., Moczek, A., Jablonka, E., Odling-Smee, J., Wray, G. A., Hoekstra, H. E., Futuyma, D. J., Lenski, R. E., Mackay, T. F. C., Schluter, D., & Strassmann, J. E. (2014). Does evolutionary theory need a rethink? *Nature*, *514*(7521), 161–164. https://doi.org/10.1038/514161a

Lanius, R. A., Vermetten, E., & Pain, C. (2010). *The impact of early life trauma on health and disease: The hidden epidemic*. Cambridge University Press. https://doi.org/10.1017/CBO9780511777042

Lee, R. B. (2018). Hunter-gatherers and human evolution: New light on old debates. *Annual Review of Anthropology*, *47*(1), 513–531. https://doi.org/10.1146/annurev-anthro-102116-041448

Lee, R. B., & Daly, R. (Eds.). (2005). *The Cambridge encyclopedia of hunters and gatherers*. Cambridge University Press.

Lewis, D. M. G., Al-Shawaf, L., Conroy-Beam, D., Asao, K., & Buss, D. M. (2017). Evolutionary psychology: A how-to guide. *American Psychologist, 72*(4), 353–373. https://doi.org/10.1037/a0040409

Lewontin, R. C. (2000). *The triple helix: Gene, organism, and environment*. Harvard University Press.

Lewontin, R. C., & Levin, R. (1997). Organism and environment. *Capitalism, Nature, Socialism, 8*(2), 95–98. https://doi.org/10.1080/10455759709358737

Lewontin, R. C., Rose, S., & Kamin, L. J. (1984). *Not in our genes: Biology, ideology, and human nature*. Pantheon Books.

Lickliter, R. (2008). The growth of developmental thought: Implications for a new evolutionary psychology. *New Ideas in Psychology, 26*(3), 353–369. https://doi.org/10.1016/j.newideapsych.2007.07.015

Lickliter, R. (2009). The fallacy of partitioning: Epigenetics' validation of the organism-environment system. *Ecological Psychology, 21*(2), 138–146. https://doi.org/10.1080/10407410902877157

Lickliter, R., & Berry, T. D. (1990). The phytogeny fallacy: Developmental psychology's misapplication of evolutionary theory. *Developmental Review, 10*(4), 348–364. https://doi.org/10.1016/0273-2297(90)90019-Z

Lickliter, R., & Honeycutt, H. (2013). A developmental evolutionary framework for psychology. *Review of General Psychology, 17*(2), 184–189. https://doi.org/10.1037/a0032932

Lickliter, R., & Witherington, D. C. (2017). Towards a truly developmental epigenetics. *Human Development, 60*(2-3), 124–138. https://doi.org/10.1159/000477996

Lloyd, E. A., & Feldman, M. W. (2002). Evolutionary psychology: A view from evolutionary biology. *Psychological Inquiry, 13*(2), 150–156. https://doi.org/10.1207/S15327965PLI1302_04

Lupien, S. J., McEwen, B. S., Gunnar, M. R., & Heim, C. (2009). Effects of stress throughout the lifespan on the brain, behaviour and cognition. *Nature Reviews Neuroscience, 10*(6), 434–445. https://doi.org/10.1038/nrn2639

Lynch, M. (2007). The frailty of adaptive hypotheses for the origins of organismal complexity. *Proceedings of the National Academy of Sciences of the United States of America, 104* (Supp. 1), 8597–8604. https://doi.org/10.1073/pnas.0702207104

Lyon, P. (2006). The biogenic approach to cognition. *Cognitive Processing, 7*(1), 11–29. https://doi.org/10.1007/s10339-005-0016-8

Margulis, L. (1998). *Symbiotic planet: A new look at evolution*. Sciencewriters.

Marlowe, F. (2004). Marital residence among foragers. *Current Anthropology, 45*(2), 277–284. https://doi.org/10.1086/382256

Mayr, E. (1961). Cause and effect in biology. *Science, 134*(3489), 1501–1506. https://doi.org/10.1126/science.134.3489.1501

Mayr, E. (1982). *The growth of biological thought: Diversity, evolution, and inheritance*. Harvard University Press.

Mayr, E. (2001). *What evolution is*. Basic Books.

Millstein, R. L. (2007). Hsp90-induced evolution: Adaptationist, neutralist, and developmentalist scenarios. *Biological Theory, 2*(4), 376–386. https://doi.org/10.1162/biot.2007.2.4.376

Montagu, A. (1986). *Touching: The human significance of the skin*. Harper & Row.

Moore, D. S. (2002). *The dependent gene: The fallacy of nature vs. nurture*. W.H. Freeman.

Moore, D. S. (2008a). Espousing interactions and fielding reactions: Addressing laypeople's beliefs about genetic determinism. *Philosophical Psychology, 21*(3), 331–348. https://doi.org/10.1080/09515080802170127

Moore, D. S. (2008b). Individuals and populations: How biology's theory and data have interfered with the integration of development and evolution. *New Ideas in Psychology, 26*(3), 370–386. https://doi.org/10.1016/j.newideapsych.2007.07.009

Moore, D. S. (2015). *The developing genome: An introduction to behavioral epigenetics*. Oxford University Press.

Moore, D. S. (2017). Behavioral epigenetics. *WIREs Systems Biology and Medicine, 9*(1), e1333. https://doi.org/10.1002/wsbm.1333

Moore, K. (2006). Biology as technology: A social constructivist framework for an evolutionary psychology. *Review of General Psychology, 10*(4), 285–301.

Narvaez, D. (2014). *Neurobiology and the development of human morality: Evolution, culture and wisdom*. Norton.

Narvaez, D., Braungart-Rieker, J., Miller-Graff, L., Gettler, L., & Hastings, P. (2016). *Contexts for young child flourishing: Evolution, family and society*. Oxford University Press. https://doi.org/10.1093/acprof:oso/9780190237790.001.0001

Narvaez, D., Four Arrows (Jacobs, D. T.), Halton, E., Collier, B. S., & Enderle, G. (Eds.). (2019). *Indigenous sustainable wisdom: First-nation knowhow for global flourishing*. Peter Lang.

Narvaez, D., Panksepp, J., Schore, A., & Gleason, T. (Eds.). (2013). *Evolution, early experience, and human development: From research to practice and policy*. Oxford University Press.

Narvaez, D., Valentino, K., Fuentes, A., McKenna, J., & Gray, P. (2014). *Ancestral landscapes in human evolution: Culture, childrearing and social well-being*. Oxford University Press. https://doi.org/10.1093/acprof:oso/9780199964253.001.0001

Narvaez, D., Wang, L., & Cheng, A. (2016). Evolved developmental niche history: Relation to adult psychopathology and morality. *Applied Developmental Science, 20*(4), 294–309. https://doi.org/10.1080/10888691.2015.1128835

Narvaez, D., & Witherington, D. (2018). Getting to baselines for human nature, development and wellbeing. *Archives of Scientific Psychology, 6*(1), 205–213. https://doi.org/10.1037/arc0000053

Narvaez, D., Woodbury, R., Cheng, Y., Wang, L., Kurth, A., Gleason, T., Deng, L., Gutzwiller-Helfenfinger, E., Christen, M., & Näpflin, C. (2019). Evolved development niche provision: Moral socialization, social maladaptation and social thriving in three countries. *SAGE Open*. Advance online publication. https://doi.org/10.1177/2158244019840123

National Human Genome Research Institute. (2018). *Genetics vs. genomics fact sheet*. https://www.genome.gov/about-genomics/fact-sheets/Genetics-vs-Genomics

Newen, A., De Bruin, L., & Gallagher, S. (2018). 4E cognition: Historical roots, key concepts, and central issues. In A. Newen, L. De Bruin, & S. Gallagher (Eds.), *The Oxford handbook of 4E cognition* (pp. 3–15). Oxford University Press. https://doi.org/10.1093/oxfordhb/9780198735410.013.1

Noble, D. (2006). *The music of life: Biology beyond genes*. Oxford University Press.

Noble, D. (2015). Evolution beyond neo-Darwinism: A new conceptual framework. *The Journal of Experimental Biology, 218*(Pt. 1), 7–13. https://doi.org/10.1242/jeb.106310

Odling-Smee, F. J. (1988). Niche constructing phenotypes. In H. C. Plotkin (Ed.), *The role of behavior in evolution* (pp. 73–132). MIT Press.

Odling-Smee, F. J., Laland, K. N., & Feldman, M. W. (2003). *Niche construction: The neglected process in evolution (MPB-37)*. Princeton University Press.

Orzack, S. H., & Forber, P. (2017). Adaptationism. In Edward N. Zalta (Ed.), *The Stanford encyclopedia of philosophy* (2017 ed.). https://plato.stanford.edu/archives/spr2017/entries/adaptationism/

Oyama, S. (1985). *The ontogeny of information: Developmental systems and evolution*. Cambridge University Press.

Oyama, S., Griffiths, P. E., & Gray, R. D. (2001). *Cycles of contingency: Developmental systems and evolution*. MIT Press.

Panksepp, J., & Panksepp, J. B. (2000). The seven sins of evolutionary psychology. *Evolution & Cognition, 6*, 108–131.

Pigliucci, M. (2007). Do we need an extended evolutionary synthesis?. *International Journal of Organic Evolution, 61*(12), 2743–2749. https://doi.org/10.1111/j.1558-5646.2007.00246.x

Ploeger, A., van der Maas, H. L. J., & Raijmakers, M. E. J. (2008a). Is evolutionary developmental biology a viable approach to the study of the human mind? *Psychological Inquiry, 19*(1), 41–48. https://doi.org/10.1080/10478400701774147

Ploeger, A., van der Maas, H. L. J., & Raijmakers, M. E. J. (2008b). Is evolutionary psychology a metatheory for psychology? A discussion of four major issues in psychology from an evolutionary developmental perspective. *Psychological Inquiry, 19*(1), 1–18. https://doi.org/10.1080/10478400701774006

Poldrack, R. A. (2018). *The new mind readers*. Princeton University Press. https://doi.org/10.2307/j.ctvc77ds2

Porges, S. W., & Carter, C. (2010). Neurobiological bases of social behavior across the life span. In M. E. Lamb, A. M. Freund. & R. M. Lerner (Eds.), *The handbook of life-span development, Vol 2: Social and emotional development* (pp. 9–50). Wiley. https://doi.org/10.1002/9780470880166.hlsd002002

Richardson, R. C. (2007). *Evolutionary psychology as maladapted psychology*. MIT Press. https://doi.org/10.7551/mitpress/7464.001.0001

Richerson, P. J., Boyd, R., & Henrich, J. (2010). Colloquium paper: Gene-culture coevolution in the age of genomics. *Proceedings of the National Academy of Sciences of the United States of America*, *107*(Suppl. 2), 8985–8992. https://doi.org/10.1073/pnas.0914631107

Rose, H., & Rose, S. P. R. (2000). *Alas poor Darwin: Arguments against evolutionary psychology*. Harmony.

Sapolsky, R. M. (2004). *Why zebras don't get ulcers* (3rd ed.). Holt.

Scher, S. J., & Rauscher, F. (Eds.). (2003). Nature read in truth or flaw: Locating alternatives in evolutionary psychology. *Evolutionary psychology: Alternative approaches* (pp. 1–29). Kluwer Academic Publishers. https://doi.org/10.1007/978-1-4615-0267-8_1

Schore, A. N. (2019). *The development of the unconscious mind*. Norton.

Sear, R., Lawson, D. W., & Dickins, T. E. (2007). Synthesis in the human evolutionary behavioural sciences. *Journal of Evolutionary Psychology*, *5*(1), 3–28. https://doi.org/10.1556/JEP.2007.1019

Shapiro, L., & Epstein, W. (1998). Evolutionary theory meets cognitive psychology: A more selective perspective. *Mind & Language*, *13*(2), 171–194. https://doi.org/10.1111/1468-0017.00072

Shonkoff, J. P., & Phillips, D. A. (2000). *From neurons to neighborhoods: The science of early childhood development*. National Academy Press.

Shonkoff, J. P., Garner, A. S., Siegel, B. S., Dobbins, M. I., Earls, M. F., McGuinn, L., Pascoe, J., Wood, D. L., The Committee on Psychosocial Aspects of Child and Family Health, The Committee on Early Childhood, Adoption, and Dependent Care, & The Section on Developmental and Behavioral Pediatrics. (2012). The lifelong effects of early childhood adversity and toxic stress. *Pediatrics*, *129*(1), e232–e246. https://doi.org/10.1542/peds.2011-2663

Small, D. L. (2008). *On deep history and the brain*. University of California Press.

Sorenson, E. R. (1998). Preconquest consciousness. In H. Wautischer (Ed.), *Tribal epistemologies* (pp. 79–115). Ashgate.

Spencer, H. (1850). *Social statics*. Robert Schalkenbach Foundation.

Stotz, K. (2014). Extended evolutionary psychology: The importance of transgenerational developmental plasticity. *Frontiers in Psychology*, *5*, 908. https://doi.org/10.3389/fpsyg.2014.00908

Stotz, K., & Griffiths, P. E. (2018). Genetic, epigenetic, and exogenetic information. In R. Joyce (Ed.), *The Routledge handbook of evolution and philosophy* (pp. 106–119). Routledge.

Stotz, K., & Narvaez, D. (2018). Niche. In V. Zeigler-Hill & T. Shackelford (Eds.), *Encyclopedia of personality and individual differences*. Springer. https://doi.org/10.1007/978-3-319-28099-8_1554-1

Tallis, R. (2004). *Why the mind is not a computer*. Imprint Academic.

Tanghe, K. B., De Tiège, A., Pauwels, L., Blancke, S., & Braeckman, J. (2018). What's wrong with the modern evolutionary synthesis? A critical reply to Welch (2017). *Biology & Philosophy*, *33*, Article 23. https://doi.org/10.1007/s10539-018-9633-3

Tarsha, M. S., Wang, L., & Narvaez, D. (2020, October). *Positive and negative first year caregiving predicts autonomic regulation five years later* [Paper presentation]. International Society for Developmental Psychobiology Virtual Conference, Rockville, MD.

Thelen, E., & Smith, L. B. (1994). *A dynamic systems approach to the development of cognition and action*. MIT Press.

Thompson, E. (2007). *Mind in life: Biology, phenomenology, and the sciences of mind*. Harvard University Press.

Thornhill, R., & Gangestad, S. W. (1996). The evolution of human sexuality. *Trends in Ecology & Evolution*, *11*(2), 98–102. https://doi.org/10.1016/0169-5347(96)81051-2

Tishkoff, S. A., Reed, F. A., Ranciaro, A., Voight, B. F., Babbitt, C. C., Silverman, J. S., Powell, K., Mortensen, H. M., Hirbo, J. B., Osman, M., Ibrahim, M., Omar, S. A., Lema, G., Nyambo, T. B., Ghori, J., Bumpstead, S., Pritchard, J. K., Wray, G. A., & Deloukas, P. (2007). Convergent adaptation of human lactase persistence in Africa and Europe. *Nature Genetics*, *39*(1), 31–40. https://doi.org/10.1038/ng1946

Tooby, J., & Cosmides, L. (2015). The theoretical foundations of evolutionary psychology. In D. M. Buss (Ed.), *The handbook of evolutionary psychology, 2nd ed. Vol 1: Foundations* (pp. 3–87). Wiley. https://doi.org/10.1002/9781119125563.evpsych101

Trevathan, W. R. (2011). *Human birth: An evolutionary perspective* (2nd ed.). Aldine de Gruyter.

Van der Kolk, B. (2014). *The body keeps the score*. Penguin.

Van Gelder, T. (1995). What might cognition be, if not computation? *The Journal of Philosophy, 91*, 345–381. https://doi.org/10.2307/2941061

Varela, F. J., Thompson, E., & Rosch, E. (1991). *The embodied mind: Cognitive science and human experience.* MIT Press. https://doi.org/10.7551/mitpress/6730.001.0001

Waddington, C. H. (1957). *The strategy of the genes.* Allen and Unwin.

West, M. J., & King, A. P. (1987). Settling nature and nurture into an ontogenetic niche. *Developmental Psychobiology, 20*(5), 549–562. https://doi.org/10.1002/dev.420200508

West-Eberhard, M. J. (2003). *Developmental plasticity and evolution.* Oxford University Press. https://doi.org/10.1093/oso/9780195122343.001.0001

Wheeler, M. (2005). *Reconstructing the cognitive world.* MIT Press. https://doi.org/10.7551/mitpress/5824.001.0001

Wheeler, M. (2014). Revolution, reform, or business as usual? The future prospects for embodied cognition. In L. Shapiro (Ed.), *The Routledge handbook of embodied cognition* (pp. 374–383). Routledge.

Wilson, D. S., & Laland, K. N. (2016). *Empowering the extended evolutionary synthesis.* The Evolution Institute.

Wilson, E. O. (1975). *Sociobiology: The new synthesis.* Harvard University Press.

Witherington, D. C., & Lickliter, R. (2016). Integrating development and evolution in psychological science: Evolutionary developmental psychology, developmental systems, and explanatory pluralism. *Human Development, 59*(4), 200–234. https://doi.org/10.1159/000450715

Witherington, D., Overton, W., Lickliter, R., Marshall, P., & Narvaez, D. (2018). Metatheories and conceptual confusions in developmental science. *Human Development, 61*(3), 181–198. https://doi.org/10.1159/000490160

Wray, G. A., Hoekstra, H. E., Futuyma, D. J., Lenski, R. E., Mackay, T. F. C., Schluter, D., & Strassmann, J. E. (2014). Comment: Does evolutionary theory need a rethink? Point: No, all is well. *Nature, 514*(7521), 161–164. https://doi.org/10.1038/514161a

13.
EVOLUTION BEYOND NEO-DARWINISM: A NEW CONCEPTUAL FRAMEWORK
*Denis Noble**

Origin of this chapter

This paper represents the culmination of ideas previously developed in a book, *The Music of Life* (Noble, 2006), and four related articles (Noble, 2011b; Noble, 2012; Noble, 2013; Noble et al., 2014). Those publications raised many questions from readers in response to which the 'Answers' pages (http://musicoflife.co.uk/Answers-menu.html) of *The Music of Life* website were drafted. Those pages, in particular the page entitled *The language of Neo-Darwinism*, were written in preparation for the present chapter. The ideas have been extensively honed in response to further questions and comments.

Introduction

The recent explosion of research on epigenetic mechanisms described in this book and elsewhere (e.g. Noble et al., 2014), and most particularly work focused on trans-generational inheritance mediated by those mechanisms (e.g. Danchin et al., 2011; Dias and Ressler, 2014; Gluckman et al., 2007; Klironomos et al., 2013; Nelson et al., 2012; Nelson and Nadeau, 2010; Nelson et al., 2010; Rechavi et al., 2011; Sela et al., 2014), has created the need to either extend or replace the Modern (neo-Darwinist) Synthesis (Beurton

* Department of Physiology, Anatomy and Genetics, Oxford, UK.

et al., 2008; Gissis and Jablonka, 2011; Noble et al., 2014; Pigliucci and Müller, 2010). This paper explains why replacement rather than extension is called for. The reason is that the existence of robust mechanisms of trans-generational inheritance independent of DNA sequences runs strongly counter to the spirit of the Modern Synthesis. In fact, several new features of experimental results on inheritance and mechanisms of evolutionary variation are incompatible with the Modern Synthesis. Fig. 13.1 illustrates the definitions and relationships between the various features of Darwinism, the Modern Synthesis and a proposed new Integrative Synthesis. The diagram is based on an extension of the diagram used by Pigliucci and Müller (Pigliucci and Müller, 2010) in explaining the idea of an extended Modern Synthesis.

The shift to a new synthesis in evolutionary biology can also be seen to be part of a more general shift of viewpoint within biology towards systems approaches. The reductionist approach (which inspired the Modern Synthesis as a gene-centred theory of evolution) has been very productive, but it needs, and has always needed, to be complemented by an integrative approach, including a new theory of causation in biology (Noble, 2008), which I have called the theory of Biological Relativity (Noble, 2012). The approach to replace the Modern Synthesis could be called the Integrative Synthesis as it would be based on the integration of a variety of mechanisms of

DOI: 10.4324/9781032702988-21

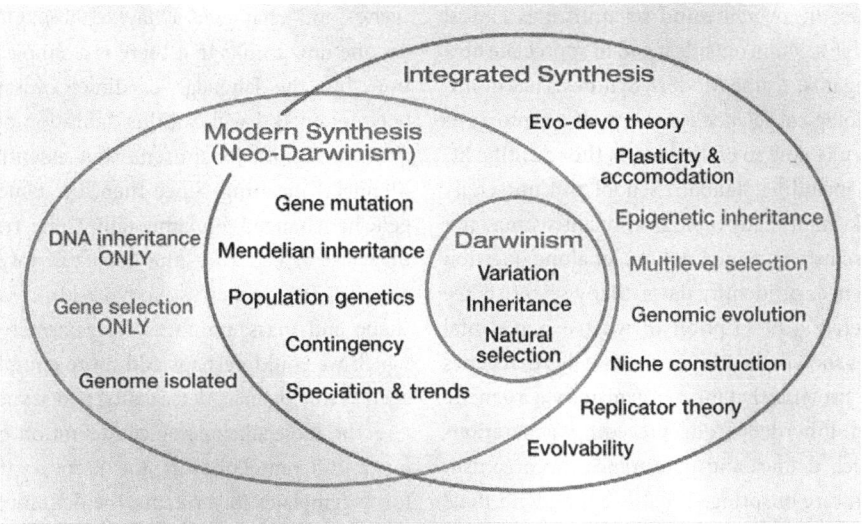

Fig. 13.1 Diagram illustrating definitions of Darwinism, Modern Synthesis (neo-Darwinism) and Integrated Synthesis. The diagram is derived from Pigliucci and Müller's (2010) presentation of an Extended Synthesis. All the elements are also present in their diagram. The differences are: (1) the elements that are incompatible with the Modern Synthesis are shown bolded on the right; (2) the reasons for the incompatibility are shown in the three corresponding bolded elements on the left. These three assumptions of the Modern Synthesis lie beyond the range of what needs to extend or replace the Modern Synthesis; (3) in consequence, the Modern Synthesis is shown as an oval extending outside the range of the extended synthesis, which therefore becomes a replacement rather than an extension.

evolutionary change that must interact, rather than the single mechanism postulated by the Modern Synthesis (Noble, 2013). We are moving to a much more nuanced multi-mechanism theory of evolution, which, interestingly, is closer to some of Darwin's ideas than to neo-Darwinism. Darwin was not a neo-Darwinist. He recognised other mechanisms in addition to natural selection and these included the inheritance of acquired characteristics.

The language of neo-Darwinism

Many of the problems with the Modern Synthesis in accommodating the new experimental findings have their origin in neo-Darwinist forms of representation rather than in experimental biology itself. These forms of representation have been responsible for, and express, the way in which 20th century biology has most frequently been interpreted. In addition, therefore, to the need to accommodate unanticipated experimental findings, we have to review the way in which we interpret and communicate experimental biology. The language of neo-Darwinism and 20th century biology reflects highly reductionist philosophical and scientific viewpoints, the concepts of which are not required by the scientific discoveries themselves. In fact, it can be shown that, in the case of some of the central concepts of 'selfish genes' or 'genetic program', no biological experiment could possibly distinguish even between completely opposite conceptual interpretations of the same experimental findings (Noble, 2006, 2011b). The concepts therefore form a biased interpretive veneer that can hide those discoveries in a web of interpretation.

I refer to a web of interpretation as it is the whole conceptual scheme of neo-Darwinism that creates the difficulty. Each concept and metaphor reinforces the overall mind-set until it is almost impossible to stand outside it and to appreciate how beguiling it is. As the Modern Synthesis has dominated biological science for over half a century, its viewpoint is now so embedded in the scientific literature, including standard school and university textbooks, that many biological scientists may not recognise its conceptual nature, let alone question incoherences or identify flaws. Many scientists see it as merely a description of what experimental work has shown: the idea in a nutshell is that genes code for proteins that form organisms via a genetic program inherited from preceding generations and which defines and determines the organism and its future offspring. What is wrong with that? This chapter analyses what I think is wrong or misleading and, above all, it shows that the conceptual scheme is neither required by, nor any longer productive for, the experimental science itself.

I will analyse the main concepts and the associated metaphors individually, and then show how they link together to form the complete narrative. We can then ask what would be an alternative approach better fitted to what we now know experimentally and to a new more integrated systems view. The terms that require analysis are 'gene', 'selfish', 'code', 'program', 'blueprint' and 'book of life'. We also need to examine secondary concepts like 'replicator' and 'vehicle'.

'Gene'

Neo-Darwinism is a gene-centred theory of evolution. Yet, its central notion, the 'gene', is an unstable concept. Surprising as it may seem, there is no single agreed definition of 'gene'. Even more seriously, the different definitions have incompatible consequences for the theory.

The word 'gene' was introduced by Johannsen (Johannsen, 1909). But the concept had already existed since Mendel's experiments on plant hybrids, published in 1866 (see Druery and Bateson, 1901), and was based on 'the silent assumption [that] was made almost universally that there is a 1:1 relation between genetic factor (gene) and character' (Mayr, 1982). Of course, no-one now thinks that there is a simple 1:1 relation, but the language of direct causation has been retained. I will call this definition of a 'gene' $gene_J$ to signify Johannsen's (but essentially also Mendel's) meaning. Since then, the concept of a gene has changed fundamentally. $Gene_J$ referred to the cause of a specific inheritable phenotype characteristic (trait), such as eye/hair/skin colour, body shape and mass, number of legs/arms/wings, to which we could perhaps add more complex traits such as intelligence, personality, and sexuality.

The molecular biological definition of a gene is very different. Following the discovery that DNA forms templates for proteins, the definition shifted to locatable DNA sequences with identifiable beginnings and endings. Complexity was added through the discovery of regulatory elements (essentially switches), but the basic cause of phenotype characteristics was still thought to be the DNA sequence as that forms the template to determine which protein is made, which in turn interacts with the rest of the organism to produce the phenotype. I will call this definition of a 'gene' $gene_M$ (see Fig. 13.2).

But unless all phenotype characteristics are attributable entirely to DNA sequences (which is false: DNA does not act outside the context of a complete cell), $gene_M$ cannot be the same as $gene_J$. According to the original view, $genes_J$ were necessarily the cause of inheritable phenotypes because that is how they were defined: as whatever in the organism is the cause of that phenotype. Johanssen even left the answer on what a gene might be vague: 'The gene was something very uncertain, "ein Etwas" ['anything'], with no connection to the chromosomes' (Wanscher, 1975). Dawkins (1982) also uses this 'catch-all' definition as 'an inheritable unit'. It would not matter whether that was DNA or something else or any combination of factors. No experiment could disprove a 'catch-all' concept as anything new discovered to be included would also be welcomed as a $gene_J$. The idea becomes unfalsifiable.

The question of causation is now an empirical investigation precisely because the modern

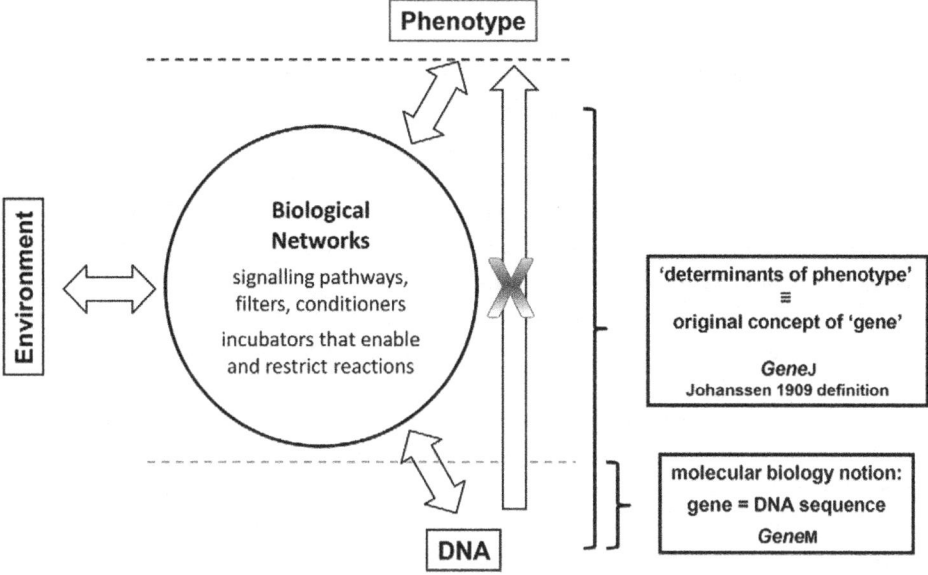

Fig. 13.2 Relationships between genes, environment and phenotype characters according to current physiological and biochemical understanding. This diagram represents the interaction between DNA sequences, environment and phenotype as occurring through biological networks. The causation occurs in both directions between all three influences on the networks. This view is very different from the idea that genes 'cause' the phenotype (right-hand arrow). This diagram also helps to explain the difference between the original concept of a gene as the cause of a particular phenotype (gene$_J$) and the modern definition as a DNA sequence (gene$_M$). For further description and analysis see Kohl et al. (2010).

definition, genes$_M$, identifies them instead with DNA sequences alone, which omits reference to all other factors. To appreciate the difference, consider Mendel's experiments showing specific phenotypes, such as smooth or wrinkled surfaces of peas. Gene$_J$ was whatever in the plant caused the peas to be smooth or wrinkled. It would not make sense to ask whether gene$_J$ was the cause. That is how it was defined. It simply is everything that determines the inherited phenotype, i.e. the trait. (Of course, different questions of an empirical nature could be asked about genes$_J$, such as whether they follow Mendel's laws. Some do; some don't.) By contrast, it makes perfect sense to ask whether a specific DNA sequence, gene$_M$, is responsible for determining the phenotype. That question is open to experimental investigation. Gene$_J$ could only be the same as gene$_M$ if DNA alone determined the phenotype.

This difference between gene$_J$ (which refers to indeterminate entities that are necessarily the cause) and gene$_M$ (whose causation is open to experimentation) is central and I will use it several times in this Chapter. The difference is in fact large as most changes in DNA do not necessarily cause a change in phenotype. Organisms are very good at buffering themselves against genomic change. Eighty per cent of knockouts in yeast, for example, are normally silent (Hillenmeyer et al., 2008), while critical biological oscillators like the cardiac pacemaker (Noble, 2011a) or circadian rhythm (Foster and Kreitzman, 2004) are buffered against genomic change through extensive back-up mechanisms.

The original concept of a gene has therefore been adopted, but then significantly changed by molecular biology. This led to a great clarification of molecular mechanisms, surely one of the greatest triumphs of 20th century biology, and widely acknowledged as such. But the more philosophical consequences of this change for higher level biology are profound and they are much less widely understood. Fig. 13.2 summarizes the difference.

Some biological scientists have even given up using the word 'gene', except in inverted commas. As Beurton et al. (Beurton et al., 2008) comment: 'It seems that a cell's enzymes are capable of actively manipulating DNA to do this or that. A genome consists largely of semi stable genetic elements that may be rearranged or even moved around in the genome thus modifying the information content of DNA.' This view is greatly reinforced by the fact that gene expression is stochastic (Chang et al., 2008) and that this itself opens the way to an extensive two-way interaction between the organism's functional networks and the structure and function of chromatin [e.g. figure 10.5 in Kupiec (Kupiec, 2014)].

The reason that the original and the molecular biological definitions have incompatible consequences for neo-Darwinism is that only the molecular biological definition, $gene_M$, could be compatible with a strict separation between the 'replicator' and the 'vehicle'. As illustrated in Fig. 13.2, a definition in terms of inheritable phenotypic characteristics (i.e., $gene_J$) necessarily includes much more than the DNA, so that the distinction between replicator and vehicle is no longer valid (Noble, 2011b). Note also that the change in definition of a gene that I am referring to here is more fundamental than some other changes that are required by recent findings in genomics, such as the 80% of 'non-coding' DNA that is now known to be transcribed (The_Encode_Project_Consortium, 2012) and which also might be included in the molecular biological definition. Those findings raise an empirical question: are those transcriptions as RNAs functional? That would extend $gene_M$ to include these additional functional sequences. The difference I refer to, by contrast, is a conceptual one. The difference between $gene_J$ and $gene_M$ would still be fundamental because it is the difference between necessary and empirically testable causality, not just an extension of the definition of $gene_M$.

'Selfish'

There is no biological experiment that could distinguish between the selfish gene theory and its opposites, such as 'imprisoned' or 'cooperative genes'. This point was conceded long ago by Richard Dawkins in his book *The Extended Phenotype*: 'I doubt that there is any experiment that could prove my claim' (Dawkins, 1982). A more complete dissection of the language and possible empirical interpretations of selfish gene theory can be found in Noble (Noble, 2011b).

'Code'

After the discovery of the double helical structure of DNA, it was found that each sequence of three bases in DNA or RNA corresponds to a single amino acid in a protein sequence. These triplet patterns are formed from any combination of the four bases U, C, A and G in RNA and T, C, A and G in DNA. They are often described as the genetic 'code', but it is important to understand that this usage of the word 'code' carries overtones that can be confusing. This section of the chapter is not intended to propose that the word 'code' should not be used. Its purpose is rather to ensure that we avoid those overtones.

A code was originally an intentional encryption used by humans to communicate. The genetic 'code' is not intentional in that sense. The word 'code' has unfortunately reinforced the idea that genes are active and even complete causes, in much the same way as a computer is caused to follow the instructions of a computer program. The more neutral word 'template' would be better. Templates are used only when required (activated); they are not themselves active causes. The active causes lie within the cells themselves because they determine the expression patterns for the different cell types and states. These patterns are communicated to the

DNA by transcription factors, by methylation patterns and by binding to the tails of histones, all of which influence the pattern and speed of transcription of different parts of the genome. If the word 'instruction' is useful at all, it is rather that the cell instructs the genome. As the Nobel-prize winner Barbara McClintock said, the genome is an 'organ of the cell', not the other way round (McClintock, 1984).

Representing the direction of causality in biology the wrong way round is confusing and has far-reaching consequences. The causality is circular, acting both ways: passive causality by DNA sequences acting as otherwise inert templates, and active causality by the functional networks of interactions that determine how the genome is activated.

'Program'

The idea of a 'genetic program' was introduced by the French Nobel laureates Jacques Monod and Francois Jacob. They referred specifically to the way in which early electronic computers were programmed by paper or magnetic tapes: 'The programme is a model borrowed from electronic computers. It equates the genetic material with the magnetic tape of a computer' (Jacob, 1982). The analogy was that DNA 'programs' the cell, tissues and organs of the body just as the code in a computer program causally determines what the computer does. In principle, the code is independent of the machine that implements it, in the sense that the code itself is sufficient to specify what will happen when the instructions are satisfied. If the program specifies a mathematical computation, for example, it would contain a specification of the computation to be performed in the form of complete algorithms. The problem is that no complete algorithms can be found in the DNA sequences. What we find is better characterised as a mixture of templates and switches. The 'templates' are the triplet sequences that specify the amino acid sequences or the RNA sequences. The 'switches' are the locations on the DNA or histones where transcription factors, methylation and other controlling processes trigger their effects. As a program, this is incomplete.

Where then does the full algorithmic logic of a program lie? Where, for example, do we find the equivalent of 'IF-THEN-ELSE' type instructions? The answer is in the cell or organism as a whole, not just in the genome.

Take as an example circadian rhythm. The simplest version of this process depends on a DNA sequence *Period* used as a template for the production of a protein PER whose concentration then builds up in the cytoplasm. It diffuses through the nuclear membrane and, as the nuclear level increases, it inhibits the transcription of *Period* (Foster and Kreitzman, 2004). This is a negative feedback loop of the kind that can be represented as implementing a 'program' like IF LEVEL X EXCEEDS Y STOP PRODUCING X, BUT IF LEVEL X IS SMALLER THAN Y CONTINUE PRODUCING X. But it is important to note that the implementation of this 'program' to produce a 24 h rhythm depends on rates of protein production by ribosomes, the rate of change of concentrations within the cytoplasm, the rate of transport across the nuclear membrane, and interaction with the gene transcription control site (the switch). All of this is necessary to produce a feedback circuit that depends on much more than the genome. It depends also on the intricate cellular, tissue and organ structures that are not specified by DNA sequences, which replicate themselves via self-templating, and which are also essential to inheritance across cell and organism generations.

This is true of all such 'programs'. To call them 'genetic programs' or 'gene networks' is to fuel the misconception that all the active causal determination lies in the one-dimensional DNA sequences. It doesn't. It also lies in the three-dimensional static and dynamic structures of the cells, tissues and organs.

The postulate of a 'genetic program' led to the idea that an organism is fully defined by its genome, whereas in fact the inheritance of cell structure is equally important. Moreover, this structure is specific to different species. Cross-species clones do not generally work. Moreover, when, very rarely, cross-species clones do work, the outcome is determined by the cytoplasmic structures and expression patterns as well as the DNA (Sun et al., 2005). In this connection it is worth noting that the basic features of structural organisation both of cells and of multicellular organisms must have been determined by physical constraints before the relevant genomic information was developed (Müller and Newman, 2003; Newman et al., 2006).

As with 'code', the purpose of this section is to warn against simplistic interpretations of the implications of the word 'program'. In the extended uses to which the word has been put in biology, and in modern computing science where the concept of a distributed program is normal, 'program' can be used in many different ways. The point is that such a 'program' does not lie in the DNA alone. That is also the reason why the concept of a 'genetic program' is not testable. By necessarily including non-DNA elements, there is no way of determining whether a 'genetic program' exists. At the limit, when all the relevant components have been added in, the 'program' is the same as the function it is supposed to be programming. The concept then becomes redundant [p. 53 of Noble (Noble, 2006)]. Enrico Coen (Coen, 1999) put the point beautifully when he wrote: 'Organisms are not simply manufactured according to a set of instructions. There is no easy way to separate instructions from the process of carrying them out, to distinguish plan from execution.'

'Blueprint'

'Blueprint' is a variation on the idea of a program. The word suffers from a similar problem to the concept of a 'program', which is that it can be mistaken to imply that all the information necessary for the construction of an organism lies in the DNA. This is clearly not true. The complete cell is also required, and its complex structures are inherited by self-templating. The 'blueprint', therefore, is the cell as a whole. But that destroys the whole idea of the genome being the full specification. It also blurs and largely nullifies the distinction between replicator and vehicle in selfish gene theory.

'Book of life'

The genome is often described as the 'book of life'. This was one of the colourful metaphors used when projecting the idea of sequencing the complete human genome. It was a brilliant public relations move. Who could not be intrigued by reading the 'book of life' and unravelling its secrets? And who could resist the promise that, within about a decade, that book would reveal how to treat cancer, heart disease, nervous diseases, diabetes, with a new era of pharmaceutical targets. As we all know, it didn't happen. An editorial in *Nature* spelt this out:

> The activity of genes is affected by many things not explicitly encoded in the genome, such as how the chromosomal material is packaged up and how it is labelled with chemical markers. Even for diseases like diabetes, which have a clear inherited component, the known genes involved seem to account for only a small proportion of the inheritance …the failure to anticipate such complexity in the genome must be blamed partly on the cosy fallacies of genetic research. After Francis Crick and James Watson cracked the riddle of DNA's molecular structure in 1953, geneticists could not resist assuming it was all over bar the shouting. They began to see DNA as the "book of life," which could be read like an instruction manual. It now seems that the genome might be less like a list of parts and more like the weather system, full of complicated feedbacks and interdependencies.
>
> (Editorial, 2010)

The 'book of life' represents the high watermark of the enthusiasm with which the language of neo-Darwinism was developed. Its failure to deliver the promised advances in healthcare speaks volumes. Of course, there were very good scientific reasons for sequencing whole genomes. The benefits to evolutionary and comparative biology in particular have been immense, and the sequencing of genomes will eventually contribute to healthcare when the sequences can be better understood in the context of other essential aspects of physiological function. But the promise of a peep into the 'book of life' leading to a cure for all diseases was a mistake.

The Language of neo-Darwinism as a Whole

All parts of the neo-Darwinist forms of representation encourage the use and acceptance of the other parts. Once one accepts the idea that the DNA and RNA templates form a 'code', the idea of the 'genetic program' follows naturally. That leads on to statements like 'they [genes] created us body and mind' (Dawkins, 1976, 2006), which gets causality wrong in two ways. First, it represents genes as active causes, whereas they are passive templates. Second, it ignores the many feedbacks on the genome that contributes to circular causality, in which causation runs in both directions. Those mistakes lead to the distinction between replicators and vehicles. The problem lies in accepting the first step, the idea that there is a 'code' forming a complete program.

The distinction between the replicator and the vehicle can be seen as the culmination of the neo-Darwinist way of thinking. If all the algorithms for the processes of life lie in the genome then the rest of the organism does seem to be a disposable vehicle. Only the genome needs to replicate, leaving any old vehicle to carry it.

The distinction, however, is a linguistic confusion and it is incorrect experimentally (Noble, 2011b). The DNA passed on from one generation to the next is based on copies (though not always perfect). The cell that carries the DNA is also a copy (also not always perfect). In order for a cell to give rise to daughter cells, both the DNA and the cell have to be copied. The only difference between copying a cell and copying DNA is that the cell copies itself by growing (copying its own detailed structure gradually, which is an example of self-templating) and then dividing so that each daughter cell has a full complement of the complex cell machinery and its organelles, whereas copying DNA for the purpose of inheritance occurs only when the cell is dividing. Moreover, the complexity of the structure in each case is comparable: 'It is therefore easy to represent the three-dimensional image structure of a cell as containing as much information as the genome' (Noble, 2011a). Faithful genome replication also depends on the prior ability of the cell to replicate itself because it is the cell that contains the necessary structures and processes to enable errors in DNA replication to be corrected. Self-templating must have been prior to the development of the relevant DNA (Müller and Newman, 2003; Newman et al., 2006).

My germ line cells are therefore just as much 'immortal' (or not) as their DNA. Moreover, nearly all of my cells and DNA die with me. Those that do survive, which are the germ cells and DNA that help to form the next generation, do not do so separately. DNA does not work without a cell. It is simply an incorrect playing with words to single the DNA out as uniquely immortal.

I was also playing with words when I wrote that 'DNA alone is inert, dead' (Noble, 2011b). But at least that has a point in actual experiments. DNA alone does nothing. By contrast, cells can continue to function for some time without DNA. Some cells do that naturally, e.g. red blood cells, which live for about 100 days without DNA. Others, such as isolated nerve axons, fibroblasts (Cox et al., 1976; Goldman et al., 1973) or any other enucleated cell type, can do so in physiological experiments.

Genes$_M$ are best viewed therefore as causes in a passive sense. They do nothing until activated. Active causation lies with proteins, membranes, metabolites, organelles, etc., and the dynamic functional networks they form in interaction with the environment (Noble, 2008).

Notice also that the language as a whole is strongly anthropomorphic. This is strange, given that most neo-Darwinists would surely wish to avoid anthropomorphising scientific discovery.

An alternative form of representation

The alternative form of representation depends on two fundamental concepts. The first one is the distinction between active and passive causes. Genes$_M$ are passive causes; they are templates used when the dynamic cell networks activate them. The second concept is that there is no privileged level of causation. In networks, that is necessarily true, and it is the central feature of what I have called the theory of biological relativity, which is formulated in a mathematical context (Noble, 2012).

I will illustrate the second point in a more familiar non-mathematical way. Take some knitting needles and some wool. Knit a rectangle. If you don't knit, just imagine the rectangle. Or use an old knitted scarf. Now pull on one corner of the rectangle while keeping the opposite corner fixed. What happens? The whole network of knitted knots moves. Now reverse the corners and pull on the other corner. Again, the whole network moves, though in a different way. This is a property of networks. Everything ultimately connects to everything else. Any part of the network can be the prime mover, and be the cause of the rest of the network moving and adjusting to the tension. Actually, it would be better still to drop the idea of any specific element as prime mover. It is networks that are dynamically functional.

Now knit a three-dimensional network. Again, imagine it. You probably don't actually know how to knit such a thing. Pulling on any part of the three-dimensional structure will cause all other parts to move (cf. Ingber, 1998). It doesn't matter whether you pull on the bottom, the top or the sides. All can be regarded as equivalent. There is no privileged location within the network.

The three-dimensional network recalls Waddington's epigenetic landscape network (Fig. 13.3) and is quite a good analogy to biological networks as the third dimension can be viewed as representing the multi-scale nature of biological networks. Properties at the scale of cells, tissues and organs influence activities of elements, such as genes and proteins, at the lower scales. This is sometimes called downward causation, to

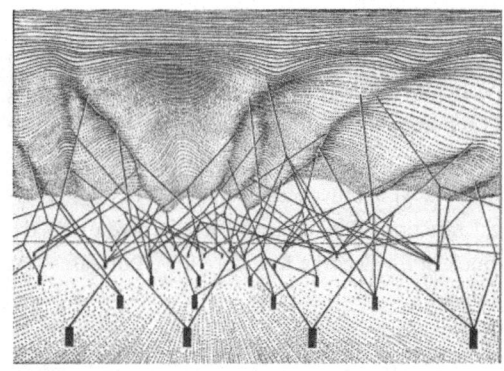

Fig. 13.3 Conrad Waddington's diagram of the epigenetic landscape. Genes (solid pegs at the bottom) are viewed as parts of complex networks so that many gene products interact between themselves and with the phenotype to produce the phenotypic landscape (top) through which development occurs. Waddington's insight was that new forms could arise through new combinations to produce new landscapes in response to environmental pressure, and that these could then be assimilated into the genome. Waddington was a system biologist in the full sense of the word. If we had followed his lead many of the more naive 20th century popularisations of genetics and evolutionary biology could have been avoided. Image taken from The Strategy of the Genes (Waddington, 1957). Reprinted (2014) by Routledge Library Editions.

distinguish it from the reductionist interpretation of causation as upward causation (Ellis et al., 2012). 'Down' and 'up' here are also metaphors and should be treated carefully. The essential point is the more neutral statement: there is no privileged scale of causality, beyond the representation of scales, perhaps. This must be the case in organisms, which work through many forms of circular causality. A more complete analysis of this alternative approach can be found in the article on Biological Relativity (Noble, 2012), from which Fig. 13.4 is taken. One of the consequences of the relativistic view is that genes$_M$ cease to be represented as active causes. Templates are passive causes, used when needed. Active causation resides in the networks, which include many components for which there are no DNA templates. It is the physics and chemistry of those dynamic networks that determine what happens.

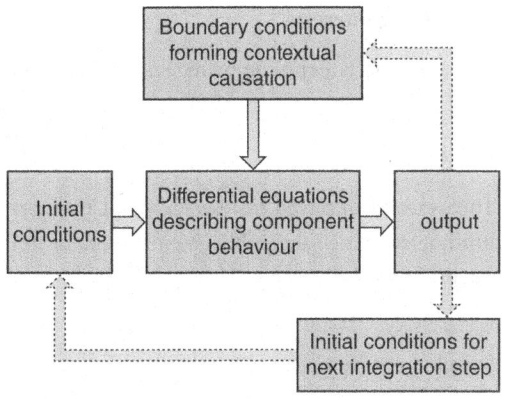

Fig. 13.4 Many models of biological systems consist of differential equations for the kinetics of each component. These equations cannot give a solution (the output) without setting the initial conditions (the state of the components at the time at which the simulation begins) and the boundary conditions. The boundary conditions define what constraints are imposed on the system by its environment and can therefore be considered as a form of contextual causation from a higher scale. This diagram is highly simplified to represent what we actually solve mathematically. In reality, boundary conditions are also involved in determining initial conditions and the output parameters can also influence the boundary conditions, while they in turn are also the initial conditions for a further period of integration of the equations. The arrows are not really unidirectional. The dotted arrows complete the diagram to show that the output contributes to the boundary conditions (although not uniquely), and determines the initial conditions for the next integration step. Legend and diagram are reproduced from Noble (Noble, 2012).

In certain respects, my chapter reflects some of the points made over 30 years ago by Ho and Saunders (Ho and Saunders, 1979), who wrote: 'The intrinsic dynamical structure of the epigenetic system itself, in its interaction with the environment, is the source of non-random variations which direct evolutionary change, and that a proper study of evolution consists in the working out of the dynamics of the epigenetic system and its response to environmental stimuli as well as the mechanisms whereby novel developmental responses are canalized.' Their ideas also owe much to those of Conrad Waddington - the term 'canalised' is one that he often used.

An important linguistic feature of the alternative, relativistic, concepts proposed here is that most or all the anthropomorphic features of the neo-Darwinist language can be eliminated, without contravening a single biological experimental fact. There may be other forms of representation that can achieve the same result. It doesn't really matter which you use. The aim is simply to distance ourselves from the biased conceptual scheme that neo-Darwinism has brought to biology, made more problematic by the fact that it has been presented as literal truth.

Conclusions

The extent to which the language of neo-Darwinism has dominated biological thought for over a century since George Romanes invented the term in a letter to *Nature* (Romanes, 1883) is remarkable. It is a tribute to the inventiveness and persuasiveness of many biologists and to their ability to communicate the original idea and its subsequent formulation as the Modern Synthesis to a very wide public. The integration of the early discoveries of molecular biology also contributed great momentum, particularly as the Central Dogma of Molecular Biology (Crick, 1970) was perceived (incorrectly as it subsequently turned out) to confirm a central assumption, which was that the genome was isolated from the lifestyle of the organism and its environment.

In retrospect, neo-Darwinism can be seen to have oversimplified biology and over-reached itself in its rhetoric. By so conclusively excluding anything that might be interpreted as Lamarckism, it assumed what couldn't be proved. As John Maynard Smith (Maynard Smith, 1998) admitted: 'It [Lamarckism] is not so obviously false as is sometimes made out', a statement that is all the more significant from being made by someone working entirely within the Modern Synthesis framework. His qualification on this statement in 1998 was that he couldn't see what the mechanism(s) might be.

We can now do so thanks to some ingenious experimental research in recent years.

Nevertheless, the dogmatism was unnecessary and uncalled for. It damaged the reputation of Lamarck, possibly irretrievably.

Lamarck should be recognised by biologists generally as one of the very first to coin and use the term 'biology' to distinguish our science, and by evolutionary biologists in particular for championing the transformation of species against some very powerful critics. Darwin praised Lamarck for this achievement: 'This justly celebrated naturalist…who upholds the doctrine that all species, including man, are descended from other species' (preface to the 4th edition of *The Origin of Species*, 1866).

Many others were damaged too, Waddington included. A little more humility in recognising the pitfalls that beset the unwary when they think they can ignore some basic philosophical principles would have been a wiser strategy. The great physicist Poincaré pointed out, in connection with the relativity principle in physics, that the worst philosophical errors are made by those who claim they are not philosophers (Poincaré, 1902, 1968). They do so because they don't even recognise the existence of the conceptual holes they fall into. Biology has its own version of those conceptual holes.

Acknowledgements

I thank Peter Hacker, Michael Joyner, Peter Kohl, Jean-Jacques Kupiec, Gerd Müller, Raymond Noble and Amit Saad for valuable discussions and comments on the paper itself, and the many correspondents who commented or asked further questions on the *Answers* pages on the Music of Life website (http://musicoflife.co.uk/Answers-menu.html). I thank Bryce Bergene, Senior Designer, Mayo Clinic Creative Media, for the design of Fig. 13.1. A video version of this figure in relation to the original extended synthesis figure can be viewed online (supplementary material Movie 1).

Competing interests

The author declares no competing financial interests.

Funding

This research received no specific grant from any funding agency in the public, commercial, or not-for-profit sectors.

Supplementary material

Supplementary material available online at http://jeb.biologists.org/lookup/suppl/doi:10.1242/jeb.106310/-/DC1

References

Beurton, P. J., Falk, R. and Rheinberger, H.-J. (2008). *The Concept of the Gene in Development and Evolution: Historical and Epistemological Perspectives.* Cambridge: Cambridge University Press.

Chang, H. H., Hemberg, M., Barahona, M., Ingber, D. E. and Huang, S. (2008). Transcriptome-wide noise controls lineage choice in mammalian progenitor cells. *Nature* **453**, 544–547.

Coen, E. (1999). *The Art of Genes.* Oxford: Oxford University Press.

Cox, R. P., Krauss, M. R., Balis, M. E. and Dancis, J. (1976). Studies on cell communication with enucleated human fibroblasts. *J. Cell Biol.* **71**, 693–703.

Crick, F. (1970). Central dogma of molecular biology. *Nature* **227**, 561–563.

Danchin, É., Charmantier, A., Champagne, F. A., Mesoudi, A., Pujol, B. and Blanchet, S. (2011). Beyond DNA: integrating inclusive inheritance into an extended theory of evolution. *Nat. Rev. Genet.* **12**, 475–486.

Dawkins, R. (1982). *The Extended Phenotype.* London: Freeman.

Dawkins, R. (1976, 2006). *The Selfish Gene.* Oxford: Oxford University Press.

Dias, B. G. and Ressler, K. J. (2014). Parental olfactory experience influences behavior and neural structure in subsequent generations. *Nat. Neurosci.* **17**, 89–96.

Druery, C. T. and Bateson, W. (1901). Experiments in plant hybridization. *Journal of the Royal Horticultural Society* **26**, 1–32.

Editorial (2010). The human genome at ten. *Nature* **464**, 649–650.

Ellis, G. F. R., Noble, D. and O'Connor, T. (2012). Top-down causation: an integrating theme within and across the sciences. *Interface Focus* **2**, 1–3.

Foster, R. and Kreitzman, L. (2004). *Rhythms of Life*. London: Profile Books.

Gissis, S. B. and Jablonka, E. (2011). *Transformations of Lamarckism. From Subtle Fluids to Molecular Biology*. Cambridge, MA: MIT Press.

Gluckman, P. D., Hanson, M. A. and Beedle, A. S. (2007). Non-genomic transgenerational Inheritance of disease risk. *BioEssays* **29**, 145–154.

Goldman, R. D., Pollack, R. and Hopkins, N. H. (1973). Preservation of normal behavior by enucleated cells in culture. *Proc. Natl. Acad. Sci. USA* **70**, 750–754.

Hillenmeyer, M. E., Fung, E., Wildenhain, J., Pierce, S. E., Hoon, S., Lee, W., Proctor, M., St Onge, R. P., Tyers, M., Koller, D. et al. (2008). The chemical genomic portrait of yeast: uncovering a phenotype for all genes. *Science* **320**, 362–365.

Ho, M. W. and Saunders, P. T. (1979). Beyond neo-Darwinism - an epigenetic approach to evolution. *J. Theor. Biol.* **78**, 573–591.

Ingber, D. E. (1998). The architecture of life. *Sci. Am.* **278**, 48–57.

Jacob, F. (1982). *The Possible and the Actual*. New York, NY: Pantheon Books.

Johannsen, W. (1909). *Elemente der Exakten Erblichkeitslehre*. Jena: Gustav Fischer.

Klironomos, F. D., Berg, J. and Collins, S. (2013). How epigenetic mutations can affect genetic evolution: model and mechanism. *BioEssays* **35**, 571–578.

Kohl, P., Crampin, E. J., Quinn, T. A. and Noble, D. (2010). Systems biology: an approach. *Clin. Pharmacol. Ther.* **88**, 25–33.

Kupiec, J.-J. (2014). Cell differentiation is a stochastic process subjected to natural selection. In *Towards a Theory of Development* (ed. A. Minelli and T. Pradeu), pp. 155–173. Oxford: OUP.

Maynard Smith, J. (1998). *Evolutionary Genetics*. New York, NY: Oxford University Press.

Mayr, E. (1982). *The Growth of Biological Thought*. Cambridge, MA: Harvard.

McClintock, B. (1984). The significance of responses of the genome to challenge. *Science* **226**, 792–801.

Müller, G. and Newman, S. A. (2003). Origination of organismal form: the forgotten cause in evolutionary theory. In *Origination of Organismal Form* (ed. G. Müller and S. A. Newman), pp. 3–10. Boston, MA: MIT Press.

Nelson, V. R. and Nadeau, J. H. (2010). Transgenerational genetic effects. *Epigenomics* **2**, 797–806.

Nelson, V. R., Spiezio, S. H. and Nadeau, J. H. (2010). Transgenerational genetic effects of the paternal Y chromosome on daughters' phenotypes. *Epigenomics* **2**, 513–521.

Nelson, V. R., Heaney, J. D., Tesar, P. J., Davidson, N. O. and Nadeau, J. H. (2012). Transgenerational epigenetic effects of Apobec1 deficiency on testicular germ cell tumor susceptibility and embryonic viability. *Proc. Natl. Acad. Sci. USA* **109**, E2766–E2773.

Newman, S. A., Forgacs, G. and Müller, G. B. (2006). Before programs: the physical origination of multicellular forms. *Int. J. Dev. Biol.* **50**, 289–299.

Noble, D. (2006). *The Music of Life*. Oxford: Oxford University Press.

Noble, D. (2008). Genes and causation. *Philos. Trans. R. Soc. A* **366**, 3001–3015.

Noble, D. (2011a). Differential and integral views of genetics in computational systems biology. *Interface Focus* **1**, 7–15.

Noble, D. (2011b). Neo-Darwinism, the modern synthesis and selfish genes: are they of use in physiology? *J. Physiol.* **589**, 1007–1015.

Noble, D. (2012). A theory of biological relativity: no privileged level of causation. *Interface Focus* **2**, 55–64.

Noble, D. (2013). Physiology is rocking the foundations of evolutionary biology. *Exp. Physiol.* **98**, 1235–1243.

Noble, D., Jablonka, E., Joyner, M. J., Müller, G. B. and Omholt, S. W. (2014). Evolution evolves: physiology returns to centre stage. *J. Physiol.* **592**, 2237–2244.

Pigliucci, M. and Müller, G. B. (2010). Elements of an extended evolutionary synthesis. In *Evolution: The Extended Synthesis* (ed. M. Pigliucci and G. B. Muller), pp. 3–17. Cambridge, MA: MIT Press.

Poincaré, H. (1902, 1968). *La Science et l'Hypothèse*. Paris: Flammarion.

Rechavi, O., Minevich, G. and Hobert, O. (2011). Transgenerational inheritance of an acquired small RNA-based antiviral response in C. elegans. *Cell* **147**, 1248–1256.

Romanes, G. J. (1883). Letter to the Editor. *Nature* **27**, 528–529.

Sela, M., Kloog, Y. and Rechavi, O. (2014). Non-coding RNAs as the bridge between epigenetic mechanisms, lineages and domains of life. *J. Physiol.* **592**, 2369–2373.

Sun, Y. H., Chen, S. P., Wang, Y. P., Hu, W. and Zhu, Z. Y. (2005). Cytoplasmic impact on cross-genus cloned fish derived from transgenic common carp (*Cyprinus carpio*) nuclei and goldfish (*Carassius auratus*) enucleated eggs. *Biol. Reprod.* **72**, 510–515.

The_Encode_Project_Consortium (2012). An integrated encyclopedia of DNA elements in the human genome. *Nature* **489**, 57–74.

Waddington, C. H. (1957). *The Strategy of the Genes*. London: Allen and Unwin.

Wanscher, J. H. (1975). An analysis of Wilhelm Johannsen's genetical term 'genotype' 1909-26. *Hereditas* **79**, 1–4.

SECTION V

BEHAVIOR GENETICS

Heritability, twin studies, adoption studies, and IQ

EDITORS' INTRODUCTION

The complexity involved in the depictions of development found in relational developmental systems (RDS)-based theories can be daunting to scholars, both in regard to the conceptual difficulties involved in integratively understanding the multiple levels of organization fused within the relational developmental system and in respect to the methodological challenges involved in using such theories as a frame for research. If challenging to scholars, such theories are often seen as virtually impossible to grasp by non-specialists (the "person in the street," to Horowitz, 2000, p. 8) and/or by media representatives. Both groups may gravitate toward "single-variable stories" (Horowitz, 2000, p. 3) about human development—such as "genes cause behavior" (e.g., see Plomin, 2018; Rushton, 1999, 2000)—in order to understand or communicate about people's lives, respectively.

Such a simplistic—a distortingly simplistic—alternative to RDS-based theories of human development is embodied in the field of behavior genetics. As explained by Horowitz (2000):

> Against the media popularity of single-variable stories, the science itself is moving inexorably toward greater and greater data-driven, integrative theoretical complexity. An exception to this is behavioral genetics. In contrast to the dynamic nonlinear interactive models full of reciprocity between and among levels and variables, behavioral genetics presents a relatively non-dynamic linear additive model that tries to assign percentages of variance in behavior and development that can be attributed to genes. The enterprise rests on the assumption that genetic influence can be expressed as a value accounting for a portion of the variance in a nondynamic linear equation for predicting behavioral functioning and furthermore, that the individual experiences of shared and nonshared environments can be assessed inferentially by the degree of biological relatedness of individuals without empirical observations of experience.
>
> Behavioral genetics involves a relatively simplistic approach when compared with the kinds of dynamic systems theories currently being elaborated. Perhaps that is why, in the mode of wanting simple answers to simple questions, behavior genetic reports are so media attracting. (p. 3)

What then is the field of behavior genetics? What is the view of human development it presents? How does it seek to support this view? Does it represent a viable, "nature" alternative to dynamic and integrative developmental systems conceptions of human development in general or of particular facets of human development (e.g., intelligence) more specifically? To begin to address these questions, we turn to a definition of this field.

According to Robert Plomin (2000, p. 30), arguably the most eminent behavior geneticist in the world, "Behavioural genetics is the genetic study of behaviour, which includes quantitative genetics (twin and adoption studies) as well as molecular genetics (DNA studies) of human and animal behaviour broadly defined to include responses of the organism from responses measured in the brain such as functional neuroimaging to self-report questionnaires." Plomin et al. (1980) add that "Behavioral genetics lies at the interface between genetics and the behavioral sciences" (p. 12), and Plomin (1986) noted that "Behavioral geneticists explore the etiology of individuality, differences among individuals in a population" (p. 5). Plomin (1986) explained that "The three basic methods used in human behavioral genetics are family, twin, and adoption studies" (p. 11).

Across all of these methods, which have remained unchanged (e.g., see Plomin, 2018) despite repeated criticism (e.g., Charney, 2017, 2022; Feldman & Riskin, 2022; Lickliter, 2016; Moore & Shenk, 2016), the goal of behavior genetic analysis is to separate (partition) the variation in a distribution of scores (e.g., for a personality attribute, temperamental characteristic, or intelligence) into the proportion due to genes and the proportion due to the environment. Although behavior geneticists admit that genes and environments may be correlated and/or may interact (and they use this term in its statistical sense), they most typically seek to compute a score (termed a *heritability coefficient*) that (in its most frequently used form) denotes the independent contribution of genetic variance to the overall differences in the distribution of scores for a given individual characteristic.

For such heritability scores to be meaningful, the methodologies of behavior genetics rest on a model of gene function that considers the possibility that genetic contributions are independent of (not correlated or interactive with) the context within which genes exist. However, genes do not work in the way that behavior geneticists imagine.

Fatal Flaws in the Behavior Genetics Model of Gene Function

Strohman (1993a, 1993b), as well as in the writings of other biologists (e.g., Feldman & Riskin, 2022; Ho, 1984, 2010, 2013; Müller-Hill, 1988), and of biological or comparative psychologists (e.g., McEwen, 1997, 1998, 1999; Meaney, 2010; Meaney et al., 1988), do not place credence in the model of genetic function involved in behavior genetics. In fact, Venter and colleagues (2001), the group that successfully mapped the sequence of the human genome, emphasize that there are two conceptual errors that should not be made in the face of the advances they and other scientists have made in understanding the structure and functional consequences of the human genome. They stress that "There are two fallacies to be avoided: determinism, the idea that all characteristics of the person are 'hard-wired' by the genome; and reductionism, the view that with complete knowledge of the human genome sequence, it is only a matter of time before our understanding of gene functions and interactions will provide a complete causal description of human variability" (Venter et al., 2001, p. 1348). Developmental scientists conducting their research within RDS-based models would make the same two criticisms (e.g., Overton, 2015).

Contemporary thought in molecular genetics thus rejects the idea that genes are structures that act *on* supragenetic levels; instead, these scientists adopt the RDS-based, probabilistic epigenetic view of the role of genes in human development that is epitomized by Gottlieb (1970, 1996, 1997). The integration— or fusion—of genes *with* the other levels of organization that comprise the person and his or her context create the individuality of behavior at and across points in ontogeny (e.g., Charney, 2017, 2022; Lickliter, 2016; Moore & Shenk, 2016)

In essence, the field of behavior genetics (e.g., Plomin, 1986, 2000, 2018; Rowe, 1994) uses a model of genetic structure and function that is specifically rejected by those scientists who study the structure and action of genes directly (e.g., Venter et al., 2001). This rejection occurs because the field of behavioral genetics not only employs a counterfactual and scientifically atavistic conception of the role of genes in human development (e.g., Ho, 1984, 2010, 2013; Strohman, 1993a, 1993b), but also because behavior genetics is a viewpoint with a conceptually flawed and empirically deficient view of the developmental process and, as well, involves the conflation of description and explanation.

For instance, in regard to process, the structural account of genetic action offered by behavior genetics suffers from the flaws of all structural accounts of development; that is, as explained by Thelen and Smith (1994, 1998, 2006; Smith & Thelen, 1993), such conceptions are inherently incomplete. These views do not explain individual behavioral performance (actions), other than to express empirically unsubstantiated confidence that in some way genetic structures translate—through the levels of cells, tissues, organs, the individual, and his or her actual context—into real-time actions.

For example, without any specification of the pathways of influence from genes to behaviors, Rowe (1994) asserted that:

> Genes can produce dispositions, tendencies, and inclinations, because people with subtly different nervous systems are differently motivated ... [and] given enough environmental opportunities [for selection of environments], the ones chosen are those most reinforcing for a particular nervous system created by a particular genotype ... the direction of the growth curve of development, and the limit ultimately attained, is set in the genes. (p. 91)

However, because behavior geneticists believe that genetic structure transcends and is independent of real-time actions, an adequate, empirically verifiable account of actual individual-in-context behavior is beyond theoretical range (Smith & Thelen, 1993). Moreover, because of the inability to explain individual performance, of actual individual-in-context behavior, behavior genetics, like other structural theories (Smith & Thelen, 1993), cannot explain the global order of behavior or developmental change itself.

In turn, in regard to the conflation of description and explanation, behavior genetics describes variability in *distributions* of a specific attribute in a specific sample, and then explains the distribution it has observed by reference to a label it has applied to one (or the other) of the "sources" of the variability genes or environment. Not only is this reification an instance of the nominal fallacy, but also—to paraphrase the parody of structural explanations presented by Smith and Thelen (1993, p. 159)—the cause of the distribution of interindividual differences in a distribution of a specific attribute is merely an abstract description of the attribute distribution itself: Behavior genetics describes the variability in a distribution, labels it with a fancy "source" term (i.e., heritability), and then imputes that there is a gene, or set of genes, that explains the distribution.

To illustrate, Rowe (1994) noted that "understanding the growth and development of a single individual has been confused with understanding the origin of different traits in a population" (p. 3). However, this confusion about the distinction between interindividual differences and intraindividual change, as well as the problem of the conflation of description and explanation, exists in behavior genetics. On the basis of heritability data, writers such as Rowe (1994) seamlessly slide from talking about descriptive "sources" of variation within a distribution into talking about the genetic basis of individual development, that is, about the "causal influence on such child outcomes as intelligence, personality, and psychopathology" (p. 1).

The logical and inferential problems with such statements are enormous. For instance, Charney (2017) explains that:

> It is important to understand that heritability estimates address *similarities/differences* in traits rather than the *causes* of traits themselves, and that the difference here is enormously consequential. The claim, e.g., that lifetime income is 50% heritable is not intended to mean that 50% of a person's income is due to her 'genes' and 50% to her 'environment.' Heritability estimates only apply to populations, not to individuals, and the word 'heritability,' in this context, must not be confused with 'genetic inheritance.' Unfortunately, although 'correlation does not entail causation,' heritability estimates are frequently misinterpreted as showing that genetic similarities *cause* trait similarities in the study population. (pp. 1–2)

We will note again the problems raised in behavior genetics by the conflation of description and explanation. Here, however, it is necessary to note that one key basis of the lack of an adequate treatment in behavior genetics of performance, developmental sequence and process, as well as the distinction between description and explanation, is that these conceptual problems are coupled in behavior genetics with a lack of an adequate theoretical understanding both of supragenetic intraorganism processes (Gottlieb, 1991, 1997, 2002/2014; Gottlieb et al., 2006) and of extraorganism contextual or ecological processes (e.g., Bronfenbrenner & Ceci, 1994; Bronfenbrenner & Morris, 2006; Horowitz, 2000; Lewis, 1997; Magnusson, 1999a, 1999b, 2000; Sameroff, 1983, 2009, 2010; Thelen & Smith, 2006). Accordingly, in behavior genetics, there is a failure to adequately measure the environment or ecology of human development. In short, to paraphrase Goldberger (1980), in his discussion of Hearnshaw's (1979) account of the scientific fraud perpetrated by behavior geneticist, Cyril Burt, in regard to the study of the heritability of intelligence, behavior geneticists have methods that give them a lot of numbers but very little sensible or useful data about human development.

Behavior Genetics as the Emperor's New Clothes

That these egregious conceptual and methodological problems exist is not news. For more than 50 years, at this writing, these problems have been identified by several scholars, including for example Bronfenbrenner and Ceci (1994), Charney (2012, 2017), Collins et al. (2000), Ford and Lerner (1992), Gottlieb (1970, 1997; Gottlieb et al., 2006), Greenberg (2004, 2011, 2013, 2015, 2016), Hirsch (1970, 1997, 2004), Horowitz (2000), Joseph (2004), Kuo (1967, 1976), Krimsky and Gruber (2013), Lehrman (1953, 1970), Lerner (1976, 1978, 2015a), Lewis (1997), Magnusson (1999a, 1999b; Magnusson & Stattin, 1998), Moore (2002, 2006, 2015a, 2016; Moore & Shenk, 2016), Overton (1973, 2015; Overton & Lerner, 2012), Panofsky (2014), Richardson (2013, 2017), Schneirla (1956, 1957), Tobach (1981; Tobach & Greenberg, 1984), Thelen and Smith (1994, 2006), Wahlsten (2012, 2013), and Witherington (2015; Witherington & Heying, 2013).

Yet, despite criticism by colleagues in the fields of psychology or developmental science, including some colleagues who themselves are leading behavior geneticists (Turkheimer, 2011), as well as by molecular geneticists (Cole, 2014; Slavich & Cole, 2013), population geneticists (e.g., Feldman, 2014; Feldman & Lewontin, 1975; Feldman & Riskin, 2022), and evolutionary biologists (e.g., Gould, 1981; Ho, 2010, 2013; Jablonka & Lamb, 2005), many proponents of behavioral genetics continue to act *as if* there is a compelling evidence base for the inheritance of behaviors as varied as academic achievement (Selzam et al., 2017), intelligence (Jensen, 1969, 1998), parenting (Scarr, 1992), morality (Wilson, 1975), temperament (Buss & Plomin, 1984), television watching (Plomin et al., 1990), and, as Charney (2017, p. 1) lists, "voting in a presidential election, creative dance performance, and utilitarian moral judgments." In addition, there are some behavior genetic accounts claiming that genes shape the role in human development of the "environment" (Harris, 1998; Plomin, 1986, 2000; Plomin & Daniels, 1987; Rowe, 1994)!

The breadth and depth of the continuing criticisms of behavior genetics have been somewhat invisible or, at least, ignored by Plomin (2000, p. 30), who claimed that:

> The controversy that swirled around behavioural genetics research during the 1970s has largely faded. During the 1980s and especially during the 1990s, the behavioural sciences became much more accepting of genetic influence.

Extending this positive view of the contributions of behavior genetics into the second decade of the 21st century, Plomin and colleagues presented in 2016 the "top ten" replicated findings from behavior genetics (Plomin et al., 2016). They note that:

> On the basis of our decades of experience in the field of behavioral genetics and our experience in writing the major textbook in the field (Plomin et al., 2013), we selected these 10 findings because in our opinion they are "big" findings, both in terms of effect size and their potential impact on psychological science. These findings are not novel precisely because we selected results that have been repeatedly verified. For this reason, each of the findings in our top 10 list has been reviewed elsewhere, and a few have been highlighted previously as "laws" of behavioral genetics, as will be noted later. Although not all of these findings are supported by formal meta-analyses, we expect that most behavioral geneticists will agree with the 10 findings on our list, but we also suspect they may wish to add to the list.
>
> (Plomin et al., 2016, pp. 3–4)

Plomin's assertions about the acceptance and contributions of behavior genetics are incorrect for several reasons.

Failures to Replicate

First, although I would not dispute the claim by Plomin et al. (2016), that behavior geneticists would applaud their Top 10 list and might even want to add additional items to it, there is strong evidence that the replicability of behavior genetics findings that Plomin et al. (2016) claim is simply not the case. Charney (2017) summarizes this evidence. The computation of heritability coefficients is not the only method used by behavior geneticists to identify the genes that (in their view) determine behavior. One additional method is termed candidate gene association (CGA) studies. This method is used to find polymorphisms associated with specific behaviors. It is useful here to note that that "When a specific variation in DNA sequence in a particular location on a chromosome occurs in >1% of the population, it is referred to as a 'polymorphism'" (Charney, 2017, p. 3). Charney goes on to explain that:

> In a typical CGA study, a researcher proposes that those individuals with a given polymorphism are more likely to exhibit a given behavior. CGA studies have typically focused upon a small number of polymorphisms on the same regions of a small number of genes (e.g., MAOA, 5-HTT, DRD2, and DRD4) for two reasons: (1) These genes are transcribed to produce proteins involved in the regulation of neurotransmitters and thus are believed to be important for human behavior; and

(2) researchers believed, on the basis of mouse and *in vitro* experiments, that they could associate polymorphic differences in regulatory regions of these genes with differences in the level of certain neurotransmitters and thereby with differences in behavior. In fact, attempts to associate these polymorphisms with actual differences in neurotransmitters in the human brain under physiological conditions have been largely unsuccessful. Nonetheless, this did not stop researchers from proposing, and claiming to have demonstrated, associations between the same polymorphisms of these neurotransmitter-related genes and a bewildering array of different behaviors.

(2017 p. 3)

In the face of such failures to replicate, Charney (2017) points out that John K. Hewitt (2012), the then editor of the journal, *Behavior Genetics*, wrote an editorial (pp. 1–2) that in part said that:

The literature on candidate gene associations is full of reports that have not stood up to rigorous replication … it now seems likely that many of the published findings of the last decade are wrong or misleading and have not contributed to real advances in knowledge.

Nevertheless, Charney (2017) points out that through the time of his writing CGA articles continued to be published.

Another method used by behavior geneticists to identify the genes that determine behavior is Genome Wide Association Studies (GWAS) (see Richardson, 2017). This method involves using, typically, thousands of research participants who either manifest a specific behavior of interest (these participants are termed *cases*) or do not manifest this behavior (these participants are termed *controls*). After such an identification is made, researchers then assess "large segments of the genome (a million or more base pairs) in an attempt to find a polymorphism that *cases*…have to a greater extent than *controls*" (Charney, 2017, p. 4). In addition, Charney notes that most GWAS "search for single nucleotide polymorphisms (SNPs), a polymorphism involving the substitution of a single base pair in a given position on the DNA molecule" (2017, p. 4).

However, here again empirical failure is the result of such research. Charney (2017, p. 4) summarized the state of this literature by explaining that these studies:

have been largely unsuccessful, leading to the so-called 'problem of missing heritability': Despite the supposed heritability of all human behavior, behavior geneticists have been unable to find any substantial associations between polymorphisms and behaviors (Turkheimer, 2011).

Despite such failures proponents of behavior genetics make claims without any independent empirical evidence that there are over nine million SNPs involved in behaviors such as educational attainment (Charney, 2017; see too Richardson, 2017). In sum, then, Plomin's (2000) claims, that 1. developmental science has accepted his view of the role of genes in human development and that, as well, 2. There are several well-replicated findings emanating from the field of behavior genetics (Plomin et al., 2016) are unfounded.

The Continuing Story of Shortcomings of Heritability Analysis

Coupled with the failures of CGA and GWAS to provide data in support of the views found in behavior genetics about the role of genes in human development are the myriad problems of heritability analysis.

Tabery (2014) explains that heritability analysis represents the essential approach of behavior genetics, that is, *variance partitioning*. As explained by Moore (2015b), this method seeks to divide (partition) the variance in some instance of the behavioral phenotype (e.g., intelligence, personality, character, voting, morality, or television watching) into the variance contributed to the phenotypic variance by genetics versus the variance contributed to the phenotypic variation by environment (typically broadly construed). Thus, the variance partitioning approach harkens back to what Anastasi (1958) termed the "How much of each?" question about the nature versus nurture controversy. Tabery (2014) contrasts the variance partitioning approach to understand the basis of human attributes with the approach associated with RDS-based research, that is, what Tabery (2014) labels *mechanism elucidation*. As explained by Moore (2015b), this approach engages the integrated, developmental system by seeking to understand the co-construction (Moore, 2015a) between the individual and context (that is, individual ⇔ context relations) involved in the *process* (note, not *mechanism*) through which all attributes of the phenotype (the whole organism) are *developed*.

This process-elucidation approach (a term that is one more clearly associated with RDS-based ideas than is the reductionist-related term "mechanism;" e.g., Lerner & Overton, 2017) is aimed at explaining how variation within the dynamic and holistic, relational development system eventuates in a specific attribute of an individual being developed. However, at best, the variance-partitioning approach can only describe purported sources of variance associated with variance across individuals in a specific attribute distribution. In addition, the variance-partitioning, which enables behavior geneticists to compute a heritability coefficient, cannot even validly fulfill its only purpose, that is, accurately partitioning variance in such a distribution. That is, in addition to ignoring that heritability is a statistic that pertains only to interindividual differences and not to intraindividual change and, as well, involves the conflation of description and explanation (e.g., describing covariation between two constructs—genes and a behavioral attribute, for instance—does not explain the basis of the covariation), the methodological problems with the methodology of heritability analysis are legion.

Calculating Heritability

For the purposes of our discussion of the computational issues associated with heritability analyses, we will symbolize heritability by the term h^2. If all the variation in an attribute (e.g., intelligence) could be attributed to the concomitant variation within the gene distribution of the sample under study, then no variation whatsoever in the attribute would be due to environmental variation. In such a case, the value for environmental variation in the population would be zero and, accordingly, h^2 would equal one, or +1.0 (the "+" sign is not actually needed here, in that heritability coefficients range only from 0.0 to 1.0). Note, however, that although there was no variation in the measure of the environment, and it thus contributed no variance to this particular heritability coefficient, this statistic does *not* mean that environment was not a basis of the attribute (Hebb, 1970; Moore, 2015b). The environment could have had an important impact on the attribute in question, even the key impact, although this influence was the same for all individuals in the sample. Donald Hebb (1970) offered a useful example of the problems associated with interpreting high heritability as reflective on low environmental influence. He did so by drawing on a "modest proposal" put forth by Mark Twain:

> Mark Twain once proposed that boys should be raised in barrels to the age of 12 and fed through the bung-hole. Suppose we have 100 boys reared this way, with a practically identical environment. Jensen agrees that environment has *some* importance (20% worth?), so we must expect that the boys

on emerging from the barrels will have a mean IQ well below 100. However, the variance attributable to the environment is practically zero, so on the "analysis of variance" argument [that is, the variance-partitioning approach], the environment is not a factor in the low level of IQ, which is nonsense. (p. 578)

In Hebb's example, environmental had no *differential* effect on the boy's IQs; presumably in all boys it has the same (severely limiting) effect. In having this same effect, environment could contribute nothing to differences between the boys. No differences—or variation—existed in the environment, and so the environment could not be said to contribute anything to differences between people. Yet, it is also obvious that environment had a major influence on each boys' IQ scores. Even with heritability equal to 1.0, the intelligence of each of the boys would have been different had he developed in an environment other than a pickle barrel.

As another example how misleading heritability coefficients can be in regard to understanding the role of environmental influences, consider the following imaginary example. Suppose a society had a law pertaining to eligibility for government office. The law was simply that men could be elected to such positions and women could not. Consider what one would need to know in order to completely and correctly divide a group of randomly chosen people from this society into one of two groups. Group 1 would consist of those who had greater than a zero percent chance of being elected to a leadership post and Group 2 would consist of those who had no chance. All that one would need to know to make this division with complete accuracy was whether a person possessed an XX pair of chromosomes or an XY pair. In the first case, the person would be a female (since possession of the XX chromosome pair leads to female development). In the second, the person would be a male. One could, thus, correctly place all possessors of the XY pair into the "greater than zero chance" group and all possessors of the XX pair into the "no chance" group.

In this example, then, *all* the differences between people with respect to the characteristic in question—eligibility for office—can be summarized by genetic differences between them, that is, possession of either the XX or XY chromosome pair. In this case the heritability of "being eligible for election" would be 1.0. In other words, in this society eligibility is 100% heritable. But, by any stretch of the imagination, does this finding mean that the eligibility characteristic is inherited, or that the differences between men and women with respect to this characteristic are genetic in nature? Is there a gene for "eligibility," one that men possess and women do not? Would behavior geneticists use the CGA method to search for evidence of an electability gene?

Of course, the answers to these questions are no, although, frankly, given the plethora of variables that have been the targets of CGA studies (Charney, 2017), we admit we are less certain about whether electability would be excluded from future instances of behavior genetics research using this method. In any case, although heritability in this imaginary case is perfect, it presents an instance analogous to the one Hebb (1970) provided. Our example indicated that contextual, social or cultural, variables—laws regarding what men and women can and cannot do—determine whether someone has a chance of being elected. Indeed, if the law in question were changed, and women were then allowed to hold office, then the heritability of the eligibility characteristic would—probably rather quickly—fall to much less than 1.0.

Of course, not all instances of the computation of heritability result in heritability coefficients at or approaching 1.0. Typically, the heritability of an attribute falls somewhere within the 0.0 to 1.0 range and, quite importantly, estimates of heritability of even the same measure of an attribute vary from sample to sample (e.g., Charney, 2017; Hirsch, 1963; Moore, 2002, 2006; Moore & Shenk, 2016). Therefore, *heritability is a property of samples and not of individual attributes* (Hirsch, 1963). Thus, an attribute of an individual cannot be appropriately spoken of as of being heritable; researchers cannot correctly speak of the trait intelligence as being heritable. Rather, heritability only refers to the extent to which genetic variation among the members of a specific sample of individuals is associated with the distribution of scores for an attribute measured in a specific sample. In fact, Jensen (1969) made this same point. In discussing the appropriateness of applying the concept of heritability to a population of people and the inappropriateness of such application to any individual within that population, Jensen (1969) stated:

> Heritability is a population statistic, describing the relative magnitude of the genetic component (or set of genetic components) in the population variance of the characteristic in question. It has no sensible meaning with reference to a measurement or characteristic in an individual. A single measurement, by definition, has no variance. (p. 42)

Hence, heritability describes something about a group and not anything about an individual. Heritability relates to the source of differences among people in a sample; it says nothing about a given attribute (e.g., intelligence, personality, etc.) within any individual in that sample. Accordingly, as Moore and Shenk (2016) explain (see too Hirsch, 1963, 1970), although the claim, that if a trait is heritable it is therefore inherited, seems so obvious as to border on a tautology, or an assertion true by definition, nothing could be further from the truth: *The demonstration of heritability says nothing about the extent to which an attribute is inherited* (Lerner, 1992; Lerner & von Eye, 1992, 1993). In fact, evidence for heritability cannot be taken as evidence for the common possession of a particular set of genes. There is no connection between the concept of heritability and the idea that a human's characteristics are caused by one or more genes (e.g., Moore, 2015b; Moore & Shenk, 2016)!

Statistical Problems Associated with Heritability Analysis: Genotype–environment Correlation

A facet of the problems associated with the calculation of heritability coefficients is the statistical problem of genotype–environment correlations. This problem is worthy of focus because it represents an Achilles Heel for researchers engaged in making estimates of the heritability of attributes of human development. For instance, David Layzer (1974, p. 1263) noted that the impact of genes (G) and environment (E) "can be unambiguously defined if, and only if, genotype–environment correlations are absent. Even then, however, a certain practical ambiguity persists. Genetic differences may influence the development of an attribute in qualitatively distinct ways. For example, [with] different thresholds, different slopes, and different final values. Heritability estimates do not take such qualitative distinctions into account."

Layzer (1974) also noted that plant and animal geneticists *can* minimize genotype–environment correlation by *randomizing environments:* thus, such researchers can take a step that is, according to Layzer, *indispensable* for the application of heritability analysis, because without such randomization there is just no means to disentangle genetic and environmental contributions to phenotypic variances. But this step is not done in IQ heritability research. Layzer (1974) explained:

> The applicability of heritability analysis does not, as is commonly assumed, hinge on the smallness of the interaction term (R) relative to the terms G and E in Fisher's decomposition of the phenotypic value. In fact, one may reasonably assume on biological grounds that genotype–environment interaction makes a substantial contribution to the phenotypic value of every phenotypically plastic attribute, except in populations where the ranges of genetic and environmental variation are severely restricted. Even so, heritability analysis can be applied to phenotypically plastic attributes, provided

that the relevant genetic and environmental variables are statistically uncorrelated. When this condition is not satisfied, the contributions of interaction to phenotypic variances and covariances cannot, in general, be separated from the contributions of genotype and environment, and heritability analysis cannot, therefore, be applied meaningfully.

In adult subpopulations, IQ and environment are well known to be more or less strongly correlated. Since differences in IQ are undeniably related to genetic differences (although not, perhaps, in a very simple way), one may safely assume that genotype–environment correlation is significant in adult subpopulations and in subpopulations composed of children reared by their biological parents or by close relatives. Hence, no valid estimate of IQ heritability can be based on data that refer to such sub-populations.

Yet data of precisely this kind make up the bulk of the available material, and many published heritability estimates have been based on them. Burt (1966), Jensen (1969), and Herrnstein (1971) for example, all cite kinship correlation data as evidence for a high value of h^2. (p. 1263)

These interpretative problems are linked to others associated with the work of hereditarians when they discuss the heritability of IQ. Here, however, we should note that the existence of genotype–environment correlation presents a seemingly insurmountable problem in studies of the heritability of any human characteristic. Analyses of the heritability of human characteristics that rely on data sets wherein such correlations exist, therefore, result in flawed estimates.

Heritability Does Not Mean Inherited

As illustrated by the problems associated with the calculation of, and the statistical analyses associated with, heritability coefficients, these methodological problems are interrelated with problems of inappropriate interpretation of the meaning of such statistics. That is, it is clear that heritability is a far less meaningful, more limited piece of information than most people seem to realize (Hirsch, 1970; Moore & Shenk, 2016). Most importantly, *heritability does not mean genetically determined.* Nevertheless, the assonance between the terms "heritability" and "inherited" (Hirsch, 1997) suggests one reason why the concept may be used in a confused and confusing manner. Indeed, for several reasons heritability *is* a difficult, confusing concept. At first blush, it would seem to pertain to the extent to which something is inherited, that is, is based in the genes. For instance, if a person was told that "intelligence is 80% heritable" (Jensen, 1969), it might be reasonable to take this statement to mean that 80% of intelligence was genetically "determined"—that a given person's intelligence was largely (80%) shaped by his or her genes and that something else (environment) shaped the small percentage remaining.

The seemingly reasonable interpretation that high heritability means to genetic reductionists such as Rushton (1999, p. 60) that, "the differences are inborn and the environment has no effect;" this claim is often made by some scientists, members of the media, and governmental policy-makers (e.g., see Horowitz, 2000). Nevertheless, this interpretation of heritability is *completely incorrect.*

However, in pertaining only to differences *between* people, heritability has absolutely nothing to do with the extent to which anything—be it genes or environment—determines characteristics *within* an individual. Heritability only refers to the extent to which differences between people in a specific characteristic can be summarized by genetic differences between these people. Lerner and von Eye (1992) provided a technical explanation of this claim, and noted that, despite the dazzling statistical pyrotechnics often involved in the computation of heritability estimates (e.g., Molenaar et al., 1990), these statistics, nevertheless, still only describe the extent to which interindividual differences in an attribute distribution measured at one point in time and under one particular set of environmental conditions are associated with interindividual differences in gene distributions. These analyses say *nothing* about the attribute per se. They say nothing about the role of genes in causing the interindividual differences in the attribute distribution. Most certainly, they say nothing about the role of genes in providing a basis for the development of the attribute within the person (e.g., see Hirsch, 1990a, 1990b, 1997). As such, Horowitz (2000) concluded that:

> The data reported in behavioral genetics studies involving degrees of relationships among twins, siblings, and biologically unrelated individuals are in themselves interesting, even if it is doubtful that these relationships tell us anything about the direct and unmediated impact of genes. (p. 3)

Nevertheless, despite the clear fact that heritability does not mean inherited (Hirsch, 1997), people building their scientific careers around the production of heritability analyses cast their work *as if* it, in fact, provided some understanding of the separate *and causal* role of genes in human behavior and development (e.g., Harden, 2021; cf. Hirsch, 1990a, 1997). This "nature as separate from nurture" work cannot achieve such a partition; indeed, the entire idea of this work—of separating the contribution of nature from nurture in the causation of an individual's behavior—is simply counterfactual.

Yet, whatever the motivations of hereditarians for perpetuating the causal misuse of the concept of heritability, they often manifest lapses in language—for example, moving from describing factors associated with interindividual differences in an attribute distribution to advancing genetically causal explanations for the behavior itself (e.g., see Rowe, 1994; Rushton, 1990). Indeed, as noted by Gottlieb (2002/2014),

> Although population thinkers tell us that, strictly speaking, h^2 and e^2 refer to sources of individual *differences* among phenotypes, as a matter of fact in actual practice these measures are often applied to a causal understanding of the outcome of individual development as well. (p. 117)

Moreover, Gottlieb goes on to note that,

> If h^2 actually was useful for estimating genetic constraints or limitations on developmental outcomes, it would have some value, but it is widely agreed by geneticists themselves that h^2 cannot be interpreted in that way (Feldman & Lewontin, 1975), although individual scientists may now and again lapse into thinking in those terms (e.g., Gottesman & Shields, 1982)."
>
> (Gottlieb, 2002/2014, p. 118)

For instance, Plomin and colleagues (1990) conducted a behavior genetic study of individual differences in television viewing in early childhood in order to "explore the etiology of individual differences" (p. 372). Plomin and colleagues contended that their data provide evidence of a "genetic influence" (p. 371) on television viewing, and used phrases that often becloud the distinction between interindividual differences in a behavior and the behavior itself and, invariably, cast their descriptive information as if genetic causality had been demonstrated:

> The remarkable result is the evidence for significant genetic influence … inherent proclivities of children are in part responsible for differences in the amount of time they choose to watch television. (p. 376)

From the vantage point of the scholarship presented by Gottlieb (1997, 2002/2014, 2004), the interpretation Plomin and colleagues (1990) gave of their data is not correct. Quite simply, it is counterfactual to contend that nature is separable from nurture, and counterproductive (to say the least) to devise statistical methods to model this imaginary (or, at best, hypothetical) situation; but, most important, genes do not, in reality, function in the manner that behavior geneticists must have them work if their "story-lines" about the influence of nature are to attain even face validity. As Gottlieb (2002/2014) pointed out:

> The actual role of genes (DNA) is not to produce an arm or a leg or fingers, but to produce protein (through the coactions inherent in the formula DNA ↔ RNA ↔ protein). The protein produced by the DNA ↔ RNA ↔ cytoplasm coaction then differentiates according to coactions with other cells in its surround. Thus, differentiation occurs according to coations *above the level of DNA ↔ RNA ↔ coactions* (i.e., at the supragenetic level). (pp. 164–165)

Criticisms of Behavior Genetics Have Increased, Not Diminished

We noted that there are other reasons why Plomin's (2000; Plomin et al., 2016) views about the status and stature of behavior genetics are incorrect. These views are contradicted by the fact that the controversy regarding the legitimacy of behavioral genetics—both as a conceptual frame for understanding the role of genes in behavioral development and as a methodology for studying the role of genes in behavioral development—has not diminished at all. Indeed, as research on epigenetics has continued to increase and be extended to behavioral development (e.g., Cole, 2014; Lester, Conradt, & Marsit, 2016; Meaney, 2010; Moore, 2015a, 2016; Slavich & Cole, 2013), the decades-long criticism of the fundamental flaws of behavior genetics have become amplified. As such, in a review of David S. Moore's (2015a) landmark book, *The developing genome. An introduction to behavioral epigenetics*, Douglas Wahlsten (2015) concluded that the most important point made by the book was that scholars

> Need to think of the gene-environment relationship differently and reassess old ideas about development and evolution. He [Moore] advocates a broadened perspective in which gene-environment co-construction of traits is seen as universal and all dreams of simple genetic determination of behavioral phenotypes are finally dismissed. (p. 421)

Indeed, these criticisms of the "dreams" (Wahlsten, 2015) of behavior geneticists raised in the early decades of the 21st century are only more current instantiations of the criticisms that occurred in the very decades that Plomin (2000) claimed were in the period within which negative critiques had abated. Scholars need only note the controversy surrounding, and the litany of criticisms about the poor reasoning and bad science found in the Herrnstein and Murry (1994) book, *The Bell Curve* (see in particular the article by Joseph and Richardson that is presented in Section VIII of this book), or the criticisms leveled at the hereditarian views of J. Philippe Rushton (1996, 1997, 1999), which rely heavily on information derived from behavior genetics, to recognize that Plomin's (2000) "declaration of victory" is an inadequate attempt to either ignore or deny the persisting flaws of behavior genetics theory and method identified by scientists from numerous disciplines.

To illustrate, in a critique of the explanatory model and method associated with behavior genetic analyses of parent behaviors and the effects of parenting on child and adolescent development, Collins and colleagues (2000) noted that:

> Large-scale societal factors, such as ethnicity or poverty, can influence group means in parenting behavior—and in the effects of parenting behaviors—in ways that are not revealed by studies of within group variability. In addition, highly heritable traits also can be highly malleable. Like traditional correlational research on parenting, therefore, commonly used behavior–genetic methods have provided an incomplete analysis of differences among individuals. (p. 220)

Accordingly, Collins and colleagues (2000) concluded:

> Whereas researchers using behavior–genetic paradigms imply determinism by heredity and correspondingly little parental influence (e.g., Rowe, 1994), contemporary evidence confirms that the expression of heritable traits depends, often strongly, on experience, including specific parental behaviors, as well as predispositions and age-related factors in the child. (p. 228)

Rewriting History

Plomin's (2000) characterization of the growing acceptance of behavior genetics view of the role of genes in human development in effect rewrites history. His assertion that it was not until the 1990s that behavioral science really came to accept the role of genes in behavioral development is incorrect. For at least a half-century prior to the period of the 1990s, genes had been accepted as part of the developmental system that propels human life across time (e.g., Anastasi, 1958; Maier & Schneirla, 1935; Novikoff, 1945a, 1945b; Schneirla, 1956, 1957). The issue is not the one that Plomin points to, then, that of accepting that genes are involved in development. Instead, the issue is *how* do genes contribute to development. Plomin's (2000) approach and that of other behavior geneticists (e.g., Harden, 2021; Rowe, 1994) involves a split, nature–reductionist treatment of this issue. Developmental scientists working with models and concepts framed by the RDS metamodel take an integrated, dynamic view of genes within the relational developmental system (e.g., Lickliter, 2016; Mascolo & Fischer, 2015; Overton, 2015; Raeff, 2016).

In fact, Plomin (2000) conceptually approaches the vacuity of the behavior genetics approach, at least as it has been pursued through the 20th century. Although he maintains that "Twin and adoption research and genetic research using nonhuman animal models will continue to thrive" in the 21st century (Plomin, 2000, p. 30), Plomin perhaps admits to the serious flaws in this approach to understanding the role of genes in behavioral development when he acknowledges that. "The greatest need is for quantitative genetic research that goes beyond heritability, that is, beyond asking whether and how much genetic factors are important in behavioral development" (Plomin, 2000, p. 31). Plomin (2000) then continued by asking a series of important questions about the role of genes in behavioral development: "How do genetic effects unfold developmentally? What are the biological pathways between genes and behaviour? How do nature and nurture interact and correlate?" (p. 31). Unfortunately, he was seeking answers to these questions through the flawed model and methods of behavior genetics and, at least through 2018 (e.g., Plomin, 2018; Plomin et al., 2016; Selzam et al., 2017) never explored the potential usefulness of an

RDS-based approach. Nevertheless, such exploration would be very useful because Plomin (2000) admitted that it would be a major mistake

> to think that genes determine outcomes in a hardwired, there's-nothing-we-can-do-about-it way. For thousands of rare single-gene disorders, such as the gene on chromosome 4 that causes Huntington's disease, genes do determine outcomes in this hardwired way. However, behavioral disorders and dimensions are complex traits influenced by many genes as well as many environmental factors. For complex traits, genetic factors operate in a probabilistic fashion like risk factors rather than predetermined programming. (p. 33)

Thus, ultimately, Plomin (2000) admitted that a probabilistic epigenetic relation is involved in accounting for the role of genes in behavioral development. Still, his views about single-gene disorders reflect an ahistorical conception of such problems of human development. That is, with respect to other such single-gene disorders (e.g., as involved with phenylketonuria [PKU]), genetic research has found means to counteract the problems produced by the genetic inheritance and has thus shown that a hardwired genetic influence is not that hard-wired after all (Scriver & Clow, 1980a, 1980b). As such, Plomin maintains a narrow view of the relational developmental system; it apparently does not include the ingenuity of scholars to capitalize on the relative plasticity within the developmental system and to demonstrate that what might seem to be hard-wired is in reality amenable to change as a consequence of its embeddedness within a dynamic system. Nevertheless, in admitting to the importance of a dynamic, relational development system in behavioral development, Plomin (2000) is, in actuality, defeating his own, split approach to the nature and nurture of behavioral development.

Moreover, other scholars are not as convinced as Plomin (2000) that the various methodologies he associates with behavior genetics will generate useful data. Consistent with the 2017 critique of Charney regarding the failures of the CGA and GWAS methods, Collins and colleagues (2000) noted that:

> One criticism is that the assumptions, methods, and truncated samples used in behavior—genetic studies maximize the effects of heredity and features of the environment that are different for different children and minimize the effects of shared family environments.... A second criticism is that estimates of the relative contributions of environment and heredity vary greatly depending on the source of data ... heritability estimates vary considerably depending on the measures used to assess similarity between children or between parents and children.... The sizable variability in estimates of genetic and environmental contributions depending on the paradigms and measures used means that no firm conclusions can be drawn about the relative strength of these influences on development. (pp. 220–221)

Similarly, and again counter to Plomin's (2000) assertion that the controversy surrounding behavior genetics faded by the 1990s, Horowitz (2000) noted that:

> One sees increasing skepticism about what is to be learned from assigning variance percentages to genes .... The skepticism is informed by approaches that see genes, the central nervous system and other biological functions and variables as contributors to reciprocal, dynamic processes which can only be fully understood in relation to sociocultural environmental contexts. It is a perspective that is influenced by the impressive recent methodological and substantive advances in the neurosciences. (p. 3)

Conclusions

Clearly, many human developmentalists do not believe in the causal "storyline" of behavior genetics. Nevertheless, "research" in behavior genetics—studies that, in effect, involve obtaining samples of people with differing degrees of biological relatedness and applying, typically, state-of-the-art measures of traits and inadequate measures of the ecology of human development (e.g., Bronfenbrenner & Ceci, 1994; Bronfenbrenner & Morris, 2006; Hoffman, 1991)—continues through this writing to be well funded and widely disseminated.

But behavior genetics is really like the story of the emperor's new clothes. Despite the positive regard some researchers hold for this area, there is actually "nothing there." The naked truth is that conceptual errors and misapplied models—no matter how often repeated or published—do not by dint of their numbers make for an adequate contribution to science (Lerner, 2015b, 2015c, 2016; Panofsky, 2014; Tabery, 2014; Moore, 2015b).

Given the myriad theoretical and methodological problems associated with behavior genetics, little can be gained either for advancing the science of human development or for adequately informing or serving Horowitz's (2000) "Person in the Street" by continuing to invest resources in the behavior genetics approach. Indeed, there seems to be compelling reasons to make human and financial investments elsewhere given, on the one hand, the counterfactual view of genetic activity inherent in behavior genetics, the several insurmountable conceptual and computational problems involved in its methods, and the lack of reliable empirical support for the claims or interpretations forwarded by behavior geneticists about the findings derived from their methods. On the other hand, the availability of the theoretically rich and empirically productive RDS-based alternatives to genetic reductionist suggest that behavior genetics work will not be useful in the developmental science in the decades following the one in which this book has been written.

An Overview of the Contributions to This Section

As was the case in Section IV, the contributions presented in this section begin with an excerpt from another important book by Gilbert Gottlieb (2002/2014), *Individual development and evolution: The genesis of novel behavior*. In his chapter entitled "From gene to organism: The developing individual as an emergent, interactional hierarchical system," Gottlieb presents a view of individual development reflecting an RDS-based view of dynamic, individual⇔context coactions, one that decries the split and reductionist approach to genes and context promoted by behavior geneticists. Gottlieb (2002/2014) defines epigenesis as individual development that is

> characterized by an increase of complexity of organization – i.e., the emergence of new structural and functional properties and competencies – at all levels of analysis (molecular, subcellular, cellular, organismic) as a consequence of horizontal and vertical coactions among its parts, including organism-environment coactions. (pp. 159–160)

He goes on to explain that neither genes nor context can separately cause development, as the purveyors of heritability coefficients as tools for understanding the putative causes of development would have it. Instead, Gottlieb (2002/2014) emphasizes that:

> The cause of development – what makes development happen – is the relationship of the two components, *not the components themselves*. Genes in themselves cannot cause development any more than

stimulation in itself in itself can cause development. When we speak of coaction as being at the heart of developmental analysis or causality what we mean is that we need to specify some relationship between at least two components of the developmental system.

(Gottlieb, 2002/2014, pp. 161–163, italics added)

He then illustrates and explains the dynamic complexity of the integrative, hierarchical developmental system wherein these coactive relationships occur through the life spans of individuals.

Moore and Shenk (2016) add to a discussion of the several shortcomings of the research disseminated by genetic reductionists by building on Moore's earlier (2002) critique of heritability research (see Section II of this book) and arguing that the notion of heritability per se is one of the most misleading concepts in the history of science. Genetic reductionists often use the results of heritability research to claim that a behavior is, at least in part (depending on the size of the heritability coefficient), to be genetically determined: For instance, to be a reflection of an individual's genetic blueprint (Plomin, 2018) or the results of what a person has gained in the individual's "genetic lottery" (Harden, 2021). Moreover, as we have noted in this section and, as well, across preceding sections, it is also often the case that popular belief in the media or in political rhetoric coincides with such interpretations of heritability. However, Moore and Shenk (2016) explain that heritability scores for a behavior (say, a facet of a cognitive skill, such as syllogistic reasoning, or working memory capacity) do not indicate anything about how genetically inheritable that behavior is and, furthermore, does not indicate what the cause of the behavior might be, or the magnitude of the contribution of genes or of the context to the development of the behavior. Moore and Shenk conclude that measures of heritability are of little value, except in very rare cases, and that continued use of the term does enormous damage to public understanding of how human beings develop any behaviors.

Among the methods that heritability researchers use is the study of twins reared apart. Certainly, the purported remarkable similarities of adult identical twins who are claimed to have been reared separately from birth or very early in infancy have received considerable attention in media. Arguably, the major study of twins reared apart, that has been seen as having actual data (as compared to the fake data reported by Sir Cyril Burt; Goldberger, 1980; Hearnshaw, 1979), is the "Minnesota Study of Twins Reared-Apart" (MISTRA) IQ Study (e.g., Bouchard et al., 1990; Bouchard & McGue, 1981). However, the methods and results of, and interpretations derived from, the MISTRA study have received strong criticism, most notably by Joseph (2004, 2010, 2013, 2015). In fact, in 2022, Joseph published a newly formulated critical analysis of the MISTRA study and concluded that the study did not provide any useful evidence that genes influenced IQ scores or specific cognitive abilities in the study participants. Not unexpectedly, the lead researcher of the MISTRA study (Bouchard, 2023) wrote a detailed response to Joseph's (2022) critique. In turn, Joseph (2023) wrote a detailed rejoinder to Bouchard's (2023) response. Joseph's rejoinder takes the form of presenting each of the points made by Bouchard in his response and then provides Joeseph's point-by-point refutation of these points.

The contest of ideas in this exchange brings into vivid relief the point made by Witherington et al. (2018), which is included as one of the contributions in Section II of this book, that is, that the debates about the role of genes in human behavior and development are embedded in perhaps irreconcilable metatheoretical differences. Accordingly, because of the rarity and importance of the illuminating paper written by Joseph (2023), we have included it in this section of the book. In addition, in Section VIII of this book, we present the 2024 critique by Joseph and Richardson of the previously noted Herrnstein and Murray (1994) book.

References

Anastasi, A. (1958). Heredity, environment, and the question "how?" *Psychological Review, 65*, 197–208.

Bouchard, T. J. (2023). The garden of forking paths; an evaluation of Joseph's 'a reevaluation of the 1990 "Minnesota Study of Twins Reared Apart" IQ Study.' *Twin Research and Human Genetics, 26*(2), 1–10.

Bouchard, T. J., Jr., Lykken, D. T., McGue, M., Segal, N. L., & Tellegen, A. (1990). Sources of human psychological differences: The Minnesota Study of Twins Reared Apart. *Science, 250*, 223–228.

Bouchard, T. J., Jr., & McGue, M. (1981). Familial studies of intelligence: A review. *Science, 212*, 1055–1059.

Bronfenbrenner, U., & Ceci, S. J. (1994). Nature-nurture reconceptualized: A bioecological model. *Psychological Review, 101*, 568–586.

Bronfenbrenner, U., & Morris, P. A. (2006). The bioecological model of human development. In W. Damon & R. M. Lerner (Eds.) & R. M. Lerner (Vol. Ed.), *Handbook of child psychology: Vol. 1. Theoretical models of human development* (6th ed., pp. 793–828). Wiley.

Burt, C. (1966). The genetic determination of differences in intelligence: A study of monozygotic twins reared together and apart. *British Journal of Psychology, 57*, 137–153.

Buss, A. H., & Plomin, R. (1984). *Temperament: Early developing personality traits*. Erlbaum.

Charney, E. (2012). Behavior Genetics and Post Genomics. *Behavioral and Brain Sciences, 35*, 331–410.

Charney, E. (2017). Genes, behavior, and behavior genetics. *WIREs Cognitive Science, 8*(1-2), e1405. https://doi.org/10.1002/wcs.1405

Charney, E. (2022). The "Golden Age" of behavior genetics? *Perspectives on Psychological Science, 17*(4), 1188–1210.

Cole, S. W. (2014). Human social genomics. *PLoS Genetics, 10*(8), 1–7.

Collins, W. A., Maccoby, E. E., Steinberg, L., Hetherington, E. M., & Bornstein, M. H. (2000). Contemporary research on parenting: The case of nature and nurture. *American Psychologist, 55*, 218–232.

Feldman, M. (2014). Echoes of the past: Hereditarianism and a troublesome inheritance. *PLoS Genetics, 10*, e1004817.

Feldman, M. W., & Lewontin, R. C. (1975). The heritability hang-up. *Science, 190*, 1163–1168.

Feldman, M. W., & Riskin, J. (2022). Why biology is not destiny. *The New York Review*. https://www.nybooks.com/online/2022/04/02/an-evolving-view-of-inheritance/

Ford, D. H., & Lerner, R. M. (1992). *Developmental systems theory: An integrative approach*. Sage.

Goldberger, A. S. (1980). Review of "Cyril Burt, psychologist." *Challenge: The Magazine of Economic Affairs, 23*, 61–62.

Gottesman, I. L., & Shields, J. (1982). *Schizophrenia: The epigenetic puzzle*. Cambridge University Press.

Gottlieb, G. (1970). Conceptions of prenatal behavior. In L. R. Aronson, E. Tobach, D. S. Lehrman, & J. S. Rosenblatt (Eds.), *Development and evolution of behavior: Essays in memory of T. C. Schneirla* (pp. 111–137). W. H. Freeman and Company.

Gottlieb, G. (1991). The experiential canalization of behavioral development: Theory. *Developmental Psychology, 27*, 4–13.

Gottlieb, G. (1997). *Synthesizing nature-nurture: Prenatal roots of instinctive behavior*. Psychology Press.

Gottlieb, G. (2002/2014). *Individual development and evolution: The genesis of novel behavior*. Psychology Press.

Gottlieb, G. (2004). Normally occurring environmental and behavioral influences on gene activity: From central dogma to probabilistic epigenesis. In C. Garcia Coll, E. Bearer, & R. M. Lerner (Eds.), *Nature and nurture: The complex interplay of genetic and environmental influences on human behavior and development* (pp. 85–106). Erlbaum.

Gottlieb, G., Wahlsten, D., & Lickliter, R. (2006). The significance of biology for human development: A developmental psychobiological systems view. In R. M. Lerner (Ed.). *Theoretical models of human development. Volume 1 of Handbook of Child Psychology* (6th ed., pp. 210–257). Editors-in-chief: W. Damon & R. M. Lerner. Wiley.

Gould, S. J. (1981). *The mismeasure of man*. Norton.

Greenberg, G. (2004). R. I. P. Genetic determinism: Please. Review of G. Kaplan and L. J. Rogers (2003). *Gene worship*. New York: Other Press. *Developmental Psychobiology, 46*, 93–96.

Greenberg, G. (2011). The failure of biogenetic analysis in psychology: Why psychology is not a biological science. *Research in Human Development, 8,* 173–191.

Greenberg, G. (2013). A long way from genes to behavior. Commentary on: With Gottlieb beyond Gottlieb: The role of epigenetics in psychobiological development. *International Journal of Developmental Science, 7,* 83–86.

Greenberg, G. (2015). The case against behavioral genetics. Review of A. Panofsky (2014). *Misbehaving science: Controversy and the development of behavior genetics.* University of Chicago Press. *Developmental Psychobiology, 57,* 854–857.

Greenberg, G. (2016). In memoriam: Ethel Tobach 1921–2015. *American Psychologist, 71,* 75.

Harden, K. P. (2021). *The genetic lottery: Why DNA matters for social equality.* Princeton University Press.

Harris, J. R. (1998). *The nurture assumption: Why children turn out the way they do.* The Free Press.

Hearnshaw, L. (1979). *Cyril Burt: Psychologist.* Cornell University Press.

Hebb, D. O. (1970). A return to Jensen and his social critics. *American Psychologist, 25,* 568.

Herrnstein, R. J. (1971). I.Q. *Atlantic Monthly, 228,* 43–64.

Herrnstein, R. J., & Murray, C. (1994). *The bell curve: Intelligence and class structure in American life.* Free Press.

Hewitt, J. K. (2012). Editorial policy on candidate gene association and candidate gene-by-environment interaction studies of complex traits. *Behavior Genetics, 42,* 1–2.

Hirsch, J. (1963). Behavior genetics and individuality understood: Behaviorism's counterfactual dogma blinded the behavioral sciences to the significance of meiosis. *Science, 142*(3598), 1436–1442.

Hirsch, J. (1970). Behavior-genetic analysis and its biosocial consequences. *Seminars in Psychiatry, 2,* 89–105.

Hirsch, J. (1990a). Correlation, causation, and careerism. *European Bulletin of Cognitive Psychology, 10,* 647–652.

Hirsch, J. (1990b). A nemesis for heritability estimation. *Behavioral and Brain Sciences, 13,* 137–138.

Hirsch, J. (1997). Some history of heredity-vs-environment, genetic inferiority at Harvard (?), and the (incredible) Bell Curve. *Genetica, 99,* 207–224.

Hirsch, J. (2004). Uniqueness, diversity, similarity, repeatability, and heritability. In C. Garcia Coll, E. Bearer, & R.M. Lerner (Eds.). *Nature and nurture: The complex interplay of genetic and environmental influences on human behavior and development* (pp. 127–138). Erlbaum.

Ho, M. W. (1984). Environment and heredity in development and evolution. In M.-W. Ho & P. T. Saunders (Eds.), *Beyond neo-Darwinism: An introduction to the new evolutionary paradigm* (pp. 267–289). Academic Press.

Ho, M. W. (2010). Development and evolution revisited. In K. E. Hood, C. T. Halpern, G. Greenberg, & R. M. Lerner (Eds.), *Handbook of developmental systems, behavior and genetics* (pp. 61–109). Wiley Blackwell.

Ho, M. W. (2013). No genes for intelligence in the fluid genome. In R. M. Lerner & J. B. Benson, (Eds.), *Advances in Child Development and Behavior: Embodiment and epigenesis: Theoretical and methodological issues in understanding the role of biology within the relational developmental system. Part B. Ontogenetic dimensions* (pp. 67–92). Elsevier.

Hoffman, L. W. (1991). The influence of family environment on personality: Accounting for sibling differences. *Psychological Bulletin, 110,* 187–203.

Horowitz, F. D. (1993). Bridging the gap between nature and nurture. A conceptually flawed issue and the need for a comprehensive and new environmentalism. In R. Plomin & G. E. McClearn (Eds.), *Nature, nurture and psychology* (pp. 341–354). APA Books.

Horowitz, F. D. (2000). Child development and the PITS: Simple questions, complex answers, and developmental theory. *Child Development, 71,* 1–10.

Jablonka, E., & Lamb, M. (2005). *Evolution in four dimensions: Genetic, epigenetic, behavioral, and symbolic variation in the history of life.* MIT Press.

Jensen, A. R. (1969). How much can we boost IQ and scholastic achievement? *Harvard Educational Review, 39,* 1–123.

Jensen, A. R. (1998). Jensen on "Jensenism." *Intelligence, 26,* 181–208.

Joseph, J. (2004). *The Gene Illusion.* Algora.

Joseph, J. (2010). Genetic research in psychiatry and psychology: A critical overview. In K. E. Hood, C. T. Halpern, G. Greenberg, & R. M. Lerner (Eds.). *Handbook of developmental systems, behavior and genetics* (pp. 557–625). Wiley Blackwell.

Joseph, J. (2013). The lost study: A 1998 adoption study of personality that found no genetic relationship between birth parents and their 240 adopted-away biological offspring. In R. M. Lerner & J. B. Benson, (Eds.), *Advances in Child Development and Behavior: Embodiment and epigenesis: Theoretical and methodological issues in understanding the role of biology within the relational developmental system. Part B. Ontogenetic dimensions.* (pp. 93–124). Elsevier.

Joseph, J. (2015). *The trouble with twin studies: A reassessment of twin research in the social and behavioral sciences.* Routledge.

Joseph, J. (2022). A reevaluation of the 1990 "Minnesota Study of Twins Reared Apart" IQ study. *Human Development, 66,* 48–65.

Joseph, J. (2023). The 1990 "Minnesota Study of Twins Reared Apart" IQ study: Ripe for retraction? *Free Associations: Psychoanalysis and Culture, Media, Groups, Politics, 89,* 1–19. http://www.freeassociations.org.uk

Krimsky S., & Gruber, J. (Eds.). (2013). *Genetic explanations: Sense and nonsense.* Harvard University Press.

Kuo, Z. Y. (1967). *The dynamics of behavior development.* Random House.

Kuo, Z.-Y. (1976). *The dynamics of behavior development: An epigenetic view.* Plenum.

Layzer, D. (1974). Heritability analyses of IQ scores: Science or numerology? *Science, 183,* 1259–1266.

Lehrman, D. S. (1953). A critique of Konrad Lorenz's theory of instinctive behavior. *Quarterly Review of Biology, 28,* 337–363.

Lehrman, D. S. (1970). Semantic and conceptual issues in the nature-nurture problem. In L. R. Aronson, E. Tobach, D. S. Lehrman, & J. S. Rosenblatt (Eds.), *Development and evolution of behavior: Essays in memory of T. C. Schneirla* (pp. 17–52). Freeman.

Lerner, R. M., & Overton, W. F. (2017). Reduction to absurdity: Why epigenetics invalidates all models involving genetic reduction. *Human Development, 60*(2–3), 107–123.

Lerner, R. M. (1976). *Concepts and theories of human development.* Addison-Wesley.

Lerner, R. M. (1978). Nature, nurture, and dynamic interactionism. *Human Development, 21,* 1–20.

Lerner, R. M. (1992). *Nature, nurture and mass murder. Readings: A Journal of Reviews and Commentary on Mental Health, 7*(3), 8–15.

Lerner, R. M. (2015a). Preface. *Handbook of child psychology and developmental science* (7th ed.). Editor-in-chief: R. M. Lerner. (pp. xv–xxi). Wiley.

Lerner, R. M. (2015b). Promoting social justice by rejecting genetic reductionism: A challenge for developmental science. *Human Development, 58,* 67–69.

Lerner, R. M. (2015c). Eliminating genetic reductionism from developmental science. *Research in Human Development, 12,* 178–188.

Lerner, R. M. (2016). Complexity embraced and complexity reduced: A tale of two approaches to human development. *Human Development, 59,* 242–249.

Lerner, R. M., & Overton, W. F. (2017). Reduction to absurdity: Why epigenetics invalidates all models involving genetic reduction. *Human Development, 60*(2–3), 107–123.

Lerner, R. M., & von Eye, A. (1992). Sociobiology and human development: Arguments and evidence. *Human Development, 35,* 12–33.

Lerner, R. M., & von Eye, A. (1993). Why Burgess and Molenaar "just don't get it." *Human Development, 36,* 55–56.

Lester, B. M., Conradt, E., & Marsit, C. (2016). Introduction to the special section on epigenetics. *Child Development, 87,* 29–37.

Lewis, M. (1997). *Altering fate.* Guilford Press.

Lickliter, R. (2016). Developmental evolution. *WIREs Cognitive Science.* https://doi.org/10.1002/wcs.1422

Magnusson, D. (1999a). Holistic interactionism: A perspective for research on personality development. In L. A. Pervin & O. P. John (Eds.), *Handbook of personality: Theory and research* (2nd ed., pp. 219–247). The Guilford Press.

Magnusson, D. (1999b). On the individual: A person-oriented approach to developmental research. *European Psychologist, 4,* 205–218.

Magnusson, D. (2000). Developmental science. In A. E. Kazdin (Ed.), *Encyclopedia of psychology* (Vol. 3, pp. 24–26). American Psychological Association and Oxford University Press.

Magnusson, D., & Stattin, H. (1998). Person-context interaction theories. In W. Damon (Series Ed.) & R. M. Lerner (Vol. Ed.), *Handbook of child psychology: Vol. 1 theoretical models of human development* (5th ed., pp. 685–759). Wiley.

Maier, N. R. F., & Schneirla, T. C. (1935). *Principles of animal behavior*. McGraw-Hill.

Mascolo, M. F., & Fischer, K. W. (2015). Dynamic development of thinking, feeling, and acting. In W. F. Overton & P. C. Molenaar (Eds.), *Theory and method. Volume 1 of the handbook of child psychology and developmental science* (7th ed.). Editor-in-chief: R. M. Lerner (pp. 113–161). Wiley.

McEwen, B. S. (1997). Possible mechanisms for atrophy of the human hippocampus. *Molecular Psychiatry, 2*, 255–262.

McEwen, B. S. (1998). Protective and damaging effects of stress mediators. *New England Journal of Medicine, 338*, 171–179.

McEwen, B. S. (1999). Stress and hippocampal plasticity. *Annual Review of Neuroscience, 22*, 105–122.

Meaney, M. (2010). Epigenetics and the biological definition of gene x environment interactions. *Child Development, 81*, 41–79.

Meaney, M., Aitken, D., Berkel, H., Bhatnager, S., & Sapolsky, R. (1988). Effect of neonatal handling of age-related impairments associated with the hippocampus. *Science, 239*, 766–768.

Molenaar, P. C. M., Boomsma, D. I., Neeleman, D., & Dolan, C. V. (1990). Using factor scores to detect G x E interactive origin of "pure" genetic or environmental factors obtained in genetic covariance structure analysis. *Genetic Epidemiology, 7*, 93–100.

Moore, D. (2006). A very little bit of knowledge: Re-evaluating the meaning of the heritability of IQD. *Human development, 49*, 347–353.

Moore, D. S. (2002). *The dependent gene: The fallacy of nature vs. nurture*. W. H. Freeman.

Moore, D. S. (2015a). *The developing genome: An introduction to behavioral epigenetics*. Oxford University Press.

Moore, D. S. (2015b). The asymmetrical bridge. Book review of James Tabery's *Beyond versus: The struggle to understand the interaction of nature and nurture*. *Acta Biotheoretica, 63*(4), 413–427.

Moore, D. S. (2016). Behavioral epigenetics. *WIREs Cognitive Science* 2016. https://doi.org/10.1002/wcs.1333

Moore, D. S., & Shenk, D. (2016). The heritability fallacy. *WIREs Cognitive Science* 2016. https://doi.org/10.1002/wcs.1400

Müller-Hill, B. (1988). *Murderous science: Elimination by scientific selection of Jews Gypsies, and others. Germany 1933–1945* (Fraser, G. R., Traps.). Oxford University.

Novikoff, A. B. (1945a). The concept of integrative levels and biology. *Science, 101*, 209–215.

Novikoff, A. B. (1945b). Continuity and discontinuity in evolution. *Science, 101*, 405–406.

Overton, W. F. (1973). On the assumptive base of the nature-nurture controversy: Additive versus interactive conceptions. *Human Development, 16*, 74–89.

Overton, W. F. (2015). Process and relational developmental systems. In W. F. Overton & P. C. M. Molenaar (Eds.), *Handbook of child psychology and developmental science, Volume 1: Theory and method* (7th ed., pp. 9–62). Editor-in-chief: R. M. Lerner. Wiley.

Overton, W. F., & Lerner, R. M. (2012). Relational developmental systems: Paradigm for developmental science in the post-genomic era. *Behavioral and Brain Sciences, 35*(5), 375–376.

Panofsky, A. (2014). *Misbehaving science: Controversy and the development of behavior genetics*. University of Chicago Press.

Plomin, R. (1986). *Development, genetics, and psychology*. Erlbaum.

Plomin, R. (2000). Behavioural genetics in the 21st century. *International Journal of Behavioral Development, 24*, 30–34.

Plomin, R. (2018). *Blueprint: How DNA makes us who we are*. Allen Lane.

Plomin, R., Corley, R., DeFries, J. C., & Faulker, D. W. (1990). Individual differences in television viewing in early childhood: Nature as well as nurture. *Psychological Science, 1*, 371–377.

Plomin, R., & Daniels, D. (1987). Why are children in the same family so different from each other? *Behavioral and Brain Sciences, 10*, 1–16.

Plomin, R., DeFries, J. C., & McClearn, G. E. (1980). *Behavioral genetics: A primer*. Freeman.
Plomin, R., DeFries, J. C., Knopik, V. S., & Neiderhiser, J. S. (2013). *Behavioral genetics* (6th ed.). Worth Publishers.
Plomin, R., DeFries, J. C., Knopik, V. S., & Neiderhiser, J. M. (2016). Top 10 replicated findings from behavioral genetics. *Perspectives on psychological science, 11*(1), 3–23.
Raeff, C. (2016). *Exploring the dynamics of human development: An integrative approach*. Oxford University Press.
Richardson, K. (2013). The evolution of intelligent systems. In R. M. Lerner & J. B. Benson, (Eds.), *Advances in child development and behavior: Embodiment and epigenesis: Theoretical and methodological issues in understanding the role of biology within the relational developmental system. Part A. Philosophical, theoretical, and biological dimensions*. (pp. 127–159). Elsevier.
Richardson, K. (2017). *Genes, brains, and human potential: The science and ideology of human intelligence*. Columbia University Press.
Rowe, D. C. (1994). *The limits of family influence: Genes, experience, and behavior*. The Guilford Press.
Rushton, J. P. (1990). Sex, ethnicity, and hormones. *Behavioral & Brain Sciences, 13*, 194, 197–198.
Rushton, J. P. (1996). Political correctness and the study of racial differences. *Journal of Social Distress & the Homeless, 5*, 213–229.
Rushton, J. P. (1997). Cranial size and IQ in Asian Americans from birth to age seven. *Intelligence, 25*, 7–20.
Rushton, J. P. (1999). *Race, evolution, and behavior* (special abridged edition). Transaction Publishers.
Rushton, J. P. (2000). *Race, evolution, and behavior* (2nd special abridged edition). Transaction Publishers.
Sameroff, A. (Ed.). (2009). *The transactional model of development: How children and contexts shape each other*. American Psychological Association.
Sameroff, A. (2010). A unified theory of development: A dialectic integration of nature and nurture. *Child Development, 81*(1), 6–22.
Sameroff, A. J. (1983). Developmental systems: Contexts and evolution. In W. Kessen (Ed.), *Handbook of child psychology: Vol. 1, history, theory, and methods* (pp. 237–294). Wiley.
Scarr, S. (1992). Developmental theories for the 1990's: Development and individual differences. *Child Development, 63*, 1–19.
Schneirla, R. C. (1956). Interrelationships of the innate and the acquired in instinctive behavior. In P. P. Grassé (Ed.), *L'instinct dans le comportement des animaux et de l'homme* (pp. 387–452). Mason et Cie.
Schneirla, T. C. (1957). The concept of development in comparative psychology. In D. Harris (Ed.), *The concept of development* (pp. 78–108). University of Minnesota Press.
Scriver, C. R., & Clow, C. L. (1980a). Phenylketonuria: Epitome of human biochemical genetics (first of two parts). *New England Journal of Medicine, 303*, 1336–1342.
Scriver, C. R., & Clow, C. L. (1980b). Phenylketonuria: Epitome of human biochemical genetics (second of two parts). *New England Journal of Medicine, 303*, 1394–1400.
Selzam, S., Krapohl, E., von Stumm, S., O'Reilly, P. F., Rimfeld, K., Kovas, Y., Dale, P. S., Lee, J. J., & Plomin, R. (2017). Predicting educational achievement from DNA. *Molecular Psychiatry, 22*, 267–272.
Slavich, G. M., & Cole, S. W. (2013). The emerging field of human social genomics. *Clinical Psychological Science, 1*, 331–348.
Smith, L. B., & Thelen, E. (Eds.). (1993). *A dynamic systems approach to development: Applications*. MIT Press.
Strohman, R. C. (1993a). Organism and experience. [Review of Lerner, R. M. Final Solutions. University Park, PA: Penn State Press, 1992.] *Journal of Applied Developmental Psychology, 14*, 147–151.
Strohman, R. C. (1993b). Book Reviews: *Final Solutions*, Richard M. Lerner, Penn State Press, 1992, and *Individual Development and Evolution: The genesis of novel behavior*, Gilbert Gottlieb, Oxford University Press, 1992. *Integrative Physiological and Behavioral Science, 28*, 99–104.
Tabery, J. (2014). *Beyond versus: The struggle to understand the interaction of nature and nurture*. MIT Press.
Thelen, E., & Smith, L. B. (1994). *A dynamic systems approach to the development of cognition and action*. MIT Press.
Thelen, E., & Smith, L. B. (1998). Dynamic systems theories. In W. Damon (Series Editor) & R.M. Lerner (Vol. Ed.), *Handbook of child psychology: Vol. I. Theoretical models of human development* (5th ed., pp. 563–633). Wiley.

Thelen, E., & Smith, L. B. (2006). Dynamic systems theories. In R. M. Lerner & W. Damon (Eds.), *Handbook of child psychology: Vol. 1. Theoretical models of human development* (6th ed., pp. 258–312). Wiley.

Tobach, E. (1981). Evolutionary aspects of the activity of the organism and its development. In R. M. Lerner & N. A. Busch-Rossnagel (Eds.), *Individuals as producers of their development: A life-span perspective* (pp. 37–68). Academic Press.

Tobach, E., & Greenberg, G. (1984). The significance of T. C. Schneirla's contribution to the concept of levels of integration. In G. Greenberg & E. Tobach (Eds.), *Behavioral evolution and integrative levels* (pp. 1–7). Erlbaum.

Turkheimer, E. (2011). Still missing. *Research in Human Development, 8*, 227–241.

Venter, J. C., Adams, M. D., Myers, E. W., Li, P. W., Mural, R. J., et al. (2001). The sequence of the human genome. *Science, 291*, 1304–1351.

Wahlsten, D. (2012). The hunt for gene effects pertinent to behavioral traits and psychiatric disorders: From mouse to human. *Developmental Psychobiology, 54*, 475–492.

Wahlsten, D. (2013). A contemporary view of genes and behavior: Complex systems and interactions. In R. M. Lerner & J. B. Benson, (Eds.), *Advances in child development and behavior: Embodiment and epigenesis: Theoretical and methodological issues in understanding the role of biology within the relational developmental system. Part A. Philosophical, theoretical, and biological dimensions* (pp. 285–306). Elsevier.

Wahlsten, D. (2015). Dynamic heredity. Review of Moore, D. S. (2015). *The developing genome: An introduction to behavioral epigenetics.* New York: Oxford University Press. *Developmental Psychobiology, 58*, 419–421.

Wilson, E. O. (1975). *Sociobiology: The new synthesis.* Harvard University Press.

Witherington, D. C. (2015). Dynamic systems in developmental science. In W.F. Overton & P.C. Molenaar (Eds.), *Handbook of child psychology and developmental science. Volume 1: Theory and method* (7th ed., pp. 63–112). Editor-in-chief: R. M. Lerner. Wiley.

Witherington, D. C., & Heying, S. (2013). Embodiment and agency: Toward a holistic synthesis for developmental science. In R. M. Lerner & J. B. Benson, (Eds.), *Advances in child development and behavior: Embodiment and epigenesis: Theoretical and methodological issues in understanding the role of biology within the relational developmental system. Part A. Philosophical, theoretical, and biological dimensions* (pp. 161–192). Elsevier.

Witherington, D. C., Overton, W. F., Lickliter, R., Marshall, P. J., & Narvaez, D. (2018). Metatheory and the primacy of conceptual analysis in developmental science. *Human Development, 61*(3), 181–198.

14.
FROM GENE TO ORGANISM
The Developing Individual as an Emergent, Interactional, Hierarchical System
Gilbert Gottlieb*

The historically correct definition of epigenesis—the emergence of new structures and functions during the course of individual development—did not specify, even in a general way, how these emergent properties come into existence. Thus, there was still room for preformation-like thinking about development, which I (Gottlieb, 1970) earlier labeled the predetermined conception of epigenesis, in contrast to a probabilistic conception (see Table 14.1 for details). That epigenetic development is probabilistically determined by active interactions among its constituent parts is now so well accepted that epigenesis itself is sometimes defined as the interactionist approach to the study of individual development (e.g., Dewsbury, 1978; Johnston, 1987). That is a fitting tribute to the career-long labors of Zing-Yang Kuo (1976), T. C. Schneirla (1961), and Daniel S. Lehrman (1970), the principal champions of the interaction idea in the field of psychology, particularly as it applies to the study of behavioral and psychological development. Thus, it seems appropriate to offer a new definition of epigenesis that includes not only the idea of the emergence of new properties but also the idea that the emergent properties arise through reciprocal interactions (coactions) among already existing constituents. Somewhat more formally expressed, the new definition of epigenesis would say that *individual development is characterized by an increase of complexity of organization—i.e., the emergence of new structural and functional properties and competencies—at all levels of analysis* (molecular, subcellular, cellular, organismic) *as a consequence of horizontal and vertical coactions among its parts, including organism-environment coactions.* Horizontal coactions are those that occur at the same level (gene-gene, cell-cell, tissue-tissue, organism-organism), whereas vertical coactions occur at different levels (gene—cytoplasm, cell—tissue, behavioral activity—nervous system) and are reciprocal, meaning that they can influence each other in either direction, from lower to higher, or from higher to lower, levels of the developing system. For example, the sensory experience of a developing organism affects the differentiation of its nerve cells, such that the more experience the more differentiation and the less experience the less differentiation. (For example, enhanced activity or experience during individual development causes more elaborate branching of dendrites and more synaptic contacts among nerve cells in the brain [Greenough & Juraska, 1979.])[1] Reciprocally, the more highly differentiated nervous system permits a greater degree of behavioral competency and the less differentiated nervous system permits a lesser degree of behavioral competency. Thus, the essence of the probabilistic conception of epigenesis is the bidirectionality of structure—function relationships, as depicted in Table 14.1. It is important to note

* University of North Carolina, Greensboro

Table 14.1 Two Versions of Epigenetic Development

Predetermined Epigenesis
Unidirectional Structure-Function Development
Genetic Activity → Structural Maturation →Function, Activity, or Experience
(DNA → RNA → Protein)

Probabilistic Epigenesis
Bidirectional Structure-Function Development
Genetic Activity ↔ Structural Maturation ↔ Function, Activity, or Experience
(DNA ↔ RNA ↔ Protein)

As applied to the nervous system, structural maturation refers to neurophysiological and neuroanatomical development, principally the structure and function of nerve cells and their synaptic interconnections. The unidirectional structure-function view assumes that genetic activity gives rise to structural maturation that then leads to function in a nonreciprocal fashion, whereas the bidirectional view holds that there are constructive reciprocal relations between genetic activity, maturation, and function. In the unidirectional view, the activity of genes and the maturational process are pictured as relatively encapsulated or insulated so that they are uninfluenced by feedback from the maturation process or function, whereas the bidirectional view assumes that genetic activity and maturation are affected by function, activity, or experience. The bidirectional or probabilistic view calls for arrows going back to genetic activity to indicate feedback serving as signals for the turning off and turning on of DNA to manufacture protein. The usual view calls for genetic activity to be regulated by the genetic system itself in a strictly feedforward manner. That the feedback (actually, feeddown) view is correct is evidenced by the experimental results of Zamenhof & van Marthens, Uphouse & Bonner, and Grouse et al. reviewed in this chapter.

Note: Throughout this work I have presented DNA → RNA → protein pathway in an oversimplified manner that, although it seems appropriate for the present purpose, does disregard the fact that a number of crucial events intervene between RNA and protein formation. In fact, according to Pritchard (1986), dozens of known factors intervene between RNA activity and protein formation! Thus, it is an oversimplification to imply that DNA and RNA alone produce specific proteins—other factors (e.g., cytoplasm) contribute to the specificity of the protein.

that this hierarchical, reciprocal, coactive definition of epigenesis holds for anatomy and physiology (cf. the embryologist P. D. Nieuwkoop's definition in Gerhart, 1987), as well as for behavior and psychological functioning. The traffic is bidirectional, neither exclusively bottom—up or top-down. The embryologists Ludwig von Bertalanffy (1933-1962) and Paul Weiss (1939-1969), and the geneticist Sewall Wright (1968) have long been championing such a systems view for developmental genetics and developmental biology. The systems view in developmental psychology is exemplified by approaches and theories that have been called ecological (Bronfenbrenner, 1979), transactional (Dewey & Bentley, 1949; Sameroff, 1983), contextual (Lerner & Kaufman, 1985), interactive (Johnston, 1987; Magnusson, 1988), probabilistic epigenetic (Gottlieb, 1970), individual—socioecological (Valsiner, 1987), structural—behavioral (Horowitz, 1987), and, most globally speaking, interdisciplinary developmental science (Cairns, 1979).

Developmental Causality (Coaction)

Behavioral (or organic or neural) outcomes of development are a consequence of *at least* (at minimum) *two* specific components of coaction (e.g., person—person, organism—organism, organism–environment, cell–cell, nucleus-cytoplasm, sensory stimulation-sensory system, activity—motor behavior). The cause of development—what makes development happen—is the relationship of the two components, not the components themselves. Genes in themselves cannot cause development any more than stimulation in itself can cause development. When we speak of coaction as being at the heart of developmental analysis or causality what we mean is that we need to specify some relationship between at least two components of the developmental system. The concept used most frequently to designate coactions at the organismic level of functioning is *experience:* experience is thus a relational term. As documented elsewhere (Gottlieb 1976a, 1976b), experience can play at least three different roles in anatomical, physiological, and behavioral development (Fig. 14.1).

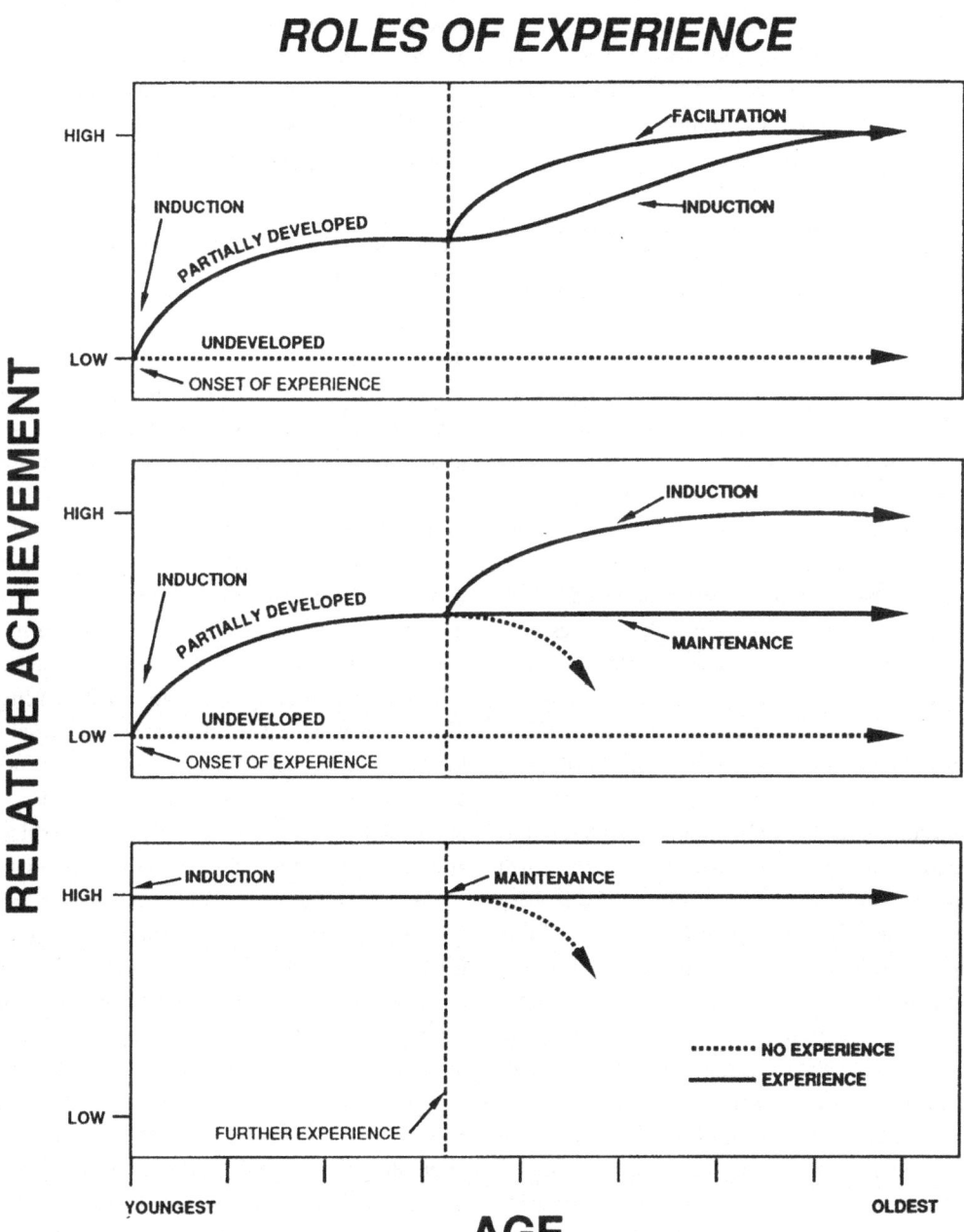

Fig. 14.1 Three roles that experience plays in the development of anatomy, physiology, and behavior.

It can be necessary to sustain already achieved states of affairs (*maintenance function*), it can temporally regulate when a feature appears during development (*facilitative function*), and it can be necessary to bring about a state of affairs that would not appear unless the experience occurred (*inductive function*).

Since first defining and formulating the various roles of experience in 1976, it has become apparent that interactive *activity* of some kind (e.g., activity—dependent regulation of gene expression) is a usual part of development at all levels from the subcellular to the organismic, so I have here (Figure 14.1) modified the early scheme to reflect that eventuality.[1] While this scheme is not exhaustive, it does call attention to the various roles that experience (interactive activity) can play in both intrinsic relations (inside the organism) and extrinsic ones (organism-organism or organism-environment relations), as Bateson (1983) and others have pointed out. In this scheme experience is defined as interactive activity, whether it is analyzed under or outside the skin of the organism, or, preferably, both under (say at the cellular level) and outside which can now be realized in research in developmental neurobiology. The techniques at hand make such thoroughgoing analyses quite feasible, especially in multidisciplinary research settings where investigators can pool their skills from the most molar levels (culture, society) to the most molecular levels of analysis. Such a multidisciplinary developmental systems approach is described and discussed in the first section of the January 1991 issue of *Developmental Psychology,* and one hopes it will become more commonplace in the years to come.

Since developing systems are by definition always changing in some way, statements of developmental causality must also include a *temporal* dimension describing when the experience occurred. For example, one of the earliest findings of experimental embryology had to do with the differences in outcome according to the time during early development when tissue was transplanted. When tissue from the head region of the embryo was transplanted to the embryo's back, if the transplantation occurred early in development the tissue differentiated according to its new surround (i.e., it differentiated into back tissue), whereas, if the transplant occurred later in development the tissue differentiated according to its previous surround so that, for example, a third eye might appear on the back of the embryo. These transplantation experiments demonstrated not only the import of time but also showed the essentially coactional nature of embryonic development.

Significance of Coaction for Individual Development

The early formulation by August Weismann of the role of the hereditary material (what came to be called genes) in individual development held that different parts of the genome or genic system caused the differentiation of the different parts of the developing organism, so there were thought to be genes for eyes, genes for legs, genes for toes, and so forth. Hans Driesch's experiment, in which he separated the first two cells of a sea urchin's development and obtained a fully formed sea urchin from each of the cells, showed that each cell contained a complete complement of genes. This means that each cell is capable of developing into any part of the body, a competency that was called *equipotentiality* or *pluripotency* in the jargon of the early history of experimental embryology and *totipotency* and *multipotentiality* in today's terms (e.g., DiBerardino, 1988). Each cell does not develop into just any part of the body even though it has the capability of doing so. Each cell develops in accordance with its surround, so cells at the anterior pole of the embryo develop into parts of the head, cells at the posterior pole develop into parts of the tail end of the body, cells in the foremost lateral region of the embryo develop into forelimbs, cells in the hindmost lateral region develop hindlimbs, the dorsal area of the embryo develops into the back, and so on. Although we do not know what actually causes cells to differentiate appropriately according to their surround, we do know that it is the cell's interaction with its surround, including other cells in that same area, that causes the cell to differentiate appropriately. The actual role of genes (DNA) is not to produce an arm or a leg or fingers, but to produce protein (through the coactions inherent in the formula DNA ↔ RNA ↔ protein). The protein produced by the DNA–RNA–cytoplasm coaction then differentiates according to coactions with other cells in its surround. Thus, differentiation occurs according to coactions *above the level of DNA—RNA coaction* (i.e., at the supragenetic level). The DNA—RNA coaction produces protein and that protein subsequently differentiates according to where it finds itself in the three-dimensional space of the embryo (the anterior- posterior, lateral, and dorsal-ventral spatial dimensions), plus the temporal dimension alluded to earlier. Thus it is the coactions during each phase of embryonic development that somehow, by means not yet understood, eventually produce a mature organism.[2]

Another demonstration that genes merely produce protein and not mature traits comes from an ingenious experiment by J. B. Gurdon (1968). If genes (DNA) in the nucleus of a cell did produce specific traits (e.g., an intestinal cell), then, should it prove possible to recover the nucleus from such a cell it would be missing that part of its genetic complement because those genes would have been used up in creating the intestinal cell. By way of proving that idea wrong, Gurdon recovered the nucleus from an intestinal cell of a tadpole and inserted that nucleus into an early embryonic (blastula) cell whose nucleus had been removed. Such nuclear transfers yield entirely normal and fertile adult male and female toads, thus proving that the differentiation of a cell does not require any loss whatsoever of genetic material and that differentiation occurs above the level of the genes after the genes, in coaction with RNA and cytoplasm, have manufactured the protein that is an essential building block of all cells in the body. When certain scientists refer to behavior or any other aspect of organismic structure or function as being "genetically determined" they are not mindful of the fact that genes synthesize protein and not fully developed features of the organism. And, as experiments on the early development of the nervous system have demonstrated, the amount of protein synthesis is regulated by neural activity itself, once again demonstrating the bidirectionality and coaction of influences during individual development (Born & Rubel, 1988).

The Hierarchical Systems View

Much has been written about the holistic or systems nature of individual development. In fact, there is no other way to envisage the manner in which development must occur if a harmoniously functioning, fully integrated organism is to be its product. In earlier chapters, figures were reproduced from the writings of the geneticist Sewall Wright and the embryologist Paul Weiss, which portray well the major components of the developing individual as an emergent, coactional, hierarchical system. So far we have dealt with the concepts of emergence and coaction as they pertain to the development of individuals. The notion of hierarchy, as it applies to individual development, simply means that coactions occur vertically as well as horizontally in all developmental systems. All the parts of the system are capable of influencing all the other parts of the system, however indirectly that influence may manifest itself. Consonant with Sewall Wright's and Paul Weiss's depiction of the developmental system, the organismic hierarchy proceeds from the lowest level, that of the genome or DNA in the nucleus, to the nucleus in the cytoplasm in the cell, to the cell in a tissue, to the tissue in an organ, the organ in an organ system, the organ system in an organism, the organism in an environment of other organisms and physical features, the environment in an ecosystem, and so on back down through the hierarchical developmental system (review by Grene, 1987). A fascinating example of the environment-to-gene hierarchical pathway is the finding that the genes that respond to heat shock in fruitfly larvae are more apt to do so if the to-be-shocked larvae have been kept at warm temperatures rather than cold ones. The characteristic heat-shock proteins that these "heat-shock genes" produce are less evident in cold-reared larvae (Singh & Lakhotin, 1988). Another dramatic developmental effect traversing the many levels from the environment back to the cytoplasm of the cell is shown by the experiments of Victor Jollos in the 1930s and Mae-Wan Ho in the 1980s. In Ho's experiment (reviewed in Chapter 12), an extraorganismic environmental event such as a brief period of exposure to ether occurring at a particular time in embryonic development can alter the cytoplasm of the cell in such a way that a different pattern of protein is produced that eventually results in a second set of wings (an abnormal "bithorax" condition) in place of the halteres (balancing organs) on the body of an otherwise normal fruitfly. Obviously, it is very likely that "signals" have been altered at various levels of the developmental hierarchy to achieve such an outcome. (Two excellent texts that describe the many different kinds of coactions that are a necessary and normal part of embryonic development are D. J. Pritchard's [1986] *Foundations of Developmental Genetics* and N. K. Wessells's [1977] *Tissue Interactions and Development*.)

It happens that when the cytoplasm of the cell is altered, as in the experiments of Jollos and Ho, the effect is transgenerational such that the untreated daughters of the treated mothers continue for a number of generations to produce bithorax offspring and do so even when mated with males from untreated lines. Such a result has evolutionary as well as developmental significance, which, to this date, has been little exploited because the neo-Darwinian, modern synthesis does not yet have a role in evolution for anything but changes in genes and gene frequencies in evolution: epigenetic development above the level of the genes has not yet been incorporated into the modern synthesis (Futuyma, 1988; Løvtrup, 1987).[3] that the first step in the evolutionary pathway may sometimes not involve a genetic change or mutation.)

Another remarkable organism—environment coaction occurs routinely in coral reef fish. These fish live in spatially well-defined, social groups in which there are many females and few males. When a male dies or is otherwise removed from the group, one of the females initiates a sex reversal over a period of about two days in which she develops the coloration, behavior, and gonadal physiology and anatomy of a fully functioning male (Shapiro, 1981). Such sex reversals keep the sex ratios about the same in social groups of coral reef fish. Apparently, it is the higher ranking females that are the first to change their sex and that inhibits sex reversal in lower ranking females in the group. Sex reversal in coral reef fish provides an excellent example of the vertical dimension of developmental causality.

The completely reciprocal or bidirectional nature of the vertical or hierarchical organization of individual development is nowhere more apparent than the responsiveness of cellular or nuclear DNA itself to events originating in the external environment of the organism. The major theoretical point of this monograph is that the genes are part of the developmental system in the same sense as other components (cell, tissue, organism), so genes must be susceptible to influence from other levels during the process of individual development. DNA produces protein, cells are composed of protein, so there must be a high correlation between the *size* of cells, amount of protein, and quantity of DNA, and there must also be a high correlation between the *number* of cells, amount of protein, and quantity of DNA, and so there is (Cavalier-Smith, 1985; Mirsky & Ris, 1951). For our behavioral-psychological purposes, it is most interesting to focus on the developing brain, where we do indeed find the expected correlation among size and/or number of brain cells, amount of protein, and quantity of DNA (Zamenhof & van Marthens, 1978, 1979). From the present point of view, it is significant that cell size, if not cell number, in the developing rodent and chick brain is responsive to two sorts of environmental input: nutrition and sensorimotor experience. Undernutrition and "supernutrition" produce newborn rats and chicks with lower and higher quantities of cerebral DNA respectively

Table 14.2 Sequence Complexities of Visual Cortex RNAs

Group	No. of Separate RNA Preparations	No. of Hybridization Reactions at $ER_o t = 211,300$	RNA Complexity as Percentage of uDNA
Unsutured	5	18	11.77 ± 0.59
Sutured	5	10	8.69 ± 0.19
Repeat-sutured	4	8	9.22 ± 0.86

Source: From Grouse et al. (1979).

(Zamenhof & van Marthens, 1978, 1979). Similar cerebral consequences are produced by extreme variations (social isolation, environmental enrichment) in sensorimotor experience during the postnatal period (Rosenzweig & Bennett, 1978).

Since the route from DNA to protein is through the mediation of RNA (DNA → RNA → protein), it is significant for the present theoretical viewpoint that social isolation and environmental enrichment produce alterations in the complexity (or diversity) of RNA sequences in the brains of rodents. (RNA complexity or diversity refers to the total number of nucleotides of individual RNA molecules.) A specific example of a change in RNA complexity as a consequence of normal and deprived visual experience is shown in Table 14.2. When the eyelids of kittens are sutured closed so they cannot receive visual stimulation, they show less RNA complexity in the visual cortex of the brain compared to normal (unsutured) kittens (Grouse et al., 1979). In general, environmental enrichment produces an increase in the complexity of expression of RNA sequences, whereas social isolation and environmental deprivation result in a significantly reduced degree of RNA complexity (Grouse et al., 1980; Uphouse & Bonner, 1975). These experientially produced alterations in RNA diversity are specific to the brain. When other organs are examined (e.g., liver), no such changes are found.

Nonlinear Causality

Because of the emergent nature of epigenetic development, another important feature of developmental systems is that causality is often not "linear" or straightforward. In developmental systems the coaction of X and Y often produces W rather than more of X or Y, or some variant of X or Y. Another, perhaps clearer, way to express this same idea is to say that developmental causality is often not obvious. In my own research, for example, I found that mallard duck embryos had to hear their own vocalizations prior to hatching if they were to show their usual highly specific behavioral response to the mallard maternal assembly call after hatching. If the mallard duck embryo was deprived of hearing its own or sib vocalizations, it lost its species-specific perceptual specificity and became as responsive to the maternal assembly calls of other species as to the mallard hen's call. To the human ear, the embryo's vocalizations sound nothing like the maternal call. It turned out, however, that there are certain rather abstract acoustic ingredients in the embryonic vocalizations that correspond to critical acoustic features that identify the mallard hen's assembly call. In the absence of experiencing those ingredients, the mallard duckling's auditory perceptual system is not completely "tuned" to those features in the mallard hen's call and they respond to the calls of other species that resemble the mallard in these acoustic dimensions. The intricacy of the developmental causal network revealed in these experiments proved to be striking. Not only must the duckling experience the vocalizations as an embryo (the experience is ineffective after hatching), the embryo must experience *embryonic* vocalizations. That is, the embryonic vocalizations change after hatching and no longer contain the proper ingredients to tune the embryo to the maternal cell (Gottlieb, 1985).

Prenatal nonlinear causality is also nonobvious because the information, outside of experimental laboratory contexts, is usually not available to us. For example, the rate of adult sexual development is retarded in female gerbils that were adjacent to a male fetus during gestation (Clark & Galef, 1988). To further compound the nonobvious, the daughters of late-maturing females are themselves retarded in that respect—a transgenerational effect!

In a very different example of nonobvious and nonlinear developmental causality, Cierpial and McCarty (1987) found that the so-called spontaneously hypertensive (SHR) rat strain employed as an animal model of human hypertension is made hypertensive by coacting with their mothers after birth. When SHR rat pups are suckled and reared by normal rat mothers after birth they do not develop hypertension. It appears that there is a "hyperactive" component in SHR mothers' maternal behavior that causes SHR pups to develop hypertension (Myers, Brunelli, Shair, Squire, & Hofer, 1989; Myers, Brunelli, Squire, Shindeldecker, & Hofer, 1989). The highly specific coactional nature of the development of hypertension in SHR rats is shown by the fact that normotensive rats do not develop hypertension when they are suckled and reared by SHR mothers. Thus, although SHR rat pups differ in some way from normal rat pups, the development of hypertension in them nonetheless requires an interaction with their mother; it is not an inevitable outcome of the fact that they are genetically, physiologically, and/or anatomically different from normal rat pups. This is a good example of the *relational* aspect of the definition of experience and developmental causality offered earlier in this chapter. The cause of the hypertension in the SHR rat strain is not in the SHR rat pups or in the SHR mothers but in the nursing relationship between the SHR rat pups and their mother.

Another example of a nonlinear and nonobvious developmental experience undergirding species-typical behavioral development is Wallman's (1979) demonstration that if chicks are not permitted to see their toes during the first two days after hatching, they do not eat or pick up mealworms as chicks normally do. Instead, the chicks stare at the mealworms. Wallman suggests that many features of the usual rearing environment of infants may offer experiences that are necessary for the expression of species-typical behavior.

The Unresolved Problem of Differentiation

The nonlinear, emergent, coactional nature of individual development is well exemplified by the phenomenon of *differentiation*, whereby a new kind of organization comes into being by the coaction of preexisting parts. If genes directly caused parts of the embryo rather than producing protein, there would be less of a problem in understanding differentiation. Since the route from gene to mature structure or organism is not straightforward, differentiation poses a significant intellectual puzzle, as recognized as early as 1962 by Ephrussi (1979), among others. The problem of differentiation also involves our limited understanding of the role of genes in development; what else, if anything, might genes do than produce protein?

It has been recognized since the time of Driesch's earth-shaking experiments demonstrating the genetic equipotentiality of all cells of the organism that the chief problem of understanding development was that of understanding why originally equipotential cells actually do become different in the course of development, i.e., how is it they differentiate into cells that form the tissues of very different organ systems. The problem of understanding development thus became the problem of understanding cellular differentiation. We still do not understand differentiation today, and it is quite telling of the immense difficulty of the problem that today's theory of differentiation is very much like the necessarily vaguer theories put forth by E. B. Wilson in 1896 and T. H. Morgan in 1934 (reviewed in Davidson, 1986), that ultimate or eventual cellular differentiation is influenced by an earlier coaction between the genetic material in the nucleus of the cell with particular regions of the cytoplasm of the cell. Some of the vagueness has been removed in recent years by the actual determination of regional differences in the cytoplasm (extensively reviewed by Davidson, 1986). Thus, the undifferentiated protein resulting from locale or regional differences of nucleo-cytoplasmic coaction is biochemically distinct, which, in some as yet unknown way, influences or biases its future course of development. For example, protein with the same or similar biochemical makeups may stay together during cellular migration during early development and thus eventually come to form a certain part of the organism by the three-dimensional spatial field considerations of the embryo mentioned earlier in this chapter. Although the actual means or mechanisms by which some cells become one part of the organism and others become another part are still unresolved, we do have a name for the essential coactions that cause cells to differentiate: they are called embryonic *inductions*. The nonlinear hallmark of developmental causality is well exemplified by embryonic induction, in which one kind of cell (*A*) coacting with a second kind of cell (*B*) produces a third kind of cell (*C*). For example, if left in place, cells in the upper one-third of an early frog embryo differentiate into nerve cells: if removed from that region, those same cells can become skin cells. Equipotentiality and the critical role of spatial position in determining differentiation in the embryo is well captured in a quotation from the autobiography of Hans Spemann, the principal discoverer of the phenomenon of embryonic induction: "We are standing and walking with parts of our body which could have been used for thinking if they had been developed in another position in the embryo" (translated by B. K. Hall, 1988, p. 174). It might have been even more striking—and equally correct—if Spemann had elected to say, "We are sitting with parts of our body which could have been used for thinking …"!

Even if we do not yet have a complete understanding of differentiation, the facts at our disposal show us that epigenetic development is correctly characterized as an emergent, coactional, hierarchical system that results in increasingly complex organization. It remains now to use that conceptual framework to fashion a developmentally based view of evolutionary change.

Notes

1. In 1976 I (Gottlieb, 1976a, 1976b) defined experience in such a way as to include spontaneous activity generated within the nervous system as well as evoked activity arising from sensory stimulation originating in the organism's environment. Since that time it has been shown that spontaneous as well as evoked activity does play an important role in the normal neuroanatomical development of the brain (e.g., Born & Rubell, 1998; Shatz & Stryker, 1988). The brain develops abnormally (deficiently) when normal spontaneous or evoked activity is curtailed by experimental means. This area of research is coming to the forefront in developmental neuroscience under the rubric *activity-dependent regulation of gene expression*, a concept that meshes extremely well with the view of the mechanisms of individual development expressed in the present work (review in Changeux & Konishi). The contribution of factors "upstream" from the genes is also being recognized by the "new look" in developmental neuroscience (e.g., Edelman, 1988).

2. With the realization that Weissmann's notion that specific genes give rise to specific parts of the body is erroneous, various biologists have been working on hierarchically organized field theory approaches to an understanding of individual morphological development. Among the current workers are Goodwin (1984), Hall (1988), and Oster, Odell, & Alberch (1980). It is curious that some ostensible field theorists such as Wolpert (1971, 1982) still hold that the morphological development of the individual is somehow preprogrammed in the genes. So, when I say "the realization that Weismann's new view is erroneous," it is obvious that the implications of that statement are not unequivocally understood by all developmental biologists, even if they happen to be working on field theories of individual development!

3. It is especially noteworthy that someone like Futuyma acknowledges the lack of developmental thinking in the neo-Darwinian concept of evolution, because he is firmly identified with the population-genetic tradition that undergirds the modern synthesis. On the other hand Futuyma seems to see no substantive difference or change in the neo-Darwinian synthesis resulting from an inclusion of developmental considerations:

"The need for a theory of development... is evident, but at this point, the need for a new evolutionary paradigm is not" (Futuyma, 1988, p. 221). Other biologists do not agree with Futuyma; they try to show why a new evolutionary paradigm is necessary when individual development is taken into account. In Chapter 14 I try to show what differences accrue to evolutionary thinking when developmental behavioral and psychological concerns are placed in the

References

Bateson, P. (1983). Genes, environment, and the development of behaviour. In T. R. Halliday & P. J. B. Slater (Eds.), *Animal behaviour, Vol. 3. Genes, development, and learning*. Oxford: Blackwell.

von Bertalanffy, L. (1962). *Modern theories of development: An introduction to theoretical biology*. New York: Harper. (Originally published in German in 1933.)

Bronfenbrenner, U. (1979). *The ecology of human development: Experiments by nature and design*. Cambridge, Massachusetts: Harvard University Press.

Cairns, R. B. (1979). *Social development: The origins and plasticity of interchanges*. San Francisco: W. H. Freeman.

Cavalier-Smith, T. (1985). Cell volume and the evolution of eukaryote genome size. In T. Cavalier-Smith (Ed.), *The evolution of genome size*. Chichester, England: Wiley.

Changeux, J.-P., & Konishi, M. (Eds.) (1987). *The neural and molecular bases of learning*. Chichester, England: Wiley.

Cierpal, M. A., & McCarty, R. (1987). Hypertension in SHR rats: Contribution of maternal environment. *American Journal of Physiology, 253*, 980–984.

Clark, N. M., & Galef, B. G. (1988). Effects of uterine position on rate of sexual development in female mongolian gerbils. *Physiology & Behavior, 42*, 15–18.

Davidson, E. H. (1986). *Gene activity in early development*. Orlando, Florida: Academic Press.

Dewey, J., & Bentley, A. F. (1949). *Knowing and the known*. Boston: Beacon.

Dewsbury, D. A. (1978). *Comparative animal behavior*. New York: McGraw-Hill.

DiBerardino, M. A. (1988). Genomic multipotentiality of differentiated somatic cells. In G. Eguchi, T. S. Okada, L. Saxén (Eds.), *Regulatory mechanisms in developmental processes*. Ireland: Elsevier.

Edelman, G. M. (1988). *Topobiology*. Basic Books.

Ephrussi, B. (1979). Mendelism and the new genetics. *Somatic Cell Genetics, 5*, 681–695.

Futuyma, D. J. (1988). *Sturm und Drang* and the evolutionary synthesis. *Evolution, 42*, 217–226.

Gehring, W. J. (1987). Homeo boxes in the study of development. *Science, 236*, 1245–1252.

Goodwin, B. C. (1984). A relational or field theory of reproduction and its evolutionary implications. In M.-W. Ho, & P. T. Saunders (Eds.), *Beyond neo-Darwinism: An introduction to the new evolutionary paradigm*. London: Academic Press.

Gottlieb, G. (1970). Conceptions of prenatal behavior. In L. R. Aronson, E. Tobach, D. S. Lehrman, & J. S. Rosenblatt (Eds.), *Development and evolution of behavior: Essays in memory of T. C. Schneirla* (pp. 111–137). San Francisco: Freeman.

Gottlieb, G. (1976a). The roles of experience in the development of behavior and the nervous system. In G. Gottlieb (Ed.), *Neural and behavioral specificity*. New York: Academic Press.

Gottlieb, G. (1976b). Conceptions of prenatal development: Behavioral embryology. *Psychological Review, 83*, 215–234.

Gottlieb, G. (1985). Development of species identification in ducklings: XI. Embryonic critical period for species-typical perception in the hatchling. *Animal Behaviour, 33*, 225–233.

Greenough, W.T., & Juraska, J. M. (1979). Experience-induced changes in brain fine structure: Their behavioral implications. In M. E. Hahn, C. Jensen, & B. C. Dudek (Eds.), *Development and evolution of brain size*. New York: Academic Press.

Grouse, L. D., Schrier, B. K., & Nelson, P. G. (1979). Effect of visual experience on gene expression during the development of stimulus specificity in cat brain. *Experimental Neurology, 64*, 354–359.

Grouse, L. D., Schrier, B. K., Letendre, C. H., & Nelson, P. G. (1980). RNA sequence complexity in central nervous system development and plasticity. *Current Topics in Developmental Biology, 16*, 381–397.

Gurdon, J. B. (1968). Transplanted nuclei and cell differentiation. *Scientific American, 219*, 24–35.

Hall, B. K. (1988). The embryonic development of bone. *American Scientist, 76*, 174–181.

Ho, M.-W. (1984). Environment and heredity in development and evolution. In M.-W. Ho & P. T. Saunders

(Eds.), *Beyond neo-Darwinism: An introduction to the new evolutionary paradigm*. London: Academic Press.

Ho, M.-W., & Fox, S. W. (Eds.) (1988). *Evolutionary processes and metaphors*. New York: Wiley.

Horowitz, F.D. (1987). *Exploring developmental theories: Toward a structural/behavioral model of development*. Hillsdale, New Jersey: Erlbaum.

Johnston, T. D. (1987). The persistence of dichotomies in the study of behavioral development. *Developmental Review, 7*, 149–182.

Kuo, Z.-Y. (1976). *The dynamics of behavior development: An epigenetic view* (enlarged ed.). New York: Plenum Press.

Lehrman, D.S. (1970). Semantic and conceptual issues in the nature-nurture problem. In L. R. Aronson, D. S. Lehrman, E. Tobach, & J. S. Rosenblatt (Eds.), *Development and evolution of behavior*. San Francisco, California: W. H. Freeman.

Lerner, R.M., & Kaufman, M. B. (1985). The concept of development in contextualism. *Developmental Review, 5*, 309–333.

Løvtrup, S. (1987). *Darwinism: Refutation of a myth*. Beckenham, Kent, England: Croom Helm.

Magnusson, D. (1988). *Individual development from an interactional perspective: A longitudinal study*. Hillsdale, New Jersey: Erlbaum.

Morgan, T. H. (1934). *Embryology and genetics*. Westport, Connecticut: Greenwood.

Myers, M. M., Brunelli, S. A., Shair, H. N., Squire, J. M., & Hofer, M. A. (1989). Relationships between maternal behavior of SHR and WKY dams and adult blood pressures of cross-fostered F1 pups. *Developmental Psychobiology, 22*, 55–67.

Myers, M. M., Brunelli, S. A., Squire, J. M., Shindeldecker, R. D., & Hofer, M. A. (1989). Maternal behavior of SHR rats and its relationship to offspring blood pressures. *Developmental Psychobiology, 22*, 29–53.

Oster, G., Odell, G., & Alberch, P. (1980). Mechanics, morphogenesis, and evolution. *Lectures on Mathematics in the Life Sciences, 13*, 165–255.

Pritchard, D. J. (1986). *Foundations of developmental genetics*. London and Philadelphia: Taylor and Francis.

Sameroff, A. J. (1983). Developmental systems: Contexts and evolution. In P. H. Mussen (Ed.), *Handbook of Child Psychology* (Vol. 1): W. Kessen (Ed.), *History, theory, and methods*. New York: Wiley.

Schneirla, T. C. (1961). Instinctive behavior, maturation—experience and development. In B. Kaplan & S. Wapner (Eds.), *Perspectives in psychological theory—Essays in honor of Heinz Werner*. New York: International Universities Press.

Shapiro, D. Y. (1981). Serial female sex changes after simultaneous removal of males from social groups of a coral reef fish. *Science, 209*, 1136–1137.

Shatz, C. J., & Stryker, M. P. (1988). Prenatal tetrodotoxin infusion blocks segregation of retinogeniculate afferents. *Science, 242*, 87–89.

Singh, A. K., &; Lakhotin, S. C. (1988). Effect of low-temperature rearing on heat shock protein synthesis and heat sensitivity in *Drosophila melanogaster*. *Developmental Genetics, 9*, 193–201.

Uphouse, L. L., & Bonner, J. (1975). Preliminary evidence for the effects of environmental complexity on hybridization of rat brain RNA to rat unique DNA. *Developmental Psychobiology, 8*, 171–178.

Valsiner, J. (1987). *Culture and the development of children's action*. Chichester, England: Wiley.

Wallman, J. (1979). A minimal visual restriction experiment: Preventing chicks from seeing their feet affects later responses to mealworms. *Developmental Psychobiology, 12*, 391–397.

Weiss, P. (1959). Cellular dynamics. *Reviews of Modern Physics, 31*, 11–20.

Wessells, N.K. (1977). *Tissue interactions and development*. Menlo Park, California: W. A. Benjamin.

Wolpert, L. (1971). Positional information and pattern formation. *Current Topics in Developmental Biology, 6*, 183–224.

Wolpert, L., & Stein, W. D. (1982). Evolution and development. In H. C. Plotkin (Ed.), *Learning, development, and culture: Essays in evolutionary epistemology*. London: Wiley.

Wright, S. (1968). *Evolution and the genetics of populations*, Vol. 1: *Genetic and biometric foundations*. Chicago: University of Chicago Press.

Zamenhof, S., & van Marthens, E. (1978). Nutritional influences on prenatal brain development. In G. Gottlieb (Ed.), *Early influences*. New York: Academic Press.

Zamenhof, S., & van Marthens, E. (1979). Brain weight, brain chemical content, and their early manipulation. In M. E. Hahn, C. Jensen, & B. C. Dudek (Eds.), *Development and evolution of brain size*. New York: Academic Press.

15.
THE HERITABILITY FALLACY
David S. Moore* and David Shenk†

The term 'heritability,' as it is used today in human behavioral genetics, is one of the most misleading in the history of science. Contrary to popular belief, the measurable heritability of a trait *does not* tell us how 'genetically inheritable' that trait is. Further, it does not inform us about what causes a trait, the relative influence of genes in the development of a trait, or the relative influence of the environment in the development of a trait. Because we already know that genetic factors have significant influence on the development of *all* human traits, measures of heritability are of little value, except in very rare cases. We, therefore, suggest that continued use of the term does enormous damage to the public understanding of how human beings develop their individual traits and identities.

© 2016 Wiley Periodicals, Inc.

Introduction

If someone were to tell you that research has proven that human intelligence is highly 'heritable,' what would you think that means? Most people would probably assume that it means people inherit a significant percentage of their intelligence directly, via their parents' genes. In fact, though, the scientific terms 'heritable' and 'heritability' actually have very little to do with genetic inheritance. This is confusing, because 'heritability' sounds like it means the same thing as 'inheritability.' The confusion about what 'heritability' actually measures significantly adds to a deep misunderstanding about how, exactly, our genomes contribute to our observable characteristics (see Charney, Genes, behavior, and behavior genetics, *WIREs Cogn Sci*, also in the collection How We Develop).

The Appropriation of 'Heritable'

For hundreds of years, the word 'heritable' was used without confusion as a synonym for 'hereditary.' But in the early 20th century, the word was repurposed to represent something new and rather narrow. At that time, geneticists had a strictly deterministic understanding of how genes influence the formation of traits. They considered the relationship between genes and the environment to be akin to the relationship between a plant seed and the rain that waters it: Genes were thought to contain specific, blueprint-like instructions for the formation of traits, whereas the environment provided the nutrients and other salubrious conditions that would allow those instructions to unfold. According to this earlier way of thinking, a person's DNA has specific instructions for blue eyes, or athletic arms, or a mathematical mind; the environment merely allows for emphasis or de-emphasis of those already-designed traits. (If this

* Pitzer College and Claremont Graduate University, Claremont, CA
† DeLTA Center, University of Iowa, Iowa City, IA

The Heritability Fallacy

sounds familiar, it's probably because it strongly resembles what many of us were taught about genetics in grade school.)

The term heritability was first given this new meaning in J. L. Lush's 1937 book *Animal Breeding Plans*.[1] In that text, Lush proposed a calculation for what he called 'heritability' that neatly codified the then-popular deterministic viewpoint. Because, Lush argued, an animal's *phenotype* (i.e., its observable traits, such as intelligence, height, eye color, etc.) is a function of genetic instructions *plus* the finishing influence of the environment, we should be able to statistically separate the influence of each.[2] Relying on mathematical guidelines from the geneticist Sewall Wright, Lush proposed that in any given group:

$$V_p(\text{phenotypic variation}) = V_g(\text{genetic variation}) + V_e(\text{environmental variation})$$

Lush asserted that the Vg portion of that total can reasonably be termed 'heritability', as it revealed the portion of the trait variation that could be accounted for by variation in genes. The intention was 'to quantify the level of predictability of passage of a biologically interesting phenotype from parent to offspring'. In this way, the new technical use of 'heritability' accurately reflected that period's understanding of genetic determinism. Still, it was a curious appropriation of the term, because—even by the admission of its proponents—it was meant only to represent how *variation* in DNA relates to *variation* in traits across a population, not to be a measure of the actual influence of genes on the development of any given trait. For example, in a large group of people with eyes of different colors (Fig. 15.1), 'Vg' only represents the extent to which *variation* in the group's DNA accounts for variation in different eye colors *in that group*—not whether or how DNA is responsible for the development of eye color. In that sense, it was a highly misleading new use of the term (even in the context of determinism) that was bound to cause confusion: And indeed it did.

Twin Studies

The new use of the term caught on. Since that time, behavioral geneticists have conducted hundreds of studies derived from the Lush-Wright concept of heritability. The most prominent method to determine heritability in human beings has been statistical comparisons of identical and fraternal twins.

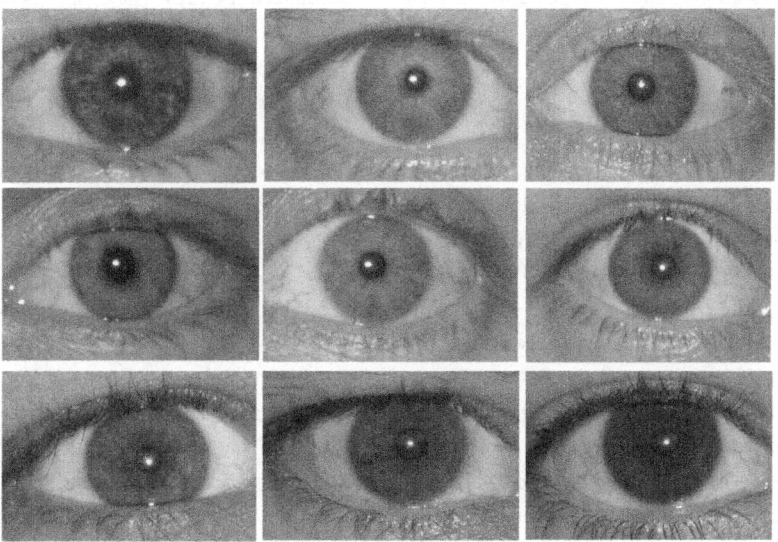

Fig. 15.1 Natural variability in human eye colors. Source: Sturm and Frudakis.[3]

In what *seems* like a straightforward approach, researchers compute correlations for a trait among identical twins and compare them to correlations for that trait among fraternal twins. Because identical twins share 100% of their DNA and fraternal twins share, on average, only 50% of their DNA, this statistical comparison yields a crisp number that *seems* to highlight that portion of a trait *caused* by genetic instruction. Even though behavioral geneticists are sometimes careful to point out that each 'heritability' measure only actually applies to variations in a specific group, they often use the term 'heritability' in a way that conveys a sense of direct genetic influence on traits. Also, because the term (if not its new-fangled scientific meaning) is so familiar to the public, the casual misinterpretation of the term's narrower meaning has been rampant in the popular press. For many decades now, we have been entertained with journalistic accounts of twin studies suggesting that 'personality is heritable',[4] 'criminals are born, not made',[5] and 'cheating genes play [a] large role in female infidelity'.[6] Twin studies have reaffirmed the strong public impression that some physical and personality traits can be passed directly from parent to child through DNA. While understandable, this impression is flatly incorrect, as brightly illustrated by three significant flaws in some scientists' use of—and thus the public's understanding of—the term 'heritability'.

The Group vs. Individual Flaw

Although the original motivation for the concept of heritability was 'to quantify the level of predictability of passage of a biologically interesting phenotype from parent to offspring', it is essential to realize that the equation Vp = Vg + Ve has no relevance at all when it comes to individual development. *The Vp in that equation refers to variation across individuals in a population, not to causal processes that occur within any individual person.* Although the term 'heritability' is often misunderstood as speaking to the degree to which a certain trait's appearance in an individual is caused by genetic factors, the logic underlying the equation nullifies that conceptualization. 'Heritability', explains author Matt Ridley, 'is a slippery concept, much misunderstood. For a start, it is a population average, meaningless for any individual person: you cannot say that Hermia has more heritable intelligence than Helena. When somebody says that heritability of height is 90 percent, he does not and cannot mean that 90 percent of my inches come from genes and 10 percent from my food.' Instead, he means that 90% of the variation in height in a group of people can be accounted for, statistically, by variations in those people's DNA. But there is no sense in which the specific DNA variations are causing 90% of a person's height. As Ridley notes, 'there is no heritability in height for the individual'.[7]

This group/individual distinction might seem like a mere mathematical technicality, but it is not. Failing to recognize the distinction introduces a critical logical flaw that, on its own, completely undermines the broad popular understanding of the term 'heritability'.

To understand the profundity of this logical flaw, consider an illustration having nothing to do with biology: One winter, in a particular neighborhood, there is a rash of house fires. Committed to fixing the problem, the city sets out to determine what caused the fires. They gather as much data as they can about all the homes in this fire-ravaged neighborhood—including those that had fires and those that did not. They find that 100% of the fire variation in the group is attributable to whether or not space heaters were present in the various homes.

Question: Is the cause of each individual fire due *only* to the presence of a space heater?

An equation analogous to Vp = Vg + Ve might, if applied to *individual* house fires, seem to suggest that the answer is a resounding yes, because the presence or absence of a space heater accounted for *all* of the variation in house fires in the neighborhood. But, actually, the answer is a resounding no, because as it turns out, every single home in this particular neighborhood was built out of highly combustible wood and painted with highly flammable paint. Importantly, that fact might not have emerged in the investigation because there was *zero variation* in paint and construction materials within this neighborhood. A statistical investigation focused on variation rather than causation is, if not very narrowly interpreted, likely to lead to a mistaken conclusion about what caused the fires and about what can be done to prevent them in the future.

To see this, consider a newer neighborhood right across the river in which every single home was built with non-flammable aluminum and painted with fire-retardant paint. Even if this neighborhood has just as many space heaters, not a single home here could ever catch fire.

So what causes individual fires? The answer is complex, and statistical variation surveys cannot address the question effectively; although all of the *variation* in fires in the fire-ravaged neighborhood can be accounted for by focusing on space heaters alone, the fact remains that multiple factors—flammable building materials, the presence of heat sources, and available oxygen, for example—are responsible for *collectively* causing fires. Thus, focusing on *variation* rather than causation can contribute to a misleading sense of how important a particular factor might be in contributing to a particular outcome.

A similar argument was advanced decades ago by Richard Lewontin,[8] who suggested a "thought experiment" that illustrates how even if a human characteristic is 100% heritable, it is still possible to easily influence the development of that characteristic by raising children in a different environment. Lewontin asked us to imagine that we plant a handful of normal seeds in an environmental context in which each developing plant receives the identical amount of light and the identical amount of nutrients, and furthermore, that the amount of light and nutrients each plant receives is adequate to support normal growth (see the left side of Fig. 15.2). In this case, the variation in height that we ultimately see in these

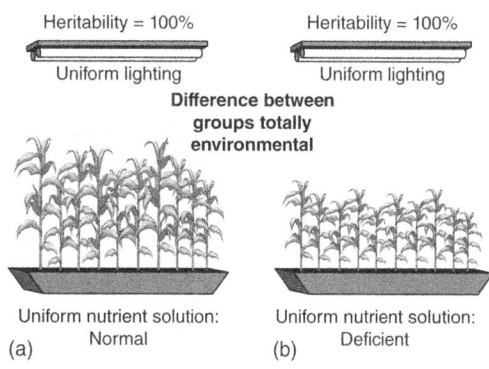

Fig. 15.2 Lewontin's thought experiment. Genetically variable seeds that develop in controlled environments grow to varying heights. The heritability of height in both the (a) and (b) panels is 100%, because all plants in each panel are exposed to the same environment; thus, all of the variation in height (within a panel) is accounted for by *genetic* variation. Despite height being 100% heritable in both the left panel *and* the right panel, plants' heights are still influenced by the quality of the nutrients they encounter in their environments; mature plants that develop in a deficient nutrient solution (b) are shorter, on average, than are mature plants that develop in a normal nutrient solution (a). Source: http://www.nyu.edu/gsas/dept/philo/faculty/block/papers/plants.gif) and Lewontin.[9]

plants can be accounted for by referencing genetic variation alone (because all of the plants are grown in the same environment, meaning there is no environmental variation at all); in this situation, the heritability of height would be calculated to be 100%. Next, imagine that we plant another handful of normal seeds in an environmental context in which each developing plant again receives the identical amount of light and the identical amount of nutrients—but now imagine that each plant is provided with a less-than-adequate amount of light and nutrients (see the right side of Fig. 15.2). In this situation, too, the heritability of height would be calculated to be 100%, because again, we have carefully controlled the environment in which this second group of plants has grown, so all of the variation in the plants' heights can be accounted for by referencing genetic variation alone. However, even though heritability in

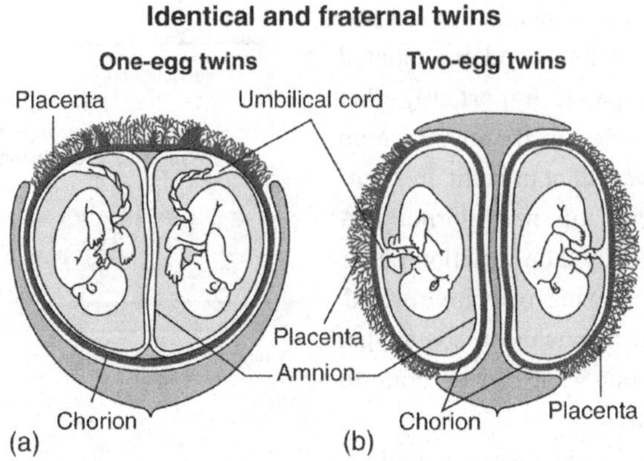

Fig. 15.3 Monozygotic ('identical') twins often share a placenta (a), whereas, dizygotic ('fraternal') twins never share a placenta (b). As a result, most monozygotic twins develop in prenatal environments that are more similar than are the prenatal environments in which dizygotic twins develop. Source: http://www.britannica.com/EBchecked/topic/217570/dizygotic-twin.

each of these subpopulations would be calculated to be 100%, we would still find that on average, the nutrient-deprived plants wind up shorter than the plants grown in adequate environmental conditions. Thus, finding that the heritability of a characteristic is 100% still does not mean that environmental factors cannot powerfully influence the development of that characteristic.

The Environmental Flaw

The math of twin studies is based on the assumption that environments encountered by identical twins are *no more similar* than environments encountered by fraternal twins. However, the formulas are not sound if identical twins, in addition to having more genetic similarity, are also exposed to more environmental similarity; in that case, increased similarities in identical twins' traits can just as easily be accounted for by environmental similarities as by genetic similarities.

In fact, identical twins are exposed to demonstrably more similar environments than are fraternal twins. It starts in the womb: while fraternal twin embryos are always connected to their mother via two *unique* placentas, identical twins most often (but not always) *share* a single placenta. This means that, beginning soon after conception,

identical twins typically have more similar access to nutrients, oxygen, and other factors than do fraternal twins (Fig. 15.3). But does this increased similarity matter? Yes. In fact, identical twins who share a placenta as fetuses have more similar IQs[10] and personalities[11] than do *identical* twins who did not share a placenta.

This environmental difference continues after birth. Scarr and McCartney argued that, because similar-looking individuals are more likely than different-looking individuals to evoke similar social responses from people around them, identical twins likely encounter more similar environments than do fraternal twins.

Twin researchers, aware of this essential conjecture underlying their work, have given it a name—the Equal Environments Assumption (EEA)—and have attempted to correct for it in their calculations *and* to demonstrate that it has no real consequences for their findings. Studies have, however, found demonstrable EEA effects[12-15] that effectively invalidate the assumptions upon which heritability claims rely.

The Biological Flaw

Most important of all is a deep flaw in an assumption that many people make about biology: That genetic influences on trait development *can be separated* from their environmental context. However, contemporary biology has demonstrated beyond any doubt that traits are produced by *interactions* between genetic and nongenetic factors that occur in each moment of developmental time (see Charney, Genes, behavior, and behavior genetics, *WIREs Cogn Sci*, also in the collection How We Develop). That is to say, *there are simply no such things as gene-only influences*. Our DNA, we now know, does not contain specific blueprint-like instructions about traits; rather, DNA segments merely contribute to the production of different kinds of RNA molecules. These RNA molecules can, in turn, regulate other DNA segments, or contribute to the production of proteins that are constituent parts of cells, cells that are assembled into systems that manifest identifiable traits. This entire process takes place in a developmental context; DNA produces its products under the influence of signals from the environment, as well as from other DNA segments (which are in turn signaled by the environment and other DNA segments, and so on). Rather than spitting out pre-determined creations front programmed instructions, 'genes' are more like customized switches that get turned on and off by particular developmental circumstances. Traits are *always* a consequence of this interactive dynamic. After determining that intelligence among poor families was nearly zero-percent 'heritable,' Eric Turkheimer and colleagues wrote: 'These findings suggest that a model of [genes plus environment] is too simple for the dynamic interaction of genes and real-world environments during development.'[16]

Patrick Bateson has echoed this sentiment, stating that heritability studies make 'the extraordinary assumption that genetic and environmental influences are independent of one another and do not interact. That assumption is clearly wrong.'[17]

The fact that DNA segments are influenced by their contexts is one reason that very highly heritable traits can nonetheless be influenced by nongenetic factors. Consider height, which has a heritability that has oftentimes been measured at close to 90% in each of numerous different populations; despite this finding, height can be drastically affected by nutrition, an environmental factor. For example, because food is not equally plentiful in North and South Korea, South Koreans are, on average, nearly five inches taller than North Koreans,[18] even though their gene pools do not substantially differ.[19] One likely reason for this is that no gene, or set of genes, transmits isolated growth instructions that directly result in an individual's height. Rather, DNA operates *only* in the context of its particular environment. Like a musician in a band, each DNA segment *only* and *always* plays its part in concert with the other players—in much the same way as a guitarist changes what she does according to the pace of the drummer, the emotion of the singer, and the charisma of the keyboardist. Similarly, when the exact same DNA segment is operating in a different environmental context, it can generate distinctly different products. It is now clear that nutritional inputs can significantly influence genetic activity, which almost certainly helps explain the height differences found among North and South Koreans. The takeaway message is clear: the now-understood model is not *genes plus environment*, but rather *genes dynamically interacting with the environment*.[20,21]

We Inherit Developmental Resources, Not Traits

Contrary to popular understanding, people *do not inherit traits* from their parents. Rather, we 'inherit' developmental resources that interact to create the person we each become. These developmental resources include DNA *as well as* nongenetic resources like RNAs, proteins, and the physical, social, and cultural environments in which we develop.[22-25] These resources range from cytoplasmic factors in the egg to the language spoken in the home.

If DNA does not, in some direct way, determine the development of traits, why do family members often strongly resemble and sound like one another? How do family members 'pass on' diseases, addictions, habits, or personality characteristics? The answer is that inherited genetic variants *do* influence every aspect of our biological and psychological identities, but those influences are mediated and modulated by all of the environmental inheritances that *constantly* interact with those genetic variants. A genetic determinist might look at siblings who share the same parents and say that their physical similarities and differences are due to a mix of shared and non-shared genes. But what we need to understand is that each individual—starting with traits as elemental as eye color—is a developmental creation of genetic and nongenetic factors that are continuously interacting with one another, and that many of the relevant nongenetic factors normally reflect the broader environment outside of our bodies.

One of the ways we know about the profound influences of environments on the formation of traits is through experiments where environmental conditions are altered. For example, when newborn rats developed for 16 days in microgravity aboard the space shuttle Columbia, they matured into adults that exhibited behavioral evidence of altered vestibular development.[26] Similarly, altering the developmental environment can change what were thought to be the instinctive behaviors of winged-queen fire ants[27] (see Blumberg, Development evolving: the origins and meanings of instinct, *WIREs Cogn Sci,* also in the collection How We Develop).

Conclusion: A True Measure of the 'Inheritability' of Complex Traits Has Never Been Developed

Although it would, of course, be useful to have a measure of the biological inheritability of complex traits, scientists have never been able to develop such a measure. The process of trait development is so dynamic and multi-causal that in natural contexts where environmental factors are not carefully controlled, accurate predictions about inheritability are effectively impossible. (This stands in contrast to the utility of heritability statistics in experimental and other artificial contexts, in which predictions about a descendant's phenotypes can sometimes be accurate because relevant environmental factors can be precisely controlled; see, for example, Gitonga et al.[28])

If a study reveals a trait to be somewhat heritable, we *can* conclude that genetic factors are associated with the trait. But this turns out to be of little value, because we already know that genetic factors influence *all* of our characteristics! At the same time, all of our characteristics are *also* influenced by nongenetic factors that *interact* with the relevant genetic factors, rendering heritability statistics nonsensical in most circumstances (Fig. 15.4).

Heritability statistics do remain useful in some limited circumstances, including selective breeding programs in which developmental environments can be strictly controlled. But in environments that are not controlled, these statistics do not tell us much. In light of this, numerous theorists have concluded that 'the term "heritability," which carries a strong conviction or connotation of something "[in]heritable" in the everyday sense, is no longer suitable for use in *human genetics,* and its use should be discontinued.'[30] Reviewing the evidence, we come to the same conclusion. Continued use of the term with respect to human traits spreads the demonstrably false notion that genes have some direct and isolated influence on traits. Instead, scientists need to help the public understand that all complex traits are a consequence of developmental processes. Without such an understanding, we are at risk of underestimating the extent to which environmental manipulations can have profoundly positive effects on development.[31,32] Thus, the way 'heritability' is used in most discussions of human phenotypes not only perpetuates false ideas; it also blinds us to steps we might otherwise take to improve the human condition.

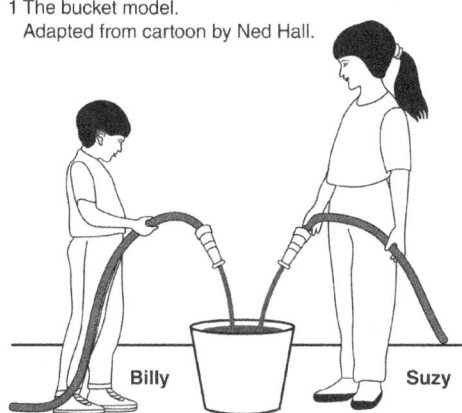
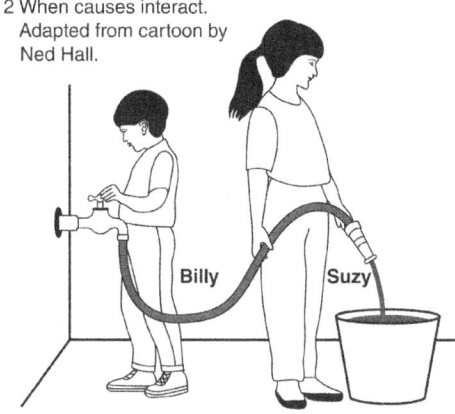

1 The bucket model. Adapted from cartoon by Ned Hall.

Here is a bucket: Billy fills it with 40L of water; then Suzy fills it with 60L of water. So, 40% of the water in the bucket is due to Billy, 60% to Suzy.

2 When causes interact. Adapted from cartoon by Ned Hall.

But suppose instead that what happened was this: Suzy brought a hose to the bucket, and then Billy turned the tap on. Now how much of the water is due to Billy and how much to Suzy?
    Answer: The question no longer makes any sense.

Fig 15.4. An illustration of why it makes little sense to attempt to quantify the relative importance of two different factors that *interact with one another to* produce an outcome. Because genetic and nongenetic factors interact with one another to produce phenotypes, it is not possible to accurately assess the relative importance of their contributions to the phenotypes they produce. Source: Keller.[29]

Further Reading

Block N. How heritability misleads about race. *Cognition* 1995, 56:99–128.

Lewontin RC, Rose S, Kamin LJ. *Not in Our Genes*. Pantheon: New York; 1984.

Michel GF, Moore CL. *Developmental Psychobiology: An Interdisciplinary Science*. Cambridge, MA: MIT; 1995.

Moore DS. Big B, little b: Myth #1 is that Mendelian genes actually exist. In: Krimsky S, ed. Genetic Explanations: Sense and Nonsense. Cambridge, MA: Harvard University Press; 2013.

Ronca AE, Alberts J. R. Effects of prenatal spaceflight on vestibular responses in neonatal rats. *J Appl Physiol* 2000, 89:2318–2324.

References

1. Lush JL. *Animal Breeding Plans*. Ames, IA: Iowa State College Press; 1937.
2. Feldman MW. Heritability: Some theoretical ambiguities. In: Keller E, Lloyd E, eds. *Keywords in Evolutionary Biology*. Cambridge, MA: Harvard University Press; 1992.
3. Sturm RA, Frudakis TN. Eye colour: portals into pigmentation genes and ancestry. *Trends Genet* 2004, 20:327–332.
4. Wade N. The twists and turns of history, and of DNA. *New York Times* 2006; 12.
5. Barr R. 'Born criminals' may be just that: Study cites inherited traits. *The Los Angeles Times* 1985.
6. Cookson C. Cheating genes play large role in female infidelity. *The Financial Times* 2004; 24.
7. Ridley M. *Nature via Nurture: Genes, Experience and What Makes Us Human*. New York: Harper Collins; 2003.
8. Lewontin RC. Annotation: the analysis of variance and the analysis of causes. *Am J Hum Genet* 1974, 26:400–411.
9. Lewontin R. Race and intelligence. *Bull At Sci* 1970, 26:2–8.
10. Melnick M, Myrianthopoulos NC, Christian JC. The effects of chorion type on variation in IQ in the NCPP twin population. *Am J Hum Genet* 1978, 30:425–433.
11. Sokol DK, Moore CA, Rose RJ, Williams CJ, Reed T, Christian JC. Intrapair differences in personality and cognitive ability among young monozygotic

twins distinguished by chorion type. *Behav Genet* 1995, 25:457-466.
12. Fosse R, Joseph J, Richardson K. A critical assessment of the equal-environment assumption of the twin method for schizophrenia. *Front Psychiatry* 2015, 6:62.
13. Hettema JM, Neale MC, Kendler KS. Physical similarity and the equal-environment assumption in twin studies of psychiatric disorders. *Behav Genet* 1995, 25:327-335.
14. Joseph J. *The Trouble with Twin Studies: A Reassessment of Twin Research in the Social and Behavioral Sciences*. New York: Routledge; 2015.
15. Richardson K, Norgate S. The equal environments assumption of classical twin studies may not hold. *Br J Educ Psychol* 2005, 75:339-350.
16. Turkheimer E, Haley A, Waldron M, D'Onofrio B, Gottesman II. Socioeconomic status modifies heritability of IQ in young children. *Psychol Sci* 2003, 14:623-628.
17. Bateson P. Behavioral development and Darwinian Evolution. In: Oyama S, Griffiths P, Gray R, eds. *Cycles of Contingency: Developmental Systems and Evolution*. Cambridge, MA: MIT Press; 2001.
18. Schwekendiek D. Height and weight differences between North and South Korea. *J Biosoc Sci* 2009, 41:51-55.
19. Johnson W, Turkheimer E, Gottesman II, Bouchard TJ Jr. Beyond heritability: twin studies in behavioral research. *Curr Dir Psychol Sci* 2010, 18:217-220.
20. Moore DS. A very little bit of knowledge: re-evaluating the meaning of the heritability of IQ. *Hum Dev* 2006, 49:347-353.
21. Moore DS. Behavioral genetics, genetics, & epigenetics. In: Zelazo PD, ed. *Oxford Handbook of Developmental Psychology*. New York: Oxford University Press; 2013.
22. Griffiths PE, Gray RD. Developmental systems and evolutionary explanation. *J Philos* 1994, 91:277-304.
23. Lickliter R, Honeycutt H. Developmental dynamics: toward a biologically plausible evolutionary psychology. *Psychol Bull* 2003, 129:819-835.
24. Odling-Smee FJ, Laland KN, Feldman MW. *Niche Construction: The Neglected Process in Evolution*. Princeton, NJ: Princeton University Press; 2013.
25. West MJ, King AP. Settling nature and nurture into an ontogenetic niche. *Dev Psychobiol* 1987, 20:549-562.
26. Walton KD, Harding S, Anschel D, Harris YT, Llinas R. The effects of microgravity on the development of surface righting in rats. *J Physiol* 2005, 565:593-608.
27. Keller L, Ross KG. Phenotypic plasticity and "cultural transmission" of alternative social organization in the fire ant *Solenopsis invicta*. *Behav Ecol Sociobiol* 1993, 33:121-129.
28. Gitonga VW, Koning-Boucoiran CF, Verlinden K, Dolstra O, Visser RG, Maliepaard C, Krens FA. Genetic variation, heritability and genotype by environment interaction of morphological traits in a tetraploid rose population. *BMC Genet* 2014, 15:146.
29. Keller E. *The Mirage of a Space Between Nature and Nurture*. Durham, NC: Duke University Press; 2010.
30. Guo SW. Gene-environment interaction and the mapping of complex traits: some statistical models and their implications. *Hum Hered* 2000, 50: 286-303.
31. Moore DS. *The Dependent Gene: The Fallacy of "nature vs. nurture"*. New York: Times Books/Henry Holt Co; 2002.
32. Moore DS. *The Developing Genome: An Introduction to Behavioral Epigenetics*. New York: Oxford University Press; 2015.

16.
THE 1990 "MINNESOTA STUDY OF TWINS REARED-APART" IQ STUDY
Ripe for Retraction?
Jay Joseph*

Reared-apart (separated) identical twins have great appeal. Though exceedingly rare, many people see them as providing the ultimate Rosetta Stone-type method of teasing apart the potential roles of nature and nurture as causes of human behavioral variation. They share a genetic identity while—in theory—growing up and living under entirely different environmental conditions. The two main ways they have come to our attention have been anecdotal stories of individual twin pairs said to share "spooky" and "eerie" behavioral similarities, and a small handful of scientific studies based on a sample of twins. Here, I focus on the latter.

The "Minnesota Study of Twins Reared Apart" (MISTRA) is one of the most famous and widely cited studies in the behavioral sciences. The study began in 1979 and ended in 2000. Many academic publications based on the MISTRA data have appeared since 2000. The MISTRA IQ study was key, due to cognitive ability's long-standing central role in the "nature-nurture debate" and the fields of psychology and behavioral genetics. The study was published in a 1990 edition of *Science*, one of the world's top scientific journals (Bouchard et al., 1990). Although the researchers assessed other psychological characteristics in that article, I refer to it here as the "IQ study" because this was the main MISTRA IQ publication.

The 1990 MISTRA sample consisted of 56 reared-apart MZ twin pairs (monozygotic, identical; 100% genetic similarity) and 30 reared-apart DZ twin pairs (dizygotic, fraternal; average 50% genetic similarity). The three MISTRA cognitive ability (IQ) measures were the "Wechsler Adult Intelligence Scale" (WAIS), the "Raven's Progressive Matrices/Mill-Hill Vocabulary Scale composite," and the "First Principal Component of Special Mental Abilities" (FPC). Reared-apart MZ twins are known as "MZA" pairs; reared-apart DZ twins are known as "DZA" pairs.

Study initiator psychologist Thomas J. Bouchard, Jr., along with David Lykken, Matthew McGue, Nancy Segal, and Auke Tellegen concluded in the 1990 *Science* article that their twin study results showed that "IQ is strongly affected by genetic factors." They estimated the heritability of IQ at "about 70%." (The use of heritability estimates in behavioral research, as well as IQ testing itself, have been disputed for decades.) MISTRA IQ heritability estimates have been cited ever since in psychology textbooks, review publications, media reports, in the social media, and in books such as the controversial 1994 *The Bell Curve* (Herrnstein & Murray, 1994; see Chapter 23 of the present volume). The *Science* article has been cited over 2,400 times since 1990 (about 72 citations per year).

* Private Practice.

My 2022 Analysis and Bouchard's 2023 Response

I have been writing about the MISTRA and other "twins reared apart" (TRA) studies for many years. In 2022, I published a newly formulated critical analysis in the journal *Human Development* (Joseph, 2022b). I concluded that contrary to most of what has been written about it, the "MISTRA IQ study failed to discover evidence that genetic factors influence IQ scores and cognitive ability across the studied population," and that the study was a replication-crisis exemplar of flawed science. I review and elaborate on these points below. The *replication crisis* has called into question the integrity of research in the behavioral sciences and science as a whole, and has been described in psychology as follows:

> The replication crisis in psychology refers to concerns about the credibility of findings in psychological science. The term, which originated in the early 2010s, denotes that findings in behavioral science often cannot be replicated: Researchers do not obtain results comparable to the original, peer-reviewed study when repeating that study using similar procedures. For this reason, many scientists question the accuracy of published findings and now call for increased scrutiny of research practices in psychology. (Psychology Today, n.d.)

In 2023, the now-retired Bouchard published a response to my article in the journal *Twin Research and Human Genetics* (Bouchard, 2023), where he claimed to have refuted my central arguments. In the present review, I respond directly to Bouchard's 2023 article and examine the question of whether the 1990 MISTRA IQ study should be added to the growing list of retracted scientific research publications. I limit my response to the most important areas of contention.

Although Bouchard concluded in 2023 that the MISTRA findings are valid in part because they should be evaluated in the context of other studies (more on this point later), he did not cite any discoveries of genes for IQ or cognitive ability at the molecular genetic level. Bouchard had recognized in 2014 that the results of gene searches beginning in the early 1990s "have been dismal in comparison with expectation" (Bouchard, 2014), and his failure to claim IQ gene associations or discoveries in 2023 suggests that he continued to hold this view.

The key points I raised in my 2022 article were as follows:

1. "Twins reared apart" (TRA) studies contain numerous non-genetic similarity-producing biases that critics have documented for over 50 years, which call into question claims that environmental confounds are minor or absent in TRA studies.
2. The MISTRA design reproduced most of these biases, including that its volunteer twin participants were not separated at birth and randomly assigned to available adoptive homes, and that most pairs grew up knowing they had a twin sibling and having had contact with each other.
3. Most pairs found in TRA studies, including the MISTRA pairs, were only *partially* reared apart.
4. DZAs (DZ or fraternal twins reared apart) were the official MISTRA control group. The MISTRA was the first TRA study to recruit DZAs as controls.
5. The MISTRA researchers failed to publish the DZA control group IQ correlations in their 1990 *Science* article, and the full-sample DZA IQ correlations remain unpublished to this day.
6. After Bouchard and colleagues omitted and bypassed their 1990 DZA control group correlations, they used their MZA (MZ or identical twins reared apart) IQ correlations alone to estimate heritability, based on their assumption that the MZA correlation "directly estimates heritability."

7. The assumption that the MZA IQ correlation directly estimates heritability is false. Even perfectly separated MZ twin pairs share many similarity-producing environmental and "cohort" influences in common. The researchers, on the other hand, assumed that similarity-producing environmental influences do not exist. They maintained this assumption by deciding to count most environmental influences as genetic influences.
8. Near-full-sample DZA IQ correlations published much later, in 2007 and 2012, show that the MISTRA MZA group and DZA group IQ correlations do not differ at a statistically significant level (see Table 16.2). This finding, by itself, leads to a conclusion that the IQ study found no evidence that hereditary factors influence IQ scores (0% IQ heritability).
9. When TRA study researchers fail to make their unpublished raw data and twins' life histories available for inspection and analysis by qualified independent reviewers, we must evaluate their findings with caution.
10. The MISTRA should be evaluated in the context of science's replication crisis, and the researchers used at least two *p-hacking* methods to arrive at a finding of above-zero IQ heritability. Strong genetic confirmation biases influenced their interpretations of the data.

A Closer Look at Bouchard's 2023 Response

Design of the Study

In my article, I quoted psychologist Raymond Fancher's (1985) description of a "definitive" or "ideal" TRA study, which would be based on randomly assigned and completely separated twin pairs representative of the total population of reared-apart twins. I concluded, "the MISTRA did not come close to meeting this standard." Bouchard responded in 2023 by explaining the difference between a true or "planned experiment" and a "natural experiment" such as the MISTRA:

> We did not conduct a true experiment. We gathered a sample of convenience and justified our conclusions on the grounds that the sample was reasonable for our purposes. The same is true for all 'natural experiments.'

I'm not opposed to natural experiments. However, I pointed out that because researchers conducting them are unable to design, control, observe, or manipulate environmental conditions and other variables, they must make *assumptions* about these conditions and variables, and these assumptions must be valid. Critics argue that many TRA study assumptions are not valid, and I quoted McGue and Bouchard's own (1989) recognition that "several" MISTRA model-fitting assumptions "are likely not to hold for cognitive abilities." Bouchard did not mention or correct this little-known 1989 statement.

Bouchard mistakenly said that Fancher described a "perfect" study, which set up his later comment that "no study is perfect, including MISTRA." Everyone understands that studies of human behavior are not perfect. Still, I quoted Fancher to show that the MISTRA and other TRA studies did not meet minimum design requirements.

How "Separated" Were the MISTRA MZA Pairs?

Five IQ studies based on supposedly "reared-apart" MZ twins (MZAs) have been published. British psychologist Cyril Burt's TRA IQ publications were discredited in the 1970s on suspicions of fraud, now established "beyond a reasonable doubt" (Tucker, 1997), and Burt's publications are no longer part of the TRA IQ literature. The authors of the first three studies, Horatio Newman and colleagues (1937), James Shields (1962), and Niels Juel-Nielsen (1965/1980) provided detailed case history information for most of the pairs they studied. The MISTRA and the authors of the final study, the Swedish Adoption/Twin Study on Aging (SATSA), produced only a few case histories. These case histories consisted mainly of cherry-picked MISTRA pairs, such as the "Jim Twins," that have been reported repeatedly in the media and MISTRA publications for decades. (For more on the genetically misleading aspects of these stories, see my review of the movie *Three Identical Strangers*, Joseph, 2019.)

I described TRA studies and their problems in detail in my 2015 book *The Trouble with Twin Studies* (Joseph, 2015). In my 2022 article, I showed that based on my analysis of the Newman, Shields, and Juel-Nielsen case histories:

- In 25/75 (33%) of the pairs, twins were separated at 12 months of age or later.
- In 56/75 (75%) of the pairs, twins had contact with each other while growing up.
- In 42/75 (56%) of the pairs, one or both twins were placed with a family member.
- In 17/75 (23%) of the pairs, twins lived together for at least 12 months after separation, or grew up next door to each other.

Bouchard didn't dispute these calculations or challenge my conclusion that "by default and until proven otherwise we must assume that the MISTRA MZA pairs were no more 'reared apart' than were…the partially reared-apart MZA samples found in the original three TRA studies." Like the earlier investigations, his was a study of twins only *partially* reared apart.

Failed Attempt to Invalidate Earlier Critics of TRA Research

As he has done since the 1980s, Bouchard spent about a quarter of his 2023 article attempting to discredit critics of the earlier studies, such as psychologist Leon Kamin in his pioneering 1974 book *The Science and Politics of I.Q.* (Kamin, 1974), sociologist Howard Taylor's 1980 *The IQ Game* (Taylor, 1980), and psychologist Susan Farber's 1981 *Identical Twins Reared Apart: A Reanalysis* (Farber, 1981).

These books appeared in the aftermath of psychologist Arthur Jensen's highly publicized promotion of TRA research (Jensen, 1969, 1970) in support of his claims in favor of high within-group IQ heritability, and at least partial genetic causes of mean IQ differences between ethnic groups. Jensen relied heavily on the subsequently discredited Burt data (discussed in my 2018 tribute to Kamin).

In the 1980s (e.g., Bouchard, 1982) and his 2023 article, Bouchard called the works of Kamin, Taylor, and Farber "pseudoanalyses" due to these authors' use of subgroup analyses in support of the importance of environmental influences on MZA behavioral and IQ resemblance. As one example of a subgroup analysis, Taylor showed that MZA pairs in the first three investigations who had been reunited before being studied correlated significantly higher for IQ than MZA pairs who had not been reunited. Bouchard claimed that he refuted these subgroup analyses long ago. Whether he succeeded in doing so is a topic for another article.

Although Bouchard's "walking in the garden of forking paths" data analysis description might have some relevance to the subgroup analyses he criticized, it has no relevance to my 2022 article, where I did not perform or endorse such analyses. In attempting to knock over the subgroup analysis straw man he created, Bouchard tried to link me to a practice I did not engage in or depend on.

Bouchard referred to Kamin, Taylor, and Farber as "discredited sources," and he attempted to paint me with the same brush by alleging that I "depended on" and "repeatedly cited" their "discredited" arguments. These were top-notch analysts whose work remains of great importance. They are not discredited sources just because Bouchard says they are, or for any other reason. He scolded journalist John Horgan (1993) for writing 30 years ago, "Kamin has shown that identical twins supposedly raised apart are often raised by members of their families or by unrelated families in the same neighborhood; some twins had extensive contact with each other while growing up." Was Bouchard implying that Kamin's descriptions as cited by Horgan were untrue, or that these factors do not bias TRA study results? Kamin's descriptions were accurate, as anyone who has read the first three studies' case descriptions can confirm (see Joseph, 2015, Tables 2.1–2.3).

My article briefly summarized the main TRA study problems and biases described by Kamin, Taylor, Farber, and others, most of which also apply to the MISTRA studies. These problems and biases related to the lack of separation described above, how twins were recruited to the studies, how they were studied, the environmental similarities they experienced, and how they were evaluated and reported. Subgroup analyses aside, all these points remain in full force.

Suppression of the MISTRA DZA Control Group Data

According to Bouchard (2023), "Joseph implies that we concealed data gathered from the dizygotic twins reared apart (DZAs)." I not only implied it, I said so directly. I continue to do so here.

The MISTRA was the first TRA study to recruit DZA pairs as controls (Bouchard et al, 1986; Segal, 2012). In my (2022b) article, I argued that Bouchard and colleagues suppressed (omitted and bypassed) their DZA control group data to arrive at desired conclusions. This allegation is serious, and I do not make such allegations lightly. Yet in his 2023 article Bouchard didn't dispute my contentions (1) that DZAs were the official MISTRA control group, (2) that in the 1990 *Science* article he and his colleagues omitted their DZA correlations and sidestepped the MISTRA "model-fitting" procedure, (3) that the *Science* article reported no control group DZA results of any kind, and (4) that the MISTRA full-sample DZA IQ correlations were never published.

My 2022 article's Table 1 (Table 16.1 here) showed the reported 1990 *Science* correlations and noted the *non-reported* DZA control group correlations. The three MZA IQ samples ranged from 42-48 pairs.

As in 1990, Bouchard wrote in 2023 that the 30 DZA pairs in the 1990 IQ study "were not included because the sample was small." He made this decision arbitrarily, presumably after reviewing the data. My article cited examples of *non-IQ* MISTRA studies appearing around the same time, where Bouchard published full-sample DZA correlations based on similar or even smaller DZA sample sizes (see Chapter 23 for examples).

I also showed that in her (2012) book *Born Together—Reared Apart: The Landmark Minnesota Twin Study*, MISTRA researcher and 1990 *Science*

Table 16.1 Reported and Nonreported Correlations in the 1990 MISTRA Science Article

Measure	MZA	MZT	DZA control group
WAIS full-scale IQ	0.69	0.88	not reported
Raven, Mill-Hill composite IQ	0.78	0.76	not reported
First principal component IQ	0.78	"not available"	not reported
Hawaii Special Mental Abilities	0.45	"not available"	not reported
Comprehensive Special Mental Abilities	0.48	"not available"	not reported
Multidimensional Personality Questionnaire	0.50	0.49	not reported
California Psychological Inventory (personality)	0.48	0.49	not reported
Strong Campbell Interest Inventory	0.39	0.48	not reported
Religiosity Scales	0.49	0.51	not reported
MPQ Traditionalism Scale	0.53	0.50	not reported

IQ measures are italicized. Intraclass correlations from Bouchard et al. (1990, p. 226), DZA, DZ twins reared apart; MZA, MZ twins reared apart; MZT, MZ twins reared together; WAIS, Wechsler Adult Intelligence Scale; MPQ, Multidimensional Personality Questionnaire. The 1990 MISTRA *Science* sample consisted of 56 MZA and 30 DZA pairs. The MISTRA-reported MZA IQ correlations were based on fewer pairs, ranging from 42 to 48 pairs. "Not available" status was reported by the researchers. No DZA correlations or results of any kind were reported in the 1990 MISTRA *Science* article.

Table 16.2 Near-full-sample MISTRA IQ Correlations at the Study's End: MZA Versus DZA Twin Pairs

	MZA pairs (experimental group)	DZA pairs (control group)	Probability value
Wechsler (WAIS) IQ correlations	0.62 (74 pairs)	0.50 (52 pairs)	$p = 0.17$ Not statistically significant at the 0.05 level
Raven's Progressive Matrices IQ correlations	0.55 (74 pairs)	0.42 (52 pairs)	$p = 0.18$ Not statistically significant at the 0.05 level

Intraclass correlations. One-tailed probability. The final 2000 MISTRA full sample consisted of 81 MZA and 56 DZA pairs. Based on calculations made at the VassarStats website. DZA, DZ twins reared apart; MZA, MZ twins reared apart; *p*, one-tailed probability; WAIS, Wechsler Adult Intelligence Scale. Sources: Wechsler (WAIS) correlations from Segal (2012, p. 286), based on the number of pairs reported on p. 284; Raven correlations from Johnson et al. (2007, p. 552), based on the number of pairs reported on p. 545. The DZA sample contained 18 opposite-sex pairs (Segal, 2012, p. 42). The DZA group correlation for the MISTRA "First Principal Component of Special Mental Abilities" measure, which was the third of three MISTRA IQ measures, has never been published.

article co-author Nancy Segal revealed that the researchers submitted an early-1980s paper to *Science* with IQ results based on only *twelve* DZA pairs. Bouchard did not dispute or comment upon any of the above examples.

I then provided Table 16.2, which included *near*-full-sample MISTRA DZA correlations published in 2007 and 2012 showing that the MISTRA IQ *r*MZA (the MZA group IQ correlation) and IQ *r*DZA (the DZA group IQ correlation) did not differ at a statistically significant level. If *r*MZA and *r*DZA do not differ significantly, we can safely conclude that non-genetic factors alone raised both IQ correlations above zero (more on this point below). As seen in Table 16.2, MZAs' greater genetic resemblance (100%) did not lead to their greater IQ behavioral resemblance versus DZAs (50% average genetic similarity).

Failure at the "Important First Step"

In *Born Together—Reared Apart*, Segal emphasized that the MZA-DZA comparison is "an important first step" in determining "whether or not" genes influence IQ and other behavioral characteristics. In my article, I quoted Segal:

The simple comparison of the MZ (or MZA) and DZ (or DZA) intraclass correlations *is an important first step* in behavioral-genetic analysis because this demonstrates *whether or not* there is genetic influence on the trait. (emphasis added)

And elsewhere in the book, Segal wrote,

Genetic effects are shown *if* the correlation for MZ or MZA twins exceeds the correlation for DZ or DZA twins. (emphasis added).

The Swedish (SATSA) researchers agreed with Segal. "When MZ correlations are not greater than DZ correlations," they wrote in 1992, "twin similarity may reflect correlated environments rather than genetic similarity" (Pedersen et al., 1992). Segal described a TRA study process in which intraclass "correlations are calculated separately for MZA and DZA twin pairs and compared." Yet in the MISTRA IQ study, MZA and DZA twin correlations *were not* compared. I showed in 2022 (Table 16.2) that the MISTRA IQ study failed at its "important first step," and for this reason alone the study found no evidence that genes influence IQ (0% heritability).

Ironically, Bouchard criticized me for supposedly "walking in the garden of forking paths." Did he notice my 2022 Fig. 16.1, entitled "The MISTRA: Two Paths to Genetic Findings"? There,

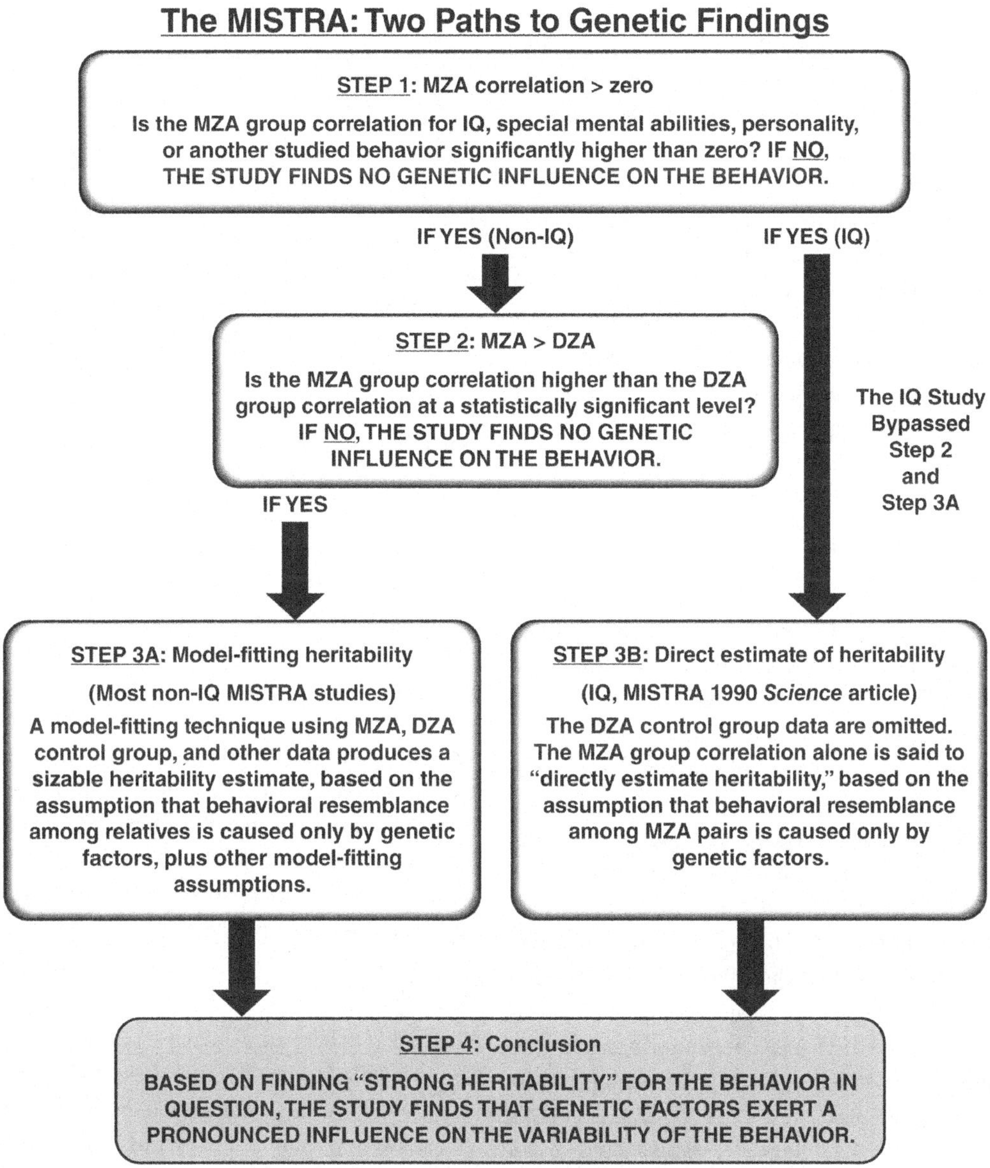

The MISTRA: Two Paths to Genetic Findings

STEP 1: MZA correlation > zero

Is the MZA group correlation for IQ, special mental abilities, personality, or another studied behavior significantly higher than zero? IF <u>NO</u>, THE STUDY FINDS NO GENETIC INFLUENCE ON THE BEHAVIOR.

→ IF YES (Non-IQ) ↓   IF YES (IQ) ↓

STEP 2: MZA > DZA

Is the MZA group correlation higher than the DZA group correlation at a statistically significant level? IF <u>NO</u>, THE STUDY FINDS NO GENETIC INFLUENCE ON THE BEHAVIOR.

IF YES ↓

The IQ Study Bypassed Step 2 and Step 3A

STEP 3A: Model-fitting heritability

(Most non-IQ MISTRA studies)

A model-fitting technique using MZA, DZA control group, and other data produces a sizable heritability estimate, based on the assumption that behavioral resemblance among relatives is caused only by genetic factors, plus other model-fitting assumptions.

STEP 3B: Direct estimate of heritability

(IQ, MISTRA 1990 *Science* article)

The DZA control group data are omitted. The MZA group correlation alone is said to "directly estimate heritability," based on the assumption that behavioral resemblance among MZA pairs is caused only by genetic factors.

STEP 4: Conclusion

BASED ON FINDING "STRONG HERITABILITY" FOR THE BEHAVIOR IN QUESTION, THE STUDY FINDS THAT GENETIC FACTORS EXERT A PRONOUNCED INFLUENCE ON THE VARIABILITY OF THE BEHAVIOR.

Fig. 16.1 Diagram of the two paths used to arrive at conclusions in favor of genetics.

I suggested that he and his colleagues arrived at their own 1990 "forking path" study decision point. I diagrammed and described how after discovering that the planned MISTRA IQ heritability path was blocked by the statistically non-significant "important first step" MZA-DZA comparison, they decided to use an IQ heritability path based only on the MZA results.

"Small Sample" and "Space Limitations"

In addition to the supposedly small size of the DZA group, Bouchard said in the 1990 *Science* article that he could not publish his DZA correlations due to "space limitations." He wrote in 2023 that the purpose of the 1990 IQ study "was to report a constructive replication of previous studies of MZA twins in the brief format provided by *Science* and explain the methodology underlying the study of MZA twins." The 1990 article ran over 6,000 words (six pages) and therefore was not "brief," and the term "constructive replication" did not appear in it.

Bouchard didn't mention or dispute the numbers I presented in Table 16.2 or provide an acceptable explanation for why his control group DZA correlations did not appear in the 1990 *Science* article. Nor did he comment on my quoting Segal's description of the MISTRA "important first step" MZA-DZA comparison, or say that Segal was wrong.

Replicating or Overturning Previous TRA Studies?

For Bouchard, the MISTRA IQ findings "replicated" those of the earlier three TRA studies. *In fact, they overturned them.* As I and others have shown (Joseph, 2015), most MZA pairs in the Newman et al., Shields, and Juel-Nielsen studies did not come close to being "reared apart" based upon most people's understanding of this term. This means *there were no valid studies of reared-apart twins to "replicate."* Moreover, these three studies did not use a DZA control group to assess the meaning of their above-zero MZA IQ correlations. For this reason, Segal (2012) saw the creation of the MISTRA DZA control group as "an important methodological improvement over past projects." Because the MISTRA MZA and DZA IQ correlations did not differ significantly, we can assume that the earlier studies, had they also used a DZA control group, would have found similar negative results.

With no apparent "space limitations," Bouchard's 2023 article provided an excellent opportunity to finally publish the MISTRA 1990 full-sample DZA IQ correlations to try to prove me wrong. Yet 33 years after the study's publication, Bouchard continued to keep his full-sample control group DZA IQ correlations secret. Why is that?

P-Hacking in the MISTRA IQ Study

A major reason why science is currently embroiled in a replication crisis is *p-hacking*, which is the practice of researchers consciously or unconsciously manipulating definitions and data, either openly or behind the scenes, to transform non-findings into "findings" that reach the conventional .05 level of statistical significance (Chambers 2017; Head et al., 2015; Stefan & Schönbrodt, 2023; see also Joseph & Baldwin, 2000). I showed in my 2023 book *Schizophrenia and Genetics: The End of An Illusion* that p-hacking, most of it out in the open, was a major aspect of the most frequently cited schizophrenia adoption studies (Joseph, 2023).

The authors of a 2015 analysis wrote that one aspect of p-hacking "occurs when researchers try out several statistical analyses and/or data eligibility specifications and then selectively report those that produce significant results." They concluded that p-hacking is "widespread throughout science" (Head et al., 2015). Bouchard wrote in 2023, "P-hacking can be defined in several different ways and Joseph provides a few. He does not provide, in a full page of text devoted to the topic (588 words), any examples of p-hacking in MISTRA." In fact, in my 2022 article, I gave two specific examples of apparent MISTRA p-hacking:

> In the area of IQ, the MISTRA researchers appear to have engaged in p-hacking (a) when they failed to publish and assess their control group DZA IQ correlations at the 1990 *Science* study stop point; and (b) when in the same article they selectively reported a method that produced statistically significant results, while failing to report the results of the planned method (MZA-DZA comparison and/or model fitting) which, the evidence suggests, produced statistically nonsignificant results.

The intended genetic natural experiment Bouchard and his colleagues described in 1986, and Segal confirmed in 2012, involved comparing rMZA versus rDZA and including these results in their model-fitting procedures, as they did in most non-IQ MISTRA studies. Three examples are a 1988 MISTRA personality study (Tellegen et al., 1988), a 1990 MISTRA study of religious interests and attitudes (Waller et al., 1990), and a 1991 MISTRA vocational interests study (Moloney et al., 1991).

Though perhaps not unusual in the academic psychology research culture of that era, where what we now call p-hacking practices were seen by some as minor violations similar to jaywalking (Simmons et al., 2018), bypassing the DZA group correlations to arrive at desired conclusions was a classic p-hacking maneuver that transformed negative MISTRA IQ results into positive ones. Bouchard's statement that I did not provide any examples of MISTRA p-hacking allowed him to bypass the two specific examples I provided.

Data Collection Stop Point

According to Bouchard, I implied that he violated the rule of "establishing a stated data collection stop point" even though I knew "full well" this "rule" was "not in place when MISTRA was conducted." If the rule was not in place in 1990, it should have been. Establishing a data-collection stop-point rule helps prevent researchers, usually working behind the scenes using their "hidden flexibility," from "peeking" at their data and stopping data collection before reaching the study's planned stop point to achieve desired and publishable results falling below the .05 level of statistical significance (Chambers, 2017). The rule also prevents researchers from collecting data *past* the stop point for the same reason.

Following the publication of their 1990 *Science* article and up to the study's 2000 endpoint, the MISTRA researchers added 25 MZA and 26 DZA twin pairs, for a final total of 81 MZA and 56 DZA pairs. Bouchard continued to withhold the full-sample DZA IQ correlations from publication and prohibited independent review of the raw data. Most likely, he did so because he was waiting for the samples to become large enough to nudge the MZA-DZA comparison under the critical .05 level of statistical significance. At that point, he could publish the full-sample DZA correlations. However, the evidence suggests that the MZA-DZA comparisons never reached the .05 significance level, which would explain why the full-sample DZA IQ correlations were never published and why Bouchard didn't allow Leon Kamin and others "anywhere near" the MISTRA raw data (Wright, 1997).

Bouchard, in 2023, quoted from his (1998) publication: "The MISTRA IQ correlations have not yet been fully analyzed. We are awaiting completion of the study before conducting a full analysis." Yet he never published this "full analysis" of the IQ data, nor did he claim to have done so.

The Replication Crisis and Academic Psychology

In a well-known 2012 article on research problems in academic psychology, Leslie John and colleagues (2012) helped develop the concept of "questionable research practices," or "QRPs." They found that the "percentage of [psychologist] respondents who have engaged in questionable practices was surprisingly high," and "that some questionable practices may constitute the prevailing research norm."

The MISTRA IQ study p-hacking behaviors I described in my 2022 article appear to match the following four (out of ten) QRPs John and colleagues described in 2012:

- **QRP #1:** "In a paper, failing to report all of a study's dependent measures [the DZA IQ correlations]."
- **QRP #2:** "Deciding whether to collect more data after looking to see whether the results were significant."
- **QRP #6:** "In a paper, selectively reporting studies that 'worked.'"
- **QRP #7:** "Deciding whether to exclude data after looking at the impact of doing so on the results."

The replication crisis has its roots in Bouchard's field of academic psychology, where shoddy, p-hacked studies based on multiple QRPs, blatantly false assumptions, researcher confirmation biases, and financial conflicts of interest were overlooked, promoted, endorsed in textbooks, and even celebrated for decades. Although many researchers did not engage in such practices and produced sound research, in some cases flawed research helped psychologists build their careers, achieve fame, and attain expert status (see Le Texier, 2019; Romm, 2015).

We have the case of the famous psychologist and IQ hereditarian Hans Eysenck. The author of a 2020 *Science* article wrote of the discovery that some of Eysenck's publications contained "suspected data manipulation" favorable to the interests of the tobacco industry that funded him (O'Grady, 2020). Eysenck was the author or co-author of over 80 books and over 1,000 scientific papers. Prior to his death in 1997, he was the most cited living psychologist and the third most cited psychologist of all time, behind Sigmund Freud and Jean Piaget (Rushton, 2001). As the *Science* article reported, the scandal has "pushed" this "psychology hero off his pedestal."

Previously, the Association for Psychological Science (APS) had given Eysenck its 1994 "William James Fellow Award" in "recognition of a lifetime of distinguished contribution to psychological science" (Association for Psychological Science, 1994). The APS statement said, approvingly, that Eysenck "has allied himself with unpopular positions, such as…the selective contribution of cigarettes to cancer based on personality. …Time and again, the accumulation of facts has vindicated him" (more on the Eysenck case and his retracted articles later).

The American Psychological Foundation (APF, affiliated with the American Psychological Association, or APA) presented Bouchard with its 2014 "Gold Medal Award for Lifetime Achievement in the Science of Psychology" (American Psychological Association, 2014). The APF ignored the non-publication of the full-sample DZA IQ data, Bouchard's circumvention of the APA's *Ethical Principles of Psychologists and Code of Conduct* data sharing requirement (see below), and many additional problems I and others have outlined. The APF Award statement described the MISTRA as "groundbreaking and inventive, exciting and controversial," and a "stunning achievement, a body of work in which all psychologists can take pride." In 2018, the APA's Division 14 awarded Bouchard its "Dunnette Prize" for the study of individual differences (Association for Psychological Science, 2018).

Small wonder that U.S. psychology's research/publication process is in a serious and long-overdue crisis. The field's leaders were far more interested in congratulating and awarding each other than looking closely at an awardee's problematic research publications.

Built-In Genetic Bias

P-hacking can also involve technology, including computer programs with built-in genetic biases. In my 2022 article, I showed that in a MISTRA publication (Johnson et al., 2007) the researchers knowingly, openly, and approvingly analyzed their data using "Mx" software containing such biases, including reducing the statistical weight given to unexpectedly high DZA correlations that didn't fit genetic models. "Genetic confirmation bias was built into the MISTRA computer software program," I wrote. Bouchard responded in 2023, accurately, that Mx "is not MISTRA software." That's like being on trial for breaking into someone's car with one's own hammer and seeking acquittal on the grounds that it was actually a neighbor's hammer.

Access to Raw Data Denied

As mentioned, Bouchard has always denied access to critically minded or independent scholars seeking to inspect and analyze the MISTRA raw data, including the DZA IQ correlations and twins' unpublished case histories and information on their degree of separation (see a 1991 letter to *Science* by geneticist Jonathan Beckwith and colleagues). According to the APA's *Ethical Principles*, however,

> After research results are published, psychologists do not withhold the data on which their

conclusions are based from other competent professionals who seek to verify the substantive claims through reanalysis and who intend to use such data only for that purpose, provided that the confidentiality of the participants can be protected and unless legal rights concerning proprietary data preclude their release. (American Psychological Association, 2017)

In 2023, Bouchard repeated his long-standing position that the "MISTRA was required by the University of Minnesota Institutional Review Board to gather informed consent from all participants and guarantee confidentiality." Confidentiality, however, can be achieved by anonymization. Providing full-sample group IQ correlations and anonymous twins' raw IQ scores would not violate twins' confidentiality. In the wake of the Cyril Burt scandal, Jensen himself wrote in 1974 that MZA data "should be published in full….so that quantitative analytical techniques other than those used by the original author can be applied to the data by anyone who wishes" (Jensen, 1974).

I argued in my 2022 article that *regardless of the reason for denying access*, when TRA researchers fail to make their unpublished raw data available, we must evaluate their findings with extreme caution, or even reject their findings outright, because TRA studies are extremely difficult to carry out due to changing policies, practices, and social conditions. Most likely, it will never again be possible to collect large enough MZA and DZA samples to conduct a new TRA replication study. Given the study's significant social, educational, and political policy implications, the MISTRA findings could be disregarded based on the researchers' "data-hoarding" practices alone.

Do Reared-Apart MZ Twins Experience Environmental Similarity Leading to IQ Similarity?

Most critics answer yes to this question. In 2023, Bouchard emphatically answered "NO." Nevertheless, in 1985, he and Segal concluded at the end of their detailed "IQ and Environment" review, "As we have found with most other environmental variables, quality of schooling, amount of schooling, and preschool enrichment do have an influence on IQ" (Bouchard & Segal, 1985).

Related to behavioral similarity in general, MZAs are always the same sex, and society socializes males and females to behave differently. For example, same-sex twins will correlate much higher for the behavioral characteristic "lipstick-wearing, yes or no?" than *opposite*-sex twins. "Genes for" lipstick-wearing behavior have nothing to do with it.

Cohort Influences

Even perfectly separated MZA pairs share many *nonfamilial* environmental influences in common. The *cohort effect* concept refers to similarities in age-matched people's IQ scores, behavior, preferences, beliefs, physical condition, and other characteristics caused not by heredity but by experiencing stages of life at the same time, in the same historical period and cultural milieu. In my 2022 article, I presented a table listing 15 shared cohort influences experienced or potentially experienced by MZA pairs separated near birth and first reunited when studied (rare even in reared-apart twin studies). These important influences included socioeconomic status (SES), gender cohort, age cohort, educational methods, oppression/racism/discrimination/privilege, national/regional/ethnic/religious/political culture, and striking physical similarity.

Conflicting Statements on Environmental (Cohort) Influences

Bouchard spent another quarter of his 2023 article addressing each of the 15 influences I listed and argued that, in most cases, research evidence suggests that these influences are "difference-producing, not similarity-producing." As if growing up experiencing several common cultural influences at the same time causes people to behave *less* alike rather than more alike.

We saw that Bouchard and Segal concluded in 1985 that quality and amount of schooling influence IQ scores. Because MZA pairs usually grow up in similar SES environments, and because the quality and amount of schooling vary depending on an adoptive family's SES, it is reasonable to conclude that MZAs' similar SES rearing environments contributed to their IQ resemblance for non-genetic reasons.

Bouchard's 2023 "shared environmental influences are difference-producing" statement conflicts with others he has made since at least the 1990 *Science* article, where he and his colleagues wrote, "The proximal cause of most psychological variance probably involves learning through experience, just as radical environmentalists have always believed." And looking back in 2016, he wrote,

> Our interpretation of the results of MISTRA was very straightforward. We expected that with regard to psychological traits, monozygotic twins reared apart were similar because their effective environments were similar. (Bouchard, 2016).

It appears that cohort and other shared environmental influences produce similar psychological characteristics after all, "just as radical environmentalists have always believed." The fact that Bouchard and colleagues (1990) decided to count most environmental influences as genetic influences is irrelevant. It was fallacious to do so, just as it was fallacious when leading behavioral geneticist and SATSA co-investigator Robert Plomin did so in his (2018) book *Blueprint: How DNA Makes Us Who We Are* (reviewed by Joseph, 2022a).

Overlooked Natural Experiments

TRA researchers and their supporters overlook countless real and potential natural experiments that are difficult to explain on genetic grounds. As one example, the American Amish (population approximately 370,000) are traditionalist Christians known for simple living, plain dress, and a reluctance to adopt many conveniences of modern technology. If pairs of separated-at-birth male MZAs (born at the same time, as are all twins) who grew up in the same Amish community were reunited for the first time at age 40, they would likely display many similarities in personality, IQ, behavior, religious beliefs and practices, sexual behavior, clothing, facial hair, and so on. The reason? Although they grew up in completely different *families*, they were raised in the same behavior-molding *culture* at the same time. For his 2023 argument to hold, Bouchard would have to conclude that growing up in an Amish community would not cause MZA pairs to behave more similarly than if they had been randomly placed into different homes spanning the entire globe.

The "Flynn Effect"

I mentioned the much-discussed "Flynn effect" in my 2022 article. Moral philosopher/IQ researcher James Flynn showed in the 1980s (Flynn, 1984, 1987, 1999), at a time when the MISTRA was well underway, that IQ scores worldwide had been increasing by about three points per decade (0.30 points per year), including supposedly "g-loaded, culture-fair" tests such as Raven's Progressive Matrices (increasing 0.50 points per year worldwide). IQ test creators periodically reform their tests and make them more difficult in order to maintain a mean of 100 and a "bell-shaped" IQ-score distribution. Flynn documented "massive IQ gains over time [which] revealed that the present generation has a huge IQ advantage over the previous generation." Genetic theories cannot explain these massive IQ gains, but environmental factors such as improved nutrition and healthcare, better teaching methods and increased spending on education, and technological advances can help explain them.

Bouchard responded in 2023 that Flynn believed in "the importance of 'heritability.'"

True enough, but Flynn also wrote in a 1987 *Psychological Bulletin* article that his findings indicate that "psychologists should stop saying that IQ tests measure intelligence. They should say that IQ tests measure abstract problem-solving ability..." (Flynn, 1987). In any case, the most relevant point is that the MISTRA twins were born at the same time and usually grew up in the same country. Therefore, their educational, learning, and cognitive skills development occurred simultaneously in similar Flynn-effect "huge IQ advantage or disadvantage" environments. Theoretically, due solely to the Flynn effect, if the MISTRA MZA pairs had been born two generations apart, the younger twins would have scored about 15 points higher than their much older yet genetically identical co-twins.

Although Flynn generally accepted twin study heritability estimates, the "massive IQ gains over time" he documented provides an additional non-genetic reason why twins will correlate higher for IQ versus randomly selected pairs of individuals spanning the entire age range. We should add the Flynn effect to the long list of environmental factors that confound genetic interpretations of MZA IQ correlations, further undermining the already extremely shaky MISTRA "MZA correlation directly estimates heritability" assumption.

In his (2009) book *What is Intelligence*, Flynn tried to "solve the paradox of how environment could appear so feeble in the twin studies and yet so potent in IQ gains over time." In my view, the "paradox" Flynn described is easily solved. Massive IQ gains over time are real, and twin study IQ genetic (heritability) findings constitute a century-long scientific illusion that Flynn was unable to recognize. Studies using "classical twin method" reared-*together* MZ-DZ comparisons help sustain this illusion. The twin method depends on the assumption that both types of twins grow up experiencing "equal environments." Critics make a compelling case that this crucial assumption is false, meaning that reared-together MZ-DZ comparisons should not be interpreted genetically (see Chapter 23 in this volume).

Were Age and Sex Influences Controlled for in the MISTRA?

In 1984, McGue and Bouchard created a formula to correct for age and sex effects on twins' behavioral correlations. "For most psychological, physiological, and medical variables," they wrote, "there are substantial age and sex effects…. Failing to correct for age and sex effects when they exist will result in overestimation of the twin intraclass correlation" (McGue & Bouchard, 1984). Bouchard thereby recognized that age and sex cohort effects alone constitute "substantial" non-genetic similarity-producing influences on behavior, and that the MISTRA needed to control for these influences. He wrote in 2023, "As Joseph acknowledges, we deal with the issue of sex differences, so that is not an issue." I said they created a "questionable and complicated statistical procedure" to adjust data for age and sex effects. I didn't say their procedure was valid or adequate, and it didn't take the Flynn effect into account. Similarity-producing age and sex influences remain an issue.

The Converging Evidence ("Triangulation") Argument

In my 2022 article I argued that we must evaluate the MISTRA IQ study based on its soundness and logic, and that it cannot be validated by previous TRA studies or other types of behavioral genetic research. "A psychological study, test, or method," I wrote, "must stand or fall on its own logic and soundness, and cannot be validated by supposed 'converging evidence' from other methods." Bouchard strongly disagreed, and argued at several points that the supposed MISTRA findings "triangulate" with or should be evaluated in the context of other types of behavioral genetic research, including animal research. Examples from his 2023 article are as follows:

> Many studies in the behavioral sciences use small samples and, consequently, are not "true experiments," and these problems bedevil all of them. This problem has been solved in

behavior genetics by using replications, multiple corroboration, constructive replications and model fitting.

No study is perfect, including MISTRA, and that is why research must rely on constructive replication and multiple corroboration (triangulation).

Studies of both genetic and environmental influences require a combination of research strategies.

We included the findings for IQ from the three previous MZA studies....We concluded that "general intelligence or IQ is strongly affected by genetic factors." These results were replicated in Sweden [SATSA] two years later... using a design that included both MZ and DZ twins reared apart and together.

Bouchard failed to mention that the SATSA researchers (Pedersen et al., 1992) defined their "reared-apart" twins as follows: "By definition, the twins reared apart were separated by the age of 11." About 44% of the SATSA twins were raised by members of the same family, usually with the mother raising one twin, and the mother's sibling or parent raising the other twin (Pedersen et al., 1984). In a 1993 publication, Bouchard saw his own investigation, which assessed and tested twins in person in Minneapolis, as far superior to the SATSA in several respects: "Their instruments are very inferior to ours....Their zygosity diagnosis is entirely by questionnaire and their data collected by mail" (Bouchard, 1993). By definition, the MISTRA twins had been separated by four years of age, not eleven. Replication means repeating a study's procedures and definitions, matching its assumptions, and observing whether the prior findings are confirmed. Clearly, the MISTRA and SATSA procedures and definitions were different.

"At no point," Bouchard wrote in 2023, "does Joseph refute the converging evidence (e.g., the animal work cited at the beginning of this manuscript) in favor of the hypothesis that genetic factors influence human traits." Bouchard thereby implied that his IQ study's findings don't hold up on their own and depend on findings from other studies, and even other species. Quite an admission.

It wasn't my task nor was it necessary to "refute" the supposed "converging evidence" that I and many other writers have critically examined in publications spanning several decades. My task was to explain in detail why the famous MISTRA IQ study—standing alone—produced no evidence that hereditary factors influence IQ score differences.

I cited the late psychologist Scott Lilienfeld and his colleagues, who observed in 2003 that the "proponents of pseudoscientific claims....typically maintain that scientific claims can be evaluated only within the context of broader claims and therefore cannot be judged in isolation" (Lilienfeld et al., 2003). Bouchard appeared to delight in his gotcha revelation that "citing Lilienfeld is ironic. As a [University of Minnesota] graduate student he gathered psychophysiological measurements from TRAs." If "he were still alive," wrote Bouchard, "he would refute Joseph" on radical environmentalism and pseudoscience.

I already knew that Lilienfeld (2010) was an admirer and former student of Bouchard, and a theme of Bouchard's 2023 article was a kind of all-or-nothing evaluation of his study's critics and supporters alike. In Bouchard's eyes, if Kamin, Taylor, Farber, and Joseph were wrong about something, everything else they said must also be wrong and they become disreputable and refuted "pseudoscientists." On the other hand, because Lilienfeld worked in TRA research and admired Bouchard, no aspects of his writings can be used to argue that Bouchard's "multiple corroboration (triangulation)" converging-evidence defense of the MISTRA work is also invoked by discredited pseudosciences in support of their claims. That's not the intellectual world most people live in.

The MISTRA IQ results can be evaluated only through a careful analysis of the study's data, methods, researcher practices, and assumptions, just as claims by the authors of a phrenology study cannot be validated by grouping their study with other methods that supposedly "converge" to predict mental characteristics.

Appeals to Authority

Bouchard began his 2023 article by quoting Charles Darwin, and ended by quoting Nobel Prize-winning physicist Richard Feynman in support of the idea that my 2022 analysis "is not science—it is, to use Feynman's term, pseudoscience." Let's look at the Feynman passage from a publication that Bouchard called an "astute critique of both physical and social science research," exactly as Bouchard quoted it:

> The first principle is that you must not fool yourself—and you are the easiest person to fool. So you have to be very careful about that. After you've not fooled yourself, it's easy not to fool other scientists… In summary, the idea is to try to give *all* of the information to help others to judge the value of your contribution; not just the information that leads to judgment in one particular direction or another. (emphasis in original)

"The idea is to try to give all of the information to help others to judge the value of your contribution; not just the information that leads to judgment in one particular direction or another." Wise words, Professor Feynman! One can assume this Nobel Prize winner would not have approved of Bouchard's decision to keep his unpublished "information" off limits to independent analysts, and to omit the control group DZA IQ correlations from his 1990 IQ study.

Since Bouchard quoted a heavyweight scientist against me, I call on another heavyweight who had doubts that a TRA study is able to disentangle potential genetic and environmental influences on behavior. In a 2007 passage, leading IQ psychologist researcher/author Robert Sternberg wrote:

> There are methods that can be helpful, such as the method of separated identical twins, but even these methods have their limitations, such as the confounding variable that identical twins tend to be placed in similar, and hence correlated, environments so that effects that may appear to be a result of genetic factors may, in fact, not be a result of such factors. (Sternberg, 2007)

Wise words, Professor Sternberg!

Conclusion: *Science* **Should Retract the MISTRA IQ Study**

Leaving aside environmental confounds, lack of proper separation, secret raw data, reliance on false or questionable assumptions, and other problem areas, I argued in my 2022 *Human Development* article (Joseph, 2022b) that the Minnesota researchers suppressed (omitted and bypassed) their DZA control group IQ correlations to arrive at desired conclusions in favor of substantial IQ heritability (70%). Had they chosen to publish their DZA correlations, they would have arrived at an undesired 0% IQ heritability finding. Nothing Bouchard wrote in his 2023 article leads me to modify this conclusion.

As seen on the Retraction Watch website and elsewhere, academic journals are retracting fraudulent and p-hacked research publications at an increasing rate. Journals have retracted at least 13 articles by Eysenck, the 13th "most eminent psychologist" of the 20th century (American Psychological Association, 2002), and dozens more have been flagged for possible retraction (Retraction Watch, 2020). The journal *Psychological Reports* alone retracted 10 Eysenck publications due to "concerns with the validity of the datasets" (Psychological Reports, 2021). *Science* unwittingly published in 2011 a subsequently retracted fraudulent study by psychologist Diederik Stapel, whose retraction count is now up to 58. *Science* reported on the Stapel fraud case in a 2012 article (Enserink, 2012).

While I was writing the original draft of the article that became this chapter, the President of Stanford University resigned from his post due to charges of research misconduct. As reported in the July 19th, 2023 edition of *Stanford Daily*, "He will also retract or issue lengthy corrections to five widely cited papers for which he was principal author after a Stanford-sponsored investigation found 'manipulation of research data'" (Baker, 2023). Two of the retracted articles appeared in *Science*. We are entering a new and long-overdue era of increased scrutiny of scientific research publications. The replication crisis has dramatically demonstrated that we cannot rely on the peer-review process to prevent the publication of methodologically unsound research based on unsupported assumptions and QRPs.

According to 2019 guidelines published by the Committee on Publication Ethics (COPE), "Retraction is a mechanism for correcting the literature and alerting readers to articles that contain such seriously flawed or erroneous content or data that their findings and conclusions cannot be relied upon." (Committee on Publication Ethics, 2019).

The current *Science* "general policies" for publication, adapted from the COPE guidelines, are found on its website. The following *Science* guidelines attempt to prevent the publication of p-hacked and fraudulent research (Science Magazine, n.d.; all emphasis in original):

> All data used in the analysis must be available to any researcher for purposes of reproducing or extending the analysis.
>
> Authors should present results in a complete and transparent fashion so that stated conclusions are backed by appropriate statistical evaluation and limitations of the study are frankly discussed.
>
> *Rules for stopping data collection*. Did you define rules for stopping data collection in advance (for example, specific intermediary and final endpoints)?
>
> *Data inclusion/exclusion criteria*. What criteria did you apply for inclusion and exclusion of data? Were these criteria established prospectively?
>
> *Research objectives*. State the objectives of the research, clearly distinguishing pre-specified hypotheses from hypotheses suggested after initiation of the data analyses.

The *Science* journals generally require all data underlying the results in published papers to be publicly and immediately available. Post-publication embargoes are not permitted, nor are stipulations for readers to contact the authors.

I doubt these *Science* policies were in force in 1990, but whether Bouchard and colleagues played by the rules as they stood then, long before the replication crisis, has no relevance to whether *Science* should now retract the MISTRA IQ study because its authors violated its policies. Regardless of the researchers' intent and the dysfunctional research/publication culture in which they were required to operate, there can be no statute of limitations for false-conclusion p-hacked studies carrying huge social, educational, and political implications.

The *Science* policies section addresses the issue of study retraction directly:

> In cases of identified errors or irreproducibility of research findings reported in a *Science* journal paper, a retraction is likely if the core conclusions are thereby invalidated. An accumulation of errors identified in a paper may cause the editors to lose confidence in the integrity of the data presentation, and the paper may be retracted.

Because the "accumulation of errors identified" in the virtually irreproducible 1990 MISTRA IQ article resulted in its "core conclusions" being "invalidated," the Editors of *Science* should retract this article. They should take this step not as a form of punishment or condemnation, but to correct the scientific literature.

In a positive review of Kamin's *The Science and Politics of I.Q.* (Bouchard mentioned only negative reviews-), the late evolutionary biologist Richard Lewontin (1976) wrote that Kamin discovered "a pattern of shoddiness, carelessness, miserable experimental design, misreporting, and misrepresentation amounting to a major scandal" in IQ-hereditarian research. And Howard Taylor (1980) described what he called "The IQ game," by which he meant IQ genetic researchers' "use of assumptions that are implausible as well as arbitrary to arrive at some numerical value for the genetic heritability of human IQ scores on the grounds that no heritability calculations could be made without benefit of such assumptions." Unfortunately, not much has changed since then. IQ hereditarians and others continue to play the IQ game, and critics continue to expose this game now aided by replication-crisis-era terms, concepts, and perspectives.

***

References

American Psychological Association. (2002). *Eminent psychologists of the 20th century*. https://www.apa.org/monitor/julaug02/eminent

American Psychological Association. (2014). Gold medal award for life achievement in the science of psychology: Thomas J. Bouchard, Jr. award for distinguished scientific contributions. *American Psychologist*, 69(5), 477–479. https://doi.org/10.1037/a0036917

American Psychological Association. (2017). *Ethical principles of psychologists and code of conduct*. https://www.apa.org/ethics/code

Association for Psychological Science. (1994). *1994 William James Fellow Award: Hans J. Eysenck*. https://www.psychologicalscience.org/members/awards-and-honors/fellow-award/recipent-past-award-winners/hans_j_eysenck

Association for Psychological Science. (2018, April 9). *Bouchard receives Dunnette Prize for Study of Individual Difference*. https://www.psychologicalscience.org/publications/observer/obsonline/bouchard-receives-dunnette-prize-for-study-of-individual-difference.html

Baker, T. (2023, July 19). Stanford president resigns over manipulated research, will retract at least three papers. *Stanford Daily*. https://stanforddaily.com/2023/07/19/stanford-president-resigns-over-manipulated-research-will-retract-at-least-3-papers/#:~:text=Marc%20Tessier%2DLavigne%20failed%20to,lab%20dynamic%2C%20Stanford%20report%20says&text=July%2019%2C%202023%2C%2010%3A,by%20the%20University%20Wednesday%20morning.

Beckwith, J., Geller, L., & Sarkar, S. (1991). IQ and heredity [Letter to the editor]. *Science*, 252(5003), 191. https://www.jstor.org/stable/2875650

Bouchard, T. J., Jr. (1982). [Review of the book *The Intelligence Controversy*, by H. Eysenck versus L. Kamin]. *American Journal of Psychology*, 95(2), 346–349. https://doi.org/10.2307/1422481

Bouchard, T. J., Jr. (1993). Genetic and environmental influences on adult personality: Evaluating the evidence. In J. Hettema & I. J. Deary (Eds.), *Basic issues in personality* (pp. 15–44). Kluwer Academic Publishers.

Bouchard, T. J., Jr. (1998). Genetic and environmental influences on adult intelligence and special mental abilities. *Human Biology*, 70(2), 257–279. https://digitalcommons.wayne.edu/humbiol/vol70/iss2/7

Bouchard, T. J., Jr. (2014). Genes, evolution and intelligence. *Behavior Genetics*, 44, 549–577. https://doi.org/10.1007/s10519-014-9646-x

Bouchard, T. J., Jr. (2016). Genes and behavior: Nature via nurture. In R. J. Sternberg, S. T. Fiske, & D. J. Foss (Eds.), *Scientists making a difference: One hundred eminent behavior and brain scientists talk about their most important contributions* (pp. 73–76). Cambridge University Press.

Bouchard, T. J., Jr. (2023). The garden of forking paths; An evaluation of Joseph's "A Reevaluation of the 1990 'Minnesota Study of Twins Reared Apart' IQ Study." *Twin Research and Human Genetics*, 26(2), 133–142. https://doi.org/10.1017/thg.2023.19

Bouchard, T. J., Jr., Lykken, D. T., McGue, M., Segal, N. L., & Tellegen, A. (1990). Sources of human psychological differences: The Minnesota Study of Twins Reared Apart. *Science*, 250(4978), 223–228. 10.1126/science.2218526

Bouchard, T. J., Jr., Lykken, D. T., Segal, N. L., & Wilcox, K. J. (1986). Development in twins reared apart:

A test of the chronogenetic hypothesis. In A. Demirjian (Ed.), *Human growth: A multidisciplinary review* (pp. 299–310). Taylor & Francis.

Bouchard, T. J., Jr., & Segal, N. (1985). Environment and IQ. In B. Wolman (Ed.), *Handbook of intelligence* (pp. 391–464). Wiley.

Chambers, C. (2017). *The seven deadly sins of psychology: A manifesto for reforming the culture of scientific practice*. Princeton University Press.

Committee on Publication Ethics. (2019). *Retraction Guidelines*. https://doi.org/10.24318/cope.2019.1.4

Enserink, M. (2012, November 28). Final report: Stapel affair points to bigger problems in social psychology. *Science*. https://www.science.org/content/article/final-report-stapel-affair-points-bigger-problems-social-psychology

Fancher, R. E. (1985). *The intelligence men: Makers of the IQ controversy*. W.W. Norton.

Farber, S. L. (1981). *Identical twins reared apart: A reanalysis*. Basic Books.

Flynn, J. R. (1984). The mean IQ of Americans: Massive gains 1932 to 1978. *Psychological Bulletin*, 95(1), 29–51. https://doi.org/10.1037/0033-2909.95.1.29

Flynn, J. R. (1987). Massive IQ gains in 14 nations: What IQ tests really measure. *Psychological Bulletin*, 101(2), 171–191. https://doi.org/10.1037/0033-2909.101.2.171

Flynn, J. R. (1999). Searching for justice: The discovery of IQ gains over time. *American Psychologist*, 54(1), 5–20. https://doi.org/10.1037/0003-066X.54.1.5

Flynn, J. R. (2009, expanded paperback ed.). *What is intelligence?* Cambridge University Press.

Head, M. L., Holman, L., Lanfear, R., Kahn, A. T., & Jennions, M. D. (2015). The extent and consequences of p-hacking in science. *PLoS Biology*, 13(3). https://doi.org/10.1371/journal.pbio.1002106

Herrnstein, R. J., & Murray, C. (1994). *The bell curve: Intelligence and class structure in American life*. The Free Press.

Horgan, J. (1993, June). Eugenics revisited. *Scientific American*, 268(6), 122–131. https://www.scientificamerican.com/article/eugenics-revisited/

Jensen, A. R. (1969). How much can we boost IQ and scholastic achievement? *Harvard Educational Review*, 39(1), 1–123. https://doi.org/10.17763/haer.39.1.l3u15956627424k7

Jensen, A. R. (1970). IQ's of identical twins reared apart. *Behavior Genetics*, 1, 133–148. https://doi.org/10.1007/BF01074659

Jensen, A. R. (1974). Kinship correlations reported by Sir Cyril Burt. *Behavior Genetics*, 4(1), 1–28. https://doi.org/10.1007/BF01066704

John, L. K., Loewenstein, G., & Prelec, D. (2012). Measuring the prevalence of questionable research practices with incentives for truth telling. *Psychological Science*, 23(5), 524–532. https://doi.org/10.1177/0956797611430953

Johnson, W., Bouchard, T. J., Jr., McGue, M., Segal, N. L., Tellegen, A., Keyes, M., & Gottesman, I. I. (2007). Genetic and environmental influences on the Verbal-Perceptual-Image Rotation (VPR) model of the structure of mental abilities in the Minnesota Study of Twins Reared Apart. *Intelligence*, 35(6), 542–562. https://doi.org/10.1016/j.intell.2006.10.003

Joseph, J. (2015). *The trouble with twin studies: A reassessment of twin research in the social and behavioral sciences*. Routledge.

Joseph, J. (2018, April 4). Leon J. Kamin (1927–2017): A nemesis of genetic determinism and scientific racism. *Mad in America*. https://www.madinamerica.com/2018/04/leon-j-kamin-nemesis-genetic-determinism/

Joseph, J. (2019, June 21). Three Identical Strangers and the nature-nurture debate. *Mad in America*. https://www.madinamerica.com/2019/06/three-identical-strangers-nature-nurture-debate/

Joseph, J. (2022a). A blueprint for genetic determinism. [Review of the book *Blueprint: How DNA makes us who we are*, by Robert Plomin]. *American Journal of Psychology*, 135(4), 442–454. https://doi.org/10.5406/19398298.135.4.13

Joseph, J. (2022b). A reevaluation of the 1990 "Minnesota Study of Twins Reared Apart" IQ study. *Human Development*, 66(1), 48–65. https://doi.org/10.1159/000521922

Joseph, J. (2023). *Schizophrenia and genetics: The end of an illusion*. Routledge.

Joseph, J., & Baldwin, S. (2000). Four editorial proposals to improve social sciences research and publication. *International Journal of Risk and Safety in Medicine*, 13(2–3), 109–116. https://psycnet.apa.org/record/2005-06884-004

Juel-Nielsen, N. (1965/1980). *Individual and environment: Monozygotic twins reared apart* (revised ed.). International Universities Press.

Kamin, L. J. (1974). *The science and politics of I.Q.* Erlbaum.

Le Texier, T. (2019). Debunking the Stanford prison experiment. *American Psychologist, 74*(7), 823–839. https://doi.org/10.1037/amp0000401

Lewontin, R. C. (1976). Science and politics: An explosive mix [Review of the book *The science and politics of I.Q.*, by L. Kamin]. *Contemporary Psychology, 21*(2), 97–98. https://resolver.scholarsportal.info/resolve/00107549/v21i0002/97_sapaem.xml

Lilienfeld, S. O. (2010). Can psychology become a science? *Personality and Individual Differences, 49*(4), 281–288. https://doi.org/10.1016/j.paid.2010.01.024

Lilienfeld, S. O., Lynn, S. J., & Lohr, J. M. (2003). Science and pseudoscience in clinical psychology: Initial thoughts, reflections, and considerations. In S. O. Lilienfeld, S. J. Lynn, & J. M. Lohr (Eds.), *Science and pseudoscience in clinical psychology* (pp. 1–14). Guilford.

McGue, M., & Bouchard, T. J., Jr. (1984). Adjustment of twin data for the effects of age and sex. *Behavior Genetics, 14*(4), 325–343. https://doi.org/10.1007/BF01080045

McGue, M., & Bouchard, T. J., Jr. (1989). Genetic and environmental determinants of information processing and special mental abilities: A twin analysis. In R. J. Sternberg (Ed.), *Advances in the psychology of human intelligence* (Vol. 5, pp. 7–45). Erlbaum.

Moloney, D. P., Bouchard, T. J., Jr., & Segal, N. L. (1991). A genetic and environmental analysis of the vocational interests of monozygotic and dizygotic twins reared apart. *Journal of Vocational Behavior, 39*(1), 76–109. https://doi.org/10.1016/0001-8791(91)90005-7

Newman, H. H., Freeman, F. N., & Holzinger, K. J. (1937). *Twins: A study of heredity and environment*. University of Chicago Press.

O'Grady, C. (2020, July 15). Misconduct allegations push psychology hero off his pedestal. *Science*. https://www.sciencemag.org/news/2020/07/misconduct-allegations-push-psychology-hero-his-pedestal

Pedersen, N. L., Friberg, L., Floderus-Myrhed, B., McClearn, G. E., & Plomin, R. (1984). Swedish early separated twins: Identification and characterization. *Acta Geneticae Medicae et Gemellologiae, 33*(2), 243–250. https://doi.org/10.1017/s0001566000007285

Pedersen, N. L., Plomin, R., Nesselroade, J. R., & McClearn, G. E. (1992). A quantitative genetic analysis of cognitive abilities during the second half of the life span. *Psychological Science, 3*(6), 346–353. https://doi.org/10.1111/j.1467-9280.1992.tb00045.x

Plomin, R. (2018). *Blueprint: How DNA makes us who we are*. MIT Press.

Psychological Reports. (2021). Retraction notice. *Psychological Reports, 124*(2), 920–921. https://doi.org/10.1177/0033294120901992

Psychology Today (n.d.). *Replication crisis*. https://www.psychologytoday.com/us/basics/replication-crisis

Retraction Watch. (2020). *Journals retract 13 papers by Hans Eysenck, flag 61, some 60 years old*. https://retractionwatch.com/2020/02/12/journals-retract-three-papers-by-hans-eysenck-flag-18-some-60-years-old/

Romm, C. (2015, January 28). Rethinking one of psychology's most infamous experiments. *The Atlantic*. https://www.theatlantic.com/health/archive/2015/01/rethinking-one-of-psychologys-most-infamous-experiments/384913/

Rushton, J. P. (2001). A scientometric appreciation of H. J. Eysenck's contributions to psychology. *Personality and Individual Differences, 31*(1), 17–39. https://doi.org/10.1016/S0191-8869(00)00235-X

Science Magazine. (n.d.). *Science Journals: Editorial Policies*. https://www.science.org/content/page/science-journals-editorial-policies#authorship

Segal, N. L. (2012). *Born together—Reared apart: The landmark Minnesota twin study*. Harvard University Press.

Shields, J. (1962). *Monozygotic twins brought up apart and brought up together*. Oxford University Press.

Simmons, J. P., Nelson, L. D., & Simonsohn, U. (2018). False-positive citations. *Perspectives on Psychological Science, 13*(2), 255–259. https://doi.org/10.1177/1745691617698146

Stefan, A. M., & Schönbrodt, F. D. (2023). Big little lies: A compendium and simulation of *p*-hacking strategies. *Royal Society Open Science, 10*(220346). https://doi.org/10.1098/rsos.220346

Sternberg, R. J. (2007). Critical thinking in psychology is really critical. In R. J. Sternberg et al., (Eds.), *Critical thinking in psychology* (pp. 289–296). Cambridge University Press.

Taylor, H. F. (1980). *The IQ game: A methodological inquiry into the heredity-environment controversy*. Rutgers University Press.

Tellegen, A., Lykken, D. T., Bouchard, T. J., Jr., Wilcox, K. J., Segal, N. L., & Rich, S. (1988). Personality similarity in twins reared apart and together.

Journal of Personality and Social Psychology, 54(6), 1031–1039. https://doi.org/10.1037/0022-3514.54.6.1031

Tucker, W. H. (1997). Re-reconsidering Burt: Beyond a reasonable doubt. *Journal of the History of the Behavioral Sciences, 33*(2), 145–162. https://doi.org/10.1002/(SICI)1520-6696(199721)33:2<145::AID-JHBS6>3.0.CO;2-S

Waller, N. G., Kojetin, B. A., Bouchard, T. J., Jr., Lykken, D. T., & Tellegen, A. (1990). Genetic and environmental influences on religious interests, attitudes, and values: A study of twins reared apart and together. *Psychological Science, 1*(2), 138–142. https://doi.org/10.1111/j.1467-9280.1990.tb00083.x

Wright, L. (1997). *Twins: And what they tell us about who we are.* John Wiley & Sons.

SECTION VI

SOCIOBIOLOGY

EDITORS' INTRODUCTION

In 1975, E. O. Wilson published a book that announced the "new" scientific discipline of sociobiology. Wilson (1975a) contended that sociobiology would be the "master" synthetic discipline, a field enveloping all of behavioral and social science. The claims made by Wilson (1975a, 1975b, 1980) and others (e.g., Rushton, 1999, 2000) in support of this synthetic role for sociobiology involved ideas pertinent to features of behavior central to human reproduction, parenting, and child caregiving.

The Scientific Goals of Sociobiology

Wilson's (1975a, 1975b, 1980; Lumsden & Wilson, 1981) claims about sociobiology, as well as corresponding assertions made by others (Konner, 1982; MacDonald, 1988), have evoked both approval (e.g., Rushton, 1987, 1988a, 1988b, 1990, 1991a, 1991b, 1995, 1996, 1999) and criticism (e.g., Barlow & Silverberg, 1980; Caplan, 1978; Kitcher, 1985; Lerner, 1992; Lerner & von Eye, 1992, 1993; Lewontin, 1980), even among those who associated themselves with the "new synthesis." For instance, according to Dunbar (1987):

> Wilson created the impression that sociobiology was on the verge of replacing most of the disciplines in the social and behavioral sciences. This, of course, is arrant nonsense since sociobiology does not, of itself, deal with much of the subject matter of these disciplines. (p. 51)

However, Dunbar (1987) went on to defend the importance of Wilson's views for social and behavioral science:

> Wilson was, none the less, right to emphasize the importance of sociobiology in relation to these disciplines. What it in fact does … is to provide a unifying umbrella under which these disciplines can interact on common ground. (p. 51)

It is the process through which this unification is purported to occur that concerns us first. This process pertains to the nature–nurture controversy.

Genetic Determinism as Sociobiology's Key to Interdisciplinary Integration

In Wilson's (1975a) view, the "unifying umbrella" provided by sociobiology is the ubiquitous influence of genes on all facets of individual and social behavior and development. Indeed, Wilson's (1975a, 1975b) views exemplify a key conception of genetic determinism and genetic reductionism. The complexity of all social behavior and development and, indeed, all human culture (Lumsden & Wilson, 1981) can be reduced to a few simple genetic principles.

The core idea is one of gene reproduction. To Wilson (1975a, 1980), Dawkins (1976), and other sociobiologists (Barash, 1977; Freedman, 1979; MacDonald, 1988), the essential, core purpose of human life is *only* to reproduce genes. As Konner (1982, p. 265) put it, "A person is only a gene's way of making another gene." Similarly, Dawkins (1976, p. ix) sees humans as only "survival machines"—robot vehicles blindly programmed to preserve the selfish molecules known as genes."

Simply put, in the wake of such anthropomorphism, all of human development reduces ultimately to gene reproduction and, as such, a human organism "does not live for itself" (Wilson, 1975a, p. 3). Instead, the organism's primary function is not even to produce other organisms per se. Rather, Wilson (1975a) claimed that the primary purpose of an organism is to produce genes. In fact, the organism is seen as only the temporary carrier of genes (Wilson, 1975a, p. 3). Through reproduction, the organism "transports" genes to another organism which, in turn, transports the genes to another organism, and so on.

According to this view, humans have not evolved to produce other people, but only to replicate some of their particular complement of genes. Humans' life spans represent only a relatively short period within the vast temporal span of evolution. During this time they provide a temporary "house," or a transport, for the genes carried within them. Given this machine-like view of the "human as transport," it is clear why Dawkins (1976, p. 21) could see humans merely as "lumbering robots (housing genes that) created us, body and mind; and their preservation is the ultimate rationale for our existence."

On the basis of this core, gene reproduction principle, it is also evident why Dawkins (1976) considered genes to be selfish, although it is less evident why Dawkins decided that it was a conceptually and empirically useful scientific idea to write as if genes possessed intentionality, volition, and cognized, goal-directed purpose. Such attribution reflected the animistic thinking that Piaget (1950, 1970) discussed as prototypic of the preoperational child. Interestingly, in the wake of criticisms of the selfish gene idea and Dawkins' (1976) claims about it, Dawkins (1982) admitted that he was doubtful that there could ever be an experiment that could prove his claims (see too Noble, 2015, regarding the shortcomings of these claims). In any event, to Dawkins, genes are "concerned" with nothing other than self-replication, with reproducing themselves over and over, as many times as possible. Simply, then, in this view humans are really just (seemingly complex) duplicating machines. Their mating rituals, their family relationships, and their cultural institutions are all inventions in the service of gene reproduction.

This core genetic principle of life leads to several other ideas about how genes influence individual behavior and the social world. The more copies of one's genes one can send out into the world, then, in the terms of sociobiology, the more one is increasing one's *inclusive fitness*. In other words, the more copies of itself a genotype can transport into another generation of "gene reproducers" the greater is its inclusive fitness.

This concept derives from the sociobiological view that natural selection is the essential vehicle through which evolutionary change occurs (Dawkins, 1976; Wilson, 1975a). However, not all genes are able to compete equally in the face of the fierce and rigorous challenges imposed by the natural environment; only the most aggressive genotypes will succeed in this struggle for survival. It is here, in the link between sociobiological ideas regarding inclusive fitness and aggression, that the ideas of Konrad Lorenz (1940a, 1940b, 1965, 1966, 1974,1975) play an important role in the shaping of sociobiological thinking.

Sociobiology and Human Aggression

A parallel arises between the views of sociobiologists (Dawkins, 1976; Konner, 1982) and the ideas of Konrad Lorenz. Lorenz (1966) saw human aggression as both inevitable and inherent in the human genome—to the point of providing for innate "militant enthusiasm." To Lorenz, not only do the genes of humans make them aggressive, but also the "highest form" of humans should be the most aggressive—the most militaristic—because such action reflects the presence of genes that have succeeded most in the struggles of natural selection. Similarly, in the sociobiological world of selfish genes, and the robotic "survival machines" that house them, aggression functions to allow genes to enhance their inclusive fitness; "blind" (i.e., unthinking, and machinelike) aggression allows genes to eliminate anything in the environment that interferes with their reproduction.

Thus, to Lorenz (1966) and to sociobiologists (e.g., Konner, 1982), selfish—indeed, ruthless—human aggression is the cornerstone of genes' control over human functioning. The most successful genes—the "best" of evolution, if you will (Lorenz, 1940b)—will be the most successful at ruthless aggression. They will be the most selfishly directed to maximizing their presence in the gene pool and, at the same time, to minimizing the genes of others. Of this blind, ruthless, militant aggression, Dawkins (1976) said:

To a survival machine, another survival machine (which is not its own child or another close relative) is part of its environment, like a rock or a river or a lump of food. It is something that gets in the way, or something that can be exploited. It differs from a rock or a river in one important respect: it is inclined to hit back. This is because it too is a machine which holds its immortal genes in trust for the future, and it too will stop at nothing to preserve them. Natural selection favours genes which control their survival machines in such a way that they make the best use of their environments. This includes making the best use of other survival machines, both of the same and other species. (p. 71)

Similarly, and redolent of Lorenz's (1966) views in his book, *On Aggression*, Konner (1982) explained:

I believe in the existence of innate aggressive tendencies in humans (p. 203) ... if we are ever to control human violence we must first appreciate that humans have a natural, biological tendency to react violently as individuals or as groups, in certain situations. (p. xviii)

According to Lorenz (1966), these innate violent reactions are elicited in a reflex-like manner among either individuals or groups. The reflex occurs when members of an in-group are threatened by members of an out-group. Similarly, as we have noted, Dawkins (1976) contended that selfish genes impel humans to act aggressively against survival machines other than those of their own, close genetic group, and Wilson (1975a; see also Flohr, 1987, p. 199) likewise believes that fear of (and hatred toward) an out-group (xenophobia) is innate in humans. For example, in his book, *On Human Nature*, Wilson (1978) wrote of "hidden biological prime movers," and contended that:

In all periods of life there is (a) ... powerful urge to classify other human beings into two artificially sharpened categories. We seem able to be fully comfortable only when the remainder of humanity can be labeled as members versus nonmembers, kin versus nonkin, friend versus foe. (p. 70)

In short, all genotypes must struggle arduously to include as many copies of themselves as possible in the gene pool. However, within sociobiological thinking all genotypes are not "created equal." That is, whereas all genotypes strive to maximize their inclusive fitness, genotypes differ in what is termed in sociobiology *gametic potential*, (i.e., the potential of a genotype to replicate itself). Differences in gametic potential are associated with the differences that exist between males and females.

Sex Differences in Gametic Potential

Within sociobiology, it is held that men and women differ in their potential for transmitting copies of their genes into the future. For example, as claimed by Konner (1982, p. xviii), "as now seems clearly demonstrated, there are biological reasons why women, like other primate females, have a weaker aggressive tendency than males." According to sociobiologists, because aggression is the key to getting one's genotype reproduced maximally, it follows that women, lacking an aggressive ability sufficient to compete with men, must evolve some other strategy to enhance their inclusive fitness. Sociobiologists contend that the strategy that women use derives completely from the nature of the specialized cells used by women to transmit copies of their genes to future generations.

Genotype copies are contained in gametes, that is, sperms and ova. Both types of gametes function to maximize the inclusive fitness of the genotypes they carry. However, the two types of gametes have a different potential for such reproduction—due to the anatomical and physiological differences between the "lumbering robots"—men and women—housing these gametes. Men, who—it is of more than passing interest to note—were the founders and leading proponents of sociobiology, can generate a large number of genotype copies. Their gametes can be "sent forth to multiply" quite readily—millions can be sent out with each ejaculation. Thus, in the terms of sociobiology, their "gametic potential" is great, given that there is—at least theoretically—a ready, large pool of recipients of their gametes. Freedman (1979) put this idea as follows:

> Since mammalian males produce many more sperm than females produce ova, any given male has far greater potential for producing offspring. He is also more inclined to compete with other males over the "scarce" resource, females. (p. 2)

Simply put, then, any male has a greater potential for enhancing his inclusive fitness than any female, given males' greater gametic potential. Moreover, males must have, in general, a more aggressive genotype than females, because they must compete for access to the female gametes, viewed as a "resource" for the deposit of the males' sperm. Such competition is, of course, highly desirable in the view of sociobiologists, because it ensures that the most aggressive genotypes—those best suited to succeed against the struggles of natural selection—will reproduce most often.

In turn, the genotypes of females impel women to try to reproduce in quite a different way. One might understand the origin and development of this "alternative," female reproductive strategy if one asks these questions:

> Given the vast difference in reproductive potential, and if the point of life is to actualize such potential, is it not reasonable to expect that on the average the male pattern of courtship will differ from the female? Might nature not have arranged it so that men are ready to fecundate almost any female and that selectivity of mates has become the female prerogative?
>
> (Freedman, 1979, p. 12)

In answer to such questions, van den Berghe and Barash (1977) noted:

> Human females, as good mammals who produce a few, costly and therefore precious, offspring, are choosy about picking mates who will contribute maximally to their offspring's fitness, whereas males, whose production of offspring is virtually unlimited, are much less picky. (p. 813)

Gametic Potential and Social and Sexual Development

What does the sexes' different gametic potential imply for understanding male and female social behavior and development? Given the selfishness of genes and the single-minded direction of the duplicating machines housing them, men develop sexual mores dictating the acceptability (if not the appropriateness) of multiple sexual partners. Indeed, van den Berghe and Barash (1977) argued that the different gametic potential of men and women explains

> the widespread occurrence in human societies of polygamy, hypergamy, and double standards of sexual morality. There is another related reason for the sexual double standard in such things as differential valuation of male and female virginity and differential condemnation of adultery: "marital infidelity of the spouse can potentially reduce the fitness of the husband more than that of the wife. Women stand to lose much less if their husbands have children out of wedlock than vice versa (p. 813).... In addition, a woman will, at a maximum, produce some 400 fertile eggs in her lifetime, of which a dozen at most will grow up to reproductive age, while a man produces millions of sperm a day and can theoretically sire hundreds of children. Not surprisingly, females tend to go for quality, and males for quantity." (p. 814)

Moreover, given the large number of offspring they can potentially produce, a male's parental investment in any one offspring is quite small. Unfortunately for the recipients of males' genetic copies—women—their gametic potential is quite different, and so, too, is their parental investment. They can replicate themselves at most every nine months. Even with multiple births, a woman cannot replicate her genes as much in a lifetime as a man can in a short period of time. Therefore, according to sociobiologists, a woman's investment in her offspring is much greater than is a man's. Moreover, sociobiologists also contend that because women cannot reproduce very frequently, women will not be motivated toward frequent copulation with multiple partners.

Instead, it is believed that women need to protect their offspring and assure their survival, and that this need should motivate them to keep their impregnators bound to them. As a consequence, the view of sociobiologists is that females develop monogamous sexual behaviors and a devotion to childbearing and rearing. Van den Berghe and Barash (1977) argued:

> For a woman, the successful raising of a single infant is essentially close to a full-time occupation for a couple of years, and continues to claim much attention and energy for several more years. For a man, it often means only a minor additional burden.... (M)ost societies make no attempt to equalize parental care; they leave women holding the babies. (p. 813)

Lest anyone contend that the different moral, sexual, and social developments of men and women are merely products of socialization, Barash (1977, p. xv) argued that the sex differences in gene reproduction strategy explain "why women have almost universally found themselves relegated to the nursery, while men derive the greatest satisfaction from their jobs." Van den Berghe and Barash (1977, p. 815) further noted that "ethnographic evidence points to different reproductive strategies on the part of men and women, and to a remarkable consistency in the institutionalized means of accommodating these biological predispositions." Van den Berghe and Barash (1977), therefore, concluded:

> Men are selected for engaging in male–male competition over resources appropriate to reproductive success, and women are selected for preferring men who are successful in that endeavor.

Any genetically influenced tendencies in these directions will necessarily be favored by natural selection. (p. 814)

Dawkins (1976) embellished these ideas by contending that women's exploitation by men is biologically determined. He argued that the sexes' behavioral developments are differentiated not only by the different number of sex cells that can be used for genotype reproduction but also by the different size of their respective sex cells:

> The sex cells or "gametes" of males are much smaller and more numerous than the gametes of females ... it is possible to interpret all the other differences between the sexes as stemming from this one basic difference. (p. 152)

and

> Sperms and eggs ... contribute equal numbers of genes, but eggs contribute far more in the way of food reserves: indeed sperms make no contribution at all, and are simply concerned with transporting their genes as fast as possible to an egg .... Female exploitation begins here. (p. 153)

Sex differences in the gametic potential and size of the gametes result not only in female exploitation in general but, in particular, in the legitimization of extramarital sex for males, but not for females, and of the use of violence toward wives who have extramarital sexual relations. To explain these sex differences, Freedman (1979) argued:

> We have to assume that cultural universals reflect those aspects of our species that were evolutionarily derived (evolved). Male promiscuity is universally winked at because there is nothing much we can do about it, and Kinsey's (Kinsey et al., 1953) main findings appear to be descriptions of the species: males must have "frequent outlets" for sex, whether heterosexual or homosexual; whereas many females can go for long periods without copulation or masturbation .... And this difference appears to hinge on the difference in gametic potential that we have been discussing. (p. 19)...As in the gelada baboon, in humans female jealousy is based not on the male's sex act with another woman but on his potential attachment to the latter .... Male jealousy is rather different .... It does not make evolutionary sense for the male to invest in a child not possessing his genes and the murderous jealousy exhibited by a cuckolded male is biologically sensible. Furthermore, the cuckold's retribution can strike either the female or the male cheater ... and most legal systems (perhaps all patrilineal systems) wink at the ensuing violence. (pp. 20–21)

Dawkins (1976) extended across the life span the idea of the biological basis of men's promiscuous sexual interests. He offered both "a possible explanation of the evolution of the menopause in females" (p. 136) and, at the same time, an account of the sociobiological basis of the existence of what are colloquially (and pejoratively) termed "dirty old men":

> The reason why the fertility of males tails off gradually rather than abruptly is probably that males do not invest so much as females in each individual child anyway. Provided he can sire children by young women, it will always pay even a very old man to invest in children rather than in grandchildren. (p. 136)

Conclusions about Genetic Determinism and Human Development

In the theory of sociobiology advanced by Wilson (1975a, 1975b), men—impelled mechanistically by their genes—are oriented to seek sexual relations with as many women as possible, to achieve more and more copies of their genes, and to not be overly devoted to or concerned with any one or any few given "replicates." Women in contrast, are oriented to remain monogamous in order to maximize the probability that their relatively few replicates will survive. In essence, then, men and women are genetically impelled to differ in ways that are consistent with traditional (i.e., stereotypic) sex-role patterns.

As did Freud (1923) and later Erikson (1968), Wilson's (1975a, 1975b) sociobiology in effect holds that "anatomy is destiny" regarding key features of behavioral development—ones involving reproduction, parenting, child caregiving, and sexuality. In other words, Wilson (1975a, 1975b), Dawkins (1976), Freedman (1979), and other sociobiologists (e.g., Barash, 1977; Konner, 1982; MacDonald, 1988; Rushton, 1999, 2000) built a genetic reductionist-based edifice encompassing the very core of all human behavior and development—the reproduction of men and women, the character of the family, and survival of the species. Any notions of nurture or of nature–nurture fusion as sources of key features of human behavior are mere fictions, if genes work in the way that sociobiology requires, that of selfish, goal-directed, and intentional agents. According to sociobiology, after other, more superficial "causes" of human behavior are stripped away (e.g., "causes" involved in an individual's development such as character or moral values), genes provide the ultimate basis for human functioning: the replication of genotypes.

According to this conception, the social world does not coact in a fundamental causal manner with humans' genes, much less act as an alternative source for human development. Instead, according to sociobiologists, the social world—human mores (e.g., regarding sexual permissiveness or monogamy), social institutions (such as marriage and the family), and, indeed, all of human culture—is nothing other than the outcome of strategies laid down by humans' genes for their own replication. In short, sociobiologists have complete faith in the inevitable reducibility of human behavior to the functioning of selfish genes.

Akin to the ideas of Lorenz (1966), this genetic–determinist view has necessarily xenophobic and ruthlessly (if not militantly) selfish implications for society. The faith in genetic determinism and reductionism maintained by sociobiologists is expressed by Dawkins (1976) in his claim that "It can be perfectly proper to speak of 'a gene for behavior so-and-so' even if we haven't the faintest idea of the chemical chain of embryonic causes leading from gene to behaviour" (pp. 65–66). Dawkins (1976) also stated "Be warned that if you wish, as I do, to build a society in which individuals cooperate generously and unselfishly towards a common good, you can expect little help from biological nature" (p. 3).

To what extent is this sociobiological view of human development, and of society, supported by scientific evidence? Asked another way, what scientific evidence do sociobiologists draw on to legitimate their claims, and how adequate is this evidence?

Evaluating Sociobiological Claims

Given Wilson's (1975a, p. 4) original definition of sociobiology as "the systematic study of the biological basis of all social behavior," it may seem surprising, and perhaps contradictory, to learn that Wilson (1980, p. 296) also contended that "contrary to an impression still widespread among social scientists, sociobiology is not the theory that human behavior has a genetic basis." Perhaps, Wilson (1980) was just playing with words. Perhaps, he meant that sociobiology is not a "theory" but only a "perspective," or merely a rather general framework within which to systematically study the biological and, therefore, ultimately, the genetic basis of all social behavior. Whether his statement pertaining to social scientists' mistaken impressions about sociobiology rests on a difference in meaning between the phrases "the theory that ..." and "the systematic study of ...," Wilson's own words show that sociobiology is the study of the role of the connection between genes and human social behavior. He (1980), in fact, used the term *sociobiological theory* to represent this linkage. Wilson (1980) claimed:

> Real sociobiological theory allows no less than three possibilities concerning the present status of human social behavior: (a) During the rapid evolution of the human brain, natural selection exhausted any genetic variability of the species affecting social behavior, so that today virtually all human beings are identical with respect to behavioral potential. In addition, the brain has been "freed" from these genes in the sense that all outcomes are determined by culture. The genes, in other words, merely prescribe the capacity for culture. (b) Genetic variability has been exhausted, as in (a). But the resulting uniform genotype predisposes psychological development toward certain outcomes as opposed to others. In an ethological sense, species-specific human attributes exist and, as in animal repertories, they have a genetic foundation. (c) Genetic variability still exists, and, as in (b), at least some human behavioral attributes have a genetic foundation.
>
> Having identified these alternatives, and stressed the freedom of the discipline of sociobiology from the necessity of any particular outcome, I can now add that the evidence appears to lean heavily in favor of alternative (c). (p. 296)

In the case of each of the given options—(a), (b), and (c)—stress is given to the links among evolution, genetic variability, and human development and society. However, if sociobiologists have spent a good deal of time exploring the first two of the three options, such work has not found its way into the published literature. Hence, Wilson is correct in asserting that to the extent that "evidence" exists in support of any of the three options, it does so in regard to option (c). Yet, support for (c) does not exist because the three options have been repeatedly subjected to comparative scientific analyses. Rather, the preponderance of published sociobiological work—at least insofar as the human literature is concerned—has taken as its "working assumption" option (c). The "evidence" derived from such work constitutes not a test of competing hypotheses but, rather, an attempt to bring empirical observations to bear on a demonstration of a guiding presupposition.

That is, given what are quite well-known facts of genetic variability (e.g., McClearn, 1981), it would be nothing short of preposterous to conduct a scientific investigation predicated on the idea that genetic variability does not exist. As a consequence, we do not believe it plausible that either Wilson or other sociobiologists are not fully aware of this quite basic evidence about the existence of human genetic variability. Consequently, it is equally difficult to envision that any serious scientific attention could be paid by sociobiologists to options (a) or (b). Therefore, these two options cannot be, and, as we indicated, are not treated as, viable counters to (c). Instead, this last conception is the only one actually pursued scientifically by sociobiologists. But, given that no alternatives are really comparatively tested, such pursuit is more a demonstration of how empirical phenomena coincide with a conceptual presupposition than a critical test of theoretical options. How do such demonstrations proceed? Three types of evidence have been invoked.

Comparisons of Humans and Nonhumans: The Concept of Homology

One way in which sociobiologists demonstrate that human social behavior is constrained by evolutionarily shaped genes is to draw parallels between the behaviors of humans and nonhuman animals. If the behaviors of distinct species can be described similarly, it is argued that there must be some evolutionary connection, or continuity, between them. A common evolutionary pathway for a physical structure or a behavioral function in distinct species is termed a *homology*. Simply, then, sociobiologists argue that if the characteristics of two species can be described in a common way, evidence is present of homologous evolution. The positing of such homology is offered as proof that the characteristics in question are controlled, or constrained, by evolutionarily shaped genes.

The use of such "evidence" is exemplified in the writing of Freedman (1979) and Rushton (1999, 2000). For instance, Freedman (1979) attempted to document his views that human males' gametic potential gives rise to sexually promiscuous behavior—in order to increase their opportunities to garner the "scarce resource" of females' ova—whereas human females' gametic potential makes them more monogamous. In support of his idea, Freedman claimed that he found homologies between fruit flies, rhesus monkeys, and South American jungle-dwelling, polygynous humans. Freedman (1979) argued that in all species:

> Females tend to cluster about an average number of young whereas males form a greatly skewed curve, some very successful, many not successful at all. And, since most mammals are polygynous ... this tendency may characterize the entire class Mammalia. (p. 13)

Freedman (1979, p. 14) carried his argument one step further. By again using what he regarded as common behavioral descriptions across species, he attempted to provide an evolutionary and genetic account for inevitable human male promiscuity and also for the genetically preordained urge to seek sexual relations with other females even to the point of forcing oneself onto them (i.e., committing rape) (Freedman, 1979). First, Freedman (1979) cites the work of Grzimek (1972) that:

> In spring, when the gonads are at the peak of their development, there are attempts to "rape" strange females in the mallard and pintail and a few other species (Grzimek, 1972, p. 270). (p. 14)

Second, Freedman made an inference about the "promiscuous, polygynous intentions" of ducks and, finally, drew a conclusion about the insatiable, continuous, and carnal search by human males for females with whom to copulate. Freedman (1979) contended:

> It would appear that if the mallard drake had his way his would be a polygynous species and, in fact, one does occasionally see a consortship of two females and a male. In our own species and our own culture, I am asserting nothing startling when I point out that with sexual maturity, most heterosexual males are in constant search of females, and if inhibited about sexual contact, they fantasize almost continuously and fairly indiscriminately about such contact ... adolescent males in our culture frequently experience life as a nearly continuous erection—spaced by valleys of depression that accompany sexual disappointment. (p. 14)

Are these descriptions, and those by other sociobiologists (Barash, 1977; Wilson, 1975a), of purportedly comparable human and nonhuman social behavior, satisfactory proof of the evolutionary and genetic bases of human behavior? Does apparent descriptive similarity establish evolutionary homology?

The answer to both of these questions is no, for several reasons, not the least of which is the difficulty of accumulating sound scientific evidence of common evolutionary descent when only physical attributes are being considered (Atz, 1970; Gould, 1980). The task is even more problematic in the case of behavioral characteristics, as even very similar behaviors (a) may be manifestations of quite different processes, and/or (b) may serve different functions (Bitterman, 1965, 1975).

In regard to (a), it is a truism that one can describe similar behaviors across even vastly different species. For instance, a conclusion that can be drawn from the work of Bitterman (1965, 1975) is that insects, fish, rats, and humans all "learn;" that is, in members of each of these species, systematic and relatively permanent changes in behavior occur in relation to experience. Nevertheless, the ways in which these species learn—the processes of learning—vary considerably. For example, it would be difficult to contend that thought processes play a part in the learning of insects at any point in their lives. In turn, it would be equally difficult to argue that cognition does not enter into human learning perhaps for anything other than the earliest weeks of life span; however, even in this portion of ontogeny cognition may play a role (Piaget, 1970).

Accordingly, although experience-based changes occur in all animals' adjustment to the environment, this similarity is at best evidence for an analogy, not a homology (Atz, 1970; Schneirla, 1957). In other words, different processes may subserve analogous functions. But to claim that such descriptive analogies are indicative of common evolutionary histories is, at best, naive, and, at worst, poor scholarship. Dunbar (1987) was frank in admitting this limitation in sociobiological scholarship:

> Many of those who were influential in promoting the sociobiological perspective ... (e.g., E. O. Wilson) tend to be unaware of the more sophisticated nature of the behaviour of higher organisms and are apt to regard even advanced mammals simply as scaled-up insects. (p. 53)

In turn, and in regard to the aforementioned point (b), the presence of identical behaviors in different organisms does not constitute proof for even common function or purpose. To illustrate, the reasons that male mallard ducks might force copulation upon a female of their species are certainly distinct from those involved when a human male rapes a human female. Indeed, to label both the male duck's behavior and actions of the human male with the same term (rape) seems to trivialize, through biological reductionism, what is certainly a complex and violent human act, one that may not even be a behavior predicated in any way on sexuality or sexual feelings (Sunday & Tobach, 1985).

Can Freedman (1979), Barash (1977), or other sociobiologists who argue for homology on the basis of such cross-species descriptions, contend that the devaluation of women in many sectors of modern society, and the legitimization of violence as a means of exercising social (and political control), do not enter into the primary causation of forced copulation by human males and/or that they enter as well into the basis of such behaviors in ducks? We think not. Simply stated, the mere portrayal of behaviors in two species as appearing comparable is no proof at all of their common evolutionary heritage. Nor is it any proof at all regarding the extent to which such behaviors are genetically constrained or produced. Indeed, this conclusion seems to have been reached by Wilson (1980) himself. He noted that:

> We cannot rest the hypothesis of genetic constraint in human social behavior on the indirect evidences of homology. (p. 297)

If the sociobiologists' behavioral homologies do not constitute adequate proof for the genetic basis of human social behavior, what then does? Two other types of evidence have been offered, ones pertaining to the concepts of heritability and adaptation. We consider first heritability.

Sociobiology and Heritability Analyses

The myriad conceptual and methodological problems associated with heritability analyses of human behavior have been discussed in Section V of this book. We need not reiterate here the information we discussed about the counterfactual view of genetic functioning, the flawed reasoning, and the methodological shortcomings associated with heritability research. We may simply note that, in relying on heritability as a source of support for their hereditarian views, sociobiologists are, in effect, relying on no evidence at all (Hirsch, 1997).

Nevertheless, Wilson (1980) argued that data from heritability research supported the one of the three possible explanations upon which sociobiology rests. The notion was that genetic variability exists and, as such, that at least some human behavioral attributes have a genetic foundation. Accordingly, Wilson (1980) saw that heritability research not only supported the presence of genetic variability but also that it did so in a manner supporting the hereditarian claims of sociobiology. This seemingly straightforward perspective evokes, in actuality, a thicket of conceptual confusion.

First, sociobiologists do not have to look to behavior genetics to document the clear fact that genetic variability exists. In fact, behavior genetics and its use of research about heritability does not provide proof about the presence of human genetic variability—molecular genetics and population genetics provide this information. Heritability analysis capitalizes on (begins with the acknowledged fact of) genetic variability and then seeks to partition variability into hereditary and environmental sources of variance.

Second, however, sociobiologists' reliance on the findings of heritability research as offering support for their views is completely ill-conceived. Sociobiologists wish to talk about behavioral attributes that are common to a species. The task of the sociobiologist is to show scientifically that such attributes uniformly and unequivocally characterize the subgroups of humans in question (e.g., males and females), and do so because of the possession of evolutionarily based genetic "directives" for genotype reproduction.

Stated simply, sociobiologists wish to demonstrate that some human attributes (i.e., ones common to a given group and dealing with that group's reproductive strategy and, hence, inclusive fitness), have a genetic basis. In other words, sociobiologists want to demonstrate the common, or invariant, inheritance of these attributes but, in relying on evidence from the study of heritability, they are using information that capitalizes not on commonality of inheritance but on its variability!

In essence, then, sociobiologists are trying to claim support for the importance of invariant heredity for human characteristics by pointing to evidence that shows there is variation in heredity. Hirsch (1997) made a similar point. Hirsch noted (1997, p. 210) that "The misleading picture that emerged in *Sociobiology* was that heritability is the very essence of evolution." This depiction of the connection between sociobiology and evolution was flawed, Hirsch (1997) argued, because:

> Wilson was downright irresponsible in his failure to emphasize the inherent contradiction in this picture, namely that the important characters have the lowest heritabilities. In the words of his own source "characters with the lowest heritabilities are those most closely connected with reproductive fitness, while the characters with the highest heritabilities are those that might be judged on biological grounds to be the least important determinants of natural fitness" (Falconer, 1960, p. 167) (p. 210) … Similarly, Collins and colleagues (2000) pointed out that "genetic factors that are highly important in a behavior do not show up in a study of heritability of that behavior because the genetic factor is uniform for all members of a population. Thus, analyzing the variation of a factor within a population does not provide exhaustive information concerning either the genetic or the environmental contributions to the factor. (p. 220)

Accordingly, there are insurmountable logical, conceptual, methodological, and empirical problems involved in sociobiologists' reliance on data derived from heritability research for evidence in support of their version of hereditarian claims. As such, neither this line of "evidence" nor that provided by work associated with the concept of homology can be used by sociobiologists to support their ideas about human behavior and development. There is, then, only one possible line of evidence left for them to use to establish the validity of their ideas: adaptation.

Are Adaptations Everywhere?

A cornerstone of the sociobiological "method" is to offer explanations in the vein of Rudyard Kipling's (1933) "Just-So Stories" of how particular social behaviors, or differences among people in their social status or roles, came to be (Gould, 1980). As recounted by Gould (1980):

> Rudyard Kipling asked how the leopard got its spots, the rhino its wrinkled skin. He called his answers "just-so stories." When evolutionists try to explain form and behavior, they also tell just-so stories—and the agent is natural selection. Virtuosity in invention replaces testability as the criterion for acceptance. (p. 258)

According to Gould (1980), this unacceptable scientific procedure led the biologist von Bertalanffy (1969) to complain:

> If selection is taken as an axiomatic and a priori principle, it is always possible to imagine auxiliary hypotheses—unproved and by nature unprovable—to make it work in any special case .... Some adaptive value ... can always be construed or imagined.... I think the fact that a theory so vague, so insufficiently verifiable and so far from the criteria otherwise applied in "hard" science, has become a dogma, can only be explained on sociological grounds. Society and science have been so steeped in the ideas of mechanism, utilitarianism, and the economic concept of free competition, that instead of God, selection was enthroned as ultimate reality. (p. 11)

According to both Gould (1980) and von Bertalanffy (1969), the key feature of sociobiological "just-so stories" is that these current arrangements in society are adaptations, Adaptations are changes that enhance fitness, that have been shaped by natural selection, over the eons of human evolution to have this function, and that are now represented in the human genotype. Yet, it is the key element in these arguments—the presence of an adaptation, of a change in fitness—that all too often remains a scientifically unverified, post hoc story.

Indeed, as admitted by Dunbar (1987):

> A simple statement that X increases the fitness of those that perform it explains nothing: it is strictly tautologous for improving fitness is what every sociobiological explanation implicitly assumes. What we need to know—and this is the heart of any sociobiological explanation—is: How does it increase fitness?

> It is the transparent failure to answer this question that has left so many sociobiologists open to criticisms of "Just-So" story-telling and unscientific practice. Since we necessarily have to rely on comparative observations rather than experimental manipulation when tackling evolutionary problems, we are particularly exposed to this kind of accusation. The only way to avoid it is to provide as

watertight a case as is possible by showing that proximate problems of survival or reproduction are in fact resolved when individuals behave in a specified way, and that efficient solutions to these problems will result in increased contributions to the species' future gene pool. This will not always be easy, but, unless it can be done, sociobiological explanations will always be open to skeptical doubts, particularly where these doubts are fuelled by political or religious conviction. (p. 50)

Despite these explanatory difficulties, sociobiologists see adaptations—changes in fitness "designed" by (or, actually, "resulting" from) natural selection—as being everywhere. In the view of sociobiologists, these changes in fitness, because they are adaptations, are optimizations. That is, as argued as well by 19th-century social Darwinists (Tobach et al., 1974), natural selection results in genetically based features that are the "time-tested," best possible outcomes of humans' evolutionary history.

According to sociobiologists, then, that which exists is an adaptation: Humans' social behaviors and the niches they occupy in the social hierarchy have been shaped by natural selection to take their present form. As claimed succinctly by Konner (1982, p. 18), "An organism has characteristics, they must have been selected for or they wouldn't be here now."

Given this centrality of the concept of adaptation in sociobiologists' thinking, is there the direct, uniform, and singular pathway that sociobiologists infer from evolution, through natural selection, to adaptation and the present character of people and society? In addition, why is presenting a story—which is a possible scenario of the way natural selection *could have* resulted in a given feature of human behavior—not sufficient to establish scientifically that just such a history transpired?

The Concept of "exaptation"

The work of Gould and Vrba (1982) is quite relevant to these questions. They provided a new term in evolutionary biology in order to clarify some important, but confusing, uses of the term *adaptation*. Gould and Vrba noted that one meaning of adaptation is the shaping of a feature of the organism (e.g., a physical attribute or behavior) by natural selection for the function it now performs. A second meaning is a more static one, and refers to the immediate way in which a physical feature or a behavior enhances the organism's current ability to fit its context. This second meaning does not take into account the historical origin of the feature, but only whether the organism's physical or behavioral characteristics help it to meet the current demands of its environment.

Gould and Vrba (1982) cited Williams (1966) as adhering to the first definition of adaptation. Williams (1966, p. 6) contended that one should speak of adaptation only when one can "attribute the origin and perfection of this design to a long period of selection for effectiveness in this particular role." Bock's (1979) views illustrate the second definition of adaptation. Bock indicated that "an adaptation is ... a feature of the organism ... which interacts operationally with some factor of its environment so that the individual survives and reproduces" (p. 39).

Gould and Vrba (1982) claimed that a confusion, therefore, exists regarding a central concept in evolutionary theory—adaptation. This conflict exists because the single term "adaptation" has been used when, in fact, there are different criteria for the historical basis of a given organism's feature and for its current use. Darwin (1859) himself may have seen this potential confusion:

> The sutures in the skulls of young mammals have been advanced as a beautiful adaptation for aiding parturition, and no doubt they facilitate, or may be indispensable for this act; but as sutures occur in the skulls of young birds and reptiles, which have only to escape from a broken egg, we may infer that

this structure has arisen from the laws of growth, and has been taken advantage of in the parturition of the higher animals. (p. 197)

In other words, although Darwin saw the necessity of unfused sutures in the skulls of young mammals, he was uncertain about labeling the unfused sutures as adaptations. This uncertainty occurred because the unfused sutures were not built by selection to function as they now do in mammals (Gould & Vrba, 1982). But if the unfused sutures are not adaptations, if they were not shaped by natural selection, what are they and where did they come from? Clearly, a new term must be used to rectify the confusion, and Gould and Vrba (1982, p. 6) provided one. They suggested that such characters evolved for other usages (or for no function at all), and were later "coopted" for their current role. They termed such characters *exaptations*. The characters are fit for their current role (i.e., they are *aptus*), but they were not designed by natural selection for this role, therefore, they are not *ad aptus* (pushed toward fitness by natural selection).

To illustrate, it is useful to consider exaptation pertinent to the features of microevolution. This illustration of exaptation indicates how this concept may account for a feature of the genome that, to those committed to an adaptationist program, might appear anomalous. Gould and Vrba (1982) pointed out that:

For a few years after Watson and Crick elucidated the structure of DNA, many evolutionists hoped that the architecture of genetic material might fit all their presuppositions about evolutionary processes. The linear order of nucleotides might be the beads on a string of classical genetics: one gene, one enzyme; one nucleotide substitution, one minute alteration for natural selection to scrutinize. We are now, not even 20 years later, faced with genes in pieces, complex hierarchies of regulation and, above all, vast amounts of repetitive DNA. High repetitive, or satellite, DNA can exist in millions of copies: middle-repetitive DNA, with its tens to hundreds of copies, forms about one quarter of the genome in both *Drosophila* and *Homo*. What is all the repetitive DNA for (if anything)? How did it get there? (p. 101)

Some of the repeated DNA may be conventional adaptations, selected for a role in regulation (e.g., the repeated copies may bring previously separated parts of the genome into new, aptative interrelation). However, there is too much repetitive DNA for such direct adaptation to account for all of it. A second, traditional (i.e., adaptationist program-oriented) basis for the presence of so much repeated DNA has been forwarded. This suggestion is that repetitive DNA exists because it is needed for *future* evolution, that is, it exists to provide for a "flexible future;" for instance, non-used, redundant copies are free to alter because their adaptive product is still being produced by the remaining DNA copies (e.g., Cohen, 1976; Kleckner, 1977). However, this second argument is teleological because it permits future needs to determine present circumstances.

Whereas Gould and Vrba (1982) claimed that future uses are quite significant consequences of repeated DNA, the potential future use cannot be held to empirically determine the prior status of the genome. In turn, the concept of exaptation capitalizes on the idea that repeated DNA may, indeed, have a significant future use but does so without recourse to teleological, "final cause" explanations. In making these contributions, the concept of exaptation furthers understanding of how features of the genome provide a basis for plastic microevolutionary processes. Gould and Vrba (1982) explained that:

Defenders of the second tradition understand how important repetitive DNA is to evolution, but only know the conventional language of adaptation for expressing this conviction. But since utility is a future condition (when the redundant copy assumes a different function or undergoes secondary adaptation for a new role), an impasse in expression develops. To break this impasse, we might suggest that repeated copies are nonapted features available for cooptation later, but not serving any direct function at the moment. When coopted, they will be exaptations in their new role (with secondary adaptive modifications if altered).

What then is the source of these exaptations? According to the first tradition, they arise as true adaptations and later assume their different function. The second tradition, we have argued, must be abandoned. A third possibility has recently been proposed (or rather, better codified after previous hints): perhaps repeated copies can originate for no adaptive reason that concerns the traditional Darwinian level of phenotypic advantage (Doolittle & Sapienza, 1980; Orgel & Crick, 1980). Some DNA elements are transposable: if these can duplicate and move, what is to stop their accumulation as long as they remain invisible to the phenotype (if they become so numerous that they begin to exert energetic constraint upon the phenotype, then natural selection will eliminate them)? Such "selfish DNA" may be playing its own Darwinian game at a gene level, but it represents a true nonaptation at the level of the phenotype. Thus, repeated DNA may often arise as a nonaptation. Such a statement in no ways argues against its vital importance for evolutionary futures. When used to great advantage in that future, these repeated copies are exaptations. (p. 11)

In other words, and crucial for a synthesis of micro- and macro-evolutionary processes, Gould and Vrba (1982) claimed that there exists an "enormous pool" of nonaptations and that this pool must be the source, the "reservoir," of most evolutionary flexibility. They noted that:

We need to recognize the central role of "cooptability for fitness" as the primary evolutionary significance of ubiquitous nonaptation in organisms. In this sense, and at its level of the phenotype, this nonaptive pool is an analog of mutation—a source of raw material for further selection.

Both adaptations and nonaptations, while they may have non-random approximate causes, can be regarded as randomly produced with respect to any potential cooptation by further regimes of selection. Simply put: all exaptations originate randomly with respect to their effects. Together, these two classes of characters, adaptations and nonaptations, provide an enormous pool of variability, at a level higher than mutations, for cooptation as exaptations [and provide for] … the flexibility of phenotypic characters as a primary enhancer of or damper upon future evolutionary change. Flexibility lies in the pool of features available for cooptation (either as adaptations to something else that has ceased to be important in new selective regimes, as adaptations whose original function continues but which may he coopted for an additional role, or as nonaptations always potentially available). The paths of evolution—both the constraints and the opportunities—must be largely set by the site and nature of this pool of potential exaptations. Exaptive possibilities define the internal contribution that organisms make to their own evolutionary future. (pp. 12–13)

In sum, the concept of exaptation, and the limitations it imposes for the notion of adaptation as the sole process by which evolution occurs, present formidable conceptual and empirical problems for sociobiological thinking. The existence of exaptive processes indicates that evolution is considerably more plastic than sociobiology would imply. This plasticity is highlighted by those objections to the adaptationist program that involve the specification of the causal role played by the developing organism, and especially by its dynamic coactions with its context, in influencing the course of evolution.

The Role of the Organism in Its Own Evolution

A clear implication of Gould and Vrba's (1982) revised terminology is that not all instances of fitness are adaptations; that is, not all features of an organism's structure and function that are aptational have this character as a consequence of being shaped by natural selection. Such a possibility, if supported, would serve to weaken what Gould and Lewontin (1979) labeled the "adaptationist program," that is, the position, reflected in the earlier quote by Konner (1982, p. 18), that a feature's current aptational character implies historical shaping by natural selection for that character.

Lewontin (1981) discussed the adaptationist "program" and its conventional use of the concept of adaptation. As do Gould and Vrba (1982), Lewontin (1981) saw problems with this view of adaptation; in essence, he saw the view as deficient because it ignores the active, constructive role the organism plays in its own adaptation. The organism shapes the context to which it adapts, and, hence, there is a reciprocal, multilevel (i.e., fused) relation between organism and context, that is, there are organism ⇔ context relations (e.g., Ford & Lerner, 1992; Gottlieb, 1997; Lerner & Walls, 1999; Magnusson, 1999a, 1999b; Thelen & Smith, 1998). Thus, Lewontin's (1981) criticisms of the conventional use of the concept of adaptation is associated with a view of the organism compatible with relational developmental systems (RDS)-based conceptions of human development. Specifically, Lewontin (1981) noted:

> Organisms ... by their own life activities determine which aspects of the outer world make up their environment. Organisms change the environment by their activities ... they "construct" environments. The problem is that the concept of adaptation has been extended metaphorically from its valid domain of describing individual, short-term, goal-directed behavior to other levels ... it is pure metaphor, ideologically molded by the progressivism and optimalism of the nineteenth century, to describe numbers of chromosomes, patterns of fertility, migrations, and religious institutions as "adaptations." ... It is not simply that some evolutionary process can be described as nonadaptive, but that the entire framework is in question. Whether we look at the fossil record or at living species, we do not see them as "adapting," but as "adapted." But how can that be? How is it that, if evolution is a process of adapting, organisms always seem to be adapted. It may be more illuminating to see organisms as changing and, in the process, as reconstructing the elements of the outer world into a new environment that is sufficient for their survival. (p. 245)

For example, summarizing the literature pertaining to the character of the environment to which organisms adapt, Lewontin and Levins (1978) emphasized that reciprocal processes between organism and environment are involved in human evolution, supporting the view that human functioning is one source of its own evolutionary development. Lewontin and Levins (1978) stated that:

> The activity of the organism sets the stage for its own evolution ... the labor process by which the human ancestors modified natural objects to make them suitable for human use was itself the unique feature of the way of life that directed selection on the hand, larynx, and brain in a positive feedback that transformed the species, its environment, and its mode of interaction with nature. (p. 78)

Consistent with the position of Lewontin (1981) and Lewontin and Levins (1978) regarding the problems with the "adaptationist program," Gould and Vrba (1982) contended that recognition of the potential presence of exaptive features leads one to recognize that previously non-adaptive (note, *not* preadaptive) features may be present and may be coopted for fitness—a recognition that provides a key for plasticity in evolutionary processes and for the role of individuals' own organismic characteristics in their development. Gould and Vrba (1982) indicated:

> Flexibility lies in the pool of features available for cooptation.... The paths of evolution—both the constraints and the opportunities—must be largely set by the size and nature of this pool of potential exaptations. Exaptive possibilities define the "internal" contribution that organisms make to their own evolutionary future. (pp. 12–13)

The concept of exaptation leads to the understanding that the processes involved in evolution are plastic ones, and that plasticity involves organisms' active contributions to their own evolutionary change (e.g., Brandtstädter, 1998, 1999; Gottlieb, 1983, 1997; Lerner, 1982; Lerner & Busch-Rossnagel, 1981; Lerner & Walls, 1999). As such, exaptation is a concept consistent with a key theme in the developmental systems "alternative" to a hereditarian view of the role of biology in human development (e.g., Ford & Lerner, 1992; Gottlieb, 1997; Lerner & Walls, 1999; Magnusson, 1999a, 1999b; Thelen & Smith, 1998). According to this alternative, it is possible to envision how processes exist that contribute to the plasticity of people's functioning, processes that allow people to play a role in the ontogeny—and, through a concept introduced by Gottlieb (1983, 1991, 1997), in the phylogeny as well—of their own flexible characteristics.

The Relational Developmental System and the Role of the Concept of "Behavioral Neophenotypes" in Evolutionary Change

As described by Lewontin (1981), it is possible to view the organism as other than just the host of its evolutionarily provided genes and, as we have emphasized throughout this book, it is likewise possible to view the importance of the organism's activity across ontogeny as more than just the maturationally predetermined unfolding of hereditarily fixed progressions. The key alternative view is one that sees biological and contextual factors as reciprocally coactive. As such, developmental changes are probabilistic in respect to normative outcomes due to variation in the timing of the biological, psychological, and social factors that provide interactive bases of ontogenetic progressions (e.g., Gottlieb, 1970, 1996, 2004; Schneirla, 1957; Tobach, 1981).

As discussed in previous sections of this book, this view has been labeled as probabilistic epigenetic by Gottlieb (1970), and developed by him (Gottlieb, 1976a, 1976b, 1991, 1992, 1997, 2004; Gottlieb et al., 2006) and earlier by Schneirla (1956, 1957) and Tobach and Schneirla (1968). Probabilistic epigenesis constitutes a defining feature of RDS-based theories, and the fusions among levels of organizations within the system that it reflects provides the basis of plasticity in development across the human life span (e.g., Ford & Lerner, 1992; Gottlieb, 1997; Magnusson, 1999a, 1999b; Thelen & Smith, 2006). As we have just noted, Lewontin (1981) indicated what such plasticity in development may mean for altering the course of evolution. In turn, Gottlieb (2002/2014) also provided a quite intriguing discussion of the role of plastic developmental functioning in shaping evolutionary change.

Although biologists such as Garstang (1922), de Beer (1930), and Goldschmidt (1933) previously argued that developmental changes may lead to evolution, they also believed that a genetic change or a mutation was necessary to create the developmental changes. Gottlieb (2002/2014), however, argued for an evolutionary pathway in which ontogenetic development leads to evolutionary change and, quite significantly, where genetic change occurs as a secondary or tertiary result of behavioral changes shaped t by nongenetic changes in species-typical development. Gottlieb's conception draws on a notion introduced by Kuo (1967), of behavioral neophenotype, that is, a behavioral innovation, or ontogenetic novelty, made possible by the plasticity of the organism and its probabilistic, dynamic coactions with its context.

Gottlieb contended that a behavioral neophenotype is likely the first step in an evolutionary sequence that proceeds from behavioral change, to morphological change, to genetic change (see too Jablonka & Lamb, 2005). More specifically, the emergence of a behavioral neophenotype encourages new environmental relationships that, in turn, bring out latent possibilities for morphological and physiological changes. Gottlieb (2002/2014) noted that somatic mutation, cytoplasmic alteration, or change in gene regulation may also take place at this point, however, an alteration of structural genes need not take place in this secondary stage of the process. However, a change in genes or in gene frequency does occur in the third stage, wherein as a consequence of long-term geographic or behavioral isolation (i.e., separate breeding populations), such alteration takes place.

Because of the plasticity that exists in organisms (and especially ones with larger relative brain size such as humans; Gottlieb, 1987), a plasticity textured by the probabilistic, dynamic coactions they have across ontogeny with their context, an evolutionary pathway is created that is inconsistent with the conception of evolutionary change found in evolutionary epistemology and in the associated predetermined–epigenetic view of organism change. Gottlieb's ideas about behavioral neophenotypes have been incorporated by Jablonka and Lamb (2005) as one of the four systemically integrated dimensions of evolution. They note (2005, p. 1) that "many animals transmit information to others by behavioral means, which gives then a third hereditary system," along with genetics, epigenetics, and culture.

Thus, on the basis of both this contribution to evolution by organism development, and the implications of the concept of exaptation, it is possible to conclude that the key features of sociobiological thinking are severely scientifically limited. Evolution processes are not, therefore, just comprised of phylogenetically continuous changes that, by virtue of the antecedent and independent effects of the physical world, shape via natural selection particular cognitive structures, reproductive strategies, or parent–child relations. The particular set of behavioral or social features present in a person, social group, or culture cannot be judged as contributing to or diminishing the survival of the human species by virtue of the adaptationist assertion that the features have or have not been shaped and selected for fitness.

Conclusions about the Presence of Evidence in Support of the Sociobiological View of Human Development

As was the case in regard to the lines of evidence relating to the concept of homology and to the use of heritability research data, the third line of evidence relied on by sociobiologists—an adaptationist storyline to explain what are purported to be genetically based differences in individual and social development—fails. "Just-so stories" (Gould, 1980) about human evolutionary history are used to substitute superficial descriptions for in-depth explanations. Alternative paths to current fitness (or aptation) are excluded from scientific consideration or analysis.

Equally serious problems arise in regard to the other two lines of evidence relied on by sociobiologists—involving the inappropriate postulation of homologies between nonhuman and human animals and the misuse of the concept of heritability, which of course is a fatally flawed idea in regard to understanding human development. Indeed, as noted by Moore (2015a, 2015b), the variance-partitioning approach epitomized by heritability analyses "black boxes" the study of development. The logical and empirical problems of sociobiology reveal the weak scientific basis of this theory. The severity of these problems suggests that sociobiological thinking has little relevance for the understanding of human behavior and development in general, or for individual or group differences in particular.

Nevertheless, the scientific vacuity of sociobiological ideas about human development has not deterred some writers from using these ideas to propose theories about the evolutionary basis of individual and/or group differences in numerous features of human development (e.g., MacDonald, 1988, 1994; Rushton, 1988a, 1988b, 1996, 1999, 2000), for example, sexuality, intelligence, criminality, and parenting. It is important to evaluate an example of this type of work in order to illustrate the quality of the evidence that sociobiologists use to make pronouncements about the hereditary basis of group differences in human development.

To provide this illustration, we focus on the work of arguably the most visible of the hereditarian writers who used sociobiology to explain group differences in human behaviors: J. Philippe Rushton (1999, 2000). We focus on his views about human evolution and the quality of the scientific work he does to support his ideas.

The Work of J. Philippe Rushton

Rushton's (1997) work rested on a split view of the nature–nurture issue. In fact, he not only splits genes from context in his attempts to explain human development but also he saw a split between the people whose work is associated with hereditarian versus developmental systems conceptions, a split that divides—in his view—good from poor scientists. That is, Rushton (1997) noted that:

> Most of those engaged in the serious study of race today do so from either the "hermeneutical" or the "race-realist" perspective. At one extreme, those I have termed "hermeneuticists" approach race as an epiphenomenon, a mere social construction, with political and economic forces as the real causal agents worthy of study. Rather than research race, hermeneuticists research those who do. At the other end of the forum, those I term the "race-realists" view race as a natural phenomenon to be observed, studied, and explained. Alternative and intermediate positions certainly exist, but the most heated debate currently takes place between advocates of the two polar positions. The hermeneutical approach relies on textual, historical, and political analysis; the race-realist approach is empirical and employs a panoply of scientific methodologies, including surveys, psychometrics, and genetics. Because the hermeneutical viewpoint sees inexorable links between theory and practice, its writings are often prescriptive and assume an advocacy position. The race-realist viewpoint is descriptive and typically avoids prescribing policy. To their opposite numbers, hermeneuticists come across as muddled, heated, and politically committed to "antiracism"; the race-realists come across to their opponents as cold, detached, and suspect of hiding a "racist" agenda. (p. 78)

In effect, this instance of Rushton's split conception is actually one of labeling hermeneuticists as "obfuscating politically correct ad hominemists" and seeing race realists as "objective crusaders for scientific and social truth." It may be, however, that this characterization hoists Rushton "on his own petard." To see if it does, we turn to a discussion of the ideas and methods Rushton used to seek and present "truth."

Rushton's Tripartite Theory of Race, Evolution, and Behavior

Rushton (1999, 2000) proposed a tripartite racial view of human evolution, one which purports to show that in regard to characteristics of human functioning linked to successful development (e.g., high intelligence and occupational achievement, good parenting and caregiving skills, and low criminality), the three racial groups he identified (what he terms "Orientals," "Whites," and "Blacks") differ significantly. Although Rushton (1999) never defined the concept of "race," he noted that there are:

> Three major cases: *Orientals* (East Asians, Mongoloids), *Whites* (Europeans, Caucasoids), and *Blacks* (Africans, Negroids). To keep things simple, I will use these common names instead of scientific ones and will not discuss subgroups within the races.
>
> On average, Orientals are slower to mature, less fertile, and less sexually active, have larger brains and higher IQ scores. Blacks are at the opposite end in each of these areas. Whites fall in the middle, often close to the Orientals. (p. 18)

There are numerous, and well-known data sets contradicting Rushton's all-too-facile divisions. For instance, consider the variable that Rushton (1999) considered to be the most clearly linked to the biological and, hence, evolutionary differences between the racial groups he described, that is, reproductive maturation. For example, he noted that "races tend to differ in the age when they reach milestones such as the end of infancy, the start of puberty, adulthood, and old age" (Rushton, 1999, pp. 27–28) and that "Blacks reach sexual maturity sooner than Whites, who in turn mature sooner than Orientals. This is true for things like age at first menstruation, first sexual experience, and first pregnancy" (Rushton, 1999, p. 30). Rushton failed to attend to abundant information which indicates unequivocally that his assertion is simply incorrect.

To illustrate, Rushton (1999) ignored Hiernaux's (1968) data showing that pubertal maturation (i.e., age of menarche, the age of the first menstrual cycle) among Africans can vary from as low as 12.4 years to as high as 18.8 years; as such, the ontogenetic rate of maturation of some Africans is substantially slower than those of many groups of Asians and Europeans studied by Hiernaux (1968). In turn, Tanner (1973, 1991) reported a secular trend wherein the time of pubertal maturation decreased over the course of the 20th century for numerous groups of Europeans, for European Americans, and for Asians (e.g., Japanese). In fact, the latter group showed the most dramatic decrease in time of maturation for all groups studied by Tanner (1991). In describing trends after World War II, Tanner noted that "in improving postwar conditions, there was a decline of some 11 months per decade until 1975, when the trend leveled out to practically zero" (1991, p. 638). Thus, pubertal maturation is a quite plastic phenomenon, responsive to the nutritional and medical resources present in the ecology of developing individuals.

The data reflecting such plasticity directly contradict the tripartite differences specified by Rushton (1999). As such, *either* Rushton is open to criticism for weak and inadequate scholarship as a consequence of his not knowing of data sets that had been quite prominent in the biological and human development literatures for several decades at the time of his writing, *or* he is open to criticism for biased and inadequate scholarship as a consequence of failing to acknowledge that his ideas are convincingly contradicted by strong, countervailing data. In either case, Rushton's (1999, p. 96) self-congratulatory assertion that "I have not ignored any important studies" is simply incorrect. Indeed, Winston (1997a, 1997b) explained that Rushton makes similar "errors" in regard to his claiming support for his tripartite racial theory from data about the brain size of the three "racial groups" Rushton (1997, 1999) described (cf. Peters, 1995a, 1995b; Winston, 1996).

Rushton's Ideas about Different Reproductive Strategies across Race Groups

Despite the inadequate scholarship that characterizes the evidentiary basis for his claims, Rushton (1999) went on to propose that the bases for the reproductive and associated behavioral differences he associated with the three racial groups he discussed lies in the different "reproductive strategies" characterizing them. He described a continuum of reproductive strategies wherein "At one end of this scale are r-strategies that rely on high reproductive rates. At the other end are K-strategies that rely on high levels of parental care" (p. 24).

The different strategies depicted across this continuum are useful in biology to depict the reproductive rates of separate species (that are trying to survive and reproduce in diverse ecological niches (Johanson & Edey, 1981). For instance, a sponge, living and reproducing on the ocean floor, will produce literally thousands of offspring during a given reproductive cycle, and this level of reproduction will increase the probability of a few offspring withstanding the harsh currents and otherwise dangerous ecology of the ocean bottom for a period sufficient for their survival and eventual perpetuation of the species. In turn, given elephants' enormous nutritional needs during their lengthy prenatal gestation period and postnatal years, the probability of offspring survival is enhanced when a small number, most typically one, offspring, is produced during a reproductive cycle.

Thus, the r–K distinction is useful for describing *differences* between species in how their rate of reproduction fits the ecological niche within which they live. However, there is no validity for applying this concept to differences *within* a species in the reproductive rates of different individuals or groups. Yet, this is an error that Rushton (1999) made, and, in fact, admitted that he did! He noted that the r–K "scale is generally used to compare the life histories of different species of animals. I have used it to explain the small but real differences between the human races" (Rushton, 1999, p. 24).

Hence, Rushton (1999) misapplied the r–K distinction in two ways. First, he took a concept that describes differences between species and applied it to differences within a species *without any biological evidence of the validity of such an application*. Nevertheless, in response to the question of whether his r–K concept applied only to differences between species and not to within-species differences, Rushton (1999, p. 103) asserted without any documentation that, "It applies to both."

Second, Rushton used a descriptive concept to explain differences within a species—and his explanation was that, basically, the group he called "Blacks" represent an evolutionarily less-advanced form of organism, in that their reproductive strategy is more closely aligned with more "primitive," r-like organisms. Indeed, Rushton (1999) used his r–K explanation to account for purported differences between "Orientals" and "Whites," who he claimed are more "K-selected" and "Blacks," who he contends are more r-selected," in their investment in their children.

He indicated that, "Highly K-selected men invest time and energy in their children rather than the pursuit of sexual thrills. They are "dads" rather than "cads"" (Rushton, 1999, p. 24). Moreover, Rushton (1999, pp. 35–36) asserted—*without any citation whatsoever to bolster his statements*—that "In Africa, the female-headed family is part of an overall social pattern. It consists of early sexual union and the procreation of children with many partners. It includes fostering children away from home, even for several years, so mothers remain sexually active.... In Black Africa and the Black Caribbean, as in the American underclass ghetto, groups of pre-teens and teenagers are left quite free of adult supervision." The ideas of Belsky, Steinberg, and Draper (1991) and of Ellis, Schlomer, Tilley, and Butler (2012) are redolent of these unfounded ideas presented by Rushton (1999).

Amazingly, Rushton (1999) showed no awareness (e.g., through discussion or even mere citation) of the rich literature pertinent to the African American family (e.g., Demo, Allen, & Fine, 2000; McAdoo, 1977, 1991, 1993a, 1993b, 1995, 1998, 1999; McCubbin et al., 1998). This literature presents data providing a point-for-point contradiction of Rushton's undocumented assertions. Accordingly, when Rushton (1999, p. 105) asserted that "scientists have a special duty to examine the facts and tell the truth," one may wonder whether he included himself within the group held to this standard. In any case, it seems clear, from the evaluations that have been made of the quality of the "data" Rushton forwarded regarding his ideas, that the "truth" was not being told by either the data he presented or the interpretations he made of the data.

Evaluations of Rushton's Evidence

It is useful to consider a critique of the breadth of the evidence Rushton has presented in regard to his ideas. Cernovsky (1997) noted that Rushton's studies of racial differences (e.g., Rushton, 1998a, 1998b, 1990, 1991a, 1991b, 1995), as well as those of other researchers working to support his findings (e.g., Lynn, 1993)

> are noteworthy for their excessive reliance on very low correlation coefficients from obsolete data sets to postulate causal relationships. When a given method produces findings inconsistent with their ... views, they conveniently switch to a different method. An independent statistical re-examination of the same source of data by others may produce dramatically different results. (p. 1)

To illustrate, Cernovsky and Litman (1993) reanalyzed the data that Rushton (1990) used to demonstrate that there were significant race differences involving what Rushton termed "Mongoloid," "Caucasoid," and "Negroid" groups across nations in crime rates (e.g., involving homicide, rape, and serious assault). The data, Rushton (1990) claimed, indicated that the Negroid group had higher rates of crime than did either of the other two groups. However, Cernovsky and Litman (1993) found that the race differences reported by Rushton (1990) were not strong and, in fact, were largely weak and inconsistent. Not only did Rushton (1990) *not* present any evidence why these small differences among races should be considered genetic in origin but also Cernovsky and Litman (1993) found that, in Rushton's own data, reliance on race to predict an individual's likelihood of committing a crime "would result in an absurdly high rate (99.9%) of false positives" (p. 31).

Similarly, Gorey and Cryns (1995) reassessed some of the data that Rushton used (1988a, 1998b 1990, 1991, 1995) to illustrate the evolutionary and genetic deficits of "Negroids" in regard to intelligence, rate of physical maturation, personality and temperament, sexuality, and social/familial organization. The results of Gorey and Cryns' (1995) independent analysis of these data contradicted Rushton's characterization of the support provided for his hereditarian views of race differences. Gorey and Cryns found that the "relationships are very close to zero and some are in the opposite direction than postulated by Rushton" (Cernovsky, 1997, p. 2).

To illustrate some of the problems with Rushton's interpretation of the literature, we may note that Rushton (1990) cited the assessment of Beals, Smith, and Dodd (1984) of the relations between brain weight and race as providing support for his contention that "Negroids" have lower brain weights than do the other two race groups Rushton considers. Yet, Cernovsky and Litman (1993) noted that the statistical conclusions of Beals and colleagues (1984) "are the opposite of his own: brain weight is *not* primarily related to race" (p. 35). In addition, Cernovsky and Litman (1993) indicated as well that Rushton "selectively reports data confirming his theory ... this renders the data reported in (his work) *worthless* for generalization" (p. 35).

In addition, Cernovsky (1992) noted that Rushton's (1988a, 1998b, 1990, 1991) information suffers from conceptual and methodological flaws, for example, relating to his ignoring environmental effects such as secular trends (e.g., as in Tanner, 1991); statistical problems, associated with interpreting data with restricted ranges or with the overinterpretation of low correlations; and either omitting contradictory information from the literature he reviewed or, as illustrated by the above-noted work of Gorey and Cryns (1995), interpreting contradictory information as supportive of his ideas.

Conclusions about the Quality of Rushton's Hereditarian Views of Race Differences

Given the numerous dimensions of critical scientific problems associated with Rushton's work, we agree with Cernovsky's (1995) view that:

> Although Rushton's writings and public speeches instill the vision of Blacks as small-brained, oversexed criminals who multiply at a fast rate and are afflicted with mental disease, his views are neither based on a bona fide scientific review of literature nor on contemporary scientific methodology. His dogma of bioevolutionary inferiority of Negroids is not supported by empirical evidence. (p. 677)

In sum, given this quality of work that Rushton (1999) employed to document his views, we believe that an appropriate conclusion about Rushton's scholarship is reflected in Cernovsky's (1997) view, that:

> Rushton's racial theory is logically inconsistent, built on methodologically obsolete procedures, and is not supported by credible data sets selected in an objective manner. (p. 4)

As is the case with the other lines of evidence that intend to provide sociobiological evidence in support of the genetic basis of human behavior and development (relating to homology, heritability, and adaptation), Rushton's work reduces to no evidence at all.

Conclusions: Why Isn't Nativism "dead?"

We have reviewed several different approaches to nature/hereditarian conceptions of human development. All approaches—behavior genetics theory, the assessment of the heritability of intelligence, the study of instincts, and sociobiology—involve nature–nurture split concepts. Each has been seen to have critical conceptual, methodological, and empirical problems. Although the work of Rushton (1999, 2000) may be an exemplar of the bad science associated with these views—combining conceptual problems, including counterfactual assertions with poor methodology and misinterpretations or misrepresentations of data—instances of such scholarly shortcomings abound in the literatures associated with these instances of genetic reductionism.

Why? With so many conceptual, methodological, and empirical problems associated with these instances of genetic reductionism, why are there still examples of these approaches being presented at this writing? Why, as well, are these approaches still being given scientific attention, for example, through the awarding of research funds or by publication of this work in good scientific journals?

Tabery (2014) offers one answer. An exemplar of genetic reductionist models, he suggests that the variance-partitioning approach of behavior genetics and the process-elucidation approach of research associated with RDS-based theories (e.g., Moore, 2015b) address different questions. As a consequence, the data from one approach are not appreciated or seen as relevant to proponents of the other approach. Although this idea has merit—proponents of the two approaches do address different questions—we think that this answer is not compelling for at least two reasons.

First, as Moore (2015b) explains, the flaws of the variance-partitioning approach are so numerous that it cannot be fairly said that the two approaches are scientifically commensurate in regard to their respective capacities to describe, explain, and optimize the diverse intraindividual change trajectories comprising human development within the relational developmental system. Second, Tabery's (2014) answer ignores the psychology of the "Person in the Street" discussed by Horowitz.

As Horowitz (2000) explained, the simplicity of the hereditarian answer to the questions of the "Person In the Street" about human development (i.e., "The answer is that it is in your genes") continues to be attractive to people and seen as newsworthy to the media (e.g., see Wade, 2014; but see too Feldman, 2014, for a critique of Wade, 2014). Often, neither the "Person in the Street" nor the media have patience for more complex answers (e.g., "The answer depends on the particular history of fusions within the developmental system").

Bleier (2008) extends the points made by Horowitz (2000). Bleier emphasizes that one reason for interest in sociobiology is that such purportedly scientific accounts of genetic determinism align with political ideas that account for inequitable treatment of women (or, we would add, other groups marginalized because of race, ethnicity, religion, or other demographic or behavioral characteristics) on the basis of unalterable genetic differences and not because of social prejudice or discrimination. Bleier explains that this ideological alignment between, on the one hand, sociobiological assertions about genetic determinism and, on the other, marginalizing and minoritizing political ideology is ill conceived because of the inadequacies and distortions inherent in the science that sociobiologists use to support their claims.

A key contribution of Bleier's presentation is that the flawed science involved in sociobiology provides no credible justification for the scientific validity of political assertions such as "we had best resign ourselves to the fact that the more unsavory aspects of human behavior, like wars, racism, and class struggle, are inevitable results of evolutionary adaptations based in our genes" or that "the particular roles performed by women and men in society are also biologically, genetically determined; in fact, civilization as we know it, or perhaps any at all, could not have evolved in any other way" (Bleier, 2008, p. 185).

Accordingly, given both the theoretical and practical reasons that are associated with attraction to hereditarian ideas, versions of such formulations are likely to continue to be forwarded. As in the children's game, Whack-A-Mole, as soon as the failures of one instantiation of genetic reductionism are compellingly refuted, other instances of this problem-riddled conception, and another version of this idea, pops up. As a consequence, and as Lewontin cautioned, the "price" developmental scientists must pay for the continued possible use of such conceptions is the need to remain vigilant about their appearance. Developmental scientists must be prepared to discuss the poor science involved in genetic reductionist ideas and to point out the inadequate bases they provide for public policy and for applications pertinent to improving human life (Lerner, 2015a, 2015b; see, too, Schneirla, 1966; Tobach, 1994). Developmental scientists must be ready to suggest alternatives, such as RDS-based views, to hereditarian views of research about and applications for human development.

An Overview of the Contributions to This Section

Writing in the penultimate decade of the 20th century, just five years after the publication of Wilson's (1975) *Sociobiology: The New Synthesis*, Gould (1980/2019) presented a prescient critique of the future scientific standing of sociobiology and explained why the interest in sociobiology that existed at the time of his writing would dissipate in future decades. Gould noted that the eminent biologist, Ludwig von

Bertalanffy (1933), a scientist who may be credited as a key contributor to the creation of general systems theory (von Bertalanffy, 1968), perceptively predicted that Darwinian natural selection would fail as a comprehensive theory because it explained too much. Gould explains that, in attempting to substantiate its claims about the explanatory hegemony of sociobiology, through a reduction of the human sciences to biology via Darwinism and natural selection, much of sociobiology argumentation has relied on storytelling about adaptation arising through the purported powerful and pervasive optimizing character of natural selection (e.g., see Gould, 1978). Gould explains that, whereas humans are certainly adaptive, evolution involves much more than genetic transmission and these other dimensions of evolution also contribute to adaption.

Akin to the work presented in earlier sections by Gottlieb (1997, 2002/2014), Jablonka and Lamb (2007), Lickliter (2016), Narvaez et al. (2022), and Noble (2015), there are multiple dimensions of evolution, including epigenetics, behavior, and culture. The latter two dimensions—which are nongenetic—function to support and transmit information about adaptation. As such, Gould correctly predicted that the stories presented by sociobiologists would wane in interest among scientists, and that the sociobiological attempt to reduce all of human behavior and development to natural selection would fail.

Lerner and von Eye (1992) also consider the evidence base of the claims of sociobiologists They note that the genetic reductionism of sociobiologists are predicated on three arguments: (1) humans are impelled to act to increase replications of their genes in future generations (i.e., the ideas of inclusive fitness); (2) The gametes of males and females differ in their potential for replication; and (3) The differences in gametic potential result in sex differences in reproductive strategies that are reflected in different social behavior and development (most importantly in differences in parental investment in offspring). Lerner and von Eye then note that sociobiologists use three lines of research to support these ideas: (1) Interspecies comparisons and the concept of homology; (2) Heritability estimates of human behaviors; and (3) The idea that the evolution of human behavior invariably involves adaptation. The authors present logical and empirical shortcomings of all these lies of work and they conclude that sociobiological ideas are of little use in accounting either for general features of human development or for the nature or presence of individual differences.

References

Atz, J. W. (1970). The application of the idea of homology to behavior. In L. R. Aronson, E. Tobach, D. S. Lehrman, & J. S. Rosenblatt (Eds.), *Development and evolution of behavior: Essays in memory of T. C. Schneirla* (pp. 53–74). W. H. Freeman.

Barash, D. P. (1977). *Sociobiology and behavior*. Elsevier.

Barlow, G. W., & Silverberg, J. (Eds.). (1980). *Sociobiology: Beyond nature/nurture? Reports, definitions, and debate*. Westview.

Beals, K. L., Smith, C. L., & Dodd, S. M. (1984). Brain size, cranial morphology, climate, and time machines. *Current Anthropology*, 25, 301–330.

Belsky, J., Steinberg, L., & Draper, P. (1991). Childhood experience, interpersonal development, and reproductive strategy: An evolutionary theory of socialization. *Child Development*, 62, 647–670.

Bitterman, M. E. (1965). Phyletic differences in learning. *American Psychologist*, 20, 396–410.

Bitterman, M. E. (1975). The comparative analysis of learning. *Science*, 188, 699–709.

Bleier, R. (2008). Sociobiology, Biological Determinism, and Human Behavior. In M. Wyer, M. Barbercheck, D. Cookmeyer, H. Ozturk, & M. Wayne (Eds.), *Women, science, and technology: A reader in feminist science studies* (2nd ed., pp. 185–204). Routledge.

Bock, W. J. (1979). A synthetic explanation of macroevolutionary change – A reductionistic approach. *Bulletin of the Carnegie Museum of Natural History*, 13, 20–69.

Brandtstädter, J. (1998). Action perspectives on human development. In W. Damon (Series Ed.) & R. M. Lerner (Vol. Ed.), *Handbook of child psychology: Vol. 1. Theoretical models of human development* (5th ed., pp. 807–863). Wiley.

Brandtstädter, J. (1999). The self in action and development: Cultural, biosocial, and ontogenetic bases of intentional self-development. In J. Brandtstädter & R.M. Lerner (Eds.), *Action and self-development: Theory and research through the life-span* (pp. 37–65). Sage.

Caplan, A. L. (1978). *The sociobiology debate*. Harper & Row.

Cernovsky, Z. Z. (1992). J. P. Rushton on Negroids and Caucasoids: Statistical concepts and disconfirmatory evidence. *The International Journal of Dynamic Assessment and Instruction, 2*(2), 55–67.

Cernovsky, Z. Z. (1995). On the similarities of American blacks and whites. *Journal of Black Studies, 25*(6), 672–679.

Cernovsky, Z. Z. (1997). *Statistical methods and behavioral similarities of blacks and whites*. Paper presented at the 58th Annual Convention of the Canadian Psychological Association, Toronto, Canada.

Cernovsky, Z. Z., & Litman, L. C. (1993). Re-analyses of J. P. Rushton's crime data. *Canadian Journal of Criminology, 35*(1), 31–36.

Cohen, S. (1976). Transposable genetic elements and plasmid evolution. *Nature, 263*, 731–738.

Collins, W. A., Maccoby, E. E., Steinberg, L., Hetherington, E. M., & Bornstein, M. H. (2000). Contemporary research on parenting: The case of nature and nurture. *American Psychologist, 55*, 218–232.

Darwin, C. (1859). *The origin of species*. John Murray.

Dawkins, R. (1976). *The selfish gene*. Oxford University.

Dawkins, R. (1982). *The extended phenotype*. Freeman.

de Beer, G. R. (1930). *Embryology and evolution*. Clarendon Press.

Demo, D. H., Allen, K. R., & Fine, M. A. (2000). *Handbook of family diversity*. Oxford University Press.

Doolittle, W. F., & Sapienza, C. (1980). Selfish genes, the phenotype paradigm, and genome evolution. *Nature, 284*, 601–603.

Dunbar, R. I. M. (1987). Sociobiological explanations and the evolution of ethnocentrism. In V. Reynolds, V. Falger, & I. Vine (Eds.), *The sociobiology of ethnocentrism* (pp. 48–59). Croom Helm.

Ellis, B. J., Schlomer, G. L., Tilley, E. H., & Butler, E. A. (2012). Impact of fathers on risky sexual behavior in daughters: A genetically and environmentally controlled sibling study. *Development and Psychopathology, 24*, 317–332.

Erikson, E. H. (1968). *Identity, youth, and crisis*. Norton.

Falconer, D. S. (1960). *Quantitative genetics*. Oliver & Boyd.

Feldman, M. (2014). Echoes of the past: Hereditarianism and a troublesome inheritance. *PLoS Genetics, 10*, e1004817.

Flohr, H. (1987). Biological bases of social prejudices. In V. Reynolds, V. Falger, & I. Vine (Eds.), *The sociobiology of ethnocentrism* (pp. 190–207). Croom Helm.

Ford, D. H., & Lerner, R. M. (1992). *Developmental systems theory: An integrative approach*. Sage.

Freedman, D. G. (1979). *Human sociobiology: A holistic approach*. Free Press.

Freud, S. (1923). *The ego and the id*. Hogarth Press.

Garstang, W. (1922). The theory of recapitulation: A critical re-statement of the biogenetic law. *Journal of the Linnean Society of London, Zoology, 35*, 81–101.

Goldschmidt, R. (1933). Some aspects of evolution. *Science, 78*, 539–547.

Gorey, K. M., & Cryns, A. G. (1995). Lack of racial differences in behavior: A quantitative replication of Rushton's (1988) review and an independent meta-analysis. *Personality and Individual Differences, 19*, 345–353.

Gottlieb, G. (1970). Conceptions of prenatal behavior. In L. R. Aronson, E. Tobach, D. S. Lehrman, & J. S. Rosenblatt (Eds.), *Development and evolution of behavior: Essays in memory of T. C. Schneirla* (pp. 111–137). Freeman.

Gottlieb, G. (1976a). The roles of experience in the development of behavior and the nervous system. In G. Gottlieb (Ed.), *Neural and behavioral specificity* (pp. 25–54). Academic Press.

Gottlieb, G. (1976b). Conceptions of prenatal development: Behavioral embryology. *Psychological Review, 83*, 215–234.

Gottlieb, G. (1983). The psychobiological approach to developmental issues. In M. M. Haith & J. Campos (Eds.), *Handbook of child psychology: Infancy and biological bases* (Vol. 2, pp. 1–26). Wiley.

Gottlieb, G. (1991). The experiential canalization of behavioral development: Theory. *Developmental Psychology, 27*, 4–13.

Gottlieb, G. (1997). *Synthesizing nature-nurture: Prenatal roots of instinctive behavior.* Psychology Press.
Gottlieb, G. (2002/2014). *Individual development and evolution: The genesis of novel behavior.* Psychology Press.
Gottlieb, G. (2004). Normally occurring environmental and behavioral influences on gene activity: From central dogma to probabilistic epigenesis. In C. Garcia Coll, E. Bearer, & R. M. Lerner (Eds.), *Nature and nurture: The complex interplay of genetic and environmental influences on human behavior and development* (pp. 85–106). Erlbaum.
Gottlieb, G., Wahlsten, D., & Lickliter, R. (2006). The significance of biology for human development: A developmental psychobiological systems view. In R. M. Lerner (Ed.), *Theoretical models of human development. Volume 1 of handbook of child psychology* (6th ed., pp. 210–257). Editors-in-chief: W. Damon & R. M. Lerner. Wiley.
Gould, S. J. (1978). Sociobiology: The art of storytelling. *New Scientist,* November 16, 530–533.
Gould, S. J. (1980). Jensen's last stand. *New York Review of Books, 27,* 38–44.
Gould, S. J., & Lewontin, R. C. (1979). The spandrels of San Marco and the panglossian paradigm: A critique of the adaptationist programme. In J. Maynard Smith & R. Holliday (Eds.), *The evolution of adaptation by natural selection* (pp. 581–598). Royal Society of London.
Gould, S., & Vrba, E. (1982). Exaptation: A missing term in the science of form. *Paleobiology, 8,* 4–15.
Grzimek, B. (Ed.). (1972). *Animal life encyclopedia.* Van Nostrand Reinhold.
Hiernaux, J. (1968). Ethnic differences in growth and development. *Eugenics Quarterly, 15,* 12–21.
Hirsch, J. (1997). Some history of heredity-vs-environment, genetic inferiority at Harvard (?), and The (incredible) Bell Curve. *Genetica, 99,* 207–224.
Horowitz, F. D. (2000). Child development and the PITS: Simple questions, complex answers, and developmental theory. *Child Development, 71,* 1–10.
Jablonka, E., & Lamb, M. (2005). *Evolution in Four Dimensions: Genetic, epigenetic, behavioral, and symbolic variation in the history of life.* MIT Press.
Jablonka, E., & Lamb, M. (2007). Précis of evolution in four dimensions. *Behavioral and Brain Sciences, 30*(4), 353–365. https://doi.org/10.1017/S0140525X07002221
Johanson, D. C., & Edey, M. A. (1981). *Lucy: The beginnings of humankind.* Simon & Schuster.
Kinsey, A. C., Pomeroy, W. B., Martin, C. E., & Gebhard, P. H. (1953). *Sexual behavior in the human female.* Saunders.
Kipling, R. (1993). *How the leopard got his spots and other Just so stories.* Barefoot Books.
Kitcher, P. (1985). *Vaulting ambition.* MIT.
Kleckner, J. H. (1977). Alcoholics can drink again, or can they? *Psychology, 14*(2), 6–8.
Konner, M. (1982). *The tangled wing.* Holt, Rinehart & Winston.
Kuo, Z. Y. (1967). *The dynamics of behavior development.* Random House.
Lerner, R. M. (1982). Children and adolescents as producers of their own development. *Developmental Review, 2,* 342–370.
Lerner, R. M. (1992). *Final solutions: Biology, prejudice, and genocide.* Penn State Press.
Lerner, R. M. (2015a). Promoting social justice by rejecting genetic reductionism: A challenge for developmental science. *Human Development, 58,* 67–69.
Lerner, R. M. (2015b). Eliminating genetic reductionism from developmental science. *Research in Human Development, 12,* 178–188.
Lerner, R. M., & Busch-Rossnagel, N. A. (1981). Individuals as producers of their development: Conceptual and empirical bases. In R. M. Lerner & N. A. Busch-Rossnagel (Eds.), *Individuals as producers of their development: A life-span perspective* (pp. 1–36). Academic Press.
Lerner, R. M., & von Eye, A. (1992). Sociobiology and human development: Arguments and evidence. *Human Development, 35,* 12–33.
Lerner, R. M., & von Eye, A. (1993). Why Burgess and Molenaar "just don't get it." *Human Development, 36,* 55–56.
Lerner, R. M., & Walls, T. (1999). Revisiting individuals as producers of their development: From dynamic interactionism to developmental systems. In J. Brandtstädter & R. M. Lerner (Eds.), *Action and self-development: Theory and research through the life-span* (pp. 3–36). Sage.

Lewontin, R. C. (1980). Sociobiology: Another biological determinism. *International Journal of Health Sciences, 10*(3), 347–363.
Lewontin, R. C. (1981). On constraints and adaptation. *The Behavioral and Brain Sciences, 4*, 244–245.
Lewontin, R. C., & Levins, R. (1978). Evolution. *Encyclopedia Einaudi* (Vol. 5). Einaudi.
Lickliter, R. (2016). Developmental evolution. *WIREs Cognitive Science* 2016. doi: 10.1002/wcs.1422.
Lorenz, K. (1940a). Durch Domestikation verursachte Störungen arteigenen Verhaltens. *Zeitschrift für angewandte Psychologie und Charakterkunde, 59*, 2–81.
Lorenz, K. (1940b). Systematik und Entwicklungsgedanke im Unterricht, *Der Biologe 9*, 24–36.
Lorenz, K. (1965). *Evolution and modification of behavior.* University of Chicago Press.
Lorenz, K. (1966). *On aggression.* Harcourt, Brace & World.
Lorenz, K. (1974). Letter: Lorenz clarifies ideas. *Human Behavior*, September 6.
Lorenz, K. (1975). Konrad Lorenz responds to Donald Campbell. In R. I. Evans (Ed.), *Konrad Lorenz: The man and his ideas* (pp. 119–128). Harcourt Brace Jovanovich.
Lumsden, C. J., & Wilson, E. O. (1981). *Genes, mind, and culture.* Harvard University.
Lynn, R. (1993). Further evidence for the existence of race and sex differences in cranial capacity. *Social Behavior and Personality, 21*, 89–92.
MacDonald, K. (Ed.). (1988). *Sociobiological perspectives on human development.* Springer-Verlag.
MacDonald, K. (1994). *A people that shall dwell alone: Judaism as an evolutionary group strategy.* Greenwood.
Magnusson, D. (1999a). Holistic interactionism: A perspective for research on personality development. In L. A. Pervin & O. P. John (Eds.), *Handbook of personality: Theory and research* (2nd ed., pp. 219–247). The Guilford Press.
Magnusson, D. (1999b). On the individual: A person-oriented approach to developmental research. *European Psychologist, 4*, 205–218.
McAdoo, H. (1991). Family values and outcomes for children. *Journal of Negro Education, 60*, 361–365.
McAdoo, H. (1993a). Family equality and ethnic diversity. In K. Altergott (Ed.), *One world, many families* (pp. 52–55). National Council on Family Relations.
McAdoo, H. (1993b). The social cultural contexts of ecological developmental family models. In P. Boss, W. Doherty, & W. Schyumm (Eds.), *Sourcebook of family theories and methods: A contextual approach* (pp. 298–301). Plenum.
McAdoo, H. P. (1977). A review of the literature related to family therapy in the Black community. *Journal of Contemporary Psychotherapy, 9*, 15–19.
McAdoo, H. P. (1995). Stress levels, family help patterns, and religiosity in middle- and working-class African American single mothers. *Journal of Black Psychology, 21*, 424–449.
McAdoo, H. (1998). African American families: Strength and realities. In H. I. McCubbin, E. A. Thompson, A. I. Thompson, & J. E. Fromer, (Eds.), *Resiliency in ethnic minority families: African American families* (pp. 17–30). Sage.
McAdoo, H. P. (1999). Diverse children of color. In H. E. Fitzgerald, B. M. Lester, & B. S. Zuckerman (Eds.), *Children of color: Research, health, and policy issues* (pp. 205–218). Garland Publishing.
McClearn, G. E. (1981). Evolution and genetic variability. In E. S. Gollin (Ed.), *Developmental plasticity: Behavioral and biological aspects of variations in development* (pp. 3–31). Academic Press.
McCubbin, H. I., Thompson, E. A., Thompson, A. I., & Futrell, J. A. (Eds.). (1998). *Resiliency in ethnic minority families: African-American families.* Sage.
Moore, D. S. (2015a). *The developing genome: An introduction to behavioral epigenetics.* Oxford University Press.
Moore, D. S. (2015b). The asymmetrical bridge. Book review of James Tabery's Beyond versus: The struggle to understand the interaction of nature and nurture. *Acta Biotheoretica, 63*(4), 413–427.
Narvaez, D., Moore, D. S., Witherington, D. C., Vandiver, T. I., & Lickliter, R. (2022). Evolving evolutionary psychology. *American Psychologist, 77*(3), 424–438.
Noble, D. (2015). Evolution beyond neo-Darwinism: A new conceptual framework. *The Journal of Experimental Biology, 218*, 7–13.

Orgel, L. E., & Crick, F. H. C. (1980). Selfish DNA: The ultimate parasite. *Nature, 284,* 604–607.
Peters, M. (1995a). Does brain size matter? A reply to Rushton and Ankney. *Canadian Journal of Experimental Psychology, 47,* 751–756.
Peters, M. (1995b). Race differences in brain size: Things are not as clear as they seem to be. *American Psychologist, 49,* 570–576.
Piaget, J. (1950). *The psychology of intelligence.* Harcourt Brace.
Piaget, J. (1970). Piaget's theory. In P. H. Mussen (Ed.), *Carmichael's manual of child psychology* (3rd ed., Vol. 1, pp. 703–723). Wiley.
Rushton, J. P. (1987). An evolutionary theory of health, longevity, and personality: Sociobiology, and r/K reproductive strategies. *Psychological Reports, 60,* 539–549.
Rushton, J. P. (1988a). Do r/K reproductive strategies apply to human differences? *Social Biology, 35,* 337–340.
Rushton, J. P. (1988b). Race differences in behavior: A review and evolutionary analysis. *Personality and Individual Differences, 9,* 1009–1024.
Rushton, J. P. (1990). Sex, ethnicity, and hormones. *Behavioral & Brain Sciences, 13,* 194, 197–198.
Rushton, J. P. (1991a). Do r-K strategies underlie human race differences? A reply to Weizmann et al. *Canadian Psychology, 32,* 29–42.
Rushton, J. P. (1991b). Race, brain size, and intelligence: Another reply to Cernovsky. *Psychological Reports, 66,* 659–666.
Rushton, J. P. (1995). *Race, evolution, and behavior.* Transaction.
Rushton, J. P. (1996). Political correctness and the study of racial differences. *Journal of Social Distress & the Homeless, 5,* 213–229.
Rushton, J. P. (1997). Cranial size and IQ in Asian Americans from birth to age seven. *Intelligence, 25,* 7–20.
Rushton, J. P. (1999). *Race, evolution, and behavior* (Special Abridged ed.). Transaction Publishers.
Rushton, J. P. (2000). *Race, evolution, and behavior* (2nd Special Abridged ed.). Transaction Publishers.
Schneirla, R. C. (1956). Interrelationships of the innate and the acquired in instinctive behavior. In P. P. Grassé (Ed.), *L'instinct dans le comportement des animaux et de l'homme.* Mason et Cie.
Schneirla, T. C. (1957). The concept of development in comparative psychology. In D. Harris (Ed.), *The concept of development* (pp. 78–108). University of Minnesota Press.
Schneirla, T. C. (1966). Instinct and aggression: Reviews of Konrad Lorenz, *Evolution and modification of behavior* (Chicago: The University of Chicago Press, 1965), and *On aggression* (New York: Harcourt, Brace & World, 1966). *Natural History, 75,* 16.
Sunday, S. R., & Tobach, E. (Eds.). (1985). *Violence against women: A critique of the sociobiology of rape.* Gordian.
Tabery, J. (2014). *Beyond versus: The struggle to understand the interaction of nature and nurture.* MIT Press.
Tanner, J. M. (1973). Growing up. *Scientific American, 229,* 34–43.
Tanner, J. (1991). Menarche, secular trend in age of. In R. M. Lerner, A. C. Petersen, & J. Brooks-Gunn (Eds.), *Encyclopedia of adolescence* (Vol. 1, pp. 637–641). Garland.
Thelen, E., & Smith, L. B. (1998). Dynamic systems theories. In W. Damon (Series Editor) & R. M. Lerner (Vol. Ed.), *Handbook of child psychology: Vol. I Theoretical models of human development* (5th ed., pp. 563–633). Wiley.
Thelen, E., & Smith, L. B. (2006). Dynamic systems theories. In R. M. Lerner & W. Damon (Eds.), *Handbook of child psychology: Vol. 1. Theoretical models of human development* (6th ed., pp. 258–312). Wiley.
Tobach, E. (1981). Evolutionary aspects of the activity of the organism and its development. In R. M. Lerner & N. A. Busch-Rossnagel (Eds.), *Individuals as producers of their development: A life-span perspective* (pp. 37–68). Academic Press.
Tobach, E. (1994). ...Personal is political is personal is political... *Journal of Social Issues, 50,* 221–224.
Tobach, E., Gianutsos, J., Topoff, H. R., & Gross, C. G. (1974). *The four horses: Racism, sexism, militarism, and social Darwinism.* Behavioral Publications.
Tobach, E., & Schneirla, T. C. (1968). The biopsychology of social behavior of animals. In R. E. Cooke & S. Levin (Eds.), *Biologic basis of pediatric practice* (pp. 68–82). McGraw-Hill.

van den Berghe, P. L., & Barash, D. P. (1977). Inclusive fitness and human family structure. *American Anthropologist, 79,* 809–823.

von Bertalanffy, L. (1933). *Modern theories of development.* Oxford University Press.

von Bertalanffy, L. (1968). *General systems theory.* Braziller.

von Bertalanffy, L. (1969). Chance or law. In A. Koestler (Ed.), *Beyond reductionism* (pp. 59–84). Hitchinson.

Wade, N. (2014). *A troublesome inheritance: Genes, race and human history.* Penguin Books.

Williams, G. C. (1966). *Adaptation and natural selection.* Princeton University.

Wilson, E. O. (1975a). *Sociobiology: The new synthesis.* Harvard University Press.

Wilson, E. O. (1975b, December 11). For sociobiology. *New York Review of Books.*

Wilson, E. O. (1978). *On human nature.* Harvard University Press.

Wilson, E. O. (1980). A consideration of the genetic foundation of human social behavior. In G. W. Barlow & J. Silverberg (Eds.), *Sociobiology: Beyond nature/nurture* (pp. 295–305). Westview.

Winston, A. S. (1996). The context of correctness: A comment on Rushton. *Journal of Social Distress and the Homeless, 5,* 231–250.

Winston, A. S. (1997a). Genocide as a scientific project. *American Psychologist, 52,* 182–183.

Winston, A. S. (1997b). Rushton and racial differences: Further reasons for caution. *Journal of Social Distress and the Homeless, 6,* 199–202.

17.
SOCIOBIOLOGY AND THE THEORY OF NATURAL SELECTION
*Stephen Jay Gould**

Natural Selection as Storytelling

Ludwig von Bertalanffy, a founder of general systems theory and a holdout against the neo-Darwinian tide, often argued that natural selection must fail as a comprehensive theory because it explains too much—a paradoxical, but perceptive statement. He wrote (1969:24, 11):

> If selection is taken as an axiomatic and a priori principle, it is always possible to imagine auxiliary hypotheses—unproved and by nature unprovable—to make it work in any special case... Some adaptive value... can always be construed or imagined.

I think the fact that a theory so vague, so insufficiently verifiable and so far from the criteria otherwise applied in "hard" science, has become a dogma, can only be explained on sociological grounds. Society and science have been so steeped in the ideas of mechanism, utilitarianism, and the economic concept of free competition, that instead of God, Selection was enthroned as ultimate reality.

Similarly, the arguments of Christian fundamentalism used to frustrate me until I realized that there are, in principle, no counter cases and that, on this ground alone, literal bibliolatry is bankrupt. The theory of natural selection is, fortunately, in much better straits. It could be invalidated as a general cause of evolutionary change. (If, for example, Lamarckian inheritance were true and general, then adaptation would arise so rapidly in the Lamarckian mode that natural selection would be powerless to create and would operate only to eliminate.) Moreover, its action and efficacy have been demonstrated experimentally by 60 years of manipulation within Drosophila bottles—not to mention several thousand years of success by plant and animal breeders.

Yet in one area, unfortunately a very large part of evolutionary theory and practice, natural selection has operated like the fundamentalist's God—he who maketh all things. Rudyard Kipling asked how the leopard got its spots, the rhino its wrinkled skin. He called his answers "just-so stories." When evolutionists try to explain form and behavior, they also tell just-so stories—and the agent is natural selection. Virtuosity in invention replaces testability as the criterion for acceptance. This is the procedure that inspired von Bertalanffy's complaint. It is also the practice that has given evolutionary biology a bad name among many experimental scientists in other disciplines. We should heed their disquiet, not dismiss it with a claim that they understand neither natural selection nor the special procedures of historical science.

* Harvard University, Cambridge, MA

This style of storytelling might yield acceptable answers if we could be sure of two things: (1) that all bits of morphology and behavior arise as direct results of natural selection and (2) that only one selective explanation exists for each bit. But, as Darwin insisted vociferously, and contrary to the mythology about him, there is much more to evolution than natural selection. (Darwin was a consistent pluralist who viewed natural selection as the most important agent of evolutionary change, but who accepted a range of other agents and specified the conditions of their presumed effectiveness. In Chapter 7 of the *Origin* (6th ed.), for example, he attributed the cryptic coloration of a flat fish's upper surface to natural selection and the migration of its eyes to inheritance of acquired characters. He continually insisted that he wrote his 2-volume *Variation of Animals and Plants Under Domestication* (1868), with its Lamarckian hypothesis of pangenesis, primarily to illustrate the effect of evolutionary factors other than natural selection. In a letter to *Nature* in 1880, he used the sharpest and most waspish language of his life to castigate Sir Wyville Thomson for caricaturing his theory by ascribing all evolutionary change to natural selection.)

Since God can be bent to support all theories, and since Darwin ranks closest to deification among evolutionary biologists, panselectionists of the modern synthesis tended to remake Darwin in their image. But we now reject this rigid version of natural selection and grant a major role to other evolutionary agents—genetic drift, fixation of neutral mutations, for example. We must also recognize that many features arise indirectly as developmental consequences of other features subject to natural selection—see classic (Huxley 1932) and modern (Gould 1966 and 1975; Cock 1966) work on allometry and the developmental consequences of size increase. Moreover, and perhaps most importantly, there are a multitude of potential selective explanations for each feature. There is no such thing in nature as a self-evident and unambiguous story.

When we examine the history of favored stories for any particular adaptation, we do not trace a tale of increasing truth as one story replaces the last, but rather a chronicle of shifting fads and fashions. When Newtonian mechanical explanations were riding high, G.G. Simpson wrote (1961:1686):

> The problem of the pelycosaur dorsal fin... seems essentially solved by Romer's demonstration that the regression relationship of fin area to body volume is appropriate to the functioning of the fin as a temperature regulating mechanism.

Simpson's firmness seems almost amusing since now—a mere 15 years later with behavioral stories in vogue—most paleontologists feel equally sure that the sail was primarily a device for sexual display. (Yes, I know the litany: It might have performed both functions. But this too is a story.)

On the other side of the same shift in fashion, a recent article on functional endothermy in some large beetles had this to say about the why of it all (Bartholomew and Casey 1977: 883):

> It is possible that the increased power and speed of terrestrial locomotion associated with a modest elevation of body temperatures may offer reproductive advantages by increasing the effectiveness of intraspecific aggressive behavior, particularly between males.

This conjecture reflects no evidence drawn from the beetles themselves, only the current fashion in selective stories. We may be confident that the same data, collected 15 years ago, would have inspired a speculation about improved design and mechanical advantage.

Sociobiological Stories

Most work in sociobiology has been done in the mode of adaptive storytelling based upon the optimizing character and pervasive power of natural selection. As such, its weaknesses of methodology are those that have plagued so much of evolutionary theory for more than a century.

Sociobiologists have anchored their stories in the basic Darwinian notion of selection as individual reproductive success. Though previously underemphasized by students of behavior, this insistence on selection as individual success is fundamental to Darwinism. It arises directly from Darwin's construction of natural selection as a conscious analog to the laissez-faire economics of Adam Smith with its central notion that order and harmony arise from the natural interaction of individuals pursuing their own advantages (see Schweber 1977).

Sociobiologists have broadened their range of selective stories by invoking concepts of inclusive fitness and kin selection to solve (successfully I think) the vexatious problem of altruism—previously the greatest stumbling block to a Darwinian theory of social behavior. Altruistic acts are the cement of stable societies. Until we could explain apparent acts of self-sacrifice as potentially beneficial to the genetic fitness of sacrificers themselves—propagation of genes through enhanced survival of kin, for example—the prevalence of altruism blocked any Darwinian theory of social behavior.

Thus, kin selection has broadened the range of permissible stories, but it has not alleviated any methodological difficulties in the process of storytelling itself. Von Bertalanffy's objections still apply, if anything with greater force because behavior is generally more plastic and more difficult to specify and homologize than morphology. Sociobiologists are still telling speculative stories, still hitching without evidence to one potential star among many, still using mere consistency with natural selection as a criterion of acceptance.

David Barash (1976), for example, tells the following story about mountain bluebirds. (It is, by the way, a perfectly plausible story that may well be true. I only wish to criticize its assertion without evidence or test, using consistency with natural selection as the sole criterion for useful speculation.) Barash reasoned that a male bird might be more sensitive to intrusion of other males before eggs are laid than after (when he can be certain that his genes are inside). So Barash studied two nests, making three observations at 10-day intervals, the first before the eggs were laid, the last two after. For each period of observation, he mounted a stuffed male near the nest while the male occupant was out foraging. When the male returned he counted aggressive encounters with both model and female. At time one, males in both nests were aggressive toward the model and less, but still substantially, aggressive toward the female as well. At time two, after eggs had been laid, males were less aggressive to models and scarcely aggressive to females at all. At time three, males were still less aggressive toward models, and not aggressive at all toward females. Barash concludes that he has established consistency with natural selection and need do no more (1976: 1099-1100):

> These results are consistent with the expectations of evolutionary theory. Thus aggression toward an intruding male (the model) would clearly be especially advantageous early in the breeding season, when territories and nests are normally defended... The initial, aggressive response to the mated female is also adaptive in that, given a situation suggesting a high probability of adultery (i.e., the presence of the model near the female) and assuming that replacement females are available, obtaining a new mate would enhance the fitness of males.... The decline in male-female aggressiveness during incubation and fledgling stages could be attributed to the impossibility of being cuckolded after the eggs have been laid... The results are consistent with an evolutionary interpretation. In addition, the term "adultery" is unblushingly employed in this letter without quotation marks, as I believe it reflects a true analogy to the human concept, in the sense of Lorenz. It may also be prophesied that continued application of a similar evolutionary approach will eventually shed considerable light on various human foibles as well.

Consistent, yes. But what about the obvious alternative, dismissed without test in a line by Barash: male returns at times two and three, approaches the model a few times, encounters no reaction, mutters to himself the avian equivalent of "it's that damned stuffed bird again," and ceases to bother. And why not the evident test: expose a male to the model for the *first* time *after* the eggs are laid.

We have been deluged in recent years with sociobiological stories. Some, like Barash's are plausible, if unsupported. For many others, I can only confess my intuition of extreme unlikeliness, to say the least—for adaptive and genetic arguments about why fellatio and cunnilingus are more common among the upper classes (Weinrich 1977), or why male panhandlers are more successful with females and people who are eating than with males and people who are not eating (Lockard et al. 1976).

Not all of sociobiology proceeds in the mode of storytelling for individual cases. It rests on firmer methodological ground when it seeks broad correlations across taxonomic lines, as between reproductive strategy and distribution of resources, for example (Wilson 1975), or when it can make testable, quantitative predictions as in Trivers and Hare's work on haplodiploidy and eusociality in Hymenoptera (Trivers and Hare 1976). Here sociobiology has had and will continue to have success. And here I wish it well. For it represents an extension of basic Darwinism to a realm where it should apply.

Special Problems for Human Sociobiology

Sociobiological explanations of human behavior encounter two major difficulties, suggesting that a Darwinian model may be generally inapplicable in this case.

Limited Evidence and Political Clout

We have little direct evidence about the genetics of behavior in humans; and we do not know how to obtain it for the specific behaviors that figure most prominently in sociobiological speculation—aggression, conformity, etc. With our long generations, it is difficult to amass much data on heritability. More important, we cannot (ethically, that is) perform the kind of breeding experiments, in standardized environments, that would yield the required information. Thus, in dealing with humans, sociobiologists rely even more heavily than usual upon speculative storytelling.

At this point, the political debate engendered by sociobiology comes appropriately to the fore. For these speculative stories about human behavior have broad implications and proscriptions for social policy—and this is true quite apart from the intent or personal politics of the storyteller. Intent and usage are different things; the latter marks political and social influence, the former is gossip or, at best, sociology.

The common political character and effect of these stories lies in the direction historically taken by innatist arguments about human behavior and capabilities—a defense of existing social arrangements as part of our biology.

In raising this point, I do not act to suppress truth for fear of its political consequences. Truth, as we understand it, must always be our primary criterion. We live, because we must, with all manner of unpleasant biological truth—death being the most pervasive and ineluctable. I complain because sociobiological stories are not truth but unsupported speculations with political clout (again, I must emphasize, quite apart from the intent of the storyteller). All science is embedded in cultural contexts, and the lower the ratio of data to social importance, the more science reflects the context.

In stating that there is politics in sociobiology, I do not criticize the scientists involved in it by claiming that an unconscious politics has intruded into a supposedly objective enterprise. For they are behaving like all good scientists—as human beings in a cultural context. I only ask for a more explicit recognition of the context and, specifically, for more attention to the evident impact of speculative sociobiological stories. For example, when the *New York Times* runs a weeklong front page series on women and their rising achievements and expectations, spends the first four days documenting progress toward social equality, devotes the last day to potential limits upon this progress, and advances sociobiological stories as the only argument for potential limits—then we know that these are stories with consequences:

> Sociologists believe that women will continue for some years to achieve greater parity with men, both in the work place and in the home. But an uneasy sense of frustration and pessimism is growing among some advocates of full female equality in the face of mounting conservative opposition. Moreover, even some staunch feminists are reluctantly reaching the conclusion that women's aspirations may ultimately be limited by inherent biological differences that will forever leave men the dominant sex.
>
> (New York Times, Nov. 30, 1977)

The article then quotes two social scientists, each with a story.

> If you define dominance as who occupies formal roles of responsibility, then there is no society where males are not dominant. When something is so universal, the probability is—as reluctant as I am to say it—that there is some quality of the organism that leads to this condition.
>
> It may mean that there never will be full parity in jobs, that women will always predominate in the caring tasks like teaching and social work and in the life sciences, while men will prevail in those requiring more aggression—business and politics, for example—and in the 'dead' sciences like physics.

Adaptation in Humans Need Not Be Genetic and Darwinian

The standard foundation of Darwinian just-so stories does not apply to humans. That foundation is the implication: if adaptive, then genetic—for the inference of adaptation is usually the only basis of a genetic story, and Darwinism is a theory of genetic change and variation in populations.

Much of human behavior is clearly adaptive, but the problem for sociobiology is that humans have so far surpassed all other species in developing an alternative, nongenetic system to support and transmit adaptive behavior—cultural evolution. An adaptive behavior does not require genetic input and Darwinian selection for its origin and maintenance in humans; it may arise by trial and error in a few individuals that do not differ genetically from their groupmates, spread by learning and imitation, and stabilize across generations by value, custom, and tradition. Moreover, cultural transmission is far more powerful in potential speed and spread than natural selection—for cultural evolution operates in the "Lamarckian" mode by inheritance through custom, writing, and technology of characteristics acquired by human activity in each generation.

Thus, the existence of adaptive behavior in humans says nothing about the probability of a genetic basis for it, or about the operation of natural selection. Take, for example, Trivers' (1971) concept of "reciprocal altruism." The phenomenon exists, to be sure, and it is clearly adaptive. In honest moments, we all acknowledge that many of our "altruistic" acts are performed in the hope and expectation of future reward. Can anyone imagine a stable society without bonds of reciprocal obligation? But structural necessities do not imply direct genetic coding. (All human behaviors are, of course, part of the potential range permitted by our genotype—but sociobiological speculations posit direct natural selection for specific behavioral traits.) As Benjamin Franklin said: "We must all hang together, or assuredly we shall all hang separately."

Failure of the Research Program for Human Sociobiology

The grandest goal—I do not say the only goal—of human sociobiology must fail in the face of these difficulties. That goal is no less than the reduction of the behavioral (indeed most of the social) sciences to Darwinian theory.

Wilson (1975) presents a vision of the human sciences shrinking in their independent domain, absorbed on one side by neurobiology and on the other by sociobiology.

But this vision cannot be fulfilled, for the reason cited above. Although we can identify adaptive behavior in humans, we cannot tell thereby if it is genetically based (while much of it must arise by fairly pure cultural evolution). Yet the reduction of the human sciences to Darwinism requires the genetic argument, for Darwinism is a theory about genetic change in populations. All else is analogy and metaphor.

My crystal ball shows the human sociobiologists retreating to a fallback position—indeed it is happening already. They will argue that this fallback is as powerful as their original position, though it actually represents the unravelling of their fondest hopes. They will argue: yes, indeed, we cannot tell whether an adaptive behavior is genetically coded or not. But it doesn't matter. The same adaptive constraints apply whether the behavior evolved by cultural or Darwinian routes, and biologists have identified and explicated the adaptive constraints. (Steve Emlen (this volume) reports, for example, that some Indian peoples gather food in accordance with predictions of optimal foraging strategy, a theory developed by ecologists. This is an exciting and promising result within an anthropological domain—for it establishes a fruitful path of analogical illumination between biological theory and non-genetic cultural adaptation. But it prevents the assimilation of one discipline by the other and frustrates any hope of incorporating the human sciences under the Darwinian paradigm.)

But it does matter. It makes all the difference in the world whether human behaviors develop and stabilize by cultural evolution or by direct Darwinian selection for genes influencing specific adaptive actions. Cultural and Darwinian evolution differ profoundly in the three major areas that embody what evolution, at least as a quantitative science, is all about:

1. Rate. Cultural evolution, as a "Lamarckian" process, can proceed orders of magnitude more rapidly than Darwinian evolution. Natural selection continues its work within *Homo sapiens,* probably at characteristic rates for change in large, fairly stable populations, but the power of cultural evolution has dwarfed its influence (alteration in frequency of the sickling gene vs. changes in modes of communication and transportation). Consider what we have done with ourselves in the past 3000 years, all without the slightest evidence for any biological change in the size or power of the human brain.
2. Modifiability. Complex traits of cultural evolution can be altered profoundly all at once (social revolution, for example). Darwinian change is much slower and more piecemeal.
3. Diffusibility. Since traits of cultural evolution can be transmitted by imitation and inculcation, evolutionary patterns include frequent and complex anastomosis among branches. Darwinian evolution in sexually reproducing animals is a process of continuous divergence and ramification with few opportunities for coming together (hybridization or parallel modification of the same genes in independent groups).

I believe that the future will bring mutual illumination between two vigorous, independent disciplines—Darwinian theory and cultural history. This is a good thing, joyously to be welcomed. But there will be no reduction of the human sciences to Darwinian theory and the research program of human sociobiology will fail. The name, of course, may survive. It is an irony of history that movements are judged successful if their label sticks, even though the emerging content of a discipline may lie closer to what opponents originally advocated. Modern geology, for example, is an even blend of

Lyell's strict uniformitarianism and the claims of catastrophists (Rudwick 1972; Gould 1977). But we call the hybrid doctrine by Lyell's name and he has become the conventional hero of geology.

I welcome the coming failure of reductionistic hopes because it will lead us to recognize human complexity at its proper level. For consumption by Time's millions, my colleague Bob Trivers maintained: "Sooner or later, political science, law, economics, psychology, psychiatry, and anthropology will all be branches of sociobiology" (*Time*, Aug. 1, 1977:54. It is one thing to conjecture, as I would allow, that common features among independently developed legal systems might reflect adaptive constraints and might be explicated usefully with some biological analogies. It is quite another to state, as Trivers did, that the mores of the entire legal profession will be subsumed, along with a motley group of other disciplines, as mere epiphenomena of Darwinian processes.

I read Trivers' statement the day after I had sung in a full production of Berlioz' *Requiem*. And I remembered the visceral reaction I had experienced upon hearing the 4 brass choirs, finally amalgamated with the 10 tympani in the massive din preceding the great *Tuba mirum*—the spine tingling and the involuntary tears that almost prevented me from singing. I tried to analyze it in the terms of Wilson's conjecture—reduction of behavior to neurobiology on the one hand and sociobiology on the other. And I realized that this conjecture might apply to my experience. My reaction had been physiological and, as a good mechanist, I do not doubt that its neurological foundation can be ascertained. I will also not be surprised to learn that the reaction has something to do with adaptation (emotional overwhelming to cement group coherence in the face of danger, to tell a story). But I also realized that these explanations, however "true," could never capture anything of importance about the meaning of that experience.

And I say this not to espouse mysticism or incomprehensibility, but merely to assert that the world of human behavior is too complex and multifarious to be unlocked by any simple key. I say this to maintain that this richness—if anything—is both our hope and our essense.

Summary

Even since Darwin proposed it, the theory of natural selection has been marred by an uncritical style of speculative application to the study of individual adaptations: one simply constructs a story to explain how a shape, function, or behavior might benefit its possessor. Virtuosity in invention replaces testability and mere consistency with evolutionary theory becomes the primary criterion of acceptance. Although this dubious procedure has been used throughout evolutionary biology, it has recently become the primary style of explanation in sociobiology.

Human sociobiology presents two major problems related to this tradition. First, evidence is so poor or lacking that speculative storytelling assumes even greater importance than usual. Secondly, the existence of behavioral adaptation does not imply the operation of Darwinian processes at all—for non-genetic cultural evolution, working in the Lamarckian mode, dwarfs by its rapidity the importance of slower Darwinian change. The sociobiological vision of a reduction of the human sciences to biology via Darwinism and natural selection will fail. Instead, I anticipate fruitful, mutual illumination by analogy between independent theories of the human and biological sciences.

Literature Cited

Barash, D. 1976. Male response to apparent female adultery in the mountain bluebird (*Sialia currucoides*): An evolutionary interpretation. *American Naturalist* 110:1097–1101.

Bartholomew, G.A. and T.M. Casey. 1977. Endothermy during terrestrial activity in large beetles. *Science* 195:882–883.

Bertalanffy, L. von. 1969. Chance or law. *In* A. Koestler (ed.). *Beyond reductionism*. Hutchinson, London.

Cock, A.G. 1966. Genetical aspects of metrical growth and form in animals. *Quarterly Review of Biology* 4l:131–190.

Darwin, C. 1868. *The variation of animals and plants under domestication*. John Murray, London.

Darwin, C. 1880. Sir Wyville Thomson and natural selection. *Nature* 23:32.

Gould, S.J. 1966. Allometry and size in ontogeny and phylogeny. *Biological Reviews* 41:587–640.

Gould, S.J. 1975. Allometry in primates, with emphasis on scaling and the evolution of the brain. In Approaches to primate paleobiology. *Contributions to Primatology* 5:244–292.

Gould, S.J. 1977. Eternal metaphors of paleontology. In A. Hallam (ed.). *Patterns of evolution*. Elsevier, Amsterdam, pp. 1–26.

Huxley, J. 1932. *Problems of relative growth*. MacVeagh, London.

Lockard, J.S., L.L. McDonald, D.A. Clifford, and R. Martinez. 1976. Panhandling: Sharing of resources. *Science* 191:406–408.

Rudwick, M.J.S. 1972. *The meaning of fossils*. Macdonald, London.

Schweber, S.S. 1977. The origin of the *Origin* revisited. *Journal of the History of Biology* 10:229–316.

Simpson, G.G. 1961. Some problems of vertebrate paleontology. *Science* 133:1679–1689.

Trivers, R. 1971. The evolution of reciprocal altruism. *Quarterly Review of Biology* 46:35–57

Trivers, R. and H. Hare. 1976. Haplodiploidy and the evolution of the social insects. *Science* 191:249–263.

Weinrich, J.D. 1977. Human sociobiology: Pair-bonding and resource predictability (effects of social class and race). *Behavioral Ecology and Sociobiology* 2:91–118.

Wilson, E.O. 1975. *Sociobiology: The New Synthesis*. Harvard University Press, Cambridge, Massachusetts.

18.
SOCIOBIOLOGY AND HUMAN DEVELOPMENT
Arguments and Evidence
Richard M. Lerner[*] and Alexander von Eye[†]

In 1975, E.O. Wilson published a book announcing the "new" scientific discipline of sociobiology. Wilson (1975a) contended that sociobiology would be the "master" synthetic discipline, a field enveloping all of behavioral and social science. In this article, arguments forwarded by Wilson and others in support of this synthetic role for sociobiology are evaluated, especially as these ideas have been advanced in regard to understanding human behavior and development. In particular we focus on sociobiological claims pertinent to features of behavior central to human reproduction, parenting, and child caregiving. Our assessment draws on and extends prior discussions of sociobiological ideas (Barlow and Silverberg, 1980; Caplan, 1978; Kitcher, 1985; Lewontin et al., 1984), particularly in our concern with issues relevant to human development and in our provision of an alternative to sociobiological views of the role of biology in human development, one based on a developmental contextual (Lerner, 1984, 1986, 1991) perspective regarding human development.

The analysis undertaken here is, we believe, timely. Wilson's (1975a, b, 1980; Lumsden and Wilson, 1981) claims about sociobiology, as well as corresponding assertions made by others (Konner, 1982; MacDonald, 1988), have both found support and application (Rushton, 1987, 1988a, b) and evoked strong criticism (Barlow and Silverberg, 1980; Caplan, 1978; Kitcher, 1985; Lerner, in press), even among those associating themselves with the "new synthesis." For instance, according to Dunbar (1987, p. 51):

> Wilson created the impression that sociobiology was on the verge of replacing most of the disciplines in the social and behavioral sciences. This, of course, is arrant nonsense since sociobiology does not, of itself, deal with much of the subject matter of these disciplines.

However, Dunbar (1987, p. 51) went on to defend the importance of Wilson's views for social and behavioral science:

> Wilson was, none the less, right to emphasize the importance of sociobiology in relation to these disciplines. What it in fact does ... is to provide a unifying umbrella under which these disciplines can interact on common ground.

It is the mechanism through which this unification is purported to occur that concerns us first. This mechanism bears on a key conceptual issue in the study of human development, the nature-nurture controversy.

---

[*] The Pennsylvania State University is in State College, PA
[†] MIchigan State University, East Lansing, MI

Genetic Determinism as Sociobiology's Key to Interdisciplinary Integration

In Wilson's (1975a) view, the "unifying umbrella" provided by sociobiology is the ubiquitous influence of genes on all facets of individual and social behavior and development. Indeed, Wilson's (1975a, b) views exemplify a key conception of genetic determinism—thorough genetic reductionism. The complexity of all social behavior and development and, indeed, all human culture (Lumsden and Wilson, 1981) can be reduced to a few simple genetic principles. The key one is the idea of "gene reproduction." To Wilson (1975a, 1980), Dawkins (1976), and other sociobiologists (Barash, 1977; Freedman, 1979; MacDonald, 1988), the essential, core purpose of human life is *only* to reproduce genes. As Konner (1982, p. 265) puts it, "A person is only a gene's way of making another gene." Similarly, Dawkins (1976, p. ix) sees humans as only "survival machines—robot vehicles blindly programmed to preserve the selfish molecules known as genes."

Simply put, all of human development reduces ultimately to gene reproduction and, as such, a human organism "does not live for itself" (Wilson, 1975a, p. 3). Instead, the organism's "primary function is not even to produce other organisms; it produces genes, and it serves as their temporary carrier" (Wilson, 1975a, p. 3). According to this view, humans have not evolved to produce other people, but only to replicate their particular complement of genes. Humans' life spans represent only a relatively short period within the vast temporal span of evolution. During this time, they provide a temporary "house," or a transport, for the genes carried within them. Given this machine-like view of the "human as transport," it is clear why Dawkins (1976, p. 21) can see humans merely as "lumbering robots [housing genes that] created us, body and mind; and their preservation is the ultimate rationale for our existence."

Given this core, "gene reproduction" principle, it is evident also why Dawkins (1976) sees genes as selfish. Genes are "concerned" with nothing other than self-replication, with reproducing themselves over and over, as many times as possible. Simply, then, in this view humans are really just (seemingly complex) duplicating machines. Their mating rituals, their family relationships, and their cultural institutions are all inventions in the service of gene reproduction.

This core genetic principle of life leads to several other ideas about how genes influence individual behavior and the social world. The more copies of one's genes one can send out into the world, then, in terms of sociobiology, the more one is increasing one's *inclusive fitness*. This concept derives from the sociobiological view that natural selection is the essential vehicle through which evolutionary change occurs (Dawkins, 1976; Wilson, 1975a). However, not all genes are able to compete equally in the face of the fierce and rigorous challenges imposed by the natural environment; only the most *aggressive* genotypes will succeed in this struggle for survival.

Sociobiology and Human Aggression

A parallel arises between the views of sociobiologists (Dawkins, 1976; Konner, 1982) and the ideas of Konrad Lorenz. Lorenz (1966) saw human aggression as both inevitable and inherent in the human genome—to the point of providing for innate "militant enthusiasm." To Lorenz, not only do the genes of humans make them aggressive, but the "highest form" of humans should be the most aggressive—the most militaristic—since such action reflects the presence of genes that have succeeded most in the struggles of natural selection. Similarly, in the sociobiological world of selfish genes, and the robotic "survival machines" that house them, aggression functions to allow genes to enhance their inclusive fitness; "blind" (that is to say, unthinking, machine-like) aggression allows genes to eliminate anything in the environment that interferes with their reproduction.

Thus, to Lorenz (1966) and to sociobiologists (Konner, 1982), selfish— indeed ruthless—human aggression is the cornerstone of genes' control over human functioning. The most successful genes—the "best" of evolution, if you will (Lorenz, 1940)—will be the most successful at ruthless aggression. They will be the most selfishly directed to maximizing their presence in the gene pool and, at the same time, to minimizing the genes of others. Similarly, and redolent of Lorenz's (1966) views in his book, *On Aggression*, Konner (1982) explains:

> I believe in the existence of innate aggressive tendencies in humans [p. 203] … if we are ever to control human violence we must first appreciate that humans have a natural, biological tendency to react violently, as individuals or as groups, in certain situations.
>
> (p. xviii)

To Lorenz (1966), these innate violent reactions are elicited in a reflex-like manner among either individuals or groups. The reflex occurs when members of an in-group are threatened by members of an out-group. Similarly, Dawkins (1976) contends that selfish genes impel humans to act aggressively against survival machines not of their own, close genetic group (i.e., a child or another close relative), and Wilson (1975a; see also Flohr, 1987, p. 199) likewise believes that fear of (and hatred toward) an out-group (xenophobia) is innate in humans. For example, in his book, *On Human Nature*, Wilson (1978, p. 70) writes of "hidden biological prime movers," and contends:

> In all periods of life there is [a] … powerful urge to … classify other human beings into two artificially sharpened categories. We seem able to be fully comfortable only when the remainder of humanity can be labeled as members versus nonmembers, kin versus nonkin, friend versus foe.

In sum, all genotypes must struggle arduously to include as many copies of themselves as possible in the gene pool. However, within sociobiological thinking all genotypes are *not* "created equal"; that is, whereas all genotypes strive to maximize their inclusive fitness, genotypes differ in what is termed in sociobiology *gametic potential*, that is, in the potential of a genotype to replicate itself. Differences in gametic potential are associated with the differences that exist between males and females.

Sex Differences in Gametic Potential

Within sociobiology, it is held that men and women differ in their potential for transmitting copies of their genes into the future. For example, as claimed by Konner (1982, p. xviii), "as now seems clearly demonstrated, there are biological reasons why women, like other primate females, have a weaker aggressive tendency than males." Since aggression is the key, according to sociobiologists, to getting one's genotype reproduced maximally, it follows that women, lacking an aggressive ability sufficient to compete with men, must evolve some other strategy to enhance their inclusive fitness. The strategy that women use derives completely, sociobiologists contend, from the nature of the specialized cells used by women to transmit copies of their genes to future generations.

Genotype copies are contained in gametes, that is, sperms and ova. Both types of gametes function to maximize the inclusive fitness of the genotypes they carry. However, the two types of gametes have a different *potential* for such reproduction—due to the anatomical and physiological differences between the "lumbering robots"—men and women—housing these gametes. Men, who—it is of more than passing interest to note—were the founders and leading proponents of sociobiology, can generate a large number of genotype copies. Their gametes can be "sent forth to multiply" quite readily—millions can be sent out with each ejaculation. Thus, in the terms of sociobiology, their "gametic potential" is great, given that there is—at least theoretically—a ready, large pool of recipients of their gametes. Freedman (1979, p. 2) puts this idea as follows:

> Since mammalian males produce many more sperm than females produce ova, any given male has far greater potential for producing offspring. He is also more inclined to compete with other males over the "scarce" resource, females.

Simply put, then, any male has a greater potential for enhancing his inclusive fitness than any female, given males' greater gametic potential. Moreover, males *must* have in general a more aggressive genotype than females, since they must compete for access to the female gamete, viewed as a "resource" for the deposit of the males' sperm. Such competition is, of course, highly desirable in the view of sociobiologists, since it ensures that the most aggressive genotypes—those best suited to succeed against the struggles of natural selection—will reproduce most often.

In turn, the genotypes of females impel women to try to reproduce in quite another way. One might understand the origin and development of this "alternative," female reproductive strategy if one asks these questions:

> Given the vast difference in reproductive potential, and if the point of life is to actualize such potential, is it not reasonable to expect that on the average the male pattern of courtship will differ from the female? Might nature not have arranged it so that men are ready to fecundate almost any female and that selectivity of mates has become the female prerogative?
>
> (Freedman, 1979, p. 12)

In answer to such questions, van den Berghe and Barash (1977, p. 813) note:

> Human females, as good mammals who produce a few, costly and therefore precious, offspring, are choosy about picking mates who will contribute maximally to their offspring's fitness, whereas males, whose production of offspring is virtually unlimited, are much less picky.

Gametic Potential and Social and Sexual Development

What does the sexes' different gametic potential imply for understanding male and female social behavior and development? Given the selfishness of genes, and the single-minded direction of the duplicating machines housing them, men develop sexual mores dictating the acceptability (if not the appropriateness) of multiple sexual partners. Indeed van den Berghe and Barash (1977) argue that the different gametic potential of men and women explains:

> the widespread occurrence in human societies of polygamy, hypergamy, and double standards of sexual morality. There is another related reason for the sexual double standard in such things as differential valuation of male and female virginity and differential condemnation of adultery: marital infidelity of the spouse can potentially reduce the fitness of the husband more than that of the wife. Women stand to lose much less if their husbands have children out of wedlock than vice versa [p. 813] ... In addition, a woman will, at a maximum, produce some 400 fertile eggs in her lifetime, of which a dozen at most will grow up to reproductive age, while a man produces millions of sperm a day and can theoretically sire hundreds of children. Not surprisingly, females tend to go for quality, and males for quantity.
>
> (p. 814)

Moreover, given the large number of offspring they can potentially produce, a male's parental investment in any one is quite small. Unfortunately for the recipients of males' genetic copies—women, their gametic potential is quite different, and so too is their parental investment. They can replicate themselves at most every 9 months. Even with multiple births, a woman cannot replicate her genes as much in a lifetime as a man can in a short period of time. Therefore, a woman's investment in her offspring is much greater than is a man's. Moreover, since they cannot reproduce very frequently, women will not be motivated toward frequent copulation with multiple partners. Instead, women need to protect their offspring and assure their survival, and this need should motivate them to keep their impregnators bound to them. As a consequence, females develop monogamous sexual behaviors and a devotion to childbearing and rearing. Van den Berghe and Barash (1977, p. 813) argue:

> For a woman, the successful raising of a single infant is essentially close to a full-time occupation for a couple of years, and continues to claim much attention and energy for several more years. For a man, it often means only a minor additional burden ... [M]ost societies make no attempt to equalize parental care; they leave women holding the babies.

Lest anyone contend that the different moral, sexual, and social developments of men and women are merely products of socialization, Barash (1977, p. xv) argues that the sex differences in gene reproduction strategy explain "why women have almost universally found themselves relegated to the nursery, while men derive the greatest satisfaction from their jobs." van den Berghe and Barash (1977, p. 815) note further that "ethnographic evidence points to different reproductive strategies on the part of men and women, and to a remarkable consistency in the institutionalized means of accommodating these biological predispositions." They therefore conclude:

> Men are selected for engaging in male-male competition over resources appropriate to reproductive success, and women are selected for preferring men who are successful in that endeavor. Any genetically influenced tendencies in these directions will necessarily be favored by natural selection.

(van den Berghe and Barash, 1977, p. 814)

Dawkins (1976) embellishes these ideas by contending that women's *exploitation* by men is biologically determined. He argues that the sexes' behavioral developments are differentiated not only by the different number of sex cells that can be used for genotype reproduction but, as well, by the different size of their respective sex cells:

> the sex cells or "gametes" of males are much smaller and more numerous than the gametes of females ... it is possible to interpret all the other differences between the sexes as stemming from this one basic difference [p. 152] ... Sperms and eggs ... contribute equal numbers of genes, but eggs contribute far more in the way of food reserves: indeed sperms make no contribution at all, and are simply concerned with transporting their genes as fast as possible to an egg ... Female exploitation begins here.

(p. 153)

Sex differences in the gametic potential and size of the gametes result not only in female exploitation in general but, in particular, in the legitimation of extramarital sex for males, but not for females, and of the use of violence toward wives who have extramarital sexual relations. To explain these sex differences, Freedman (1979) argues:

> [W]e have to assume that cultural universals reflect those aspects of our species that were evolutionarily derived (evolved). Male promiscuity is universally winked at because there is nothing much we can do about it, and Kinsey's [Kinsey et al., 1953] main findings appear to be descriptions of the species: males must have frequent "outlets" for sex, whether heterosexual or homosexual; whereas many females can go for long periods without copulation or

masturbation ... And this difference appears to hinge on the difference in gametic potential that we have been discussing.

(p. 19)

As in the gelada baboon, in humans female jealousy is based not on the male's sex act with another woman but on his potential attachment to the latter ... Male jealousy is rather different. ... It does not make evolutionary sense for the male to invest in a child not possessing his genes, and the murderous jealousy exhibited by a cuckolded male is biologically sensible ... Furthermore, the cuckold's retribution can strike either the female or the male cheater ... and most legal systems (perhaps *all* patrilineal systems) wink at the ensuing violence

(pp. 20–21)

Dawkins (1976) extends across the life span the idea of the biological basis of men's promiscuous sexual interests. He offers both "a possible explanation of the evolution of the menopause in females" (p. 136) and, at the same time, an account of the sociobiological basis of the existence of what are colloquially (and pejoratively) termed "dirty old men":

The reason why the fertility of males tails off gradually rather than abruptly is probably that males do not invest so much as females in each individual child anyway. Provided he can sire children by young women, it will always pay even a very old man to invest in children rather than in grandchildren.

(p. 136)

Conclusions: Genetic Determinism and Human Development

In the sociobiology advanced by Wilson (1975a, b), men—impelled mechanistically by their genes—are oriented to seek sexual relations with as many women as possible, to achieve more and more copies of their genes, and to not be overly devoted to or concerned with any one or any few given "replicates." Women, in contrast, are oriented to remain monogamous in order to maximize the probability that their relatively few replicates will survive. In essence, then, men and women are genetically impelled to differ in ways that are consistent with traditional, that is, stereotypic, sex role patterns.

As did Freud (1923), and later Erikson (1968), Wilson's (1975a, b) sociobiology in effect holds that "anatomy is destiny" regarding key features of behavioral development—ones involving reproduction, parenting, child caregiving, and sexuality. In other words, Wilson (1975a, b), Dawkins (1976), Freedman (1979), and other sociobiologists (Barash, 1977; Konner, 1982; MacDonald, 1988) have built a natural edifice encompassing the very core of all human behavior and development—the reproduction of men and women, the character of the family, and the survival of the species. Any notions of nurture or of nature–nurture fusion as sources of key features of human behavior are mere fictions if genes work in the way that sociobiology requires—that is, as selfish, goal-directed, intentional agents. According to this view, after other, more superficial "causes" of human behavior are stripped away (for instance, "causes" involved in an individual's development, such as learning or social values), genes provide the ultimate basis for human functioning, the replication of genotypes.

According to this conception, the social world does not interact with humans' genes, much less act as an alternative source for human development. Instead, to sociobiologists, our social world—human mores (e.g., regarding sexual permissiveness or monogamy), social institutions (such as marriage and the family), and indeed all of human culture—is nothing other than the outcome of strategies laid down by humans' genes for their own replication. Sociobiologists have complete faith in the inevitable reducibility of human behavior to the functioning of selfish genes. Akin to Lorenz (1966), this genetic determinism view has necessarily xenophobic and ruthlessly (if not militantly) selfish implications for society. The faith in genetic determinism and reductionism maintained by sociobiologists is expressed by Dawkins (1976) in his claims that "it can be perfectly proper to speak of 'a gene for behavior so-and-so' even if we haven't the faintest idea of the chemical chain of embryonic causes leading from gene to behavior" (pp. 65-66) and that "Be warned that if you wish, as I do, to build a society in which individuals cooperate generously and unselfishly towards a common good, you can expect little help from biological nature" (p. 3).

To what extent is this sociobiological view of human development, and of society, supported by scientific evidence? Asked another way, what scientific evidence do sociobiologists draw on to legitimate their claims, and how adequate is this evidence? We turn now to examination of these key questions.

Evaluating Sociobiological Claims

Given Wilson's (1975a, p. 4) original definition of sociobiology as "the systematic study of the biological basis of all social behavior," it may seem surprising, and perhaps contradictory, to learn that Wilson (1980, p. 296) also contends that "contrary to an impression still widespread among social scientists, sociobiology is not the theory that human behavior has a genetic basis." Perhaps Wilson (1980) is just playing with words. Perhaps he means that sociobiology is not a "theory" but only a "perspective," or merely a rather general framework within which to study systematically the biological and, therefore, ultimately, the genetic basis of all social behavior. Whether or not his statement pertaining to social scientists' mistaken impressions about sociobiology rests on a difference in meaning between the phrases "the theory that ..." and "the systematic study of ...," Wilson's own words show that sociobiology *is* the study of the role of the connection between genes and human social behavior. Wilson (1980, p. 296) in fact uses the term "sociobiology theory" to represent this linkage. He claims:

> Real sociobiological theory allows no less than three possibilities concerning the present status of human social behavior: (a) During the rapid evolution of the human brain, natural selection exhausted any genetic variability of the species affecting social behavior, so that today virtually all human beings are identical with respect to behavioral potential. In addition, the brain has been "freed" from these genes in the sense that all outcomes are determined by culture. The genes, in other words, merely prescribe the capacity for culture. Or, (b) genetic variability has been exhausted, as in (a). But the resulting uniform genotype predisposes psychological development toward certain outcomes as opposed to others. In an ethological sense, species-specific human traits exist and, as in animal repertories, they have a genetic foundation. Or, (c) genetic variability still exists, and, as in (b), at least some human behavioral traits have a genetic foundation. Having identified these alternatives, and stressed the freedom of the discipline of sociobiology from the necessity of any particular outcome, I can now add that the evidence appears to lean heavily in favor of alternative (c).

In the case of each of the given options—(a), (b), and (c)—stress is given to the links among evolution, genetic variability, and human development and society. However, if sociobiologists have spent a good deal of time exploring the first two of the three options, such work has not found its way into the published literature. Hence, Wilson is correct in asserting that to the extent that "evidence" exists in support of any of the three options, it does so in regard to option (c). Yet, support for (c) does not exist because the three options have repeatedly been subjected to comparative scientific analyses. Rather, the preponderance of published sociobiological work—at least insofar as the human literature is concerned—has taken as its "working assumption" option (c). The "evidence" derived from such work constitutes not a test of competing hypotheses, but, rather, an attempt to bring empirical observations to bear on a demonstration of a guiding presupposition.

Indeed, given what are quite well-known facts of genetic variability (McClearn, 1981), it would be nothing short of preposterous to conduct a scientific investigation predicated on the idea that genetic variability does *not* exist. As a consequence, we do not believe it plausible that either Wilson or other sociobiologists are not fully aware of this quite basic evidence about the existence of immense human genetic variability. As a consequence, it is equally difficult to envision that any serious scientific attention could be paid by sociobiologists to options (a) or (b). Therefore, these two options cannot be, and, as we have indicated, are not, treated as viable counters to (c). Instead, this last conception is the only one actually pursued scientifically by sociobiologists. But, given that no alternatives are really comparatively tested, such pursuit is more a demonstration of how empirical phenomena coincide with a conceptual presupposition than a critical test of theoretical options.

How do such demonstrations proceed? Three types of evidence have been invoked. It is useful to examine each type separately.

Comparisons of Humans and Nonhumans: The Concept of Homology

One way in which sociobiologists demonstrate that human social behavior is constrained by evolutionarily shaped genes is to draw parallels between the behaviors of humans and of nonhuman animals. If the behaviors of distinct species can be described similarly, it is argued that there must be some evolutionary connection, or continuity, between them. A common evolutionary pathway for a physical structure or a behavioral function in distinct species is termed a *homology*. Simply, then, sociobiologists argue that if the characteristics of two species can be described in a common way, evidence is present of homologous evolution. The positing of such homology is offered as proof that the characteristics in question are controlled, or constrained, by evolutionarily shaped genes.

The use of such "evidence" is exemplified in the writing of Freedman (1979). In attempting to document his views that human males' gametic potential gives rise to sexually promiscuous behavior—in order to increase their opportunities to garner the "scarce resource" of females' ova—while human females' gametic potential makes them more monogamous, Freedman finds homologies between fruit flies, rhesus monkeys, and South American jungle-dwelling, polygynous humans. Freedman (1979, p. 13) argues that in all species

> females tend to cluster about an average number of young whereas males form a greatly skewed curve, some very successful, many not successful at all. And, since most mammals are polygynous ... this tendency may characterize the entire class Mammalia.

Freedman (1979, p. 14) carries his argument one step further. By again using what he regards as common behavioral descriptions across species, he attempts to provide an evolutionary and genetic account not only for inevitable human male promiscuity but also for the genetically preordained urge to seek sexual relations with other females even to the point of forcing oneself onto them, that is, committing rape (Freedman, 1979). First, citing the work of Grzimek (1972, p. 270), Freedman notes that "[i]n spring, when the gonads are at the peak of their development, there are attempts to 'rape' strange females in the mallard and pintail and a few other species." Second, Freedman makes an inference about the "promiscuous, polygynous intentions" of ducks and, finally, draws a conclusion about the insatiable, continuous, and carnal search by human males for females with whom to copulate. Freedman (1979, p. 14) contends:

> It would appear that if the mallard drake had his way his would be a polygynous species and,

in fact, one does occasionally see a consortship of two females and a male. ... In our own species and our own culture, I am asserting nothing startling when I point out that with sexual maturity, most heterosexual males are in constant search of females, and if inhibited about sexual contact, they fantasize almost continuously and fairly indiscriminately about such contact ... [A]dolescent males in our culture frequently experience life as a nearly continuous erection—spaced by valleys of depression that accompany sexual disappointment.

Are these descriptions, and those by other sociobiologists (Barash, 1977; Wilson, 1975a), of purportedly comparable human and nonhuman social behavior, satisfactory proof of the evolutionary and genetic bases of human behavior? Does apparent descriptive similarity establish evolutionary homology?

The answer to both of these questions is no, for several reasons, not the least of which is the difficulty of accumulating sound scientific evidence of common evolutionary descent when only physical attributes are being considered (Atz, 1970; Gould, 1980). The task is even more problematic in the case of behavioral characteristics, as even very similar behaviors (a) may be manifestations of quite different processes, and/or (b) may serve different functions (Bitterman, 1965, 1975).

In regard to (a), it is a truism that one can describe similar behaviors across even vastly different species. For instance, insects, fish, rats, and humans all "learn"; that is, in members of each of these species systematic and relatively permanent changes in behavior occur in relation to experience. Nevertheless, the ways in which these species learn—the processes of learning—vary considerably. For example, it would be difficult to contend that thought processes play a part in the learning of insects at any point in their lives. In turn, it would be equally difficult to argue that cognition does not enter into human learning for anything other than the earliest years of the life span, and even in infancy cognition may play a role (Piaget, 1970).

Accordingly, although experience-based changes occur in all animals' adjustment to the environment, this similarity is at best evidence for an *analogy*, not a homology (Atz, 1970; Schneirla, 1957). In other words, different processes may subserve analogous functions. But to claim that such descriptive analogies are indicative of common evolutionary histories is, at best, naive, and, at worst, poor scholarship. Dunbar (1987) is frank in admitting this limitation in sociobiological scholarship:

[M]any of those who were influential in promoting the sociobiological perspective ... (e.g., E.O. Wilson) ... tend to be unaware of the more sophisticated nature of the behaviour of higher organisms and are apt to regard even advanced mammals simply as scaled-up insects.

(p. 53)

In turn, and in regard to point (b) above regarding multifunctionality, the presence of identical behaviors in different organisms does not constitute proof for even common function or purpose. To illustrate, the reasons that male mallard ducks might force copulation upon a female of their species are certainly distinct from those involved when a human male rapes a human female. Indeed, to label both the male duck's behavior and the actions of the human male with the same term (rape) seems to trivialize, through biological reductionism, what is certainly a complex and violent human act, one that current scholars point out may not even be a behavior predicated in any way on sexuality or sexual feelings (Sunday and Tobach, 1985).

Can Freedman (1979), Barash (1977), or other sociobiologists who argue for homology on the basis of such cross-species descriptions, contend that the devaluation of women in many sectors of modern society, and the legitimation of violence as a means of exercising social (and political control), enter *not* into the primary causation of forced copulation by human males *and/or* enter as well into the basis of such behaviors in ducks? We think not. Simply stated, the mere portrayal of behaviors in two species as appearing comparable is no proof at all of their common evolutionary heritage. Nor is it any proof at all regarding the extent to which such behaviors are genetically constrained or produced.

Indeed, this conclusion seems to have been reached by Wilson (1980) himself. He notes that, "We cannot rest the hypothesis *of genetic* constraint in human social behavior on the indirect evidences of homology" (Wilson, 1980, p. 297). If the sociobiologists' behavioral homologies do not constitute adequate proof for the genetic basis *of* human social behavior, what then does? Two other types of evidence have been offered, pertaining to the concepts of heritability and adaptation. We first consider heritability.

The Concept of Heritability

Wilson (1980) has argued that the third of the three possible theoretical options upon which sociobiology rests is the one that current evidence favors heavily. This is the notion stressing that genetic variability exists and, as such, that at least some human behavioral traits have a genetic foundation. This seemingly straightforward statement evokes, in actuality, a thicket of conceptual confusion.

Sociobiologists wish to talk about behavioral characteristics—traits—that are common to a species. The task of the sociobiologist is to show scientifically that such traits uniformly and unequivocally characterize the subgroups of humans in question (e.g., males and females), and do so because of the possession *of* evolutionarily based genetic "directives" for genotype reproduction. Stated simply, sociobiologists wish to demonstrate that some human traits—that is, ones common to a given group and dealing with that group's reproductive strategy and hence inclusive fitness—have a genetic basis.

To do so—to demonstrate the common, or invariant, inheritance of these traits—sociobiologists rely on the concept of *heritability*. The line of argument is that, if it can be shown that the trait in question is heritable, it must be the case that the trait is commonly inherited. The claim that if a trait is heritable it is therefore inherited seems so obvious as to border on a tautology or an assertion true by definition. However, nothing could be further from the truth: *The demonstration of heritability says virtually nothing about the extent to which a trait is commonly inherited.* In fact, evidence for heritability cannot be taken as evidence for the common possession of a particular set of genes.

Indeed, quite the opposite is the case. Heritability is "the percentage of variability attributable to genotype" (Pianka, 1978, p. 11). It is an estimate of variation in genes, not of commonality. In fact, we will show that if a human trait were underlain by genes common to all people in a group (as sociobiologists need to establish), the estimated heritability index for that trait would be zero.

If this preamble to a discussion of heritability suggests that sociobiologists are using the concept in a confused and confusing manner, this inference is, from our perspective, entirely warranted. Heritability is a difficult, confusing concept. At first blush it would seem to pertain to the extent to which something is inherited, that is, is based in the genes. For instance, if one were told that "intelligence is 80% heritable" (Jensen, 1969), it might be reasonable to take this to mean that 80% of intelligence was genetically "determined"—that is, that a given person's intelligence was largely (80%) shaped by his or her genes and that something else (environment) shaped the small percentage remaining.

This reasonable interpretation—one often made by some scientists, members of the media, and governmental policymakers—is *completely incorrect*. Technically, heritability pertains only to differences *between* people and has absolutely nothing to do with the extent to which anything—be it genes or environment—determines characteristics *within* an individual. What heritability does refer to is the extent to which differences between people in a specific characteristic can be summarized by genetic differences between these people.

For a technical explanation of this claim, consider the index of "broad heritability" (a term we discuss again later) as a sample case of the usual variance decomposition approach. The coefficient is

$$H^2 = \frac{s^2(G)}{s^2(P)} \quad (1)$$

where H^2 = broad heritability, $s^2(G)$ denotes the genetically determined variance and $s^2(P)$ the phenotypical variance. As is well known, the variance of a variable is defined as

$$s = \left(\sum_i (x_i - x)^2\right) \Big/ (n-1). \quad (2)$$

From this expression we see that (1) relates the individuals' genetic differences to their phenotypical differences. The coefficient H^2 measures the portion of phenotypical variability that can be accounted for by genetic variability.

The coefficient in (1) has been criticized for many reasons such as, for instance, lack of inclusion of possible homogamy or the (dis)similarity of environments. Therefore, more sophisticated approaches have been proposed, such as Cattell's (1960) multiple abstract variance analysis approach and, more recently, the structural equation models presented by Chipuer et al. (1990). However, these approaches have in common the expression of heritability in terms of variance components. Thus, the interpretational basis is, as for H^2, a variability rather than a commonality measure.

To give an example of how misleading heritability interpretations can be, suppose a society had a law pertaining to eligibility for government office. The law was simply that men could be elected to such positions and women could not. Consider what one would need to know in order to divide completely correctly a group of randomly chosen people from this society into one of two groups. Group 1 would consist of those who had greater than a zero percent chance of being elected to a leadership post and group 2 would consist of those who had no chance. All that one would need to know to make this division with complete accuracy was whether a person possessed an XX pair of chromosomes or an XY pair. In the first case, the person would be a female (since possession of the XX chromosome pair leads to female development). In the second, the person would be a male. One could thus correctly place all possessors of the XY pair into the "greater than zero chance" group and all possessors of the XX pair into the "no chance" group.

In this example, then, *all* the differences between people with respect to the characteristic in question—eligibility for office—can be summarized by genetic differences between them, that is, possession of either the XX or the XY chromosome pair. In this case the heritability of "being eligible" would be 1.0. In other words, in this society eligibility is 100% heritable. But, by any stretch of the imagination, does this mean that the eligibility characteristic is inherited, or that the differences between men and women with respect to this characteristic are genetic in nature?

Is there a gene for "eligibility," one that men possess and women do not? Of course, the answer to these questions is no. Although heritability in this case is perfect, it is social (environmental) variables—laws regarding what men and women can and cannot do—that determine whether or not someone has a chance of being elected. Indeed, if the law in question were changed, and women were now allowed to hold office, then the heritability of the eligibility characteristic would—probably rather quickly—fall to much less than 1.0.

Hebb (1970) offers another useful example, one drawing on a "modest proposal" put forth by Mark Twain:

> Mark Twain once proposed that boys should be raised in barrels to the age of 12 and fed through the bung-hole. Suppose we have 100 boys reared this way, with a practically identical environment. Jensen agrees that environment has *some* importance (20% worth?), so we must expect that the boys on emerging from the barrels will have a mean IQ well below 100. However, the variance attributable

to the environment is practically zero, so on the "analysis of variance" argument, the environment is not a factor in the low level of IQ, which is nonsense.

(p. 578)

In Hebb's example, environment had no *differential* effect on the boys' IQs; presumably in all boys it has the same (severely limiting) effect. In having this same effect, environment could contribute nothing to differences between the boys. No difference—or variation—existed in the environment, and so the environment could not be said to contribute anything to differences between people. Yet, it is also obvious that environment had a major influence on the boys' IQ scores. Even with IQ heritability equal to + 1.0, the intelligence of each of the boys would have been different had he developed in an environment other than a barrel.

A third example is based on actual empirical research. Partanen et al. (1966) analyzed data from 172 monozygotic and 557 dizygotic male twin pairs. All participants were alcohol users. The aim of the study was to estimate the degree to which *alcohol abuse* is genetically determined. When measured by frequency of alcohol consumption, alcohol abuse seems to have at least a modest genetic component (heritability = 0.40). However, if one uses the amount of alcohol consumed on each occasion, the heritability estimate drops considerably (to 0.27). A third measure of alcohol abuse, the number of citations and other social conflicts resulting from drinking, yields a heritability estimate of 0.02.

Thus, judgments concerning heritability can depend largely on the definition and operationalization of the behavior under study. In addition, the confusion between commonality and variability can lead to misinterpretation. Accordingly, *high heritability does not mean developmental fixity*. A high estimate of heritability means that environment does not contribute very much to *differences* among people in their expression of a trait; yet environment may still provide an important (although invariant) source of the expression of that trait, for instance in determining the average level of a trait shown by people in a given group.

From Hebb's example we see clearly that although heritability may be high, the characteristic in question may still be influenced by the environment. Even when environment contributes nothing to *differences* between people in a population, this fact does not mean that the population characteristic is fixed by heredity or that it is unavailable to environmental influence. As Hebb well points out, while contributing nothing to differences between people, environment can still be a uniformly potent source of behavioral development and functioning within each of the people in a group.

A related point has been made by Lehrman (1970). When geneticists speak of a trait as heritable, all they mean is that one is able to predict the trait distribution in the offspring of a group on the basis of knowing the trait distribution in the parent group. One can predict the distribution of eye color in the offspring generation merely by knowing the distribution of eye color among the parents. Thus, while geneticists may use the term *hereditary* or *inherited* as interchangeable with the term *heritable*, they are not, by such usage, making any statements about the *process* involved in the development of this trait. In other words, the geneticist is not saying anything at all about the way that nature and nurture serve as sources of a heritable trait. Thus, the geneticist is not saying anything about the extent to which the expression of the trait may change in response to environmental modification. In short, a geneticist would not say that a highly heritable trait cannot be influenced by the environment. Rather, the geneticist would probably recognize, as we now must, that even if the heritability *of* a trait is +1.0, an almost infinite number of *expressions* (phenotypes) of that trait may be expected to develop as a result of an interaction with the almost infinite number of environments to which any one genotype may be exposed.

Those who equate heritability with genetic determination assume that as the magnitude of heritability increases from zero to +1.0, less and less can be done through environmental modifications to alter the expression of the trait. Correspondingly, they assume that if heritability is low, more room is left for alteration of the trait by means of environmental manipulation. This argument is fallacious. As noted by Scarr-Salapatek (1971, p. 1128):

> The most common misunderstanding of the concept "heritability" relates to the myth of fixed intelligence: if h^2 [heritability] is high, this reasoning goes, then intelligence is genetically fixed and unchangeable at the phenotypic level. This misconception ignores the fact that h^2 is a population statistic, bound to a given set of environmental conditions at a given point in time. Neither intelligence nor h^2 estimates are fixed.

In short, whatever the level reached by an estimate of heritability, environmental variation may be a (or the) key causative factor. In addition, it is clear that high heritability does not mean developmental fixity. If the social context changes, one cannot be certain if any information one has about heritability still applies.

Several other statistical and methodological problems associated with the determination of heritability are important to note (Hirsch, 1970, 1976, 1990a, b; Hirsch et al., 1980; Lerner, 1986; McGuire and Hirsch, 1977; Walsten, 1990). A key problem arises in regard to differences between the concept of broad heritability (H^2), mentioned earlier, and the additional concept of narrow heritability (h^2). To understand these concepts, we should recognize that the contributions of variation in heredity and environment to the variance in a given behavior such as general intelligence or personality might be expressed in several ways.

Hereditary and environmental variance can relate to behavioral variance separately and independently. In such a case, what one contributes is unrelated to what the other contributes. Each may be labeled as a main effect, in the conventional analysis of variance sense of the term. Alternatively, the contributions of hereditary variance and environmental variance may interact, again in the conventional analysis of variance meaning of the term. Finally, heredity-environment (or genotype-environment) correlation occurs when hereditary (genotype) and environmental differences covary. Together, the concepts of main, interactional, and correlational contributions of hereditary and environmental variation allow the concepts of broad heritability (H^2) and narrow heritability (h^2) to be distinguished and the statistical problems associated with them to be noted.

H^2 is the proportion of the total variation in a given behavior that may be attributed to the sum of (a) the independent genetic variation (i.e., the main effect of hereditary variance), and (b) the variance due to genotype-environment correlation (McGuire and Hirsch, 1977, p. 46). H^2 is of little interest in genetics (McGuire and Hirsch, 1977), primarily because it does not allow the contribution of genotype variance per se to be disentangled from environmental variance; and, of course, such separation is an objective of heritability analyses in the first place.

Narrow heritability (h^2) is simply the proportion of variance in behavior that may be attributed solely to the main effect of hereditary (genotypic) variance; that is, h^2 is the hereditary variance that exists independent of, and thus merely separately adds to, environmental variance (McGuire and Hirsch, 1977). It is the determination of h^2 toward which most heritability analysis is aimed. But there, especially in the case of the analysis of such variation in human behavior, lies the rub. It is in human heritability analysis, and the calculation of h^2, that the statistical and methodological problems inherent in this work arise.

It makes sense to attribute through heritability analysis (and the calculation of h^2) variation in some behavior to variation in heredity *only* if the contributions of heredity and of the environment are additive (Walsten, 1990). If heredity and environment interact, the calculation of heritability is quite problematic (Bullock, 1990;

Feldman and Lewontin, 1975; McGuire and Hirsch, 1977; Walsten, 1990). However, in assessing heritability among humans, the statistical techniques used to determine if there is evidence for heredity—environment interaction (analysis of variance or its equivalent, multiple regression) are not as sensitive to the presence of interactions as they are to the independent (and hence additive) influences of heredity and environment. In other words, these statistical techniques cannot detect as readily the presence of interactions as they can the presence of main effects (Walsten, 1990).

This problem is especially apparent when these statistical tests are used with relatively small numbers of observations, and it is unfortunately the case that such small samples are generally the rule in social and behavioral science studies involving the calculation of heritability (Walsten, 1990). Thus, the smaller the sample used in a study, the greater the likelihood of the statistical tests being unable to identify an interaction that is actually present. In such situations, the inference that an interaction between heredity and environment does not exist, and that in turn the contributions of these two factors involve only main, and therefore only additive, effects, is incorrect. As a consequence, heritability estimates, and any conclusions based on them, are similarly misconceived.

If a large sample of observations exists, other methodological problems occur. If a scientist is interested in determining how much variance in a behavior is accounted for by variance in what people inherit, versus what they experience in their environment, it is imperative that the nature of the specific heredity involved in the group under study be completely certain. If one wants to determine the extent to which variation in children's intelligence is accounted for by the genes provided to children by their mothers and fathers, as compared to the environments provided by these parents, one must be certain to measure the intelligence of children and their actual biological mothers and fathers. As Hirsch et al. (1980, p. 236) have argued, "A sine qua non for the study of heredity is proof positive of the presumed biological relationship, i.e., ascertainment of the biological validity of the designated kinships, such as parent-offspring, sibling, etc."

Unfortunately, however, few studies of heritability have included controls for presumed biological relatedness. This absence makes one uneasy about presuming that in all studies involving estimates of the contribution of genes to behavioral resemblance between, for instance, children and parents, accurate designations of biological relatedness have occurred. Such concerns are magnified when, in the few studies including such controls, evidence emerges that substantial proportions of the people labeled as "biological parent" are in fact unrelated biologically to the children in question. For instance, Hirsch et al. (1980) were able to determine, through blood testing, the actual biological relationships between parents and children within a subsample of 38 of the 112 families they studied. In 13% of the families in this subsample, there were children who could not have been the biological offspring of at least one of the putative parents in the family.

Similarly, Philipp (1973; Hirsch, 1990a; Hirsch et al., 1980), reporting evidence derived from comparable analyses done on samples from the U.K., noted that there were data disqualifying as the presumed fathers 30% of the husbands within the families being studied.

When correct information regarding biological relatedness is unavailable to the researcher who is appraising heritability, literal miscalculations and inferential errors abound. Such concerns have led Kempthorne (1990, p. 139), a quantitative genetic analyst, to the view that "most of the literature on heritability in species that cannot be experimentally manipulated, for example, in mating, should be ignored." In a similar vein, population geneticists Feldman and Lewontin (1975, p. 116B) have concluded that "Certainly the sample estimate of heritability, either in the broad or narrow sense, but most especially in the broad sense, is nearly equivalent to no information at all for any serious problem of human genetics."

Thus, what heritability estimates at best provide (and *only* in the case of additive genotypic variation only, large sample sizes, compelling evidence for no genotype-environment interaction or correlation, and valid evidence for the relevant biological relatedness) is an estimate of the extent to which genetic differences (variation) within a given group are associated with differences (variation) in the scores for a trait measured among people in that group at a specific time in their lives.

In sum, then, heritability estimates describe only characteristics of a distribution of scores; they describe only a feature of differences between people. Such estimates say nothing about the trait itself. Such estimates, in particular, say nothing about the genetic and/or environmental determination (or cause) of the trait *within* any person in a group. Certainly, from such an estimate of *between*-people differences (which is what heritability is), one cannot legitimately make any statements about how humans have been selected for the homogeneous presence of a trait (Lewontin et al., 1984). Indeed, to the extent such homogeneity exists, heritability must be low, due to lack of variation.

Sociobiologists, in using the presence of heritability to support their claim for the genetic determination of universal characteristics of human social functioning, are, in effect, "shooting themselves in the foot" every time they argue that the presence of heritability supports this claim. As we have seen, heritability data, to the extent that they are useful at all, imply just the opposite. Any attempt to use heritability data to support a claim for the universal, evolutionarily based genetic determination of behavior is thus an argument based on a misunderstanding and misapplication of the heritability concept.

There is, then, only one line of argument left to support sociobiological claims about the evolutionarily based, genetic source of human social behavior. This line of argument involves the concept of adaptation and the view that the patterns of human development seen in society reflect evolutionarily based, and therefore naturally selected, genetically influenced adaptations to the pressures of humans's context.

Are Adaptations Everywhere?

A cornerstone of the sociobiological "method" is to offer explanations in the vein of Kipling's *Just-So Stories* of how particular social behaviors, or differences among people in their social status or roles, came to be (Gould, 1980). As recounted by Gould (1980, p. 258):

> Rudyard Kipling asked how the leopard got its spots, the rhino its wrinkled skin. He called his answers "just-so stories." When evolutionists try to explain form and behavior, they also tell just-so stories—and the agent is natural selection. Virtuosity in invention replaces testability as the criterion for acceptance.

According to Gould (1980), this unacceptable scientific procedure led the biologist von Bertalanffy (1969, p. 11) to complain:

> If selection is taken as an axiomatic and a priori principle, it is always possible to imagine auxiliary hypotheses—unproved and by nature unprovable—to make it work in any special case ... Some adaptive value ... can always be construed or imagined. ... I think the fact that a theory so vague, so insufficiently verifiable and so far from the criteria otherwise applied in "hard" science, has become a dogma, can only be explained on sociological grounds. Society and science have been so steeped in the ideas of mechanism, utilitarianism, and the economic concept of free competition, that instead of God, Selection was enthroned as ultimate reality.

According to both Gould (1980) and von Bertalanffy (1969), the key feature of sociobiological "just-so stories" is that these current arrangements in society are *adaptations*; that is, they are changes that enhance fitness, that have been shaped by natural selection over the eons of human evolution to have this function, and that are now represented in the human genotype. Yet, it is the key element in these arguments—the presence of an adaptation, of a change in fitness—that all too often remains a scientifically unverified, post hoc story.

Indeed, as admitted by Dunbar (1987, p. 50):

A simple statement that "X increases the fitness of those that perform it" explains nothing: it is strictly tautologous, for improving fitness is what every sociobiological explanation implicitly assumes. What we need to know—and this is the heart of any sociobiological explanation—is: *How* does it increase fitness? ... It is the transparent failure to answer this question that has left so many sociobiologists open to criticisms of "Just-So" story-telling and unscientific practice. Since we necessarily have to rely on comparative observations rather than experimental manipulation when tackling evolutionary problems, we are particularly exposed to this kind of accusation. The only way to avoid it is to provide as watertight a case as is possible by showing that proximate problems of survival or reproduction are in fact resolved when individuals behave in a specified way, and that efficient solutions to these problems will result in increased contributions to the species' future gene pool. This will not always be easy, but, unless it can be done, sociobiological explanations will always be open to skeptical doubts, particularly where these doubts are fuelled by political or religious conviction.

Despite these explanatory difficulties, sociobiologists see adaptations—changes in fitness "designed" by (or, actually, "resulting" from) natural selection—as everywhere. And, in the view of sociobiologists, these changes in fitness, since they are adaptations, are *optimizations*. That is, as argued as well by 19th-century social Darwinists (Tobach et al., 1974), natural selection results in genetically based features that are the "time-tested," best possible outcomes of humans' evolutionary history.

To sociobiologists, then, that which exists is an adaptation: Humans' social behaviors and the niches they occupy in the social hierarchy have been shaped by natural selection to take their present form. As claimed succinctly by Konner (1982, p. 18), "An organism has characteristics; they must have been selected for or they wouldn't be here now." Given the centrality of the concept of adaptation in sociobiologists' thinking, we may ask whether there is the direct, uniform, and singular pathway that sociobiologists infer from evolution, through natural selection, to adaptation and the present character of people and society. We may ask also precisely why presenting a story—which is a possible scenario of the way natural selection *could have* resulted in a given feature of human behavior—is not sufficient to establish scientifically that just such a history transpired.

The work of Gould and Vrba (1982) is quite relevant to these issues. They try to provide a new term in evolutionary biology in order to clarify some important, but confusing, uses of the term "adaptation." Gould and Vrba note that one meaning of adaptation is the shaping of a feature of the organism (a physical attribute or a behavior, for instance) by natural selection for the function it now performs. A second meaning is a more static one, referring to the immediate way in which a physical feature or a behavior enhances the organism's current ability to fit its context. This second meaning does not take into account the historical origin of the feature, but only whether the organism's physical or behavioral characteristics help it to meet the current demands of its environment.

Gould and Vrba (1982) cite Williams (1966) as adhering to the first definition of adaptation. Williams (1966, p. 6) contended that one should speak of adaptation only when one can "attribute the origin and perfection of this design to a long period of selection for effectiveness in this particular role." Bock's (1979) views illustrate the second definition of adaptation. Bock indicates that "an adaptation is ... a feature of the organism ...

which interacts operationally with some factor of its environment so that the individual survives and reproduces" (p. 39).

Gould and Vrba (1982) believe that a confusion therefore exists regarding a central concept in evolutionary theory, adaptation. This conflict exists because a single term has been used, despite the fact that different criteria for the historical basis of a given organism feature and for its current use are involved in the two meanings of the term. Darwin (1859, p. 197) himself may have seen this potential confusion:

> The sutures in the skulls of young mammals have been advanced as a beautiful adaptation for aiding parturition, and no doubt they facilitate, or may be indispensable for this act; but as sutures occur in the skulls of young birds and reptiles, which have only to escape from a broken egg, we may infer that this structure has arisen from the laws of growth, and has been taken advantage of in the parturition of the higher animals.

In other words, while Darwin saw the necessity of unfused sutures in the skulls of young mammals, he was uncertain about labeling the unfused sutures as adaptations, because the unfused sutures were not built by selection to function as they now do in mammals (Gould and Vrba, 1982). But if the unfused sutures are not adaptations, if they were not shaped by natural selection, what are they and where did they come from? Clearly, a new term must be used to rectify the confusion, and Gould and Vrba (1982, p. 6) provide one. They suggest that such characters evolved for other usages (or for no function at all), and were later "coopted" for their current role. They term such characters "exaptations." The characters are fit for their current role (i.e., they are aptus), but they were not designed by natural selection for this role, therefore, they are not *ad aptus* (pushed toward fitness by natural selection).

A clear implication of Gould and Vrba's revised terminology is that not all instances of fitness are adaptations; that is, not all features of an organism's structure and function that are aptational have this character as a consequence of being shaped by natural selection. Such a possibility, if supported, would serve to weaken what Gould and Lewontin (1979) have labeled the "adaptationist program"—the position, reflected in the earlier quote by Konner (1982, p. 18) that a feature's current aptational character implies historical shaping by natural selection for that character.

Lewontin (1981) has discussed the adaptationist "program" and its conventional use of the concept of adaptation. As do Gould and Vrba (1982), Lewontin (1981) sees problems with this view of adaptation; in essence, he sees the view as deficient because it ignores the active, constructive role the organism plays in its own adaptation. The organism shapes the context to which it adapts, and hence there is a reciprocal, multilevel (i.e., fused) relation between organism and context (Lerner, 1978, 1986, 1991; Lerner and Kauffman, 1985). Lewontin's (1981) criticisms of the conventional use of the concept of adaptation lead to a view *of* the organism compatible within a developmental contextual conception of human development (Lerner, 1991). Specifically, Lewontin (1981, p. 245) notes:

> Organisms ... by their own life activities determine which aspects of the outer world make up their environment. Organisms change the environment by their activities ... they "construct" environments. The problem is that the concept of adaptation has been extended metaphorically from its valid domain of describing individual, short-term, goal-directed behavior to other levels ... [I]t is pure metaphor, ideologically molded by the progressivism and optimism of the nineteenth century, to describe numbers of chromosomes, patterns of fertility, migrations, and religious institutions as "adaptations" ... It is not simply that some evolutionary process can be described as nonadaptive, but that the entire framework is in question. Whether we look at the fossil record or at living species, we do not see them

as "adapting," but as "adapted". But how can that be? How is it that, if evolution is a process of *adapting,* organisms always seem to be *adapted?* It may be more illuminating to see organisms as *changing* and, in the process, as reconstructing the elements of the outer world into a new environment that is sufficient for their survival.

Consistent with the position of Lewontin (1981) regarding the problems with the "adaptationist program," Gould and Vrba (1982) contend that recognition of the potential presence of exaptative features leads one to recognize that previously nonaptative (note, *not* preadaptive) features may be present and may be coopted for fitness—a recognition that provides a key for plasticity in evolutionary processes and for the role of individuals' own organismic characteristics in their development. Gould and Vrba (1982, pp. 12–13) indicate:

> Flexibility lies in the pool of features available for cooptation ... The paths of evolution—both the constraints and the opportunities—must be largely set by the size and nature of this pool of potential exaptations. Exaptive possibilities define the "internal" contribution that organisms make to their own evolutionary future.

The concept of exaptation leads to the understanding that the processes involved in evolution are plastic ones. The concept is consistent with a key theme in the developmental contextual alternative to a biological determinist view of the role of biology in human development (Lerner, 1991, in press). According to this alternative, processes exist that contribute to the plasticity of people's functioning—that allow them to play a role in the development of their own flexible characteristics.

Conclusions

Hence, the third line of evidence relied on by sociobiologists—an adaptationist story line to explain what are purported to be genetically based differences in individual and social development—fails. "Just-so stories" (Gould, 1980) about human evolutionary history are used to substitute superficial descriptions for in-depth explanations (Piaget, 1979); alternative paths to current fitness (or aptation) are excluded from scientific consideration or analysis.

Equally serious problems arise in regard to the other two lines of evidence relied on by sociobiologists—involving the inappropriate postulation of homologies between nonhuman and human animals and the misuse of the concept of heritability. These logical and empirical problems reveal the weak scientific basis of the sociobiological viewpoint. The severity of these problems suggests that sociobiological thinking has little relevance for the understanding of human behavior and development in general, or individual differences (and, most specifically, sex differences) in particular.

The limited scientific utility of the sociobiological concept of genetic determinism is consistent with the shortcomings of other reductionistic views of the role of genes in human behavior (Lewontin et al., 1984; Tobach et al., 1974). However, one need not eschew the role of genes in human behavior and development because of the failure of genetic reductionistic models such as sociobiology. Developmental contextual views (Lerner, 1986, 1991) and developmental systems models (Gottlieb, 1991b) incorporate gene structure and function within a "fusion," or a reciprocal integration, of variables from multiple levels of analysis.

In developmental contextualism, the unit of analysis is the changing *relation* between organism (or gene) and context; it is not either element of this relation alone. There is compelling evidence that genes do not directly—that is, in and of themselves—produce any structural or functional characteristics of an organism (Gottlieb, 1991a, b). Genes are not independently acting sources of development. Instead, the action of genes (i.e., genetic expression) is "affected by events at other levels of the [developmental] system, including the environment of the organism" (Gottlieb, 1991b, p. 5), and all levels of organization within this integrated, developmental system "may be considered as potentially equal" (Gottlieb, 1991b, p. 6). The literature pertinent to developmental contextualism indicates that (a) intraorganism variables making up to the proximal context of the gene, and (b) extraorganism contextual variables, exist in a reciprocally influential relation with genes. If, as seems to be the case (Gottlieb, 1991a, b; Lerner, 1984, 1991), increasing theoretical and empirical attention is paid to synthetic viewpoints such as developmental contextualism, it may be that the future will bring greater understanding of the ubiquitous but dynamically interactive role of genes in human behavior and development.

Acknowledgment

Supported in part by NICHD grant HD23229.

References

Atz, J.W. (1970). The application of the idea of homology to behavior. In L.R. Aronson, E. Tobach, D.S. Lehrman, & J.S. Rosenblatt (Eds.), *Development and evolution of behavior: Essays in memory of T.C. Schneirla* (pp. 53–74). San Francisco, CA: Freeman.

Barash, D.P. (1977). *Sociobiology and behavior*. New York: Elsevier.

Barlow, G.W., & Silverberg, J. (Eds.). (1980). *Sociobiology: Beyond nature/nurture? Reports, definitions and debate*. Boulder, CO: Westview Press.

Bitterman, M.E. (1965). Phyletic differences in learning. *American Psychologist, 20*, 396–410.

Bitterman, M.E. (1975). The comparative analysis of learning. *Science, 18*, 699–709.

Bock, W. (1979). A synthetic explanation of macroevolutionary change—a reductionistic approach. *Bulletin of the Carnegie Museum of Natural History, 13*, 20–69.

Bullock, D. (1990). Methodological heterogeneity and the anachronistic status of ANOVA in psychology. *Behavioral and Brain Sciences, 13*, 122–123.

Caplan, A.L. (Ed.). (1978). *The sociobiology debate*. New York: Harper & Row.

Cattell, R.B. (1960). The multiple abstract variance analysis equations and statistics. *Psychological Review, 67*, 122–146.

Chipuer, H.M., Rovine, M.J., & Plomin, R. (1990). LISREL modeling: Genetic and environmental influences on IQ revisited. *Intelligence, 14*, 11–29.

Darwin, C. (1859). *On the origin of species by means of natural selection or the preservation of favored races in the struggle for life*. London: John Murray.

Dawkins, R. (1976). *The selfish gene*. New York: Oxford University Press.

Dunbar, R.I.M. (1987). Sociobiological explanations and the evolution of ethnocentrism. In V. Reynolds, V. Falger, & I. Vine (Eds.), *The sociobiology of ethnocentrism* (pp. 48–59). London: Croom Helm.

Erikson, E. (1968). *Identity, youth, and crisis*. New York: Norton.

Feldman, M.W., & Lewontin, R.C. (1975). The heritability hang-up. *Science, 190*, 1163–1168.

Flohr, H. (1987). Biological bases of social prejudices. In V. Reynolds, V. Falger, & I. Vine (Eds.), *The sociobiology of ethnocentrism* (pp. 190–207). London: Croom Helm.

Freedman, D.G. (1979). *Human sociobiology; A holistic approach*. New York: Free Press.

Freud, S. (1923). *The ego and the id*. London: Hogarth.

Gottlieb, G. (1991a). The experiential canalization of behavioral development: Theory. *Developmental Psychology, 27*, 4–13.

Gottlieb, G. (1991b). The experiential canalization of behavioral development: Research. *Developmental Psychology, 27*, 35–39.

Gould, S.J. (1980). Sociobiology and the theory of natural selection. In G.W. Barlow & J. Silverberg (Eds.), *Sociobiology: Beyond nature/nurture* (pp. 257–269). Boulder, CO: Westview Press.

Gould, S.J., & Lewontin, R.C. (1979). The spandrels of San Marco and the Panglossian paradigm: A critique of the adaptationist programme. In J. Maynard Smith & R. Holliday (Eds.), *The evolution of adaptation by natural selection*. London: Royal Society of London.

Gould, S.J., & Vrba, E. (1982). Exaptation: A missing term in the science of form. *Paleobiology, 8*, 4–15.

Grzimek, B. (Ed.). (1972). *Animal life encyclopedia*. Vol. 7. New York: Von Nostrand Reinhold.

Hebb, D.O. (1970). A return to Jensen and his social critics. *American Psychologist, 25*, 568.

Hirsch, J. (1970). Behavior-genetic analysis and its biosocial consequences. *Seminars in Psychiatry, 2*, 89–105.

Hirsch, J. (1976). Review of *Sociobiology* by E.O. Wilson. *Animal Behavior, 24*, 707–709.

Hirsch, J. (1990a). Correlation, causation, and careerism. *European Bulletin of Cognitive Psychology, 10,* 647–652.

Hirsch, J. (1990b). A nemesis for heritability estimation. *Behavioral and Brain Sciences, 13,* 137–138.

Hirsch, J., McGuire, T.R., & Vetta, A. (1980). Concepts of behavior genetics and misapplications to humans. In J. Lockard (Ed.), *The evolution of human social behavior* (pp. 215–238). New York: Elsevier.

Jensen, A.R. (1969). How much can we boost IQ and scholastic achievement? *Harvard Educational Review, 19,* 1–123.

Keinpthorne, O. (1990). How does one apply statistical analysis to our understanding of the development of human relationships? *Behavioral and Brain Sciences, 13,* 138–139.

Kinsey, A.C. et al. (1953). *Sexual behavior in the human female.* Philadelphia: Saunders.

Kitcher, P. (1985). *Vaulting ambition.* Cambridge, MA: MIT Press.

Konner, M. (1982). *The tangled wing.* New York: Holt, Rinehart and Winston.

Lehrman, D.S. (1970). Semantic and conceptual issues in the nature-nurture problem. In L.R. Aronson, E. Tobach, D.S. Lehrman, & J.S. Rosenblatt (Eds.), *Development and evolution of behavior: Essays in memory of T.C. Sclmeirla.* San Francisco, CA: Freeman.

Lerner, R.M. (1978). Nature, nurture and dynamic interactionism. *Human Development, 21,* 1–20.

Lerner, R.M. (1984). *On the nature of human plasticity.* New York: Cambridge University Press.

Lerner, R.M. (1986). *Concepts and theories of human development* (2nd ed.). New York: Random House.

Lerner, R.M. (1991). Changing organism-context relations as the basic process of development: A developmental contextual perspective. *Developmental Psychology, 27,* 27–32.

Lerner, R.M. (in press). *Final solutions: Biology, prejudice, and genocide.* University Park, PA: Pennsylvania State University Press.

Lerner, R.M., & Kauffman, M.B. (1985). The concept of development in contextualism. *Developmental Review, 5,* 309–333.

Lewontin, R.C. (1981). On constraints and adaptation. *Behavioral and Brain Sciences, 4,* 244–245.

Lewontin, R.C., Rose, S., & Kamin, L.J. (1984). *Not in our genes: Biology, ideology, and human nature.* New York: Pantheon Books.

Lorenz, K. (1940). Durch Domestikation verursachte Störungen arteigenen Verhaltens. *Zeitschrift für angewandte Psychologie und Charakterkunde, 59,* 2–81.

Lorenz, K. (1966). *On aggression.* New York: Harcourt, Brace & World.

Lumsden, C.J., & Wilson, E.O. (1981). *Genes, mind, and culture.* Cambridge, MA: Harvard University Press.

MacDonald, K.B. (Ed.). (1988). *Sociobiological perspectives on human development.* New York: Springer-Verlag.

McClearn, G.E. (1981). Evolution and genetic variability. In E.S. Golin (Ed.), *Developmental plasticity: Behavioral and biological aspects of variations in development* (pp. 3–31). New York: Academic Press.

McGuire, T.R., & Hirsch, J. (1977). General intelligence (g) and heritability (H^2, h^2). In I.C. Uzgiris & F. Weizmann (Eds.), *The structuring of experience* (pp. 25–72). New York: Plenum.

Partanen, J., Brunn, K., & Markkanen, T. (1966). Inheritance of drinking. *The Finnish Foundation for Alcohol Studies, 14.*

Philipp, E.E. (1973). Discussion in *Law and ethics of AID and embryo transfer* (p. 66). Ciba Foundation Symposium 17 (new series). Amsterdam: Elsevier, Excerpta Medica, North Holland.

Piaget, J. (1970). Piaget's theory. In P.H. Müssen (Ed.), *Carmichael's manual of child psychology* (3rd ed., Vol. 1, pp. 703–732). New York: Wiley.

Piaget, J. (1979). Relations between psychology and other sciences. *Annual Review of Psychology, 30,* 1–8.

Pianka, E.R. (1978). *Evolutionary ecology.* New York: Harper & Row.

Rushton, J.P. (1987). An evolutionary theory of health, longevity, and personality: Sociobiology and r/k reproductive strategies. *Psychological Reports, 60,* 539–549.

Rushton, J.P. (1988a). Race differences in behavior: A review and evolutionary analysis. *Personality and Individual Differences, 9,* 1009–1024.

Rushton, J.P. (1988b). Do r/k reproductive strategies apply to human differences? *Social Biology, 35,* 337–340.

Scarr-Salapatek, S. (1971). Unknowns in the IQ equation. *Science, 174,* 1223–1228.

Schneirla, T.C. (1957). The concept of development in comparative psychology. In D.B. Harris (Ed.), *The concept of development* (pp. 78–108). Minneapolis: University of Minnesota Press.

Sunday, S.R., & Tobach, E. (Eds.). (1985). *Violence against women: A critique of the sociobiology of rape.* New York: Gordian Press.

Tobach, E., Gianutsos, J., Topoff, H.R., & Gross, C.G. (1974). *The four horses: Racism, sexism, militarism, and social Darwinism.* New York: Behavioral Publications.

van den Berghe, P.L., & Barash, D.P. (1977). Inclusive fitness and human family structure. *American Anthropologist, 79,* 809–823.

von Bertalanffy, L. (1969). Chance or law. In A. Koestler (Ed.), *Beyond reductionism.* London: Hitchinson.

Walsten, D. (1990). Insensitivity of the analysis of variance to heredity-environment interaction. *Behavioral and Brain Sciences, 13,* 109–120.

Williams, G.C. (1966). *Adaptation and natural selection.* Princeton, NJ: Princeton University Press.

Wilson, E.O. (1975a). *Sociobiology: The new synthesis.* Cambridge, MA: Harvard University Press.

Wilson, E.O. (1975b). Human decency is animal. *The New York Times Magazine,* October 12.

Wilson, E.O. (1978). *On human nature.* Cambridge, MA: Harvard University Press.

Wilson, E.O. (1980). A consideration of the genetic foundation of human social behavior. In G.W. Barlow & J. Silverberg (Eds.), *Sociobiology: Beyond nature/nurture* (pp. 295–305). Boulder, CO: West View Press.

SECTION VII

EPIGENETICS

EDITORS' INTRODUCTION

According to the concept of embodiment associated with the relational developmental systems (RDS) metamodel, morphological, physiological, psychological, and behavioral attributes of the person, in fusion with culture, have a temporal (historical) parameter (Marshall et al., 2021; Overton, 2008, 2013, 2015). According to Overton, 2013, (p. 103):

> Embodiment includes not merely the physical structures of the body but the body as a form of lived experience, actively engaged with the world of socio-cultural and physical objects. The body as form references the biological point-of-view, the body as lived experience references the psychological subject standpoint, and the body actively engaged with the world represents the socio-cultural point-of-view.

Peter Marshall and colleagues (2021) note that embodiment explains how Overton's (2008, 2013, 2015) tripartite conception of embodiment accounts for how a dynamic system can construct itself, that is, possess the attribute of autopoiesis involved in the processes of a living, open, and dynamic system. Marshall et al. (2021, p. 4), explain that:

> living things actively self-maintain themselves through the constant regeneration of the conditions that are necessary to sustain their material existence" (Marshall et al., 2021, p. 3) ... [and thus reflect] "constitutive autonomy ... in contrast to behavioral autonomy, where the identity of the system is imposed externally by an operator or observer.

David Witherington (2014, p. 27) makes a similar point in noting that the tripartite conception of embodiment enables each person to act "as its own cause, organizing and producing itself such that it causes and results from itself. In this way, living systems constitute natural ends or purposes." Embodiment, then, depicts the ongoing coaction between these systems and levels of organization, both internal and external to the person, and conveys the inter-penetrating and bidirectional nature of experience at every level of development of a human being (e.g., Marshall et al., 2021).

As such, embodiment has implications across ontogeny and phylogeny (Ho, 2010; Jablonka & Lamb, 2005). One key implication of embodiment involves the idea that qualitative changes emerge across the life span through the integration of organisms and contextual levels of organization (Lerner, 1984, 2002). A second key implication is the creation of relative plasticity in phylogeny and ontogeny occurring because of embodied actions resulting in autopoietic, or self-constructing, change in the developmental system (Witherington, 2014, 2015). Relative plasticity characterizes the relations between organisms and contexts that, across time, create qualitative change in developmental processes within and across generations (Lerner, 1984). This qualitative discontinuity involves what developmental scientists have termed "epigenetic (emergent) change" (e.g., Gottlieb, 1997, 1998; Werner, 1957) in ontogeny. In turn, the action of genetic ⇔ context processes, which are instances of embodied change within the developmental system, are the focus of study in the field of epigenetics (e.g., Meaney, 2010; Misteli, 2013; Moore, 2015a; Slavich & Cole, 2013).

It is important to distinguish the differences in denotation for these two uses of the term *epigenesis*. As we have noted, within the description of developmental change across the life span, the term *epigenesis* refers to the emergence of qualitatively discontinuous characteristics (e.g., developmental stages) across ontogeny (see Gottlieb, 1997, 1998; Lerner, 1984; Lerner & Benson, 2013a, 2013b, 2013c; Werner, 1957). In turn, Misteli (2013), noting that the term *epi* comes from the Greek and means "over" or "above," indicated that epigenetic effects are effects that are ones "beyond" the effects of genes. Accordingly, in the literature of evolutionary biology and molecular biology, the term *epigenetics* refers to a process involving gene ⇔ context relations resulting in the modification of information transmitted by DNA (through messenger RNA, or mRNA) across long, even multigenerational time scales (e.g., Meaney, 2010; Misteli, 2013; Slavich & Cole, 2013).

The two concepts (of epigenetic/emergent change across ontogeny and changes in the information transmitted by DNA through epigenetics) may pertain of course to interrelated phenomena. Emergent change across the life span is explained, within theories associated with the RDS metamodel, by systems changes involving mutually influential relations among levels of organizations, which would include the gene ⇔ context relations involved in epigenetics (e.g., Lerner & Benson, 2013a, 2013b, 2013c; Witherington, 2014). As well, contemporary scholarship about the features of epigenetics and evolution reflects the concept of embodied change (of fusion, or integration, of changes at all levels of organization within the developmental system). The embodiment of biological change within the RDS means that the impact of an individual's biology on his or her developmental change can be altered (enhanced) through autopoietic-based changes or through planned applications of developmental science in the service of promoting individual thriving or social justice.

Bateson and Gluckman (2011) observed that gene expression is fundamentally shaped by variables external to the cell nucleus (where deoxyribonucleic acid, DNA, is located). They stressed, therefore, that "A willingness to move between different levels of analysis has become essential for an understanding of development and evolution" (Bateson & Gluckman, 2011, p. 5). Similarly, Keller (2010) explained that it is erroneous either to conceptualize development as involving separate causal influences or to posit that attributes of the person develop as an outcome of the interaction of causal elements. Indeed, she noted that the concept of interaction is itself flawed, in that its use is predicated on the idea that there exists attributes that are at least conceptually separate. Keller explained that the concept of developmental dynamics precludes such separation. She emphasized that,

> From its very beginning, development depends on the complex orchestration of multiple courses of action that involve interactions among many different kinds of elements—including not only preexisting elements (e.g., molecules) but also new elements (e.g., coding sequences) that are formed out of such interactions, temporal sequences of events, dynamical interactions, etc.
>
> (Keller, 2010, pp. 6–7)

Keller (2010), in discussing the elements of the epigenetic system, reflects the idea of the research moment of the opposites of identity, discussed by Overton (2015); but, as well, by pointing to the presence of dynamic coactions, her ideas reflect also the moment of the identity of opposites. Moreover, Pigliucci and Muller (2010) noted that genes are not as much generators of evolutionary change as they are followers in the evolutionary process. They explained that "evolution progresses through the capture of emergent interactions into genetic-epigenetic circuits, which are passed to and elaborated on in subsequent generations" (Pigliucci & Muller, 2010, p. 14). Similarly, West-Eberhard (2003) connected evolution and the presence of relative plasticity across development. She explained that environmental variables are a major basis of adaptive evolutionary change. As also pointed out by Pigliucci and Muller (2010), West-Eberhard (2003) noted that genetic mutation does not provide either the origin or the evolution of novel adaptive characteristics because "genes are followers not leaders, in evolution" (p. 20). In addition, she explained that the relative plasticity of the phenotype can facilitate evolution by providing immediate changes in the organism (West-Eberhard, 2003). Similarly, Gissis and Jablonka (2011) noted that plasticity "is … a large topic, but, just as Lamarck anticipated, an understanding of plasticity is now recognized as being fundamental to an understanding of evolution" (p. xiii).

Crystallizing the embodiment of variables from all levels of organization within the relational developmental system that create epigenetic change across generations, Jablonka and Lamb (2005) presented evidence demonstrating that human evolution involves four interrelated dimensions: genes, epigenetics, behavior, and culture. They explained that contemporary research in molecular biology indicates clearly that current, neo-Darwinian assumptions about the role of genes in evolution are mistaken. This research demonstrates that cells can transmit information to daughter cells through non-DNA, epigenetic means. Therefore, genetic and epigenetic processes constitute two dimensions of evolution. In addition, animals can transmit information across generations through their behavior, which constitutes a third dimension of evolution. A fourth dimension of evolution is constituted by culture, in that humans "inherit" from their parent's symbols and, in particular, language. As such, Jablonka and Lamb (2005) concluded that "It is therefore quite wrong to think about heredity and evolution solely in terms of the genetic system. Epigenetic, behavioral, and symbolic inheritance also provide variation on which natural selection can act" (p. 1).

These epigenetic effects referred to by Jablonka and Lamb (2005) occur because chemicals in the cell either allow or do not allow DNA to be transcribed into mRNA (Misteli, 2013). For example, acetyl groups, when linked with one of the four base chemicals comprising DNA, that is, to cytosine, allow DNA transcription; this process is termed acetylation. In turn, when methyl groups are linked to cytosine, then there is no transcription of DNA into mRNA. This process is termed "methylation." In short, the acetylation process allows DNA to be transcribed into mRNA (and therefore play a role in producing proteins), and methylation processes silence DNA transcription.

If DNA is not transcribed into mRNA, then this DNA cannot play a role in the production of proteins for use by the cell. Because this silencing of gene transcription can persist (can remain stable) across generations (Meaney, 2010; Misteli, 2013; Roth, 2012; Slavich & Cole, 2013), epigenetic influences constitute heritable changes explained by processes other than DNA. Indeed, Gissis and Jablonka (2011), in a book discussing the transformations of Lamarckian theory that have arisen in relation to the increasingly more active focus on epigenetic processes in the study of evolution and development (Meaney, 2010), noted that a form of inheritance of acquired characteristics does exist in the form of epigenetic inheritance systems.

This system of epigenetic effects involves chemicals within the cell, within the internal milieu of the body, and within the external ecology within which the body is embedded (Misteli, 2013; Roth, 2012; Slavich & Cole, 2013) or embodied, in the terms used by Overton (2013, 2015). For instance, Roth (2012) noted that the genome of infants is modified by epigenetic changes involving experiential and environmental variables. She explained that parental stress, infant separation, or caregiver nurturance or maltreatment can alter methylation patterns that affect neurobiology and behavior across the life span. Similarly, Slavich and Cole (2013) discussed evidence that changes in the expression of hundreds of genes occurs as a function of the physical and social environments inhabited by humans, and they noted that "external social conditions, especially our subjective perceptions of these conditions, can influence our most basic internal biological processes—namely, the expression of our genes" (p. 331)—a view that again highlights the implications of embodied biological changes as a focus of actions aimed at enhancing positive human development and, ultimately, social justice (Lerner, 2018; Lerner & Overton, 2008).

We return in the concluding portion of Section VIII of this book to the implications of attributes of dynamic relational development. As such embodiment and gene ⇔ context coactions, such as those associated with epigenetics, may promote health and positive human development and, as well, social justice. Here we note, however, that the evidence concerning epigenetics, embodied action, and plasticity that today is understood as accounting for the features of evolutionary and developmental change necessarily leads to deep skepticism about the "extreme nature" (Rose & Rose, 2000) of the claims of biological reductionists, for example, evolutionary psychology (Rose & Rose, 2000), sociobiology (Lerner, 2002; Lerner & von Eye, 1992), and behavior genetics (Molenaar, 2014). Clearly, the claims of such reductionists are inconsistent with the now quite voluminous evidence in support of the role of epigenetics in the multiple, integrated dimensions of human evolution, discussed above (Bateson, 2015; Coall et al., 2015; Gissis & Jablonka, 2011; Jablonka & Lamb, 2005; Lickliter & Honeycutt, 2015). Moreover, these claims run counter to research that has importantly begun focusing on the role of the organism's active agency (McClelland et al., 2015), and of culture (Mistry & Dutta, 2015), in creating change within and across generations.

In contrast to the claims of biological reductionists, a process-relational paradigm and concepts associated with the RDS metamodel (Overton, 2015) suggest that transmission across generations is accounted for by the plastic embodied processes of the individual functioning in a reciprocal, that is, bidirectional (⇔), relation with his/her physical and cultural context. Thus, within the dynamic, RDS perspective, and in the context of contemporary evolutionary scholarship, (e.g., Gissis & Jablonka, 2011; Ho, 2010; Keller, 2010; Lickliter & Honeycutt, 2015; Meaney, 2010; Moore, 2015a, 2015b, 2016), the "Just So" stories (Gould, 1981) of evolutionary psychology are conceptually and empirically flawed. Furthermore, embodiment constitutes the basis for epigenesis within the person's life span (Gottlieb, 1997, 1998), including qualitative discontinuity across ontogeny in relations among biological, psychological, behavioral, and social-cultural variables. Evidence for the relative plasticity of human development within the integrated levels of the ecology of human development makes biologically reductionist accounts (or, equally, completely sociogenic accounts) of parenting, offspring development, or sexuality implausible, at best, and entirely fanciful, at worst (Lerner, 1984, 2002, 2006).

In sum, the RDS metamodel provides an approach to the study of evolutionary and ontogenetic change that capitalizes on the dynamic, mutually influential relations between developing individuals and their complex and changing ecology. These "strands" of theory merged in the 1970s, 1980s, and 1990s and created a focus on models emphasizing that time and place matter in regard to shaping the course of life (Bronfenbrenner, 2005; Elder, 1998; Elder & Shanahan, 2006; Elder et al., 2015). A key emphasis in these ideas is that the scientific study of human development needs to involve the individual and diversity of people to understand human development. The process-relational paradigm that framed conceptions of the bases of human development is associated with the generation of several RDS models of human development (Overton, 2013; Overton & Müeller, 2013), conceptions that were used to guide the study of individuals, contexts, and their dynamic interrelations across the life span.

An Overview of the Contributions to This Section

Cole (2009) contrasts the biologically determinist conception that genes function as the unidirectional case of social behavior (i.e., genes → social behavior) with the evidence that variation in social and other contextual factors alter the expression of genes (i.e., context → gene expression and, as well, genes ⇔ context relations). Cole discusses progress in the burgeoning field of social genomics, which includes study of the types of genes that are able to be socially regulated and, as well, the mutually influential pathways between genes and social behavior.

In addition, Cole (2014) explains that gene expression can vary in relations to the vicissitudes of everyday life, noting that differences in social circumstances associated with urbanity, socioeconomic status, social status social isolation, or social threat are associated with variation in gene transcription in leukocytes and diseased tissues such as metastatic cancers. Cole also discusses the genomics of optimal health and thriving.

Consistent with the presentations in this section of Cole (2009, 2014; see too Slavich & Cole, 2013), Moore (2016) notes that research in the study of behavioral epigenetics has demonstrated that human development rests on the coaction of the contexts encountered across life (e.g., the nature of parenting experienced by a child) and a person's lifestyle choices (e.g., diet, exercise, and sleep) as well as on the complement of genes received at conception. These data indicate that genes alone do not determine a person's intelligence, temperament, or likelihood of having cardiovascular disease or cancer. Also reflecting agreement with Slavich and Cole (2013), Moore explains the importance of developmental epigenetics for medical practice and, as well, because some of the epigenetic changes occurring across a person's life factors appear to be transmissible across generations, for a revised, Lamarckian-type of inheritance of acquired characteristics (see too Jablonka & Lamb, 2007, in Section IV of this book).

Wahlsten (2016) discussed the important contributions made in Moore's (2015a) landmark book on behavioral epigenetics, which framed much of the information Moore presented in his contribution appearing in this section of the book. Wahlsten compliments the erudition involved in Moore's (2015a) work and agrees that, given the still early phase in this field of research, caution is needed in interpreting findings because there remains so much additional work to do to understand known and to-be-discovered factors that may moderate or mediate extant findings. Wahlsten also notes some of the important directions to be included in future research, including the nature of allelic differences in a gene and parameters of gene–environment interactions, which means that different genotypes respond to environmental changes differentially and therefore must involve allelic variation in a population. Overall, Wahlsten—in agreement with Moore, Cole, and Slavich—concludes that work in behavior epigenetics or social genomics constitutes a different, a dynamic, view of heredity. The dynamics of heredity necessitate a revised conceptualization of the gene-environment relationship, development, and evolution. Wahlsten (2016) sums up his discussion of dynamic heredity with the eloquent and succinct admonition that this reality requires that genetic reductionists simple genetic determination of behavior need, finally, to be dismissed.

References

Bateson, P. (2015). Ethology and human development. In W. F. Overton & P. C. Molenaar (Eds.), *Theory and method. Volume 1 of the handbook of child psychology and developmental science* (7th ed.). Editor-in-chief: R. M. Lerner. (pp. 208–243). Wiley.

Bateson, P., & Gluckman, P. (2011). *Plasticity, development and evolution*. Cambridge University Press.

Bronfenbrenner, U. (2005). *Making human beings human: Bioecological perspectives on human development*. Sage.

Coall, D. A., Callan, A., Dickins, T. E., & Chisholm, J. S. (2015). Evolution and prenatal development. *Social, Emotional, and Personality Development, 3*, 57–105.

Cole, S. W. (2009). Social regulation of human gene expression. *Current Directions in Psychological Science, 18*(3), 132–137.

Cole, S. W. (2014). Human social genomics. *PLoS Genetics, 10*(8), e1004601.

Elder, G. H. (1998). The life course and human development. In W. Damon (Series Ed.) & R. M. Lerner (Vol. Ed.), *Handbook of child psychology: Vol. 1 theoretical models of human development* (5th ed., pp. 939–991). Wiley.

Elder, G. H., Jr., & Shanahan, M. J. (2006). The life course and human development. In R. M. Lerner (Ed.). *Theoretical models of human development. Volume 1 of handbook of child psychology* (6th ed., pp. 665–715). Editors-in-chief: W. Damon & R. M. Lerner. Wiley.

Elder, G. H., Shanahan, M. J., & Jennings, J. A. (2015). Human development in time and place. In M. H. Bornstein and T. Leventhal (eds), *Handbook of child psychology and developmental science, vol. 4: Ecological settings and processes in developmental systems* (7th ed., pp. 6–54). Wiley.

Gissis, S. B., & Jablonka, E. (Eds.). (2011). *Transformations of Lamarckism: From subtle fluids to molecular biology*. MIT Press.

Gottlieb, G. (1997). *Synthesizing nature-nurture: Prenatal roots of instinctive behavior*. Erlbaum.

Gottlieb, G. (1998). Normally occurring environmental and behavioral influences on gene activity: From central dogma to probabilistic epigenesis. *Psychological Review, 105*, 792–802.

Gould, S. J. (1981). *The mismeasure of man*. Norton

Ho, M. W. (2010). Development and evolution revisited. In K. E. Hood, C. T. Halpern, G. Greenberg, & R. M. Lerner (Eds.), *Handbook of developmental systems, behavior and genetics* (pp. 61–109). Wiley Blackwell.

Jablonka, E., & Lamb, M. (2005). *Evolution in Four Dimensions: Genetic, epigenetic, behavioral, and symbolic variation in the history of life*. MIT Press.

Jablonka, E., & Lamb, M. (2007). Précis of evolution in four dimensions. *Behavioral and Brain Sciences, 30*(4), 353–365. https://doi.org/10.1017/S0140525X07002221

Keller, E. F. (2010). *The mirage of a space between nature and nurture*. Duke University Press.

Lerner, R. M. (1984). *On the nature of human plasticity*. Cambridge University Press.

Lerner, R. M. (2002). *Adolescence: Development, diversity, context, and application*. Prentice-Hall.

Lerner, R. M. (2006). Developmental science, developmental systems, and contemporary theories of human development. In R. M. Lerner (Ed.), *Theoretical models of human development. Volume 1 of handbook of child psychology* (6th ed., pp. 1–17). Editors-in-chief: W. Damon & R. M. Lerner. Wiley.

Lerner, R. M. (2018). *Concepts and theories of human development* (4th ed.). Routledge.

Lerner, R. M., & Benson, J. B. (Eds.). (2013a). *Embodiment and Epigenesis: Theoretical and Methodological Issues in Understanding the Role of Biology within the Relational Developmental System*. Volume 1: Philosophical, Theoretical, and Biological Dimensions. Advances in Child Development and Behavior (Vol. 44). Elsevier.

Lerner, R. M., & Benson, J. B. (Eds.). (2013b). *Embodiment and Epigenesis: Theoretical and Methodological Issues in Understanding the Role of Biology within the Relational Developmental System*. Volume 2: Ontogenetic Dimensions. Advances in Child Development and Behavior (Vol. 45). Elsevier.

Lerner, R. M. & Benson, J. B. (2013c). Introduction: Embodiment and epigenesis: A view of the issues. In R. M. Lerner & J. B. Benson (Eds.), *Advances in child development and behavior: Embodiment and epigenesis: Theoretical and methodological issues in understanding the role of biology within the relational developmental system* (pp. 1–20). Elsevier.

Lerner, R. M., & Overton, W. F. (2008). Exemplifying the integrations of the relational developmental system: Synthesizing theory, research, and application to promote positive development and social justice. *Journal of Adolescent Research, 23*(3), 245–255.

Lerner, R. M., & von Eye, A. (1992). Sociobiology and human development: Arguments and evidence. *Human Development, 35*, 12–33.

Lickliter, R., & Honeycutt, H. (2015). Biology, development, and human systems. In W. F. Overton & P. C. M. Molenaar (Eds.), *Handbook of child psychology and developmental science, Volume 1: Theory and method* (7th ed., pp. 162–207). Editor-in-chief: R. M. Lerner. Wiley.

Marshall, P. J., Houser, T. M., & Weiss, S. M. (2021). The shared origins of embodiment and development. *Frontiers in Systems Neuroscience, 15*, 726403. https://doi.org/10.3389/fnsys.2021.726403

McClelland, M. M., Geldhof, G. J., Cameron, C. E., & Wanless, S. B. (2015). Development and self-regulation. In W. F. Overton & P.C. Molenaar (Eds.), *Handbook of child psychology and developmental science. Volume 1: Theory and method* (7th ed., pp. 523–565). Editor-in-chief: R. M. Lerner. Wiley.

Meaney, M. (2010). Epigenetics and the biological definition of gene x environment interactions. *Child Development, 81*, 41–79.

Misteli, T. (2013). The cell biology of genomes: Bringing the double helix to life. *Cell, 152*, 1209–1212.

Mistry, J., & Dutta, R. (2015). Human development and culture: Conceptual and methodological issues. In W. F. Overton & P. C. Molenaar (Eds.), *Handbook of child psychology and developmental science. Vol. 1: Theory and method* (7th ed., pp. 369–406). Editor-in-chief: R. M. Lerner. Wiley.

Molenaar, P. C. M. (2014). Dynamic models of biological pattern formation have surprising implications for understanding the epigenetics of development. *Research in Human Development, 11*, 50–62.

Moore, D. S. (2015a). *The developing genome: An introduction to behavioral epigenetics*. Oxford University Press.

Moore, D. S. (2015b). The asymmetrical bridge. Book review of James Tabery's beyond versus: The struggle to understand the interaction of nature and nurture. *Acta Biotheoretica, 63*(4), 413–427.

Moore, D. S. (2016). Behavioral epigenetics. *WIREs Cognitive Science* 2016. https://doi.org/10.1002/wcs.1333

Overton, W. F. (2008). Embodiment from a relational perspective. In W. F. Overton, U. Mueller & J. L. Newman (Eds.), *Developmental perspective on embodiment and consciousness* (pp. 1–18). Erlbaum Associates

Overton, W. F. (2013). A New Paradigm for Developmental Science: Relationism and Relational-Developmental Systems. *Applied Developmental Science, 17*(2), 94–107.

Overton, W. F. (2015). Process and relational developmental systems. In W. F. Overton & P. C. M. Molenaar (Eds.), *Handbook of child psychology and developmental science, Volume 1: Theory and method* (7th ed., pp. 9–62). Editor-in-chief: R. M. Lerner. Wiley.

Overton, W.F., & Müller, U. (2013). Metatheories, theories, and concepts in the study of development. In R. M. Lerner, M.A. Easterbrooks, & J. Mistry (Eds.), *Handbook of psychology. Vol. 6: Developmental psychology* (2nd ed., pp. 19–58). Editor-in-chief: I. B. Weiner. Wiley.

Pigliucci, M., & Mueller, G. B. (2010). Elements of an extended evolutionary synthesis. In M. Pigliucci & G. B. Mueller (Eds.), *Evolution – The extended synthesis* (pp. 3–17). MIT Press.

Rose, H., & Rose, S. (2000). Introduction. In. H. Rose & S. Rose (Eds.), *Alas poor Darwin: Arguments against evolutionary psychology* (pp. 1–13). Vintage.

Roth, T. L. (2012). Epigenetics of neurobiology and behavior during development and adulthood. *Developmental Psychobiology, 54*, 590–597.

Slavich, G. M., & Cole, S. W. (2013). The emerging field of human social genomics. *Clinical Psychological Science, 1*, 331–348.

Wahlsten, D. (2016). Dynamic heredity. Review of Moore, D. S. (2015). *The developing genome: An introduction to behavioral epigenetics*. Oxford University Press. *Developmental Psychobiology, 58*(3), 419–451.

Werner, H. (1957). The concept of development from a comparative and organismic point of view. In D. B. Harris (Ed.), *The concept of development* (pp. 125–148). University of Minnesota Press.

West-Eberhard, M. J. (2003). *Developmental plasticity and evolution*. Oxford University Press.

Witherington, D. C. (2014). Self-organization and explanatory pluralism: Avoiding the snares of reductionism in developmental science. *Research in Human Development, 11*(1), 22–36.

Witherington, D. C. (2015). Dynamic systems in developmental science. In W. F. Overton & P. C. Molenaar (Eds.), *Handbook of child psychology and developmental science. Volume 1: Theory and method* (7th ed., pp. 63–112). Editor-in-chief: R. M. Lerner. Wiley.

19.
SOCIAL REGULATION OF HUMAN GENE EXPRESSION
*Steve W. Cole**

Genetics vs. Genomics

The conceptual relationship between genes and social behavior has shifted significantly during the past 20 years. As genes have come to be understood as concrete DNA sequences, rather than abstractions inferred from inheritance, it has become increasingly clear that social factors can play a significant role in regulating the activity of human genes. DNA encodes the potential for cellular behavior, but that potential is only realized if the gene is expressed - if its DNA is transcribed into RNA and translated into protein (Fig. 19.1). Proteins shape the structure of a cell, and endow its characteristic behaviors such as movement, metabolism, and biochemical response to external stimuli (e.g., neurotransmission). Absent their transcription, DNA genes have no effect on health or behavioral phenotypes. With the advent of a sequenced human genome and the emergence of DNA microarray technologies, scientists can now survey the expression of all human genes simultaneously and map the specific subset of genes that are active in a given cell at a given point in time. One surprising finding from the field of "functional genomics" is that the expression of a specific gene is often more an exception than the rule. Cells are highly selective about which genes they express, and our DNA encodes a great deal more genetic potential than is realized in RNA and protein. Even more striking has been the discovery that the social world outside our bodies influences which genes are transcribed within the nuclei of our cells (the RNA "transcriptome").

Social Regulation of Gene Expression

The possibility that social factors might regulate gene expression first emerged in the context of bio-behavioral health research. Social stress and isolation have long been known to affect the onset and progression of disease (Seeman, 1996). That effect is particularly strong for viral infections, where social factors have been linked to increased replication of cold-causing rhinoviruses (Cohen, Doyle, Skoner, Rabin, & Gwaltney. 1997), the AIDS virus, HIV-1 (Cole, 2008), and several cancer-related viruses (Antoni et al., 2006). Viruses are little more than small packages of 10-100 genes that hijack the protein production machinery of their host cells (us) to make more copies of themselves. As obligate parasites of our living cells, viruses have evolved within a micro-environment structured by our own genome. If social factors can regulate the expression of viral genes, that suggests that our own complement of ~22,000

---

* Department of Medicine, Division of Hematology-Oncology, UCLA School of Medicine, Cousins Center for PNI, UCLA AIDS Institute, Jonsson Comprehensive Cancer Center, the UCLA Molecular Biology Institute, and the HopeLab Foundation

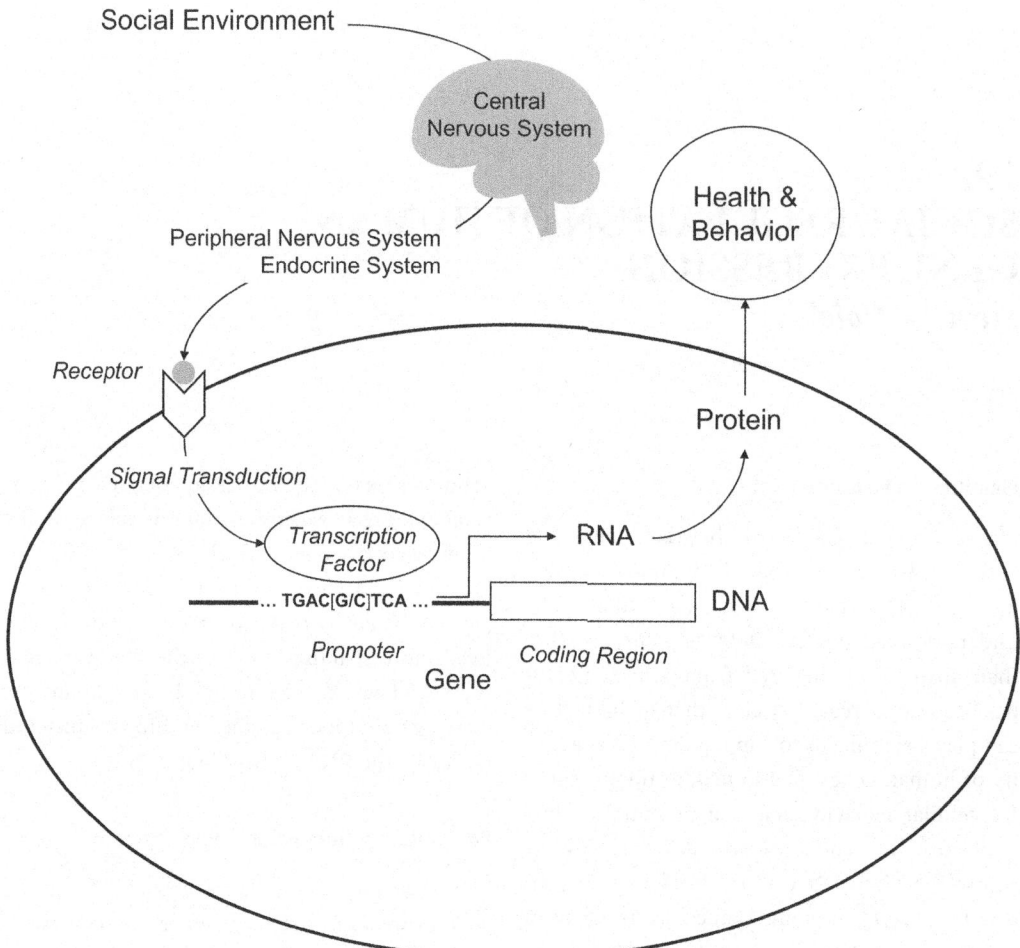

Fig. 19.1 Social signal transduction. Socio-environmental processes regulate human gene expression by activating central nervous system processes that subsequently influence hormone and neurotransmitter activity in the periphery of the body. Peripheral signaling molecules interact with cellular receptors to activate transcription factors, which bind to characteristic DNA motifs in gene promoters to initiate (or repress) gene expression. Only genes that are transcribed into RNA actually impact health and behavioral phenotypes. Individual differences in promoter DNA sequences (e.g., the [G/C] polymorphism shown here) can affect the binding of transcription factors, and thereby influence genomic sensitivity to socio-environmental conditions.

genes is likely to be regulated in biologically significant ways as well.

One of the first studies to analyze the relationship between social factors and human gene expression surveyed transcriptional profiles in white blood cells (leukocytes) from healthy older adults who differed in the extent to which they felt socially connected to others (Cole et al., 2007). Among the 22,283 genes assayed, 209 showed systematically different levels of expression in people who reported feeling lonely and distant from others consistently over the course of 4 years (Fig. 19.2). These effects did not involve a random smattering of all human genes, but focally impacted three specific groups of genes. Genes supporting the early "accelerator" phase

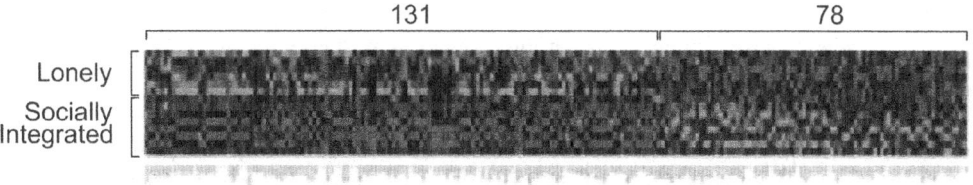

Fig. 19.2 Social regulation of gene expression in human immune cells. Expression of 22,283 human gene transcripts was assayed in 10 million blood leukocytes sampled from each of 14 older adults who showed consistent differences over 4 years in their level of subjective social isolation. 209 gene transcripts showed ≥ 30% difference in average expression level in leukocytes from 6 people experiencing chronic social isolation vs. 8 experiencing consistent social integration. In the heat-plot above, each row represents data from 1 of the 14 study participants, each column contains expression values for one of the 209 differentially active genes (Adapted from Cole, et al., 2007).

of the immune response - inflammation - were selectively up- regulated. However, two groups of genes involved in the subsequent "steering" of immune responses were down-regulated: genes involved in responses to viral infections (particularly Type I interferons), and genes involved in the production of antibodies by B lymphocytes. These results provided a molecular framework for understanding why socially isolated individuals show heightened vulnerability to inflammation-driven cardiovascular diseases (i.e., excessive non-specific immune activity) and impaired responses to viral infections and vaccines (i.e., insufficient immune responses to specific pathogens). A major clue about the psychological pathways mediating these effects came from the observation that differential gene expression profiles were most strongly linked to a person's subjective sense of isolation, rather than their objective number of social contacts.

Additional studies have identified transcriptional correlates of other socio-environmental conditions such as low socio-economic status (SES) (Chen et al., 2008) and the chronic threat of social loss (e.g., having a spouse with cancer) (Miller et al., 2008). These analyses also found up-regulated expression of leukocyte inflammatory genes, and identified specific psychological processes that appeared to contribute to those dynamics. For example, among children with asthma, those from a low-SES background tended to interpret ambiguous situations as threatening, and that perception of threat was more strongly linked to differential gene expression than was SES *per se* (Chen et al., 2008).

Several studies have shown that social influences can penetrate remarkably deeply into our bodies. The nervous system plays a key role in perceiving and responding to social stimuli, and social conditions have been found to regulate the expression of neural genes such as the Nerve Growth Factor *NGF* gene (Sloan et al., 2007) and the glucocorticoid receptor gene (Zhang et al., 2006). More surprising is the discovery that key immune system genes are also sensitive to social conditions (Sloan et al., 2007). Immune cells exert selective pressure on the evolution of viral genomes, and many viruses also appear to have developed a genomic sensitivity to our social conditions (as reviewed above). However, even pathogens that escape our immune system may still modulate gene transcription in response to host stress and social conditions. Most human cancers are invisible to the immune system, but some still change gene expression patterns in response to social stress (Antoni et al., 2006). One recent study of women with ovarian cancer found more than 220 genes to be selectively up-regulated in tumors from women with low levels of social support and high depressive symptoms (Lutgendorf et al., 2008). If our socially sensitive immune system is not conveying those effects, how do social influences reach into the damaged genome of a cancer cell? New insights have come from bioinformatic analyses of "social signal transduction."

Social Signal Transduction

Molecular biologists construe signal transduction as a local process by which signaling molecules outside the cell interact with cellular receptors to initiate a cascade of biochemical reactions inside the cell, ultimately stimulating a protein "transcription factor" to activate gene expression (Fig. 19.1). Transcription factors flag a particular stretch of DNA (the "coding region" of a gene) for transcription into RNA. Which genes can be activated by a given transcription factor is determined by the nucleotide sequence of the gene's promoter - the stretch of DNA lying upstream of the coding region. For example, the transcription factor NF-κB binds to the nucleotide motif GGGACTTTCC, whereas CREB/ATF transcription factors target the motif TGACGTCA. These two transcription factors are activated by different receptor-mediated signal transduction pathways, providing distinct molecular channels by which extra-cellular events can regulate intracellular genomic response. The distribution of transcription factor-binding motifs across our ~22,000 gene promoters constitutes a "wiring diagram" that maps micro-environmental processes onto genome-wide transcriptional responses.

Transduction of socio-environmental influences into functional genomic responses is mediated by the brain's perception of social conditions, and its subsequent regulation of hormones, neurotransmitters, and other signaling molecules that disseminate throughout the body to activate cellular receptors and transcription factors. For example, the sympathetic nervous system (SNS) and the hypothalamic-pituitary-adrenal (HPA) axis represent two major pathways by which central nervous system (CNS) perceptions of negative social conditions can regulate gene transcription in a wide array of somatic cells (Sapolsky, 1994). Positive psychological states may also regulate human gene expression (Dusek et al., 2008), although their molecular mediators are less well understood.

Links between social experiences and neural/endocrine responses have long been recognized, but the breadth of their impact on gene expression has only recently become apparent following the sequencing of the human genome. Early computational analyses of the human genome sequence suggested that promoter DNA sequences might provide for psychologically specific transcriptional responses. For example, any gene bearing the motif GGTACAATCTGTTCT in its promoter might potentially be stimulated by severe, overwhelming stress experiences that release cortisol, because the cortisol-stimulated glucocorticoid receptor (GR) binds specifically to that DNA motif. In contrast, genes bearing the CREB/ATF promoter motif TGACGTCA would be predicted to activate in response to active-coping, fight-or-flight stress responses associated with catecholamine release and beta-adrenergic receptor signaling. Based on the distribution of these promoter motifs across the human genome, it appears that these two distinct psychological stress experiences may trigger very different transcriptional responses. Genes predicted to be cortisol-responsive disproportionately encode receptors and other molecules involved in a cell's "perception" of its local environment. In contrast, putative catecholamine-responsive genes include few receptors, but a high concentration of signal transduction molecules and transcription factors involved in cellular "decision-making" (converting receptor-mediated perception into changes in gene expression and cellular behavior). Thus, severe, overwhelming stress may trigger a cellular form of "denial" (altering perception), whereas active-coping challenges induce something more akin to "sublimation" (altering responses to perceptions).

A sequenced human genome also provided new analytic infrastructure for mapping the molecular signaling pathways that convert socio-environmental conditions into differential gene expression. One approach reverses the normal

Social Regulation of Human Gene Expression

flow of biological information from the environment, through transcription factor activity, and into gene expression (Fig. 19.1). This analysis scans the promoters of differentially expressed genes to identify transcription factor-binding motifs that are over-represented in activated promoters, and thus reflect which specific transcription factors drove the observed differences in gene expression (Cole, Yan, Galic, Arevalo, & Zack, 2005). This approach has uncovered some surprising differences between the transcriptional signals "sent" by the brain, and the transcriptional signals "heard" by the human genome. In studies of chronic loneliness and threat of social loss (Cole et al., 2007; Miller et al., 2008), analyses indicated that the inflammation-driving NF-κB transcription factor played a key role in orchestrating both patterns of differential gene expression. Results also suggested that the GR was failing to inhibit NF-κB's pro-inflammatory activity as it should. Neither study found decreases in circulating cortisol levels that might explain the reduced GR activity. If the HPA axis were sending the proper anti-inflammatory cortisol signal, why would stressed people's leukocytes not down-regulate NF-κB transcription of inflammatory genes? The answer appears to involve a reduction in the GR's sensitivity to cortisol – rendering the leukocyte transcriptome deaf to the brain's request to down-regulate pro-inflammatory genes (Cole et al., 2007; Miller et al., 2008). Both chronic loneliness and threat of social loss appear to disconnect this key physiologic feedback system, and may thereby increase the risk of inflammation-related disease (Seeman, 1996). Similar analyses have identified other alterations in transcription factor activity that may connect low SES to inflammatory gene expression in asthma (Chen et al., 2008), and connect low social support and depression to altered gene expression in ovarian cancer (Lutgendorf et al., 2008).

Remodeling the Body

Because RNA transcription shapes the protein complement of our cells, and those proteins mediate cellular function (Fig. 19.1), psychological regulation of gene expression implies that the social world can remodel the functional characteristics of the human body. Consider the ability of chronic social stress to increase *NGF* gene expression, and thereby enhance the growth of SNS neural fibers in the lymph node tissues that structure immune responses (Sloan et al., 2007). This socio-environmental remodeling of lymph node innervation at $Time_1$ can persist, providing a denser neural network through which subsequent stressful exposure at $Time_2$ distributes SNS neurotransmitters into the lymph node. As a consequence, the immune system mounts a poorer response to viral infection at $Time_2$ solely because lymph node innervation was remodeled by differing social conditions at $Time_1$. In the model of Fig. 19.3, social stress at $Time_1$ ($Environment_1$) is transmitted through the nervous system ($Body_1$) into behavioral stress responses ($Behavior_1$) and increased *NGF* gene expression (RNA_1). Up-regulated *NGF* increases SNS innervation of the lymph node, and thereby alters the functional relationship between the nervous and immune systems ($Body_2$). When that functionally remodeled $Body_2$ encounters a new viral infection in $Environment_2$, increased SNS neurotransmitter release can inhibit transcription of Type I interferon genes (RNA_2). As a consequence of that impaired anti-viral response, intensified disease alters physical tissue characteristics and behavioral capacities in the future (RNA_3 and $Behavior_3$). In this way, the experience of $Environment_1$ not only "gets inside the body" but "stays there" in a concrete molecular sense that propagates through multiple gene transcriptional responses, physiologic systems, and time epochs.

Socio-environmental conditions can also regulate the molecular composition of CNS cells, and thereby alter psychological and behavioral responses to future environments (Zhang et al., 2006). Because the molecular composition of our cells constitutes the physical machinery by which we perceive and respond to the world around us ("Body" in Fig. 19.3), and that molecular composition is itself subject to remodeling by socio-environmental influences,

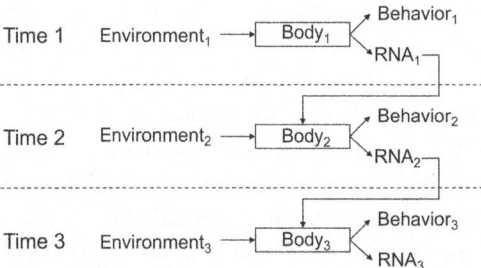

Fig. 19.3 RNA as a molecular medium of recursive development. Social conditions at one point in time (Environment$_1$) are transduced into changes in Behavior (Behavior$_1$) and gene expression (RNA$_1$) via CNS perceptual processes that trigger systemic neural and endocrine responses (mediated by Body$_1$). Those RNA transcriptional dynamics may alter molecular characteristics of cells involved in environmental perception or response, resulting in a functionally altered Body$_2$. Body$_2$ may respond differently to a given environmental challenge that would the previous Body$_1$, resulting in different Behavioral (Behavior$_2$) and RNA transcriptional responses (RNA$_2$). The persisting effect of RNA transcriptional dynamics on cellular protein and functional characteristics provides a molecular framework for understanding how socioenvironmental conditions in the past may continue to affect current behavior and health, and how those historical conditions interact with current environments to shape our future trajectories (e.g., Body$_3$, Behavior$_3$, RNA$_3$). Because gene transcription serves as both a cause of social behavior (by shaping Body) and a consequence of social behavior (a product of Environment x Body), RNA can serve as the physical medium for a recursive developmental trajectory that integrates genetic characteristics and historical-Environmental regulators to understand individual biological and behavioral responses to current Environmental conditions.

gene expression constitutes both a cause and a consequence of behavior. RNA can be construed as the physical medium of a recursive developmental system in which social, behavioral, and health *outcomes* at one point in time also constitute *inputs* that shape our future responses to the environment (e.g., as in Heckman's model of human capability development, which analyzes how capacities developed at Time$_1$ impact our ability to capitalize on environmental opportunities at Time$_2$ (Heckman, 2007)).

Future research will push these models out of accessible immune cells and into the more sensitive CNS structures that shape social, cognitive, and affective processes. It will also be critical to define the particular features of social environments that trigger transcriptional remodeling of specific cells. Given the key role of neuroendocrine responses in mediating these effects, the most decisive influences may involve our psychological reactions to social conditions rather than the properties of the external condition *per se*. After all, it is the subjective perception of conditions as threatening or uncertain that directly triggers SNS and HPA responses (Sapolsky, 1994). Our genome's social sensitivity ultimately stems from the capacity of social conditions to affect CNS perceptions of safety vs. threat (Dickerson & Kemeny, 2004), and thereby trigger biological stress responses that alter gene transcription.

The New Genetics

With genes and environments now operating in parallel to shape our RNA-driven bodies, the integration of those two streams of influence has become a central challenge in biological analyses of human health and behavior. The regulatory paradigm outlined in Fig. 19.1 provides a framework for analyzing their interplay in the context of Gene x Environment interactions. For example, variations in the DNA sequence of the promoter regulating the serotonin transporter gene (5HTT) can affect the binding of environmentally sensitive transcription factors, and thereby buffer the effects of adverse social environments on the risk of depression and other affective behaviors (Caspi et al., 2003; Champoux et al., 2002). These effects extend into the realm of immune response and survival (Capitanio et al., 2008), and thus may also shape the evolutionary trajectory of our DNA genome at a population level. In integrating the molecular biology of gene structure, the environmental control of gene expression, and the social biology of individual behavior and survival, the *5HTT* promoter polymorphism exemplifies the new "environmentally conscious" conception of genetics in which cellular and organismic behavior constitute the fundamental units of evolutionary selection, and genes and environments depend mutually on one another to shape that behavior by structuring our brains and bodies.

Research in social genomics has now clearly established that our interpersonal world exerts biologically significant effects on the molecular composition of the human body. These effects typically target a non-random ~1% of the human genome (though often a different 1% depending upon the social circumstances and cell type studied). Major topic for future exploration will involve determining which particular genes are subject to social regulation, what types of social conditions elicit such dynamics, which psychological and biological pathways mediate those effects, and which DNA polymorphisms moderate or intensify their impact.

Acknowledgments

Supported by R01 CA116778 from the National Cancer Institute.

References

Antoni MH, Lutgendorf SK, Cole SW, Dhabhar FS, Sephton SE, McDonald PG, et al. The influence of bio-behavioural factors on tumour biology: pathways and mechanisms. Nat Rev Cancer. 2006; 6(3):240–248. [PubMed: 16498446]

Capitanio JP, Abel K, Mendoza SP, Blozis SA, McChesney MB, Cole SW, et al. Personality and serotonin transporter genotype interact with social context to affect immunity and viral set-point in simian immunodeficiency virus disease. Brain Behav Immun. 2008; 22(5):676–689. Epub 2007 Aug 2023. [PubMed: 17719201]

Caspi A, Sugden K, Moffitt TE, Taylor A, Craig IW, Harrington H, et al. Influence of life stress on depression: moderation by a polymorphism in the 5-HTT gene. Science. 2003; 301(5631): 386–389. [PubMed: 12869766] A landmark paper identifying a transcription-regulating DNA polymorphism as a molecular vulnerability factor for the development of depression in response to psychological stress.

Champoux M, Bennett A, Shannon C, Higley JD, Lesch KP, Suomi S.J. Serotonin transporter gene polymorphism, differential early rearing, and behavior in rhesus monkey neonates. Mol Psychiatry. 2002; 7(10): 1058–1063. [PubMed: 12476320]

Chen E, Miller GE, Walker HA, Arevalo JM, Sung CY, Cole SW. Genome-wide transcriptional profiling linked to social class in asthma. Thorax. 2008 doi: 10.1136/thx.2007.095091.

Cohen S, Doyle WJ, Skoner DP, Rabin BS, Gwaltney JM. Social ties and susceptibility to the common cold. Journal of the American Medical Association. 1997; 227:1940–1944. [PubMed: 9200634]

Cole SW. Psychosocial influences on HIV-1 disease progression: neural, endocrine, and virologic mechanisms. Psychosom Med. 2008; 70(5):562–568. [PubMed: 18541906]

Cole SW, Hawkley LC, Arevalo JM, Sung CY, Rose RM, Cacioppo JT. Social regulation of gene expression in human leukocytes. Genome Biol. 2007; 8(9):R189. [PubMed: 17854483] This study provided the first indication that social factors might systematically regulate human genome activity in the immune system, and it pioneered novel strategies for identifying the social signal transduction pathways involved.

Cole SW, Yan W, Galic Z, Arevalo J, Zack JA. Expression-based monitoring of transcription factor activity: The TELiS database. Bioinformatics. 2005; 21(6):803–810. [PubMed: 15374858]

Dickerson SS, Kemeny ME. Acute stressors and cortisol responses: a theoretical integration and synthesis of laboratory research. Psychol Bull. 2004; 130(3): 355–391. [PubMed: 15122924]

Dusek JA, Otu HH, Wohlhueter AL, Bhasin M, Zerbini LF, Joseph MG, et al. Genomic counter-stress changes induced by the relaxation response. PLoS ONE. 2008; 3(7):e2576. [PubMed: 18596974]

Heckman JJ. The economics, technology, and neuroscience of human capability formation. Proc Natl Acad Sci USA. 2007; 104(33): 13250–13255. Epub 12007 Aug 13258. [PubMed: 17686985]

Lutgendorf SK, Degeest K, Sung CY, Arevalo JM, Penedo F, Lucci J 3rd, et al. Depression, social support, and beta-adrenergic transcription control in human ovarian cancer. Brain Behav Immun. 2008; 10:10. [PubMed: 18638541]

Miller GE, Chen E, Sze J, Marin T, Arevalo JM, Doll R, et al. A functional genomic fingerprint of chronic stress in humans: blunted glucocorticoid and increased NF-kappaB signaling. Biol Psychiatry. 2008; 64(4):266–272. Epub 2008 Apr 2028. [PubMed: 18440494]

Sapolsky, RM. Why zebras don't get ulcers: A guide to stress, stress-related diseases, and coping. Freeman; New York: 1994.

Seeman TE. Social ties and health: the benefits of social integration. Ann Epidemiol. 1996; 6(5):442–451. [PubMed: 8915476]

Sloan EK, Capitanio JP, Tarara RP, Mendoza SP, Mason WA, Cole SW. Social stress enhances sympathetic innervation of primate lymph nodes: mechanisms and implications for viral pathogenesis. J Neurosci. 2007; 27(33):8857–8865. [PubMed: 17699667].

Zhang TY, Bagot R, Parent C, Nesbitt C, Bredy TW, Caldji C, et al. Maternal programming of defensive responses through sustained effects on gene expression. Biol Psychol. 2006; 73(1):72–89. Epub 2006 Feb 2028. [PubMed: 16513241]

Miller G, Chen E, Cole SW. Health psychology: Developing biologically plausible models linking the social world and physical health. Ann. Rev. Psychol. 2009; 60:501–524. [PubMed: 19035829].

20.
HUMAN SOCIAL GENOMICS
Steven W. Cole*

Introduction

The spectacular adaptive success of *Homo sapiens* is attributable in large part to our capacity to self-organize into complex social systems or "metaorganisms" [1–3]. Research in human social genomics has begun to clarify how these extraorganismic social systems reciprocally regulate our intraorganismic physiologic function by modulating tissue-specific programs of gene expression [3–5]. Social regulation of gene expression has long been observed in animal models of morphological plasticity such as worker bee maturation into guards and scouts, cichlid sex switching, and status-dependent changes in body size, coloring, brain development, immune response, and reproductive capacity [6–9]. However, scientists, policy makers, and the general public have long wondered how such animal dynamics might pertain to everyday human life. Studies of human social genomics are now clarifying which specific types of human genes are subject to social regulation and mapping the social signal transduction pathways that mediate these effects. The results of these analyses are shedding new light on the molecular basis for social influences on individual heath, the genomic basis for human thriving, and the metagenomic capabilities that emerge from networked communities of socially sensitive genomes and underpin human group selection and the evolution of our hypersocial life history strategy [2, 10].

Human Social Genomics

Initial indications that social environments might significantly affect the functional activity of the human genome came from studies dissecting leukocyte gene expression profiles into genetic and environmental components [4, 11]. Gibson and colleagues found that ~5% of genes expressed in leukocytes showed appreciable genetic regulation (e.g., via expression quantitative trait loci), whereas >50% showed significant differences in expression across pastoral, rural, and urban social environments [11]. These results documented a substantial relationship between general social context and genome function, and motivated further analysis of the specific features of the social environment that drive the observed differences in gene expression (e.g., physicochemical stimuli, microbial exposures, and social/psychological influences on physiology). Parallel studies on the transcriptional correlates of social disparities in health subsequently suggested that both physical and psychological processes contribute to the net effect of a given social environment, with each mechanism activating some distinct gene modules as well as

---

* Department of Medicine, Division of Hematology-Oncology, UCLA School of Medicine, University of California, Los Angeles, CA

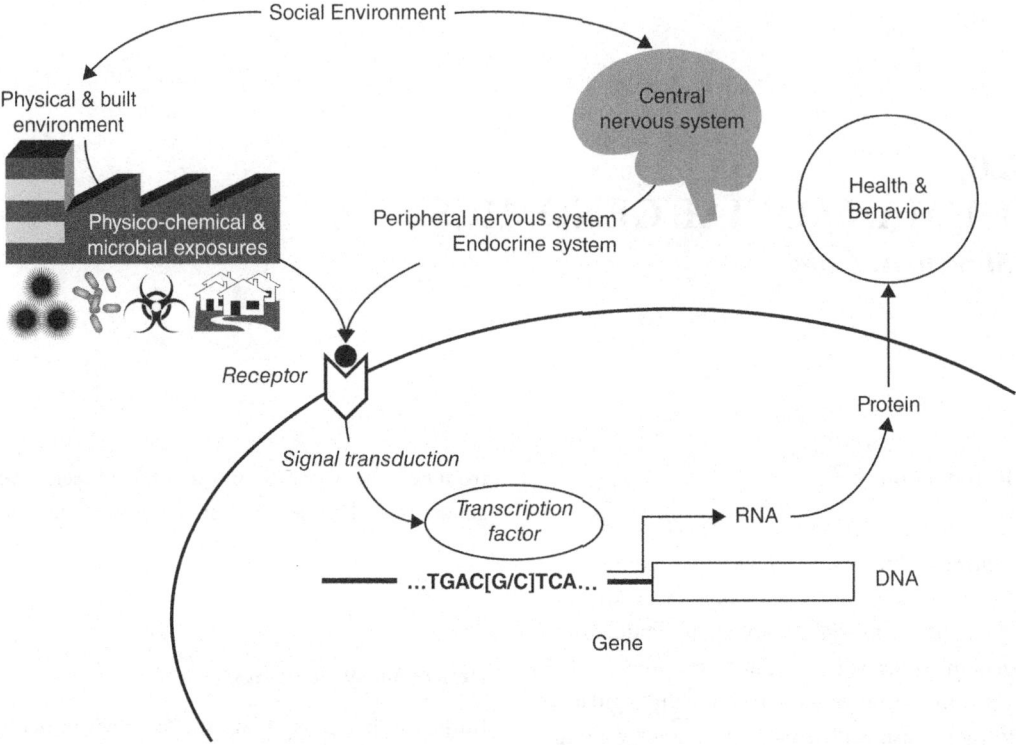

Fig. 20.1 Social regulation of human gene expression. Social environments can influence human gene expression via physicochemical processes (e.g., toxins, pollutants, and microbes) and psychological processes (e.g., experiences of threat or uncertainty) that trigger neural and endocrine responses (e.g., activation of the sympathetic nervous system). In both cases, biochemical mediators engage cellular receptor systems, which activate intracellular signal transduction pathways culminating in the activation (or repression) of transcription factors that proximally regulate the transcription of genes bearing response elements for that particular factor. The gene regulatory "wiring diagram" that maps specific biochemical signals to specific gene expression responses represents an evolved genomic program that was presumably adaptive under ancestral conditions but may have distinct maladaptive effects in the very different social environments of contemporary human life. doi: 10.1371/journal.pgen.1004601.g001.

a conserved generalized response to adverse life circumstances (Fig. 20.1) [3, 5, 12].

A prototypic example comes from studies of social isolation, which is one of the most robust epidemiologic risk factors for chronic illness and mortality [13]. Genome-wide transcriptional profiling of leukocytes from people experiencing chronic social isolation identified >200 genes that showed >50% difference in average expression levels relative to those observed in socially integrated people [14, 15]. Genes up-regulated in socially isolated individuals included a set of transcripts that play a central role in inflammation (e.g., *IL1B*, *IL8*, *PTGS2*), whereas down-regulated transcripts were involved in Type I interferon innate antiviral responses (e.g., *ISG*, *IFI*, *MX*, and *OAS* family genes) and in antibody production (e.g., *IGL*, *IGH*, *IGJ*, and *IGK*) [5, 14, 15]. Epidemiologists have long debated whether the health effects of social isolation stem predominately from a lack of social contact per se (i.e., reduced network size, economic opportunity, personal assistance, and interpersonal contact) or from the subjective experience of being lonely and disconnected from the rest of society and the threat/stress reactions that ensue [1, 13, 16]. These transcriptome analyses suggested that both subjective and objective isolation likely play a role but do so through distinct gene regulatory pathways. Objective isolation was associated with reduced expression of antibody synthesis genes (perhaps due to reduced exposure to socially transmitted microbes), whereas subjective isolation associated with increased expression of proinflammatory genes and reduced expression of Type I interferon genes and transcripts specifically involved in synthesis of immunoglobulin G$_1$ (IgG$_1$) antibodies (a pattern subsequently linked to fight-or-flight threat responses from the sympathetic nervous system) [3, 5].

A Conserved Transcriptional Response to Adversity

Following the initial analyses of social isolation, a diverse array of studies has begun to document similar leukocyte transcriptome shifts in other adverse social conditions, including low socioeconomic status (SES) [9, 17–20], chronic stress (e.g., care-giving for a dying spouse) [21, 22], bereavement [23], posttraumatic stress disorder (PTSD) [24, 25], and cancer diagnosis [26, 27]. Across these diverse forms of adversity, a common pattern of conserved transcriptional response to adversity (CTRA) has emerged, including increased expression of proinflammatory genes and decreased expression of genes involved in Type I interferon innate antiviral responses and IgG antibody synthesis [3, 5, 28].

Subsequent studies using experimental animal models have shown that CTRA gene expression profiles can be induced in leukocytes by repeated social threat [9], unstable social hierarchies [8, 29], and low social status [7]. Randomized controlled studies in humans have also shown that CTRA gene expression profiles can be suppressed by interventions such as cognitive behavioral stress management [26], meditation [30, 31], yoga [32], and Tai Chi [33].

Mechanistic studies in animal and cell culture systems have also shown that activation of fight-or-flight signaling pathways in the sympathetic nervous system (SNS) plays a major role in evoking CTRA gene expression profiles (Fig. 20.2A). SNS activation of the CTRA is mediated in large part by β-adrenergic receptors, which stimulate transcription factors such as nuclear factor kappa-light-chain-enhancer of activated B cells (NF-κB), GATA, and cAMP response element-binding protein (CREB) to selectively up-regulate transcription of proinflammatory genes (e.g., *IL6* [19]) while simultaneously inhibiting the activity of transcription factors, such as the interferon response factor family, that control transcription of Type I interferon genes (e.g., *IFNB* [34]). This pleiomorphic modulation of multiple transcription control pathways allows for rapid and relatively focal changes in leukocyte transcriptomes during extended periods of organismic threat and SNS fight-or-flight responses.

A second pathway inducing the CTRA involves broader and more durable changes in the transcriptional underpinnings of immune cell growth and development. Bioinformatic decomposition of transcriptome shifts in the heterogenous leukocyte population initially identified the primary cellular mediators of CTRA transcriptome shifts as myeloid lineage immune cells such as monocytes and dendritic cells [15]. Subsequent analysis of physically isolated leukocyte subpopulations confirmed that monocytes mediate many of the transcriptional effects of human social adversity [21, 25], and mechanistic analyses in animal models determined that the SNS can up-regulate the production of a distinct subpopulation of immature, proinflammatory monocytes by altering hematopoietic processes in the bone marrow [9]. These effects are again mediated by β-adrenergic signaling, which up-regulates transcription of the myelopoietic growth factor granulocyte-macrophage colony-stimulating factor (GM-CSF) (*Csf3*) and thereby enhances monocyte differentiation and development [9]. The resulting proinflammatory skew in the composition of the whole body monocyte pool persists for the life of the cells, which can sequester in reservoirs such as the spleen and reemerge months later in response to social threat [35].

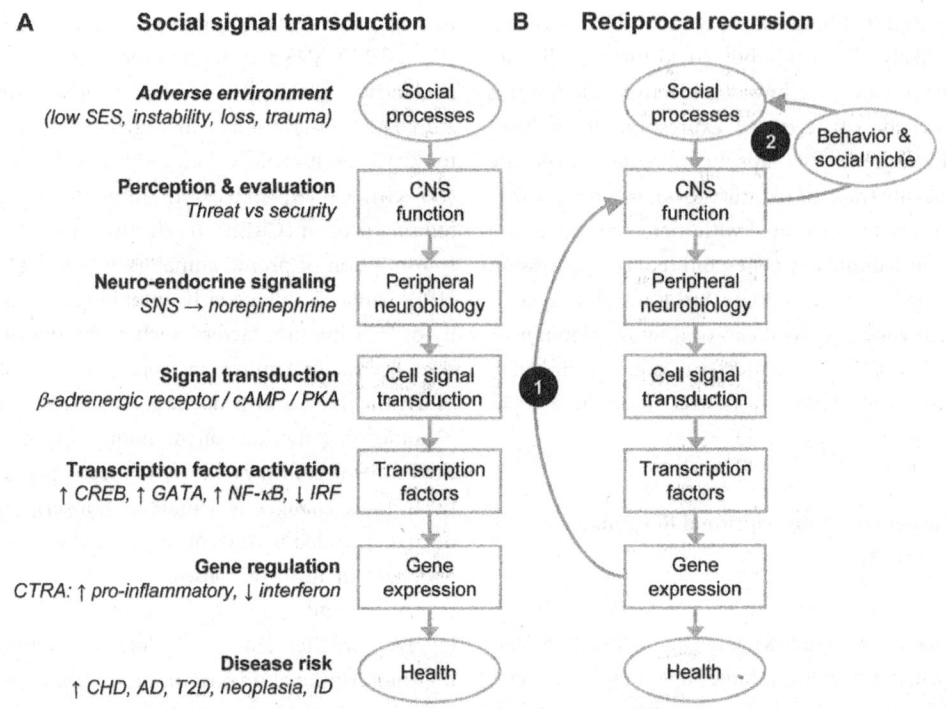

Fig. 20.2 **Social signal transduction and recursive network genomics.** (A) A simple (acyclic) social signal transduction pathway maps adverse social conditions onto activation of the conserved transcriptional response to adversity (CTRA) in leukocytes. Brain-mediated perceptions of social threat activate the sympathetic nervous system (SNS), leading to release of norepinephrine (NE) at SNS nerve terminals, activation of β-adrenergic receptors on adjacent cells, and stimulation or repression of specific transcription factors In response to the cyclic 3′-5′ adenosine monophosphate/protein kinase A (cAMP/PKA) signaling pathway, β-adrenergic-responsive transcription factors Induce the CTRA gene expression program by stimulating transcription of genes encoding proinflammatory cytokines and suppressing transcription of genes encoding Type I Interferons and IgG antibodies. CTRA gene expression programs prepare the body to respond to wounding injury and bacterial infections but may promote chronic illnesses such as cardiovascular disease (CVD), Alzheimer disease (AD), Type II diabetes (T2D), and metastatic cancer while undermining host resistance to virally mediated infectious diseases (ID). (B) Superimposed effects of reciprocal endogenous and exogenous recursive feedback on the social signal transduction pathway can propagate the impact of a transient adverse environmental shock. In this system, a transient environmental threat activates the core social signal transduction pathway to stimulate transcription of proinflammatory cytokine genes, as part of the CTRA. Arc 1 shows an endogenous biological feedback loop in which the proinflammatory gene products signal the brain to activate a programmed set of sickness behaviors that include reduced social motivation, fatigue, anhedonia, and negative emotional states. Arc 2 shows how the resulting reductions in individual social behavior and altered social niche selection evoke less supportive and more hostile responses from the surrounding social network and thereby create a more adverse social environment. Effects of the exogenous social recursion loop (Arc 2) are propagated via the core social signal transduction cascade into continued CTRA activation and continued endogenous biological recursion (Arc 1). Reciprocating feedback may thus maintain the system in a new dynamic equilibrium that maintains altered endogenous inflammation and exogenous social influence long after the initiating transient shock has passed. Similar recursive feedback can occur at every level of the social signal transduction cascade, resulting in complex systems dynamics that can trigger persistent sequelae such as PTSD and biological embedding of early life social conditions without requiring any durable genomic modification (e.g., mutation or epigenetic marking). Abbreviations: CHD, coronary heart disease; CNS (central nervous system); IRF, interferon regulatory factor. doi:10.1371/journal.pgen.1004601.g002.

A third pathway by which adverse social conditions can induce persistent transcriptional alterations in immune cells involves up-regulation of the *NGF* gene that supports the growth and differentiation of the SNS nerve fibers innervating lymph nodes [29]. The resulting increase in neurotransmitter delivery to the lymph node increases throughput from the brain to the immune system and thereby induces a persistent shift in the gene regulatory program of the tissue-resident pool of cells (e.g., down-regulating Type I interferon transcription and host resistance to viral infection [29]). In combination with SNS/β-adrenergic regulation of focal transcriptional programs in existing monocytes and de novo monocyte production through myelopoiesis, social threat-induced neoinnervation of lymphoid tissues provides three distinct gene regulatory dynamics that converge to up-regulated CTRA gene expression in immune cells during extended periods of social threat or adversity [36].

There has been great interest in the possibility that epigenetic marks such as histone acetylation or DNA methylation might mediate social influences on human gene expression. Environmental conditions clearly influence epigenetic profiles in human immune cells [37], and some studies have begun to link immune cell DNA methylation profiles to specific social environmental conditions such as low social status [7] and early life stress exposure [38–41]. However, those epigenetic profiles are only weakly correlated with differences in immune cell gene expression [7, 41], and much remains to be learned about what role epigenetic dynamics play in social genomic effects such as the CTRA.

Beyond the immune system, studies have recently begun to map relationships between social conditions and gene expression in diseased tissues such as breast, ovarian, and prostate cancers [42]. In these solid epithelial tumors, low social support is associated with increased expression of many gene modules that promote cancer progression and metastasis, including transcripts involved in inflammation, wound repair (epithelial-mesenchymal transition), blood vessel growth (angiogenesis), and resistance to programmed cell death (anoikis and chemotherapy resistance) [42]. Parallel effects have been observed in experimental animal models of cancer, and those analyses again identify a prominent role for SNS activation of β-adrenergic signaling [43, 44]. The specific genes regulated by β-adrenergic signaling in tumor tissues differ in tissue-specific ways from those engaged by β-adrenergic signaling in leukocytes, but they share some common teleological principles in their distinct cellular contexts. In both, chronic threat and SNS signaling activate tissue-specific transcriptional defense programs that enhance organismic mobility and prime wound healing and antimicrobial responses to tissue injury [3, 5, 36]. Some SNS-induced transcriptional responses play central roles in both the systemic leukocyte pool and the localized context of a growing tumor. In the context of metastatic breast cancer, for example, stress effects on tumor metastasis are mediated by β-adrenergic up-regulation of monocytes and macrophages within the tumor microenvironment [44] (i.e., the same dynamic that drives the CTRA transcriptome shift in circulating blood [9]). Other neuroendocrine signaling pathways such as glucocorticoid release from the hypothalamic-pituitary-adrenal system may also play a role in mediating relationships between social conditions and cancer biology [45]. For example, socially isolated animals show glucocorticoid-related increases in breast cancer burden [46, 47], and molecular analyses find glucocorticoid signaling to promote several cancer-related processes including mammary adipocyte metabolism [47, 48], inhibition of tumor suppressor genes such as *TP53* [49], and resistance to chemotherapy-mediated programmed cell death [50, 51]. Interestingly, a key gene involved in mediating these effects shows evidence of recent adaptive variation across human populations [52].

Genomics of Human Thriving

Transcriptional defense programs such as the CTRA presumably evolved to help adapt molecular physiology to the types of sporadic and transient threats that generally characterized our ancestral environments. In the contemporary human social ecology, however, chronic activation of these transcriptional defense programs by purely symbolic or anticipated threats likely acts to undermine health by promoting inflammation-related chronic diseases such as Type II diabetes, atherosclerosis, neurodegeneration, and metastatic cancer while simultaneously undermining host resistance to viral infections [3, 5, 19, 36]. However, genomes evolve to help us thrive and proliferate, not to generate disease [53]. As such, mapping the evolved regulatory logic of the human genome should have more to tell us about human well-being and the biology of thriving than it does about disease per so. If we seek to avoid the chronic activation of costly molecular defense programs such as the CTRA, what is the best way for us to lead our lives?

One recent study by Fredrickson and colleagues took this approach to ask which kind of "happiness" might best oppose the CTRA [54]. This was a more complex question than it might seem because several different forms of human happiness exist, and they imply very different approaches to promoting human health. Philosophers have long distinguished between a hedonic form of well-being generated by the pursuit of positive emotional experiences and consummatory self-gratification (i.e., self-focused happiness) and a more eudaimonic form of well-being that stems from devoting one's efforts to a noble cause or purpose outside the self (i.e., self-transcendent happiness). Fredrickson and colleagues found that high levels of eudaimonic well-being were associated with significantly lower levels of CTRA-related gene expression (i.e., a more favorable, or less threatened, molecular profile) [54]. In contrast, people who showed comparatively high levels of hedonic well-being relative to their level of eudaimonic well-being showed significantly elevated CTRA gene expression (above and beyond any effects of hedonistic behavioral factors such as smoking, alcohol consumption, or adiposity). Many psychological theories propose that high levels of life satisfaction and positive affect and low levels of experienced stress (i.e., hedonic well-being) should promote physical health. This genomics-based analysis implies that positive affect alone may not be sufficient and that interventions targeting the self-transcendent approach to psychological well-being may provide the greatest parallel increments to molecular well-being.

What else might molecular indicators of well-being tell us about the best way for humans to live? Several randomized controlled experimental studies have shown that contemplative practices such as mindfulness meditation, physical practices such as yoga and Tai Chi, and cognitive-behavioral stress management programs can all reduce pro-inflammatory gene expression in people confronting significant life adversity [26, 30–33]. Analyses comparing long-term meditators to novice controls have also shown such differences [55–58] and imply that even under favorable life circumstances there may exist practical opportunities to enhance molecular well-being. Transcriptomic measures of human well-being are just beginning to enter behavioral science, but their use should spread rapidly because they yield information that is qualitatively distinct from that available through other well-being assessments such as conscious introspection. Fredrickson et al., for example, found striking differences in the gene transcriptional correlates of eudaimonic and hedonic happiness despite the fact that these two types of happiness did not differ at all in self-reported measures of emotional well-being.

Network Genomics and Social Recursion

Although studies have begun to examine both the positive and negative impacts of broad social climates such as SES and social stability, little is presently known about how these influences are transmitted via concrete social networks. It should be possible to map the flow of social genomic influences through networks of individuals and observe the contagious development of coordinated "cultures" of gene expression. Network genomics analyses could clarify a wide range of issues ranging from the distributed nature of individual resilience to basic questions regarding human evolution and gene-culture interaction. Emergent systems-level capacities lie at the heart of individual well-being, group selection, and our hypersocial life history strategy [2, 10]. Network genomics provides a concrete framework for analyzing how individual socially sensitive human genomes spontaneously self-assemble into coherent metagenomic networks with new systems-level capabilities.

Network genomics analyses are complicated by the fact that people actively shape and select their social environments. These active network dynamics are mediated by brains and nervous systems that are themselves transcriptionally plastic and thus sensitive to both biological and social influences [59, 60]. Fig. 20.2B shows how this kind of transcriptional plasticity in the neural underpinnings of social signal transduction can create recursive feedback systems that can persist over time in stable attractor states and can occasionally shift rapidly to new stable equilibrium in response to a transient environmental shock such as a traumatic event. Fig. 20.2A depicts the simple acyclic social signal transduction pathway that translates social threat into *CTRA gene* expression. Fig. 20.2B superimposes two empirically documented positive feedback loops—one endogenous biological loop in which proinflammatory gene products from the CTRA signal back to the brain to reduce prosocial motivation, perception, and behavior (Arc 1) [61–63] and a second exogenous social loop (Arc 2) in which altered behaviors evoke more suspicious, distant, and hostile responses from the surrounding social environment [1]. Adverse social responses propagate the individual's experience of social threat and thus promote continued CTRA signaling, continued proinflammatory feedback to the brain, continued suppression of prosocial behavior, continued adverse social reactions, and so forth. Absent either one of these feedback loops, the effects of a transient environmental shock would generally damp and decay over time. When both loops are present, however, the output of one becomes an input to the other, and they can reciprocally propagate defensive gene expression responses long after the precipitating shock has passed. This provides a purely functional mechanism by which transient environmental conditions can produce persistent biological, psychological, and social sequelae (e.g., as in PTSD or biological embedding of adverse early life social conditions) without involving any persistent DNA modification (e.g., epigenetic marking).

Prescriptive Genomics

Functional genomics occupies a unique nexus between basic human nature, as reflected in the evolved regulatory programming of our genome to generate successful human life, and the environmentally contingent realization of that innate potential for thriving [53]. As we learn more about how everyday life circumstances influence the transcriptional realization of our genomic potential, human social genomics comes to acquire a certain prescriptive aspect. Disease, death, threat, resilience, thriving, and generativity are not value-neutral in evolutionary terms nor are they in humanitarian terms. As such, molecular indicators of human well-being such as the CTRA and blood-informative transcripts (BIT) [12, 64] may add a new dimension to social, cultural, and political discourse by providing objective leading indicators of human biological well-being that can help gauge the extent to which particular social conditions are congruent with basic human nature [10] and help realize our genomically endowed potential for thriving.

Perhaps the most transformative implication of prescriptive social genomics lies in the new opportunities it provides to us as individuals as we seek to optimize our own personal well-being. Given the diverse array of genetic, developmental, historical, and socioenvironmental factors that shape our realized transcriptomes, there may not exist any single intervention that enhances biological well-being for everyone. However, we should still be able to empirically optimize our personal well-being by selecting the lifestyle components that best work for us in the context of our own particular life circumstances. Determining what works at the individual level has historically been complicated by the fact that the primary objective indicators of well-being involve disease and mortality events that generally occur decades after their environmental precipitants and leave little room for a do-over. Our emerging ability to monitor the molecular underpinnings of human well-being in everyday life using portable real-time RNA sequencing interpreted through evaluated molecular biomarkers such as the CTRA and BIT gene sets provides a wholly new opportunity to discover what works for us personally, long before those molecular dynamics coalesce into overt disease. Continuous real-time molecular biofeedback opens up a new era of human molecular self-awareness and biological self-determination. If we judge societies at least in part by the extent to which they help each of us realize our genomically endowed potential for wellbeing, then the molecularly quantified self could represent one of our own society's most significant achievements. After all, each of our human genomes is fundamentally a system for converting environmental information into molecular resources according to the accumulated wisdom of 4 million years of hominid evolution, and those who stand to gain most from the insights they afford are we whose lives they create.

References

1. Cacioppo JT, Hawkley LC (2009) Perceived social isolation and cognition. Trends Cogn Sci 13: 447–454.
2. Wilson EO (2012) The Social Conquest of Earth. New York: Liveright Publishing, W. W. Norton.
3. Cole SW (2013) Social regulation of human gene expression: mechanisms and implications for public health. Am J Public Health 103 Suppl 1: S84–S92.
4. Gibson G (2008) The environmental contribution to gene expression profiles. Nat Rev Genet 9: 575–581.
5. Irwin MR, Cole SW (2011) Reciprocal regulation of the neural and innate immune systems. Nat Rev Immunol 11: 625–632.
6. Robinson GE, Fernald RD, Clayton DF (2008) Genes and social behavior. Science 322: 896–900.
7. Tung J, Barreiro LB, Johnson ZP, Hansen KD, Michopoulos V, et al. (2012) Social environment is associated with gene regulatory variation in the rhesus macaque immune system. Proc Natl Acad Sci U S A 109: 6490–6495.
8. Cole SW, Arevalo JM, Ruggerio AM, Heckman JJ, Suomi S (2012) Transcriptional modulation of the developing immune system by early life social adversity. Proc Natl Acad Sci U S A 109: 20578–20583.
9. Powell ND, Sloan EK, Bailey MT, Arevalo JM, Miller GE, et al. (2013) Social stress up-regulates inflammatory gene expression in the leukocyte transcriptome via beta-adrenergic, induction of myelopoiesis. Proc Natl Acad Sci U S A 110: 16574–16579.
10. Churchland PS (2011) Braintrust: what neuroscience tells us about morality. Princeton (New Jersey): Princeton University Press.
11. Idaghdour Y, Czika W, Shianna KV, Lee SH, Visscher PM, et al. (2010) Geographical genomics of human leukocyte gene expression variation in southern Morocco. Nat Genet 42: 62–67.
12. Nath AP, Arafat D, Gibson G (2012) Using blood informative transcripts in geographical genomics: impact of lifestyle on gene expression in fijians. Front Genet 3: 243.
13. Holt-Lunstad J, Smith TB, Layton JB (2010) Social relationships and mortality risk: a meta-analytic review. PLoS Med 7: e1000316.
14. Cole SW, Hawkley LG, Arevalo JM, Sung CY, Rose RM, et al. (2007) Social regulation of gene expression in human leukocytes. Genome Biol 8: 1–13.
15. Cole SW, Hawkley LC, Arevalo JM, Cacioppo JT (2011) Transcript origin analysis identifies antigen-presenting cells as primary targets of socially regulated gene expression in leukocytes. Proc Natl Acad Sci USA 108: 3080–3085.
16. Seeman TE (1996) Social ties and health: the benefits of social integration. Ann Epidemiol 6: 442–451.
17. Chen E, Miller GE, Walker HA, Arevalo JM, Sung CY, et al. (2009) Genome-wide transcriptional profiling linked to social class in asthma. Thorax 64: 38–43.
18. Miller GE, Chen E, Fok AK, Walker H, Lim A, et al. (2009) Low early-life social class leaves a biological residue manifested by decreased glucocorticoid and increased proinflammatory signaling. Proc Natl Acad Sci USA 106: 14716–14721.
19. Cole S, Arevalo J, Takahashi R, Sloan EK, Lutgendorf S, et al. (2010) Computational identification of

19. gene-social environment interaction at the human IL6 locus. Proc Natl Acad Sci U S A 107: 5681–5686.
20. Chen E, Miller GE, Kobor MS, Cole SW (2011) Maternal warmth buffers the effects of low early-life socioeconomic status on pro-inflammatory signaling in adulthood. Mol Psychiatry 16: 729–737.
21. Miller GE, Chen E, Sze J, Marin T, Arevalo JM, et al. (2008) A functional genomic fingerprint of chronic stress in humans: blunted glucocorticoid and increased NF-kappaB signaling. Biol Psychiatry 64: 266–272.
22. Miller GE, Murphy MLM, Cashman R, Ma R, Arevalo JMG, et al. (2014) Greater inflammatory activity and blunted glucocorticoid signaling in monocytes of chronically stressed caregivers. Brain, Behavior, and Immunity. In press. doi: 10.1016/j.bbi.2014.05.016
23. O'Connor MF, Schultze-Florey CR, Irwin MR, Arevalo JM, Cole SW (2014) Divergent gene expression responses to complicated grief and non-complicated grief. Brain Behav Immun 37: 78–83.
24. Segman RH, Shefi N, Goltser-Dubner T, Friedman N, Kaminski N, et al. (2005) Peripheral blood mononuclear cell gene expression profiles identify emergent post-traumatic stress disorder among trauma survivors. Mol Psychiatry 10: 500–513, 425.
25. O'Donovan A, Sun B, Cole S, Rempel H, Lenoci M, et al. (2011) Transcriptional control of monocyte gene expression in post-traumatic stress disorder. Dis Markers 30: 123–132.
26. Antoni MH, Lutgendorf SK, Blomberg B, Stagl J, Carver CS, et al. (2012) Transcriptional modulation of human leukocytes by cognitive-behavioral stress management in women undergoing treatment for breast cancer. Biological Psychiatry 71: 366–372.
27. Cohen L, Cole SW, Sood AK, Prinsloo S, Kirschbaum C, et al. (2012) Depressive symptoms and cortisol rhythmicity predict survival in patients with renal cell carcinoma: role of inflammatory signaling. PLoS ONE 7: e42324.
28. Cole SW (2010) Elevating the perspective on human stress genomics. Psychoneuroendocrinology 35: 955–962.
29. Sloan EK, Capitanio JP, Cole SW (2008) Stress-induced remodeling of lymphoid innervation. Brain Behav Immun 22: 15–21.
30. Creswell JD, Irwin MR, Burklund LJ, Lieberman MD, Arevalo JM, et al. (2012) Mindfulness-Based Stress Reduction training reduces loneliness and pro-inflammatory gene expression in older adults: A small randomized controlled trial. Brain Behav Immun 26: 1095–1101.
31. Black DS, Cole SW, Irwin MR, Breen E, St Cyr NM, et al. (2012) Yogic meditation reverses NF-kappaB and IRF-related transcriptome dynamics in leukocytes of family dementia caregivers in a randomized controlled trial. Psychoneuroendocrinology 38: 348–55.
32. Bower JE, Greendale G, Crosswell AD, Garet D, Stemlieb B, et al. (2014) Yoga reduces inflammatory signaling in fatigued breast cancer survivors: A randomized controlled trial. Psychoneuroendocrinology 43: 20–29.
33. Irwin M, Olmstead R, Breen E, Witarama T, Carrillo C, et al. (2014) Tai Chi Chih reduces cellular and genomic markers of inflammation in breast cancer survivors with insomnia. J Natl Cancer Inst. In press.
34. Collado-Hidalgo A, Sung C, Cole S (2006) Adrenergic inhibition of innate antiviral response: PKA blockade of Type I interferon gene transcription mediates catecholamine support for HrV-1 replication. Brain Behav Immun 20: 552–563.
35. Wohleb ES, McKim DB, Shea DT, Powell ND, Tarr AJ, et al. (2014) Reestablishment of Anxiety in Stress-Sensitized Mice Is Caused by Monocyte Trafficking from die Spleen to the Brain. Biol Psychiatry 75: 970–981.
36. Slavich GM, Cole SW (2013) The emerging field of human social genomics. Clin Psychol Sci 1: 331–348.
37. Fraga MF, Ballcstar E, Paz MF, Ropcro S, Scticn F, et al. (2005) Epigenetic differences arise during the lifetime of monozygotic twins. Proc Nad Acad Sci USA 102: 10604–10609.
38. Borghol N, Sudennan M, McArdle W, Racine A, Hallett M, et al. (2011) Associations with early-life socio-economic position in adult DNA methylation. Int J Epidemiol 41: 62–74.
39. Essex MJ, Thomas Boyce W, Hertzman C, Lam LL, Armstrong JM, et al. (2011) Epigenetic Vestiges of Early Developmental Adversity: Childhood Stress Exposure and DNA Mediylation in Adolescence. Child Dev 84: 58–75.
40. Naumova OY, Lee M, Koposov R, Szyf M, Dozier M, et al. (2012) Differential patterns of whole-genome. DNA methylation in institutionalized children and children raised by their biological parents. Dev Psychopathol 24: 143–155.
41. Lam LL, Emberly E, Fraser HB, Neumann SM, Chen E, et al. (2012) Factors underlying variable DNA

methylation in a human community cohort. Proc Nad Acad Sci U S A 109 Suppl 2: 17253–17260.
42. Cole SW (2013) Nervous system regulation of the cancer genome. Brain Behav Immun 30 Suppl: S10–S18.
43. Thaker PH, Han LY, Kamat AA, Arevalo JM, Takahashi R, et al. (2006) Chronic stress promotes tumor growth and angiogenesis in a mouse model of ovarian carcinoma. Nat Med 12: 939–944.
44. Sloan EK, Priceman SJ, Cox BF, Yu S, Pimentel MA, et al. (2010) The sympathetic nervous system induces a metastatic switch in primary breast cancer. Cancer Res 70: 7042–7052.
45. Volden PA, Conzen SD (2013) The influence of glucocorticoid signaling on tumor progression. Brain Behav Immun 30 Suppl: S26–31.
46. Hermes GL, Delgado B, Tretiakova M, Cavigelli SA, Krausz T, et al. (2009) Social isolation dysregulates endocrine and behavioral stress while increasing malignant burden of spontaneous mammary tumors. Proc Nad Acad Sci USA 106: 22393–22398.
47. Williams JB, Pang D, Delgado B, Kocherginsky M, Tretiakova M, et al. (2009) A model of gene-environment interaction reveals altered mammary gland gene expression and increased tumor growth following social isolation. Cancer Prev Res (Phila) 2: 850–861.
48. Volden PA, Wonder EL, Skor MN, Carmean CM, Patel FN, et al. (2013) Chronic, social isolation is associated with metabolic gene expression changes specific to mammary adipose tissue. Cancer Prev Res (Phila) 6: 634–645.
49. Feng Z, Liu L, Zhang C, Zheng T, Wang J, et al. (2012) Chronic restraint stress attenuates p53 function and promotes tumorigenesis. Proc Natl Acad Sci USA 109: 7013–7018.
50. Pang D, Kocherginsky M, Krausz T, Kim SY, Conzen SD (2006) Dexamethasone decreases xenograft response to Paclitaxel through inhibition of tumor cell apoptosis. Cancer Biol Titer 5: 933–940.
51. Wu W, Pew T, Zou M, Pang D, Conzen SD (2005) Glucocorticoid receptor-induced MAPK phosphatase-1 (MPK-1) expression inhibits paclitaxel-associated MAPK activation and contributes to breast cancer cell survival. J Biol Chem 280: 4117–4124.
52. Luca F, Kashyap S, Southard C, Zou M, Witonsky D, et al. (2009) Adaptive variation regulates the expression of the human SGK1 gene in response to stress. PLoS Genet 5: el000489.
53. Fox Keller E (2012) Genes, genomes, and genomics. Biological Theory 6: 132–140.
54. Fredrickson BL, Grewen KM, Coffey KA, Algoe SB, Firestine AM, et al. (2013) A functional genomic perspective on human well-being. Proc Nad Acad SciU S A 110: 13684–13689.
55. Dusek JA, Otu HH, Wohlhueter AL, Bhasin M, Zerbini LF, et al. (2008) Genomic counter-stress changes induced by the relaxation response. PLoS ONE 3: e2576.
56. Bhasin MK, Dusek JA, Chang BH, Joseph MG, Denninger JW, et al. (2013) Relaxation response induces temporal transcriptome changes in energy metabolism, insulin secretion and inflammatory pathways. PLoS ONE 8: e62817.
57. Qu S, Olafsrud SM, Meza-Zepeda LA, Saatcioglu F (2013) Rapid gene expression changes in peripheral blood lymphocytes upon practice of a comprehensive yoga program. PLoS ONE 8: e61910.
58. Kaliman P, Alvarez-Lopez MJ, Cosin-Tomas M, Rosenkranz MA, Lutz A, et al. (2014) Rapid changes in histone deacetylases and inflammatory gene expression in expert meditators. Psychoneuroendocrinology 40: 96–107.
59. Cole SW (2005) The complexity of dynamic host networks. In: Deisboeck TS, Kresh JY, editors. Complex Systems Science in BioMedicine. New York: Kluwer Academic - Plenum Publishers. pp. 605–629.
60. Fowler JH, Dawes CT, Christakis NA (2009) Model of genetic variation in human social networks. Proc Natl Acad Sci U S A 106: 1720–1724.
61. Dantzer R, O'Connor JC, Freund GG, Johnson RW, Kelley KW (2008) From inflammation to sickness and depression: when the immune system subjugates the brain. Nat Rev Neurosci 9: 46–56.
62. Inagaki TK, Muscatell KA, Irwin MR, Cole SW, Eisenberger NI (2012) Inflammation selectively enhances amygdala activity to socially threatening images. Neuroimage 59: 3222–3226.
63. Eisenberger NI, Inagaki TK, Mashal NM, Irwin MR (2010) Inflammation and social experience: an inflammatory challenge induces feelings of social disconnection in addition to depressed mood. Brain Behav Immun 24: 558–563.
64. Preminger M, Arafat D, Kim J, Nath AP, Idaghdour Y, et al. (2013) Blood-informative transcripts define nine common axes of peripheral blood gene expression. PLoS Genet 9: el003362.

21.
BEHAVIORAL EPIGENETICS
*David S. Moore**

Why do we grow up to have the traits we do? Most 20th century scientists answered this question by referring only to our genes and our environments. But recent discoveries in the emerging field of behavioral epigenetics have revealed factors at the interface between genes and environments that also play crucial roles in development. These factors affect how genes work; scientists now know that what matters as much as which genes you *have* (and what environments you encounter) is how your genes are *affected* by their contexts. The discovery that what our genes *do* depends in part on our experiences has shed light on how Nature and Nurture interact at the molecular level inside of our bodies. Data emerging from the world's behavioral epigenetics laboratories support the idea that a person's genes alone cannot determine if, for example, he or she will end up shy, suffering from cardiovascular disease, or extremely smart. Among the environmental factors that can influence genetic activity are parenting styles, diets, and social statuses. In addition to influencing how doctors treat diseases, discoveries about behavioral epigenetics are likely to alter how biologists think about evolution, because some epigenetic effects of experience appear to be transmissible from generation to generation. This domain of research will likely change how we think about the origins of human nature.

Introduction

How do we come to have our characteristics, from eye color to height to personality? Although theorists traditionally referred to Nature and Nurture when answering this question—or, more recently, genes and environments—we now know that a third factor, 'epigenetics,' plays a central role in trait development. Serving as the interface between genetic and environmental factors, epigenetic processes illustrate how Nature and Nurture work together in development. The implication of this discovery is that genes cannot have effects that are independent of context. Just as all of our phenotypes depend on our genes for their development, they also depend on non-genetic factors; thus, there really cannot be any traits that are *strictly* genetic, because a gene's context always matters (see Chapter 11). Epigenetic processes have been linked to diverse phenomena, including memory and learning, cancer, addiction, diabetes, aging, and the effects of such factors as exercise, nutrition, environmental toxins, and early-life experiences...and this is just a partial list.[1]

Defining Epigenetics

'Epigenetics' is an old word—Aristotle believed all of our characteristics arise in a process he called 'epigenesis'—and it has had several definitions

* Pitzer College & Claremont Graduate University, Claremont, CA

through the years. In the 1940s, the biologist Conrad Waddington began using the word in a modern way, to refer to how genes interact with their local environments to build organisms.[2] Waddington understood that we each begin life as a fertilized egg, and that as development unfolds, this egg divides into trillions of cells that each contain the same genetic information. Therefore, in order to develop into *different* kinds of cells—blood cells and brain cells and bone cells—there has to be a way for *nongenetic* factors to turn different genes on or off. Although different contemporary theorists define epigenetics in different ways, scientists agree that epigenetics is fundamentally about development. And one way in which development can be influenced is via the effects of contextual factors on genetic activity. In fact, genetic activity can be regulated somewhat like a light bulb controlled by a *dimmer* switch: it can be turned off completely, or can be turned on to greater or lesser degrees. And because contexts influence genetic activity by physically attaching various chemicals to our genes, these chemicals are literally 'epi'-genetic: they are '*on*' genes. What has made recent research on epigenetics so exciting is the discovery that the environments in which we develop—how we are nurtured—can influence the epigenetic state of our DNA, and, therefore, who we become.[3]

It's Not Just about the Genes You *Have*, It's about What Your Genes Are *Doing*

Because of how most parents, teachers, and the media talk about genes, it is easy to imagine that genes are agents that make active decisions about how to build bodies and minds. But the fact is that the DNA that contains our genes is not capable of independent action; if you put a bowl of naked DNA on your desk, it will just sit there, inert (see Ref 4 for additional information). This is why genetic *regulation* is so important. What your genes are induced to *do* by their contexts—which include other genes, of course—is just as important as what genes you *have*; a gene you have inherited from your parents is of no consequence if that gene remains turned off by epigenetic processes, so you might as well not have the gene at all.

Epigenetic Mechanisms

The genetic information in our DNA is unavailable in certain circumstances. For instance, if a DNA segment is bunched up very tightly, the cellular 'machinery' that ordinarily accesses that information will be unable to make adequate contact with the segment. When that happens, the bunched-up DNA segment will be unavailable for use, and the proteins normally produced using that segment will not be produced.

Epigenetic factors can alter access to genetic information in several ways (Fig. 21.1). Large molecules called 'histones' are associated with DNA, and these can be modified through a variety of processes. One of these processes is called 'acetylation'; histone acetylation generally causes DNA segments to become *more* accessible, and thereby leads to increased gene expression and ultimately to the production of more of the proteins associated with those DNA segments. In contrast, chemicals called 'methyl groups' can attach to DNA segments and thereby decrease access to those segments; when this happens, the segments are said to be 'methylated,' leading to decreased gene expression and Ultimately to reduced production of proteins associated with those DNA segments. Thus, histone acetylation effectively turns genes on (or speeds up rates of protein production), whereas DNA methylation effectively turns genes off (or slows down rates of protein production). Other epigenetic processes—such as histone methylation, phosphorylation, or ubiquitination, to name a few—can likewise promote or suppress gene expression, but these processes have not yet received as much research attention as DNA methylation or histone acetylation. Regardless, in each of these cases, the genome is 'marked' by the addition of chemicals that influence genetic activity without actually changing the coded information in the DNA.

Behavioral Epigenetics

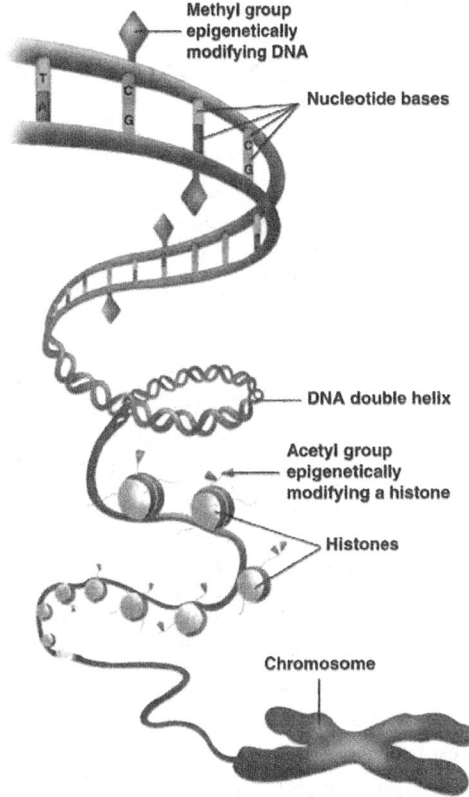

Fig. 21.1 A schematic diagram of DNA pulled from a chromosome, showing the double helix wrapped around histones, and some epigenetic modifications to both the DNA and the histones.

Several factors contribute to cells' epigenetic states. First, different cell types are epigenetically different. For instance, blood cells have certain genes activated and other genes deactivated, and this combination is what gives those cells the distinctive characteristics of blood cells. In contrast, neural cells have *other* combinations of activated and deactivated genes, and this combination is what makes these cells neurons. Second, additional epigenetic marks are located at random throughout the genome.[5] Third, the locations of some epigenetic marks are related to variations in individuals' DNA sequences.[6] And finally, the epigenetic state of particular DNA segments in particular cells can sometimes reflect an individual's prior *experiences*.[7,8]

An Early Insight into Epigenetic Regulation

In the 1960s, scientists were grappling with a perplexing question: Why do a woman's *two* X chromosomes not produce twice the number of proteins as a man's single X chromosome? We now know that early in embryonic development, epigenetic processes disable one of the X chromosomes in each cell of female mammals, and whichever one is inactivated stays inactivated for a lifetime.[9] X-Inactivation occurs at random, so some of a woman's cells have an inactivated X chromosome originally provided by her father, but others have an inactivated X chromosome originally provided by her mother; so, women are 'epigenetic mosaics,' because the X chromosomes in different cells are in different epigenetic states.[10] When genes that contribute to *coloration* are located on X chromosomes, we can actually *see* this mosaicism; a calico cat has multiple fur colors arranged in patches in her coat, and this patterning reflects the random inactivation of one X chromosome in some cells and the other X chromosome in other cells (Fig. 21.2). Two important lessons emerge from considering calicos. First, epigenetic effects can be bold; even individuals with identical genomes can have very different appearances owing to epigenetics. Second a particular calico's coat-color pattern doesn't change as she ages, so epigenetic modifications can be extremely stable.

Fig. 21.2 A photograph of a calico cat. Image courtesy of Howard Cheng.

Epigenetic Effects of Experience

Similarly bold effects of epigenetics occur in other species, too, and the effects can sometimes be traced to experiences. For example, genetically identical female honeybees can develop into either infertile workers with small bodies and short lifespans or fecund queens with larger bodies and longer lifespans; these stark differences reflect epigenetic effects of their distinctive diets.[11] Likewise, mice with identical genomes can nonetheless have very different characteristics. In one well-studied strain, some of the mice are yellow, obese, and at risk of developing tumors and diabetes, while others are brownish, thin, and healthy; the rest fall along continua between these extremes.[12] The differences between these mice are epigenetic, reflecting DNA methylation (Fig. 21.3). Although these differing epigenetic states result from random processes to some extent, they can also be influenced during embryonic development by dietary factors. Specifically, providing pregnant mice of this strain with a diet containing supplemental methyl groups causes pups developing *in utero* to be more likely to develop brownish coats and healthy constitutions; the supplemented diet contributes to DNA methylation in the not-yet-born offspring, which silences DNA that contributes to a yellow coat, obesity, and cancer.[14]

Studies of people, too, have revealed epigenetic effects of experience. Groundbreaking research revealed that identical twins—who share identical genomes—have differing epigenetic profiles. Importantly, although epigenetic profiles for pairs of identical twins were *similar* when they were young, the twins' profiles diverged as they had unique experiences. Specifically, older twins who spent more of their lives apart had more epigenetic differences throughout their genomes.[15] Thus, experiences can leave 'marks' on DNA that affect how genes are expressed.[16]

Some of the most compelling data on the epigenetic effects of experience have come from the labs of Michael Meaney, Moshe Szyf, and their colleagues at McGill University. These researchers discovered that natural variations in behaviors of mother rats affect how newborn offspring react to stress later in life. Specifically, although all normal mother rats lick and groom their newborn pups, mothers that lick and groom their pups a *lot* have offspring that grow up better able to tolerate mild

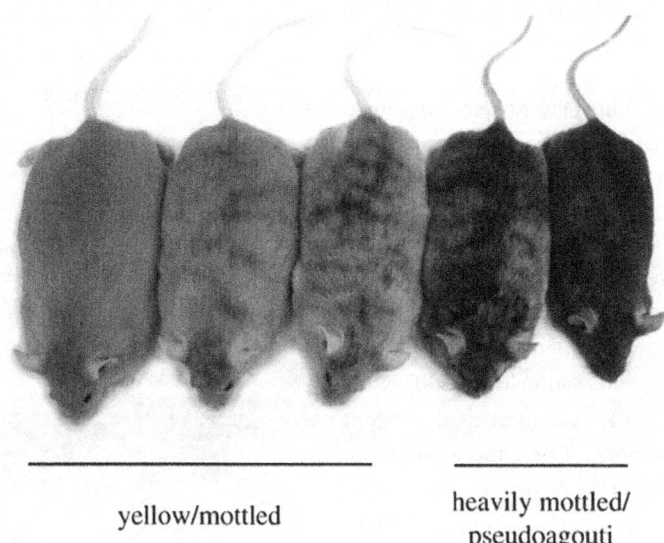

yellow/mottled   heavily mottled/pseudoagouti

Fig. 21.3 A photograph of mice that are genetically identical, but nonetheless have a spectrum of coat colors. (Reprinted with permission from Ref 13. Copyright 2012).

stressors in adulthood; pups that are not licked and groomed as much ultimately behave more fearfully when stressed.[17] When the McGill researchers looked for epigenetic effects of these early experiences, they discovered that reduced exposure to postnatal licking and grooming leaves particular DNA segments highly methylated in within the brain's hippocampus. Therefore, these DNA segments are downregulated in the affected hippocampal cells, which ultimately leads to reduced production of a particular protein—called the glucocorticoid receptor, or GR—that helps regulate stress.[18] Subsequent work with rodents revealed similar epigenetic effects of early experiences on other DNA segments in other brain areas as well.[19,20]

Analogous effects occur in humans and other primates. For instance, in one study, when compared to people who did not experience child abuse, adults abused as children had hippocampal cells with DNA that was more methylated in a region associated with the GR.[7] Thus, in this correlational study, there was reduced expression of a stress-moderating protein in the brains of people who experienced bad parenting years earlier. Similarly, an *experimental* investigation examined the epigenetic effects of maternal deprivation on newborn monkeys. When these monkeys were adults, they had hundreds of locations in blood DNA—and thousands of locations in brain DNA—where the patterns of methylation varied depending on how the monkeys were reared.[8] Another experimental study of monkeys revealed that subtler manipulations can also produce epigenetic changes in blood-derived DNA.[21] This study found that the stress experienced by low-ranking animals in a troop's dominance hierarchy affects DNA methylation. Thus, in our primate relatives, methylation—and hence gene expression—is responsive to changes in social status. Finally, there appears to be a correlation between one's socioeconomic status in childhood and one's epigenetic profile decades later; among 45-year-olds currently in various socioeconomic conditions, those who experienced poverty in childhood had different amounts of methylation (compared to unimpoverished children) in over one thousand regions of blood-derived DNA.[22] All of these studies suggest that some epigenetic alterations can serve as *records* of previous experiences.[23] Epigenetics helps explain how our experiences literally get under our skin.

Epigenetics, All Day, Every Day

Epigenetic processes remain important beyond our formative years, and they do not only register exceptional events. In fact, epigenetic events are unfolding in your body right now: if you retain any information after reading this essay, it will be because your brain changed in a physical way that allowed you to store the information in your long-term memory.[24] And these kinds of physical changes require the construction of new proteins in your brain's neurons, a process that requires genes to be epigenetically 'turned on.'[25]

Other behaviors also influence our epigenetic states. For example, exercise causes changes in DNA methylation in muscle cells.[26] Likewise, what we eat and drink can have epigenetic effects.[27] One reason is because methyl groups are present in food, specifically foods that contain B-complex vitamins or a nutrient called choline; B_9, for example, is present in asparagus, eggs, and dark green leafy vegetables, whereas choline is present in cauliflower, meats, and milk. These methyl-rich foods can influence DNA methylation. In addition, some foods influence histone acetylation. For example, broccoli sprouts or turmeric—a spice related to ginger—can affect histone acetylation in some cell types.[28] Consequently, some theorists think these effects explain how certain foods lower the risk of developing colon cancer.[29] Even simply providing people with a daily nutritional supplement containing methyl groups has been found in some experiments to reduce symptoms of both clinical depression and degenerative arthritis (see Refs 30 and 31, respectively).

Some research has examined how the diets of *pregnant* animals influence the epigenetic states of their offspring. Just as supplementing the diet of some pregnant mice can affect the coat colors and constitutions of the next generation, manipulating the diet of pregnant sheep can influence DNA methylation in offspring and leave them obese, resistant to insulin, and suffering from high blood pressure.[32] Correlational studies have also uncovered epigenetic abnormalities in 60-year-old *people* who were exposed to famine while developing *in utero*.[33] The discovery that prenatal experience can be recorded in the epigenome has stoked interest in the developmental origins of health and disease,[34] and contributed to dawning comprehension about how prenatal experience can produce long-term effects on phenotypes, from body size,[35] to metabolic functioning,[36] to a person's likelihood of developing schizophrenia.[37]

This Is Not Your Father's Biology ... but You Might Still Carry His Epigenetics

DNA methylation is relatively stable; once a segment of DNA is methylated, it typically stays that way. In addition, even when a cell in a body divides, it generally gives rise to two 'daughter' cells that each have the same DNA methylation profile that characterized the 'parent' cell; this is known as 'cellular inheritance' of epigenetic marks. However, things are very different when we consider *transgenerational* epigenetic inheritance—transmission of epigenetic marks not from cell to cell during an organism's life, but from organism to organism when a new generation is produced. Normally, most epigenetic marks are 'erased' between generations, shortly after conception of a new embryo (Fig. 21.4); this process ensures that new embryos will be composed of stem cells able to become any of the different cell types in a mature body.[38] This process is consistent with biology's dominant theory, which holds that an animal's experiences cannot affect the traits inherited by its offspring. Nonetheless, there is now compelling evidence that DNA methylation in mammalian sperm or eggs can sometimes elude between-generation 'erasure,' and thereby be transmitted to offspring.[12,38,39]

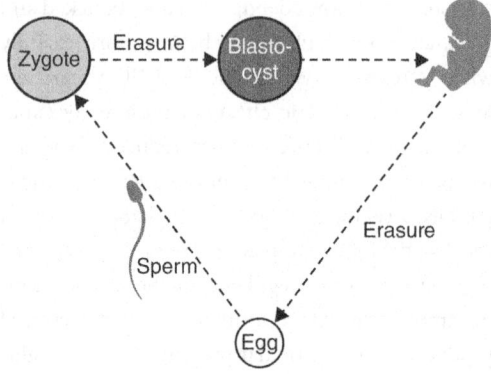

Fig. 21.4 A schematic diagram of the 'life cycle' of methylation under normal circumstances. DNA methylation is 'erased' once shortly after a new organism is conceived (a zygote is a newly conceived organism and a blastocyst is an early embryo), and once more in the primordial germ cells that will become that new organism's sperm or egg cells. (Reprinted with permission from Ref 1. Copyright 2015 Oxford University Press).

Some theorists argue that transgenerational epigenetic inheritance via sperm or eggs is rare but others believe 'epigenetic inheritance is ubiquitous' (see Refs 38 and 40, p. 131, respectively). Why the discrepancy?

One reason transgenerational epigenetic inheritance is controversial is because theorists have different ideas about what should count as 'inheritance.'[1,4] Some writers insist that transgenerational epigenetic inheritance requires transmission of epigenetic marks that are on sperm or egg cell DNA. But a phenomenon described by the McGill researchers suggests that transgenerational transmission of epigenetic marks can sometimes occur independently of genome transmission. In this phenomenon, female rat pups that were licked and groomed a lot grew up to be adults that—because of their epigenetic state—were less reactive in stressful situations. In addition, these rats matured into more attentive mothers, licking and grooming *their* offspring in ways that produced the same distinctive epigenetic characteristic found in the mothers. Thus, even though the epigenetic marks were not actually transmitted with the genome in sperm or eggs, they were nonetheless reliably reproduced in one generation after another, via a behavioral mechanism.[3,41]

Regardless of *how* it occurs, it is now clear that in some species, ancestors can beget descendants that share the ancestors' distinctive epigenetic states. For instance, epigenetic phenomena are responsible for so-called 'parent-of-origin' effects, in which a gene's activity depends on whether it originated in a father or a mother.[42] In addition, some fascinating circumstantial evidence suggestive of epigenetic inheritance across multiple generations has recently been uncovered in a human population.[43,44] Studies of the transgenerational inheritance of epigenetic marks will continue to draw researchers' attention for at least two reasons. First, such effects have evolutionary significance.[45] In fact, one research team has already described how exposure to a pesticide produces epigenetic effects in rats that are transmitted across generations in a way that these investigators believe provides 'direct experimental evidence for a role of epigenetics as a determinant factor in evolution.'[46,47] Second, because the transgenerational transmission of epigenetic marks provides a mechanism by which individuals' experiences can affect their *descendants*' characteristics, this phenomenon permits a quasi-Lamarckian mode of inheritance that would have been unthinkable to most 20th century biologists.[5] Incorporating transgenerational epigenetic inheritance into a comprehensive theory of biology will be a challenge, as the reigning theory of evolution—the neo-Darwinian Modern Synthesis—cannot accommodate this sort of phenomenon.[48] Therefore, new ideas about inheritance will be an area of active exploration in coming years.

Conclusion

Research on behavioral epigenetics has helped explain some long-standing enigmas, and driven home some important points. By revealing how gene expression is controlled, this research has reemphasized that genetic and nongenetic factors always collaborate to produce our characteristics; there are no features of our bodies or minds that are determined strictly by our DNA. By revealing how contexts can influence epigenetics, this research has explained how early-life experiences can produce lifelong effects, and has clarified how 'identical' twins come to differ. And by revealing how exposure to novel events can lead to the epigenetic initiation of structural changes in our brains, this research has illuminated how learning is implemented in the nervous system, thereby further clarifying the relationship between mind and body.

Behavioral epigenetics has the potential to have an enormous impact on our world. This research will likely alter how psychiatrists treat people with psychological disorders, how physicians practice their art, how public health organizations promote healthy lifestyles, and how government agencies evaluate the risk of pesticides, nutritional supplements, and medications.[1] Research on epigenetics will have consequences that are this wide-ranging because the human epigenome is central to human nature.

Further Reading

"Genetics" (http://www.nature.com/scitable/topic/genetics-5).

"Translation/RNA Translation" (http://www.nature.com/scitable/definition/translation-173).

"Epigenetics" (http://www.nature.com/scitable/spotlight/epigenetics-26097411).

"Early concepts of evolution: Jean Baptiste Lamarck" (http://www.blackwellpublishing.com/ridley/a-z/Lamarckian_inheritance.asp).

References

1. Moore DS. *The Developing Genome: An Introduction to Behavioral Epigenetics.* New York: Oxford University Press; 2015.
2. Waddington CH. The basic ideas of biology. Waddington CH. *Towards a Theoretical Biology, Vol. 1: Prolegomena*, 1–32. Edinburgh: Edinburgh University Press; 1968.
3. Meaney MJ. Epigenetics and the biological definition of gene x environment interactions. *Child Dev* 2010, 81:41–79.

4. Moore DS. *The Dependent Gene: The Fallacy of Nature vs. Nurture.* New York: W.H. Freeman; 2002.
5. Richards EJ. Inherited epigenetic variation—revisiting soft inheritance. *Nat Rev Genet* 2006, 7:395–401.
6. Kerkel K, Spadola A, Yuan E, Kosek J, Jiang L, Hod E, Li K, Murty VV, Schupf N, Vilain E, et al. Genomic surveys by methylation-sensitive SNP analysis identify sequence-dependent allele-specific DNA methylation. *Nat Genet* 2008, 40: 904–908.
7. McGowan PO, Sasaki A, D'Alessio AC, Dymov S, Labonte B, Szyf M, Turecki G, Meaney MJ. Epigenetic regulation of the glucocorticoid receptor in human brain associates with childhood abuse. *Nat Neurosci* 2009, 12:342–348.
8. Provencal N, Suderman MJ, Guillemin C, Massart R, Ruggiero A, Wang D, Bennett AJ, Pierre PJ, Friedman DP, Cote SM, et al. The signature of maternal rearing in the methylome in rhesus macaque prefrontal cortex and T cells. *J Neurosci* 2012, 32:15626–15642.
9. van den Berg IM, Laven JSE, Stevens M, Jonkers I, Galjaard RJ, Gribnau J, van Doorninck JH. X Chromosome inactivation is initiated in human preimplantation embryos. *Am J Hum Genet* 2009, 84:771–779.
10. Lyon MF. Sex chromatin and gene action in the mammalian X-chromosome. *Am J Hum Genet* 1962, 14:135–148.
11. Kucharski R, Maleszka J, Foret S, Maleszka R. Nutritional control of reproductive status in honeybees via DNA methylation. *Science* 2008, 319:1827–1830.
12. Morgan HD, Sutherland HG, Martin DI, Whitelaw E. Epigenetic inheritance at the agouti locus in the mouse. *Nat Genet* 1999, 23:314–318.
13. Cropley JE, Dang THY, Martin DIK, Suter CM. The penetrance of an epigenetic trait in mice is progressively yet reversibly increased by selection and environment. *Proc R Soc Lond B Biol Sci* 2012, 279:2347–2353.
14. Waterland RA, Jirtle RL. Transposable elements: targets for early nutritional effects on epigenetic gene regulation. *Mol Cell Biol* 2003, 23:5293–5300.
15. Fraga MF, Ballestar E, Paz MF, Ropero S, Setien F, Ballestar ML, Heine-Suner D, Cigudosa JC, Urioste M, Benitez J, et al. Epigenetic differences arise during the lifetime of monozygotic twins. *Proc Natl Acad Sci USA* 2005, 102: 10604–10609.
16. Roth TL. Epigenetics of neurobiology and behavior during development and adulthood. *Dev Psychobiol* 2012, 54:590–597.
17. Francis D, Diorio J, Liu D, Meaney MJ. Nongenomic transmission across generations of maternal behavior and stress responses in the rat. *Science* 1999, 286:1155–1158.
18. Weaver IC, Cervoni N, Champagne FA, D'Alessio AC, Sharma S, Seckl JR, Dymov S, Szyf M, Meaney MJ. Epigenetic programming by maternal behavior. *Nat Neurosci* 2004, 7:847–854.
19. Murgatroyd C, Patchev AV, Wu Y, Micale V, Bockmuhl Y, Fischer D, Holsboer F, Wotjak CT, Almeida OF, Spengler D. Dynamic DNA methylation programs persistent adverse effects of early-life stress. *Nat Neurosci* 2009, 12:1559–1566.
20. Roth TL, Lubin FD, Funk AJ, Sweatt JD. Lasting epigenetic influence of early-life adversity on the BDNF gene. *Biol Psychiatry* 2009, 65:760–769.
21. Tung J, Barreiro LB, Johnson ZP, Hansen KD, Michopoulos V, Toufexis D, Michelini K, Wilson ME, Gilad Y. Social environment is associated with gene regulatory variation in the rhesus macaque immune system. *Proc Natl Acad Sci USA* 2012, 109:6490–6495.
22. Borghol N, Suderman M, McArdle W, Racine A, Hallett M, Pembrey M, Hertzman C, Power C, Szyf M. Associations with early-life socio-economic position in adult DNA methylation. *Int J Epidemiol* 2012,41:62–74.
23. Heijmans BT, Tobi EW, Lumey LH, Slagboom PE. The epigenome: archive of the prenatal environment. *Epigenetics* 2009, 4:526–531.
24. Zhang TY, Meaney MJ. Epigenetics and the environmental regulation of the genome and its function. *Annu Rev Psychol* 2010, 61:439–466, C431-433.
25. Mattick JS. RNA as the substrate for epigenome-environment interactions. *Bioessays* 2010, 32: 548–552.
26. Lindholm ME, Marabita F, Gomez-Cabrero D, Rundqvist H, Ekstrom TJ, Tegner J, Sundberg CJ. An integrative analysis reveals coordinated reprogramming of the epigenome and the transcriptome in human skeletal muscle after training. *Epigenetics* 2014, 9:1557–1569.
27. Choi S-W, Friso S. Epigenetics: a new bridge between nutrition and health. *Adv Nutr* 2010, 1:8–16.
28. Dashwood RH, Ho E. Dietaryhistone deacetylase inhibitors: from cells to mice to man. *Semin Cancer Biol* 2007, 17:363–369.

29. Carey N. *The Epigenetics Revolution*. London: Icon Books; 2011.
30. Papakostas GI, Mischoulon D, Shyu I, Alpert JE, Fava M. S-Adenosyl methionine (SAMe) augmentation of serotonin reuptake inhibitors for antidepressant nonresponders with major depressive disorder: a double-blind, randomized clinical trial. *Am J Psychiatry* 2010, 167:942–948.
31. Najm WI, Reinsch S, Hoehler F, Tobis JS, Harvey PW. S-Adenosyl methionine (SAMe) versus celecoxib for the treatment of osteoarthritis symptoms: a doubleblind cross-over trial [ISRCTN36233495]. *BMC Musculoskelet Disord* 2004, 5:6.
32. Sinclair KD, Allegrucci C, Singh R, Gardner DS, Sebastian S, Bispham J, Thurston A, Huntley JF, Rees WD, Maloney CA, et al. DNA methylation, insulin resistance, and blood pressure in offspring determined by maternal periconceptional B vitamin and methionine status. *Proc Natl Acad Sci USA* 2007, 104:19351–19356.
33. Heijmans BT, Tobi EW, Stein AD, Putter H, Blauw GJ, Susser ES, Slagboom PE, Lumey LH. Persistent epigenetic differences associated with prenatal exposure to famine in humans. *Proc Natl Acad Sci USA* 2008, 105:17046–17049.
34. Gluckman PD, Hanson MA, Buklijas T. A conceptual framework for the developmental origins of health and disease. *J Dev Orig Health Dis* 2010, 1:6–18.
35. Junien C, Nathanielsz P. Report on the IASO Stock Conference 2006: early and lifelong environmental epigenomic programming of metabolic syndrome, obesity and type II diabetes. *Obes Rev* 2007 8:487–502.
36. Lumey LH, Stein AD, Susser E. Prenatal famine and adult health. *Annu Rev Public Health* 2011, 32:237–262.
37. Susser ES, Lin SP. Schizophrenia after prenatal exposure to the Dutch Hunger Winter of 1944-1945. *Arch Gen Psychiatry* 1992, 49:983–988.
38. Daxinger L, Whitelaw E. Understanding transgenerational epigenetic inheritance via the gametes in mammals. *Nat Rev Genet* 2012,13: 153–162.
39. Franklin TB, Russig H, Weiss IC, Graff J, Linder N, Michalon A, Vizi S, Mansuy IM. Epigenetic transmission of the impact of early stress across generations. *Biol Psychiatry* 2010, 68:408–415.
40. Jablonka E, Raz G. Transgenerational epigenetic inheritance: prevalence, mechanisms, and implications for the study of heredity and evolution. *O Rev Biol* 2009, 84:131–176.
41. Champagne FA, Weaver IC, Diorio J, Dymov S, Szyf M, Meaney MJ. Maternal care associated with methylation of the estrogen receptor-alpha 1b promoter and estrogen receptor-alpha expression in the medial preoptic area of female offspring. *Endocrinology* 2006, 147:2909–2915.
42. Sapienza C, Peterson AC, Rossant J, Balling R. Degree of methylation of transgenes is dependent on gamete of origin. *Nature* 1987, 328:251–254.
43. Kaati G, Bygren LO, Pembrey M, Sjostrom M. Transgenerational response to nutrition, early life circumstances and longevity. *Eur J Hum Genet* 2007, 15:784–790.
44. Pembrey ME, Bygren LO, Kaati G, Edvinsson S, Northstone K, Sjostrom M, Golding J, Team AS. Sex-specific, male-line transgenerational responses in humans. *Eur J Hunt Genet* 2006, 14:159–166.
45. Gottlieb G. *Individual Development and Evolution: The Genesis of Novel Behavior*. New York: Oxford University Press; 1992.
46. Anway MD, Cupp AS, Uzumcu M, Skinner MK. Epigenetic transgenerational actions of endocrine disruptors and male fertility. *Science* 2005, 308:1466–1469.
47. Skinner MK, Savenkova MI, Zhang B, Gore AC, Crews D. Gene bionetworks involved in the epigenetic transgenerational inheritance of altered mate preference: environmental epigenetics and evolutionary biology. *BMC Genomics* 2014, 15:377.
48. Moore DS. Individuals and populations: How biology's theory and data have interfered with the integration of development and evolution. *New Ideas Psychol* 2008, 26:370–386.

22.
DYNAMIC HEREDITY
*Douglas Wahlsten**

The DNA in the fertilized egg is faithfully replicated through mitosis, so almost every cell in the adult has the same DNA sequence as the zygote. Differentiation into various kinds of cells involves expression of a limited set of genes in the DNA and silencing of many others. Which genes are switched on and which are turned off is apparent in the RNA transcripts in the cell (Edgar, Domrachev, & Lash, 2002). Some genes are shut off in a particular cell type for the remainder of the organism's life, while others are very dynamic and change expression by the hour or even within a few minutes as a result of brain activity. Given what has been known for several decades, it is clear that nervous system development as well as learning and memory must involve changes in gene expression.

Cancer researchers documented how DNA methylation and histone acetylation are involved in turning specific genes off or on. Pioneering studies of licking by mother mice and rats showed that normal experiences can also alter those processes and the effects can be transmitted across generations without changing the DNA itself (Francis, Diorio, Liu, & Meaney, 1999; Meaney, 2001). At the time, it was not apparent whether those spectacular findings were rare exceptions, but it is now recognized that DNA methylation sites are widely distributed across the entire genome and methylation/acetylation are involved in the expression of virtually all genes (Laird, 2010; Pelizzola & Ecker, 2011; Pelizzola & Molinaro, 2011). Thousands of methylation sites have been located and are now said to constitute the "methylome" of an individual. Different individuals can have different profiles of which sites are methylated and therefore differ in the extent of expression of various genes (Isles & Wilkinson, 2008; Sun, 2014). Transmission of methylation state across generations appears to be an integral part of heredity (Champagne & Meaney, 2001; Champagne, 2008; Meaney, 2001). Methylation effects also are involved in dynamic psychological processes such as learning and memory (Day et al., 2013; Levenson & Sweatt, 2005). (Yu)

The recent book by David Moore reviews what is now a large and diverse literature about epigenetics in relation to early experience of abuse, learning and memory, drug addiction, mental disorders, and other phenotypes of interest to psychobiologists. He examines recent findings with a critical mind, noting the great promise of the findings for treating recalcitrant behavioral problems, while urging caution in interpreting research that involves so many factors that are difficult to control. He notes that many kinds of investigations in this domain are still in the early stages and need confirmation in other labs. He cites numerous relevant studies that can provide a convenient portal into the current and future literature for the reader.

* *University of Alberta, Edmonton, Alberta, Canada*

This is especially true for studies of transgenerational influences that appear to show Lamarckian inheritance of acquired characters. It is widely accepted dogma that formation of germ cells through meiosis erases all epigenetic marks, so that the zygote starts its life as a blank slate of unregulated DNA, but some studies (Barros & Offenbacher, 2009) are now calling this creed into question. In some instances the methylation state of an early embryo appears to be copied from one of its parents, a methylation state that in some cases arose from a parent's experience. Moore does not cast his vote unequivocally in favour of Lamarckism, but he notes that, if the recent findings are substantiated, our conceptions of heredity will need to be reformed and modernized. Implications will be profound for the theory of evolution because epigenetic processes could allow for much faster and more adaptive hereditary change than those that require us to wait patiently for a rare and fortuitous mutation in the DNA.

Moore also places the story in its historical context. The recent discoveries in epigenetics can be seen as outgrowths and confirmation of conceptual foundations laid by decades of careful experimentation and theorizing in developmental psychobiology. Scientists in this field knew that something akin to methylation and acetylation just had to be involved (Gottlieb, 1998, 2007). Developmental studies of early environment effects on later brain structure and behavior required it. Thus, the recent molecular studies substantiate the earlier literature on development of phenotypes. Perhaps the studies now have a greater impact on a broader spectrum of researchers because some find so-called "hard," molecular evidence to be more persuasive than what many biologists see as "soft," more labile behavioral data. Today, even the more reductionistic-minded investigators are being persuaded of the importance of epigenetics because we can point to a biological mechanism, not just some psychological test scores.

Although the book is long and profusely documented, there are aspects of epigenetics that need further elucidation. One pertains to the nature of allelic differences in a gene. This is the stuff of behavior genetics, an important source of individual differences in phenotypes in a population. Most of the developmental studies cited by Moore do not involve allelic genetic differences. Textbook examples of genetic differences often involve dramatic phenotypic effects of an amino acid substitution that alters the structure and function of a protein. Recent data from genome-wide scans for genetic variants relevant to behavioral phenotypes within the normal range, however, indicate that DNA variants with major effects on phenotypes are extremely rare or not detectable at all (Bondy, 2011; Wahlsten, 2012). This frustrating reality contradicts the compelling evidence from twin and family studies that suggests a substantial influence of hereditary variation, albeit one that might account for only half or less of statistical variance in many interesting phenotypes (Polderman et al., 2015). It is conceivable that epigenetic contributions to parent-offspring and twin resemblance involve variation in methylation sites that are numerous and can influence gene expression by degree without any change in protein encoding DNA sequence (Lopizzo et al., 2015). If some variants affect the relative extent of methylation of a specific gene, the consequences for development could be very real but also very subtle and undetectable with statistical methods. Mere correlation between a DNA marker and a phenotype cannot distinguish between an effect arising from a nearby DNA sequence variation that affects protein structure directly and a more labile transgenerational effect involving DNA methylation.

Gene-environment interaction is mentioned in several places in the book, but the implications of epigenetics for this phenomenon need further discussion. Gene-environment interaction is not simply a matter of environments having differential effects on gene expression. Interaction means that different genotypes respond to environmental changes differentially (Bagot & Meaney, 2010). To have such an interaction, there must be some kind of allelic differences in a gene in a population. A few dramatic examples of such interactions are well documented in the literature, but recent findings from epigenetic research suggest these kinds of interactions may be much more widespread than previously imagined (Kyburz, Karouzakis, & Ospelt, 2014; Nishioka, Bundo, Kasai, & Iwamoto, 2012). If an allele of a gene affects the extent or ease of methylation, this could have an important influence on how that gene responds to altered environments, even though the protein product derived from the gene is unaltered. The interaction could change the quantity of gene product without affecting its qualitative structure. Subtle interactions of this kind may be ubiquitous.

Moore points out that recent research has uncovered hitherto unimagined complexity at the genetic/molecular level. One such challenge to our comprehension is alternative splicing of RNA transcripts. In vertebrate animals, most genes are present as several physically separate chunks of DNA (exons), each of which codes for just part of a protein molecule, and those chunks are separated by long sequences of DNA that do not code directly for protein. The non-coding portion of a gene usually accounts for more than 90% of its DNA, and much of this must be involved in regulation of gene action through methylation and other processes. The transcribed RNA fragments can be spliced together in many different ways before the RNA molecule is translated into protein, which gives rise to a situation where a single gene can provide the base sequences needed to construct dozens of different proteins. It appears that variation in methylation can influence which specific RNA segments are transcribed and then spliced (Luco et al., 2010; Luco, Allo, Schor, Kornblihtt, & Misteli, 2011). Environmentally induced changes might silence not an entire gene but just a segment that is needed for one or a few kinds of proteins. Thus, epigenetic phenomena could play an important role is which specific protein is indeed derived from a single gene in a particular kind of cell under a given kind of environmental condition. Effects of this nature could help an organism to adapat rapidly to an altered environment without committing it to long term, multigenerational change arising from a novel mutation in the DNA of an exon.

At the dawn of the era of molecular genetics, there was hope that knowing the exact nucleotide sequence of the DNA would finally answer many questions that have vexed developmental biologists and psychologists for decades. That hope has been shattered. Instead of answering big questions, recent findings have uncovered a daunting degree of complexity at the molecular level and suggested a vast array of subtle variations that, taken together, can exert important influences on development, yet be virtually undetectable as distinct variants in heredity. Where all this recent research is leading developmental psychobiology is difficult to imagine. Moore discusses prospects for intervention using novel drugs that utilize epigenetic processes to treat consequences of early abuse, drug addiction, and other challenging behavioral conditions, but he is suitably circumspect and cautions that, at this early stage in the investigations, much depends on conjecture. Perhaps his most important point is that we need to think of the gene-environment relationship differently and reassess old ideas about development and evolution. He advocates a broadened perspective in which gene-environment *co-construction* of traits is seen as universal and all dreams of simple genetic determination of behavioral phenotypes are finally dismissed.

References

Bagot, R. C., & Meaney, M. J. (2010). Epigenetics and the biological basis of gene x environment interactions. The Journal of the American Academy of Child and Adolescent Psychiatry, 49, 752–771.

Barros, S. P, & Offenbacher, S. (2009). Epigenetics: connecting environment and genotype to phenotype and disease. Journal of Dental Research, 88, 400–408.

Bondy, B. (2011). Genetics in psychiatry: are the promises met? The World Journal of Biological Psychiatry, 12, 81–88.

Champagne, F., & Meaney, M. J. (2001). Like mother, like daughter: evidence for non-genomic transmission of parental behavior and stress responsivity. Progress in Brain Research, 133, 287–302.

Champagne, F. A. (2008). Epigenetic mechanisms and the transgenerational effects of maternal care. Frontiers in Neuroendocrinology, 29, 386–397.

Day, J. J., Childs, D., Guzman-Karlsson, M. C., Kibe, M., Moulden, J., Song, E., & Sweatt, J. D. (2013). DNA methylation regulates associative reward learning. Nature Neuroscience, 16, 1445–1452.

Edgar, R., Domrachev, M., & Lash, A. E. (2002). Gene Expression Omnibus: NCBI gene expression and hybridization array data repository. Nucleic Acids Research, 30, 207–210.

Francis, D., Diorio, J., Liu, D., & Meaney, M. J. (1999). Nongenomic transmission across generations of maternal behavior and stress responses in the rat. Science, 286, 1155–1158.

Gottlieb, G. (1998). Normally occurring environmental and behavioral influences on gene activity: From central dogma to probabilistic epigenesis. Psychological Review, 105, 792–892.

Gottlieb, G. (2007). Probabilistic epigenesis. Developmental Science, 10, 1–11.

Isles, A. R., & Wilkinson, L. S. (2008). Epigenetics: what is it and why is it important to mental disease? British Medical Bulletin, 85, 35–45.

Kyburz, D., Karouzakis, E., & Ospelt, C. (2014). Epigenetic changes: the missing link. Best Practice & Research Clinical Rheumatology, 28, 577–587.

Laird, P. W. (2010). Principles and challenges of genome-wide DNA methylation analysis. Nature Reviews Genetics, 11, 191–203.

Levenson, J. M., & Sweatt, J. D. (2005). Epigenetic mechanisms in memory formation. Nature Reviews Neuroscience, 6, 108–118.

Lopizzo, N., Bocchio, C. L., Cattane, N., Plazzotta, G., Tarazi, F. I., Pariante, C. M., & Cattaneo, A. (2015). Gene-environment interaction in major depression: focus on experience-dependent biological systems. Frontiers in Psychiatry, 6, 68.

Luco, R. F., Allo, M., Schor, I. E., Kornblihtt, A. R., & Misteli, T. (2011). Epigenetics in alternative pre-mRNA splicing. Cell, 144, 16–26.

Luco, R. F., Pan, Q., Tominaga, K., Blencowe, B. J., Pereira-Smith, O. M., & Misteli, T. (2010). Regulation of alternative splicing by histone modifications. Science, 327, 996–1000.

Meaney, M. J. (2001). Maternal care, gene expression, and the transmission of individual differences in stress reactivity across generations. Annual Review of Neuroscience, 24, 1161–1192.

Nishioka, M., Bundo, M., Kasai, K., & Iwamoto, K. (2012). DNA methylation in schizophrenia: progress and challenges of epigenetic studies. Genome Medicine, 4, 96.

Pelizzola, M., & Ecker, J. R. (2011). The DNA methylome. FEBS Lettres, 585, 1994–2000.

Pelizzola, M., & Molinaro, A. (2011). Methylated DNA immunoprecipitation genome-wide analysis. Methods in Molecular Biology, 791, 113–123.

Polderman, T. J., Benyamin, B., de Leeuw, C. A., Sullivan, P. F., van, B. A., Visscher, P. M., & Posthuma, D. (2015). Meta-analysis of the heritability of human traits based on fifty years of twin studies. Nature Genetics, 47, 702–709.

Sun, Y. V. (2014). The influences of genetic and environmental factors on methylome-wide association studies for human diseases. Current Genetic Medicine Reports, 2, 261–270.

Wahlsten, D. (2012). The hunt for gene effects pertinent to behavioral traits and psychiatric disorders: from mouse to human. Developmental Psychobiology, 54, 475–492.

SECTION VIII

IMPLICATIONS FOR PROGRAMS AND POLICIES

EDITORS' INTRODUCTION

Among the many split conceptions maintained by viewing the study of development through a Cartesian lens (Overton, 2015) was the split between basic and applied research. However, within models of human development derived from the ideas of the process-relational paradigm, this split joins other ones (e.g., nature–nurture or continuity–discontinuity) in being rejected. When one studies the embodied individual within the developmental system, then explanations of how changes in the individual ⇔ context relation at Time 1 may eventuate in subsequent changes in this relation at Time 2, Time 3, etc., are tested by altering the Time 1 person ⇔ context relation. When such alterations are conducted in the ecologically valid setting of the individual, these assessments constitute tests of the basic, relational process of human development and, at the same time, applications—interventions—into the course of human development (Lerner, 2002). Indeed, depending on the level of analysis, aggregation, and time scale at which these interventions are implemented, such changes in the ecology of the individual ⇔ context relation may involve relationships between individuals (e.g., mentoring relationships), community-based programs, or social policies (e.g., Bronfenbrenner, 2005).

As we have explained, the rationale for applying developmental science to enhance the lives of individuals or groups is predicated on the presence of relative plasticity in human development, a concept that is derived from relational developmental systems (RDS) based ideas, such as bidirectionally influential individual ⇔ context relations and embodiment. The relative plasticity of human development is a fundamental strength in, and the basis of optimism about, human development. Developmental scientists can be hopeful that there are combinations of person and context that can be identified or created (through programs or policies) to enhance the lives of all individuals and groups. In other words, developmental scientists may act to change the course of developmental regulations, of individual ⇔ context relations, in manners aimed at optimizing the opportunities for individual and group trajectories across life to reflect health and thriving.

As such, developmental scientists have in the repertoire of models and methods in their intellectual "toolbox" the means to work to promote a better life for all people, to give diverse individuals the requisite chances needed to maximize their aspirations and actions aimed at being active producers of their positive development and to promote a more socially just world (Lerner, 2002, 2004a, 2004b; Lerner & Overton, 2008). In this regard, Lerner and Overton (2008) noted that theoretically predicated changes in the dynamic, relational developmental system need to be evaluated in regard to how more positive development may be promoted among individuals whose ecological characteristics (e.g., socioeconomic circumstances or educational opportunities) lower the probability of such development. To contribute significantly to creating a developmental science aimed at promoting social justice, scholars need to identify the means to change individual ⇔ context relations in manners that enhance the probability that all individuals, no matter their individual characteristics or contextual circumstances, have a greater opportunity to experience positive development (e.g., see Fisher et al., 2013).

Indeed, Fisher and Lerner (2013) noted that social justice focuses on the rights of all groups in a society to have fair access to and a voice in policies governing the distribution of resources essential to their physical and psychological well-being. Social justice focuses also on social inequities, characterized as avoidable and unjust social structures and policies that limit access to resources based solely on group or individual characteristics such as race/ethnicity, age, gender, sexual orientation, physical or developmental ability status, and/or immigration status, among others.

Developmental science framed by the process-relational paradigm has a clear agenda involving such scholarship. For instance, Fisher et al. (2013) provided a vision for social justice-relevant research in developmental science. Some of the research foci they discuss include, addressing the pervasive systemic disparities in opportunities for development; investigating the origins, structures, and consequences of social inequities in human development; identifying societal barriers to health and well-being; identifying barriers to fair allocation and access to resources essential to positive development; identifying how racist and other prejudicial ideologies and behaviors develop in groups; studying how racism, heterosexism, classism, and other forms of chronic and acute systemic inequities and political marginalization may have a "weathering" effect on physical and mental health across the life span; enacting evidence-based prevention and policy research aimed at demonstrating if systemic oppression can be diminished and psychological and political liberation can be promoted; taking a systems-level approach to reducing unjust institutional practices and to promoting individual and collective political empowerment within organizations, communities, and local and national governments; evaluating programs and policies that alleviate developmental harms caused by structural injustices; and, creating and evaluating empirically based interventions that promote a just society that nurtures life-long healthy development in all of its members.

The Necessity of a Scientifically Useful and Socially Just Developmental Science

Developmental science may be at a crossroad. Given the irreparable logical and empirical shortcomings of reductionist and essentialist approaches to human development, especially those involving genes, there is no scientific value in the continued theoretical or empirical use of these ideas, whether we are discussing past instantiations of them, such as those forwarded by Skinner (1971) or Lorenz (1965), or examples of them present in the essentialist and reductionist literature at the time of this writing, for instance, EDP (e.g., Bjorklund, 2015; Del Giudice & Ellis, 2016), sociobiological neo-eugenics (e.g., Belsky, 2014) or behavior genetics (e.g., Plomin et al., 2016). Focusing on these flawed ideas as a basis for research or as a means to formulate applications to social policies or programs is a waste of valuable scholarly resources and has the potential to foster applications of developmental science that are derived from seriously mistaken ideas (Lerner, 2015).

Developmental scientists enacting many of the roles associated with their work—for example, faculty members participating in hiring, tenure, and promotion decisions, teachers, mentors, peer reviewers, editors, and of course researchers—are faced, then, with a decision. Do they embrace the complexity of human development in the enactment of all of their roles and, as such, articulate that essentialist approaches are no longer acceptable frames for the conduct of developmental science (e.g., they likely would if they were faced with evaluating work that used phrenology as the frame for scholarship), or do they allow egregiously flawed thinking and associated work to fill the minds of their students and the pages of their journals (in the name, perhaps, of academic freedom)?

We articulate the dimensions of this decision with more than a little trepidation, given the range of responses we expect it will elicit. However, more than the quality of developmental science is at stake. Developmental scientists should also recognize that civil society may hang in the balance, given the repeated applications of essentialist thinking finding its way into public policy discourse in the United States and internationally at this writing, for instance, regarding political nativist ideas about racial, ethnic, and religious diversity and about immigration and immigrants (e.g., see Feldman, 2014; Feldman & Riskin, 2022). The quality of life and the welfare of millions of people may be affected by where developmental scientists stand in regard to these issues and what they may be willing to say publicly about them.

What then might developmental scientists do? As citizens, developmental scientists can act by exercising the rights of any citizen in a socially-just democracy (Lerner, 1976, 2004a, 2004b). However, as scientists, developmental scientists can act to describe, explain, and optimize human development across the lifespan (of course). In such efforts, developmental scientists would do well to heed the advice of Horowitz (2000) in regard to how, in the face of the simplistically seductive ideas of hereditarianism, they must find the will to act in a manner supportive of social justice. Horowitz (2000) noted that:

> If we accept as a challenge the need to act with social responsibility then we must make sure that we do not use single-variable words like genes or the notion of innate in such a determinative manner as to give the impression that they constitute the simple answers to the simple questions asked by the Person in the Street lest we contribute to belief systems that will inform social policies that seek to limit experience and opportunity and, ultimately, development, especially when compounded by racism and poorly advantaged circumstances. Or, as Elman and Bates and their colleagues said in the concluding section of their book *Rethinking Innateness* (Elman et al., 1998), "If our careless, under-specified choice of words inadvertently does damage to future generations of children, we cannot turn with innocent outrage to the judge and say 'But your Honor, I didn't realize the word was loaded.'" (p. 8)

What developmental scientists can do, then, is to frame their work with terms that reflect the holism, plasticity, and optimism about the possibility of promoting positive human development by theory-predicated integration of individuals and contexts, by promoting *adaptive* individual ⇔ context relations. In addition, they can then couple this language with the appropriate methodology in order to conduct research that provides evidence about the dynamics of the relational developmental system and the health and thriving that can be promoted among diverse individuals by appropriately engaging this system. With such evidence in hand, developmental scientists can then offer evidence-based applications that contribute to policies and programs promoting positive human development.

Conclusions: Moving Beyond the Heredity Hoax

The epigenetic and embodied developmental changes that characterize individual ⇔ context relations within the autopoietic relational developmental system, and that provide a rationale for and optimism about applying developmental science in the service of promoting thriving and social justice for all people, requires "a theoretical framework more akin to current dynamic systems models than to traditional conceptions of either behavioral development or evolution" (Harper, 2005, p. 352).

Overton (2013, 2015) provided this theoretical framework. Derived from a process-relational paradigm, the RDS metamodel that he has forwarded has explained why the Cartesian-split-mechanistic scientific paradigm that until recently functioned as the standard conceptual framework for subfields of developmental science (including inheritance, evolution, and organismic—prenatal, cognitive, emotional, motivational, sociocultural—development) has been progressively failing as a scientific research program. (Overton, 2013, p. 22)

He noted:

An alternative scientific paradigm composed of nested metatheories with relationism at the broadest level and relational developmental systems as a midrange metatheory is offered as a more progressive conceptual framework for developmental science. Termed broadly the *relational developmental systems paradigm*, this framework accounts for the findings that are anomalies for the old paradigm; accounts for the emergence of new findings; and points the way to future scientific productivity.

(Overton, 2013, p. 22)

And so where do we go from here? What is the future trajectory of developmental sciences in the second quarter of the 21st century and beyond?

The theoretical orientations and interests of contemporary cohorts of developmental scientists, the aspiration to produce scholarship that matters in the real world, and the need for evidence-based means to address the challenges of the 21st century have coalesced to make Kurt Lewin's (1952, p. 169) quote, that "There is nothing so practical as a good theory," an oft-proven empirical reality. The scientific and societal value on which the developmental science of the succeeding decades of the 21st century will be judged is whether its theoretical and methodological tools accurately reflect the diversity and dynamism of human development and are centered on promoting thriving across the life span. Therefore, we believe that promoting social justice is, and will be, the most significant lens through which the contributions of developmental science will be viewed.

Why? The first quarter of the 21st century began with democracies found in WEIRD (Western, Educated, Industrialized, Rich, and Democratic) nations having broad international respect that had been continuing on a higher and higher path that was largely uninterrupted since the end of World War II. However, the world began to change beginning at least with 9/11 in 2001, and ensuing terrorist atrocities around the world. These changes have been coupled in the ensuing years of the first quarter of the 21st century with the rise of autocratic and even fascist-leaning regimes, even in NATO countries, including the United States; the emergence of within-nation insurrections, civil wars, and international wars that threatened to bring Western and non-Western nations into conflict either by proxy or directly; the growth of gun violence and mass murders, especially in the United States (where, at this writing, gun murders are the leading cause of death among U.S. children); global climate changes that threaten the viability of continued life on the planet; a global pandemic that resulted in millions of deaths and illuminated the inequities around the globe in healthcare resources; and relentless racism and disparities in economic, educational, safety, and health imposed on people of color in the United States and for youth in low and middle-income countries around the world and, as well, the marginalization and minoritization of individuals not only because of their race, but because of their ethnicity, gender identity, and sexual orientation. These injustices were coupled with inequities within and across healthcare systems, elementary and secondary education, higher education, business and industry, and politics.

Given the course of changes across these years, it might be easy, and even quite reasonable, to discuss our RDS-based approach to developmental science as seeking to eliminate one set of ideas that arise in relation to these inequities by addressing absences of socially just and democratic opportunities through research that is based on or at least uses the results of good science, as either a Sisyphean task or, even worse, as too small or too limited to be useful. However, with recognition of the challenges and horrors visited upon the diverse individuals of our world and, as well, humbled by the limited influence developmental science can hope to have on helping address these challenges, we believe there is no alternative that is reasonable to pursue other than to try to contribute. The goals of this book, and of the final material included in this last section, is to indicate how good science can legitimately and usefully help "heal the world." We believe that inequities and the rejection of socially just approaches to equitable commitments to the health, education, and well-being of diverse people has been rationalized by egregiously flawed ideas about genetically fixed and irreparable attributes of people of specific races, religions, ethnicities, life choices, political beliefs, etc. Eliminating genetic reductionism-based falsehoods will not totally heal the word. However, it can be a part of such healing. Good science enacted by scholars believing in the possibilities for thriving for everyone that derive from a dynamic, relational developmental system can make this important, albeit limited, contribution to the damages of humans that continue to be done by humans.

In the context of this hope for the future, a hope based on our understanding of the theoretical and methodological progress being made in advancing understanding of the dynamics of human development, we believe it is possible that the second quarter of the 21st century may be a tipping point in the description, explanation, *and optimization* of human development across the life span. If our hope is actualized, then future developmental science authors will be able to note that the end of the first quarter of the 21st century was the beginning of an explicit integration of developmental science scholarship with social justice and, perhaps as well, the launching of sustained and meaningful contributions by developmental scientists to liberty and justice for all.

An Overview of the Contributions to This Section

Many of the books published by genetic reductionists have gained at least their "15 minutes of fame" within national and international scientific discussions, media presentations, and political debates about whether social policies or programs are evidence based or only manifestations of political ideology. In Section V, we noted that the 1994 book by Herrnstein and Murray, *The Bell Curve: Intelligence and Class Structure in American Life*, is certainly one of these books. Herrnstein and Murry presented the book as providing the evidentiary base for the influence of genes on intellectual abilities and the social and economic attainments in American society and, as a consequence of the validity of the view that racial differences in the mean IQs of Black and White Americans played at least some role in explaining racial differences in the social, economic, health, educational, etc., standing within the United States.

As we noted in Section V, there were several criticisms of the scientific quality of the work that Herrnstein and Murray (1994) used as the evidentiary basis for this analysis of the implications of genes and intelligence test scores on equity and social justice in American life. Nevertheless, the debate regarding the use of genetic reductionist methods, such as heritability research designs and data analysis, continues through this writing (e.g., see Joseph, 2022 and, in Section V, Joseph, 2023). As such, Joseph and Richarson used the 2024, 30th "anniversary" of the publication of *The Bell Curve* to assess the status of the use of these scientific methods to study the purported link between genes and intelligence in different racial groups.

We open this section with this important contribution to science and to society. Joseph and Richardson evaluate both the claims of Herrnstein and Murray (1994) and the research they used to support their claim of substantial IQ heritability of intelligence. They explain how this knowledge base is egregiously flawed because of findings from contemporary results of molecular genetic research and from estimates of IQ heritability derived from twins-reared-apart designs; they note that this twins-reared-apart research suffered from several biases, including omitted control groups and reliance on false assumptions. They conclude that Herrnstein and Murray (1994) did not present any valid evidence of genetic causes of either between-ethnic-group or social-class IQ score differences *or* of IQ score differences within groups. They emphasize that, as was argued in 1994, the social and political policy recommendations derived from *The Bell Curve* should be rejected because of the scientific inadequacy of evidence in support of these ideas.

How, then, might the sort of dynamic, RDS-based models discussed across the section of this book be used to enable scientifically sound evidence to be applied to policies and programs in specific sectors of society? Darling-Hammond, et al. (2020) provide a sample case in regard to the educational ecosystem in the United States. They present the implications for school and classroom practices of an emerging consensus about the science of learning and development (e.g., Cantor & Osher, 2021), an approach to educational and developmental science that is explicitly derived from RDS-based models and methods (e.g., Cantor & Osher, 2021; Immordino-Yang et al., 2019).

Darling-Hammond et al. use the RDS-based framework to integrate evidence from the learning sciences and from several areas of educational research to suggest the use of teaching strategies that support the types of developmentally nurturant relationships and student-specific and diversity-sensitive learning opportunities needed to promote children's well-being, healthy development, and transferable learning. Moreover, because of meaningful variation across students, time, and place, they discuss evidence-based approaches that provide useful means to help educators respond to the individual characteristics and contexts creating students' variability, addressing adversity, and supporting resilience. Overall, they provide a dynamic, RDS-based vision for schools and educators to devise proven means for all youth to develop along positive pathways to a flourishing adulthood.

In their 2021 edited book, Cantor and Osher brought together scholars and practitioners from the biological/medical sciences, the social and behavioral sciences, educational science, and fields of law and social and educational policy in an attempt to understand the bases and status of an integrated science of learning and development and to use this knowledge to present a vision for progress that would contribute to enhancing the educational and life opportunities of all young people but, particularly, youth who have been marginalized and minoritized because of racism, poverty, and inequities associated with many of the other pernicious "isms" existing in society. The Darling-Hammond et al. (2020) essay is an example of the efforts made by Cantor, Osher, and their colleagues in advancing such work.

In their last chapter in the Cantor and Osher (2021) book, which serves as a coda for their work and the work of contributors to the book, they conclude that the scholarship presented in their book constitutes a compelling evidence base for the fact that human development involves a system of dynamic relations across the integrated levels of organization comprising the ecology of behavior and development. As such, they contend that human thriving should be understood as a dynamic process that goes beyond individual well-being to include individual and collective growth that is strength-based and multidimensional, reflecting positive growth in and across multiple domains, from physical to emotional, cognitive, and social, wherein the concept of collective thriving becomes relevant. They point out that it is the responsibility of all people interested in human thriving to use this knowledge to design a system characterized by robust equity, one wherein all people are able to take advantage of high-quality opportunities for learning and positive development.

In the last contribution in Section VIII, Lerner and Greenberg also offer a coda as well, but here for the present book. We appeal to readers to reject genetic reductionism and embrace relationism and complexity of dynamic systems. Noting that this contribution and the book as a whole are aimed at countering the bad science involved in genetic reductionism that continues, in various guises, to appear, they offer one possible call to action based on the superordinate concepts associated with, and the sound and, often, compelling data derived from research associated with, dynamic, RDS-based approaches to human development. Accordingly, they present what may be learned about dynamic systems approaches to human development from the contributions found in this book. They close the contribution by discussing some implications of dynamic systems ideas for the promotion of social justice and authentic equity for all people.

References

Belsky, J. (2014, November 30). The downside of resilience. *New York Times, Sunday Review*, p. SR4.

Bjorklund, D. F. (2015). Developing adaptations. *Developmental Review, 38*, 13–35.

Bronfenbrenner, U. (2005). *Making human beings human: Bioecological perspectives on human development*. Sage.

Cantor, P., & Osher, D. (2021). Whole-child development, learning, and thriving in an era of collective adversity, disruptive change, and increasing inequality. In P. Cantor and D. Osher (Eds.), *The science of learning and development: Enhancing the lives of all young people* (pp. 233–254). Routledge.

Darling-Hammond, L., Flook, L., Cook-Harvey, C., Barron, B., & Osher, D. (2019). Implications for educational practice of the science of learning and development. *Applied Developmental Science, 24*(2), 97–140. https://doi.org/10.1080/10888691.2018.1537791

Del Giudice, M., & Ellis, B. J. (2016). Evolutionary foundations of developmental psychopathology. In D. Cicchetti (Ed.), *Developmental psychopathology, Vol. 12: Developmental neuroscience* (3rd ed., pp. 1–58). Wiley.

Elman, J. L., Bates, E. A., Johnson, M. H., Karmiloff-Smith, A., Parisi, D., & Plunkett, K. (1998). *Rethinking innateness*. MIT Press.

Feldman, M. (2014). Echoes of the past: Hereditarianism and a troublesome inheritance. *PLoS Genetics, 10*(12), e1004817.

Feldman, M. W., Riskin, J. (2022). Why biology is not destiny. [Review of the book *The Genetic Lottery*, by K. Harden]. *The New York Review*. https://www.nybooks.com/articles/2022/04/21/why-biology-is-not-destiny-genetic-lottery-kathryn-harden/

Fisher, C. B., Busch, N. A., Brown, J. L., & Jopp, D. S. (2013). Applied developmental science: Contributions and challenges for the 21st century. In R. M. Lerner, M. A. Easterbrooks, & J. Mistry (Eds.), *Handbook of psychology. Vol. 6: Developmental psychology* (2nd ed., pp. 516–546). Editor-in-chief: I. B. Weiner. Wiley.

Fisher, C. B., & Lerner, R. M. (2013). Promoting positive development through social justice: An introduction to a new ongoing section of applied developmental science. *Applied Developmental Science, 17*(2), 57–59.

Harper, L. (2005). Epigenetic Inheritance and the Intergenerational Transfer of Experience. *Psychological Bulletin, 131*(3), 340–360. https://doi.org/10.1037/0033-2909.131.3.340

Herrnstein, R. J., & Murray, C. (1994). *The bell curve: Intelligence and class structure in American life*. The Free Press.

Horowitz, F. D. (2000). Child development and the PITS: Simple questions, complex answers, and developmental theory. *Child Development, 71*, 1–10.

Immordino-Yang, M. H., Darling-Hammond, L., & Krone, C. R. (2019). Nurturing nature: How brain development is inherently social and emotional, and what this means for education. *Educational Psychologist, 54*(3), 185–204.

Lerner, R. M. (1976). *Concepts and theories of human development*. Addison-Wesley.

Lerner, R. M. (2002). *Concepts and theories of human development* (3rd ed.). Erlbaum.

Lerner, R. M. (2004a). Genes and the promotion of positive human development: Hereditarian versus developmental systems perspectives. In C. G. Coll, E. L. Bearer, & R. M. Lerner (Eds.) *Nature and nurture: The complex interplay of genetic and environmental influences on human behavior and development* (pp. 1–33). Lawrence Erlbaum Associates, Inc.

Lerner, R. M. (2004b). *Liberty: Thriving and civic engagement among America's youth.* Sage.

Lerner, R. M. (2015). Eliminating genetic reductionism from developmental science. *Research in Human Development, 12,* 178–188.

Lerner, R. M., & Overton, W. F. (2008). Exemplifying the integrations of the relational developmental system: Synthesizing theory, research, and application to promote positive development and social justice. *Journal of Adolescent Research, 23,* 245–255.

Lewin, K. (1952). *Field theory in social science: Selected theoretical papers.* Tavistock.

Lorenz, K. (1965). *Evolution and modification of behavior.* University of Chicago Press.

Overton, W. F. (2013). Relationism and relational developmental systems: A paradigm for developmental science in the post-Cartesian era. *Advances in child development and behavior, 44,* 21–64. https://doi.org/10.1016/b978-0-12-397947-6.00002-7

Overton, W. F. (2015). Process and relational developmental systems. In W. F. Overton & P. C. M. Molenaar (Eds.), *Handbook of Child Psychology and Developmental Science, Volume 1: Theory and method* (7th ed., pp. 9–62). Editor-in-chief: R. M. Lerner. Wiley.

Plomin, R., DeFries, J. C., Knopik, V. S., & Neiderhiser, J. M. (2016). Top 10 replicated findings from behavioral genetics. *Perspectives on Psychological Science, 11*(1), 3–23.

Skinner, B. F. (1971). *Beyond freedom and dignity.* Knopf.

23.
THE BELL CURVE AT 30
A Closer Look at the Within- and Between-Group IQ Genetic Evidence*
Jay Joseph† and Ken Richardson‡

Part I: Introduction

The year 2024 marks 30 years since the publication of *The Bell Curve: Intelligence and Class Structure in American Life* (Herrnstein & Murray, 1994). The book's authors were Harvard psychologist Richard Herrnstein and conservative U.S. political scientist Charles Murray. Herrnstein and Murray (hereafter, "H & M") argued that human intelligence, as supposedly measured by IQ tests, is substantially determined by genetic variations influencing individual and group differences. The most controversial aspect of *The Bell Curve* (hereafter, "*TBC*") was its authors' claim that it seems "highly likely to us that both genes and the environment have something to do with [IQ] racial differences" between U.S. Whites and Blacks (African Americans; p. 311).

"The inheritance of intelligence is probably the most controversial topic in the whole of science," wrote geneticist Adam Rutherford, "and when it is combined with the study of population differences, evolution, and race, there we have the prospect of a perfect storm" (Rutherford, 2021, p. 132). Ten years prior to entering the storm, Murray published *Losing Ground: American Social Policy, 1950–1980*, where he proposed "scrapping the entire [U.S.] federal welfare and income-support structure for working-aged persons, including AFDC [welfare], Medicaid, Food Stamps, Unemployment Insurance, Worker's Compensation, subsidized housing, disability insurance, and the rest" (Murray, 1984, pp. 227-228). Like Murray, Herrnstein did not conduct human genetic or IQ research. A former student of B. F. Skinner, his area of expertise was animal learning and studying the learned response of pigeons (Baum, 2002), at times for use by the U.S. military during the Vietnam era (Herrnstein, 1965; Herrnstein & Loveland, 1964; Psych-Agitator, 1979). Herrnstein, who we will see became involved in the IQ genetics debate in the 1970s, passed away shortly before *TBC* was published in 1994. Murray has been active since then, including in the social media.

The 1994 publication of *TBC* (over 300,000 copies sold) was reported widely in major U.S. media outlets, largely uncritical of the book's methods and assumptions (and, at times, its major conclusions). Subsequent reviews were mostly negative, and many can be found in edited collections drawn from various sources (e.g., Fraser, 1995; Jacoby & Glauberman, 1995). Entire books were devoted to refuting the main *TBC* arguments (e.g., Devlin et al., 1997; Fischer et al., 1996; Kincheloe et al., 1996). Here, we build on previous reviews in light of more recent critiques, publications, discoveries, and conceptual advances. An important question is why, despite the criticism, *TBC* remains "alive."

---

* This chapter is published for the first time in this book.
† Private Practice, Oakland, CA.
‡ Open University (UK), Milton Keynes, Buckinghamshire, England

DOI: 10.4324/9781032702988-39

H & M's within-group "IQ heritability" arguments were based on behavioral genetic methods, theories, and concepts. The field of *human behavioral genetics* investigates the causes of individual differences in psychological/behavioral characteristics such as IQ and personality, and emphasizes the calculation of heritability estimates. The field's main research methods are family, twin, adoption, and molecular genetic studies.

After enormous worldwide protests in 2020 against racism and other forms of oppression and injustice, the Behavior Genetics Association (BGA) issued a "Diversity and Inclusion Action Plan" calling for a "renewed urgency for scientific organizations worldwide to acknowledge the role science has played in racism." It candidly stated, "The history of our field is inextricably linked with racism, including the misuse of behavior genetic research to support violent eugenic policies" (Behavior Genetics Association, 2021).

Nevertheless, the *Bell Curve* message continues to maintain and support institutional racism and inspires "race scientists" and powerful networks of groups in government and institutions influencing policies on "race" and class (Evans, 2018; Gillborn et al., 2022; Panofsky et al., 2021). In 2023, a former U.S. President running for reelection spoke of immigrants as supposedly "poisoning the blood of our country" (Rosen et al., 2023). A few months later, a major U.S. poll found that almost half (47%) of registered voters endorsed this statement (Khanna et al., 2024). Even more reason to embark on yet another in-depth critical examination of this book.

No Gene Discoveries Were Cited in The Bell Curve

Because IQ molecular genetic research was at an early stage in 1994, H & M did not cite any IQ gene (genetic variant) associations or discoveries. The behavioral "candidate gene" era, also at an early stage in 1994, turned out to be an expensive and embarrassing bust (Chabris et al., 2012; Charney, 2012; Farrell et al., 2015; Plomin, 2018; Sullivan, 2017; Yong, 2019). The title of Chabris and colleagues' 2012 article read, "Most Reported Genetic Associations with General Intelligence Are Probably False Positives."

Subsequent *genome-wide association* (GWAS) and *polygenic score studies* (PGS) found *single-nucleotide polymorphisms* (SNPs) "associated with" intelligence (reviewed in Deary et al., 2022; Plomin & von Stumm, 2018). SNPs number in the millions and are considered common minority variants of genes present in at least 1% of the population. SNPs can be used as probabilistic markers or associative "tags" for identifying nearby genetic loci that may harbor actual causative variants when the genetic profiles of cases are compared to controls.

GWAS and PGS SNP associations may be spurious or non-causal, as, like the earlier candidate gene studies, they are subject to potential environmental confounds and other problems. GWAS/PGS problems and potential biases include population stratification, a lack of individual predictive value, the practice of increasing the sample size to find statistically significant gene associations, that "variation explained by" does not mean "caused by," and the potential fishing expedition aspect of "hypothesis-free" studies in which researchers base their conclusions on statistically significant yet chance associations (see Baverstock, 2019; Burt, 2023; Charney, 2022; Coop & Przeworski, 2022; Gusev, 2023; Joseph & Boyle, 2024; Meyer et al., 2023; Non & Cerdeña, 2023; Richardson & Jones, 2019; Turkheimer, 2016, 2019; Veller & Coop, 2024).

Main Themes of The Bell Curve

As summarized by psychologist Richard Nisbett and colleagues, the authors of *The Bell Curve*

> argued that IQ tests are an accurate measure of intelligence; that IQ is a strong predictor of school and career achievement; that IQ is highly heritable; that IQ is little influenced by environmental factors; that racial differences in IQ are likely due at least in part,

and perhaps in large part, to genetics; that environmental effects of all kinds have only a modest effect on IQ; and that educational and other interventions have little impact on IQ and little effect on racial differences in IQ. The authors were skeptical about the ability of public policy initiatives to have much impact on IQ or IQ-related outcomes.

(Nisbett et al., 2012, p. 130)

H & M revived the old argument that the supposed "dysgenic" effects of higher birth rates among people and groups with lower IQ threaten to reduce intelligence levels in the U.S. (Chapter 15). "Something worth worrying about is happening to the cognitive capital of the country," they wrote (H & M, 1994, p. 341). Without referring to them as eugenic, they supported policies encouraging the reproduction of people with high IQ scores, discouraging the reproduction of people with low IQ scores, and basing U.S. immigration laws on "competency rules" (pp. 548-549). H & M predicted that U.S. society would become increasingly polarized and that a genetic "cognitive elite" would eventually be isolated in enclaves protected from the genetically low-IQ "underclass" (Chapter 21). They believed that differential reproduction patterns among the "dull" and the "cognitive elite" would lead to the creation of a U.S. genetic "custodial state," which they imagined as a "high-tech and more lavish version of the Indian reservation for some substantial minority of the nation's population, while the rest of America tries to go about its business" (p. 526). For H & M, such an outcome would result from heredity, not political decisions and policies causing greater economic inequality and poverty. They supported modifying or scrapping U.S. affirmative action programs in education and the workplace.

Firmly believing that IQ is a valid measure of human intelligence, the key evidence H & M cited in favor of substantial IQ heritability was produced by twin studies, which we put under the microscope in the present review.

The Replication Crisis in Science and Psychology

There is a growing realization that the scientific research/publication process is in crisis. It is known as the *replication crisis* (also known as the replicability crisis, or the reproducibility crisis), brought about by independent analysts' discovery that they could not replicate some key psychology research findings (Open Science Collaboration, 2015). The original findings were probably non-findings resulting from poorly performed research, at times based on vaguely defined characteristics and using make-do measures, including data and conclusions manipulated by researchers to match their own or their funding sources' expectations. Research publications are being retracted at an increasing rate (Oransky & Redman, 2024; see also the Retraction Watch website), and there are continuing concerns that "academia's reward system incentivizes researchers to publish flashy, poor quality, or even fraudulent work" (O'Grady, 2023; see also McKie, 2024).

An unnamed wise person once said, "Data don't tell stories; scientists tell stories" (quoted in Chambers, 2017, p. 175). Previously accepted stories researchers told their colleagues and the public about their data are receiving increasing replication crisis attention and scrutiny, and we can now evaluate behavioral research using replication crisis concepts and terms such as the *questionable research practices* (QRPs) described by behavioral scientist Leslie John and colleagues in 2012 (John et al., 2012). These researchers found that, as reported anonymously by U.S. academic psychologists, the use of QRPs to obtain desired results was "surprisingly high" (p. 524).

Although many researchers did not engage in unethical/questionable practices and produced sound research, the replication crisis is rooted in Herrnstein's U.S. academic psychology field. For decades, field leaders overlooked and even celebrated shoddy, QRP-ridden studies based on blatantly false assumptions, researcher confirmation biases, and financial conflicts of interest (Chambers, 2017; Le Texier, 2019; O'Grady, 2020; Ritchie, 2020). Every behavioral and social science research publication, including those cited in *TBC*, must now be viewed through the lens of replication crisis terms, concepts, and perspectives. Top behavioral geneticist Robert Plomin and colleagues have argued that behavioral genetic studies replicate well (Plomin et al., 2016). However, although this claim has been challenged (e.g., Lerner, 2018), the most important question one should ask about a behavioral genetic study is not whether its results replicate, but whether its authors' conclusions were based on sound assumptions.

IQ and Intelligence Are Not the Same

We focus on IQ genetic research while noting that "IQ" is not "intelligence." IQ is a rating/count of correct answers to an assemblage of questions and tasks, the cognitive demands of which are vague. The most important property of any test is its *construct validity*, the degree to which it measures what it is supposed to measure (Schimmack, 2021). Establishing IQ test construct validity is difficult because there is still no certainty about what intelligence is. The intelligence concept is based largely on intuition and physical metaphors such as "strength," "power," or "capacity." A group of American Psychological Association (APA) experts (Nisbett et al., 2012, p. 131) avoided the issue by quoting Linda Gottfredson (1994/1997) on how intelligence "involves…catching on," "making sense" of things, or "figuring out" what to do. The *Cambridge Handbook of Intelligence* editors noted, "There has never been much agreement on what intelligence is" (Sternberg & Kaufman, 2011, p. xv).

Although H & M assumed that IQ is a valid measure of a well-defined function, IQ is not intelligence (i.e., cognitive ability, learning ability) in any general sense. Several lines of research suggest that "real life" cognition is much more complex than that required by items found in an IQ test (see Richardson & Norgate, 2014).

The Psychometric Approach in Dispute

Psychometrics is the branch of psychology that develops, administers, and interprets quantitative tests to measure individual differences in psychological characteristics (traits) such as intelligence, aptitude, and personality. The psychometric claim that standardized IQ tests such as the Stanford-Binet and Wechsler scales, in addition to the supposedly "culture-fair" Raven's Progressive Matrices, measure innate (native) intelligence (or general intelligence, represented as "g") has been the subject of intense debate (Gould, 1981; Sternberg, 1995). The critics' objections to this claim include the following points:

- The psychometric emphasis on individual differences in abilities and variance partitioning is only one of many approaches to studying human intelligence. Variance partitioning cannot unambiguously identify sources of variance and, therefore, truly distinguish potential genetic (G) from environmental (E) influences.
- General intelligence (g) is the product of a mathematical formula. It has no physical reality. Its identity is imputed, not revealed.
- There is no consensus definition of "intelligence." The psychometric approach finesses this problem because it is about differences, not the nature of the function.
- IQ tests measure school learning more than innate intelligence. By definition, test items are selected according to how well they correlate with school performance.
- IQ tests are biased against non-Caucasians and the working class, in part because school performance is closely related to class/cultural background.
- IQ tests measure narrow abilities and de-emphasize "real world" problem-solving. So, associations with job performance are weak and uncertain (Richardson & Norgate, 2014).
- It cannot be assumed that intelligence is "normally distributed" in a bell curve. IQ tests' construct validity is questionable, although their (built-in) predictive validity is often passed off as such, quite unscientifically. A constant validity crisis surrounds the tests (Richardson, 2022).

As the authors of an APA task force report acknowledged in 1996, "the psychometric approach [to cognitive ability] is the oldest and best established, but others also have much to contribute" (Neisser et al., 1996, p. 80). And it is self-evident that "oldest and best established" does not equal valid.

In their (1996) book *Inequality by Design: Cracking the Bell Curve Myth*, sociologist Claude Fischer and colleagues described the contrasting psychometric and information processing approaches to human intelligence (see also Tabery, 2014). *Information processing* seeks to understand how humans learn and process information, leading to behavioral outcomes. Psychometrics, as we have seen, is concerned with differences among people on how they score on psychological tests. Fischer and colleagues observed that psychometric IQ tests "are refined to magnify differences," and that the famous bell-shaped curve is an invention of psychometrists who assume that intelligence is normally distributed. As psychologist Donald Dorfman (1995, p. 420) pointed out, "The distribution of IQ test scores cannot be expected to follow a bell curve unless it is constructed by the tester to do so." There is strong evidence that brain and other physiological and psychological parameters are far from "normal" in distribution (Blanca et al., 2013; Micceri, 1989).

H & M based many of their claims about the importance of cognitive ability on the "Armed Forces Qualification Test" (AFQT), part of the more extensive U.S. National Longitudinal Survey of Youth (NLSY) data collected from over 11,000 participants aged 15-23. They called the NLSY "an expensive, well-designed set of cognitive and aptitude tests…The measure of cognitive ability extracted from this test battery was the U.S. Armed Forces Qualification Test, the AFQT" (H & M, 1994, p. 120). To be precise, H & M used four AFQT subtests as measures of cognitive ability: "Word Knowledge," "Paragraph Comprehension," "Arithmetic Reasoning," and "Mathematics Knowledge." All four are based on learning (Fischer et al., 1996), but H & M called participants' subtest scores "highly *g* loaded" and a "measure of general cognitive ability." They referred to AFQT scores as "IQ scores" (H & M, 1994, p. 120).

Fischer and colleagues showed that the AFQT data used by H & M did not actually produce a bell curve. Only after a "good deal of statistical mashing and stretching," as Fischer and colleagues put it, were H & M able to create a bell-shaped distribution (Fischer et al., 1996, p. 32). Consumers of psychometric data and claims should be aware of the field's tendency to generate bell curves that don't exist in reality. "Our concern here," wrote Fischer and colleagues, "is with the psychometric insistence that there *must* be a bell curve" (p. 32, emphasis in original). Creating artificial bell-shaped distributions is one aspect of what critics have called "the grand illusion of psychometry" (Lewontin et al., 1984, p. 92).

In sum, our aim in this chapter is to assess the genetic evidence H & M cited in support of their IQ heritability claims, including some relevant history of the IQ genetics debate. In Part III, we address their argument that genetic influences explain a portion of ethnic-group and social-class IQ score differences. The larger question of whether or how heredity influences human intelligence is beyond the scope of our analysis.

We begin by challenging the practice of creating heritability estimates as a nature/nurture ratio in behavioral research, a practice both widely used and widely disputed for decades.

The "Heritability Fallacy"

Heritability (narrow-sense heritability, or h^2) is an essential behavioral genetic concept. According to field leaders, heritability is "the proportion of phenotypic variance that can be accounted for by genetic differences among individuals" (Plomin et al., 2013, p. 87). Heritability estimates range from 0% (0.0) to 100% (1.0).

Critics, however, argue that heritability estimates in behavioral research depend on several questionable or unsupported theoretical assumptions (see Block, 1995; Burt & Simons, 2014; Gusev, 2023; Hirsch, 1997; Ho, 2013; Keller, 2010;

Lerner, 2018; Lewontin, 1974; McGuire & Hirsch, 1977; Moore & Shenk, 2016; Taylor, 1980; Wahlsten & Gottlieb, 1997). In a 2022 article, Nicolas Robette, Emmanuelle Génin, and Françoise Clerget-Darpoux examined heritability assumptions and concluded, "None of the hypotheses inherent in heritability estimates are verified in humans" (Robette et al., 2022). One heritability assumption is that genetic and environmental factors are separate (additive) and do not interact. Stanford University professors Marcus Feldman and Jessica Riskin (2022) objected: "We can no more unbraid genetics and environment than we can unbraid history and culture, or climate and landscape, or language and thought."

Additional points raised by critics of using heritability estimates in social and behavioral science research include (adapted from Joseph, 2023b, Table 2.1):

- Heritability estimates were developed as an animal breeding statistic where genetic and environmental conditions can be controlled. They do not measure the strength of genetic influences on psychiatric conditions or behavioral characteristics such as IQ, nor do they measure the relative importance of genetic and environmental influences.
- Heritability is about variation, not so much about causes. If a study of the factors that sustain human life found that everyone in England drank 6,000 ml of water each day (no variation), a researcher could conclude that water plays no role in sustaining human life in England (0% of the variance).
- High heritability does not mean low malleability. Even when heritability is high, a simple environmental change or intervention can have a substantial impact (for example, the diet that prevents phenylketonuria, or PKU).
- Because it is a population statistic, heritability does not describe the importance of genetic factors as they relate to an individual.
- Heritability estimates apply only to a specific population, at a particular time, and in a particular set of environments. Estimates can change substantially under different environmental conditions.
- *Within*-group behavioral heritability implies nothing about *between*-group genetic differences, such as between ethnic groups or economic classes.
- Gene expression switches on and off "epigenetically" in response to environmental events and challenges, providing additional evidence that genetic and environmental influences are interactive, not additive.
- Heritability estimates depend on researchers' acceptance of a string of disputed assumptions about people, genetics, behavioral characteristics (traits), and psychiatric conditions, including the reliability and validity of the latter.

Although they sound alike, "heritability" and "heredity/inherited" are different concepts (Fischbach & Niggeschmidt, 2021; Hirsch, 1997; Keller, 2010). Following the lead of many behavioral geneticists, H & M blurred the distinction between the common (folk) and technical meanings of the word "heritable." The "complex mechanisms of heredity are little appreciated by non-specialists," wrote psychologist Scott Stoltenberg, "because of misunderstandings that are perpetuated when words used for technical terms have other, more widely understood, folk meanings" (Stoltenberg, 1997, p. 89).

Developmental psychologist David Moore concluded in "The Heritability Fallacy" that the "term 'heritability,' as it is used today in human behavioral genetics, is one of the most misleading in the history of science" (Moore & Shenk, 2016). We concur.

Herrnstein and Murray Estimated "Substantial" Genetic Influences

Although heritability is a flawed and misleading concept, the main arguments in *TBC* rested on the claim that IQ is roughly 60% heritable—a long-disputed yet standard conclusion found in many psychology textbooks. H & M wrote, "Cognitive ability is substantially heritable, apparently no less than 40 percent and no more than 80 percent" (H & M, 1994, p. 23). They settled on a "middling" 60% estimate, although "the balance of the evidence suggests that 60 percent may err on the low side" (p. 105). They saw heritability as "a ratio that ranges between 0 and 1 and measures the relative contribution of genes to the variation observed in a trait" (p. 106), and they believed that substantial IQ heritability was "beyond significant technical dispute" (p. 22). In Part II, we examine the evidence H & M cited in support of their 60% or higher IQ heritability estimate.

Subsequent Molecular Genetic Researchers Estimated "Very Small" Genetic Influences

Problem areas and potential confounds notwithstanding, subsequent molecular genetic GWAS and PGS results do not support H & M's heritability estimate. A 2023 study based on the UK Biobank (Williams et al., 2023, p. 85) found that genes "explained 29% of the variance in cognitive test performance" and produced a general intelligence PGS that "explained 7.6% of the variance in g." In the UK Biobank dataset, "fluid intelligence" is estimated from a two-minute, time-limited, 13-item test administered on a computerized touchscreen at initial reception or from a web-based application at a later "moment" at home (UK Biobank, n.d.).

A 2022 GWAS of "educational attainment" (EA, or years of schooling) based on a large sample of 3 million individuals found a "genome-wide polygenic predictor, or polygenic index (PGI), explain[ing] 12–16% of educational attainment variance" (Okbay et al., 2022, p. 437). Researchers often see educational attainment as a "proxy" for cognitive ability (e.g., Chen et al., 2024; Davies et al., 2016; Plomin & von Stumm, 2018). In *TBC*, H & M expressed their belief that intelligence "is roughly measured by educational attainment" (p. 739) and that educational attainment and IQ are "closely linked" (p. 111). In a later publication, Murray called educational attainment "a rough proxy measure for IQ" (Murray, 2020, p. 279).

In a 2023 review article, Robert Plomin concluded that based on a 2018 GWAS by Jeanne Savage and colleagues, and other analyses, the "current polygenic score for intelligence…predicted 4% of the variance…The polygenic scores for educational attainment and intelligence predicted 10% of the variance of intelligence test scores" (Plomin, 2023, p. 80; see also von Stumm et al., 2023). The Savage and colleagues study obtained data from 14 cohorts using anything that looked vaguely like "intelligence," including SAT scores, math and reading tests, "verbal fluency," digit span, and "immediate and delayed recall tests" (Savage et al., 2018, p. 913).

Harvard statistical geneticist Alexander Gusev (2023, "Heritability of Educational Attainment, 5.0 Summary" section) showed that the corrected (nearly) total direct SNP heritability estimate for educational attainment is 9-16% (Young et al., 2018), and that the common direct SNP effect on educational attainment is just 4% (Howe et al., 2022). Gusev concluded that the total effect of genetic influences on educational attainment is "very small." A subsequent 2024 educational attainment "cognitive ability proxy" GWAS based on East Asian populations found SNP heritabilities below 10% (Chen et al., 2024).

While the GWAS/PGS problem areas and potential environmental confounds we mentioned earlier remain in full force, these modest molecular genetic results contrast sharply with H & M's 60%+ IQ heritability figure. Large differences between earlier twin-study-based and later molecular-genetic-based heritability estimates in behavioral and psychiatric genetic research have led to the still-unresolved *missing heritability problem* (Joseph, 2012; Maher, 2008; Manolio et al., 2009).

The Bell Curve's genetic and social policy arguments depended on establishing "substantial" within-group IQ heritability. The above molecular genetic results, however, converge on a finding of *low* IQ heritability. If these GWAS and PGS estimates are roughly accurate, even as critics might counter that heritability is a misleading concept or that the true estimates are lower or even zero, the *Bell Curve* argument falls apart for this reason alone.

Twin Studies, Arthur Jensen, Cyril Burt, Herrnstein's 1970s IQ Writings, and Leon Kamin

Although genetic ethnic (racial) differences in intelligence claims go back to the beginning of psychology (Chase, 1980; Kamin, 1974; Tucker, 1994b), the modern controversy over the role of genetic influences on IQ and the causes of ethnic differences in IQ scores was initiated by University of California, Berkeley psychologist Arthur Jensen in a 1969 article published in the academic journal *Harvard Educational Review*, "How Much Can We Boost IQ and Scholastic Achievement?" (Jensen, 1969). This highly controversial and widely publicized 1969 article, which appeared in the wake of the U.S. Civil Rights Movement, was followed by Jensen's 1970 article in the journal *Behavior Genetics*, which focused more closely on "twins reared apart" (TRA) IQ studies (Jensen, 1970). Jensen argued in 1969 that monozygotic (identical) twins reared apart (MZAs) provide "the conceptually simplest estimate of heritability…since, if their environments are uncorrelated, all they have in common are their genes" (1969, p. 51; MZAs share a 100% genetic resemblance). He believed that MZA and other biological relative IQ correlations showed that intelligence is "highly heritable" (1969, p. 46). Jensen did not directly study twins. Instead, he based his arguments on his interpretation of TRA study data collected by earlier researchers, from which he estimated "upper-bound" IQ heritability at 82% while claiming that post-natal "influences in the social-psychological environment" of MZAs were negligible (1970, p. 146).

Jensen also argued that the observed mean IQ score gap between U.S. Whites and Blacks was likely caused, at least in part, by hereditary factors. In his view, because within-group "intelligence variation has a large genetic component," it was a "not unreasonable hypothesis that genetic factors are strongly implicated in the average Negro-white intelligence difference" of about 15 IQ points (1969, p. 82; in that era, the term "Negro" was still used in the U.S.). Jensen argued that what he saw as high within-group IQ heritability coupled with genetic ethnic IQ differences posed a problem for society. "Current welfare policies, unaided by eugenic foresight," he wrote, "could lead to the genetic enslavement of a substantial segment of our population" (1969, p. 95).

IQ hereditarianism is the belief that IQ tests measure human intelligence and that intelligence is largely, though not entirely, determined at conception by hereditary factors passed from parent to offspring. IQ hereditarians believe that genetic influences largely determine individual IQ score differences and often see ethnic group differences as partially resulting from such influences. A concise description of IQ hereditarian views is found in a statement supporting *TBC* published in a 1994 edition of the *Wall Street Journal* (Gottfredson, 1994/1997). This statement was endorsed by many of the era's top IQ hereditarians.

"Twins Reared Apart" (TRA) Studies as the Key Evidence Cited by IQ Hereditarians

MZAs have great appeal. Though extremely rare, many people see them as providing the ultimate Rosetta Stone-type method of teasing apart the potential roles of nature and nurture as causes of human behavioral differences. MZAs share a genetic identity while—in theory—growing up and living under entirely different environmental conditions. Nevertheless, the theorized separation of genes and environment in TRA studies has never been achieved (Joseph, 2015). Not even close.

Five TRA IQ studies have been published as of 2024. British psychologist Cyril Burt's (1966 and earlier publications) IQ TRA study was discredited in the 1970s on suspicions that he invented data, twins, and research assistants, suspicions now established "beyond a reasonable doubt"

(Tucker, 1997. See also Dorfman, 1978; Gould, 1981; Hearnshaw, 1979. For defenses of Burt, see Fletcher, 1991; Joynson, 1989). By consensus, Burt's twin study publications were removed from the TRA IQ literature decades ago (Bouchard & McGue, 1981). Unlike Burt, the authors of the first three studies, Horatio Newman and colleagues (1937), James Shields (1962), and Niels Juel-Nielsen (1965/1980) provided detailed case history information for most of the pairs they studied. TRA studies are extremely difficult to perform because reared-apart twins are rare. Due to changing policies, practices, and social conditions, collecting a large enough sample of reared-apart twins to conduct a new TRA study will likely never again be possible.

Cyril Burt was the product of a broken research/publication system that allowed (and still allows) researchers to collect, analyze, rework, and even fabricate data behind the scenes (Carey, 2011) to arrive at desired yet false conclusions that make it past the academic peer-review process and subsequently get cited as findings in textbooks and review articles. (The analyses found in TBC were not peer-reviewed prior to the book's publication [Dorfman, 1995].) At the time of the "Burt Affair," many saw Burt as a bad apple in an otherwise healthy behavioral science/psychology research barrel. The replication crisis has forever changed our perception of the barrel and has led to increasing calls to require or encourage *preregistration* in behavioral research (Chambers, 2017; Joseph & Baldwin, 2000; Nosek et al., 2018).

Richard Herrnstein's 1970s Writings on IQ

In the wake of the "firestorm" ignited by Jensen's racial differences claims (Tucker, 1994b), Herrnstein jumped into the debate by publishing a controversial 1971 article about IQ in the U.S. magazine *Atlantic Monthly* (now *The Atlantic*; Herrnstein, 1971). Two years later, he published *I.Q. in the Meritocracy*, expanding on his views and placing IQ heritability in the 70-85% range (Herrnstein, 1973). Herrnstein argued that IQ scores had a strong genetic component and were an important determinant of people's success in life, that economic classes differed in their innate levels of intelligence, and that society would see increasing stratification and the creation of high and low castes based on inherited intelligence. He returned to these themes two decades later in *TBC*. Herrnstein believed that TRA studies provided the most convincing evidence that human intelligence is "highly heritable" (1971, p. 57). He discussed but did not arrive at any firm conclusions about the causes of White-Black IQ differences in either 1971 or 1973.

A striking feature of Harvard scholar Herrnstein's *Atlantic Monthly* article was his reliance on *Jensen's* evaluation of the then-four existing TRA studies. Jensen regarded Burt's subsequently discredited publications as the "most interesting," in part because, based on Burt's (made-up) data, "the separated twins were spread over the entire range of socioeconomic levels" (Jensen, 1969, p. 52). Following Jensen, Herrnstein pointed to Burt as providing the most convincing evidence because "there was literally no general correlation in the occupational levels of the homes into which the pairs were separated" (Herrnstein, 1971, p. 55). There was no indication that Herrnstein was familiar with the four original TRA study publications on which his arguments rested, including the detailed case histories. He elaborated on TRA study results in *I.Q. in the Meritocracy*. He now cited these studies in his book's reference section, but again there was no indication that he had a good understanding of how they were performed. There was little discussion of the numerous problems found in research on reared-apart twins (see Joseph, 2015).

IQ hereditarians are concerned with human reproduction patterns and preventing what they fear is a dysgenic decline of "national intelligence" (Cattell, 1937). Whereas many developmental psychologists see talents and skills as abundant and believe that "education should be designed to reveal the talents and skills in each child" (Cantor et al., 2021, p. 7), IQ hereditarians believe that talents and skills are scarce. Clearly unhappy that U.S. society was "stuck" with uncontrolled IQ "mating patterns" leading to a "shortage of high-grade intellect," in his 1971 *Atlantic Monthly* article, Herrnstein wrote:

> We are not, for example, on the verge of Galton's vision of eugenics, even though we now have the mental test that he thought was the crucial prerequisite. For good or ill, and for some time to come, we are stuck with mating patterns as people determine them for themselves. No sensible person would want to entrust state-run human breeding to those who control today's states. There are, however, practical corollaries of this knowledge, more humble than eugenics, but ever more salient as the growing complexity of human society makes acute the shortage of high-grade intellect.
>
> (Herrnstein, 1971, pp. 57-58)

Herrnstein seemed to favor some type of future eugenic interventions on human reproduction, even though they would not be based on *Galton's* vision of eugenics or administered by those who control *today's* states. The 1971 form of government in the United States was a democratic republic, and perhaps Herrnstein believed that a more authoritarian political system would be needed to enforce a compulsory eugenic program against those who surely would resist it. IQ hereditarian psychologist/eugenicist Richard Lynn implicitly promoted this view (Lynn, 2001). H & M discussed and cited Lynn's work favorably throughout *TBC* (Kamin, 1995), referring to Lynn as "a leading scholar of racial and ethnic differences" in IQ (H & M, 1994, p. 272). For his part, Lynn (2001, p. 275) believed that H & M's "analysis is implicitly eugenic because, as they identified the underclass as a genetic problem, it calls implicitly for a eugenic solution."

Leon Kamin Enters the Debate

Student activists at Princeton University approached psychology professor Leon Kamin and urged him to read Herrnstein's *Atlantic Monthly* article. In a 1973 interview, Kamin recalled, "I got interested in the problem of intelligence testing about a year ago, largely as a result of reading Professor Herrnstein's article in the *Atlantic*" (Egerton, 1973). Unlike Herrnstein, Kamin carefully examined the original books and articles describing the four existing TRA studies and, in 1974, published *The Science and Politics of I.Q.* (Kamin, 1974). Long before the replication crisis shined a long-overdue critical light on psychological research, Kamin's analysis began the process of discrediting the Burt study (Joseph, 2018).

Although the "Burt data" played a central role in Jensen's 1969 and 1970 IQ heritability estimates, by 1974, he was compelled to recognize that Burt's IQ "correlations are useless for hypothesis testing" (Jensen, 1974, p. 24). Jensen was familiar with Kamin's circulating pre-publication analysis of Burt's IQ TRA articles and the troubling problems Kamin found in them. Jensen's 1974 *Behavior Genetics* article was a fast-tracked attempt to get out in front of (scoop) Kamin by creating the impression that IQ genetic researchers could independently police their own research and then publish their findings in an academic journal (Hirsch, 1981; Tucker, 1994a). Were it not for Kamin's meticulous detective work, psychology textbooks might still cite Burt's TRA publications as providing conclusive evidence in favor of high IQ heritability.

Major problems Kamin (1974) found in the three remaining TRA studies included (1) that most MZA pairs were separated late or incompletely; (2) that many pairs grew up near each other and had substantial contact; (3) that researchers' assessment of the degree of twins' separation and behavioral similarity frequently depended on trusting the (potentially unreliable) accounts of the twins themselves; (4) that MZA pairs were placed and grew up in similar or "correlated" environments; (5) that there were problems with the IQ tests used, including a lack of test standardization in some cases; (6) that MZA IQ correlations were subject to age- and sex-confounds, which inflated these correlations for non-genetic reasons; and (7) that "unconscious experimenter bias" may have led researchers to score MZAs similarly. Leon Kamin can be considered the father of the replication crisis in U.S. academic psychology.

Kamin showed that the Newman et al. and Shields TRA studies recruited volunteer twins using media appeals and that the twins had to have been aware of each other's existence to be able to respond. They may have responded to the appeal, or discovered each other, *because of* their behavioral similarities. Based on a 2022 analysis (Joseph, 2022b, p. 49) of the Newman et al., Shields, and Juel-Nielsen case histories, in addition to sharing an intrauterine environment:

- In 25/75 (33%) of the pairs, twins were separated at 12 months of age or later.
- In 56/75 (75%) of the pairs, twins had contact with each other while growing up.
- In 42/75 (56%) of the pairs, one or both twins were placed with a family member.
- In 17/75 (23%) of the pairs, twins lived together for at least 12 months after separation or grew up next door to each other.

In his 1970 article, Jensen (p. 135) wrote that "all of Shields' twins were separated before six months." This statement was a "gross error of fact," as sociologist Howard Taylor put it (1980, p. 87). By our count (and Kamin's [1974, p. 143]), only 28/44 (64%) of the Shields MZA pairs were separated at or before six months of age, and all had various degrees of contact after separation (see Joseph, 2015, Table 2.2). Six of the 44 (14%) Shields MZA pairs were separated between four and nine years of age. Many more errors can be found in Jensen's 1969 and 1970 publications (see Hirsch, 1975; Kamin, 1974, Chapter 6).

A close look at Shields' 1962 book describing his study shows that this *was not* a study of reared-apart twins using any reasonable definition of the term (Joseph, 2015, Chapter 2). To cite just one example (also cited by Kamin in 1974), the following passages are found in Shields' case description of a 17-year-old male MZA pair, "Bertram and Christopher":

> Separated at birth. ...The Mother died the day after the twins were born. The paternal aunts decided to take one twin each and they have brought them up amicably, living next-door to one another. ...They are constantly in and out of each other's houses. ...Both twins became garage mechanics. ...They have always been closely attached to each other, and now go out courting together. ...When they were younger, Christopher used to follow Bertram around "as if he were a younger brother."
>
> (Shields, 1962, pp. 164-165)

Something to consider: IQ hereditarians, including Jensen and Herrnstein in the 1970s right up to those currently posting on the social media, count Bertram and Christopher as a pair of "reared-apart monozygotic twins."

Herrnstein built his case in the 1970s for high IQ heritability, and what he saw as its accompanying social, political, and reproductive implications on massively flawed TRA studies he appeared to be largely unfamiliar with. As we will show in Part II, he repeated this folly two decades later in *The Bell Curve*.

Part II: Evidence Cited in *The Bell Curve* in Favor of Substantial Within-Group IQ Heritability

We now begin our analysis of the evidence H & M cited in support of genetic influences on within-group IQ score differences (IQ heritability). Like Jensen before them, H & M's argument in favor of 60% or higher within-group IQ heritability was based (1) on assuming that IQ tests measure human intelligence, (2) on accepting a heritability estimate as an indicator of the magnitude (strength) of genetic influences on IQ score differences, and (3) on genetic interpretations of IQ score correlations among family members and twins (reared together or reared apart). Under the heading "How Much is

IQ a Matter of Genes?" (pp. 105-109), H & M presented the evidence they believed showed that IQ is a "substantially heritable" trait. Let's take a look at their evidence.

Reared-Apart Relatives Do Share Common Environmental Influences

Based on the "plausible assumption" that "blood relatives who were raised apart. ...do not, in fact, share environments" (H & M, 1994, p. 107), H & M promoted TRA and other studies of supposedly reared-apart relatives as providing "direct" estimates of heritability. This "no shared environment" assumption, however, is false. Even perfectly separated reared-apart relatives share many common non-familial "cohort" influences (Elder & Shanahan, 2006; Joseph, 2022b; Rose, 1982; Schaie, 1965). Although they do not share the same family home, they share similar behavior-shaping environmental influences such as extended family, cultural, religious, SES, educational, technological, legal, mass media, regional, political, and oppression/privilege based on ethnicity and gender.

The *cohort effect* concept refers to the resemblance in similar-age people's IQ scores (see the "Flynn effect" discussion later in Part II), abilities, behavior, preferences, beliefs, physical condition, and other characteristics caused not by heredity but by experiencing stages of life at the same time in the same historical period and cultural milieu. MZA pairs are born on the same day and, therefore, experience many of these behavior-shaping cohort influences at the same time during the same developmental stages. MZAs will also elicit more similar treatment from their social environments due to their substantial physical resemblance, including physical attractiveness and height (Cropanzano & James, 1990; Gugushvili, A., & Bulczak, 2023). Because in most societies men and women are treated differently and are socialized from birth to behave, think, and feel in differing gender-specific ways, members of an MZA pair will behave more similarly for this reason alone.

The fallacious *TBC* claim that reared-apart relatives "do not, in fact, share environments" is based on defining the environment as mainly the family environment, thereby overlooking or denying the impact of the above-mentioned *non-familial* cohort influences. For example, the world's most famous TRA researcher defended his studies on the grounds that such influences are "difference-producing, not similarity-producing" (Bouchard, 2023, p. 137). As if growing up experiencing several common cultural influences at the same time causes people to behave *less* alike rather than more alike.

H & M believed that TRA studies provide the "most unambiguous" evidence of genetic influences on IQ, which "produce some of the highest estimates of heritability" (H & M, 1994, p. 105). They cited two modern TRA studies (Bouchard et al., 1990; Pedersen et al., 1992), and two review articles providing pooled IQ correlations for the three earlier studies. Apart from Burt, the Bouchard et al. investigation was the only TRA study discussed by H & M. "Identical twins," they wrote,

> share all their genes, and if they have been raised apart since birth, then the only environment they shared was that in the womb. Except for the effects on their IQs of the shared uterine environment, their IQ correlation directly estimates heritability.
>
> (H & M, 1994, p. 107)

TRA study critics have made a compelling case that this statement, including its implication that most studied pairs were "raised apart since birth," is false (see Farber, 1981; Joseph, 2015, 2022b, 2023a; Kamin, 1974; Kamin & Goldberger, 2002; Lewontin et al., 1984; Rose, 1982; Rutter, 2006; Taylor, 1980). The critics' findings undercut what H & M saw as the "purest of the direct comparisons" (p. 107) in support of substantial IQ heritability. In addition to shared cohort influences and the lack of true separation we documented earlier, fewer than 3% of the MZA pairs in the TRA studies Herrnstein cited in the 1970s (excluding Burt) were separated during the first year of life, were reared with no knowledge of their twinship, and were studied at the time of their first meeting (Farber, 1981, p. 60).

We have shown that in TRA studies, about half the time, one or both twins were placed with a family member. This trend continued in the "Swedish Adoption/Twin Study of Aging" TRA study cited by H & M (SATSA; Pedersen et al. 1992), where the researchers obtained their initial data by mailed questionnaires. According to co-investigators Nancy Pedersen, Robert Plomin, and colleagues, twins (average age 65.6 when tested) often were reared in the same extended family. "For 44% of the pairs," they wrote, "the rearing parents of one twin were biologically related to rearing parents of the other twin. In most of these cases, the biological mother reared one twin and her sibs or parents reared the other" (Pedersen et al., 1984, p. 244). Presumably, many of these pairs grew up near each other in similar cultural and SES families, and attended family events together. Pedersen and colleagues defined twins as "reared apart" if they had been "separated *by the age of 11*" (Pedersen et al., 1992, p. 347, emphasis added). The SATSA supplied additional evidence that TRA studies' "no shared postnatal environment" claim is utterly false.

The "Minnesota Study of Twins Reared Apart" (MISTRA)

The other and much more well-known TRA study published between Herrnstein's 1970s and 1990s writings on IQ was the "Minnesota Study of Twins Reared Apart" (MISTRA). Study initiator psychologist Thomas J. Bouchard, Jr. and colleagues concluded that the results of their 1990 IQ study published in *Science*, one of the world's top scientific journals, showed that "general intelligence or IQ is strongly affected by genetic factors" (Bouchard et al., 1990, p. 227) and that IQ heritability is "about 70%" (p. 223). The editor of *Science*, Daniel Koshland, had invited Bouchard to submit an article about the study and its results (Segal, 2012, p. 13). Bouchard and colleagues used three different IQ measures: the "Wechsler Adult Intelligence Scale" (WAIS), the "Raven's Progressive Matrices/Mill-Hill Vocabulary Scale composite," and the "First Principal Component of Special Mental Abilities." The 1990 MISTRA sample consisted of 56 MZA and 30 DZA pairs (p. 223; DZAs = dizygotic or fraternal twins reared apart, sharing a 50% average genetic resemblance). The 1990 *Science* article was the main MISTRA IQ heritability publication. According to Google Scholar, it has been cited over 2,400 times.

The study was paralleled by MISTRA-supplied or promoted cherry-picked stories of individual MZA pairs, such as the "Jim Twins," "Fireman Twins," "Giggle Twins," and Oskar Stöhr and Jack Yufe. Stories about their supposed genetically caused "eerie" or "spooky" behavioral similarities have repeatedly appeared in popular books and the mainstream and social media since 1979 (e.g., Barba, 2023; Rawson, 1979; Segal, 1999; Wright, 1998). Far more than simple human-interest stories, many writers and media outlets have used these pairs to sell genetic determinism to the public in ways that academic research results never could (Horgan, 1999; Joseph, 2016). Although the twins' stories are often interesting, for reasons that include the cohort effects we described earlier, these selectively reported anecdotal reports provide no evidence that genetic factors influence human behavioral characteristics (Dusek, 1987). This conclusion also applies to the famous set of identical triplets whose, at times, tragic stories were the subject of the popular 2018 movie *Three Identical Strangers* (reviewed by Joseph, 2019). We do take note, however, that a 2022 report on a Korean female MZA pair "raised in very different environments" (one in South Korea, the other in the U.S.) found a "substantial" 16-point IQ score difference between the twins (Segal & Hur, 2022, p. 7)—roughly the difference one would expect to find between genetically unrelated persons reared apart (Plomin & DeFries, 1980).

H & M saw the MISTRA as the foundation of their claim that IQ is substantially heritable, and implied that the study's results might even vindicate Cyril Burt:

An ironic afterword centers on Burt's claim that the correlation between the IQs of identical twins reared apart is +.77. A correlation this large almost irrefutably supports a large genetic influence on IQ. Since the attacks on Burt began, it had been savagely derided as fraudulent, the product of Burt's fiddling with the data to make his case. In 1990, the Minnesota twin study, accepted by most scholars as a model of its kind, produced its most detailed estimates of the correlation of IQ between identical twins reared apart. The procedure that most closely paralleled Burt's yielded a correlation of +.78.

(H & M, 1994, p. 12)

The most modern study of identical twins reared in separate homes [the MISTRA] suggests a heritability for general intelligence between .75 and .80, a value near the top of the range found in the contemporary technical literature.

(p. 107)

Although journalists often told a version of the story that Bouchard was "boggled" (Wright, 1998, p. 41) or "flabbergasted" (Rawson, 1979) by MZAs' supposed behavioral similarities, Bouchard held strong IQ hereditarian beliefs prior to initiating the MISTRA in 1979. In a 1976 publication, he argued that "human intelligence is largely under genetic control" and that "class differences in intelligence have an appreciable genetic component" (Bouchard, 1976, p. 193). Bouchard later recalled how his "teaching research by Jensen" drew student protests at the University of Minnesota in the 1970s (Holden & Bouchard, 2009, p. 27). In keeping with these views, he published a positive review of *TBC* (Bouchard, 1995) and a mostly positive review of Richard Lynn's *Dysgenics: Genetic Deterioration in Modern Populations* (Bouchard, 1999). Bouchard's top MISTRA associate, University of Minnesota psychologist David Lykken, held similar IQ hereditarian views (Lykken, 1995, pp. 216-217; Lykken, 2004). Lykken went further and called for the "licensure of parenthood [as] the only real solution to the problem of sociopathy and crime" (1995, p. 231). The replication crisis has led to a much greater understanding that strong confirmation biases often influence how researchers gather and interpret their data.

H & M wrote that in TRA studies, MZAs live their lives "often not knowing of each other's existence" (H & M, 1994, p. 107). This statement is misleading, especially so for the MISTRA, which depended on recruiting volunteer twin pairs who, in most cases, had to have been aware of each other's existence to participate in the study. Like the earlier Newman et al. and Shields TRA studies, the MISTRA sample was a "collection of cases" based mainly on volunteer twin participants who responded to media appeals or media attention (Segal, 2012, p. 36). How could they have responded if they didn't already know they had a twin sibling? Like Herrnstein's handling of TRA research in the 1970s, H & M gave little indication that they had a firm understanding of how the MISTRA researchers performed their study, analyzed their data, and reached their conclusions. Having relied second-hand on Jensen's faulty analyses and the Burt "data" in the 1970s, Herrnstein now relied on conclusions found in what we will show was Bouchard and colleagues' problematic 1990 *Science* IQ study.

Many well-intentioned critics of *TBC* in the 1990s avoided discussing the MISTRA and its claims of high IQ heritability, most likely because it was difficult to challenge Bouchard and colleagues' conclusions. However, recent analyses suggest that the Minnesota researchers' findings in favor of above-zero IQ heritability should be disregarded.

The key points raised in Joseph's (2022b) analysis of the 1990 MISTRA IQ study are summarized below (adapted from Joseph, 2023a):

- TRA studies contain numerous non-genetic similarity-producing biases, including cohort influences, that scholars have documented for over 50 years. These influences cast doubt on claims that environmental confounds are minor or absent in such studies.
- Although the researchers took some steps to eliminate biases found in the older investigations, their study design reproduced many of these biases, including the fact that MISTRA volunteer twin participants were not separated at birth and then randomly assigned to available adoptive homes. By definition, MISTRA twin pairs had been separated by four years of age (Bouchard et al., 1990, p. 224; Segal, 2012, p. 37). Most twins knew they had a twin sibling, and most had contact with each other. Like the earlier studies, the MISTRA pairs were only *partially* reared apart.
- DZAs were the official MISTRA control group, designated as such by Bouchard at the beginning of the study in 1979 (Segal, 2012, p. 12, 343 Note 69). The MISTRA was the first TRA study to recruit DZAs as controls, which the researchers saw as an important improvement over the earlier studies. As Bouchard and colleagues described it four years prior to the *Science* publication, "Our study is the first to have included a control group of dizygotic twins reared apart (DZA). ... DZA twins allow us to test the two most common competing hypotheses proposed as alternatives to the genetic hypothesis as an explanation of the similarity between MZA twins: placement bias and recruitment bias" (Bouchard et al., 1986, p. 300).
- In the 1990 *Science* article's opening paragraph (p. 223), Bouchard and colleagues continued their theme of using MZA and DZA twins to test hypotheses. "MZA and DZA twin pairs," they wrote, "are a fascinating experiment of nature. They also provide the simplest and most powerful method for disentangling the influence of environmental and genetic factors on human characteristics." However, in the same article, they failed to publish the results of this "powerful experiment of nature" (see Table 16.1 in this book), or any other DZA (control group) data, due to what they called "space limitations and the smaller size of the DZA sample (30 sets)" (p. 223). To the best of our knowledge, the full-sample DZA IQ correlations for the three IQ measures they used remain unpublished to this day.
- Near-full-sample DZA IQ correlations published many years later (with little discussion) show that the MISTRA MZA group and DZA group Wechsler (WAIS; Segal, 2012, p. 286) and Raven (Johnson et al., 2007, p. 552) IQ correlations do not differ at a statistically significant level (see Table 16.2 in this book). This finding leads to a conclusion that the study found no evidence that hereditary factors influence IQ scores (0% IQ heritability). If the MZA group (100% genetic resemblance) and DZA group (50% average genetic resemblance) IQ correlations do not differ significantly, we can safely conclude that non-genetic factors alone raised both twin-group IQ correlations above zero (Pedersen et al., 1992). The MISTRA researchers never published their "First Principal Component" DZA IQ correlations. On the "personality" side of the study, they never published the MISTRA *Sixteen Personality Factor Questionnaire* results (16PF; listed as a MISTRA-administered personality test in Bouchard, 1984, p. 160), even though the 16PF had been used extensively in personality research for decades.

- After Bouchard and colleagues decided to omit and bypass their 1990 DZA control group correlations, most likely because including them would have led to an unwanted 0% heritability finding, they used their MZA IQ correlations alone to estimate heritability (see Figure 16.1). They claimed that an MZA IQ correlation "directly estimates heritability," meaning they assumed that pre- and post-natal environmental influences played no role in elevating their MZA IQ correlations above zero. (Contrary to some accounts, the researchers did not compare their MZA and MZT [monozygotic twins reared together] IQ correlations for the purpose of establishing or calculating heritability.)
- The MISTRA assumption that the MZA IQ correlation directly estimates heritability is false. Even perfectly separated MZ twin pairs share many similarity-producing environmental and cohort influences in common. The "Flynn effect" of "massive" IQ score gains over time also inflates MZA IQ correlations because twins are born at the same time (more on this point later).
- Bouchard and colleagues attempted to shore up the implausible "MZA IQ correlation directly estimates heritability" assumption by deciding to count most environmental influences as genetic influences. They wrote that MZA pairs' "identical genomes make it probable that their effective environments are similar. ... MZA twins tend to elicit, select, seek out, or create very similar effective environments, and, to that extent, the impact of these experiences is counted as a genetic influence" (Bouchard et al., 1990, pp. 227-228). In so doing, the researchers arrived at conclusions in favor of strong genetic influences on IQ and other types of human behavior by circularly assuming the very same thing.
- The researchers failed to make their unpublished raw data and twins' life histories available for inspection and analysis by qualified independent reviewers. For this reason alone, we must evaluate their findings with extreme caution because they are based on virtually irreproducible sample populations.
- The MISTRA should be evaluated in the context of science's replication crisis, and the investigators' finding of above-zero IQ heritability was based on p-hacked research. Strong genetic confirmation biases influenced their interpretations of the data.

Invalid Findings

P-hacking is the practice of researchers consciously or unconsciously manipulating definitions and data, either openly or behind the scenes, to transform non-findings into publishable "findings" that reach the conventional .05 level of statistical significance (Head et al., 2015; Stefan & Schönbrodt, 2023). If we use language developed in the current replication crisis in science, the MISTRA IQ study was p-hacked through Bouchard and colleagues' use of at least four of the ten "questionable research practices" (QRPs) described by Leslie John and colleagues in 2012 (John et al., 2012, p. 525):

- **QRP #1:** "In a paper, failing to report all of a study's dependent measures" (the DZA IQ correlations).
- **QRP #2:** "Deciding whether to collect more data after looking to see whether the results were significant" (by continuing to recruit, add, and test new twin pairs between 1990 and 2000, most likely in the [unfulfilled] hope of creating a large enough sample to nudge the MZA-DZA comparison under the critical .05 level of statistical significance).
- **QRP #6:** "In a paper, selectively reporting studies that 'worked'" (by dropping the biometrical model-fitting procedure used in most MISTRA non-IQ publications).

- **QRP #7:** "Deciding whether to exclude data after looking at the impact of doing so on the results" (the DZA IQ correlations).

According to MISTRA researcher and 1990 *Science* article co-author Nancy Segal, "The simple comparison of the MZ (or MZA) and DZ (or DZA) intraclass correlations is an important first step in behavioral-genetic analysis because this demonstrates *whether or not* there is genetic influence on the trait" (Segal, 2012, p. 62, emphasis added). The evidence strongly suggests that the never-disclosed 1990 MISTRA DZA IQ results showed that the MZA and DZA correlations did not differ significantly ($rMZA \ngtr rDZA$), which led the researchers to suppress their DZA IQ correlations to avoid a conclusion of "not."

Now basing their genetic "findings" on the MZA IQ correlations and accompanying assumptions alone, the researchers submitted their p-hacked study manuscript to *Science*. Unfortunately, it appears that MISTRA-friendly editor Daniel Koshland and the peer reviewers who greenlighted the article for publication (1) didn't notice that the MISTRA control group correlations were missing, or (2) accepted the researchers' weak explanation that they omitted their DZA correlations due to "space limitations and the smaller size of the DZA sample," even though we showed that the manuscript opened with the statement, "MZA and DZA twin pairs...provide the simplest and most powerful method for disentangling the influence of environmental and genetic factors on human characteristics."

Bouchard responded to Joseph's (2022b) analysis in a 2023 article, where he wrote,

> The DZAs were not included because the sample was small; the purpose of the paper was to report a constructive replication of previous studies of MZA twins in the brief format provided by *Science* and explain the methodology underlying the study of MZA twins.
>
> (Bouchard, 2023, p. 136)

The "brief format" *Science* article ran six pages and said nothing about being a "constructive replication" of previous studies of MZA twins. Moreover, Bouchard and colleagues published and analyzed *smaller* full-sample DZA correlations in several non-IQ MISTRA studies appearing around the same time (e.g., Bouchard, Segal, & Lykken, 1990, "special mental abilities" = 25 DZA pairs; Moloney et al., 1991, "vocational interests" = 27 DZA pairs; Tellegen et al., 1988, "personality" = 27 DZA pairs). Joseph subsequently published a reply to Bouchard's article (Joseph, 2023a), where he called on *Science* to retract the 1990 MISTRA IQ study article because it violated the journal's guidelines designed to prevent the publication of p-hacked research.

We have shown that the key TRA study H & M cited in favor of substantial IQ heritability failed to produce valid evidence supporting the existence of genetic variants that influence differences in IQ scores. A process whereby academic psychologists and others could accept the MISTRA "as a model of its kind" (H & M, 1994, p. 12) makes it easier to understand why science is now embroiled in a replication crisis that began in Herrnstein's and Bouchard's chosen field of U.S. academic psychology.

A final evaluation of the three most well-known TRA IQ studies finds that one was fraudulent (Burt), another defined "reared apart" twin pairs as those first separated by the age of eleven (SATSA), and the other was based on p-hacked findings (MISTRA). By promoting the last two studies and implying that the first might be acceptable after all, H & M helped sustain a long-running scientific smoke-and-mirrors show otherwise known as "reared-apart" or "separated-at-birth" twin studies.

Other Evidence Cited by Herrnstein and Murray in Support of Within-Group IQ Heritability

In addition to what we have shown are invalid "direct" IQ heritability estimates produced by TRA research, and the (non-twin) adoption methods we will soon describe, H & M cited "indirect" estimates derived from reared-together sibling correlations, including twin studies that compare the behavioral resemblance of reared-together MZ pairs (MZTs) versus reared-together DZ pairs (DZTs). They did not cite any original-data studies, and described these indirect methods as follows:

> Indirect methods compare the IQ correlations between people with different levels of shared genes growing up in comparable environments—siblings versus half-siblings or versus cousins, for example, or MZ twins versus fraternal (dizygotic, DZ) twins, or nonadoptive siblings versus adoptive siblings. The underlying idea is that, for example, if full siblings raised in the same home and half-siblings raised in the same home differ in their IQ correlations, it is because they differ in the proportion of genes they share: full siblings share about 50 percent of genes, half siblings about 25 percent.
>
> (H & M, 1994, p. 107)

Most behavioral geneticists recognize that non-twin family comparisons are vulnerable to environmental influences shared by members of the same or extended family and, therefore, do not provide evidence that genetic factors influence behavior. According to Plomin and colleagues (2013, p. 191), "Family studies by themselves cannot disentangle genetic and environmental influences." They continued that although "first-degree relatives living together" correlate for general intelligence, "this resemblance could be due to genetic or to environmental influences because such relatives share both" (p. 194). For this reason, Plomin and other influential reviewers usually cite twin and adoption studies, which they claim provide the main "quantitative genetic" evidence in favor of substantial IQ heritability (e.g., Plomin & Deary, 2015; Plomin & Kawakami, 2024; Ritchie, 2015).

The Classical Twin Method

The *classical twin method* (or "twin method") compares the behavioral resemblance of MZTs versus DZTs. Like reared-apart pairs, MZTs are assumed to share 100% of their segregating genes, whereas DZT pairs share on average 50%. Twin method results usually show that MZT pairs behave more alike or correlate higher on psychological tests versus same-sex DZT pairs at a statistically significant level. We designate this finding "$rMZT > rDZT$" (with r representing the twin-group IQ score correlation). A behavioral twin study finding of $rMZT > rDZT$ is rarely disputed. What *is* disputed is how this finding should be interpreted.

The Equal Environment Assumption (EEA)

The MZT-DZT *equal environment assumption* (EEA) has always been the most controversial twin method assumption (Richardson & Norgate, 2005). According to the EEA, MZT and same-sex DZT pairs grow up experiencing roughly equal environments, and the only behaviorally relevant factor distinguishing these pairs is their differing degree of *genetic* relationship to each other. As one group of twin researchers emphasized, the EEA "is crucial to everything that follows from twin research" (Alford et al., 2005, p. 155). Twin researchers interpret $rMZT > rDZT$ genetically by assuming that the EEA is valid, or at least that it "appears reasonable" (Plomin et al., 2013, p. 81).

Critics, on the other hand, have argued since the 1930s (reviewed in Joseph, 2004, Chapter 2) that the EEA as it relates to behavioral and psychiatric twin studies *is not* valid or even "reasonable" because, when compared with same-sex DZT pairs, MZT pairs grow up experiencing (1) more similar treatment by parents, teachers, and others; (2) more similar physical and social environments; (3) more similar treatment by society due to sharing a very similar physical appearance; (4) a much greater tendency to spend time together, to have common friends and peer influences, and to model their behavior on each other (DZs more often than MZs strive to make themselves different); and (5) identity confusion and emotional attachment to each other to a far greater extent than DZTs (Cohen et al., 1975; Kringlen, 1967; Torgersen, 1979). Due to these obvious environmental confounds, some analysts argue that the results of twin method MZT-DZT comparisons should not be interpreted genetically (Charney, 2013; Fosse et al., 2015; Jackson, 1960).

When challenged to defend genetic interpretations of *r*MZT > *r*DZT, modern twin researchers and their supporters usually concede the point that MZT pairs grow up experiencing more similar environments than DZTs (e.g., Bouchard, 1997; DeFries & Plomin, 1978; Derks et al., 2006; Evans & Martin, 2000; Kendler, 1983). As criminology researchers J. C. Barnes and colleagues acknowledged, ironically, since they were engaged in a lengthy 2014 defense of twin research in their field,

> Critics of twin research have correctly pointed out that MZ twins tend to have more environments in common relative to DZ twins, including parental treatment...closeness with one another...belonging to the same peer networks...being enrolled in the same classes... and being dressed similarly.
>
> (Barnes et al., 2014, p. 597)

Instead of abandoning the twin method in the 1960s due to a growing understanding that MZT and DZT environments are in fact *un*equal, which meant that the EEA as it was then defined was false (Joseph, 2004, 2013), some twin researchers of that era chose to subtly redefine the EEA to keep their primary research method alive (e.g., Gottesman, 1966). They now recognized that MZT environments were more similar, but continued to defend genetic interpretations of *r*MZT > *r*DZT by arguing (a) that due to their greater genetically caused behavioral similarity, MZTs "create" or "elicit" more similar environments for themselves when compared with DZTs, which should be counted as a genetic effect; or (b) that critics must show that MZTs grow up experiencing more similar "trait-relevant" environments than DZTs.

However, the (a) "twins create their environments" EEA defense does not hold because, among other reasons, it circularly assumes in advance the genetic basis of behavioral differences in the process of testing for it (for more problems with the "a" defense, see Joseph, 2015, Chapter 7). Turning to the (b) "trait-relevant" EEA defense, twin researchers (not their critics) bear the burden of proof for identifying *specific and exclusive* trait-relevant environmental influences on IQ scores. However, as Bouchard acknowledged in 1997, "in spite of years of concerted effort by psychologists, there is very little knowledge of the trait-relevant environments that influence IQ" (Bouchard, 1997, p. 145). Because researchers are unable to identify specific and exclusive IQ-relevant environmental influences, followed by demonstrating that MZTs and DZTs were equally exposed to these influences, the "trait-relevant" defense fails. Overall, the "twins create their environments," "trait-relevant," and other arguments twin researchers have invoked in defense of the EEA and the twin method do not hold up under critical examination (Burt & Simons, 2014; Joseph, 2006, 2023b Chapter 4; Pam et al., 1996).

Twin researchers often point to a series of "EEA-test" studies they and their colleagues have performed since the late 1960s (e.g., Loehlin & Nichols, 1976; Scarr & Carter-Saltzman, 1979). In these studies, researchers "tested" the validity of the assumption in ways other than the only way it can be tested, which is a straightforward determination of whether MZTs and same-sex DZTs grow up experiencing roughly equal environments—yes or no. If the answer is no, as it almost always is, genetic interpretations of *r*MZT > *r*DZT are not supported (Joseph, 2023b). Although many twin researchers (e.g., Plomin et al., 2013, p. 81) argue that discovering "violations" of the EEA in a particular twin study indicates only that they may have "overestimated" heritability, the proper conclusion is that *r*MZT > *r*DZT should not be interpreted genetically because, like family studies, potential

Murray Reverses Course

As opposed to what he and Herrnstein wrote in *TBC*, in his 2020 book *Human Diversity: The Biology of Gender, Race, and Class*, Charles Murray downplayed the importance of TRA studies due to their "limited potential" (Murray, 2020, p. 215) and instead promoted the "analytic power of comparing [reared-together] MZ and DZ twins" as supplying the best evidence in support of behavioral and IQ heritability (p. 213). A stunning turnaround, yet Murray did not acknowledge that his new emphasis on reared-*together* twin studies and an accompanying deemphasis of the now "limited potential" TRA studies contrasted sharply with his *TBC* promotion of the latter as providing the "purist" and "most unambiguous" evidence in favor of high IQ heritability. After ignoring the EEA in *TBC*, in *Human Diversity*, Murray briefly addressed the assumption and the debate surrounding it. He concluded, mistakenly, "overall, heritability as estimated by twin studies appears to be accurate" (p. 217).

Murray spent several pages of *Human Diversity* singing the praises of Tinca Polderman and colleagues' frequently cited 2015 meta-analysis of 2,748 twin-method studies performed in 39 countries between 1958 and 2012. These studies assessed more than 17,000 physical, medical, and psychological traits (characteristics) and found $rMZT > rDZT$ for 84% of the traits. Polderman and colleagues concluded that their "results provide compelling evidence that all human traits are heritable: not one trait had a weighted heritability estimate of zero" (Polderman et al., 2015, p. 708). However, like H & M in 1994, they failed to defend or even mention the EEA, that is, the assumption upon which genetic interpretations of $rMZT > rDZT$ are based. Without the EEA in behavioral twin research, $rMZT > rDZT$ can be interpreted as showing that greater environmental similarity causes greater behavioral similarity, and nothing more.

Conclusions: IQ rMZT > rDZT Should Not Be Interpreted Genetically

As most twin researchers now understand, the evidence shows overwhelmingly that MZT childhood, adolescent, and adult environments (including levels of identity confusion and attachment) are much more similar than DZT environments. Unfortunately, twin researchers continue to misinterpret IQ, behavioral, and psychiatric $rMZT > rDZT$ genetically by using illogical arguments, while conceding only that unequal environments might lead to "overestimating heritability" in a particular study. Textbooks, review articles, and media reports continue to endorse these unsupported interpretations, in the process misleading new generations of students, researchers, and the public at large.

Based on twin method results, Plomin and colleagues calculated a general intelligence heritability estimate of 52% (2013, pp. 194-195). All well and good, but they and other twin method supporters must also defend comparable or even higher twin method heritability estimates for behaviors that critics point to as examples of the folly of twin research. These behaviors include "dog ownership" (55% heritability; Fall et al., 2019), "problematic child masturbatory behavior" (77% heritability; Långström et al., 2002), "adult loneliness" (48% heritability; Boomsma et al., 2005), "vegetarianism" (77% heritability; Wesseldijk et al., 2023), "tea and coffee drinking preferences in males" (46% heritability; Luciano et al., 2005), and "health insurance coverage" (40%-50% heritability; Wehby & Shane, 2019).

Plomin and colleagues were spot on when they concluded that IQ "family studies by themselves cannot disentangle genetic and environmental influences." We are simply concluding that this statement also holds true for twin method MZT-DZT IQ comparisons.

We are unaware of any valid argument supporting genetic interpretations of IQ studies using the classical twin method. Combined with our TRA study analysis, we conclude that no twin study published to date has produced valid evidence that genetic factors influence within-group IQ score differences (IQ heritability). Although twin research and its underlying assumptions have thus far avoided mainstream replication crisis scrutiny, such scrutiny is long overdue and may lead to finally solving the "missing heritability" problem.

Adoption Studies

In their "How Much is IQ a Matter of Genes" section (pp. 105-108), H & M mentioned IQ adoption research and the lower IQ heritability estimates these studies produced. They did not cite a specific adoption study, and in their view, "estimating heritabilities from any relationship other than for identical twins is inherently more uncertain" (p. 710; we briefly discuss "transracial" adoption studies in Part III). A more fitting name for a behavioral adoption study is *a study of abandoned and rejected children*. This term emphasizes the life conditions children experience between birth and their eventual placement with an adoptive family and the psychological impact these experiences may have had throughout a person's life (Cassou et al., 1980).

Correlational IQ adoption studies typically find that the IQs of adopted-away children correlate more with their biological parents than their adoptive parents, which behavioral genetic researchers interpret as evidence in favor of genetic influences on IQ (e.g., Loehlin et al., 1989). Based on a different model and different research questions, *IQ-gain adoption studies* have shown that the adopted-away biological children of poor or working-class parents achieve a roughly 12-point IQ increase when raised in the homes of families in the upper ranges of the socioeconomic (SES) scale (Capron & Duyme, 1989; Duyme et al., 1999; Schiff et al., 1982; Schiff & Lewontin, 1986). IQ-gain study researchers interpreted their findings as evidence in favor of important environmental influences on IQ. Genetically oriented commentators usually focus on correlational studies, whereas environmentally oriented commentators usually focus on IQ-gain studies.

For both types of IQ adoption studies, potential environmental confounds include that most adopted children (1) shared a prenatal environment with their often-stressed birthmother during sensitive developmental periods, and transgenerational stress effects can be important (Nugent & McCarthy, 2015); (2) were reared for a certain period by their biological parent(s); (3) suffered a disruption of attachment bonds with the biological parent(s) who gave them up for adoption; (4) may have been placed between separation and adoption into unstable or psychologically/developmentally harmful environments, such as foster homes and orphanages; and (5) potentially share with birthparents similar physical appearance, ethnicity, culture, religion, and so on.

Another major adoption study confound, more relevant to the correlational studies, is *selective placement*. Couples wishing to adopt children are carefully screened by adoption agencies for emotional and financial stability, and children are often intentionally placed into adoptive homes matched on some of the characteristics, including SES and presumed genetic status, of their biological family background. "Fitting the home to the child," wrote behavioral genetic adoption researcher Harry Munsinger, "has been the standard practice in most adoption agencies, and this selective placement can confound genetic endowment with environmental influence to invalidate the basic logic of an adoptive study" (Munsinger, 1975, p. 627). Behavioral geneticists recognize that selective placement "could cloud the separation of nature and nurture by placing adopted-apart 'genetic' relatives into correlated environments," and that "some adoption studies show selective placement for IQ" (Plomin et al., 2013, p. 79). TRA studies, also known as "adopted-twin studies," are similarly vulnerable to selective placement bias.

Many commentators have described these and other problems and potential environmental confounds in adoption studies of IQ, psychiatric conditions, and other types of behavior (e.g., Bouchard & McGue, 2003; Burt & Simons, 2014; Cross, 1996; Joseph, 2006, 2023b; Kamin, 1974, 1981; Kamin, in Eysenck vs. Kamin, 1981; Lewontin et al., 1984; Moore, 2007; Nisbett, 2009; Plomin et al., 2013; Richardson, 2022; Richardson & Norgate, 2006; Rutter, 2006; Schiff & Lewontin, 1986; Sham & Kendler, 2008; Stoolmiller, 1999; Thomas, 2017). These problems are magnified when studies are carried out in societies containing discrimination and oppression based on ethnicity, gender, nationality, religion, language, or class.

Family Correlations

For the sibling correlations they cited, H & M relied on an analysis by Heather Chipuer, Michael Rovine, and Robert Plomin (Chipuer et al., 1990), who used a behavioral genetic *biometrical model-fitting* procedure to estimate IQ heritability based on family IQ correlations previously compiled by Bouchard and McGue (1981). The assumptions underlying model-fitting techniques have been challenged by many commentators (e.g., Goldberger, 1979; Lewontin, 1974; McGuire & Hirsch, 1977; Taylor, 1980; Zuk et al., 2012).

Nancy Segal described the behavioral genetic biometrical (ACE) model-fitting assumptions used in the non-IQ MISTRA studies as follows:

> Biometrical methods can test how well a model fits a given set of data. Model fit refers to the discrepancy between the actual data (observed values) and the results implied by the model (expected values). The simplest such models assume that shared genes underlie similarity between relatives, mating occurs at random (is not assortative), genetic effects are additive, genetic and environmental effects are independent from each other, and genetic and environmental effects combine additively.
>
> (Segal, 2012, p. 63)

Not one of the above assumptions is true for human populations (only under a "no IQ genetics" null hypothesis would the random mating assumption be true), and even leading behavioral geneticists have recognized that some model-fitting assumptions might not be valid (e.g., Eaves et al., 1989; Loehlin, 1978; Rutter, 2006). Segal did not mention that in 1989, Bouchard and his MISTRA colleague Matt McGue recognized that in relation to their study's model-fitting assumptions, "several of these assumptions are likely not to hold for cognitive abilities" (McGue & Bouchard, 1989, p. 23).

In their various models fitted to existing data, Chipuer and colleagues also assumed that reared-together relatives share their environments completely, that reared-apart relatives share none of their environments, and that MZT and DZT environments are equal (Chipuer et al., 1990, p. 13, Figure 1, Table 1). The last two assumptions are false. MZT environments are much more similar than DZT environments, and extended families potentially share the cohort and other cultural influences we described earlier.

Are Family Environments Largely Irrelevant for IQ?

Following the questionable behavioral genetic distinction between "shared" and "non-shared" environments (Plomin & Daniels, 1987), H & M believed that the IQ environmental variance is due to "unknown" influences and that "little can be traced" to the family:

> The evidence is growing that whatever variation is left over for the environment to explain (i.e., 40 percent of the total variation, if the heritability of IQ is taken to be .6), relatively little can be traced to the shared environments created by families. It is, rather, a set of environmental influences, mostly unknown at present, that are experienced by individuals as individuals. The fact that family members resemble each other in intelligence in adulthood as much as they do is very largely

explained by the genes they share rather than the family environment they shared as children. These findings suggest deep roots indeed for the cognitive stratification of society.

(H & M, 1994, p.108)

H & M's conclusion that the shared family environment accounts for only a small portion of the IQ environmental variance was largely based on twin studies, where researchers calculate the portion of variance remaining after calculating the supposed genetic variance. As we have seen, model-fitting variance-partitioning calculations, especially those based on twin research, are suspect. The mistaken claim that parents, schools, and life experiences "don't make a difference" was a major theme of Robert Plomin's book *Blueprint: How DNA Makes Us Who We Are* (Plomin, 2018, Chapter 8; see the critical reviews of *Blueprint* by Comfort, 2018a, 2018b; Joseph, 2022a; Kaufman, 2019; Richardson & Joseph, 2020). The behavioral genetic methods and theories H & M relied on could lead to the absurd conclusion that, due to a lack of environmental variation, "water has no influence on a fish's development because all fish live in water" (Kaufman & Moore, 2008, Section 2).

In their discussion of the IQ-gain adoption studies, H & M recognized that findings of a 12-point IQ increase "implicate the home environment as a factor in the development of cognitive ability" (H & M, 1994, p. 412). Similarly, after reviewing the IQ adoption study literature, Nisbett and colleagues concluded in 2012:

> We can be confident that the environmental differences that are associated with social class have a large effect on IQ. We know this because adopted children typically score 12 points or more higher than comparison children (e.g., siblings left with birth parents or children adopted by lower SES parents), and adoption typically moves children from lower to higher SES homes.

(Nisbett et al., 2012, p. 136)

H & M's and Nisbett and colleagues' evaluations of the IQ-gain adoption studies shared common features, which did not square with H & M's earlier conclusion that little IQ variability "can be traced to the shared environments created by families."

The "Flynn Effect"

H & M (1994, p. 307) addressed what they called the "Flynn effect." Moral philosopher and IQ researcher James Flynn showed in the 1980s and 1990s that IQ scores worldwide had been increasing by about three points per decade (0.30 points per year; Flynn, 1984, 1999), including supposedly "*g*-loaded, culture-fair" tests such as Raven's Progressive Matrices (increasing 0.50 points per year worldwide; Flynn, 2009). Due to these gains, at one point Flynn concluded that "psychologists should stop saying that IQ tests measure intelligence. They should say that IQ tests measure abstract problem-solving ability" (Flynn, 1987, p. 188).

To maintain a mean score of 100 and a bell-shaped score distribution, IQ test creators (psychometrists) periodically revise their tests to make them more difficult. Similarly, to keep golf scores comparable over time, professional golf tournaments make courses longer and more difficult (such as by narrowing fairways and adding lakes, trees, and sand traps) as players and equipment get better over time (for one example, see Whitten & O'Riley, 2023). Professional golfers need much greater skills to shoot a round of 70 in 2024 than were needed in 1964, and IQ test takers need much greater skills to score 110 on an IQ test in 2024 than were needed in 1964.

Psychometrists were aware of rising IQ test scores, but it was left to a scholar outside the field to bring the issue to the world's attention. Flynn documented "massive IQ gains over time [which] revealed that the present generation has a huge IQ advantage over the previous generation" (Flynn, 1999, p. 5). Genetic theories cannot explain these massive IQ gains over time, but gains can be explained by environmental factors such as improved healthcare, better teaching methods, greater complexity in society, improved problem-solving skills, increased spending on education, and technological advances. "We have every reason to believe," wrote Nisbett (2009, p. 50), "that the culture is producing superior executive-control functions than were found for earlier eras," which has led to higher scores on fluid-intelligence tasks.

An additional explanation lies in the demographic swelling of the U.S. and European middle classes over the period in question, accompanied by greater use in families of test-relevant cultural learning such as text and number literacy and an improved sense of place and self-confidence in the power structure (Richardson & Norgate, 2014). These factors are relevant to MZT twins, who, much less often than DZT twins, strive for distinct identities, and to children adopted into middle-class homes.

Implications for Twins Reared Apart (TRA) Studies

MZA pairs found in the MISTRA and other TRA studies were born at the same time and usually grew up in the same country. Therefore, their educational, learning, and cognitive skills development occurred simultaneously in similar Flynn-effect "huge IQ advantage or disadvantage" environments, constituting yet another major TRA study environmental confound.

Bouchard acknowledged that the Flynn effect is real, and that "environmental influences" accounted for IQ gains over time (Bouchard & McGue, 2003, p. 17). Yet he overlooked the following crucial theoretical point: Using the same IQ test version and due solely to the Flynn effect, if studied MZA pairs had been born two generations apart, the younger twins would have scored about 15 points higher than their genetically identical co-twins (Joseph, 2023a. The MISTRA researchers adjusted their correlations for "age and sex effects" [McGue & Bouchard, 1984], but the adjustments were inadequate and did not account for Flynn's findings). In addition to the shared birth-cohort environmental influences we described earlier, Flynn effect IQ gains suggest that a sizable chunk of the MZA IQ correlations Herrnstein and Murray cited as the "most unambiguous" evidence in favor of strong genetic influences on IQ can be explained not by the genes twins share, but by the *birth date* they share

"Significant Flynn effects" were also found in Sweden, where the SATSA was performed (Rönnlund et al., 2013, p. 19). The SATSA researchers based their IQ heritability conclusions on twin pairs only partially reared apart, and on what we have shown are problematic model-fitting techniques and the unsupported assumption that the "MZA correlation directly estimates heritability" (Pedersen et al., 1992).

In his book *What is Intelligence*, Flynn tried to "solve the paradox of how environment could appear so feeble in the twin studies and yet so potent in IQ gains over time" (Flynn, 2009, p. 83). In our view, the "paradox" Flynn described is easily solved. Massive IQ gains over time are real, and twin study IQ genetic (heritability) findings constitute a century-long scientific illusion that Flynn was unable to recognize (Joseph, 2023a). In Part III, we return to the Flynn effect in the context of genetic ethnic differences in IQ claims.

Conclusions: No Valid Evidence in Support of Within-Group Genetic Influences on IQ

We have now reviewed the evidence H & M cited in support of substantial within-group genetic influences on IQ and found it wanting. Their evidence consisted of TRA studies and, to a lesser extent, family correlations, reared-together twin studies, and adoption studies. We showed that the genetic findings reported in these studies do not hold up under critical examination due to environmental confounding, a reliance on questionable or false assumptions and concepts, the use of questionable research practices, and other problem areas. In the language of psychometrics and behavioral genetics, H & M presented no valid evidence in favor of above-zero within-group IQ heritability.

Bouchard (2023, p. 139) and other behavioral geneticists have argued that although a particular method or study may contain flaws, the "converging evidence" from the behavioral genetic family, twin, and adoption studies cited by H & M and others points to substantial IQ heritability. But if each study or method fails, the entire argument fails. Bad science combined does not add up to good science. As the esteemed developmental psychologist Richard Lerner once concluded, "Behavior genetics is really like the story of the emperor's new clothes. Despite the positive regard some researchers hold for this area, there is actually nothing there" (Lerner, 2004, p. 10).

Part III: Evidence Cited in *The Bell Curve* in Favor of Genetic Ethnic (Racial) and Class Differences in IQ

We now arrive at *The Bell Curve*'s most disputed claim—that because IQ is supposedly substantially heritable within groups, some portion of the mean IQ score gap *between* groups, such as between social classes and ethnic groups, is caused by genetic factors. H & M wrote, "Ethnic differences in measured cognitive ability have been found since intelligence tests were invented" (p. 270). Indeed, leading U.S. psychologists of the 1920s "found" that a sizable portion of Jewish, Eastern European, and Southern European immigrants were genetically "feeble-minded" (Chase, 1980; Kamin, 1974; Tucker, 1994b) and that their intelligence tests demonstrated, as a top psychologist of that time put it, the "definite…superiority of the Nordic [race] type" (Brigham, 1923, p. 182). We now understand that such claims were based on elitism, prejudice, U.S. political leaders' desire to keep "non-Nordic" potentially politically radical "foreigners" out of the United States in the aftermath of the Russian Revolution, and the way the tests were constructed to match individual and group difference expectations for what was pre-conceived as "intelligence." These claims were backed by what most observers now understand to be junk science, which leads to the question of whether claims about genetic "ethnic differences" in IQ found in *TBC* were merely an updated version of such "science."

IQ hereditarians believe that establishing substantial within-group IQ heritability is a prerequisite for claiming between-group genetic IQ differences. Because H & M provided no valid evidence in favor of within-group IQ genetic influences ("substantial" or otherwise), **their claim that genes for cognitive ability contribute to ethnic (racial) and class differences in IQ was invalid on its face.** Part III, therefore, is intended mainly for readers who remain convinced that H & M provided solid evidence in favor of substantial within-group IQ genetic influences (heritability).

Ethnic Differences and "Race"

In several places, H & M strongly implied that some portion of the supposed U.S. White-Black IQ mean difference "gap" of 15 IQ points (hereafter "W-B") can be attributed to genetics. As they wrote:

> If the reader is now convinced that either the genetic or environmental explanation [for W-B] has won out to the exclusion of the other, we have not done a sufficiently good job of presenting one side or the other. It seems highly likely to us that both genes and the environment have something to do with racial differences. What might the mix be? We are resolutely agnostic on that issue; as far as we can determine, the evidence does not yet justify an estimate.
>
> (H & M, 1994, p. 311)

The above statement is sometimes misinterpreted as H & M saying they were "agnostic" about whether genetic influences explain a portion of W-B. However, they were only agnostic about how much ("the mix"), not whether genes for cognitive ability contribute to W-B. Murray subsequently tried to absolve *TBC* of genetic ethnic-differences-in-IQ accusations by referencing the "agnosticism" clause in *TBC*, but then betrayed himself in several places (see Tucker, 2024, Chapter 1). Elsewhere in the book, H & M engaged in what Harvard psychologist Howard Gardner called "scholarly brinksmanship," meaning they encouraged their readers to "draw the strongest conclusions, while allowing the authors to disavow this intention" (Gardner, 1995, p. 64).

Many critics of *TBC* accepted that W-B exists while arguing that environmental factors explain the gap entirely. This view represents a misunderstanding of the biology of diversity and human history. There are no human "races" in the sense usually intended and promoted in *TBC*. Humans share at least 99.5% of their genes. Genes that vary are those peripheral to the most crucial functions, such as human mental ability. For historical/migration reasons, those that vary tend to cluster as visible but superficial group differences, sometimes associated with cultural characteristics (Gusev, 2023; Nature, 2023). Such peripheral clusters are often taken as markers for innate (but invisible) differences, such as with intelligence, and are used to rationalize the social-inequality view implicit in *TBC* and Murray's later book *Human Diversity* (Duncan et al., 2024; Marks, 2017; Norton et al., 2019; Rutherford, 2021).

In 2022, a group of scholars published an article in *Science* where they concluded, "After a long history of race being treated as a biological variable, there is now broad agreement that racial classifications are a product of historically contingent social, economic, and political processes" (Lewis et al., 2022, p. 250). Genetic racial difference claims in *TBC* are predicated on pseudoscientific categories of race. Modern genetics shows that the term "race" should be abolished as a category or descriptor of humans. Still, the lingering influence of *TBC* contributes to its durability.

IQ, Gender, and Ethnicity

"It is claimed," wrote Jensen, "that the psychometrist can make up a test that will yield any kind of score distribution he pleases. This is roughly true, but some types of distributions are much easier to obtain than others" (Jensen, 1980, p. 71). Although Jensen believed that IQ ethnic differences are at least partly explained by genetics, he implied that psychometrists *could* create tests in which Blacks and Whites scored equally, or even tests in which Blacks outscore Whites (Matarazzo & Wiens, 1977; Williams, 1972). However, most choose not to.

When psychometrists assume cognitive ability equality between groups, they create tests to reflect this assumption. For example, males and females score roughly the same on IQ tests because test creators design the tests to produce this result based on their assumption that males and females are equally intelligent. Based on other assumptions, test creators allow ethnic differences to remain. Test items that are obviously biased are only one type of bias contributing to W-B. Another aspect is the test creators' selection of items that maintain observed differences between ethnic groups, even if these items are not obviously ethnically or class-biased (Mensh & Mensh, 1991). According to evolutionary biologist Richard Lewontin, Leon Kamin, and Steven Rose in *Not in Our Genes*, from the beginning, IQ

> test items that differentiated boys from girls… were removed, since the tests were not meant to make that distinction; differences between social classes, or between ethnic groups or races, however, have not been massaged away, precisely because it is these differences that the tests are *meant* to measure.
>
> (Lewontin et al., 1984, p. 89, emphasis in original)

And as one of us wrote,

> In effect, then, Galton's aim, and that of his followers, became simply an attempt to reproduce an existing set of ranks (social class) in another, the test scores, and pretend that the

latter is a measure of something else. This is, and remains, the fundamental strategy of the intelligence-testing movement.

(Richardson, 2000, p. 27)

Terman and Wechsler

In test pioneer psychologist Lewis Terman's 1937 revision of the Stanford-Binet IQ test, he wrote that a "few tests in the trial batteries which yielded [the] largest sex differences were early eliminated as probably unfair" (Terman & Merrill, 1937, p. 34). Because he assumed that males and females are equally intelligent, to avoid "unfairness," Terman created a test whose results reflected this assumption.

On the other hand, for Terman, it was fair and even axiomatic that ethnic groups and "races" differed in innate intelligence, with Whites on top. He wrote in *The Measurement of Intelligence* (Terman, 1916) of "the dullness" of "Spanish-Indian and Mexican families" and "negroes" (p. 91). He predicted that after "experimental methods" were performed, "there will be discovered enormously significant racial differences in general intelligence, differences that cannot be wiped out by any scheme of mental culture" (p. 92).

Psychologist David Wechsler, who developed the most widely used IQ tests, followed Terman in deciding to eliminate sex (but not ethnic) differences in his IQ tests (Wechsler, 1944). Wechsler described the question of sex differences as follows:

> In trying to arrive at an answer as to whether there are sex differences in intelligence much depends upon how one defines intelligence, and on the practical side, on the types of tests one uses in measuring it. The contemporary approach, contrary to the historical point of view, adopts a sort of null hypothesis. Unfortunately *this procedure turns out to be a circular affair* since the nature of the tests selected can prejudice or determine in advance what the findings will be.
>
> (Wechsler, 1958, p. 144, emphasis added)

As Wechsler acknowledged, his "null hypothesis" for male-female differences in intelligence was that the sexes are equal, and he designed tests "circularly" to produce this result. However, following Terman, he chose to reject a similar null hypothesis for U.S. ethnic groups due to his belief that there is "little doubt" that "ethnic and cultural differences in intelligence" exist (Wechsler, 1958, p. 90), and he designed tests circularly to produce *this* result.

Pioneers of IQ testing such as Wechsler and Terman could have created separate norms for groups experiencing vastly different social and educational environments (such as oppressed Blacks in the U.S. Jim Crow South). Wechsler, however, wrote in 1944 that it was not "possible to do this at present," and he standardized his test on White test-takers only (p. 107). In a published 1972 statement in support of Herrnstein's and Jensen's IQ genetics publications (though not specifically in support of the latter's ethnic differences claims), Wechsler and others endorsed the view that "hereditary influences" on human ability and behavior "are very strong" (Page, 1972, p. 660). The statement, which compared IQ hereditarians to Galileo, Darwin, and Einstein, cited no specific evidence or research methods in support of the supposed "great role played by heredity in human behavior" (p. 660).

According to Jensen (1980, p. 623), "The practice of eliminating and counterbalancing items to minimize sex differences is based on the assumption that the sexes do not really differ in general intelligence." What Jensen failed to articulate was the assumption upon which ethnic and class differences are allowed to remain, which could be stated, "Test creator's failure to remove and counterbalance items to eliminate ethnic and class differences is based on their assumption that ethnicities and classes really do differ in general intelligence." As described earlier, test items are selected according to how well they correlate with school performance, which is closely related to class and cultural background.

If IQ tests are viewed as the product of wealthy and powerful White men who encouraged

Terman, Wechsler, and others to create tests to "scientifically" validate their positions in society and the world (domestically and through colonialism/neo-colonialism), their bloodlines and positions required (1) the assumption of female-male genetic IQ equality (the wives, mothers, and daughters of the wealthy and powerful), (2) the assumption of the genetic IQ inferiority of the working class and people of color (the employees of the wealthy and powerful, and the lower classes they share a country with), and (3) the genetic IQ inferiority of the Global South (the colonial and neo-colonial "subjects" of the wealthy and powerful). Regarding the latter, Richard Lynn and other IQ hereditarians created a "National IQ" dataset, accompanied by maps frequently appearing in the social media, showing supposedly low national IQ scores in Africa and other areas of the Global South (e.g., Lynn & Becker, 2019). Several analysts have demonstrated that these datasets are based on highly flawed analyses (e.g., Kamin, 2006; Sear, 2022), with population and health professor Rebecca Sear concluding, "No future research should use this dataset, and published papers which have used the dataset should be corrected or retracted" (2022, p. 11).

IQ hereditarians often say the existence of W-B is undisputed and that the debate centers only on the causes of the gap (e.g., Rushton & Jensen, 2005; Warne, 2020; Winegard, 2023). However, the existence of the gap *is* disputed by those who argue that race is invalid as a biological variable and that IQ tests are designed to reproduce assumed ethnic and class differences in intelligence.

The Flynn Effect in the Ethnic-Differences Context

Claims that IQ tests such as Raven's Progressive Matrices are "culture-free" and are "among the two or three tests having the highest *g* loadings" (Jensen, 1998, p. 38) are refuted by the Flynn effect. Raven scores are "absolutely drenched in culture"; otherwise, the observed five-IQ-points-per-decade Raven gains would not have occurred (Nisbett, 2009, pp. 47-48).

James Flynn and his colleague Wiliam Dickens found that Blacks gained 4 to 7 IQ points on non-Hispanic Whites between 1972 and 2002 (Dickens & Flynn, 2006). The IQs of African Americans may have lagged behind White IQs in 1994, but they were not inferior. As Flynn wrote:

> Since 1945, Blacks have gained at an average rate of over 0.30 points per year and gained a total of 16 points over 50 years. So the Blacks of 1995 should have matched the mean IQ of the Whites of 1945.
>
> (Flynn, 1998, p. 40)

Flynn later calculated that U.S. Blacks of 2002 would have outscored U.S. Whites of 1947-48 by 4.3 IQ points (Full-Scale Black IQ 104.3 versus White IQ 100; Flynn, 2008, pp. 72-73). To confirm his calculation, Flynn proposed selecting a representative sample of contemporary U.S. Blacks and giving them the old IQ test. "Frantic appeals," he discovered, "have not located anyone interested" in such a project (p. 73).

Imagine a time machine transporting 1,000 randomly selected 2002 U.S. Blacks into a large auditorium with 1,000 randomly selected 1947-48 U.S. Whites. Both groups are tested using 1948-era IQ tests. Based on Flynn's findings, one would expect the Black group's mean IQ to be 4.3 points *higher* than the White group's mean IQ. (Using the same time machine and continuing our earlier golf analogy, champion golfers of 1964 probably wouldn't make the cut in 2024 tournaments.) So much for claims that IQ tests measure innate intelligence, and that Whites are more intelligent than Blacks.

H & M failed to recognize how the Flynn effect demolished their arguments favoring genetic ethnic differences, genetic class differences, and "dysgenic trends" (p. 355). IQ hereditarian theories predict that national IQ scores will remain static or more likely decline, as opposed to the undisputed "massive" gains Flynn documented. To some degree, H & M were aware of how this contradiction played out in the ethnic-differences context:

> The Flynn effect. Indirect support for the proposition that the observed B/W [Black-White] difference could be the result of environmental factors is provided by the worldwide phenomenon of rising test scores.
>
> (H & M, 1994, p. 307)

Furthermore:

> On the average, whites today may differ in IQ from whites, say, two generations ago as much as whites today differ from blacks today. Given their size and speed, the shifts in time necessarily have been due more to changes in the environment than to changes in the genes.
>
> (p. 308)

The Flynn effect sent a wrecking ball into the already crumbling within-group IQ genetic evidence and supplied the final nails in the coffin of *TBC* genetic ethnic and class differences in cognitive ability claims. Psychologist Ulric Neisser was the head of the APA task force on the state of intelligence research, which was created during the *Bell Curve* controversy (Neisser et al., 1996). "Herrnstein and Murray were aware of these [Flynn-effect] gains," he wrote, "but gave them short shrift—an understandable decision, considering how profoundly they undermine many of the claims of *The Bell Curve*" (Neisser, 1998, p. 4).

Herrnstein and Murray Refuted Their Own Genetic Ethnic-Group Differences Argument

Ironically, H & M emphasized that the existence of between-group mean IQ score differences proves nothing about possible between-group genetic differences, even if IQ is "genetically transmitted in individuals":

> A good place to start is by correcting a common confusion about the role of genes in individuals and in groups. As we discussed in Chapter 4, scholars accept that IQ is substantially heritable, somewhere between 40 and 80 percent, meaning that much of the observed variation in IQ is genetic. And yet this information tells us nothing for sure about the origin of the differences between races in measured intelligence. This point is so basic, and so commonly misunderstood, that it deserves emphasis: *That a trait is genetically transmitted in individuals does not mean that group differences in that trait are also genetic in origin.* Anyone who doubts this assertion may take two handfuls of genetically identical seed corn and plant one handful in Iowa, the other in the Mojave Desert, and let nature (i.e., the environment) take its course. The seeds will grow in Iowa, not in the Mojave, and the result will have nothing to do with genetic differences. The environment for American blacks has been closer to the Mojave and the environment for American whites has been closer to Iowa.
>
> (H & M, 1994, p. 298, emphasis in original)

H & M acknowledged that Richard Lewontin, a leading critic of IQ hereditarianism (e.g., Lewontin et al., 1984), had proposed a version of the "sack of seeds" thought experiment 24 years earlier (Lewontin, 1970). Computational biologists Joshua Schraiber and Michael Edge reassessed and confirmed Lewontin's argument that within-group heritability implies nothing about between-group heritability. They showed that "even if the between-group heritability were estimable, it is consistent with potentially infinitely many configurations of genetic differences among populations and cannot be used to infer the direction(s) of group differences in genetic contributions to a trait" (Schraiber & Edge, 2023, p. 13). Schraiber and Edge concluded, "In line with Lewontin's intuition, the heritabilities of phenotypic differences between groups are not constrained by heritability within groups: perfect knowledge of within-group heritability provides no information about between-group heritability" (p. 2; see also Roseman & Bird, 2023).

Quickly forgetting their "Iowa/Mohave desert" example, where they recognized that White environments are like Iowa and Black environments are like the Mohave desert, H & M produced several pages of statistical speculation supposedly showing, under the assumption "that IQ is 60 percent heritable," that a purely environmental explanation of W-B is "implausible" (H & M, pp. 298-303). They should have stuck with their Iowa/Mohave desert example, showing that the environment can entirely and easily explain group differences. "Environmental explanations may successfully circumvent" seemingly "implausible" non-genetic explanations of B-W, they wrote, "but the explanations have to be formulated rather than simply assumed" (p. 299). The Flynn effect provides an excellent formulation because it shows—by itself—that environmental influences on IQ test scores are "massive."

Transracial Adoption Studies

In their "Ethnic Differences in Cognitive Ability" chapter, H & M discussed "transracial adoption studies," the most famous of which was conducted by researchers at the University of Minnesota. The "Minnesota Transracial Adoption Study" (Scarr & Weinberg, 1976; not to be confused with the separate Minnesota Study of Twins Reared Apart) assessed the IQ scores of 130 Black/interracial children who had been adopted by "advantaged white families," in addition to IQ scores for other adoptees and the adoptive parents' biological children. H & M believed that although Scarr and Weinberg's 1976 results were inconclusive regarding the source of W-B, the results of their 1992 follow-up report (Weinberg et al., 1992) did "not favor the no-genetics case." They concluded that a "mixed gene and environmental source" of W-B "seems to us the most plausible conclusion" (H & M, 1994, p. 310).

We discussed major IQ adoption study problems in Part II, and some critically minded authors specifically addressed issues in the transracial adoption studies (e.g., Cross, 1996; Kamin, 1981; Thomas, 2017). Especially in the replication crisis context, the transracial studies that H & M examined provide no valid evidence in favor of genetic influences on either within- or between-group IQ score differences.

Caste-like Minority Groups

Anthropologist John Ogbu (1978) documented the fate of caste-like minority groups such as the Māori in New Zealand, the Burakumin of Japan, Aborigines in Australia, low castes in India, and the Irish in Great Britain. He believed that the lower performance of minority children was the result of their status as members of an oppressed group.

Koreans in Japan provide an additional example. Korean immigrants, whose nation was a colony of Japan for 35 years, formed an oppressed lower-caste group in Japan (Moon, 2010). They also scored lower than members of the dominant culture on IQ tests. "Koreans," wrote Fischer and colleagues in 1996,

> who are of the same "racial" stock as Japanese and who in the United States do about as well academically as Americans of Japanese origin (that is, above average), are distinctly "dumb" in Japan. The explanation cannot be racial, nor even cultural in any simple way.
>
> (Fischer et al., 1996, p. 172)

Fischer and colleagues' thesis was as follows: "A racial or ethnic group's position in society determines its measured intelligence rather than vice versa" (1996, p. 173).

The psychological impact of oppression (a word rarely found in the behavioral genetic and IQ hereditarian literature) gives rise to negative self-views, can depress earlier test-relevant learning, motivation, and concentration, and can create increased anxiety in test situations. Brummelman and Sedikides (2023) concluded that children from low-SES backgrounds, apart from their actual abilities and achievements, perceive themselves as less intelligent, less able to grow their intelligence, less deserving, and less worthy. "Subjective Socioeconomic Status" research has identified SES-related adverse effects on views of the self (Kraft et al., 2022). Family stress can affect a child's health and vitality even before birth (Skinner, 2014).

Ireland won hard-fought independence from the United Kingdom in the lower counties in 1921. Irish neuroscientist Kevin Mitchell, while otherwise believing that human behavioral differences are "innate," described the rise in Irish IQ (Mitchell, 2018, p. 167). Though disputed by some IQ hereditarians (e.g., Warne, 2022), Mitchell explained that the average IQ in Ireland was 85 in the 1970s, compared with the UK average of 100. "This was taken that the Irish were constitutionally stupid," he wrote, "not just ignorant and poorly educated, but irredeemably simple." IQ scores in Ireland are now on par with UK scores, which Mitchell attributed to improved education and better environmental circumstances that allowed the Irish "potential to flourish."

The same potential to flourish exists in all human skin shades, ethnicities, classes, castes, genders, regions, societies, countries, latitudes, and continents. It is only necessary to unlock it by eliminating racism, neo-colonialism, and other forms of oppression. Clearly, the United States is not the only country in which oppressed ethnic groups score lower on standardized IQ tests. They score lower as an aspect of their oppression, not because they are less intelligent.

Part IV: General Conclusions

The Bell Curve was published in 1994 amid great controversy due to its themes of IQ hereditarianism and genetic ethnic and class differences in IQ, and what H & M saw as the social and political policy implications that followed from these positions. Although widely criticized, the book has been cited favorably by IQ hereditarians and others in publications (e.g., Bouchard, 1995; Cofnas, 2020; Jensen, 1998; Rushton, 2000, featuring a recommendation of Rushton's book by Bouchard, p. 2; Sesardic, 2005; Warne, 2020) and in the social media. The damage created by this massively flawed book goes well beyond academia, as its implicit message extends worldwide to politicians, the media, institutional and military leaders, educators, and so on.

As mentioned in Part I, we focused on the supposed IQ genetic evidence cited in *TBC*. The larger question of whether or how heredity influences human intelligence was beyond the scope of our analysis. We showed that H & M's IQ-hereditarian argument was based on disputed concepts and methods, including IQ, heritability, race, twin studies, general intelligence, and variance-partitioning model-fitting techniques. H & M cited no IQ gene discoveries or associations at the molecular genetic level. The polygenic score studies Murray later promoted in *Human Diversity* are subject to low polygenic scores (weak associations) as well as environmental confounding and other problems described by the scholars we cited in Part I. Like many others before us, we conclude that Herrnstein and Murray presented no valid evidence in support of genetic causes of between-ethnic-group or social-class IQ score differences. And like few others before us, we also conclude that they produced no valid evidence that genes influence IQ score differences *within* groups.

It is not the task of critics to establish "true" twin-study-based or molecular-genetic-based IQ heritability estimates, or to demonstrate that they are zero. The only valid use of a heritability estimate is to predict the results of a selective breeding program under controlled conditions (Rose, 1997; Wahlsten, 1990), and heritability estimates should have been abandoned in the social and behavioral sciences decades ago. For people critical of genetic interpretations of twin study results, of IQ testing, and of heritability estimation, the "heritability of IQ" concept is not valid—even under the assumption that genetic influences play a role in causing individual differences in IQ scores. The IQ-hereditarian argument is as false today as it was in the 1920s, in 1969, and in 1994.

As many earlier critics noted, *TBC* continued the long tradition of using questionable or junk science to promote conservative and racial political agendas, accompanied by disingenuous claims that "we are just following the science." The book "was clearly a social policy document that aimed to reshape society and eliminate social programs" (Winston, 2024, pp. viii-ix).

In light of the Behavior Genetics Association's commendable (2021) "Diversity and Inclusion Action Plan" we mentioned in Part I, in addition to previous personal attacks on critics of scientific racism (including one by former BGA president Sandra Scarr [1998, p. 231], who called Leon Kamin, Marcus Feldman, and Stephen J. Gould "politically driven liars" and "despicable" "thugs with pens instead of microphones"), the BGA should consider officially honoring these and other opponents of scientific racism.

In 1976, long before the replication crisis began, Richard Lewontin wrote that in *The Science and Politics of I.Q.*, Leon Kamin discovered in IQ-genetic research "a pattern of shoddiness, carelessness, miserable experimental design, misreporting, and misrepresentation amounting to a major scandal" (Lewontin, 1976, p. 97). Four years later, Howard Taylor (1980, p. 7) described "The IQ game," meaning IQ genetic researchers' "use of assumptions that are implausible as well as arbitrary to arrive at some numerical value for the genetic heritability of human IQ scores on the grounds that no heritability calculations could be made without benefit of such assumptions." As we have attempted to show, *The Bell Curve* continued both dreadful traditions and helped disguise politically and even militarily imposed social, political, economic, and international inequality as biological inequality.

Abbreviations

16PF = Sixteen Personality Factor Questionnaire;
AFQT = Armed Forces Qualification Test;
APA = American Psychological Association;
BGA = Behavior Genetics Association;
DZA = dizygotic (fraternal) twins reared apart;
DZT = dizygotic twins reared together;
EA = educational attainment;
EEA = equal environment assumption;
g = general intelligence;
GWAS = genome-wide association study;
H & M = Herrnstein & Murray;
IQ = intelligence quotient score(s);
MISTRA = Minnesota Study of Twins Reared Apart;
MZA = monozygotic (identical) twins reared apart;
MZT = monozygotic twins reared together;
NLSY = National Longitudinal Survey of Youth;
PGS = polygenic score (study);
QRP = questionable research practice;
SAT = Scholastic Aptitude Test;
SATSA = Swedish Adoption/Twin Study of Aging;
SES = socioeconomic status;
SNP = single nucleotide polymorphism;
TBC = The Bell Curve;
TRA = twins reared apart (study);
WAIS = Wechsler Adult Intelligence Scale;
W-B = U.S. White-Black (African American) mean IQ score difference (gap).

References

Alford, J. R., Funk, C. L., & Hibbing, J. R. (2005). Are political orientations genetically transmitted? *American Political Science Review, 99*(2), 153–167. https://doi.org/10.1017/S0003055405051579

Barba, V. (2023, January 15). Twins separated at birth find each other and discover they've led identical lives. *The*

Mirror. https://www.mirror.co.uk/news/us-news/twins-separated-birth-find-each-28960150

Barnes, J. C., Wright, J. P., Boutwell, B. B., Schwartz, J. A., Connolly, E. J., Nedelec, J. L., & Beaver, K. M. (2014). Demonstrating the validity of twin research in criminology. *Criminology, 52*, 588–626. https://doi.org/10.1111/1745-9125.12049

Baum, W. M. (2002). The Harvard Pigeon Lab under Herrnstein. *Journal of the Experimental Analysis of Behavior, 77*(3), 347–355. https://doi.org/10.1901/jeab.2002.77-347

Baverstock, K. (2019). Polygenic scores: Are they a public health hazard? *Progress in Biophysics and Molecular Biology, 149*, 4–8.

Behavior Genetics Association (BGA). (2021, April). *BGA diversity and inclusion plan*. https://www.bga.org/content.aspx?page_id=22&club_id=971921&module_id=567723#:~:text=BGA%20members%20reflect%20the%20diversity,and%20belonging%20through%20their%20actions.

Blanca, M. J., Arnau, J., López-Montiel, D., Bono, R., & Bendayan, R. (2013). Skewness and kurtosis in real data samples. *Methodology: European Journal of Research Methods for the Behavioral and Social Sciences, 9*(2), 78–84. https://doi.org/10.1027/1614-2241/a000057

Block N. (1995). How heritability misleads about race. *Cognition, 56*(2), 99–128. https://doi.org/10.1016/0010-0277(95)00678-r

Boomsma, D. I., Willemsen, G., Dolan, C. V., Hawkley, L. C., & Cacioppo, J. T. (2005). Genetic and environmental contributions to loneliness in adults: The Netherlands Twin Register study. *Behavior Genetics, 35*(6), 745–752. https://doi.org/10.1007/s10519-005-6040-8

Bouchard, T. J., Jr. (1976). Genetic factors in intelligence. In A. R. Kaplan (Ed.), *Human behavior genetics* (pp. 164–197). Charles C. Thomas.

Bouchard, T. J., Jr. (1984). Twins reared together and apart: What they tell us about human diversity. In S. W. Fox (Ed.), *Individuality and determinism: Chemical and biological bases* (pp. 147–184). New York: Plenum Press.

Bouchard, T. J., Jr. (1995). Breaking the last taboo. [Review of the book *The bell curve*, by R. J. Herrnstein & C. Murray]. *Contemporary Psychology, 40*, 415–421. https://doi.org/10.1037/003626

Bouchard, T. J., Jr. (1997). IQ similarity in twins reared apart: Findings and responses to critics. In R. Sternberg & E. Grigorenko (Eds.), *Intelligence, heredity, and environment* (pp. 126–160). Cambridge University Press.

Bouchard, T. J., Jr. (1999). [Review of the book *Dysgenics: Genetic deterioration in modern populations*, by Richard Lynn]. *American Journal of Human Biology, 11*(2), 272–274.

Bouchard, T. J., Jr. (2023). The garden of forking paths; An evaluation of Joseph's 'A reevaluation of the 1990 "Minnesota Study of Twins Reared Apart" IQ Study.' *Twin Research and Human Genetics, 26*(2), 133–142. https://doi.org/10.1017/thg.2023.19

Bouchard, T. J., Jr., Lykken, D. T., McGue, M., Segal, N. L., & Tellegen, A. (1990). Sources of human psychological differences: The Minnesota Study of Twins Reared Apart. *Science, 250*(4978), 223–228. https://doi.org/10.1126/science.2218526

Bouchard, T. J., Jr., Lykken, D. T., Segal, N. L., & Wilcox, K. J. (1986). Development in twins reared apart: A test of the chronogenetic hypothesis. In A. Demirjian (Ed.), *Human growth: A multidisciplinary review* (pp. 299–310). Taylor & Francis.

Bouchard, T. J., Jr., & McGue, M. (1981). Familial studies of intelligence: A review. *Science, 212*(4498), 1055–1059. https://doi.org/10.1126/science.7195071

Bouchard, T. J., Jr., & McGue, M. (2003). Genetic and environmental influences on human psychological differences. *Journal of Neurobiology, 54*(1), 4–45. https://doi.org/10.1002/neu.10160

Bouchard, T. J., Jr., Segal, N. L., & Lykken, D. T. (1990). Genetic and environmental influences on special mental abilities in a sample of twins reared apart. *Acta Geneticae Medicae et Gemellologiae (Twin Research), 39*(2),193–206. https://doi.org/10.1017/s0001566000005420

Brigham, C. C. (1923). *A study of American intelligence*. Princeton University Press.

Brummelman, E., & Sedikides, C. (2023). Unequal selves in the classroom: Nature, origins, and consequences of socioeconomic disparities in children's self-views. *Developmental Psychology, 59*(11), 1962–1987. https://doi.org/10.1037/dev0001599

Burt, C. H. (2023). Challenging the utility of polygenic scores for social science: Environmental confounding, downward causation, and unknown biology.

Behavioral and Brain Sciences, 1–36. https://doi.org/10.1017/S0140525X22001145

Burt, C. H., & Simons, R. L. (2014). Pulling back the curtain on heritability studies: Biosocial criminology in the postgenomic era. *Criminology*, 52(2), 223–262. https://doi.org/10.1111/1745-9125.12036

Burt, C. L. (1966). The genetic determination of differences in intelligence: A study of monozygotic twins reared together and apart. *British Journal of Psychology*, 57(1), 137–153. https://doi.org/10.1111/j.2044-8295.1966.tb01014.x

Cantor, P., Lerner, R. M., Pittman, K. J., Chase, P. A., & Gomperts, N. (2021). *Whole-child development, learning, and thriving*. Cambridge University Press.

Capron, C., & Duyme, M. (1989). Assessment of effects of socio-economic status on IQ in a full cross-fostering study. *Nature*, 340(6234), 552–554. https://doi.org/10.1038/340552a0

Carey, B. (2011, November 2). Fraud case seen as a red flag for psychology research. *New York Times*. https://www.nytimes.com/2011/11/03/health/research/noted-dutch-psychologist-stapel-accused-of-research-fraud.html?

Cassou, B., Schiff, M., & Stewart, J. (1980). Génétique et schizophrénie: Réévaluation d'un consensus [Genetics and schizophrenia: Reevaluation of a consensus]. *Psychiatrie de l'Enfant*, 23(1), 87–201.

Cattell, R. B. (1937). *The fight for our national intelligence*. P. S. King & Son.

Chabris, C. F., Hebert, B. M., Benjamin, D. J., Beauchamp, J., Cesarini, D., van der Loos, M., Johannesson, M., Magnusson, P. K., Lichtenstein, P., Atwood, C. S., Freese, J., Hauser, T. S., Hauser, R. M., Christakis, N., & Laibson, D. (2012). Most reported genetic associations with general intelligence are probably false positives. *Psychological Science*, 23(11), 1314–1323. https://doi.org/10.1177/0956797611435528

Chambers, C. (2017). *The seven deadly sins of psychology: A manifesto for reforming the culture of scientific practice*. Princeton University Press.

Charney, E. (2012). Behavior genetics and postgenomics. *Behavioral and Brain Sciences*, 35(5), 331–358. https://doi.org/10.1017/S0140525X11002226

Charney, E. (2013). Nature and nurture. [Review of the Book *Man is by nature a political animal*, by P. Hatemi & Rose McDermott (Eds.)], *Perspectives on Politics*, 11, 558–561. doi:10.1017/S1537592713000893

Charney, E. (2022). The "golden age" of behavior genetics? *Perspectives on Psychological Science*, 17(4), 1188–1210. https://doi.org/10.1177/17456916211041602

Chase, A. (1980). *The legacy of Malthus: The social costs of the new scientific racism*. University of Illinois Press. (Originally published in 1977)

Chen, T. T., Kim, J., Lam, M., Chuang, Y. F., Chiu, Y. L., Lin, S. C., Jung, S. H., Kim, B., Kim, S., Cho, C., Shim, I., Park, S., Ahn, Y., Okbay, A., Jang, H., Kim, H. J., Seo, S. W., Park, W. Y., Ge, T., Huang, H., … Won, H. H. (2024). Shared genetic architectures of educational attainment in East Asian and European populations. *Nature Human Behaviour*. https://doi.org/10.1038/s41562-023-01781-9

Chipuer, H. M., Rovine, M. J., & Plomin, R. (1990). LISREL modeling: Genetic and environmental influences on IQ revisited. *Intelligence*, 14(1), 11–29. https://doi.org/10.1016/0160-2896(90)90011-H

Cofnas, N. (2020). Research on group differences in intelligence: A defense of free inquiry. *Philosophical Psychology*, 33(1), 125–147. https://doi.org/10.1080/09515089.2019.1697803

Cohen, D. J., Dibble, E., Grawe, J., & Pollin, W. (1975). Reliably separating identical from fraternal twins. *Archives of General Psychiatry*, 32(11), 1371–1375. https://doi.org/10.1001/archpsyc.1975.01760290039004

Comfort, N. (2018a). Genetic determinism rides again. *Nature*, 561(7724), 461–463. https://doi.org/10.1038/d41586-018-06784-5

Comfort, N. (2018b, October 5th). Lies, damn lies, and GWAS. *Genotopia*. https://genotopia.scienceblog.com/506/lies-damned-lies-and-gwas/

Coop, G., & Przeworski, M. (2022). Lottery, luck, or legacy [Review of the book *The genetic lottery: Why DNA matters for social equality*, by K. P. Harden]. *Evolution*, 76(4), 846–853. https://doi.org/10.1111/evo.14449

Cropanzano, R., & James, K. (1990). Some methodological considerations for the behavioral genetic analysis of work attitudes. *The Journal of Applied Psychology*, 75(4), 433–439. https://doi.org/10.1037/0021-9010.75.4.433

Cross, R. W. (1996). The bell curve and transracial adoption studies. In J. L. Kincheloe, S. R. Steinberg, & A. D. Gresson III (Eds), *Measured lies: The bell curve examined* (pp. 332–342). St. Martin's Press.

Davies, G., Marioni, R. E., Liewald, D. C., Hill, W. D., Hagenaars, S. P., Harris, S. E., Ritchie, S. J., Luciano, M., Fawns-Ritchie, C., Lyall, D., Cullen, B., Cox, S. R., Hayward, C., Porteous, D. J., Evans, J., McIntosh, A. M., Gallacher, J., Craddock, N., Pell, J. P., Smith, D. J., … Deary, I. J. (2016). Genome-wide association study of cognitive functions and educational attainment in UK Biobank (N=112 151). *Molecular Psychiatry 21*(6), 758–767. https://doi.org/10.1038/mp.2016.45

Deary, I. J., Cox, S. R., & Hill, W. D. (2022). Genetic variation, brain, and intelligence differences. *Molecular Psychiatry, 27*(1), 335–353. https://doi.org/10.1038/s41380-021-01027-y

DeFries, J. C., & Plomin, R. (1978). Behavioral genetics. *Annual Review of Psychology, 29*, 473–515. https://doi.org/10.1146/annurev.ps.29.020178.002353

Derks, E. M., Dolan, C. V., & Boomsma, D. I. (2006). A test of the equal environment assumption (EEA) in multivariate twin studies. *Twin Research and Human Genetics, 9*(3), 403–411. https://doi.org/10.1375/183242706777591290

Devlin, B., Fienberg, S. E., Resnick, D. P., & Roeder, K. (Eds.). (1997). *Intelligence, genes, and success: Scientists respond to the bell curve.* Springer.

Dickens, W. T., & Flynn, J. R. (2006). Black Americans reduce the racial IQ gap: Evidence from standardization samples. *Psychological Science, 17*(10), 913–920. https://doi.org/10.1111/j.1467-9280.2006.01802.x

Dorfman D. D. (1978). The Cyril Burt question: New findings. *Science, 201*(4362), 1177–1186. https://doi.org/10.1126/science.201.4362.1177

Dorfman, D. D. (1995). Soft science with a neoconservative agenda. [Review of the book *The Bell Curve*, by R. J. Herrnstein & C. Murray]. *Contemporary Psychology, 40*(5), 418–421. https://doi.org/10.1037/003627

Duncan, R. G., Krishnamoorthy, R., Harms, U., Haskel-Ittah, M., Kampourakis, K., Gericke, N., Hammann, M., Jimenez-Aleixandre, M., Nehm, R. H., Reiss, M. J., & Yarden, A. (2024). The sociopolitical in human genetics education. *Science, 383*, 826–828. DOI: 10.1126/science.adi8227

Dusek, V. (1987). Bewitching science. *Science for the People, 19*(6),19–22. https://science-for-the-people.org/wp-content/uploads/2015/07/SftPv19n6s.pdf

Duyme, M., Dumaret, A. C., & Tomkiewicz, S. (1999). How can we boost IQs of "dull children"?: A late adoption study. *Proceedings of the National Academy of Sciences of the United States of America, 96*(15), 8790–8794. https://doi.org/10.1073/pnas.96.15.8790

Eaves, L. J., Eysenck, H. J., & Martin, N. G. (1989). *Genes, culture, and personality: An empirical approach.* Academic Press.

Egerton, J. (1973). The misuse of IQ testing: Interview with Leon Kamin. *Change: The Magazine of Higher Learning, 5*(8), 40–43.

Elder, G. H., Jr., & Shanahan, M. J. (2006). The life course and human development. In R. M. Lerner (Ed.), *Theoretical models of human development. Volume 1 of handbook of child psychology* (6th ed., pp. 665–715). Editors-in-chief: W. Damon & R. M. Lerner. Wiley.

Evans, G. (2018, March 2). The unwelcome revival of "race science." *The Guardian.* www.theguardian.com/news/2018/mar/02/the-unwelcome-revival-of-race-science

Evans, D. M., & Martin, N. G. (2000). The validity of twin research. *GeneScreen, 1*, 77–79. https://doi.org/10.1046/j.1466-9218.2000.00027.x

Eysenck, H. J., vs. Kamin, L. J. (1981). *The intelligence controversy.* John Wiley & Sons.

Fall, T., Kuja-Halkola, R., Dobney, K., Westgarth, C., & Magnusson, P. K. E. (2019). Evidence of large genetic influences on dog ownership in the Swedish Twin Registry has implications for understanding domestication and health associations. *Scientific Reports, 9*(1), 7554. https://doi.org/10.1038/s41598-019-44083-9

Farber, S. L. (1981). *Identical twins reared apart: A reanalysis.* Basic Books.

Farrell, M. S., Werge, T., Sklar, P., Owen, M. J., Ophoff, R. A., O'Donovan, M. C., Corvin, A., Cichon, S., & Sullivan, P. F. (2015). Evaluating historical candidate genes for schizophrenia. *Molecular Psychiatry, 20*(5), 555–562. https://doi.org/10.1038/mp.2015.16

Feldman, M. W., & Riskin, J. (2022, April 21). Why biology is not destiny [Review of the Book *The genetic lottery: Why DNA matters for social equality,* by K. P. Harden]. *The New York Review of Books.* https://www.nybooks.com/articles/2022/04/21/why-biology-is-not-destiny-genetic-lottery-kathryn-harden/

Fischbach, M., & Niggeschmidt, K. F. (2021). *Heritability of intelligence: A clarification from a biological point of view.* Springer.

Fischer, C. S., Hout, M., Sánchez Jankowski, M., Lucas, S. R., Swidler, A., & Voss, K. (1996). *Inequality by design: Cracking the bell curve myth.* Princeton University Press.

Fletcher, R. (1991). *Science, ideology, and the media: The Cyril Burt scandal.* Transaction Publishers.

Flynn, J. R. (1984). The mean IQ of Americans: Massive gains 1932 to 1978. *Psychological Bulletin, 95*(1), 29–51. https://doi.org/10.1037/0033-2909.95.1.29

Flynn, J. R. (1987). Massive IQ gains in 14 nations: What IQ tests really measure. *Psychological Bulletin, 101*(2), 171–191. https://doi.org/10.1037/0033-2909.101.2.171

Flynn, J. R. (1998). IQ gains over time: Toward finding the causes. In U. Neisser (Ed.), *The rising curve: Long-term gains in IQ and related measures* (pp. 25–66). American Psychological Association.

Flynn, J. R. (1999). Searching for justice: The discovery of IQ gains over time. *American Psychologist, 54*(1), 5–20. https://doi.org/10.1037/0003-066X.54.1.5

Flynn, J. R. (2008). *Where have all the liberals gone? Race, class, and ideals in America.* Cambridge University Press.

Flynn, J. R. (2009). *What is intelligence?* (Expanded paperback ed.). Cambridge University Press.

Fosse R., Joseph J., & Richardson, K. (2015). A critical assessment of the equal environment assumption of the twin method for schizophrenia. *Frontiers in Psychiatry, 6*:62, 1–10. https://www.frontiersin.org/articles/10.3389/fpsyt.2015.00062

Fraser, S. (Ed.). (1995). *The bell curve wars: Race, intelligence, and the future of America.* Basic Books.

Gardner, H. (1995). Scholarly brinksmanship. In R. Jacoby & N. Glauberman (Eds.), *The bell curve debate: History, documents, opinions* (pp. 61–72). Times Books.

Gillborn, D., McGimpsey, I., & Warmington, P. (2022). The fringe is the centre: Racism, pseudoscience and authoritarianism in the dominant English education policy network. *International Journal of Educational Research, 115*, Article 102056. https://doi.org/10.1016/j.ijer.2022.102056.

Goldberger, A. S. (1979). Heritability. *Economica, 46*(184), 327–347. https://doi.org/10.2307/2553675

Gottesman, I. I. (1966). Genetic variance in adaptive personality traits. *Journal of Child Psychology and Psychiatry, 7*(3–4), 199–208. https://doi.org/10.1111/j.1469-7610.1966.tb02246.x

Gottfredson, L. S. (1994, December 13; 1997). Mainstream science on intelligence. *Wall Street Journal,* A18. Republished in 1997 in *Intelligence, 24*(1), 13–23.

Gould, S. J. (1981). *The mismeasure of man.* W. W. Norton.

Gugushvili, A., & Bulczak, G. (2023). Physical attractiveness and intergenerational social mobility. *Social Science Quarterly, 104*(7). 1360–1382. https://doi.org/10.1111/ssqu.13320

Gusev, A. (2023). A molecular genetics perspective on the heritability of human behavior and group differences. http://gusevlab.org/projects/hsq/

Head, M. L., Holman, L., Lanfear, R., Kahn, A. T., & Jennions, M. D. (2015). The extent and consequences of p-hacking in science. *PLoS Biology, 13*(3): e1002106. https://doi.org/10.1371/journal.pbio.1002106

Hearnshaw, L. S. (1979). *Cyril Burt: Psychologist.* Cornell University Press.

Herrnstein, R. J. (1965, September). In defense of bird brains. *The Atlantic.* https://www.theatlantic.com/magazine/archive/1965/09/in-defense-of-bird-brains/659179/

Herrnstein, R. J. (1971, September). I.Q. *Atlantic Monthly, 228,* 43–64. https://cdn.theatlantic.com/assets/media/files/sept_1971_-_herrnstein_-_i.q.pdf

Herrnstein, R. J. (1973). *I.Q. in the meritocracy.* Little, Brown and Company.

Herrnstein, R. J., & Loveland, D. H. (1964, October 23). Complex visual concept in the pigeon. *Science, 146*(3643), 549–551. https://doi.org/10.1126/science.146.3643.549

Herrnstein, R. J., & Murray, C. (1994). *The bell curve: Intelligence and class structure in American life.* The Free Press.

Hirsch, J. (1975). Jensenism: The bankruptcy of "science" without scholarship. *Educational Theory, 25*(1), 3–26. https://doi.org/10.1111/j.1741-5446.1975.tb00663.x

Hirsch, J. (1981). To "unfrock the charlatans." *SAGE Race Relations Abstracts, 6*(2), 1–65. Sage Publications.

Hirsch, J. (1997). Some history of heredity-vs-environment, genetic inferiority at Harvard (?), and the (incredible) bell curve. *Genetica, 99*(2–3), 207–224. https://doi.org/10.1007/BF02259524

Ho, M. W. (2013). No genes for intelligence in the fluid genome. *Advances in Child Development and Behavior, 45,* 67–92. https://doi.org/10.1016/b978-0-12-397946-9.00004-x

Holden, C., & Bouchard, T. (2009). Newsmaker interview: Behavioral geneticist celebrates twins, scorns PC science. *Science, 325* (5936), new series, 27–27. https://doi.org/10.1126/science.325_27

Horgan, J. (1999). *The undiscovered mind.* The Free Press.

Howe, L. J., Nivard, M. G., Morris, T. T., Hansen, A. F., Rasheed, H., Cho, Y., Chittoor, G., Ahlskog, R., Lind, P. A., Palviainen, T., van der Zee, M. D., Cheesman, R., Mangino, M., Wang, Y., Li, S., Klaric, L., Ratliff, S. M., Bielak, L. F., Nygaard, M., Giannelis, A., … Davies, N. M. (2022). Within-sibship genome-wide association analyses decrease bias in estimates of direct genetic effects. *Nature Genetics, 54*(5), 581–592. https://doi.org/10.1038/s41588-022-01062-7

Jackson, D. D. (1960). A critique of the literature on the genetics of schizophrenia. In D. Jackson (Ed.), *The etiology of schizophrenia* (pp. 37–87). New York: Basic Books.

Jacoby, R., & Glauberman, N. (Eds.). (1995). *The bell curve debate.* Times Books.

Jensen, A. R. (1969). How much can we boost IQ and scholastic achievement? *Harvard Educational Review, 39*(1), 1–123. https://doi.org/10.17763/haer.39.1.l3u15956627424k7

Jensen, A. R. (1970). IQ's of identical twins reared apart. *Behavior Genetics, 1*(2), 133–148. https://doi.org/10.1007/BF01071829

Jensen, A. R. (1974). Kinship correlations reported by Sir Cyril Burt. *Behavior Genetics, 4*(1), 1–28. https://doi.org/10.1007/BF01066704

Jensen, A. R. (1980). *Bias in mental testing.* Free Press.

Jensen, A. R. (1998). *The g factor: The science of mental ability.* Praeger.

John, L. K., Loewenstein, G., & Prelec, D. (2012). Measuring the prevalence of questionable research practices with incentives for truth telling. *Psychological Science, 23*(5), 524–532. https://doi.org/10.1177/0956797611430953

Johnson, W., Bouchard, T. J., Jr., McGue, M., Segal, N. L., Tellegen, A., Keyes, M., & Gottesman, I. I. (2007). Genetic and environmental influences on the Verbal-Perceptual-Image Rotation (VPR) model of the Structure of Mental Abilities in the Minnesota Study of Twins Reared Apart. *Intelligence, 35,* 542–562. https://doi.org/10.1016/j.intell.2006.10.003

Joseph, J. (2004). *The gene illusion: Genetic research in psychiatry and psychology under the microscope.* Algora.

Joseph, J. (2006). *The missing gene: Psychiatry, heredity, and the fruitless search for genes.* New York: Algora.

Joseph, J. (2012). The "missing heritability" of psychiatric disorders: Elusive genes or non-existent genes? *Applied Developmental Science, 16*(2), 65–83. https://doi.org/10.1080/10888691.2012.667343

Joseph, J. (2013). The use of the classical twin method in the social and behavioral sciences: The fallacy continues. *Journal of Mind and Behavior, 34*(1), 1–39. https://psycnet.apa.org/record/2013-26783-001

Joseph, J. (2015). *The trouble with twin studies: A reassessment of twin research in the social and behavioral sciences.* Routledge.

Joseph, J. (2016, March 6). "Bewitching Science" revisited: Tales of reunited twins and the genetics of behavior. *Mad in America.* https://www.madinamerica.com/2016/03/bewitching-science-revisited-tales-of-reunited-twins-and-the-genetics-of-behavior/

Joseph, J. (2018, April 4th). Leon J. Kamin (1927-2017): A nemesis of genetic determinism and scientific racism. *Mad in America.* https://www.madinamerica.com/2018/04/leon-j-kamin-nemesis-genetic-determinism/

Joseph, J. (2019, June 21). Three identical strangers and the nature-nurture debate. *Mad in America.* https://www.madinamerica.com/2019/06/three-identical-strangers-nature-nurture-debate/

Joseph, J. (2022a). A blueprint for genetic determinism. [Review of the book *Blueprint: How DNA makes us who we are,* by R. Plomin]. *American Journal of Psychology, 135*(4), 442–454. https://doi.org/10.5406/19398298.135.4.13

Joseph, J. (2022b). A reevaluation of the 1990 "Minnesota Study of Twins Reared Apart" IQ study. *Human Development, 66*(1), 48–65. https://doi.org/10.1159/000521922

Joseph, J. (2023a, Fall). The "Minnesota Study of Twins Reared Apart" IQ study: Ripe for retraction? *Free Associations: Psychoanalysis and Culture, Media, Groups, Politics, 89.* https://freeassociations.org.uk/FA_New/OJS/index.php/fa/article/view/458/719

Joseph, J. (2023b). *Schizophrenia and genetics: The end of an illusion* (1st English ed.). Routledge.

Joseph, J., & Baldwin, S. (2000). Four editorial proposals to improve social sciences research and publication. *International Journal of Risk and Safety in Medicine, 13*(2–3), 109–116. https://psycnet.apa.org/record/2005-06884-004

Joseph, J., & Boyle, M. (2024). An alternative framework for assessing psychiatric genetics research. In A. Cantú, E. Maisel, & C. Ruby (Eds.), *Theoretical alternatives to the psychiatric model of mental disorder labeling: Contemporary frameworks, taxonomies, and models* (pp. 464–486). Ethics International Press.

Joynson, R. B. (1989). *The Burt affair*. RKP.

Juel-Nielsen, N. (1965/1980). *Individual and environment: Monozygotic twins reared apart* (rev. ed.). International Universities Press.

Kamin, L. J. (1974). *The science and politics of I.Q.* Erlbaum.

Kamin, L. J. (1981). Commentary. In S. Scarr (Ed.), *Race, social class, and individual differences in I. Q.* (pp. 467–482). Erlbaum.

Kamin, L. J. (1995). Lies, damn lies, and statistics. In R. Jacoby & N. Glauberman (Eds.), *The bell curve debate: History, documents, opinions* (pp. 81–105). Times Books.

Kamin, L. J. (2006). African IQ and mental retardation. *South African Journal of Psychology, 36*(1), 1–9. https://doi.org/10.1177/008124630603600101

Kamin, L. J., & Goldberger, A. S. (2002). Twin studies in behavioral research: A skeptical view. *Theoretical Population Biology, 61*(1), 83–95. https://doi.org/10.1006/tpbi.2001.1555

Kaufman, S. B. (2019, January 19). There is no nature-nurture war. *Scientific American*. https://blogs.scientificamerican.com/beautiful-minds/there-is-no-nature-nurture-war/

Kaufman, S. B., & Moore, D. S. (2008, October 24). What makes us who we are? Straight talk about twin studies, genes, and parenting. *Psychology Today*. https://www.psychologytoday.com/us/blog/beautiful-minds/200810/what-makes-us-who-we-are

Keller, E. F. (2010). *The mirage of a space between nature and nurture*. Duke University Press.

Kendler, K. S. (1983). Overview: A current perspective on twin studies of schizophrenia. *American Journal of Psychiatry, 140*(11), 1413–142. https://doi.org/10.1176/ajp.140.11.1413

Khanna, K., Salvanto, A., De Pinto, J., & Backus, F. (2024, January 14). CBS News poll finds Trump's national lead grows as GOP nominating contests kick off. *CBS News*. https://www.cbsnews.com/news/trump-national-lead-grows-opinion-poll-republican-primary/?ftag=CNM-00-10aab7e&linkId=263914723

Kincheloe, J. L., Steinberg, S. R., & Gresson III, A. D. (Eds.). (1996). *Measured lies: The bell curve examined*. St. Martin's Press.

Kraft, P., Kraft, B., Hagen, T., & Espeseth, T. (2022). Subjective socioeconomic status, cognitive abilities, and personal control: Associations with health behaviours. *Frontiers in Psychology, 12*, 784758. https://doi.org/10.3389/fpsyg.2021.784758.

Kringlen, E. (1967). *Heredity and environment in the functional psychoses: An epidemiological-clinical study*. Oslo: Universitetsforlaget.

Långström, N., Grann, M., & Lichtenstein, P. (2002). Genetic and environmental influences on problematic masturbatory behavior in children: A study of same-sex twins. *Archives of Sexual Behavior, 31*(4), 343–350. https://doi.org/10.1023/a:1016224326301

Le Texier, T. (2019). Debunking the Stanford Prison Experiment. *American Psychologist, 74*(7), 823–839. https://doi.org/10.1037/amp0000401

Lerner, R. M. (2004). Genes and the promotion of positive human development: Hereditarian versus developmental systems perspectives. In C. Garcia Coll, E. L. Bearer, & R. M Lerner (Eds.), *Nature and nurture: The complex interplay of genetic and environmental influences on human behavior and development* (pp. 1–33). Erlbaum.

Lerner, R. M. (2018). *Concepts and theories of human development* (4th Ed.). Routledge.

Lewis, A. C. F., Molina, S. J., Appelbaum, P. S., Dauda, B., Di Rienzo, A., Fuentes, A., Fullerton, S. M., Garrison, N. A., Ghosh, N., Hammonds, E. M., Jones, D. S., Kenny, E. E., Kraft, P., Lee, S. S., Mauro, M., Novembre, J., Panofsky, A., Sohail, M., Neale, B. M., & Allen, D. S. (2022). Getting genetic ancestry right for science and society. *Science, 376*(6590), 250–252. https://doi.org/10.1126/science.abm7530

Lewontin, R. C. (1970). Race and intelligence. *Bulletin of the Atomic Scientists, 26*(3) 2–8. https://doi.org/10.1080/00963402.1970.11457774

Lewontin, R. C. (1974). Annotation: The analysis of variance and the analysis of causes. *American Journal of Human Genetics, 26*(3), 400–411. https://www.ncbi.nlm.nih.gov/pmc/articles/PMC1762622/

Lewontin, R. C. (1976). Science and politics: An explosive mix [Review of the book *The science and politics of I.Q., by L. J. Kamin], *Contemporary Psychology*, 21(2), 97–98. https://resolver.scholarsportal.info/resolve/00107549/v21i0002/97_sapaem.xml

Lewontin, R. C., Rose, S., & Kamin, L. J. (1984). *Not in our genes*. Pantheon.

Loehlin, J. C. (1978). Heredity-environment analyses of Jencks's IQ correlations. *Behavior Genetics*, 8(5), 415–436. https://doi.org/10.1007/BF01067938

Loehlin, J. C., Horn, J. M., & Willerman, L. (1989). Modeling IQ change: Evidence from the Texas Adoption Project. *Child Development*, 60(4), 993–1004. https://doi.org/10.1111/j.1467-8624.1989.tb03530.x

Loehlin, J. C., & Nichols, R. C. (1976). *Heredity, environment, and personality: A study of 850 pairs of twins*. University of Texas Press.

Luciano, M., Kirk, K. M., Heath, A. C., & Martin, N. G. (2005). The genetics of tea and coffee drinking and preference for source of caffeine in a large community sample of Australian twins. *Addiction*, 100(10), 1510–1517. https://doi.org/10.1111/j.1360-0443.2005.01223.x

Lykken, D. T. (1995). *The antisocial personalities*. Erlbaum.

Lykken, D. T. (2004). The new eugenics [Review of the book *Eugenics: A reassessment*, by R. Lynn]. *Contemporary Psychology*, 49(6), 670–672. https://access.portico.org/stable?au=phzmhc725

Lynn, R. (2001). *Eugenics: A reassessment*. Praeger.

Lynn, R., & Becker, D. (2019). *The intelligence of nations*. Ulster Institute for Social Research.

Maher, B. (2008). Personal genomes: The case of the missing heritability. *Nature*, 456(7218), 18–21. https://doi.org/10.1038/456018a

Manolio, T. A., Collins, F. S., Cox, N. J., Goldstein, D. B., Hindorff, L. A., Hunter, D. J., McCarthy, M. I., Ramos, E. M., Cardon, L. R., Chakravarti, A., Cho, J. H., Guttmacher, A. E., Kong, A., Kruglyak, L., Mardis, E., Rotimi, C. N., Slatkin, M., Valle, D., Whittemore, A. S., Boehnke, M., ... Visscher, P. M. (2009). Finding the missing heritability of complex diseases. *Nature*, 461(7265), 747–753. https://doi.org/10.1038/nature08494

Marks, J. (2017). *Is science racist?* Polity Press.

Matarazzo, J. D., & Wiens, A. N. (1977). Black Intelligence Test of Cultural Homogeneity and Wechsler Adult Intelligence Scale scores of black and white police applicants. *Journal of Applied Psychology*, 62(1), 57–63. https://doi.org/10.1037/0021-9010.62.1.57

McGue, M., & Bouchard, T. J., Jr. (1984). Adjustment of twin data for the effects of age and sex. *Behavior Genetics*, 14(4), 325–343. https://doi.org/10.1007/BF01080045

McGue, M., & Bouchard, T. J., Jr. (1989). Genetic and environmental determinants of information processing and special mental abilities: A twin analysis. In R. J. Sternberg (Ed.), *Advances in the psychology of human intelligence* (Vol. 5, pp. 7–45). Hillsdale, NJ: Erlbaum.

McGuire, T. R. & Hirsch, J. (1977). General intelligence (g) and heritability (H^2, h^2). In I. Uzgiris & F. Weitzmann (Eds.), *The structuring of experience* (pp. 25–72). Plenum Press.

McKie, R. (2024, February 3). "The situation has become appalling": Fake scientific papers push research credibility to crisis point. *The Guardian*. https://www.theguardian.com/science/2024/feb/03/the-situation-has-become-appalling-fake-scientific-papers-push-research-credibility-to-crisis-point

Mensh, E., & Mensh, H. (1991). *The IQ mythology: Class, race, gender, and inequality*. Southern Illinois Press.

Meyer, M. N., Appelbaum, P. S., Benjamin, D. J., Callier, S. L., Comfort, N., Conley, D., Freese, J., Garrison, N. A., Hammonds, E. M., Harden, K. P., Lee, S. S., Martin, A. R., Martschenko, D. O., Neale, B. M., Palmer, R. H. C., Tabery, J., Turkheimer, E., Turley, P., & Parens, E. (2023). Wrestling with social and behavioral genomics: Risks, potential benefits, and ethical responsibility. *The Hastings Center Report*, 53 Suppl 1(Suppl 1), S2–S49. https://doi.org/10.1002/hast.1477

Micceri, T. (1989). The unicorn, the normal curve, and other improbable creatures. *Psychological Bulletin*, 105(1), 156–166. https://doi.org/10.1037/0033-2909.105.1.156

Mitchell, K. J. (2018). *Innate: How the wiring of our brains shapes who we are*. Princeton University Press.

Moloney, D. P., Bouchard, T. J., Jr., & Segal, N. L. (1991). A genetic and environmental analysis of the vocational interests of monozygotic and dizygotic twins reared apart. *Journal of Vocational Behavior*, 39(1), 76–109. https://doi.org/10.1016/0001-8791(91)90005-7

Moon, R. (2010, Fall). Pre-war Japan and the origins of the Korean diaspora. *Stanford Spice Digest*. https://fsi9-prod.s3.us-west-1.amazonaws.com/s3fs-public/Koreans_inJapan.pdf

Moore, D. S. (2007). A very little bit of knowledge: Re-evaluating the meaning of the heritability of IQ.

Human Development, 49(6), 347–353. https://doi.org/10.1159/000096534

Moore, D. S., & Shenk, D. (2016). The heritability fallacy. WIREs Cognitive Science. https://doi.org/10.1002/wcs.1400

Munsinger, H. (1975). The adopted child's IQ: A critical review. Psychological Bulletin, 82(5), 623–659. https://doi.org/10.1037/0033-2909.82.5.623

Murray, C. (1984). Losing ground: American social policy, 1950–1980. Basic Books.

Murray, C. (2020). Human diversity: The biology of gender, race, and class. Twelve.

Nature. (2023, April 11). Why Nature is updating its advice to authors on reporting race or ethnicity. Nature, 616(7956), 219. https://doi.org/10.1038/d41586-023-00973-7

Neisser, U. (1998). Introduction: Rising test scores and what they mean. In U. Neisser (Ed.), The rising curve: Long-term gains in IQ and related measures (pp. 3–22). American Psychological Association.

Neisser, U., Boodoo, G., Bouchard, T. J., Jr., Boykin, A. W., Brody, N., Ceci, S. J., Halpern, D. F., Loehlin, J. C., Perloff, R., Sternberg, R. J., & Urbina, S. (1996). Intelligence: Knowns and unknowns. American Psychologist, 51(2), 77–101. https://doi.org/10.1037/0003-066X.51.2.77

Newman, H. H., Freeman, F. N., & Holzinger, K. J. (1937). Twins: A study of heredity and environment. University of Chicago Press.

Nisbett, R. E. (2009). Intelligence and how to get it. W. W. Norton.

Nisbett, R. E., Aronson, J., Blair, C., Dickens, W., Flynn, J., Halpern, D. F., & Turkheimer, E. (2012). Intelligence: New findings and theoretical developments. American Psychologist, 67(2), 130–159. https://doi.org/10.1037/a0026699

Non, A. L., & Cerdeña, J. P. (2023). Considerations, caveats, and suggestions for the use of polygenic scores for social and behavioral traits. Behavior Genetics, 10.1007/s10519-023-10162-x. https://doi.org/10.1007/s10519-023-10162-x

Norton, H. L., Quillen, E. E., Bigham, A.W., Pearson, L. N., & Dunsworth, H. (2019). Human races are not like dog breeds: Refuting a racist analogy. Evolution: Education and Outreach, 12, 17. https://doi.org/10.1186/s12052-019-0109-y

Nosek, B. A., Ebersole, C. R., DeHaven, A. C., & Mellor, D. T. (2018). The preregistration revolution. Proceedings of the National Academy of Sciences of the United States of America, 115(11), 2600–2606. https://doi.org/10.1073/pnas.1708274114

Nugent, B. M., & McCarthy, M. M. (2015). Epigenetic influences on the developing brain: Effects of hormones and nutrition. Advances in Genomics and Genetics, 2015(5), 215–225.

Ogbu, J. U. (1978). Minority education and caste: The American system in cross-cultural perspective. Academic Press.

O'Grady, C. (2020, July 15). Misconduct allegations push psychology hero off his pedestal. Science. https://www.sciencemag.org/news/2020/07/misconduct-allegations-push-psychology-hero-his-pedestal

O'Grady, C. (2023, October 13). How the reform-minded new editor of psychology's flagship journal will shake things up. Science. https://www.science.org/content/article/how-reform-minded-new-editor-psychology-s-flagship-journal-will-shake-things

Okbay, A., Wu, Y., Wang, N., Jayashankar, H., Bennett, M., Nehzati, S. M., Sidorenko, J., Kweon, H., Goldman, G., Gjorgjieva, T., Jiang, Y., Hicks, B., Tian, C., Hinds, D. A., Ahlskog, R., Magnusson, P. K. E., Oskarsson, S., Hayward, C., Campbell, A., Porteous, D. J., ... Young, A. I. (2022). Polygenic prediction of educational attainment within and between families from genome-wide association analyses in 3 million individuals. Nature Genetics, 54(4), 437–449. https://doi.org/10.1038/s41588-022-01016-z

Open Science Collaboration. (2015). Psychology: Estimating the reproducibility of psychological science. Science, 349(6251), aac4716-1–aac47168. https://doi.org/10.1126/science.aac4716

Oransky, I., & Redman, B. (2024, January 11). Rooting out scientific misconduct. Science, 383. https://www.science.org/doi/10.1126/science.adn9352

Page, E. B. (1972). Behavior and heredity. American Psychologist, 27(7), 660–661. https://doi.org/10.1037/h0038215

Pam, A., Kemker, S. S., Ross, C. A., & Golden, R. (1996). The "equal environment assumption" in MZ-DZ comparisons: An untenable premise of psychiatric genetics? Acta Geneticae Medicae et Gemellologiae (Twin Research), 45(3), 349–360. https://doi.org/10.1017/S0001566000000945

Panofsky, A., Dasgupta, K., & Iturriaga, N. (2021). How white nationalists mobilize genetics: From genetic ancestry and human biodiversity to counterscience and metapolitics. *American Journal of Physical Anthropology*, *175*, 387–398. https://doi.org/10.1002/ajpa.24150

Pedersen, N. L., Friberg, L., Floderus-Myrhed, B., McClearn, G. E., & Plomin, R. (1984). Swedish early separated twins: Identification and characterization. *Acta Geneticae Medicae et Gemellologiae*, *33*(2), 243–250. https://doi.org/10.1017/s0001566000007285

Pedersen, N. L., Plomin, R., Nesselroade, J. R., & McClearn, G. E. (1992). A quantitative genetic analysis of cognitive abilities during the second half of the life span. *Psychological Science*, *3*, 346–353. https://doi.org/10.1111/j.1467-9280.1992.tb00045.x

Plomin, R. (2018). *Blueprint: How DNA makes us who we are*. MIT Press.

Plomin R. (2023). Celebrating a century of research in behavioral genetics. *Behavior Genetics*, *53*(2), 75–84. https://doi.org/10.1007/s10519-023-10132-3

Plomin, R., & Daniels, D. (1987). Why are children in the same family so different from one another? *Behavioral and Brain Science*, *10*(1), 1–16. https://doi.org/10.1017/S0140525X00055941

Plomin, R., & Deary, I. (2015). Genetics and intelligence differences: Five special findings. *Molecular Psychiatry*, *20*(1), 98–108. https://doi.org/10.1038/mp.2014.105

Plomin, R., & DeFries, J. C. (1980). Genetics and intelligence: Recent data. *Intelligence*, *4*(1), 15–24. https://doi.org/10.1016/0160-2896(80)90003-3

Plomin, R., DeFries, J. C., Knopik, V. S., & Neiderhiser, J. M. (2013). *Behavioral genetics* (6th ed.). New York: Worth.

Plomin, R., DeFries, J. C., Knopik, V. S., & Neiderhiser, J. M. (2016). Top ten replicated findings from behavioral genetics. *Perspectives on Psychological Science*, *11*, 3–23. https://doi.org/10.1177/1745691615617439

Plomin, R., & Kawakami, K. (2024). Nature, nurture and nonshared environment in cognitive development. *PsyArXiv.* https://doi.org/10.31234/osf.io/qndj6

Plomin, R., & von Stumm, S. (2018). The new genetics of intelligence. *Nature Reviews Genetics*, *19*, 148–159. https://doi.org/10.1038/nrg.2017.104

Polderman, T. J. C., Benyamin, B., de Leeuw, C. A., Sullivan, P. F., van Bochoven, A., Visscher, P. M., & Posthuma, D. (2015). Meta-analysis of the heritability of human traits based on fifty years of twin studies. *Nature Genetics*, *47*, 702–709. https://doi.org/10.1038/ng.3285

Psych-Agitator. (1979). Another true tale from the annals of orthodoxy: Richard Herrnstein, the Limited War Laboratory, pigeons, and the Vietnam War. *Behavior Analysis and Social Action*, *1*(4), 23–25. https://doi.org/10.1007/BF03406125

Rawson, R. (1979, May 7th). Two Ohio strangers find they're twins at 39—and a dream to psychologists. *People.* http://www.people.com/people/archive/article/0,20073583,00.html

Richardson, K. (2000). *The making of intelligence*. Columbia University Press.

Richardson, K. (2022). *Understanding intelligence*. Cambridge University Press.

Richardson, K., & Jones, M. C. (2019). Why genome-wide associations with cognitive ability measures are probably spurious. *New Ideas in Psychology*, *55*, 35–41. https://doi.org/10.1016/j.newideapsych.2019.04.005

Richardson, K., & Joseph, J. (2020). Hail to the polygenic republic. [Review of the book *Blueprint: How DNA makes us who we are*, by Robert Plomin]. *British Journal of Psychology*, *111*(1), 148–150. https://doi.org/10.1111/bjop.12422

Richardson, K., & Norgate, S. (2005). The equal environments assumption of classical twin studies may not hold. *British Journal of Educational Psychology*, *75*(3), 339–350. https://doi.org/10.1348/000709904X24690

Richardson, K., & Norgate, S. (2006). A critical analysis of IQ studies of adopted children. *Human Development*, *49*, 319–335. https://doi.org/10.1159/000096531

Richardson, K., & Norgate, S. H. (2014). Does IQ measure ability for complex cognition? *Theory & Psychology*, *24*(6), 795–812. https://doi.org/10.1177/0959354314551163

Ritchie, S. (2015). *Intelligence: All that matters*. John Murray Learning.

Ritchie, S. (2020). *Science fictions: How fraud, bias, negligence, and hype undermine the search for truth*. Henry Holt and Co.

Robette, N., Génin, E., & Clerget-Darpoux, F. (2022). Heritability: What's the point? What is it not for? A human genetics perspective. *Genetica*, *150*(3–4), 199–208. https://doi.org/10.1007/s10709-022-00149-7

Rönnlund, M., Carlstedt, B., Blomstedt, Y., Nilsson, L., & Weinehall, L. (2013). Secular trends in cognitive test performance: Swedish conscript data 1970–1993.

Intelligence, 41(1), 19–24. https://doi.org/10.1016/j.intell.2012.10.001

Rose, R. J. (1982). Separated twins: Data and their limits. [Review of the book *Identical twins reared apart: A reanalysis*, by S. Farber]. *Science, 215*(4535), 959–960. https://doi.org/10.1126/science.215.4535.959

Rose, S. (1997). *Lifelines: Life beyond the genes*. Oxford University Press.

Roseman, C. C., & Bird, K. A. (2023). Between group heritability and the status of hereditarianism as an evolutionary science. *BioRxiv*. https://doi.org/10.1101/2023.12.18.572247

Rosen, J., Watson, K., & Rinaldi, O. (2023, December 18). Trump blasted for saying immigrants are "poisoning the blood of our country." *CBS News*. https://www.cbsnews.com/news/trump-immigrants-poisoning-the-blood-of-our-country-reaction/

Rushton, J. P. (2000). *Race, evolution, and behavior: A life history perspective* (2nd Special Abridged Ed.). Charles Darwin Research Institute.

Rushton, J. P., & Jensen, A. R. (2005). Thirty years of research on race differences in cognitive ability. *Psychology, Public Policy, and Law, 11*(2), 235–294. https://doi.org/10.1037/1076-8971.11.2.235

Rutherford, A. (2021). *How to argue with a racist* (paperback ed.). Experiment Publishing.

Rutter, M. (2006). *Genes and behavior: Nature-nurture interplay explained*. Blackwell.

Savage, J. E., Jansen, P. R., Stringer, S., Watanabe, K., Bryois, J., de Leeuw, C. A., Nagel, M., Awasthi, S., Barr, P. B., Coleman, J. R. I., Grasby, K. L., Hammerschlag, A. R., Kaminski, J. A., Karlsson, R., Krapohl, E., Lam, M., Nygaard, M., Reynolds, C. A., Trampush, J. W., Young, H., … Posthuma, D. (2018). Genome-wide association meta-analysis in 269,867 individuals identifies new genetic and functional links to intelligence. *Nature Genetics, 50*(7), 912–919. https://doi.org/10.1038/s41588-018-0152-6

Schaie, K. W. (1965). A general model for the study of developmental problems. *Psychological Bulletin, 64*(2), 92–107. https://doi.org/10.1037/h0022371

Scarr, S. (1998). On Arthur Jensen's integrity. *Intelligence, 26*(3), 227–232. https://doi.org/10.1016/S0160-2896(99)80005-1

Scarr, S., & Carter-Saltzman, L. (1979). Twin method: Defense of a critical assumption. *Behavior Genetics, 9*(6), 527–542. https://doi.org/10.1007/BF01067349

Scarr, S., & Weinberg, R. A. (1976). IQ test performance of black children adopted by white families. *American Psychologist, 31*(10), 726–739. https://doi.org/10.1037/0003-066X.31.10.726

Schiff, M., Duyme, M., Dumaret, A., & Tomkiewicz, S. (1982). How much *could* we boost scholastic achievement and IQ scores? A direct answer from a French adoption study. *Cognition, 12*(2), 165–196. https://doi.org/10.1016/0010-0277(82)90011-7

Schiff, M., & Lewontin, R. C. (1986). *Education and class: The irrelevance of IQ genetic studies*. Clarendon Press.

Schimmack, U. (2021). The validation crisis in psychology. *Meta-Psychology, 5*, 1–9. https://doi.org/10.15626/MP.2019.1645

Schraiber, J. G., & Edge, M. D. (2023). Heritability within groups is uninformative about differences among groups: Cases from behavioral, evolutionary, and statistical genetics. *BioRxiv*. https://doi.org/10.1101/2023.11.06.565864

Sear, R. (2022). "National IQ" datasets do not provide accurate, unbiased or comparable measures of cognitive ability worldwide. *PsyArXiv*. https://doi.org/10.31234/osf.io/26vfb

Segal, N. L. (1999). *Entwined lives: Twins and what they tell us about human behavior*. Dutton.

Segal, N. L. (2012). *Born together—Reared apart: The landmark Minnesota twin study*. Harvard University Press.

Segal, N. L., & Hur, Y. M. (2022). Personality traits, mental abilities and other individual differences: Monozygotic female twins raised apart in South Korea and the United States. *Personality and Individual Differences, 194*, 1–9. https://doi.org/10.1016/j.paid.2022.111643

Sesardic, N. (2005) *Making sense of heritability*. Cambridge University Press.

Sham, P. C., & Kendler, K. S. (2008). Genetic etiology. In R. Murray et al. (Eds.), *Essential psychiatry* (4th ed., pp. 80–94). Cambridge University Press.

Shields, J. (1962). *Monozygotic twins brought up apart and brought up together*. Oxford University Press.

Skinner, M. K. (2014). Environmental stress and epigenetic transgenerational inheritance. *BMC Medicine 12*, 153. https://doi.org/10.1186/s12916-014-0153-y

Stefan, A. M. & Schönbrodt, F. D. (2023). Big little lies: A compendium and simulation of *p*-hacking strategies. *Royal Society Open Science, 10*(220346). https://doi.org/10.1098/rsos.220346

Sternberg, R. J. (1995). For whom the bell curve tolls: A review of the bell curve. [Review of the book *The bell curve: Intelligence and class structure in American life*, by R. J. Herrnstein & C. Murray]. *Psychological Science*, 6(5), 257–261. https://doi.org/10.1111/j.1467-9280.1995.tb00508.x

Sternberg, R. J., & Kaufman, S. B. (2011). Preface. In R. J. Sternberg & S. B. Kaufman (Eds.), *The Cambridge handbook of intelligence* (pp. xv–xx). Cambridge University Press.

Stoltenberg S. F. (1997). Coming to terms with heritability. *Genetica*, 99(2–3), 89–96. https://doi.org/10.1007/BF02259512

Stoolmiller, M. (1999). Implications of the restricted range of family environments for estimates of heritability and nonshared environment in behavior-genetic adoption studies. *Psychological Bulletin*, 125(4), 392–409. https://doi.org/10.1037/0033-2909.125.4.392

Sullivan P. F. (2017). How good were candidate gene guesses in schizophrenia genetics? *Biological Psychiatry*, 82(10), 696–697. https://doi.org/10.1016/j.biopsych.2017.09.004

Tabery, J. (2014). *Beyond versus: The struggle to understand the interaction of nature and nurture*. MIT Press.

Taylor, H. F. (1980). *The IQ game: A methodological inquiry into the heredity-environment controversy*. Rutgers University Press.

Tellegen, A., Lykken, D. T., Bouchard, T. J., Jr., Wilcox, K. J., Segal, N. L., & Rich, S. (1988). Personality similarity in twins reared apart and together. *Journal of Personality and Social Psychology*, 54(6), 1031–1039. https://doi.org/10.1037/0022-3514.54.6.1031

Terman, L. M. (1916). *The measurement of intelligence*. Houghton Mifflin.

Terman, L. M., & Merrill, M. A. (1937). *Measuring intelligence*. Houghton Mifflin.

Thomas, D. (2017). Racial IQ differences among transracial adoptees: Fact or artifact? *Journal of Intelligence*, 5(1), 1. https://doi.org/10.3390/jintelligence5010001

Torgersen, S. (1979). The determination of twin zygosity by means of a mailed questionnaire. *Acta Geneticae Medicae et Gemellologiae*, 28(3), 225–236. https://doi.org/10.1017/s0001566000009077

Tucker, W. H. (1994a). Fact and fiction in the discovery of Sir Cyril Burt's flaws. *Journal of the History of the Behavioral Sciences*, 30(4), 335–347. https://doi.org/10.1002/1520-6696(199410)30:4<335::AID-JHBS2300300403>3.0.CO;2-5

Tucker, W. H. (1994b). *The science and politics of racial research*. University of Illinois Press.

Tucker, W. H. (1997). Re-reconsidering Burt: Beyond a reasonable doubt. *Journal of the History of the Behavioral Sciences*, 33(2),145–162. https://doi.org/10.1002/(SICI)1520-6696(199721)33:2<145::AID-JHBS6>3.0.CO;2-S

Tucker, W. H. (2024). *"The Bell Curve" in perspective: Race, meritocracy, inequality, and politics*. Palgrave Macmillan. https://link.springer.com/book/10.1007/978-3-031-41614-9

Turkheimer, E. (2016). Weak genetic explanation 20 years later: Reply to Plomin et al. *Perspectives on Psychological Science*, 11(1), 24–28. https://doi.org/10.1177/1745691615617442

Turkheimer, E. (2019). The social science blues. *Hastings Center Report*, 49(3), 45–47. https://doi.org/10.1002/hast.1008

UK Biobank. (n.d.). Category 100027, fluid intelligence/reasoning https://biobank.ndph.ox.ac.uk/ukb/label.cgi?id=100027

Veller C., & Coop, G. M. (2024). Interpreting population- and family-based genome-wide association studies in the presence of confounding. *PLoS Biology*, 2(4): e3002511. https://doi.org/10.1371/journal.pbio.3002511

von Stumm, S., Kandaswamy, R., & Maxwell, J. (2023). Gene-environment interplay in early life cognitive development. *Intelligence*, 98, Article 101748. Advance online publication. https://doi.org/10.1016/j.intell.2023.101748

Wahlsten, D. (1990). Insensitivity of the analysis of variance to heredity-environment interaction. *Behavioral and Brain Sciences*, 13(1), 109–120. https://doi.org/10.1017/S0140525X00077797

Wahlsten, D., & Gottlieb, G. (1997). The invalid separation of effects of nature and nurture: Lessons from animal experimentation. In R. J. Sternberg & E. Grigorenko (Eds.), *Intelligence, Heredity, and Environment* (pp. 163–192). Cambridge University Press.

Warne, R. T. (2020). *In the know: Debunking 35 myths about human intelligence*. Cambridge University Press.

Warne, R. T. (2022, December 17). Irish IQ: The massive rise that never happened. *Warne Blogpost*.

https://russellwarne.com/2022/12/17/irish-iq-the-massive-rise-that-never-happened/

Wechsler, D. (1944). *The measurement of adult intelligence* (3rd ed.). Williams and Wilkins.

Wechsler, D. (1958). *The measurement and appraisal of adult intelligence* (4th ed.). Williams and Wilkins.

Wehby, G. L., & Shane, D. (2019). Genetic variation in health insurance coverage. *International Journal of Health Economics and Management, 19*(3–4), 301–316. https://doi.org/10.1007/s10754-018-9255-y

Weinberg, R. A., Scarr, S., & Waldman, I. D. (1992). The Minnesota Transracial Adoption Study: A follow-up of IQ test performance at adolescence. *Intelligence, 16*(1), 117–135. https://doi.org/10.1016/0160-2896(92)90028-P

Wesseldijk, L. W., Tybur, J. M., Boomsma, D. I., Willemsen, G., & Vink, J. M. (2023). The heritability of pescetarianism and vegetarianism. *Food Quality and Preference, 103*, 1–5. https://doi.org/10.1016/j.foodqual.2022.104705.

Whitten, C., & O'Riley, C. (2023, March 16). A comprehensive history of every change made to Augusta National Golf Club. *Golf Digest.* https://www.golfdigest.com/story/complete-changes-to-augusta-national

Williams, C. M., Labouret, G., Wolfram, T., Peyre, H., & Ramus, F. (2023). A general cognitive ability factor for the UK Biobank. *Behavior Genetics, 53*(2), 85–100. https://doi.org/10.1007/s10519-022-10127-6

Williams, R. L. (September 1972). *The BITCH-100: A Culture-Specific Test.* American Psychological Association Annual Convention. Honolulu, Hawaii.

Winegard, B. (2023, May 18). Why talk about race differences? *Aporia.* https://www.aporiamagazine.com/p/why-talk-about-race-differences

Winston, A. S. (2024). Forward. W. H. Tucker, *"The bell curve" in perspective: Race, meritocracy, inequality, and politics* (pp. vii–ix). Palgrave Macmillan. https://link.springer.com/book/10.1007/978-3-031-41614-9

Wright, W. (1998). *Born that way.* Knopf.

Yong, E. (2019, May 17). A waste of 1,000 research papers: Decades of early research on the genetics of depression were built on nonexistent foundations. How did that happen? *The Atlantic.* https://www.theatlantic.com/science/archive/2019/05/waste-1000-studies/589684/

Young, A. I., Frigge, M. L., Gudbjartsson, D. F., Thorleifsson, G., Bjornsdottir, G., Sulem, P., Masson, G., Thorsteinsdottir, U., Stefansson, K., & Kong, A. (2018). Relatedness disequilibrium regression estimates heritability without environmental bias. *Nature Genetics, 50*(9), 1304–1310. https://doi.org/10.1038/s41588-018-0178-9

Zuk, O., Hechter, E., Sunyaev, S. R., & Lander, E. S. (2012). The mystery of missing heritability: Genetic interactions create phantom heritability. *PNAS, 109*(4), 1193–1198. https://doi.org/10.1073/pnas.1119675109

24.
USING THE SCIENCE OF LEARNING AND DEVELOPMENT TO TRANSFORM EDUCATIONAL PRACTICE

Linda Darling-Hammond[*], Lisa Flook[†], Channa Cook-Harvey[‡], Brigid Barron[§], and David Osher[**]

Using the Science of Learning and Development to Transform Educational Practice

In recent years, a great deal has been learned about how biology and environment interact to produce human learning and development. In this chapter, we examine how the sciences of learning and development (Figure 241), in combination with decades of educational research, can inform more productive educational settings in which the design of classrooms and the school as a whole supports student's thriving. Recent syntheses of research (Cantor, Osher, Berg, Steyer, & Rose, 2018; Osher, Cantor, Berg, Steyer, & Rose, 2018) from neuroscience and the developmental and learning sciences point to the following foundational principles for education:

1. Brain Development Is Malleable

Development is a lifelong process that begins before birth and continues throughout the life span. People can always learn new skills from birth through adulthood because the brain never stops growing and changing in response to environments, experiences, and relationships. The ability to learn is dynamic, not fixed.

2. Contexts Determine Development

The nature of these experiences and relationships matters greatly to the growth of the brain and the development of skills. Neuronal connections that are part of brain development are enhanced by good nutrition; positive, affirming interactions and responses; experiences that support a sense of safety and trust that enables healthy attachment; and experiences that allow for exploration of language and the physical world. This wiring of the brain establishes a foundation for building more complex skills and abilities in later years that are important for academics and life more generally.

3. Human Relationships Catalyze Healthy Development and Learning

Supportive, responsive relationships with caring adults from birth into adulthood provide the

---

[*] Learning Policy Institute, Palo Alto, VA
[†] Learning Policy Institute, Palo Alto, VA
[‡] Learning Policy Institute, Palo Alto, VA
[§] Stanford University, Stanford, CA
[**] American Institutes for Research, Arlington VA

foundation for healthy development and learning. Optimal brain architecture is developed by the presence of warm, empathetic, consistent relationships; interactive communication; positive experiences; and positive perceptions of these experiences. Supportive, responsive relationships in childhood and adolescence also have an important protective effect. A stable relationship with at least one committed adult can buffer the potentially negative effects of even serious adversity.

4. Learning Is Social, Emotional, and Academic

Emotions and social relationships affect learning. Positive relationships, including trust in the teacher, and positive emotions, such as interest and excitement, open up the mind to learning. Negative emotions, such as fear of failure, anxiety, and self-doubt, reduce the capacity of the brain to process information and to learn. Engagement and effort are supported in classrooms where children feel they are not stigmatized or stereotyped, where they see that they can improve with effort (for example, by revising their work), where they are respected and valued by their teachers and peers, and where they are working on things that matter to themselves and others.

5. Culture Grounds Learning

Because learning occurs as we build on experiences that are, in turn, rooted in our cultural contexts, effective learning environments are responsive to culture (Nasir, Lee, Pea, & de Royston, 2020). And because sensitivity to children's cues is so important to productive relationships, cultural competence is a critical component of the learning environment. Adults who have the cultural competence to appreciate and understand children's experiences and communications are better able to get in sync with the child and to respond affirmatively and appropriately in ways that support both relationships and the cognitive learning process.

6. Children Actively Construct Knowledge

Students dynamically shape their own learning based on their experiences, relationships, and social contexts. Learners compare new information to what they already know to create mental models that enable them to connect new learning to their past experiences and draw inferences about new situations. This process works best when students have multiple opportunities to connect the knowledge to personally relevant topics and lived experiences and when they are active participants in constructing and applying their knowledge to situations that are meaningful to them.

7. Variability in Human Development Is the Norm, Not the Exception

The pace and profile of each child's development is unique. Although development generally progresses in somewhat predictable stages, children begin at different starting points and learn and acquire skills at different rates and in different ways. Rather than assuming all children will respond to the same teaching approaches equally well, effective teachers personalize supports and intervention for different children. Supportive schools avoid attaching labels to children or designing learning experiences around a mythical average. When educators try to force all children to fit one sequence or pacing guide, they miss the opportunity to nurture the individual potential of each child, and they can cause children to adopt counterproductive views about themselves and their own learning potential that undermine progress.

Implications for Schools

We now know from the science of learning and development that environments that are relationship-rich and attuned to students' learning and developmental needs can buffer students' stress, foster engagement, and support learning. To support student achievement, attainment, and behavior, research suggests that schools should attend to four major domains, shown in Figure 5.1 and described next:

1. Building a supportive environment in both classrooms and the school as a whole
2. Explicitly nurturing social and emotional learning
3. Developing productive instructional strategies that enable motivation, competence, and self-directed learning
4. Creating individualized and integrated supports that address student needs, including the effects of trauma and adversity.

1. Building Positive Classroom and School Environments

Warm, caring, supportive student-teacher relationships, as well as other child-adult relationships, are linked to better school performance and engagement, greater social competence, and willingness to take on challenges (Osher et al., 2018). Students who are at higher levels of risk for poor outcomes can benefit especially from nurturing relationships with teachers and other adults, especially when these relationships are culturally sensitive and responsive.

Students need a sense of physical and psychological safety for learning to occur because fear and anxiety undermine cognitive capacity and short-circuit the learning process. Students learn best when they can connect what happens in school to their cultural contexts and experiences, when their teachers are responsive to their strengths and needs, and when their environment is "identity safe" (Steele & Cohn-Vargas, 2013), reinforcing their value and belonging. This is especially important given the societal and school-based aggressions many children, especially those living under adverse conditions, experience. For all these reasons, and because children develop through individual trajectories shaped by their unique traits and experiences, teachers need to know them well to create productive learning opportunities.

Creating Structures That Support Strong, Continuous Relationships

Personalizing the educational setting so that it responds to individual students' interests and needs, as well as their home and community contexts, is one of the most powerful levers to support attachment to and success in school. Students are more likely to attend and succeed academically when they have strong, trusting, and supportive connections to adults, including at least one committed relationship with a close advisor or mentor (Friedlaender et al., 2014).

Developing these relationships can be difficult in schools where organizational structures minimize opportunities for personalized relationships, as is often the case in "factory-model" schools designed a century ago for efficient batch processing of masses of students (Tyack, 1974). At that time, U.S. schools adopted the Prussian age grading model that typically moves students to another teacher each year and to as many as seven or eight teachers daily in secondary schools. Secondary teachers may see 150 to 200 students per day in short 45-minute blocks, and, despite their best efforts, are unable to know all of their students or their families well. Counselors are assigned to attend to the personal needs of hundreds of students, also an unmanageable task, and students who experience adversity may have no one to turn to for support.

The design of most U.S. secondary schools is particularly at odds with the needs of adolescents as it de-emphasizes personal connections with adults and focuses on competitive ranking of students (e.g., in academic tracking and ranking; in tryouts for activities) just as young people most need to develop a strong sense of belonging, connection, and personal identity (Eccles & Roeser, 2009). Depersonalized contexts are most damaging when students are also experiencing the effects of poverty, trauma, and discrimination without supports to enable them to become resilient.

Environments that provide opportunities for stronger relationships among adults and students create more productive contexts for learning. For

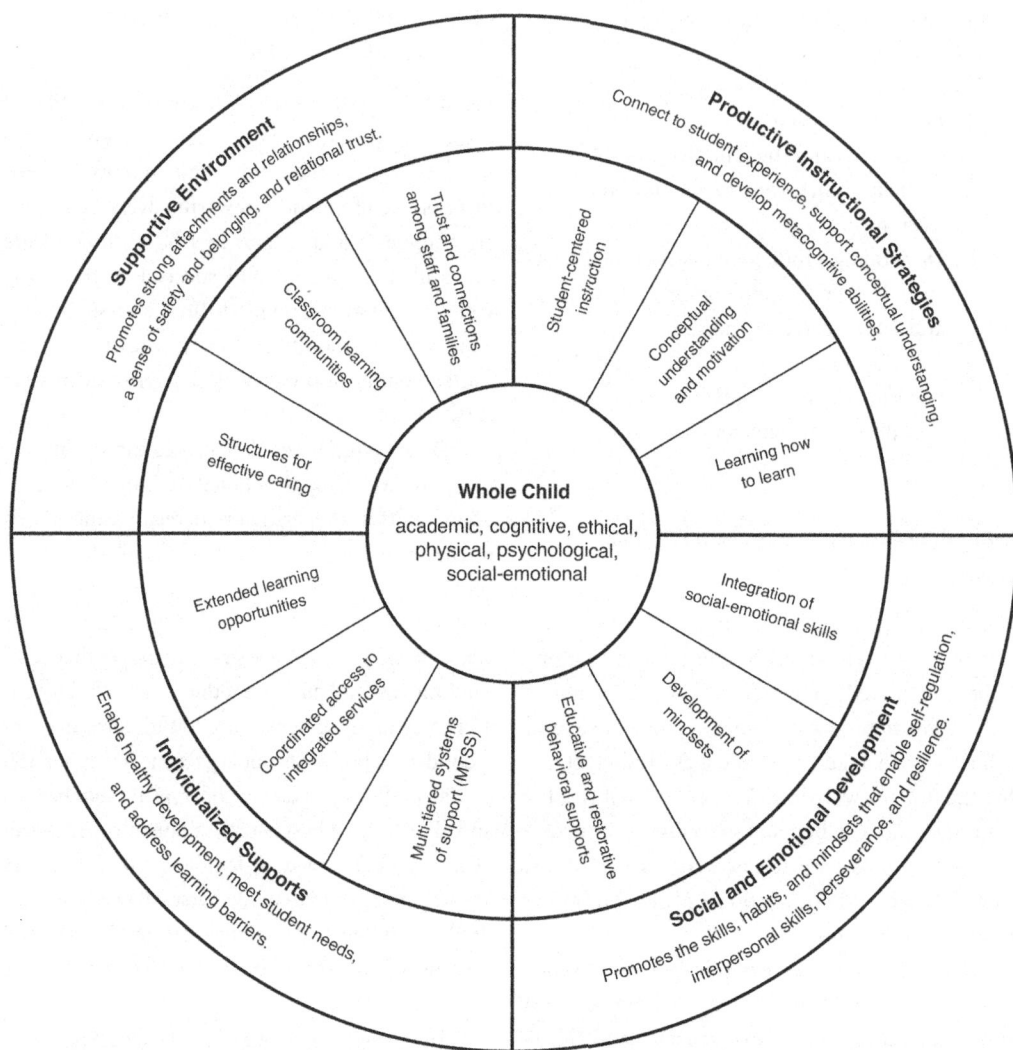

Fig. 24.1 SoLD principles of practice.

example, small schools or small learning communities with personalizing structures—such as advisory systems in which advisors work with a small group of students over multiple years, teaching teams that share students, or looping with the same teachers over two years or more—have been found to improve student achievement, attachment, attendance, attitudes toward school, behavior, motivation, and graduation rates (Bloom & Unterman, 2014; Darling-Hammond, 2006; Felner, Seitsinger, Brand, Burns, & Bolton, 2007). Teachers in such personalized settings report a heightened sense of efficacy, while parents report feeling more comfortable reaching out to the school for assistance.

Structures are important to set the stage for the relationships children need to support their development, but the nature of those relationships and the resulting educational experiences depend on the attitudes, skills, and capacity of staff; the school climate; and the practices that are adopted for instruction, classroom management, school discipline, and more. We turn to these practices next.

Creating Classroom Communities

Learning is a transactional process in which both students and teachers must learn how to understand and communicate with each other in order to produce learning. This requires an intentional community that ensures a sense of belonging and safety. In addition, a culture of participation encourages student agency in the context of a culturally responsive approach that values diverse experiences and involvement in the community (Hamedani, Zheng, Darling-Hammond, Andree, & Quinn, 2015; Noguera, Darling-Hammond, & Friedlaender, 2017).

In developmentally grounded schools, classroom management is approached as something that is done *with* students and not *to* them. Productive classrooms are organized not around compliance emphasizing the punishment of misbehavior, but on the promotion of student responsibility through their participation in developing common norms (LePage, Darling-Hammond, & Akar, 2005). This enables the development of respectful relationships not only between teachers and students, but also among the students themselves, as students are taught to develop social competencies, such as making friends, managing conflict, and caring for others. The teacher's active role in co-regulating children's behavior helps scaffold the child's development toward self-regulation by providing the child a repertoire of words and strategies to use to manage different situations.

One example of such a developmentally grounded approach found to be effective in a number of urban schools builds shared responsibility for learning and classroom organization between teachers and students. The teacher creates a consistent learning environment by working with students to establish a cooperative plan for classroom rules, procedures, use of time, and academic learning. Students shift from being "tourists" to being "citizens" as they create a classroom constitution and take responsibility for dozens of activities in the classroom that teachers might otherwise take on themselves. As they are taught citizenship skills and given multiple chances for leadership in small and large ways, students gain the experiences necessary to become self-disciplined. All adults in the school learn to work with children in consistent ways, and home/community involvement is encouraged. In evaluations of this approach, researchers found that student and teacher attendance improved; discipline referrals declined; and classroom climate and long-term student achievement improved (Freiberg, Huzinec, & Templeton, 2009).

The development of a learning community helps teachers to manage the classroom, not only because children feel more connected, but because it allows for greater collaboration among peers, who gain in competence and agency.

Building Relational Trust and Family Engagement

Relational trust among teachers, parents, and school leaders is another key resource that predicts the likelihood of gains in achievement and other student outcomes where instructional expertise is also present. Trust derives from an understanding of one another's efforts and goals, along with a sense of obligation toward each other, grounded in a common mission. As Bryk and Schneider put it: "Trust is the connective tissue that holds improving schools together" (2002).

Building strong relationships between the school and the family improves academic outcomes for students. One study found parent involvement to be key to the steep achievement gains in 100 Chicago elementary schools: controlling for other variables, students were ten times more likely to achieve substantial gains in mathematics and have increased student motivation and participation in schools with strong parental involvement (Bryk, Sebring, Allensworth, Easton, & Luppescu, 2010).

In another series of meta-analyses, researchers consistently found significant positive effects of parental involvement on academic achievement for children of color in all grades, pre-k through 12th grade (Jeynes, 2012, 2017). The largest effect sizes were for programs that supported parents' direct engagement with their children, including those that

- encouraged parents to engage in shared reading with their children;
- involved parents and teachers working together as partners to develop common strategies, guidelines, and expectations for children;
- increased communication between parents and teachers; and
- involved parents in checking students' homework.

Schools can nurture strong staff–parent relationships by building in time and supports for teachers and advisors to engage parents as partners with valued expertise. They can do this by planning teacher time for home visits, positive phone calls and text or email messages home, school meetings and student–teacher–parent conferences scheduled flexibly around parents' availability, and regular exchanges between home and school (Osher & Osher, 2002).

Enabling Culturally Competent Classrooms

Just as relational trust is key to strong learning contexts, the lack of trust within schools or between schools and families can inhibit learning. When children feel a lack of safety or belonging, or when they experience troubling inconsistencies, their cognitive load is increased, which negatively affects cognition and learning.

If students are to feel safe and have a sense of belonging, they must be understood and respected by their teachers. One aspect of this understanding derives from an appreciation of culture; that is, the shared cultural practices, norms, and belief systems that humans construct in a range of communities defined by family, religion, region, activities or interests, ethnic group membership, or other bonds that are associated with "repertoires of practice" (Gutierrez & Rogoff 2003). At its root, culturally sensitive teaching must appreciate the complexity of individuals' multiple contexts for development, as these provide insights for how to help students make connections among ideas.

Culturally responsive teaching uses "the cultural knowledge, prior experiences, frames of reference, and performance styles of ethnically diverse students to make learning encounters more relevant to and effective for them. It teaches *to and through* the strengths of students" (Gay, 2000). This includes recognizing students' culturally grounded experiences as a foundation on which to build knowledge, exhibiting cultural competency in interacting with students and families, and demonstrating an ethic of deep care while transmitting a sense of efficacy to students (Carter & Darling-Hammond, 2016).

Teachers can learn about the strengths and needs of students as well as their families' funds of knowledge through regular check-ins and class meetings, conferencing, journaling, close observation of students and their work, and connections to parents. These practices can foster developmentally informed relationships among students, parents, and staff.

Eliminating Stereotype Threat

Teachers' perceptions about their students shape expectations that often predict student achievement apart from prior ability. Unfortunately, many teachers attribute inaccurate characterizations of academic ability and behavior to students based on race and ethnicity, have lower expectations of black and Latinx students, and interact with them less positively than with white students (Irvine, 2003; Tenenbaum & Ruck, 2007). Schools reinforce these beliefs to the extent that they group or track students in ways that convey messages about perceived ability, deliver stereotypic messages associated with group status, and emphasize ability rather than effort in their judgments about students and their attributions of causes of success.

The way students are treated outside of and within school can trigger social identity threat, which affects students who are members of groups that have been evaluated negatively in society as a function of race, ethnicity, language, income, disability status, gender, sexual orientation, or other traits. Social identity threat leads to significant stress, symptoms of anxiety and depression, and, sometimes, challenging behavior that results from an attempt to protect one's identity from perceived attack (Major & Schmader, 2018).

Students who have received societal and/or school-delivered messages that they are less capable because of their identities will often translate those views into self-perceptions of ability affecting their performance on school tasks or tests. Stereotype threat, the "social identity threat that occurs when one fears being judged in terms of a group-based stereotype" (Murphy, Steele, & Gross, 2007), induces stress and reduction in working memory and focus, leading to impaired performance.

For many students who carry social identity threats, schools may be viewed as inherently unsafe spaces. Many students of color, LGBTQ students, and others who experience intense societal discrimination are keenly aware of the fact that schools have often been the focus of legal battles focused on their status, such as disputes about segregation and access to school facilities or activities. The heightened assumption that they are not welcome can be exacerbated if they don't see themselves or their identities represented in the curriculum, faculty, staff, policies, practices, or school climate in general. Among the psychic costs of social identity threat in school are feelings of marginalization causing students to "disengage or disidentify with the setting" (Steele, Spencer, & Aronson, 2002).

Stereotype threat can be mitigated in the classroom through teachers' explicit affirmations that students are seen as competent and valued. Many dozens of studies have shown that when students receive such affirmations, performance on tests, grades, and other academic measures improve significantly in ways that are frequently maintained over time (Steele, 2011). Teachers can convey affirming attitudes by exposing students to an intellectually demanding curriculum and supporting them in mastering it, conveying their confidence that students can learn; teaching students strategies they can use to monitor and manage their own learning; encouraging students to excel; and building on the individual and cultural resources they bring to the school.

2. Developing Social, Emotional, and Cognitive Habits, Skills, and Mindsets

Crafting school and classroom environments that support and encourage positive student behavior as well as learning requires recognizing that academic, social, and emotional learning are interconnected—and that they can be explicitly taught. A "key task for educators becomes the intentional development of these skills, traits, strategies, and attitudes in conjunction with the development of content knowledge and academic skills" (Farrington et al., 2012).

Educators have long known that students' academic learning and social-emotional learning (SEL) go hand in hand and that the development of prosocial mindsets, skills, and habits gives students the capacity to persist through challenging work, collaborate with others, take risks while learning, think critically, and communicate effectively. Social, emotional, and other conditions of cognitive engagement affect how the nervous system responds and the degree to which students tap their cognitive and psychological resources, influencing how students can focus their attention and make decisions.

The Collaborative for Academic, Social, and Emotional Learning (CASEL) identifies five main areas of competence: self-awareness, self-management, social awareness, relationship skills, and responsible decision making (Collaborative for Academic, Social, and Emotional Learning,

2013). Researchers at the University of Chicago have developed a comprehensive framework that describes how these and related "co-cognitive" factors are interconnected and jointly provide the foundation for academic learning. The habits, skills, and mindsets are reflected in:

- academic behaviors, such as going to class, completing homework, studying, staying organized, and participating in class;
- academic perseverance, which refers to how well a student completes school assignments to the best of his or her ability despite challenges or obstacles;
- academic mindsets, or a student's attitudes or beliefs about himself or herself in relation to academic work;
- learning strategies, the processes and tactics one employs to aid in the work of thinking, remembering, or learning; and
- social skills, those acceptable behaviors that improve social interactions, such as cooperation, assertion, responsibility, and empathy.

(Farrington et al., 2012)

Formal programs teaching SEL have shown considerable success. A metaanalysis of more than 200 controlled studies of SEL programs representing more than 270,000 students from urban, rural, and suburban schools found that these students showed greater improvements than comparison students in their social and emotional skills; attitudes about themselves, others, and school; social and classroom behavior; and test scores and school grades. They also experienced reductions in stress and depression (Durlak, Weissberg, Dymnicki, Taylor, & Schellinger, 2011). Benefits of these experiences were found to endure and to serve as a protective factor (e.g., preventing conduct problems and drug use) on follow-up measures collected 6 months to 18 years later (Taylor, Oberle, Durlak, & Weissberg, 2017).

Integration of SEL into other courses capitalizes on teachable moments, reinforcing skills throughout the school day (Jones & Bouffard, 2012). In a set of studies of urban high schools that specifically organize their efforts to support social and emotional learning, researchers found that infusion of SEL opportunities in every aspect of the schools produced positive outcomes for student engagement, achievement, attainment, and behavior (being collaborative and supportive of their peers, resilient, employing a growth mindset). SEL infusion ranged from curricula focused on perspective-taking and empathy in history and English language arts to community and social problem solving in social studies, math, and science; community service projects; and the teaching of specific conflict resolution strategies and the use of restorative practices (Hamedani et al., 2015). A whole-school approach imbued with an equity-focused lens and a social justice orientation enabled students to act as agents of change, gaining a sense of efficacy, which contributed to their academic engagement and success.

Educative and Restorative Approaches to Discipline

Social-emotional learning also informs a developmentally appropriate approach to schoolwide discipline, which recognizes students' behaviors as demonstrations of their needs pointing to skills that need to be taught and developed. Explicit teaching of self-regulation, conflict resolution, and other skills can create a virtuous circle of responsible behavior. Studies have found, for example, that even in elementary school, when students learn and practice skills of conflict resolution, they become more inclined to work out problems among themselves before the problems escalate (Johnson, Johnson, Dudley, & Acikgoz, 1994).

Research also finds that coercive discipline, in which teachers manage student behavior largely through punishments, inhibits the students' development of responsibility, ultimately increasing misbehavior as students abandon self-responsibility for their learning and behavior and develop resistance and opposition to school (Lewis, 2001). By contrast, an educative approach supports learning, as teachers' proactive and positive responses create a safe and empowering classroom environment when they use reinforcing and reminding language (including verbal and nonverbal cues); approach students in a nonthreatening manner; present students with problem-solving options as a means of de-escalating potentially explosive situations; and use nonpunitive, restorative consequences (Turnaround for Children, 2016). Students who learn in such communities have higher levels of self-understanding, commitment, performance, and belongingness, and fewer discipline problems.

Educative approaches are important for addressing the excessive reliance on exclusionary discipline in many schools, which persists in spite of evidence that punishment and exclusion do not work and often have harmful effects (Osher, Bear, Sprague, & Doyle, 2010). Student suspension and expulsion produce a range of dysfunctional consequences: the more time students spend out of the classroom, the more their sense of connection to the school wanes, both socially and academically. This distance promotes disengaged behaviors, such as truancy, chronic absenteeism, and antisocial behavior (Hemphill, Toumbourou, Herrenkohl, McMorris, & Catalano, 2006), which in turn exacerbate a widening achievement gap. The frequency of student suspensions is linked to academic declines and an increased likelihood of dropping out.

This is particularly the case for many students of color, who are disproportionately removed from class and school and are removed for longer terms, especially for minor offenses. Exclusionary discipline does not teach students new strategies they can use to solve problems, nor does it help teachers understand how they may unintentionally trigger or escalate problem behavior (Losen, 2015). Extensive use of exclusionary discipline undermines school climate overall by degrading the sense of community while exacerbating misbehavior, as students who are suspended often return frustrated and angry, further behind academically, and more likely to disrupt others.

Many schools have decreased their suspension and expulsion rates by adopting social-emotional learning and restorative practices (Skiba, Arredondo, & Rausch, 2014). Restorative discipline provides systems for students to reflect on any mistakes, repair damage to the community, and get counseling when needed. Relationships and trust are supported through universal interventions such as daily classroom meetings, community-building circles, and conflict resolution strategies. When challenging events occur, schools use restorative conferences or peer mediation, which bring together the parties involved in conflict, with the support of a facilitator, to talk about what happened, the impact, and how to repair any harm to the individual or the community.

Syntheses of research suggest that restorative practices result in fewer and less racially disparate suspensions and expulsions, fewer disciplinary referrals, improved school climate, higher quality teacher-student relationships, and improved academic achievement (Fronius, Persson, Guckenburg, Hurley, & Petrosino, 2016). Creating an environment in which students learn to be responsible and have opportunities to contribute can transform social, emotional, and academic behavior and outcomes.

3. Providing Supports for Student Motivation and Learning

Having created a supportive environment for learning, what are the curriculum designs, instructional approaches, and assessment practices that will enable students to deeply understand disciplinary content and develop skills that will allow them to solve complex problems, communicate effectively, and ultimately manage their own learning?

Children are natural learners and inherently seek to learn things that matter in their immediate everyday world. To support children's learning, adults make connections between new situations and familiar ones, focus children's attention, structure experiences, and organize the information children receive, while helping them develop strategies for intentional learning and problem solving (National Research Council, 2000).

Mindsets, Motivation, and Learning

Learning is a function both of teaching—what is taught and how it is taught—and student perceptions about the material being taught and about themselves as learners. Students' beliefs and attitudes have a powerful effect on their learning and achievement.

Four key mindsets have been identified as important for perseverance and academic success for students. They include:

1. Belief that one belongs at school
2. Belief in the value of the work
3. Belief that effort will lead to increased competence
4. A sense of self-efficacy and the ability to succeed (Farrington, 2013).

Shaping productive mindsets stimulates a cascade of effects that accumulate over time to result in higher levels of academic engagement that become self reinforcing. For example, a growth mindset—the belief that effort will lead to increased competence—ultimately heightens mental and affective abilities and perseverance (Dweck, 2000). The core principle that skills and capacities can be developed is consistent with evidence that the brain is constantly growing and changing in response to experience.

Closely related to these developmental and cognitive processes is the issue of motivation for learning. Students will work harder to achieve understanding and will make greater progress when they are motivated to learn. However, motivation is not just inherent in the individual; it can be developed by skillful teaching.

Researchers have found that student motivation in the classroom is fostered by three major considerations: (1) the nature of the *task* and its value to the student; (2) the nature of the *learner* and his or her expectations of success; and (3) the nature of the *learning environment* and the extent to which it emphasizes learning goals and provides support (Blumenfeld, Puro, & Mergendoller, 1992). The learner's perceptions of the task matter greatly. As Carol D. Lee notes, the learner implicitly asks: "What am I being asked to do?" "Am I capable of tackling these tasks?" "Is this task meaningful to me?" "What supports are available to me to wrestle with this task?" "Do I feel safe in attempting to wrestle with this task?" and "How do I weigh any risks or competing goals?" (Lee, 2017).

A *learning task* has more value to students if it is relevant to their lives, can be connected to events they have experienced or care about, and offers choices of topics, research strategies, or modes of presentation that allow students to make connections to their interests and talents. The task should also be approachable (i.e., within the zone of proximal development), and it should be structured to provide evidence of progress along the way, so that it offers ongoing incentives to continue. Students are more likely to value learning when intrinsic reasons for learning are emphasized, as when the task potentially benefits others and/or results in products or performances that have an audience beyond the teacher.

To be motivated, students need to believe they can be successful. Their *expectations for success* influence their willingness to engage in learning tasks. These expectations depend on their perceptions of the task and their own abilities, as well as on their inclinations to undertake new learning, tackle difficult tasks, and take risks. These inclinations, in turn, are related to self-perceptions of ability and mindsets as well as teachers' actions. To enhance expectancies of success, teachers can actively guide student efforts, scaffold instruction, and provide multiple pathways to learning.

The *learning environment* supports motivation when learning and mastery goals are emphasized, rather than grades or performance goals. Learning goals are encouraged when scaffolding and support are provided; effort and improvement are recognized; mistakes are treated as learning opportunities; students have the opportunity to revise their work; evaluation emphasizes learning; individual competition and comparison are minimized; and students are grouped by topic, interest, or choice rather than by their performance.

Teaching for Understanding

The expectations that graduates have the problem-solving and interpersonal skills needed for 21st-century success require a focus on instruction designed to foster outcomes such as critical thinking, communication, collaborative problem solving, and high-level reasoning. These abilities cannot be developed through passive, rote-oriented learning focused on the memorization of disconnected facts. They require paths to deeper understanding supporting the transfer of skills and use of knowledge in new situations (Goldman & Pellegrino, 2015).

These goals point us to some important insights from the learning sciences. For example, the development of neural pathways is associated with exposure to and generation of language (Kuhl. 2000), which implies that students must be active generators of content in a classroom and not just receivers. Furthermore, emotion triggers learning as it affects excitement and attention (Immordino-Yang & Damasio, 2007) and thus must be a consideration in designing instruction that is mentally engaging. At the same time, consistent structures that allow the student to know what to expect and how to be successful reduce cognitive load and free up the mind for learning other challenging material (Paas, Renkl, & Sweller, 2003).

With these goals and insights in mind, specific pedagogical moves that support the learning process and increase intrinsic motivation include:

- choice of tasks that have the right amount of challenge, such as demanding analysis to answer a question or develop a product, with supportive guidance and feedback;
- well-designed questions to stimulate inquiry and engagement as well as to support students putting information together to find answers and consolidate understanding;
- use of multiple and varied representations of concepts that allow students to "hook into" understanding in different ways;
- design of instructional conversations and "joint productive activity" (Tharp, Estrada, Dalton, & Yamaguchi, 2000) that allow students to discuss their emerging thinking and hear other ideas, developing concepts, language, and further questions in the process;
- encouragement for students to elaborate, question, and self-explain; and
- instruction and curriculum that use apprentice-style relationships in which knowledgeable practitioners or older peers facilitate students' ever-deeper participation in a particular field or domain.

(Bransford & Donovan, 2005)

The importance of these principles has been documented in schools with successful outcomes serving large numbers of students of color and students from low-income families (Noguera et al., 2017). In these schools, teachers organize inquiry in ways that help students develop an understanding of their own learning and how to continually improve, teaching and reinforcing cognitive strategies.

Part of effective teaching is learning what students already know and how they can bring that knowledge into the classroom context. As Nasir and colleagues point out, "Often, people can competently perform complex cognitive tasks outside of school, but may not display these skills on school-type tasks" (2006). For example, complex statistical calculations used on the basketball court may not initially carry into the mathematics class, unless teachers are alert to support the transfer by building on this kind of real-world knowledge.

The bridge between students' experiences and school content can be built using a cultural modeling approach that draws on the familiar to make the structure of a domain visible and explicit to students (Lee, 2007). Carol D. Lee, for example, writes about how she illustrated symbolic meanings in literature by beginning with rap songs and texts the students knew and carrying their insights into study of more formal canonic texts. Similarly, Jo Boaler's study of the outcomes of inquiry-based instructional practices in mathematics classrooms serving students from low-income families found that inequalities were reduced when teachers contextualized problems and made them relevant to students' lives, introducing new concepts through discussion and asking students to explain and discuss their thinking (2002).

Such inquiry-based learning typically takes place in collaborative groups that provide opportunities to collaborate with peers in ways that support the development of self-regulation, executive function, and social skills. Extensive research identifies significant achievement benefits of social learning in well-managed groups (Ginsburg-Block et al., 2006; Barron & Darling-Hammond. 2008). Collaborative learning can provide students with learning assistance from peers within their zone of proximal development and opportunities to strengthen metacognitive skills as they share insights, resolve differing perspectives through argument, explain their thinking about a phenomenon, observe the strategies of others, and listen to explanations (Barron & Darling-Hammond, 2008).

While group work can enhance student learning, teachers must know how to structure this work. In successful use of cooperative approaches, such as Complex Instruction, teachers help students structure roles within the group and provide questions and tasks that guide the group's discussion. Teachers create group-worthy tasks in which all must engage for the work to be successfully accomplished and they scaffold the material to be learned. They play an active role in constructing the tasks and questions that help students learn to coordinate their work and frame their ideas in terms that reflect the modes of inquiry in the discipline. These efforts produce strong learning gains and reduce achievement gaps among student groups (Cohen & Lotan, 2014).

Mastery-Oriented Assessment

Finally, a mastery-focused approach to assessment that emphasizes learning goals has been found to help sustain achievement-directed behavior over time and to orient learners toward a focus on improving competence and deeply understanding the work they produce. In addition, assessments that place value on growth rather than on scores earned at one discrete moment have been found to create higher motivation and higher levels of cognitive engagement. In contrast, evaluative, comparison-oriented testing leads to students' decreased interest in school, distancing from the learning environment, and a lowered sense of self-confidence and personal efficacy (Eccles & Roeser, 2009).

Many schools that have been particularly successful in reducing opportunity and achievement gaps for traditionally marginalized students have adopted performance-based assessments that build higher-order thinking and performance skills; collaboration and communication skills; motivation and engagement; and co-cognitive skills such as self-regulation, executive function, resilience, perseverance, and growth mindset (Noguera et al., 2017). In these schools, projects, papers, portfolios, and other products are evaluated through rubrics that clearly describe dimensions of quality, and students have the opportunity to revise their work in response to feedback. These kinds of performance assessments encourage higher-order thinking and reasoning while requiring students to demonstrate understanding. The assessments themselves are learning tools that also build students' co-cognitive skills such as planning, organizing, and other aspects of executive functioning; resilience and perseverance in the face of challenges; and a growth mindset (Darling-Hammond & Adamson, 2014).

4. Creating Integrated Systems of Support to Address Student Needs

To enable this kind of learning, many students need additional supports. Each year in the United States, more than 40 million children are exposed to violence, crime, abuse, or psychological trauma. These types of adverse childhood experiences (ACEs) are connected to poor health and educational outcomes, including absenteeism and changes in school performance as trauma affects focus, attention, and anxiety (National Child Traumatic Stress Network, 2018).

Adversity happens in all communities, and healthy development does as well. However, inequality creates increased risks. Poverty and racism, together and separately, make the experience of chronic stress and adversity more likely as they manifest in food and housing insecurity and survival concerns. In schools where students encounter implicit bias and punitive discipline tactics rather than supports for handling adversity, their stress is magnified.

School practices can exacerbate or buffer the effects of childhood adversity. The Center on the Developing Child at Harvard University has identified a common set of actions schools, families, and communities can take to make it more likely that children will experience positive outcomes in the face of significant adversity (Center on the Developing Child, 2016). These include

- facilitating supportive adult-child relationships that extend over time;
- building a sense of self-efficacy and control by teaching and reinforcing social and emotional skills that help children handle adversity, such as the ability to calm emotions and manage responses; and
- creating strong, dependable, supportive routines for both managing classrooms and checking in on student needs.

Effective school environments take a systematic approach to promoting children's development. Environments that are trauma-sensitive provide children with structure, psychological safety, adult alertness, and responsiveness. This responsiveness requires a personalized approach to identify and address each child's full range of needs.

A key aspect of creating a supportive environment is a shared developmental framework among all of the adults in the school, coupled with procedures for ensuring that students receive additional help for health, mental health, social, emotional, or academic needs when they need them, without costly and elaborate labeling procedures standing in the way. Multi-tiered systems of support (MTSS) that organize multidisciplinary student support teams to focus on individual student needs are one increasingly widespread approach.

Most systems include three tiers of support (Adelman & Taylor, 2008). The first tier is universal—everyone experiences it. Ideally, it uses teaching strategies grounded in universal designs for learning that are broadly successful with children who learn in different ways, as well as use of explicit social-emotional learning models and positive behavioral support strategies that are culturally and linguistically responsive. Tier 2 services and supports address the needs of students who are at some elevated level of risk or who need some additional support in particular areas. The risk may be demonstrated by behavior (e.g., number of absences) or may be due to having experienced a known risk factor (e.g., the loss of a parent). These services may include academic supports (e.g., Reading Recovery, mathematics tutoring, extended learning time) or family outreach, counseling, and behavioral supports. Schools may operate counseling groups to support students who have experienced loss or violence, who are managing traumatic events, and who need mental health supports. They may use social workers to help students—and sometimes their families—access supports and services. Tier 3 services involve intensive interventions for students who are at particularly high levels of risk or whose needs are not sufficiently met by tier 2 interventions. Tier 3 services might include wraparound services, one-on-one mental health supports, and effective special education (Osher et al., 2016). These supports often benefit from collaboration with local service agencies and community-based organizations with communication feedback loops to school-based staff.

A whole-child approach that deals with students in connected rather than fragmented ways is critically important. Substantial improvements can occur for children when staff and parents work together from a shared understanding of child development. The School Development Program (SDP) is an example of this approach. Building a school culture that addresses all the developmental pathways—social, emotional, cognitive, physical, and moral/ethical—the program establishes collaborative teams among principals, parents, teachers, community leaders, superintendents, and health care workers, teaching them about child development and grounding collective action in a shared framework. Research on the SDP across many cities and schools shows that it has helped reduce absenteeism and suspension, improve school climate, increase student self-concept, and strengthen student achievement (Lunenburg, 2011).

Integrated Student Services

Awareness of the pervasiveness of student toxic stress across the income spectrum and the growth of child poverty in economically traumatized communities have created demands for integrated student services, also called wraparound services, which link schools to a range of academic, health, and social services. Integrated student support (ISS) programs address the reality that children whose families are struggling with poverty—and the housing, health, and safety concerns that often go with it—cannot learn most effectively unless their non-academic needs are also met. The goal is to remove barriers to school success by connecting students and families to service providers in the community, or bringing those services into the school.

Models like the Children's Aid Society in New York City, the West Philadelphia Improvement Corps, and Communities in Schools have brought social services to schools through community partnerships for over 30 years, offering on-site child care and early childhood development; job training and literacy learning for parents; health care and mental health services; and food assistance. A research synthesis that examined 11 experimental studies of ISS models found significant positive effects on student progress in school, attendance, achievement, and grades. These studies also found measurable decreases in grade retention, dropout rates, and absenteeism (Moore & Emig, 2014). Similar findings emerged from a more recent study of the Massachusetts Wraparound Zones program (Gandhi et al., 2015).

These features come together in community school models that partner with community agencies and resources to provide an integrated focus on academics, health and social services, youth and community development, and community engagement. Many operate year-round, from morning to evening, serving both children and adults. A recent review of 125 studies of community schools found significant evidence for the benefits of these approaches for student outcomes ranging from attendance and behavior to learning and educational attainment (Oakes, Maier, & Daniel, 2017).

Summary

The emerging science of learning and development makes it clear that a whole-child approach to education, which begins with strong relationships and a positive school climate that affirms and supports all students, is essential to support academic achievement as well as healthy development. Classrooms that build on these assets to engage students in motivating and meaningful ways, that build on prior knowledge and cultural experience, and that support student agency and metacognition leverage significantly greater learning. Research and the wisdom of practice offer significant insights for policy makers and educators about how to develop such rich learning environments. The challenge ahead is to assemble the whole village—schools, health care organizations, youth and family serving agencies, state and local governments, philanthropists, and families—to work together to ensure that every young person receives the benefit of what is known about how to support his or her healthy path to a productive future.

References

Adelman, H. S., & Taylor, L. (2008). School-wide approaches to addressing barriers to learning and teaching. In B. Doll & J. Cummings (Eds.), *Transforming school mental health services: Population-based approaches to promoting the competency and wellness of children*. Thousand Oaks, CA: Corwin Press.

Barron, B., & Darling-Hammond, L. (2008). How can we teach for meaningful learning? In L. Darling-Hammond (Ed.), *Powerful learning: What we know about teaching for understanding*. San Francisco, CA: Jossey-Bass.

Bloom, H. S., & Unterman, R. (2014). Can small high schools of choice improve educational prospects for disadvantaged students? *Journal of Policy Analysis and Management, 33(2),* 290–319.

Blumenfeld, P. C., Puro, P., & Mergendoller, J. (1992). Translating motivation into thoughtfulness. In H. H. Marshall (Ed.), *Redefining student learning* (pp. 207–241). New York, NY: Ablex Publishing Corporation.

Boaler, J. (2002). Learning from teaching: Exploring the relationship between reform curriculum and equity. *Journal for Research in Mathematics Education, 33(4),* 239–258.

Bransford, J. D., & Donovan, M. S. (2005). Scientific inquiry and how people learn. In *Plow students learn: History, mathematics, and science in the classroom* (pp. 397–420). Washington, DC: The National Academies Press.

Bryk, A., & Schneider, B. (2002). *Trust in schools: A core resource for improvement*. New York: Russell Sage Foundation.

Bryk, A. S., Sebring, P. B., Allensworth, E., Easton, J. Q., & Luppescu, S. (2010). *Organizing schools for improvement: Lessons from Chicago*. Chicago, IL: University of Chicago Press.

Cantor, P., Osher, D., Berg, J., Steyer, L., & Rose, T. (2018). Malleability, plasticity, and individuality: How children learn and develop in context. *Applied Developmental Science*. doi:10.1080/10888691.2017.1398649

Carter, P., & Darling-Hammond, L. (2016). Teaching diverse learners. In D. H. Gitomer & C. Bell (Eds.), *Handbook of research on teaching* (5th ed., pp. 593–638). Washington, DC: American Educational Research Association.

Center on the Developing Child. (2016). *From best practices to breakthrough impacts: A science-based approach to building a more promising future for young children and families*. Cambridge, MA: Harvard University, Author.

Cohen, E. G., & Lotan, R. A. (2014). *Designing groupwork: Strategies for the heterogeneous classroom*. New York, NY: Teachers College Press.

Collaborative for Academic, Social, and Emotional Learning. (2013). *2013 CASEL guide: Effective social and emotional learning programs: Preschool and elementary school edition*. Chicago, IL: Collaborative for Academic, Social, and Emotional Learning.

Darling-Hammond, L., & Adamson, F. (2014). *Beyond the bubble test: How performance assessments support 21st century learning*. San Francisco, C.A: John Wiley & Sons.

Darling-Hammond, L., Ross, P., & Milliken, M. (2006). High school size, organization, and content: What matters for student success? *Brookings Papers on Education Policy, 2006/2007* (9), 163–203.

Durlak, J. A., Weissberg, R. P., Dymnicki, A. B., Taylor, R. D., & Schellinger, K. B. (2011). The impact of enhancing students' social and emotional learning: A metaanalysis of school-based universal interventions. *Child Development, 82*(1), 405–432.

Dweck, C. S. (2000). *Self-theories: Their role in motivation, personality, and development*. London, UK: Psychology Press.

Eccles, J. S., & Roeser, R. W. (2009). Schools, academic motivation, and stage-environment fit. In R. M. Lerner & L. Steinberg (Eds.), *Handbook of adolescent psychology*. Hoboken, NJ: John Wiley & Sons.

Farrington, C. A. (2013). *Academic mindsets as a critical component of deeper learning*. Chicago, IL: Consortium on Chicago School Research.

Farrington, C. A., Roderick, M., Allensworth, E., Nagaoka, J., Keyes, T. S., Johnson, D. W., & Beechum, N. O. (2012). *Teaching adolescents to become learners: The role of non-cognitive factors in shaping school performance: A critical literature review* (p. 5). Chicago, IL: Consortium on Chicago School Research.

Felner, R. D., Seitsinger, A. M., Brand, S., Burns, A., & Bolton, N. (2007). Creating small learning communities: Lessons from the project on high-performing learning communities about "what works" in creating productive, developmentally enhancing, learning contexts. *Educational Psychologist, 42*(4), 209–221.

Freiberg, H. J., Huzinec, A. C., & Templeton, S. M. (2009). Classroom management: A pathway to student achievement: A study of fourteen inner-city elementary schools. *Elementary School Journal, 110*(1), 63–80.

Friedlaender, D., Burns, D., Lewis-Charp, H., Cook-Harvey, C. M., Zheng, X., & Darling-Hammond, L. (2014). *Student-centered schools: Closing the opportunity gap*. Stanford, CA: Stanford Center for Opportunity Policy in Education.

Fronius, T., Persson, H., Guckenburg, S., Hurley, N., & Petrosino, A. (2016). *Restorative justice in U.S. schools: A research review*. San Francisco, CA: WestEd.

Gandhi, A., Slama, R., Park, S., Russo, P., Bzura, R., & Williamson, S. (2015). *Focusing on the whole student: Final report on the Massachusetts Wraparound Zones*. Waltham, MA: American Institutes for Research.

Gay, G. (2000). *Culturally responsive teaching: Theory, research, and practice*. New York, NY: Teachers College Press.

Ginsburg-Block, M. D., Rohrbeck, C. A., & Fantuzzo, J. W. (2006). A meta-analytic review of social, self-concept, and behavioral outcomes of peer-assisted learning. *Journal of Educational Psychology, 98*, 732–749.

Goldman, S., & Pellegrino, J. (2015). Research on learning and instruction: Implications for curriculum, instruction, and assessment. *Policy Insights from the Behavioral and Brain Sciences, 2*(1), 33–41.

Gutierrez, K., & Rogoff, B. (2003). Cultural ways of learning: Individual traits or repertoires of practice. *Educational Researcher, 32*(5), 19–25.

Hamedani, M. G., Zheng, X., Darling-Hammond, L., Andree, A., & Quinn, B. (2015). *Social emotional learning in high school: How three urban high schools engage, educate, and empower youth*. Stanford, CA: Stanford Center for Opportunity Policy in Education.

Hemphill, S. A., Toumbourou, J. W., Herrenkohl, T. I., McMorris, B. J., & Catalano, R. F. (2006). The effect of school suspensions and arrests on subsequent adolescent antisocial behavior in Australia and the United States. *Journal of Adolescent Health, 39*(5), 736–744.

Immordino-Yang, M. H., & Damasio, A. (2007). We feel, therefore we learn: The relevance of affective and social neuroscience to education. *Mind, Brain, and Education, 1*(1), 3–10.

Irvine, J. J. (2003). *Educating teachers for diversity: Seeing with a cultural eye*. New York, NY: Teachers College Press.

Jeynes, W. H. (2012). A meta-analysis of the efficacy of different types of parental involvement programs for urban students. *Urban Education, 47*(4), 706–742.

Jeynes, W. H. (2017). A meta-analysis: The relationship between parental involvement and Latino student outcomes. *Education and Urban Society, 49*(1), 4–28.

Johnson, D. W., Johnson, R., Dudley, B., & Acikgoz, K. (1994). Effects of conflict resolution training on elementary school students. *The Journal of Social Psychology, 134*(6), 803–817.

Jones, S. M., & Bouffard, M. B. (2012). Social and emotional learning in schools: From programs to strategies. *Social Policy Report, 26*(4), 1–33.

Kuhl, P. (2000). *A new view of language acquisition*. Washington, DC: National Academy of Sciences.

Lee, C. D. (2007). *Culture, literacy, and learning: Taking bloom in the midst of the whirlwind*. New York, NY: Teachers College Press.

Lee, C. D. (2017). Integrating research on how people learn and learning across settings as a window of opportunity to address inequality in educational processes and outcomes. *Review of Research in Education*, 1–24.

LePage, P., Darling-Hammond, L., & Akar, H. (2005). Classroom management. In L. Darling-Hammond & J. Bransford (Eds.), *Preparing teachers for a changing world: What teachers should learn and be able to do*. San Francisco, CA: Wiley.

Lewis, R. (2001). Classroom discipline and student responsibility: The students' view. *Teaching and Teacher Education, 17*(3), 307–319.

Losen, D. J. (2015). *Closing the discipline gap*. Columbia, NY: Teachers College Press.

Lunenburg, F. C. (2011). The Comer School Development Program: Improving education for low-income students. *National Forum of Multicultural Issues Journal, 8*(1), 1–14.

Major, B., & Schmader, T. (2018). Stigma, social identity threat, and health. In B. Major, J. F. Dovidio, & B. G. Link (Eds.), *The Oxford handbook of stigma, discrimination, and health*. New York, NY: Oxford University Press.

Moore, K. A., & Emig, C. (2014). *Integrated student supports: A summary of the evidence base for policymakers*. White paper #2014-05. Bethesda, MD: Child Trends.

Murphy, M. C., Steele, C. M., & Gross, J. J. (2007). Signaling threat: How situational cues affect women in math, science, and engineering settings. *Psychological Science, 18*(10), 879–885.

Nasir, N. S., Lee, C. D., Pea, R., & de Royston, M. M. (2020). *Handbook of the cultural foundations of learning*. New York: Routledge.

Nasir, N. S., Rosebery, A. S., Warren, B., & Lee, C. D. (2006). Learning as a cultural process: Achieving equity through diversity. In *The Cambridge handbook of the learning sciences* (pp. 489–504). New York, NY: Cambridge University Press.

National Child Traumatic Stress Network. (2018). *Creating, supporting, and sustaining trauma informed schools: A systems framework*. Retrieved from https://www.nctsn.org/resources/creating-supporting-and-sustaining-trauma-informed-schools-system-framework

National Research Council. (2000). *How people learn: Brain, mind, experience, and school: Expanded edition*. Washington, DC: National Academies Press.

Noguera, P., Darling-Hammond, L., & Friedlaender, D. (2017). Equal opportunity for deeper learning. In R. Heller, R. Wolfe, & A. Steinberg (Eds.), *Rethinking readiness: Deeper learning for college, work, and life*. Cambridge, MA: Harvard Education Press.

Oakes, J., Maier, A., & Daniel, J. (2017). *Community schools: An evidence-based strategy for equitable school improvement*. Boulder, CO: National Education Policy Center and Palo Alto, CA: Learning Policy Institute.

Osher, D., Bear, G., Sprague, J., & Doyle, W. (2010). How we can improve school discipline. *Educational Researcher, 39*(1), 48–58.

Osher, D., Cantor, P., Berg, J., Steyer, L., & Rose, T. (2018). Drivers of human development: How relationships and context shape learning and development. *Applied Developmental Science*. doi: 10.1080/10888691.2017.1398650

Osher, D., Kidron, Y., Brackett, M., Dymnicki, A., Jones, S., & Weissberg, R. P. (2016). Advancing the science and practice of social and emotional learning: Looking back and moving forward. *Review of Research in Education, 40*(1), 644–681.

Osher, T. W., & Osher, D. M. (2002). The paradigm shift to true collaboration with families. *Journal of Child and Family Studies, 11*(1), 47–60.

Paas, F., Renkl, A., & Sweller, J. (2003). Cognitive load theory: Instructional implications of the interaction between information structures and cognitive architecture. *Instructional Science, 32*(1–2), 1–8.

Skiba, R. J., Arredondo, M. I., & Rausch, M. K. (2014). *New and developing research on disparities in discipline*. Bloomington, IN: The Equity Project at Indiana University.

Steele, C. M. (2011). *Whistling Vivaldi: How stereotypes affect us and what we can do*. New York, NY: W.W. Norton & Company.

Steele, C. M., Spencer, S. J., & Aronson, J. (2002). Contending with group image: The psychology of stereotype and social identity threat. In *Advances in experimental social psychology* (Vol. 34, pp. 379–440). Cambridge, MA: Academic Press.

Steele, D. M., & Cohn-Vargas, B. (2013). *Identity safe classrooms: Places to belong and learn*. Thousand Oaks, CA: Corwin Press.

Taylor, R. D., Oberle, E., Durlak, J. A., & Weissberg, R. P. (2017). Promoting positive youth development through school-based social and emotional learning interventions: A meta-analysis of follow-up effects. *Child Development, 88*(4), 1156–1171.

Tenenbaum, H. R., & Ruck, M. D. (2007). Are teachers' expectations different for racial minority than for European American students? A meta-analysis. *Journal of Educational Psychology, 99*(2), 253–273.

Tharp, R. G., Estrada, P., Dalton, S., & Yamaguchi, L. A. (2000). *Teaching transformed: Achieving excellence, fairness, inclusion, and harmony*. Boulder, CO: Westview Press.

Turnaround for Children. (2016). *Classroom and Behavior Management (CBM) unit overview*. Washington, DC: Author.

Tyack, D. B. (1974). *The one best system: A history of American urban education* (Vol. 95). Cambridge, MA: Harvard University Press.

25.
THE FUTURE OF THE SCIENCE OF LEARNING AND DEVELOPMENT
Whole-Child Development, Learning, and Thriving in an Era of Collective Adversity, Disruptive Change, and Increasing Inequality
Pamela Cantor* and David Osher[†]

We live in a country where children's life chances are defined largely by their zip code (Bronfenbrenner & Crouter, 1983; Chetty, Friedman, Hendren, Jones, & Porter, 2018; Manduca, & Sampson, 2019). Zip codes often dictate which schools children can attend and the quality of those schools; the quality of housing, water supply, and available food; access to high-quality health care; and the depth of formal community resources that exist to support families (e.g., see Sampson, 2016). Adults and children in low-income communities with large black, Latinx, and first nation's populations have experienced centuries of injustice due to systemic and institutional racism and oppression—including racist housing practices and education, health care, and economic policies; police control and brutality, segregation, minoritization; and marginalization—causing high levels of adversity and persistent stress, which can compromise community well-being (e.g., Duncan, Magnuson, & Votruba-Drzal, 2015; Galea, Merchant, & Lurie, 2020; Murry, Hill, Witherspoon, Berkel, & Bartz, 2015). COVID-19 and the economic destabilization it has caused, coupled with police brutality and societal unrest, compound existing inequities and create profound challenges to family and community stability, in particular for black and brown and first nation children and

* Turnaround for Children, New York, NY
† American Institutes for Research, Arlington VA

families (e.g., Fortuna, Tolou-Shams, Robles-Ramamurthy, & Porche, 2020). These and other prejudice-related social toxicities affect the experiences of girls and women, people with disabilities, individuals who are LGBTQ, and members of religious and ethnic minorities—worldwide. Sexism, misogyny, ableism, heterosexism, and ethnocentrism are both harmful in their own right; they also amplify and, as well, are amplified by the effects of racism and poverty (e.g., Epstein, Blake, & González, 2017).

Looked at through the lens of the concepts presented in the chapters and associated commentaries in the present book, the effects of increased adversity and inequity on development and thriving will be the defining features of the context in which many millions of young people in the United States and across the world are growing up. Although this has long been the case for many people, the new inequities are portentous: a growing percentage of the current generation of children and adolescents could now, and moving forward, be living life on an expanding economic, educational, developmental, and social precipice. Rebecca Winthrop and colleagues (Winthrop, 2018; Winthrop & McGivney, 2017) explain that, by 2030, there will be 800 million children around the globe who will have reached their adulthood without possessing basic secondary level skills. Winthrop and colleagues believe that a century of progress is needed to enable most marginalized youth to attain the

DOI: 10.4324/9781032702988-41

sort of education that the wealthiest youth experience at this writing.

Although these risks are not new, they are becoming more pronounced in the form of inequitable hardship and higher barriers to quality education and jobs, with shrinking opportunities for economic and social mobility (Ambrose, 2020; Fortuna et al., 2020) that are the product of longer-term change due in part to advances in artificial intelligence and robotics, as well as due to COVID-19. Furthermore, because of the new realities that COVID-19 has created, children and families who had surmounted barriers and secured access to high-quality education and jobs, and who had managed to move out of poverty, are being pushed back into positions of insecurity. Heightened stress, and risk, with college education, high-paying jobs, and housing security are, increasingly, more and more out of reach (e.g., Patel et al., 2020).

We are facing an impending convergence of several long-standing but now exacerbated realities that will threaten the economy of the United States and other nations for years to come and compromise the futures of specific subpopulations in particular (e.g., Blundell, Costa Dias, Joyce, & Xu, 2020). For example, in the United States there is a growing gap in opportunity and life outcomes in specific subgroups, for instance between students in wealthy, working-class communities and students in low-income urban and rural districts, especially students of color, whose circumstances have now been made significantly more challenging. The rates of poverty among U.S. children have increased and continue at alarming rates (Koball & Jiang, 2018).

The United States is seeing declining performance and opportunity at both ends of the attainment spectrum. At the higher end, U.S. students are declining in their performance on the Programme for International Student Assessment (PISA) test as compared to other industrialized nations (e.g., Darling-Hammond, 2014; Schleicher, 2019). This change reveals a risk that American students will not be as qualified for the world's most competitive job opportunities as students from other countries (OECD, 2019; Tucker, 2020a, 2020b). Looking beyond these statistics reveals that education and teaching have not been incorporating recent findings from neuroscience and the learning sciences and have not sufficiently emphasized the breadth of competencies and what are termed twenty-first century skills, that is, skills that prioritize how to use knowledge and the interpersonal and intrapersonal skills needed to thrive in work, citizenship, and life (Winthrop, 2018; Winthrop & McGivney, 2017).

The performance of U.S. students in the bottom half of the attainment distribution has also continued to decline, with clear parallels to other societal metrics including child poverty rates and the highly inequitable allocations of per-pupil funding for education across states and districts (Koball & Jiang, 2018). These trends began before COVID-19 and are anticipated to worsen given the current context. But perhaps the greatest risk to the students at this end of the distribution is that their education has never focused on the acquisition of 21st century deeper learning skills, individual and collective change-oriented skills and mastery-level competencies, which would prepare youth with the breadth of skills to adapt to a multidimensional change in our society and the demands of a changing workplace. When we ask the question—What tools in our adaptive tool kit will any of us use to navigate massive change in the way we live and work?—we realize that it is our agency, our problem-solving skills, our abilities to collaborate with others, our creativity, critical thinking, and our resilience. In other words, it is a robust 21st century tool kit of skills that we will call upon and, as well, skills not anticipated in at least the early decades of this century, that is, socially transformative skills needed to adapt successfully to a still emerging new normal.

As countries seek to expand the use of AI and to reduce health risks by further automatizing and robotizing jobs, we may face dramatic reductions in job opportunities for young and/or less skilled adults who have graduated from high school and are engaged in some postsecondary education and employment (e.g., Furman & Seamans, 2019). This trend creates a significant risk that more and more young adults will face crippling challenges due to economic hardship, a dearth of jobs—especially well-paying jobs, an inability to pursue further education, and the challenge of adequately supporting a young family's basic needs (e.g., Huang & Rust, 2018).

To the extent we believe that this set of changes will affect only some of us, but not others—we are wrong. These changes are happening quickly all around us and will affect our entire country and its prospects for growth, prosperity, and well-being for years to come. Chetty, Hendren, Jones, and Porter's (2020) analyses of mobility data showed that children born in 1940 had a 90 percent chance of earning more than their parents, but for children born four decades later, that chance had fallen to 50 percent. What would those numbers be today? Right now, we are seeing a picture where the children of the wealthy are increasingly building these adaptive 21st century skills and poor children will still be attending schools where the acquisition of literacy and math skills will be the benchmark. The magnitude of the gap that grows out of this disparity is profound in its implications.

The sociopolitical, economic, technology, and climate-change issues that affected and challenged our children's futures before the pandemic of 2020 have dramatically amplified the manifestations of inequity of educational experiences among young people across the United States and the world. These issues necessitate an examination of the specific factors that could accelerate genuine opportunity for our young people, families, and communities versus factors that will continue to condemn them to joblessness, homelessness, and stagnation. Such positive factors include educational opportunities and experiences that do not fall prey to false choices between core academic skills and competencies and the acquisition of a robust array of skills and social capital. They also include educational and community contexts designed to meet basic needs and provide the tools for individual and collective adaptation, innovation, and growth (Osher et al., 2020; Osher & Boyd-Brown, 2021).

An analogy here may be useful to underscore our point. The advent of precision medicine changed the medical field and practice in large part because it began with the acceptance of individual variability and then sought to analyze disease processes simultaneously through the specific genetic, environmental, and biologic lenses of a specific person in order to more accurately predict which combination of prevention and intervention strategies will be successful for this specific patient at a specific moment in time. Such approaches are effective when variability is understood within the context of the social determinants of health and the impact of forces such as racism and provider-driven approaches on health care delivery (e.g., Institute of Medicine, 2005; Osher & Osher, 2002).

Addressing the fundamental individuality of a person may be equally useful in the science of human learning and development. The biological, genetic, developmental, and learning principles described in these papers and commentaries, would, if enacted in research, policy, and practice, lead to a substantial redesign of developmental, learning, and measurement systems that educate and prepare individual youth for their futures, with a particular emphasis on scaling the full set of dynamic (i.e., individualócontext; Mascolo & Fischer, 2015) skills, mastery-level competencies, new measurement systems, anti-racist policies and practices, and access to social capital and opportunity that would make this possible. Indeed, the forces of 2020 can and should serve as an inflection point for instigating and empowering transformational societal, educational, and economic change, defined by the goals of social justice, multidimensional equity, economic opportunity, and human thriving for each and every young person (Osher et al., 2020).

The scholarship included in the previous chapters contributes to such transformations.

Contributions of the Chapters and Commentaries

In the chapters and the commentaries about them that are included in this book, we have seen the foundation of a comprehensive understanding of whole-child development, learning, and thriving, one that requires *a dynamic and integrated view of the journey that each child takes.* There is burgeoning scientific knowledge about the systems that govern human life, including biologic systems, social processes, and the systems of the human brain. For example, researchers can now study the brain's structure, wiring, metabolism, and connections to other systems of the body and to the external world (Immordino-Yang, Darling-Hammond, & Krone, 2019), and we know much more about the social and cultural processes that affect learning (Lee & Pea, 2020). In short, we know much more than they did when the 20th century U.S. educational systems were designed.

Together, these bodies of research and methods of analysis affirm that child and adolescent development (and human development in general) involves a system of "dynamic" (i.e., mutually influential) relations among:

1. Each child's multiple biological processes (e.g., the nervous system and epigenetic influences on the developing brain);
2. Each child's psychological processes (cognitive, affective, and behavioral) and social relational processes including how people make meaning, both individually and collectively; and
3. The collective set of coactive culturally resonant ecological contexts that each child, as well as those who are key to the child's learning and development, encounter across their life span, ranging from proximal child-parent/family relationships to more distal contexts of schools and communities, to macro contexts involving societal institutions, culture, race, the natural and designed physical ecology, and the ongoing impact of history.

(Bronfenbrenner, 2005; Bronfenbrenner & Morris, 2006; Spencer, 2006)

Current scientific understanding of these dynamic, individualized journeys must become the foundation for the beliefs and practices of practitioners, administrators, and policy makers. Specifically, *thriving* should be understood as a dynamic process that goes beyond well-being to include individual and collective growth that is strength-based and multidimensional, reflecting positive growth in and across multiple domains from physical to emotional to cognitive and social, wherein the concept of collective thriving becomes relevant (Osher et al, 2020). Practitioners, administrators, and policy makers must understand the learning processes, potentialities, and capabilities that can and will emerge in and among students across time and across dynamically coactive settings designed to promote whole-child development, learning, and thriving as well as how aversive processes, including many of our current practices, can attenuate the growth of children. These domains are linked to internal attributes and external conditions that enable favorable change (Osher et al., 2020). It is the responsibility of all people interested in the thriving of children to use this knowledge to design a system characterized by *robust* equity, "a system in which all individuals are able to take advantage of high-quality opportunities for transformative learning and development" (Osher et al., 2020, p. 3).

To apply the new knowledge of human development and learning, developmental and learning scientists, educators, and practitioners must (1) challenge the assumptions and goals that serve as the foundation for the current U.S. systems; (2) articulate and popularize a revised conception of whole-child development, learning, and thriving (which we will introduce); (3) accept—rather than seek to simplify—the complexities in how humans develop; (4) create a profound paradigm shift in the purpose of education; (5) describe a new approach to measurement of both development of core academic competencies and higher-order dynamic skills including a measurement of the settings/contexts in which they develop; and (6) do so in a manner that explicitly takes into account and addresses the impacts of institutionalized racism and white privilege.

Realizing the potential of this convergent knowledge will not be easy. There are intellectual, political, structural, ideological, and practical challenges, including broadening the diversity of voices and perceptions of those doing further research and applying the research to practice (Osher, Williamson, Kendziora, Wells, & Sarikey, 2019). This set of challenges does not mean waiting to act as the need to act is exigent, and we already know enough to start acting on this knowledge. We know enough to begin to design policy-level and practice-level actions, and to conduct research and evaluation of their efficacy with appropriate methodological approaches to measurement, design, and analysis (Cantor, Lerner, Pittman, Chase, & Gomperts, in press). Such actions must be taken in the context of a recognition of the challenges involved in generating and bringing to scale the innovations we envision (Dymnicki, Wandersman, Osher, & Pakstis, 2017). The next section discusses these challenges.

Rooting Out the Tentacles of Racism and Caste

The preceding chapters provide a picture of what the science of learning and development (SoLD) can and must do. These contributions suggest a vision of what is possible and necessary if we are to create robust equity, a nation wherein children, youth, families, and communities can thrive. This knowledge can also help us frame the significance of the events going on today, when the nature and impacts of institutionalized racism and privilege have become more evident. Racism and caste have had and continue to have a pervasive pernicious influence on every facet of American society (Spencer, 2021; Wilkerson, 2020). The discussions in this book illuminate the historical and sociological processes of systemic racism that have created inequities for people of color and privilege for white people; these processes have often been denied and are still being denied, including how these factors are embedded and continue to embed themselves in virtually every aspect of our lives, our schools, and our professions (Farrington, 2020; Nasir, Lee, Pea, & de Royston, 2020). If unchecked, these processes will continue, and perhaps amplify, and therefore not only sustain, but exacerbate, inequities.

Structural and cultural conditions of systemic racism, institutionalized prejudice, and routinized and mystified white privilege are oppressive—they limit opportunities and constrain how we think and act. While scholarship often focuses on the powerful effects of social structure and cultural factors at a societal (macro) level, it is also important to understand the mechanisms and processes through which these macro factors are crystalized, replicated, and played out (Giddens, 1991; Goffman, 1959, 1974). Routinization and mystification are two examples. Routinization (what Giddens calls "structuration") bakes-in, automatizes, and backgrounds processes; in so doing, routinization permits people to act without thinking or, even when they think, to take some things for granted, as in epistemes—unconscious' structures underlying the production of scientific knowledge (Foucault, 1970). Routinization is important to consider here because racism (and other prejudices and privileges) can be baked into law, policies, curricula, pedagogies, and ritual—they can become routine and normalized and can be reproduced and reinforced by unconscious thinking and behavior (Bordieu, 2010; Giddens, 1984; Gramsci, 1971; Guess, 2006). Mystification operates differently. It functions like a pernicious magic trick; it directs our focus and thinking by focusing us on some observable factors and hiding or obscuring others (Goffman, 1959). While demagogues and opportunists intentionally change our focus by scapegoating or manipulating pernicious anecdotes, mystification can also result from well-intended actions or what are often viewed as ideologically neutral and objective activities. For example, historian Charles Payne (2004) showed how *Brown v. Board of Education* focused attention on *de jure* segregation while ignoring *de facto* segregation (Payne, 2004). Similarly, research design, measures, and methodology can obscure how we understand the cause of problems (Ryan, 1971).

A developmental systems framework helps us see how racism and other forms of oppression and privilege exist at every ecological level and how processes (latent functions), not just measurable outcomes (manifest functions), are crystalized and how they can be oppressive (Nesselroade, 2019; Nesselroade & Molenaar, 2010). Micro-aggression and other toxic behaviors that are routinized or otherwise taken for granted ("how we do it here") can limit the ability to thrive. They can be hard to challenge in situations in which individuals feel unsafe and/or need to have to manage their public personae (Goffman, 1959). However, when not illuminated or interrogated, micro-aggressions set the stage for transactions where people coactively recreate inequities. Take the case of common assumptions about the value of punitive discipline and how the operationalization of these assumptions in both policy (e.g., codes of conduct) and teacher pedagogy contribute to problematic and racialized student-teacher-school interactions. These transactions can provoke or amplify student oppositional behavior and adult counteraggression, which, in turn, contribute to punitive, biased, and disparate disciplinary processes that contribute to suspension and dropout (Ferguson, 2020; Okonofua, Walton, & Eberhardt, 2016; Osher et al., 2004; Willis & Willis, 1981). This dynamic often starts with teacher-student disconnection and inappropriate support for student behavioral and academic engagement; this leads to a set of punitive and/or exclusionary processes (e.g., suspension) along with diminished opportunities to learn; these experiences, which are reinforced by lower teacher academic expectations and higher teacher behavioral vigilance, in turn contribute to more troubling behavior and poor academic performance, more surveillance and counteraggressive behavior, which are part of a dynamic, cascading cycle that can lead to chronic absence and dropout that can limit the ability of youth to develop the social, emotional, and cognitive skills or the certifications and social capital needed to thrive in a society wherein children of color experience bias, profiling, and mass incarceration (Alexander, 2020; Brinkley-Rubinstein & Cloud, 2020; Massey & Denton, 1993; Osher, Woodruff & Sims, 2002). Although opportunities to learn and develop not solely dependent on formal education exist outside of schools, schools are important as they are compulsory and essentially universal, and they serve as gatekeepers. Schools are in essence high-stakes; effects are powerful and long-lasting (Cairns & Cairns, 1994; Osher, Kendziora, Spier, & Garibaldi, 2014).

Racism and its embodiment in day-to-day practice is a sociocultural process that has four components: the social structuring of perception, identity formation, political legitimation, and the impacts of ideology. Social structures as well as the environments that we build affect what we see and do not see, including what aspects of our lives are visible to others. Zip codes and interstate highways, both of which reflect the historical impacts of institutionalized racism and privilege (through, for example redlining, "slum clearance," gentrification, and highway location, affect what people experience and see in their neighborhoods, (e.g., Charles, 2006; Connerly, 2002). The more segregated societies are—whether by race, social class, gender, ethnicity, or other differentiating elements—the less individuals know about members of segregated groups. This segregation can be legally mandated (Jim Crow), policy mandated (military segregation, 1862–1945), or realized by private practices such as banker and realtor redlining, which kept black people out of some areas and funneled them into other, that are now bureaucratically identified as zip codes. This segregation creates social harms. For example, consider a situation in which a white teacher who does not live in a student's community or knows about it or the student's family's strengths, suspends a black child after an interaction that the teacher had triggered by an unknowingly hurtful statement after the student had arrived late to class. Social structure and cultural lenses affect and reflect whom we interact with and under what conditions and, in the context of uninterrogated teacher bias and white privilege, may make it more likely that teachers will respond in a manner that precipitates or escalates troubling action or respond punitively to troubling behavior reactively.

Meaning making is key to how people develop and process experience, as well as how culture is shaped and shapes people (Bruner, 1990), including how a child, a child's peers, and the teacher make sense of an interaction such as the teacher suspending the student. The personal narratives that are part of identity development are one source of meaning (McAdams, 2011). Individuals develop personal narratives that provide a sense of unity and purpose, help them understand themselves, and affect their sense of righteousness and dignity. Whether these narratives are redemptive (how I grew and succeeded) or contaminated (how I was harmed), they help explain life experiences, including one's privilege or the lack thereof. These individual narratives reflect master narratives (Takaki, 1993) that also provide affirmative sense making for groups; master narratives can include a framing of history as well as of current events. Three key U.S. master narratives (each of which is contested) are American exceptionalism, the white Anglo-Saxon Protestant American identity, and individualism (including meritocracy and victim blaming); these master narratives provide a grammar of self-justifying and reinforce specific behaviors, including ones that often lead to the exclusion of alternative perspectives (e.g., Stanley, 2007).

These narratives both contribute to and are reinforced by political needs to legitimate behavior, including one's social position. Both the narratives and the related legitimation efforts are affected by dominant ideological perspectives that often infuse individual and collective sense making and have psychological ramifications for individuals' conceptions of their identity.

The Limits of Current Research

Scientific and technical solutions do not pave the way to seriously address and certainly do not eradicate the impacts of institutionalized racism, ethnocentrism, and privilege, which are both manifest (at least to those who experience the harms) and latent in all situations. In addition, the research presented across the contributions to this book do not sufficiently address two other challenges: first, aligning research, practice, and policy; and second, building upon the strengths and addressing the complex needs of linguistically and culturally diverse families and communities, with attention to individuality, intersectionality, and commonality—and doing so in a manner that is respectful of, actively engaged by, and engaging of children, youth, families, and communities (e.g., Osher, Mayer, Jagers, Kendziora, & Wood, 2019; Osher, Moroney, & Williamson, 2018). To successfully conduct research, scholars must move forward with the cultural humility and self-awareness regarding how they make sense of the world through personal and cultural lenses that have been and are socially grounded and reinforced (Francis & Osher, 2018). Further, scholars should acknowledge the ways in which scholarship has contributed to institutionalized privilege and interrogate, interrupt, and limit the ways in which privilege affects their research and work.

Prejudice and institutionalized privilege disadvantage and oppress many groups of people and have intersectional dimensions (e.g., Henry, Votruba-Drzal, & Miller, 2019; Heumann, 2020; Velez & Spencer, 2018). Although (in the words of Audre Lorde, 2005) there is no hierarchy of oppressions, and all oppression must be addressed if all children are to thrive, we will focus on racism. The structure of systemic racism that exists has the power to constrain the optimal and healthy development of youth of color, unjustly and inequitably define a young person's trajectory, and create a caste system wherein, when specific hurdles are surmounted, additional ones are encountered in the implicit service of constraining opportunity, equality, and equity (Wilkerson, 2020).

These caste-based constraints can profoundly influence what children, youth, and adults believe is possible and their sense of identity, agency, and hope, while reinforcing hegemonic master narratives that, all too often, are victim blaming and/or reductionist. The level of constraint is profound and exists in complete opposition to the inherent truths in our biologic story, our epigenetic story, and our cultural story, one that is recounted with rigorous evidence in the contributions in this book.

These ideas have been further elaborated in Osher et al. (2020), in *Thriving, Robust Equity, and Transformative Learning*. This paper broadens conceptualizations of thriving, robust equity, and transformative learning. The report, which focuses on youth, conceptualizes individual and collective thriving across all life domains and in all life spaces, and includes the individual and collective experience of voice, identity, purpose, civic engagement, and subjective well-being, along with the social, emotional, and cognitive skills that support agency, critical thinking, creativity, and the ability to address inequity. Young people can thrive even in the face of current and past adversities if they have steady exposure to contexts and relationships that are designed to provide safety, support, and challenge (Osher et al., 2020).

These ideas can be fruitfully empirically illuminated with theoretical research. The research directions we have discussed here are framed by ideas derived from dynamic, relational developmental systems-based models of human development (e.g., Fischer & Bidell, 2006; Mascolo & Bidell, 2020; Mascolo & Fischer, 2015; Overton, 2015; Spencer, 2006, 2021; Spencer, Swanson, & Harpalani, 2015). These models provide, as well, a new conceptualization of youth development, learning, and thriving that can be used to systematically address racism and caste within education and, more significantly, society as a whole.

A New Conceptualization of Whole-Child Development, Learning, and Thriving

A human embryo is formed by the combination of two people, who bring to their parenting culturally nested knowledge and experience. The embryo is nurtured in the context of the womb and emerges as an infant. The infant is born, and the parents, with the socially situated resources they possess and the knowledge they have, will pour nutrients in all forms into this unique being. They will do so without knowing how the story will unfold, but they know they must do these things for this life to unfold. They will not know the attributes of this being for a long time, or who he or she will be in the future. They only know that they must nurture their offspring with all the resources they have to enable this life to emerge and thrive. These open-ended processes, which happen every day between parents and children around the world, are what dynamic systems and web-like processes are all about. However, they are not—but should be—the processes that govern the design, structure, purpose, and goals of our education system.

The creation of one whole child is directly contingent on a set of conditions and open-ended processes that could take place in schools, homes, communities, ball fields—everywhere that children grow and learn—*if* these environments were designed to support whole-child development, learning, and thriving, by addressing the ecological factors that affect both children's development and the capacity and willingness of adults to collaborate with each other in a culturally competent manner to support a child's development. These steps are not all that different from a parent or a family or clan raising a child. When we parent, we know that what we put in shapes what will come out. Parenting is the most prevalent example we have of an open-ended system. What we take for granted in parenting—that what we put in shapes what will come out—is or could be true in educational settings, too. This fact forces us to ask—what could be true if educational and learning settings were designed to nurture children's potential? This focus is not what 20th century education was designed to do, but it is what 21st century education can become.

Scientists and educators must use the knowledge present today, about what is needed to nurture the development of a whole human being to design new education and learning systems and settings that are culturally responsive and promote robust equity. The collective implications of the defining attributes of human development and learning (i.e., the role of context, embodiment, jaggedness, developmental range, epigenetics, etc.) and their dynamic relation to one another demand the adoption of a new worldview about the purpose of education, designed as an open system to enable each child to know and reach his or her individual potential. The physicist and philosopher Thomas Kuhn (1962, 1970) termed the revisions in understanding the development education of children a *paradigm shift,* and educational historians David Tyack and William Tobin and Cuban would call this a change in the *grammar of schooling* (e.g., Tyack & Tobin, 1994). Thus far, we have created interventions and programs that have generated only incremental change, and only for some children and boutique models that neither scale nor sustain. What we need now is a new transformational paradigm shift that can, if equity is to be realized, go to scale with integrity.

Human development and the learning sciences findings converge to suggest that development is an open-ended continuum, happening over time in many different places. While development is occurring, the brain is becoming increasingly integrated and connected, acquiring the ability to perform increasingly complex skills, and improving its ability to ascribe meaning to new experiences. Human development and learning sciences show how micro-developmental processes—the construction of new skills for "proximal processes" (Vygotsky, 1978)—integrate and converge with macro-developmental processes (larger-scale processes in which many constructive activities come together to form complex skills) and then stabilize, such that they can be applied to new contexts over time. This application means that, although people develop their competencies over the long term, critical learning occurs in shorter time horizons through an accumulation of diverse, context-dependent, nonlinear growth experiences that produce increasingly important skills, competencies, identities, and knowledge.

Within our dynamic conception, whole-child development, learning, and thriving emerge from the malleability, agency, and changing as well as changeable developmental range of a child, as they draw on the available resources and build the web of relations and experiences across the settings of the child's life. The nature of these webs presents both opportunities and risks to development, but can, if well-designed and intentional, provide the foundation for the development of dynamic, complex skills that ultimately reveal the talent, passions, and potential of each child. The pathways will be diverse, and the patterns jagged, but throughout the developmental web of contexts, relationships, and experiences will be the driver of the expression of each child's genetic endowment and epigenetic attributes, will harness the malleability of the child's body and brain, and will nurture the fullest expression of what this child becomes. What is being asked of our education and learning systems is to use the knowledge we have today, which is captured in these developmental principles, to ignite these developmental and learning processes along positive trajectories for each and every child.

IF educators and youth-serving professionals understand that children and adolescents possess a broad set of potentialities across multiple domains (e.g., physical health; mental health; complex social, emotional, and cognitive development; core academic skills and knowledge; positive identity formation; agency) and recognize that each child is an integrated dynamic system with virtually infinite horizons of developmental range, AND each attribute of the child is a subsystem with its own developmental range,

THEN, educators and youth-serving professionals will make daily practice and policy decisions to ensure that *all* the adults in *all* of the environments that comprise our education and learning systems are provided with the knowledge and support so that they understand and have the capacity to build as well as build upon the strengths and potential of each child. And, they will address the sources and manifestations of institutionalized racism and privilege and establish the core features of integrated developmental and learning settings in which the primary role of the adult is not only to teach discrete skills, but to create opportunities for each young person and young people collectively to want to bring parts of their interests, passions, talents, prior experiences, culture, and existing capabilities to bear in an effort to master increasingly complex skills because they are engaged and motivated to do so.

Such settings are:

- Attuned to the presence of biological, psychological, and sociocultural (embodied) attributes of each child within each setting in his or her life;
- Able to foster positive developmental relationships in all aspects and activities;
- Integrating multidimensional practices to describe, explain, and optimize the fullest expression of developmental range of the diverse set of attributes in children who can and are acting with agency in such environments;
- Oriented to create conditions of support and opportunities for growth within and, critically, across settings that capitalize on the malleability of the child and the variable and jagged pathways through which he or she will acquire increasingly complex skills and academic competencies;
- Able to capitalize on the specific strengths and potential growth in the strengths of each youth to build the cognitive, social, emotional, metacognitive, and motivational skills and positive identities to enable the child to both influence his or her setting and positively adapt to new challenges, including transfer of skills to new settings;
- Able to address, individually and collectively, sources of institutionalized racial oppression and prejudice, white supremacy, sexism, marginalization, stereotyping, and individual bias that, taken together, diminish the opportunities for positive identity formation and the expression of an individual child's developmental range and potential; and
- Aligned with and guided by the resources for positive growth found in communities, cultural assets, families, schools, youth development programs, faith-based organizations, and athletics.

When such a web of environments and experiences is constructed by educators and community professionals, and families, then each young person's learning and healthy whole-child development will be optimized.

Conclusions

"Not everything that is faced can be changed, but nothing can be changed until it is faced."

—James Baldwin (1962)

Here is what we are facing: It is not possible to talk about the development, learning, and thriving of young people without talking about opportunity, access, resources, and social capital. And it is not possible to talk about any of those things without talking about race and institutionalized privilege. Our societal and educational structures privilege some and not others. Our education systems were designed for sorting and selecting, based on beliefs and assumptions that we now know to be false and that we have summarized in this book. But these false assumptions are not only drivers of the design of our public systems. Our public systems, including our education systems, were intentionally and systematically designed to promote, privilege, and advantage specific groups, predominantly white, middle- to upper-class males, and to oppress and marginalize other races and genders. The institutional forces that have enabled this racist system to exist can be found in every corner of our social, economic, and educational infrastructure:

> Our society was built on the racial segmentation of personhood. Some people were full humans guaranteed non-enslavement, secured from expropriation and given the protection of law, and some people—Blacks, Natives and other nonwhites—were not. That unequal distribution of personhood was an economic reality as well. It shaped your access to employment and capital; determined whether you would be doomed to the margins or given access to its elevated ranks; marked who might share in the bounty of capitalist production and who would most likely be cast out as disposable.
>
> (Bouie, 2020)

If the United States wanted to right the wrongs of today and the 400 years of policies and practices behind them, its actions would necessarily include rethinking systems based on the scientific principles presented in this book. These principles tell an optimistic story about what is possible for young people—their learning and their lives—when environments, experiences, and relationships are designed for relationship development, health, learning, and access to opportunity and resources, as well as what can happen when conditions press down upon individuals and communities and undermine the opportunity for people to thrive. Under the right conditions, the potential that exists in all young persons has a much greater chance of being expressed, no matter where their journeys begin. The principles in this book can serve as a guide not only to what we can do to optimize all young people's learning and development, but also what we must stop doing now because it is actively harmful to the learning and development of many young people. This includes dismantling the institutions that preserve and sustain harmful racist practices including tracking, harsh discipline, exclusion, shaming, and many others.

We have established that the 20th century U.S. education system was not designed based on knowledge of human development and the learning sciences. It was never designed to develop the whole child. It was not designed for equity—to see students as individuals and to unleash the potential inside them. It was designed based on the belief that some children were more deserving of opportunity than others.

Education has reached the limit of what it can achieve through standardized and outdated approaches, as has happened in other fields. These approaches work for some—mainly those born into privilege—but not for many others. When other fields, such as health or medicine, hit a point where fundamental assumptions and beliefs needed to be challenged, this led to breakthrough solutions that benefitted many more people.

For example, 25 years ago, cancer researchers were stuck using therapies that worked for some, but not others. When they asked themselves why, they recognized that they needed to understand more about the microenvironments around specific cancer cells so they could personalize individual treatments by building more targeted, context-sensitive therapies. This insight fueled new breakthroughs in cancer treatments, enabling doctors to successfully treat far more patients. It also led to the creation of a new field—immunotherapy—where an individual's own immune system is recruited to fight a cancer.

In this example, a standardized approach to treatment was disrupted and moved to a *personalized* approach. This pattern—disruption of standardization leading to personalization—has transformed other fields, too. Examples include travel through Airbnb, music through Spotify, and entertainment through Netflix. These examples of personalized approaches also enable greater access and scale, changes that have not been true of our medical and broader health systems. Although we now have a much greater array of treatments and can realize much better clinical outcomes, these treatments are not equally available, and people of color remain at greater risk of illness, something which has been profoundly unmasked by COVID-19 (Bornstein, 2020).

This kind of transformational shift must now happen in education and youth development. We envision a personalized system that sees all young people as individuals, one that recognizes their specific assets, strengths, cultures, and also their vulnerabilities, and that uses this youth-specific knowledge to influence the design of the environmental contexts and the individual-context relations that support their learning, meaning making, identity development, and their thriving as individuals (e.g., see Spencer, 2006, 2021; Spencer et al., 2015).

To achieve the transformation we need today, our education systems and community-based systems, including health, mental health, foster care, and juvenile justice, must be willing to embrace what we know about how children learn and develop. This includes strengths-based culturally responsive approaches that are child and family driven and address the individuality of each child and the elimination of traumatizing practices such as shaming, restraint, and exclusion (Center for the Developing Child at Harvard University, 2016; Osher, Penkoff, Sidana, & Kelley, 2016; Osher, Williamson et al., 2019; Rivas-Drake, Jaegers, & Martinez, 2019; Weist et al., 2019). The core message from diverse sciences is clear: the range of students' academic skills and knowledge—and, ultimately, students' potential as human beings—can be significantly influenced through exposure to highly favorable conditions, that is, learning environments and experiences that are intentionally designed to optimize student development (Bloom, 1984; Fischer & Bidell, 2006). To create this transformation, the scientific principles highlighted in this book must become the foundation for new 21st century education and learning systems designed around the acquisition of 21st century skills and mastery-level competencies.

With what we know today, we know enough to design environments that help protect children from developmental harm, including racist policies and behaviors, and promote their healthy development and success as learners. The nonnegotiable elements described in this book will simultaneously ignite brain development and learning; promote wellness and thriving; support positive identity formation; enable the acquisition of knowledge, skills, and mindsets that are critical for success in learning and work; build individual and collective competencies that foster resilience to future stresses; and provide the intellectual challenge and rigor that enable children to discover the specific personal, academic, and social achievements they desire and are capable of attaining under the right conditions. Settings designed in these ways will provide a developmental continuum of experiences, both in and out of school, and across age bands, that optimize each student's developmental range and reveal the talent, skills, and potential that all children have (Cantor et al., in press). To be successful, these settings must intentionally build and support individual, organizational, and systemic readiness. This work includes building and supporting the motivation and social, emotional, and cultural capacity of adults to attune to each child, and to each other thereby eliminating the systemic factors that undermine a person's capacity to attune, care, listen, and grow (Jennings, Minnici, & Yoder, 2019; Osher, 2018; Yoder, Holdheide, & Osher, 2018). These actions require more than muscle; they require both organizational prioritization and the readiness to be engaged by as well as to engage culturally and linguistically diverse youth, families, and community organizations.

In the design process, we can ask and answer the same question that the cancer researchers asked: What can we do that will work optimally for this specific child, in this context? This question will move scientists and educators to fundamentally different answers about the way our schools and education systems of the future must be designed—toward integrated and individualized processes and supports, using tools and platforms that enable educators to integrate academic instruction with the intentional development of the skills and mindsets possessed by all successful learners. This design work will need to address the implications of implementation science, which involve the creation of individual, collective, agency, and system-level readiness and capacity (e.g., Birken et al., 2020).

Breakthroughs do not occur when we seek to achieve them by doing what we have been doing, just a little better. The approaches we have been taking thus far to learning and schools reflect master narratives and the relative silencing of other voices. These approaches have not fully challenged our assumptions about talent—is it scarce or is it everywhere, or is it yet to be discovered; or about humans skills—are they malleable or are they fixed?; or about human potential—does it only exist in some children, or will we see what children are capable of when we design environments to reveal it?

Accordingly, in this book, we are calling for a humanistic reckoning and have presented the scientific basis for such a reckoning that will mean a profound shift in the assumptions and practices that have dominated 20th century education and perpetuated racist and white supremacist policies, biases, and practices. Building highly favorable conditions into all of the environments and supporting the capacity of adults to attune to, learn from and with, and support children in the environments in which they grow and learn will put many children and youth on the path to equity and thriving and all on the path to the fullest expression of their potential.

Indeed, the forces of 2020 can and should serve as an inflection point for instigating and empowering transformational, societal, educational, and economic change defined by the goals of social justice, multidimensional equity, economic opportunity, and human thriving for each and every young person (Osher et al., 2020). One, which will see an end to false choices between core academic competencies and the presence of deep enduring relationships, the acquisition of robust 21th century skills and the promotion of social capital. And one where we will witness the elimination of bias and stereotype emanating from privilege and hegemonic master narratives (Takaki, 1979).

Indeed, a focus on these drivers of learning and thriving will enable not only practitioners and policy makers but, as well, parents and children to imagine a world where children's lives are a succession of opportunities in which they come to know who they are and in which they discover who they could become. Imagine learning settings of all kinds where these kinds of opportunities are not only possible, but they are optimized, no matter where the children started their education journey or where they happened to live across their school years.

Imagine too that the system-wide focus on these individualócontext processes empowered educators to find how best to identify each child's specific abilities, interests, and aspirations and then align these attributes with the specific contexts that best promoted the child's talents, achievements, and successes in life. Finally, then, imagine that all children lived in a world that removed the constraints of racism, poverty, disparities, and injustices and provided them with the specific relationships and supports needed for thriving.

The message in the science is optimistic: genes are chemical followers. Context shapes the expression of our genetic attributes. This is the biologic truth. And schools designed, as Bloom (1984) suggested, using the levers of human development—so that what one child can do, nearly all children can do under highly favorable conditions—can define our new 21st century learning systems: systems designed to see and unleash talent and potential and ensure that all young people can thrive. This vision constitutes a transformational shift in the purpose and potential of education, a dismantling of the systems that constrain this vision, grounded in what we know *today* about human development, the development of the brain, and learning science.

References

Alexander, M. (2020). *The new Jim Crow: Mass incarceration in the age of colorblindness*. New York, NY: The New Press.

Ambrose, A. J. H. (2020). Inequities during COVID-19. *Pediatrics, 146*(2), e20201501. https://doi.org/10.1542/peds.2020-1501

Baldwin, J. (1962). As much truth as one can bear, New York Times Book Review, January 14, 120, 148.

Birken, S. A., Haines, E. R., Hwang, S., Chambers, D. A., Bunger, A. C., & Nilsen, P. (2020). Advancing understanding and identifying strategies for sustaining evidence-based practices: A review of reviews. *Implementation Science, 15*, Article No. 88.

Bloom, B. S. (1984). The 2 sigma problem: The search for methods of group instruction as effective as one-to-one tutoring. *Educational Researcher, 13*(6), 4–16.

Blundell, R., Costa Dias, M., Joyce, R., & Xu, X. (2020). COVID-19 and inequalities. *Fiscal Studies, 41*(2), 291–319.

Bordieu, P. (2010). *A social critique of the judgement of taste. Distinction: A social critique of the judgment of taste.* London and New York, NY: Routledge.

Bornstein, M. H. (Ed.). (2020). *Psychological insights for understanding COVID-19 and families, parents, and children.* New York: Routledge.

Bouie, J. (2020, June 26). Beyond "white fragility". *The New York Times*.

Brinkley-Rubinstein, L., & Cloud, D. H. (2020, January). Mass incarceration as a social-structural driver of health inequities: A supplement to *AJPH*. *American Journal of Public Health, 110* (S1), S14–S15. doi:10.2105/AJPH.2019.305486. PMID: 31967896. PMCID: PMC6987928

Bronfenbrenner, U. (2005). *Making human beings human: Bioecological perspectives on human development*. Thousand Oaks, CA: Sage.

Bronfenbrenner, U., & Crouter, A. C. (1983). The evolution of environmental models in developmental research. In W. Kessen (Series Ed.) & P. H. Mussen (Vol. Ed.), *Handbook of child psychology* (4th ed., pp. 357–414). Vol. 1: History, Theory, and Methods. New York: Wiley.

Bronfenbrenner, U., & Morris, P. A. (2006). The bioecological model of human development. In W. Damon & R. M. Lerner (Eds.), R. M. Lerner (Vol. Ed.), *Handbook of child psychology* (6th ed., pp. 793–828). Vol. 1: Theoretical Models of Human Development. Hoboken, NJ: Wiley.

Bruner, J. (1990). *The Jerusalem-Harvard lectures: Acts of meaning*. Cambridge, MA: Harvard University Press.

Cairns, R. B., & Cairns, B. D. (1994). *Lifelines and risks: Pathways of youth in our time*. Cambridge: Cambridge University Press.

Cantor, P., Lerner, R. M., Pittman, K., Chase, P. A., & Gomperts, N. (in press). *Whole-child development, learning, and thriving*. New York: Cambridge University Press.

Center for the Developing Child at Harvard University. (2016). *Applying the science of child development in child welfare systems*. Cambridge, MA: author.

Charles, C. Z. (2006). *Won't you be my neighbor?: Race, class, and residence in Los Angeles*. New York: Russell Sage Foundation.

Chetty, R., Friedman, J. N., Hendren, N., Jones, M. R., & Porter, S. R. (2018). *The opportunity atlas: Mapping the childhood roots of social mobility*. No. w25147. Washington, DC: National Bureau of Economic Research.

Chetty, R., Hendren, N., Jones, M. R., & Porter, S. R. (2020). Race and economic opportunity in the United States: An intergenerational perspective. *The Quarterly Journal of Economics, 135*(2), 711–783.

Connerly, C. E. (2002). From racial zoning to community empowerment: The interstate highway system and the African American community in Birmingham, Alabama. *Journal of Planning Education and Research, 22*(2), 99–114.

Darling-Hammond, L. (2014). What can PISA tell us about U.S. education policy? *New England Journal of Public Policy, 26*(1), Article No. 4, 1–14.

Duncan, G.J., Magnuson, K., & Votruba-Drzal, E. (2015). Children and socioeconomic status. In M. H. Bornstein & T. Leventhal (Eds.), *Handbook of child psychology and developmental science* (7th ed., pp. 534–573). Vol. 4: Ecological Settings and Processes. Editor-in-chief: R. M. Lerner. Hoboken, NJ: Wiley.

Dymnicki, A., Wandersman, A., Osher, D., & Pakstis, A. (2017). Bringing interventions to scale: Implications and challenges for the field of community psychology. In M. Bond, C. B. Keyes, I. Serano-Garcia, &

M. Shinn (Eds.), *APA handbook of school psychology* (Vol. II, pp. 297–310). Washington, DC: American Psychological Association.

Epstein, R., Blake, J., & González, T. (2017). *Girlhood interrupted: The erasure of Black girls' childhood.* SSRN.

Farrington, C. A. (2020). Equitable learning and development: Applying science to foster liberatory education. *Applied Developmental Science, 24*(2), 159–169.

Ferguson, A. A. (2020). *Bad boys: Public schools in the making of black masculinity.* Ann Arbor, MI: University of Michigan Press.

Fischer, K. W., & Bidell, T. R. (2006). Dynamic development of action and thought. In R. M. Lerner (Ed.), *Handbook of child psychology* (6th ed., pp. 313–399). Vol. 1: Theoretical Models of Human Development. Editors-in-chief: W. Damon & R. M. Lerner. Hoboken, NJ: Wiley.

Fortuna, L. R., Tolou-Shams, M., Robles-Ramamurthy, B., & Porche, M. V. (2020). Inequity and the disproportionate impact of COVID-19 on communities of color in the United States: The need for a trauma-informed social justice response. *Psychological Trauma: Theory, Research, Practice, and Policy, 12*(5), 443–445.

Foucault, M. (1970). The archaeology of knowledge. *Social Science Information, 9*(1), 175–185.

Francis, K., & Osher, D. (2018). The centrality of cultural competence and responsiveness. In D. Osher, D. Moroney, & S. Williamson (Eds.), *Creating safe, equitable, engaging schools: A comprehensive, evidence-based approach to supporting students* (pp. 79–86). Cambridge, MA: Harvard Education Press.

Furman, J., & Seamans, R. (2019). *AI and the economy.* Cambridge, MA: NBER working paper w24689.

Galea, S., Merchant, R. M., & Lurie, N. (2020). The mental health consequences of COVID-19 and physical distancing: The need for prevention and early intervention. *JAMA Internal Medicine, 180*(6), 817–818.

Giddens, A. (1984). *The constitution of society: Outline of the theory of structuration.* Berkley, CA: University of California Press.

Giddens, A. (1991). Structuration theory: Past, present and future. In C. Bryant & D. Jary (Eds.), *Giddens' theory of structuration: A critical appreciation* (pp. 55–66). London: Routledge.

Goffman, E. (1959). *The presentation of self in everyday life.* Harmondsworth: Penguin.

Goffman, E. (1974). *Frame analysis: An essay on the organization of experience.* Cambridge, MA: Harvard University Press.

Gramsci, A. (1971). *Selections from the prison notebooks of Antonio Gramsci.* New York: International Publishers.

Guess, T. J. (2006). The social construction of whiteness: Racism by intent, racism by consequence. *Critical Sociology, 32*(4), 649–673. doi: 10.1163/156916306779155199

Henry, D. A., Votruba-Drzal, E., & Miller, P. (2019). Child development at the intersection of race and SES: An overview. *Advances in Child Development and Behavior, 57*, 1–25.

Heumann, J. (2020). *Being Heumann large print edition: An unrepentant memoir of a disability rights activist.* Boston, MA: Beacon Press.

Huang, M-H., & Rust, R. T. (2018). Artificial intelligence in service. *Journal of Service Research, 21*(2), 155–172.

Immordino-Yang, M. H., Darling-Hammond, L., & Krone, C. R. (2019). Nurturing nature: How brain development is inherently social and emotional, and what this means for education. *Educational Psychologist, 54*(3), 185–204.

Institute of Medicine. (2005). *Building a better delivery system: A new engineering/health care partnership.* Washington, DC: The National Academies Press. https://doi.org/10.17226/11378

Jennings, P. A., Minnici, A., & Yoder, N. (2019). Creating the working conditions to enhance teacher social and emotional well-being. In D. Osher, M. J. Mayer, R. J. Jagers, K. Kendziora, & L. Wood (Eds.), *Keeping students safe and helping them thrive: A collaborative handbook on school safety, mental health, and wellness* (Vol. 1, pp. 210–239). Santa Barbara, CA: Praeger/ABC-CLIO.

Koball, H., & Jiang, Y. (2018). *Basic facts about low-income children: Children under 18 years, 2016.* New York: National Center for Children in Poverty, Columbia University Mailman School of Public Health.

Kuhn, T. S. (1962). *The structure of scientific revolutions.* Chicago, IL: University of Chicago Press.

Kuhn, T. S. (1970). *The structure of scientific revolutions* (2nd ed.). Chicago, IL: University of Chicago Press.

Lee, C., & Pea, R. (2020). *Handbook of the cultural foundations of learning.* London: Routledge.

Lorde, A. (2005). Age, race, sex and class: Women redefining difference. *Feminist Theory: A Reader*, 338–342.

Manduca, R., & Sampson, R. J. (2019). Punishing and toxic neighborhood environments independently predict the intergenerational social mobility of black and white children. *Proceedings of the National Academy of Sciences, 116*(16), 7772–7777.

Mascolo, M. F., & Bidell, T. R. (Eds.). (2020). *Handbook of integrative developmental science: Essays in honor of Kurt W. Fischer*. Abingdon, UK: Routledge.

Mascolo, M. F., & Fischer, K. W. (2015). Dynamic development of thinking, feeling, and acting. In W. F. Overton & P. C. Molenaar (Eds.), *Handbook of child psychology and developmental science* (7th ed., pp. 113–161). Vol. 1: Theory and Method. Editor-in-chief: R. M. Lerner. Hoboken, NJ: Wiley.

Massey, D., & Denton, N. A. (1993). *American apartheid: Segregation and the making of the underclass*. Cambridge, MA: Harvard University Press.

McAdams, D. P. (2011). Narrative identity. In S. Schwartz, K. Luyckx, & V. L. Vignoles (Eds.), *Handbook of identity theory and research* (pp. 99–115). New York: Springer.

Murry, V. M., Hill, N. E., Witherspoon, D., Berkel, C., & Bartz, D. (2015). Children in diverse social contexts. In M. H. Bornstein & T. Leventhal (Eds.), *Ecological settings and processes. Volume 4 of the handbook of child psychology and developmental science* (7th ed., pp. 416–454). Editor-in-chief: R. M. Lerner. Hoboken, NJ: Wiley.

Nasir, N. S., Lee, C. D., Pea, R., & de Royston, M. M. (2020). *Handbook of the cultural foundations of learning*. New York: Routledge.

Nesselroade, J. R. (2019). Developments in developmental research and theory. *Applied Developmental Science, 23*(4), 346–348.

Nesselroade, J. R., & Molenaar, P. C. M. (2010). Emphasizing intraindividual variability in the study of development over the life span. In W. F. Overton (Ed.), *Handbook of life-span development. Vol. 1: Cognition, biology, methods* (pp. 30–54). Editor-in-chief: R. M. Lerner. Hoboken: Wiley.

OECD. (2019). *PISA 2018 results (volume 1): What students know and can do*. Paris, France: PISA, OECD Publishing, https://doi.org/10.1787/5f07c754-en

Okonofua, J. A., Walton, G. M., & Eberhardt, J. L. (2016). A vicious cycle: A social-psychological account of extreme racial disparities in school discipline. *Perspectives on Psychological Science, 11*(3), 381–398.

Osher, D. (2018). Building readiness and capacity. In D. Osher, D. Moroney, & S. Williamson (Eds.), *Creating safe, equitable, engaging schools: A comprehensive, evidence-based approach to supporting students* (pp. 15–24). Cambridge, MA: Harvard Education Press.

Osher, D., & Boyd-Brown, M. J. (2021). *The intersection between thriving, equity, and human learning*. Arlington, VA: American Institutes for Research.

Osher, D., Kendziora, K., Spier, E., & Garibaldi, M. L. (2014). School influences on child and youth development. In Z. Sloboda & H. Petras (Eds.), *Advances in prevention science Vol. 1: Defining prevention science* (pp. 151–170). New York, NY: Springer.

Osher, D., Mayer, M.J., Jagers, R.J., Kendziora, K., & Wood, L. (2019). *Keeping students safe and helping them thrive: A collaborative handbook on school safety, mental health, and wellness* (2 vols.). Santa Barbara, CA: Praeger/ABC-CLIO.

Osher, D., Moroney, D., & Williamson, S. K. (2018). *Creating safe, equitable, engaging schools: A comprehensive, evidence-based approach to supporting students*. Cambridge, MA: Harvard Education Press.

Osher, T. W., & Osher, D. (2002). The paradigm shift to true collaboration with families. *Journal of Child and Family Studies, 11*(1), 47–60.

Osher, D., Penkoff, C., Sidana, A., & Kelly, P. (2016). *Improving conditions for learning for youth who are neglected or delinquent* (2nd ed.). Washington, DC: National Evaluation and Technical Assistance Center for the Education of Children and Youth Who Are Delinquent, Neglected, or at Risk (NDTAC).

Osher, D., Pittman, K., Young, J., Smith, H., Moroney, D., & Irby, M. (2020). *Thriving, robust equity, and transformative learning & development: A more powerful conceptualization of the contributors to youth success*. Washington, DC: American Institutes for Research and Forum on Youth Investment.

Osher, D., VanAker, R., Morrison, G., Gable, R., Dwyer, K., & Quinn, M. (2004). Warning signs of problems in schools: Ecological perspectives and effective practices for combating school aggression and violence. *Journal of School Violence, 2*(3), 13–37.

Osher, D., Williamson, S. K., Kendziora, K., Wells, K., & Sarikey, C. (2019). Interdisciplinary and cross-stakeholder collaboration for better outcomes. In D. Osher, M. J. Mayer, R. J. Jagers, K. Kendziora, & L. Wood (Eds.), *Keeping students safe and helping them thrive: A collaborative handbook on school safety, mental health, and wellness* (Vol. 1, pp. 389–407 and paren). Santa Barbara, CA: Praeger/ABC-CLIO.

Osher, D., Woodruff, D., & Sims, A. (2002). Schools make a difference: The relationship between education services for African American children and youth and their overrepresentation in the juvenile justice system. In D. Losen (Ed.), *Minority issues in special education* (pp. 93–116). Cambridge, MA: The Civil Rights Project, Harvard Education Publishing Group, Harvard University.

Overton, W. F. (2015). Process and relational developmental systems. In W. F. Overton & P. C. M. Molenaar (Eds.), *Handbook of child psychology and developmental science* (7th ed., pp. 9–62). Vol. 1: Theory and Method. Editor-in-chief: R. M. Lerner. Hoboken, NJ: Wiley.

Patel, J. A., Nielsen, F. B. H., Badiani, A. A., Assi, S., Unadkat, V. A., Patel, B., ... Wardle, H. (2020). Poverty, inequality and COVID-19: The forgotten vulnerable. *Public Health*, *183*, 110–111.

Payne, C. M. (2004). "The whole United States is southern!": Brown v. board and the mystification of race. *The Journal of American History*, *91*(1), 83–91.

Rivas-Drake, D., Jaegers, R. J., & Martinez, K. J. (2019). Race, ethnicity, and socio-emotional health. In D. Osher, M. J. Mayer, R. J. Jagers, K. Kendziora, & L. Wood (Eds.), *Keeping students safe and helping them thrive: A collaborative handbook on school safety, mental health, and wellness* (Vol. 1, pp. 113–141). Santa Barbara, CA: Praeger/ABC-CLIO.

Ryan, W. (1971). *Blaming the victim*. New York: Pantheon.

Sampson, R. J. (2016). The characterological imperative: On Heckman, Humphries, and Kautz's The myth of achievement tests: The GED and the role of character in American Life. *Journal of Economic Literature*, *54*(2), 493–513.

Schleicher, A. (2019). *PISA/2018: Insights and interpretations*. Paris, France: OECD. Retrieved from https://www.oecd.org/pisa/PISA%202018%20Insights%20 and%20Interpretations%20FINAL%20PDF.pdf

Spencer, M. B. (2006). Phenomenological variant of ecological systems theory (PVEST): A human development synthesis applicable to diverse individuals and groups. In W. Damon & R. M. Lerner (Eds.), R. M. Lerner (Vol. Ed.), *Handbook of child psychology* (6th ed., pp. 829–894). Vol. 6: Theoretical Models of Human Development. Hoboken, NJ: Wiley.

Spencer, M. B. (2021). Interrogating the developmental context of "we the people". In P. Cantor, R. M. Lerner, K. Pittman, P. A. Chase, & N. Gomperts (Eds.), *Whole-child development, learning, and thriving: A dynamic systems approach*. New York: Cambridge University Press.

Spencer, M. B., Swanson, D. P., & Harpalani, V. (2015). Development of the self. In M. E. Lamb (Vol. Ed.), *Handbook of child psychology and developmental science* (7th ed., pp. 750–793). Vol. 3: Socioemotional Processes. Editor-in-Chief: R. M. Lerner. Hoboken, NJ: Wiley.

Stanley, C. A. (2007). When counter narratives meet master narratives in the journal editorial-review process. *Educational Researcher*, *36*(1), 14–24.

Takaki, R. T. (1979). *Iron cages: Race and culture in nineteenth-century America*. New York: Alfred A. Knopf.

Takaki, R. T. (1993). *A different mirror: A history of multicultural America*. Boston: Little-Brown.

Tucker, M. S. (2020a, June 18). *Race in America 2020*. NCCE. Retrieved from https://ncce.org/2020/06/race-in-america-2020/

Tucker, M. S. (2020b, June 26). *COVID-19 and our schools: The real challenge*. NCCE. Retrieved from https://ncee.org/2020/06/covid-19-and-our-schools-the-real-challenge/

Tyack, D., & Tobin, W. (1994). The "grammar" of schooling: Why has it been so hard to change? *American Educational Research Journal*, *31*(3), 453–479.

Velez, G., & Spencer, M. B. (2018). Phenomenology and intersectionality: Using PVEST as a frame for adolescent identity formation amid intersecting ecological systems of inequality. *New Directions for Child and Adolescent Development*, (161), 75–90.

Vygotsky, L. (1978). *Mind in society: The development of higher psychological processes* (M. Cole, V. John-Steiner, S. Scribner, & E. Souberman, Trans.). Cambridge, MA: Harvard University Press.

Weist, M. D., Shapiro, C. J., Hartley, S. N., Bode, A. A., Miller, E., Huebner, S., ... Osher, D. (2019). Assuring strengths- and evidence-based approaches in child, adolescent and school mental health. In D. Osher, M. J. Mayer, R. J. Jagers, K. Kendziora, & L. Wood (Eds.), *Keeping students safe and helping them thrive: A collaborative handbook on school safety, mental health, and wellness*. Santa Barbara, CA: Praeger/ABC-CLIO.

Wilkerson, I. (2020). *Caste: The origins of our discontents.* New York, NY: Random House Publishing Group.

Willis, P. E., & Willis, P. (1981). *Learning to labor: How working class kids get working class jobs.* New York, NY: Columbia University Press.

Winthrop, R. (2018). *Leapfrogging inequality: Remaking education to help young people thrive.* Washington, DC: The Brookings Institution.

Winthrop, R., & McGivney, E. (2017). *Can we leapfrog? The potential of education innovations to rapidly accelerate progress skills for a changing word.* Washington, DC: Center for Universal Education, The Brookings Institution.

Yoder, N., Holdheide, L., & Osher, D. (2018). Educators matter. In D. Osher, D. Moroney, & S. Williamson (Eds.), *Creating safe, equitable, engaging schools: A comprehensive, evidence-based approach to supporting students* (pp. 223–232). Cambridge, MA: Harvard Education Press.

26.
REJECTING GENETIC REDUCTIONISM AND EMBRACING RELATIONISM AND THE COMPLEXITY OF DYNAMIC SYSTEMS[*,†]

Richard M. Lerner[‡] *and Gary Greenberg*[§]

Rejecting Genetic Reductionism and Embracing Relationism and the Complexity of Dynamic Systems

At this writing, the authors of this chapter have spent more than a combined 100 years attempting to convey to colleagues, students, and the public the egregious and irremediable logical, theoretical, and methodological errors of genetic reductionist accounts of human development (the years wherein our respective Ph.D.s were awarded are 1971 and 1970, respectively). This book continues the task we have been pursuing.

On the one hand, we marvel at the resilience of the bad science and socially dystopian policy implications of genetic reductionism. On the other hand, we are appalled by the many abhorrent ideas that are all-too-often associated with genetic reductionism, for example, racism, sexism, anti-Semitism, and religious intolerance more generally. Frankly, we are alarmed to witness the staying power of the ideas associated by with genetic reductionism used in different sectors of society to justify acts of intolerance, physical attack, and even murder. At this writing, the use of these ideas in contemporary political rhetoric or in media popularism is both unfathomable and frightening to us.

As such, this chapter and the book within which it is embedded are aimed at countering the bad science that continues, in various guises, to appear (e.g., Buss & Hippen, 2018; Harden, 2021; Plomin, 2018). Such challenges to genetic reductionism are necessary but, at the same time, insufficient (e.g., Lerner, 2015), especially at the time in history when we were completing this book, that is, in the early weeks of fall, 2024.

We recognize that we are living at a moment in time when scientific facts are being dismissed as false news or as something that authoritarian-leaning or authoritarian-embracing politicians, or seekers of power-for-the-sake-of-power, term "woke;" where the freedom to read books of one's choice and to protect one's health and plans for one's own family are being taken away; when there is a continuation of the minoritization and marginalization of people because of race, ethnicity, religion, immigrant status, identity, or sexual preferences; where education is interpreted as indoctrination; and when there is criminalization of the expression of points of view that are discrepant from the narrative of anti-democracy people (those who act to diminish or even eliminate individuals and institutions that are pillars of democracy and civil society).

What can scientists do to at least diminish the prevalence and often unrecognized power of

---

[*] This chapter is published for the first time in this book.
[†] The writing of this chapter was supported by grants from the Templeton World Charity Foundation, the National 4-H Council, and Compassion International.
[‡] *Tufts University, Medford, MA*
[§] *Wichita State University, Wichita, KS*

DOI: 10.4324/9781032702988-42

the sectors of society that are creating seemingly relentless pursual of these calamitous actions? One answer is "Nothing." However, such an answer in effect makes those who wish to see an end to the nihilism and antidemocratic features of the present time complicit in the very circumstances they deplore. Even some small, countering actions are worthy of consideration.

One Possible Call to Action

Major organizations in developmental science, and the major journals in our field, could join in formulating a counternarrative, one based on the superordinate concepts associated with, and the sound and, often, compelling data derived from research associated with, dynamic, relational developmental systems-based approaches to human development. Why should dynamic, relational developmental systems be the frame for a counternarrative to genetic reductionism?

Earlier in this book we discussed psychological science and, in particular, developmental science as fields of scholarship consistent and compatible with the principles, laws, and findings within many of the physical, life, and social/behavioral sciences (e.g., see also Lerner, 2015, 2018; Lewis, 2000; Molenaar, 2004). Indeed, these convergences have been illustrated in much of the work included in the prior sections of the present book. These examples exist along with many others. For Greenberg et al. (1999) showed how dynamic systems models can advance understanding of the emergence of complex behaviors, such as instance the evolution and development of language. Hollenstein (2007) explained how the dynamics of dyadic relationships could be modeled through the use of state space grids. Ram et al. (2014) illustrated how the dynamics among processes unfolding across different time scales enabled the study of intraindividual changes in affect, health, and interpersonal behavior. In turn, Mongin et al. (2022) explain how to ordinary differential equations used in physics can enable developmental scientists to understand the rate of change across time, changes in the rate of change across time, and changes in the changes of rates across time, etc. In short, there is a robust and growing literature illuminating the use of dynamic systems concepts to integrate conceptual and empirical links among different sciences in relational and not in reductionist manners (e.g., Hattfield, 1995, 2000; Lerner, 2015; Sanjuan, 2023).

Unfortunately, these ideas and data are often unrecognized or ignored in the writings of genetic reductionists. To illustrate the sort of authoritative, clear, and precise counternarrative material that could appear in such consensus documents, we can point to some work critiquing the genetic reductionism found in both Evolutionary Psychology (e.g., Buss, 2019; Buss & von Hipple, 2018) and Evolutionary Developmental Psychology (e.g., Bjorklund, 2015, 2016; Bjorklund & Ellis, 2005). For instance, Greenberg and Partridge (2010) have explained that:

> "Rejection of the atomistic reductionistic approach also entails a rejection of the adaptationist agenda of evolutionary psychology. As many have pointed out, including such notable evolution scientists as Gould (1997), even Darwin suggested that mechanisms other than adaptation are at work in evolution. It is a mistake and a misunderstanding of Darwinism to suppose that there is anything approaching the consensus claimed by evolutionary psychologists. Rather, pluralism of mechanisms is the rule in the still developing paradigm of evolution. For example, we now understand evolution to involve punctuated equilibrium, genetic drift, mutation, and other processes, as well as natural selection. In fact, evolution does not always involve changes in the genome. It is now recognized that not all genes of the human genome get expressed. Evolution can occur if different portions of the genome are expressed, the result perhaps of environmental impact. This would result in new phenotypes.
>
> (see Honeycutt, 2006)" (p. 122)

The ideas of Greenberg and Partridge (2010) are consistent with those of Saunders and Ho (1976), who suggested that increasing complexity is a second law of evolution after natural selection (Saunders & Ho, 1976, 1981, 1984). In addition, Greenberg (2011) goes on to note that:

"in a dynamic and changing environment, rather than genes specifying a particular developmental outcome, be it structural or behavioral, every outcome is an emergent result of the transaction between genes and their cellular, organismic, ecological, and temporal contexts. This view of epigenesis is epitomized by discoveries in biology that even identical genomes in extremely similar environments do not always follow the same developmental pathways. Ko and colleagues (Ko, Yomo, & Urabe, 1994), studying enzyme activity in bacteria, found that despite identical genomes and extremely uniform culture conditions, individual cells developed different levels of enzyme activity and grew into colonies of different size. Ko's studies showed that cell state in bacteria is determined not only by genotype and environment. Rather, "Changes of state can occur spontaneously, without any defined internal or external cause. By definition, these changes are epigenetic phenomena: dynamic processes that arise from the complex interplay of all the factors involved in cellular activities, including the genes".

(Solé & Goodwin, 2000, p. 63)" (pp. 184–185)

Using authoritative and clearly presented statements akin to the ones we have just used as illustrations, a broadly disseminated counternarrative could, perhaps, have the status of a National Academies of Science "Consensus" document about the bad science associated with past and contemporary genetic reductionist ideas. In addition, such a document could be formulated to have different versions in order to reach different sectors of society The document could explain the abundant and growing evidence (e.g., as non-exhaustively illustrated in the sections of the present book) about the plasticity of human development and present the profoundly different implications of this work for applications to policies and programs. The visibility of such a broadly disseminated collaborative consensus document might – to use the strong words employed by Jerry Hirsch in 1981 – "unfrock the charlatans." Such a statement should also explain the dangers of policy and program recommendations (e.g., neo-Eugenics policies, as in Belsky, 2014) being promulgated on the basis of such bad science, and—perhaps—enhance public discourse about how informed citizens can act to promote the positive and healthy development of all people.

Whereas such an effort would be confronted by significant pushback from some members of the scientific community and, as well, by members of both various governmental bodies and organizations within the diversity of civil society – there seems to us to be sufficient reason to embark on this idea. Nevertheless, at this writing, such a statement has not materialized. What might provide an impetus for the start of such a project?

In our view, the features of developmental science framing the contributions to the present book offer one possibility. A succinct presentation of the key ideas about dynamic developmental systems that are present in the contributions in this book are, although admittedly abstract, complex, and non-linear, a potential starting point for creation of such a consensus document. Multidisciplinary elaborations supporting the core of the consensus could be a contribution to defining a "new normal" for the scientific study of human development, one that could become a guiding document for the education of a new generation of scientists.

Such education would include the theoretical tenets of and methodological tools (i.e., measures, research designs, and data analytic techniques) involved in the use of dynamic systems approaches to the study of human development. Such an education could also inculcate within developmental scientists a dedication to application, that is, to translate their work into forms accessible to the diversity of social sectors comprising the general public. Examples are parents and children, community leaders, educators, media, and policy makers.

The essence of this translation would be that genes are plastic components of the dynamic, relational developmental system, and not blueprints of the life course of people. A person's genes, received at conception, do not design the course of human behavior and development across ontogeny, generations, and contexts (Lewontin, 2000; Lewontin et al., 1984). And, as we shall emphasize in our concluding remarks, another outcome of this new normal might be the contribution of developmental science to social justice.

We believe that it is useful, therefore, to present what may be learned about dynamic systems approaches to human development from the contributions found in this book. We will then return to some ideas about implications of dynamic systems ideas for the promotion of social justice and authentic equity for all people.

Rejecting Reductionism and Linearity and Accepting the Relationism and the Complexity of Dynamic Systems

A fair reading of the corpus of psychological science across 150 years would support the conclusion that the major emphasis in this literature has been on the continuity of psychological laws, such as they may be, across all organisms. Despite evidence to the contrary (e.g., Bitterman, 1960, 1965; Hebb, 1949; Schneirla, 1957; Tobach & Schneirla, 1968), the preponderant emphasis in this literature is on quantitative variation in the process believed to be involved in the behavior of all organisms. Any qualitative differences proposed to exist have been regarded as more apparent than real, and could be explained away through essentially linear reduction to the laws of conditioning and learning (e.g., Baer, 1976; Bijou & Baer, 1961; Skinner, 1938, 1971) or to the laws of genetics purportedly involved in sociobiology (e.g., Dawkins, 1976; Wilson, 1975), behavior genetics (e.g., Plomin, 2018), evolutionary psychology (e.g., Buss, 2019; Buss & Hippel, 2018), or evolutionary developmental psychology (Bjorklund, 2015, 2016; Bjorklund & Ellis, 2005).

Dynamic, relational developmental systems concepts offer a very different narrative about human behavior and development. The narrative rejects reductionism, in both its nature and nurture versions, and, instead, draws on concepts associated with a distinct ontology about human development, one that is qualitatively distinct from those of other living organisms. The narrative indicates that the specific instance of developmentally-nurturant relationships that can exist among humans provides bases for each individual human to thrive and, as well, for all humans to contribute to a flourishing social and physical ecology within which all people and the ecology of human life can flourish and, even more, wherein equity and social justice may be attained.

To understand human development per se, an understanding of features of dynamic systems should be discussed.

Autopoiesis

Dynamic systems are *autopoietic*, that is self-constructing. This self-construction means that each part of the dynamic system is both a product and a producer of every other part of the system. Marshall et al. (2021, p. 4) explain that:

> "living things actively self-maintain themselves through the constant regeneration of the conditions that are necessary to sustain their material existence" (Marshall et al., 2021, p. 3)… [and thus reflect] "constitutive autonomy… in contrast to behavioral autonomy, where the identity of the system is imposed externally by an operator or observer."

Witherington (2014, p. 27) makes a similar point in noting that the dynamic developmental system serves "as its own cause, organizing and producing itself such that it causes and results from itself. In this way, living systems constitute natural ends or purposes."

Each part of the system therefore *regulates* the other parts of the system *across time and* (in open- and, especially, living-systems) *place* (Brandtstädter, 1998, 2006). Such regulation may involve either *attracting* or *repelling* other parts of the system (Hollenstein, 2007) and, as well, moderating both *quantitative fluctuations* in other system components and *transformations* of short-term quantitative fluctuations *into qualitative and relatively permanent change* (Ram et al., 2014). Such qualitative transformations are understood as the features of human *development* across the life span (Overton, 2015). This idea was recognized earlier by the neuroscientist Sherrington (1951), who defined a healthy animal as a "set of organs of interlocking action regulating each other, the whole making a self regulating system" (p. 163).

In short, in a dynamic system, no one system part is a "prime mover" of other parts of the system. Each part of the system exists in a mutually-influential relation with every other part of the system. The connections among any two parts of the system must, then, be conceptually represented by a bidirectional, two-headed arrow – ⇔ – and never by a unidirectional arrow (→).

Developmental Regulation in Human Development

When human life is the dynamic system under discussion, this relation may be represented as person ⇔ context. The person in such a dynamic, mutually influential relation is a regulator of the structure, function, and changes within the context within which the person lives and develops **just as** the context is a regulator of the structure, function, and quantitative and qualitative changes in the individual living and developing within the context. In other words, every individual is a product and a producer of their own development.

The human possession of such regulatory capacity across the life span is termed *agency*. The systematic variation in a person's behavior and developmental that can derive from such agency is termed *plasticity* (Lerner, 1984, 2018). Agency and plasticity are key strengths of human development.

Relationism (and Not Reductionism)

The basic unit of analysis within dynamic systems involved in human behavior and development is the mutually-influential person ⇔ context relation or, when discussing coactions between two or more humans within the dynamic developmental system, person ⇔ person relationships. However, there is considerable variability in the substance, nature, or content of mutually-influential relations.

Indeed, it is unlikely that any two people will have the same history across their lives in these relations (Bornstein, 2019; Hirsch, 2004; Rose, 2016). Said another way, the individual relations or relationships that occur across a human life are always specific to a person (Bornstein, 2019). In addition, specific changes within each person's life during one period of life contribute to shaping subsequent, specific individual ⇔ context relations.

Mutual Beneficence

There is one instance of such within-the-person variation in the significance for ongoing exchanges with the context, including both other people and the physical ecology, that is central to understanding the dynamics of what defines optimal human development. That is, whereas developmental regulations involving an individual and a context are always mutually influential, only some of these coactions are also *mutually-beneficial*. Mutually-beneficial individual-context or individual-individual coactions are ones that promote, maintain, and sustain the well-being of all coacting individual-context or individual-individual coactions and, on a more general level, of all components of the dynamic relations within the human development system. Relations or relationships that are marked by mutuality of benefit to all facets of the system are ones that nurture the positive or healthy development of the system.

When it is the case that across time and place a person's behavior and development are marked by specific relations/relationships that are developmentally nurturing to the whole system – to the person and to the person's social and physical ecology – then a key qualitative transformation emerges in human development – an emergence that is distinct to humans; it is what developmental scientist and theologian Pamela Ebstyne King and colleagues (e.g., King et al., 2024) term a "reciprocating self." A reciprocating self is possessed by an individual who cognitively, affectively, and behaviorally is committed to live a life of mutual beneficence to self and others, who possesses a personal purpose (or a self-constructed "telos;" King et al., 2024) to "becoming one's best self *with and for others, and with a higher purpose*" (King & Mangan, 2023, p. 610).

As explained by Carpendale and Lewis (2021), this beneficent relational perspective can be found in different faith traditions and cultures: "the Japanese kanji symbol for a person is a representation of one person being supported by another person…[and] is typical in indigenous cultures such as Canadian First Nations" (p. 4). In addition, they point to the African concept of *ubuntu*, which, as derived from Zulu, means "I am because you are" or "a person is a person through other persons" (Carpendale & Lewis, 2021, p. 4).

How the dynamic, relational developmental system may promote the development of a reciprocating self is the story of what makes human beings human. There are several concepts of human ontology that are implicit in this dynamic systems model of human development: Embodiment, holism, specificity of individual ⇔ context relations and, as just noted, individual ⇔ context relations that involve developmentally-nurturant relations and the emergence of a reciprocating self.

The Ontology of Human Life and Development Within Dynamic Developmental Systems

Four interrelated ideas enable the specification of a dynamic systems-based conception of the meaning of being human. The first to consider derives from the relational character of the dynamic human system: Embodiment.

Embodiment

Overton (2015) explained that human development is embodied by 1. the physiological and morphological features of humans (e.g., neurobiology, genetics/epigenetics, and hormones), by 2. the coactions of psychological processes (e.g., involving cognitions, affect, and behaviors) with this first instance of embodiment, and by 3. the coaction of social and cultural processes with these first two instances of embodiment. Thus, Overton (2013) notes that:

> "Embodiment includes not merely the physical structures of the body but the body as a form of lived experience, actively engaged with the world of socio-cultural and physical objects. The body as form references the biological point-of-view, the body as lived experience references the psychological subject standpoint, and the body actively engaged with the world represents the socio-cultural point-of-view".
>
> (p. 103)

Embodiment, then, depicts the ongoing coaction between these systems and levels of organization, both internal and external to the person, and conveys the inter-penetrating and bidirectional nature of experience at every level of development of a human being (Cantor et al., 2021; Immordino-Yang & Yang, 2017; Schneirla, 1957).

As an example of this embodiment, Immordino-Yang and colleagues (e.g., Immordino-Yang & Yang, 2017; Immordino-Yang et al., 2019; Immordino-Yang et al., in press) point to evidence indicating that the individuals' cognitive, affective, and behavioral processes involved in meaning-making in and of their world are a product and a producer of meaningful changes in brain activity that result for the individual coacting with features of their specific contexts. Thus, developmental science requires theory-predicated and empirically systematic interrogation and, ultimately, the integration of the three domains of embodiment.

Holism

Human life and development must be understood as involving multiple domains of development that coact in the holistically integrated dynamic developmental system. The concept of holism means that the "parts of the whole" (the embodied domains) of an individual:

> "do not combine through an additive process. Instead, the combination may be better understood as a multiplicative process: When the parts combine, they produce, in combination, attributes of a novel whole that do not exist in the parts in isolation. What makes living systems unique is that they change systematically, through mutually-influential individual ⇔ context relations, into new, increasingly adaptive and complex forms (see Gould & Vrba, 1982; Jablonka & Lamb, 2005). The whole, through its self-organization, has unique systemic features that are not attributes of any part. Thus, the whole is not just quantitatively greater than the sum of its parts; it is *qualitatively different from the sum of its parts*".
>
> (Cantor et al., 2021, p. 23)

As such, the developmental process of humans: (a) dynamically links each individual and the individual's context in self-creating (autopoietic) and thus mutually influential- and, in adaptive instances, in mutually beneficial-coactions; and (b) dynamically integrates any facet of human development with other physical and physiological processes, cognitive, affective, and behavioral processes, and social and cultural processes involved in each person's developmental pathway (Lerner & Lerner, 2019). For instance, King and colleagues (in press; King & Mangan, 2023) explain that spiritual development occurs through the coactions of (embodied) individuals embedded in various contexts. Individual's experiences of transcendence result in meaning-making, purpose, fidelity, and virtue development.

Specificity of Individual ⇔ Context Relations

As we have noted, human life and development are characterized by the specificity of an individual's course of individual ⇔ context relations across time and place (e.g., Bornstein, 2019; Elder et al., 2015; Molenaar, 2004; Rose, 2016). The Specificity Principle presented by Bornstein (2017, 2019; see too Lerner & Bornstein, 2021) depicts the dimensions of biological, psychological, behavioral, relationship, and contextual individuality that must be understood to provide a complete account of human development and, as well, to maximize the chances that the application of human development theory-based research equitably promotes positive outcomes for every individual. Indeed, without recognition of specificity in models and measurement of human development (e.g., Molenaar, 2004; Rose, 2016), and in evidence-based policies and programs derived from such developmental scholarship, neither authentic equity nor social justice can be derived as contributions from developmental science (Cantor & Osher, 2021).

Rooted in the concepts associated with dynamic, RDS metatheory (Bornstein, 2019; Overton, 2015), Bornstein's Specificity Principle explains that human development always involves specific outcomes in specific individuals occurring in specific places at specific times and in specific ways. As defined by Lerner and Bornstein (2021, p. 2):

> "the Specificity Principle has five main terms: setting, person, time, process, and outcome. The Specificity Principle states that processes

of development involve mutually influential (dynamic) relations between specific individuals and their specific contexts, represented as individual ⇔ context relations. Thus, the Specificity Principle embraces whole-individual development (Cantor et al., 2021) and the uniqueness of each individual's development. The Specificity Principle is a heuristic not an analytic means. In consequence, there is no statistical test of the Specificity Principle. Rather, the Specificity Principle is meant to guide investigators in the design, report, and interpretation of research."

Of course, the specificity of individual ⇔ context relations does not negate the existence of facets of developmental processes that can be generalized across groups (i.e., differential group processes) and, as well, facets of developmental processes that are nomothetic (that are shared by all humans) (Allport, 1937, 1962; Emmerich, 1968). Kluckhohn and Murray (1948) made this point by explaining that it is simultaneously true that each human has attributes that are present in all other humans, that each human has attributes that are present in only some other humans, and that each human has attributes that are shared with no other humans (these attributes are labeled as idiographic or person-specific).

Nevertheless, the ubiquitous specificity of individual ⇔ context relations over the course of ontogeny means that, to comprehensively and holistically understand the development of human behavior across the life span, developmental scientists – concerned with the development of any construct – need to identify the idiographic, differential, and nomothetic dimensions of a target process (e.g., Cantor et al., 2021; Lerner, 2018; Molenaar, 2004; Molenaar & Nesselroade, 2015; Rose, 2016). Such identification will provide knowledge about the "valence" of the set of specific individual ⇔ context relations involved in a specific person's life. How many of these mutually-influential relations are also (1) mutually antagonistic to the components of the relation; (2) neither antagonistic nor beneficial; and (2) mutually beneficial and therefore developmentally nurturant? It is from the latter instances of the coactions between the person and the context that developmentally-nurturant relationships and a reciprocating self may develop.

Developmentally-Nurturant Relationships and the Reciprocating Self

An embodied approach to holism and specificity in human life and development suggests that any facet of the individual person involves mutually-influential coactions within the person and between the person and the context. In ideal circumstances, these inner and outer coactions are mutually beneficial. Such coactions constitute the bases of developmentally-nurturant relationships, and therefore can be the key resources for applications of RDS-based concepts to promote positive human development (Li & Julian, 2012).

However, to optimize the contributions of developmentally nurturant relationships in serving as a resource to promote positive development and thriving, another developmental transformation must emerge from such relationships. The agency of the person must be aimed at (have the purpose; e.g., Damon, 2008; Damon et al., 2003) of constituting a *reciprocating self* (Balswick et al., 2016; King, 2016; King & Defoy, 2020; King & Mangan, 2023; Schnitker et al., 2019).

As we have noted, King and Mangan (2023) explain that such a self-definition, such an identity, reflects a dynamic purpose to become one's best self with and for others. In other words, a reciprocating self provides a person-specific purpose for a human's development, one which offers the perspective that positive instantiations of human development are aligned with promoting individual, social, and physical ecological flourishing (King & Mangan, 2023; King et al., 2024; Schnitker et al., 2019).

In a sense then, in maturing instantiations of mutually-beneficial individual ⇔ individual coactions, a personally-defined purpose of each human life involves becoming a steward of the other lives within which one lives in positive, mutually-supportive relationships *and* a steward of the ecology within which human life can continue to flourish. Specific instances of such stewardship will of course vary across people, time, and place. However, in all cases, they reflect the higher purpose that King and colleagues note reflects the relational purpose of *existing with and for others* (King & Mangan, 2023, p. 610).

The adaptive significance of such mutually meaningful relationships for human and socio-ecological thriving may be found in research that documents that the absence of such mutuality can hinder healthy and positive development, across the life span, from infancy (e.g., Ourth & Brown, 1961; Spitz, 1945, 1946) through adulthood and old age (e.g., Antonucci et al., 2001; Ingersoll-Dayton et al., 1997; Slavich & Cole, 2013; Tomaka et al., 2006). However, as emphasized by Cantor (e.g., Cantor et al., 2021; Cantor & Osher, 2021) the presence of negative social relations, or even trauma, in an early period of life is not destiny. Because of the relative plasticity of human development (i.e., the potential for relatively permanent change within the autopoietic, dynamic developmental system; Lerner, 1984, 2018), providing mutually beneficial relationships that involve developmentally nurturant relationships can enhance health and positive development across the human life span (e.g., Antonucci et al., 2001; Ingersoll-Dayton et al., 1997; Lerner, 1984). This accompaniment can serve as the link between dynamic developmental systems and the promotion of social justice.

Making Humans Human: The Ontology of Dynamic, Relational Developmental Systems

The features of dynamic systems ontology that we have discussed make clear that neither nature nor nurture reduction can ever be successful in learning about what makes human beings human. Contemporary developmental science understanding of the dynamics of human life and development emphasizes that, from conception and through the entire course of life, humans are relational creations.

Human life cannot be known by reducing humans to "lumbering robots" (e.g., Dawkins, 1976) controlled completely by the DNA received at their conception. Humans also cannot be characterized as without "freedom and dignity" (e.g., Skinner, 1971) because of reduction to the stimuli they adventitiously encounter in their environment. As we have already suggested, a dynamic systems approach to human development leads us to believe that humans cannot be known by reducing to an algorithm a life-span journey toward gaining fulfillment, identity, meaning, mattering, and love, a journey that will necessarily involve hope, sacrifice, and spiritual transcendence (e.g., Balswick et al., 2016; King & Defoy, 2020) and the other coactive qualities that define each person in a specific manner. Similarly, Gastmans et al. (2023) have noted:

> "the fact that human life is fundamentally relational – from the moment of conception and throughout the life span… – means that human emotions, such as love, loyalty, fidelity, empathic concern and, as well, the need for affiliation, and mattering positively to others, are the specific features of human existence that derive from and contribute to relationships that make human beings human."

Agency, coaction, and irreducibility have been documented in voluminous empirical studies of human development across the life span (e.g., see the four volumes of the *Handbook of Child Psychology and Developmental Science*, 7th edition; Lerner, 2015). The presence of agency and self-construction means that humans have responsibility for the course and outcomes of their lives. Within developmental science, responsibility is a psychological (characterological) attribute studied as a part of moral reasoning, ethical understanding and action, and moral development (e.g., Lerner, 2018; Lerner et al., 2023; Nucci, 2019). Such aspirations, ethical mandates, and actions are not the stuff of genes, stimuli, or algorithms (Gastmans et al., 2024; Sinibaldi et al., 2022). They are the stuff of humans.

When developmental science is framed by dynamic, relational developmental systems concepts, explanations are provided of the process through which these ethical actions and relations occur, and do so without reductive reliance on to a prior-to-and outside-of-the system entity, such as a selfish gene.

Conclusions

Readers of this chapter might consider that, in the ten years preceding its writing (in the early weeks of fall, 2024), the world has seen dramatic changes that threaten the health and lives of billions of people and our planet (e.g., see Lerner et al., 2021 for a discussion). In the eyes of many around the globe, the decade began with U. S. world leadership and international respect continuing on a higher and higher path that was largely uninterrupted since the entry of the nation into World War II. However, this period also included relentless racism and disparities in economic, educational, safety, and health imposed on people of color in the United States and, as well, on individuals in low and middle-income countries around the world, that have persisted over the past century despite changes in geopolitical power. The decade ended with the world confronted with a pandemic of a magnitude not encountered for at least 100 years, with the eroding of respect for and reliance on the United States as a positive force for democracy and social justice in the world, and with world media and political attention paid to a rising tide of authoritarianism and, as well, fascism around the globe, movements associated with instances of systemic racism and the oppression and marginalization of people not only in relation to race but, as well, in regard to their ethnicity, gender identity, and sexual orientation.

These injustices were coupled with inequities within and across healthcare systems, elementary and secondary education, higher education, business and industry, and politics. Around the world, these injustices were brought into stark relief by indignities and brutalities shown to, and the murders of, children and adults because of racism and autocratic manifestations of power and self-indulgence that diminish the chances of peace, justice, and action to address the dangers of catastrophic climate change.

Given the course of changes across this 10-year period, it might be easy, and even quite reasonable, to discount scientific contributions to ameliorating this situation as a Sisyphean task. However, the vision for humanity derived from the ideas within dynamic systems approaches to human development – of the potential for all humans to thrive through developmentally-nurturant relations that contribute to the emergence of a reciprocating self – *may* make some contribution. Certainly, the potential for human thriving through instantiation of such relations may be of obvious use in countering the rising rates of poor mental health and loneliness that have become a public health crisis (Lee et al., 2023; Office of the U.S. Surgeon General, 2023; Witters, 2023; Zablotsky & Ng, 2023), a crisis exacerbated by inequities due to marginalization. Indeed, continuing to explore the fundamental significance of human "accompaniment," of reciprocating selves, may provide some impetus for recognition that the answer to the problem besetting humans at this moment in history can be addressed by nurturance of each individual to the world beyond the self.

If humanity is to flourish, then there must be strong and dynamic links between the development of a thriving individual and a thriving context, one that supports both the individual as an individual (with specific attributes and aspirations for life) and, *as well*, all specific individuals in their specific quests for health and positive lives (Lerner, 2004). The agency and relative plasticity that are defining features of human life may, when developed along with the specific mutually-influential relations within the dynamic human development system that create mutual beneficence and a reciprocation self, means that the potential for acting to promote authentic opportunities for socially-just changes in the world are defining features of a thriving human being. Such development, such flourishing, is a hope that all parents, of all nations, have for all of their children. It is a hope that bring people together with fairness of justice.

Our hope is that such a convergence of ideas and people can forge an action agenda aimed at enhancing the lives of all people and, as well, creating a socially just and equitable social world and physical ecology within which every human can thrive in community with others. Such an agenda can instantiate a genuine chance for peace on a sustainable planet and goodwill for all people.

References

Allport, G. W. (1937). *Personality: A psychological interpretation*. Holt.

Allport, G. W. (1962). The general and the unique in psychological science. *Journal of Personality, 30*(3), 405–422.

Antonucci, T. C., Lansford, J. E., & Akiyama, H. (2001). Impact of positive and negative aspects of marital relationships and friendships on well-being of older adults. *Applied Developmental Science, 5*(2), 68–75.

Baer, D. M. (1976). The organism as host. *Human Development, 19*(2), 87–98.

Balswick, J. O., King, P. E., & Reimer, K. S. (2016). *The reciprocating self: Human development in theological perspective* (2nd ed.). InterVarsity Press.

Belsky, J. (2014, November 30). The downside of resilience. *New York Times, Sunday Review*, p SR4.

Bijou, S. W., & Baer, D. M. (1961). *Child development: A systemic and empirical theory* (Vol.

Bitterman, M. E. (1960). Toward a comparative psychology of learning. *American Psychologist, 15*, 704–12.

Bitterman, M. E. (1965). Phyletic differences in learning. *American Psychologist, 20*, 396–410.

Bjorklund, D. F. (2015). Developing adaptations. *Developmental Review, 38*, 13–35.

Bjorklund, D. F. (2016). Prepared is not preformed: Commentary on Witherington and Lickliter.

Bjorklund, D. F., & Ellis, B. J. (2005). Evolutionary psychology and child development: An emerging synthesis. In B. J. Ellis & D. F. Bjorklund (Eds.), *Origins of the social mind: Evolutionary psychology and child development* (pp. 3–18). New York: Guilford.

Bornstein, M. H. (2017). The specificity principle in acculturation science. *Perspectives in Psychological Science, 12*(1), 3–45.

Bornstein, M. H. (2019). Fostering optimal development and averting detrimental development: Prescriptions, proscriptions, and specificity. *Applied Developmental Science, 23*(4), 340–345.

Brandtstädter, J. (1998). Action perspectives on human development. In W. Damon (Series Ed.) & R. M. Lerner (Vol. Ed.), *Handbook of child psychology*: Vol. 1. *Theoretical models of human development* (5th ed., pp. 807–863). New York: Wiley.

Brandtstädter, J. (2006). Action perspectives on human development. In R. M. Lerner (Ed.). *Theoretical models of human development*. Volume 1 of *Handbook of Child Psychology* (6th ed.). (pp. 516–568). Editors-in-chief: W. Damon & R. M. Lerner. Hoboken, NJ: Wiley.

Buss, D. (2019). *Evolutionary psychology: The new science of the mind*. Taylor & Francis.

Buss, D. M., & von Hippel, W. (2018). Psychological barriers to evolutionary psychology: Ideological bias and coalitional adaptations. *Archives of Scientific Psychology, 6*(1), 148–158.

Cantor, P., Lerner, R. M., Pittman, K., Chase, P. A. &, Gomperts, N. (2021). *Whole-child development, learning, and thriving: a dynamic systems approach*. Cambridge University Press.

Cantor, P., & Osher, D. (2021). The future of the science of learning and development: Whole-child development, learning, and thriving in an era of collective adversity, disruptive change, and increasing inequality. In P. Cantor and D. Osher (Eds.), *The Science of Learning and Development: Enhancing the Lives of All Young People* (pp. 233–254). Routledge.

Carpendale, J., & Lewis, C. (2021). *What makes us human: How minds develop through social interactions*. Routledge.

Damon, W. (2008). *The path to purpose: Helping our children find their calling in life*. Simon and Schuster.

Damon, W., Menon, J., & Bronk, K. C. (2003). The development of purpose during adolescence. *Applied Developmental Science, 7*(3), 119–128.

Dawkins, R. (1976). *The selfish gene*. Oxford University.

Elder, G. H., Shanahan, M. J., & Jennings, J. A. (2015). Human development in time and place. In M. H. Bornstein and T. Leventhal (eds), *Handbook of child psychology and developmental science, vol. 4: Ecological settings and processes in developmental systems* (7th ed., pp. 6–54). Hoboken, NJ: Wiley.

Emmerich, W. (1968). Personality development and concepts of structure. *Child Development 39*, 671–690.

Gastmans, C., Sinibaldi, E., Lerner, R., Yáñez, M., Kovács, L., Palazzani, L., Pegoraro, R., & Vandemeulebroucke, T. (2024). *Christian anthropological considerations on healthcare robotics*. Roboethics Working Group, Pontifical Academy for Life. Vatican City.

Gould, S. J. (1997). Darwinian fundamentalism. *New York Review of Books, 44*(10), 34–37.

Gould, S., & Vrba, E. (1982). Exaptation: A missing term in the science of form. *Paleobiology, 8*, 4–15.

Greenberg, G. (2011). The failure of biogenetic analysis in psychology: Why psychology is not a biological science. *Research in Human Development, 8*, 173–191.

Greenberg, G., & Partridge, T. (2010). Biology, evolution, and development. In W. F. Overton (Ed.), *Cognition, Biology, and Methods: Volume 1 of the Handbook of Life-span Development* (pp. 115–148). Wiley.

Greenberg, G., Partridge, T., Weiss, E., & Haraway, M. M. (1999). Integrative levels, the brain, and the emergence of complex behavior. *Review of General Psychology, 3*, 168–187.

Harden, K. P. (2021). *The genetic lottery: Why DNA matters for social equality*. Princeton University Press.

Hattfield, G. (1995). Remaking the science of mind: Psychology as natural science. In C. Fox, R. Porter, & R. Wolker (Eds.), *Inventing human science: Eigdhteenth-century domains* (pp. 184–231). Berkeley CA: University of California Press.

Hattfield, G. (2000). The brain's "new" science: Psychology, neurophysiology, and constraint. *Philosophy of Science, 67*, S388–S403.

Hebb, D. O. (1949). *The organization of behavior*. Wiley.

Hirsch, J. (1981). To "unfrock the charlatans." *Sage Race Relations Abstracts, 6*, 1–65.

Hirsch, J. (2004). Uniqueness, diversity, similarity, repeatability, and heritability. In C. Garcia Coll, E. Bearer, & R.M. Lerner (Eds.). *Nature and nurture: The complex interplay of genetic and environmental influences on human behavior and development* (pp. 127–138). Erlbaum.

Ho, M. W., & Saunders, P. T. (Eds.). (1984). Beyond neo-Darwinism: An epigenetic approach to evolution. *Journal of Theoretical Biology, 78*, 573–591

Hollenstein, T. (2007). State space grids: Analyzing dynamics across development. *International Journal of Behavioral Development, 31*(4), 384–396.

Honeycutt, H. (2006). Studying evolution in action: Foundations for a transgenerational comparative psychology. *International Journal of Comparative Psychology, 19*, 170–184.

Immordino-Yang, M. H., Darling-Hammond, L., & Krone, C. R. (2019). Nurturing nature: How brain development is inherently social and emotional, and what this means for education. *Educational Psychologist, 54*(3), 185–204.

Immordino-Yang, M. H., Nasir, N. S., Cantor, P., & Yoshikawa, H. (in press). Weaving a colorful cloth: Centering education on humans' emergent developmental potentials. In C. D. Lee, R. M. Lerner, V. L. Gadsden, & D. Osher (Eds). *The science of learning and development. Review of Research in Education, 2023*. Sage Publications.

Immordino-Yang, M. H., & Yang, X.-F. (2017). Cultural differences in the neural correlates of social–emotional feelings: An interdisciplinary, developmental perspective. *Current Opinion in Psychology, 17*, 34–40.

Ingersoll-Dayton, B., David Morgan, D., & Antonucci, T. (1997). The effects of positive and negative social exchanges on aging adults. *The Journals of Gerontology: Series B, 52*(4), S190–S199.

Jablonka, E., & Lamb, M. (2005). *Evolution in Four Dimensions: Genetic, epigenetic, behavioral, and symbolic variation in the history of life*. MIT Press.

King, P. E. (2016). The reciprocating self: Trinitarian and Christological anthropologies of being and becoming. *Journal of Psychology and Christianity, 35*(3), 215–232.

King, P. E., Baer, R. E., & Greenway, T. S. (2024). Theological perspectives on beliefs and communities of practice: Virtue systems as an integrative approach for psychologists. In M. D. Matthews & R. M. Lerner (Eds.), *Routledge International Handbooks of Multidisciplinary Perspectives on Character Development*, Volume I: *Conceptualizing and Defining Character*. (pp. 347–379). Routledge.

King, P. E., & Defoy, F. (2020). Joy as a virtue: The means and ends of joy. *Journal of Psychology and Theology, 48*(4), 308–31.

King, P. E., & Mangan, S. (2023). Hindsight in the 2020's: Looking back and forward to positive youth development and thriving. In L. Crockett, G. Carlo, & J. Schulenberg (Eds.), *APA Handbook of Adolescent and Young Adult Development*. American Psychological Association.

Kluckhohn, C., & Murray, H. (1948). Personality formation: The determinants. In C. Kluckhohn & H. Murray (Eds.), *Personality in nature, society, and culture*. New York: Knopf.

Ko, E. P., Yomo, T., & Urabe, I. (1994). Dynamic clustering of bacterial population. *Physica D: Nonlinear Phenomena, 75*(1–3), 81–88.

Lee, S., Salvador, C., Tuel, A., & Vicedo-Cabrera, A. M. (2023). Exploring the association between precipitation and hospital admission for mental disorders in Switzerland between 2009 and 2019. *PLoS One, 18*(4): e0283200

Lerner, R. M. (1984). *On the nature of human plasticity*. Cambridge University Press.

Lerner, R. M. (2004). *Liberty: Thriving and civic engagement among America's youth*. Sage.

Lerner, R. M. (Ed.). (2015). *Handbook of Child Psychology and Developmental Science*, 4th ed. Wiley.

Lerner, R. M. (2018). *Concepts and theories of human development* (4th ed.). Routledge.

Lerner, R. M., & Bornstein, M. H. (Eds.). (2021). Enriching the Study of Human Development Through the Use of the Specificity Principle: Theory, Research, and Application. *Journal of Applied Developmental Psychology, 73/74/75*.

Lerner, R. M., & Lerner, J. V. (2019). An idiographic approach to adolescent research: Theory, method, and application. In L. B. Hendry & M. Kloep (Eds.), *Reframing Adolescent Research* (pp. 25–38). Routledge.

Lerner, R. M., Lerner, J. V., & Buckingham, M. H. (2023). Prosocial behavior, positive youth development, and character virtues: A. dynamic, relational developmental systems-based model. In T. Malti & M. Davidov (Eds.), *Handbook of prosociality: Development, mechanisms, promotion* (pp. 847–865). Cambridge University Press.

Lerner, R. M., Lerner, J. V., Murry, V. M., Smith, E. P., Bowers, E. P., Geldhof, G. J., & Buckingham, M. H. (2021). Positive youth development in 2020: Theory, research, programs, and the promotion of social justice. *Journal of Research on Adolescence, 31*(4), 1114–1134.

Lewis, M. D. (2000). The promise of dynamic systems approaches for an integrated account of human development. *Child Development, 71*, 36–43.

Lewontin, R. C. (2000). *The triple helix*. Harvard University Press.

Lewontin, R. C., Rose, S., & Kamin, L. J. (1984). *Not in our genes: Biology, ideology, and human nature*. Pantheon Press.

Li, J., & Julian, M. M. (2012). Developmental relationships as the active ingredient: A unifying working hypothesis of "what works" across intervention settings. *American Journal of Orthopsychiatry, 82*(2), 157–166. DOI: 10.1111/j.1939-0025.2012.01151.x

Marshall, P. J., Houser, T. M., & Weiss, S. M. (2021). The shared origins of embodiment and development. *Frontiers in Systems Neuroscience, 15*, 726403. https://doi.org/10.3389/fnsys.2021.726403

Molenaar, P. C. M. (2004). A manifesto on psychology as idiographic science: Bringing the person back into scientific psychology, this time forever. *Measurement, 2*(4), 201–218.

Molenaar, P. C. M., & Nesselroade, J. R. (2015). Systems methods for developmental research. W. F. Overton & P. C. M. Molenaar (Eds.), Handbook of child psychology and developmental science. Vol. 1: Theory and method (7th ed., pp. 652–682). Editor-in-chief: R.M. Lerner. Wiley.

Mongin, D., Uribe, A., Cullati, S., & Courvoisier, D. S. (2022). A tutorial on ordinary differential equations in behavioral science: What does physics teach us?. *Psychological Methods*. http://dx.doi.org/10.1037/met0000517

Nucci, L. (2019). Character: A developmental system. *Child Developmental Perspectives, 13*(2), 73–78.

Office of the U.S. Surgeon General (2023). Our epidemic of loneliness and isolation: The U.S. Surgeon General's advisory on the healing effects of social connection and community. Retrieved from https://www.hhs.gov/sites/default/files/surgeon-general-social-connection-advisory.pdf

Ourth, L., & Brown, K. B. (1961). Inadequate mothering and disturbance in the neonatal period. *Child Development, 32*(2), 287–295.

Overton, W. F. (2013). A New Paradigm for Developmental Science: Relationism and Relational-Developmental Systems. *Applied Developmental Science, 17*(2), 94–107.

Overton, W. F. (2015). Process and relational developmental systems. In W. F. Overton & P. C. M.

Molenaar (Eds.), *Handbook of Child Psychology and Developmental Science*, Volume 1: *Theory and Method* (7th ed., pp. 9–62). Editor-in-chief: R. M. Lerner. Wiley.

Plomin, R. (2018). *Blueprint: How DNA makes us who we are*. London: Allen Lane.

Ram, N., Conroy, D. E., Pincus, A. L., Lorek, A., Rebar, A., & Roche, M. J., Coccia, M., Morack, J., Feldman, J., & Gerstorf, D. (2014). Examining the interplay of processes across multiple time-scales: Illustration with the intraindividual study of affect, health, and interpersonal behavior (iSAHIB). *Research in Human Development, 11*(2), 142–160.

Rose, T. (2016). *The end of average: How we succeed in a world that values sameness*. HarperCollins Publishers.

Sanjuan, M. A. F. (2023). Physics of animal navigation. *The European Physical Journal: Special topics, 232*, 231–235.

Saunders, P. T., & Ho, M.-W. (1976). On the increase in complexity in evolution. *Journal of Theoretical Biology, 63*, 375–384.

Saunders, P. T., & Ho, M. W. (1981). On the increase in complexity in evolution II. The relativity of complexity and the principle of minimum increase. *Journal of Theoretical Biology, 90*(4), 515–530.

Schneirla, T. C. (1957). The concept of development in comparative psychology. In D. B. Harris (Ed.), *The concept of development: An issue in the study of human behavior* (pp. 78–108). Minneapolis: University of Minnesota Press.

Schnitker, S. A., King, P. E., & Houltberg, B. (2019). Religion, spirituality, and thriving: Transcendent narrative, virtue, and telos. *Journal of Research on Adolescence, 29*(2), 276–290.

Sherrington, C. (1951). *Man on his nature* (revised ed.). New York: Mentor.

Sinibaldi, E., Gastmans, C., Yanez, M., Lerner, R. M., Kovacs, L., Casalone, C., Pegoraro, R., & Paglia, V. (2022). Contribution from the Catholic Church to ethical reflections in the digital era. *Nature Machine Intelligence, 2*(5), 242–244.

Skinner, B. F. (1938). *The behavior of organisms*. New York: Appleton.

Skinner, B. F. (1971). *Beyond freedom and dignity*. New York: Knopf.

Slavich, G. M., & Cole, S. W. (2013). The emerging field of human social genomics. *Clinical Psychological Science, 1*, 331–348.

Solé, R., & Goodwin, B. (2000). *Signs of life: How complexity pervades biology*. Basic Books.

Spitz, R. (1945). Hospitalism, an inquiry into the genesis of psychiatric conditions in early childhood. *Psychoanalytic Study of the Child, 1*, 53–74.

Spitz, R. (1946). Hospitalism, a follow-up report. *Psychoanalytic Study of the Child, 2*, 113–117.

Tobach, E., & Schneirla, T. C. (1968). The biopsychology of social behavior of animals. In R. E. Cooke & S. Levin (Eds.), *Biologic basis of pediatric practice* (pp. 68–82). New York: McGraw-Hill.

Tomaka, J., Thompson, S., & Palacios, R. (2006). The relation of social isolation, loneliness, and social support to disease outcome among the elderly. *Journal of Aging and Health, 18*(3), 359–384.

Wilson, E. O. (1975). *Sociobiology: The new synthesis*. Cambridge, MA: Harvard University Press.

Witherington, D. C. (2014). Self-organization and explanatory pluralism: Avoiding the snares of reductionism in developmental science. *Research in Human Development, 11*(1), 22–36.

Witters, D. (2023, May 17). *U.S. Depression Rates Reach New Highs*. Gallup. Retrieved from https://news.gallup.com/poll/505745/depression-rates-reach-new-highs.aspx

Zablotsky, B., & Ng, A. E. (2023). Mental health treatment among children aged 5-17 years: United States, 2021. *NCHS Data Brief, 472*, 1–8. PMID: 37314377.

AUTHOR INDEX

Note: Locators in *italics* represent figures and **bold** indicate tables in the text.

Abbs, B. 140
Ablah, E. 82
Acikgoz, K. 484
Acredolo, L. P. 138
Adams, M. D. 84
Adamson, F. 488
Adelman, H. S. 489
Aizawa, K. 69
Akar, H. 481
Alberch, P. 215, 287
Albers, B. 89
Alexander, C. N. 40
Alexander, M. 499
Alford, J. R. 450
Algoe, S. B. 404
Allegrucci, C. 414
Allen, K. R. 341
Allensworth, E. 481
Allo, M. 420
Allport, G. W. 519
Almeida, O. F. 413
Alpert, J. E. 413
Altmann, G. 140–141
Alvarez-Lopez, M. J. 404
Ambrose, A. J. H. 495
Amso, D. 141
Anastasi, A. 107, 263, 269
Anda, R. F. 233
Andree, A. 481
Anschel, D. 296
Anstey, M. L. 218
Antoni, M. H. 391, 393, 401, 404

Antonucci, T. C. 520
Anway, M. D. 195, 415
Arafat, D. 400, 405
Arberg, A. A. 137
Arevalo, J. 399, 400, 401, 403, 404
Armstrong, J. M. 403
Arnold, H. M. 125
Aronson, J. 483
Aronson, L. R. 8, 23, 25, 77, 84
Arredondo, M. I. 485
Asaridou, S. S. 227
Aslin, R. N. 140–141
Atz, J. W. 329–330, 367
Avital, E. 197–198, 202
Azarian, B. 46

Badyaev, A. V. 201
Bae, K. 172
Baer, D. 4, 38, 43, 44, 515
Bagot, R. C. 419
Bai, D. L. 138
Bailey, M. T. 399, 401, 403
Baker, T. 314
Baldwin, J. M. 50, 503
Baldwin, S. 306
Ballcstar, E. 403
Ballestar, E. 412
Ballestar, M. L. 412
Balling, R. 415
Balswick, J. O. 519, 520
Baltes, P. B. 3, 7, 38, 41, 44–46
Bandi Rao, S. 141

Author Index

Bandura, A. 42
Barandiaran, X. 64
Barash, D. P. 322, 324–325, 327, 329–330, 353–354, 360, 362–364, 367
Barba, V. 445
Barlow, G. W. 321, 359
Barnes, J. C. 451
Barr, R. 292
Barreiro, L. B. 399, 401, 403, 413
Barrett, K. C. 138, 183
Barrett, L. 222–223, 231
Barron, B. 488
Barros, S. P. 419
Bartholomew, G. A. 352
Bartz, D. 494
Basser, L. S. 113
Bates, E. A. 427
Bateson, P. 110–112, 119, 125, 129, 131, 136, 159–160, 174, 181, 196, 215, 217, 282, 295, 384, 386
Bateson, W. 172, 244
Baum, W. M. 433
Baverstock, K. 434
Baxter, B. 113
Bayley, N. 139
Beach, F. 105
Beals, K. L. 342
Bear, G. 485
Bearer, E. 48
Becker, D. 460
Beckwith, J. 308
Beja-Pereira, A. 226
Bell, R. Q. 36, 40
Belsky, J. 17, 19, 46, 48, 154, 158, 341, 426, 514
Belyaev, D. K. 193
Benitez, J. 412
Bennett, A. J. 411, 413
Bennett, M. R. 69
Benson, J. B. 384
Benson, P. L. 44
Bentley, A. F. 280
Berent, I. 140
Berg, J. 477
Bering, J. M. 158
Berkel, C. 494
Berkowitz, A. 29
Bernstein, R. J. 63
Bernston, G. G. 77
Berry, T. D. 224, 226
Bertenthal, B. I. 138–139, 143

Bertenthal, B. J. 183
Beurton, P. J. 242, 246
Bhasin, M. 404
Bidell, T. R. 6, 44–45, 183, 501, 505
Bijou, S. W. 4, 38, 43–44, 515
Bird, K. A. 462
Birken, S. A. 506
Bispham, J. 414
Bitterbaum, E. J. 120
Bitterman, M. E. 330, 367, 515
Bjorklund, D. F. 5, 16, 47–48, 151, 154, 158, 160, 162–164, 222, 229, 426, 513, 515
Black, D. S. 401, 404
Blake, J. 494
Blakemore, C. 113
Blanca, M. J. 437
Blauw, G. J. 414
Bleier, R. 344
Blest, A. D. 106
Block, J. 46
Block, N. 69, 437
Blomberg, B. 401, 404
Bloom, B. S. 505–506
Bloom, H. S. 480
Blumberg, M. 105–106, 114, 121–122, 135–137, 143, 162, 224
Blumenfeld, P. C. 486
Blundell, R. 495
Boaler, J. 488
Bock, W. 333, 374
Bockmuhl, Y. 413
Bohbot, V. 139
Bolhuis, J. J. 121, 136–137, 223, 225, 227, 231
Bolton, N. 480
Bonatti, L. 141
Bondy, B. 419
Bonner, J. 174, 182, 280, 285
Boomsma, D. I. 452
Boorstein, D. J. 25
Bordieu, P. 498
Borghol, N. 403, 413
Boring, E. G. 222
Bornstein, M. H. 6, 41, 45–46, 130, 134, 505, 516, 518
Borodin, P. M. 193
Bouchard, T. J., Jr. 272, 295, 299–315, 441, 444–449, 451, 454, 456–457, 463
Bouffard, M. B. 484
Bouie, J. 504
Bower, J. E. 401, 404

Bowers, K. S. 162–163
Bowlby, J. 130
Bowler, P. J. 214
Boyd, R. 199
Boyd-Brown, M. J. 496
Boyce, T. W. 403
Braida, L. 141
Brand, S. 480
Brandtstädter, J. 8, 44–46, 337, 516
Bransford, J. D. 487
Breckler, S. J. 77
Breen, E. 401, 404
Brigham, C. C. 457
Brim, O. G., Jr. 9, 38
Brinkley-Rubinstein, L. 499
Bronfenbrenner, U. 7, 40–46, 151, 160, 260, 271, 280, 386, 425, 494, 497
Brooks, D. 76
Brooks, P. 141
Brooks, R. A. 68, 229
Brown, A. A. 139
Brown, K. B. 520
Brown, R. T. 112
Brummelman, E. 463
Brunelli, S. A. 286
Bruner, J. 228–229, 500
Bryk, A. 481
Buhrmann, T. 64
Buklijas, T. 414
Bulczak, G. 444
Buller, D. J. 223, 225
Bullock, D. 371–372
Bundo, M. 420
Bunn, H. F. 192
Burghardt, G. 109
Burian, R. M. 83
Burklund, L. J. 401, 404
Burman, J. T. 227, 231
Burns, A. 480
Burrows, M. 218
Burt, C. H. 434, 437, 451, 454
Burt, C. L. 260, 266, 272, 440–442, 444–446, 449
Burtt, E. H. 120
Busch-Rossnagel, N. A. 8, 26, 40, 46, 83, 337
Bush, G. L. 181
Buss, A. H. 260
Buss, D. M. 4–5, 17–19, 157, 221–223, 225–226, 229, 232, 512, 513, 515
Buth, D. G. 181

Butler, E. A. 341
Butterfield, S. 141
Bygren, L. O. 415
Byrne, R. W 182

Cacioppo, J. T. 77, 399, 400, 401, 405
Cairns, B. D. 44–47, 50, 499
Cairns, R. B. 44–47, 50, 280, 499
Cajete, G. 72
Callina, K. S. 8, 42, 50
Campbell, D. 79
Campbell, J. H. 196
Campos, J. J. 138–139, 183
Cantor, P. 6–7, 430, 441–442, 477, 498, 505, 517–520
Capaldi, E. J. 40
Capitanio, J. P. 132, 396, 401, 403
Caplan, A. L. 321, 359
Caporale, L. 193
Capron, C. 453
Carey, B. 441
Carey, N. 413
Carey, S. 67, 142
Carmack, H. J. 5
Carmean, C. M. 403
Carpendale, J. 517
Carrillo, C. 401, 404
Carroll, S. 217, 227
Carter, C. 232
Carter, P. 482
Carter-Saltzman, L. 451
Carver, C. S. 401, 404
Casey, T. M. 352
Cashman, R. 401
Caspi, A. 396
Cassou, B. 453–454
Catalano, R. F. 485
Cattell, R. B. 369, 441
Cavalier-Smith, T. 284
Cavigelli, S. A. 403
Ceci, S. J. 260, 271
Cerdena, J. P. 434
Cervoni, N. 413
Chabris, C. F. 434
Chalmers, D. 69
Chambers, C. 306–307, 435, 441
Champagne, F. A. 214, 413, 415, 418
Champoux, M. 396
Chang, B. H. 404
Chang, H. H. 246

Changeux, J.-P. 287
Chapman, M. 40
Charles, C. Z. 499
Charney, E. 258, 260–262, 264, 270, 434, 451
Chase, A. 440, 457
Chase, P. A. 498
Chater, N. 141
Chen, E. 393, 395, 401, 403
Chen, T. T. 439
Cheng, K. 139
Chess, S. 42, 151
Chetty, R. 494, 496
Chiandetti, C. 139
Chipuer, H. M. 369, 454
Choi, S. W. 413
Chomsky, N. 139
Chow, K. L. 113
Christakis, N. A. 405
Christen, M. 71
Christian, J. C. 294
Christiansen, M. H. 140–141
Churchland, P. S. 399, 405
Cicchetti, D. 71
Cigudosa, J. C. 412
Clark, A. 69, 230
Clark, N. M. 286
Clarke, E. 5
Clayton, D. F. 399
Clearfield, M. W. 136, 138
Cloud, D. H. 499
Clow, C. L. 270
Coall, D. A. 386
Coen, E. 248
Coffey, K. A. 404
Cofnas, N. 463
Cohen, D. J. 451
Cohen, E. G. 488
Cohen, L. 401
Cohen, S. 334, 391
Cohn-Vargas, B. 479
Cole, S. W. 260, 268, 384–387, 391–392, *393*, 395, 399, 400, 401, 403, 404, 405, 520
Collado-Hidalgo, A. 401
Collins, W. A. 47, 49, 260, 269–270, 331
Colombo, J. 111–114
Colunga, E. 142
Comfort, N. 455
Confer, J. C. 222, 224
Connerly, C. E. 499

Connolly, K. 111
Conradt, E. 268
Constância, M. 195
Conway, C. M. 140
Conzen, S. D. 403
Cookson, C. 292
Coop, G. 434
Cornell, E. H. 138
Cosin-Tomas, M. 404
Cosmides, L. 222, 224, 228–229
Costa Dias, M. 495
Cote, S. M. 411, 413
Coughlin, J. 131
Cox, B. F. 403
Cox, R. P. 249
Coyne, J. A. 224
Crafts, L. W. 21
Crawford, M. L. 113
Creel, S. C. 140–141
Creswell, J. D. 401, 404
Crews, D. 218, 415
Crick, F. 28, 79, 248, 251, 335
Cronbach, L. J. 42
Cropanzano, R. 444
Cropley, J. E. *412*
Cross, R. W. 454, 462
Crosswell, A. D. 401, 404
Crouter, A. C. 494
Cryns, A. G. 342–343
Csikszentmihalyi, M. 44–46
Cupp, A. S. 415
Curtiss, S. 113
Cutler, A. 141
Cynader, M. 113
Czika, W. 399

Dale, R. A. C. 140
D'Alessio, A. C. 411, 413
Dalton, S. 487
Daly, R. 232
Damasio, A. 487
Damon, W. 37, 44, 46–47, 519
Danchin, É. 242
Dang, T. H. Y. *412*
Daniel, J. 490
Daniels, D. 261, 454
Dantzer, R. 405
Darling-Hammond, L. 430, 481, 482, 488, 495, 497
Darwin, C. 24–25, 187–188, 333, 334, 352, 357, 375

Dashwood, R. H. 413
Davidson, E. H. 286–287
Davies, G. 439
Dawes, C. T. 405
Dawkins, R. 4–6, 16, 190, 199, 232, 244, 246, 249, 322–323, 326–327, 360–361, 363–365, 515, 520
Daxinger, L. 414
Day, J. J. 418
Day, R. H. 138
Denninger, J. W. 404
Deacon, T. 141
Deary, I. J. 434, 450
de Beer, G. R. 128, 156, 215, 337
DeCasper, A. J. 171
Defoy, F. 519, 520
DeFries, J. C. 445, 451
Delgado, B. 403
Del Giudice, M. 16, 47, 158, 160–162, 426
Demo, D. H. 341
De Mol, J. 64
Denenberg, V. H. 176
Denton, N. A. 499
Derks, E. M. 451
de Royston, M. M. 478, 498
Descartes, R. 63, 69, 71
Devlin, B. 433
Dewey, J. 50, 280
Dewsbury, D. A. 279
Diamond, S. 105
Dias, B. G. 242
Diaz, N. 214
Díaz, S. 234
DiBerardino, M. A. 282
Dickens, W. T. 460
Dickerson, S. S. 396
Diennes, Z. 141
Dineva, E. 138
Diorio, J. 413, 415, 418
Di Paolo, E. 64, 69, 230
Dixon, R. A. 38, 40
Dobzhansky, T. 190, 215, 216
Dodd, S. M. 342
Dodsworth, R. O. 177
Doerr, A. 15
Dolinoy, D. C. 195
Domrachev, M. 418
D'Onofrio, B. 295
Donovan, M. S. 487
Doolittle, W. F. 335

Dor, D. 203
Dorfman D. D. 437, 441
Douglas, L. 85
Doyle, W. 391, 485
Dozier, M. 403
Draper, P. 158–159, 341
Druery, C. T. 244
Dubrova, Y. E. 201
Dudley, B. 484
Dunbar, R. I. M. 321, 330, 332, 359, 367, 374
Duncan, G. J. 494
Duncan, R. G. 458
Durham, W. H. 227
Durlak, J. A. 484
Dusek, J. A. 394, 404
Dusek, V. 445
Dutta, R. 386
Duyme, M. 453
Dweck, C. S. 486
Dymnicki, A. 484, 498
Dymov, S. 411, 413, 415

Eales, L. A. 131
Easton, J. Q. 481
Eaves, L. J. 454
Eberhardt, J. L. 499
Eccles, J. S. 479, 488
Edelman, G. M. 48, 171, 182
Edery, I. 172
Edey, M. A. 154–155, 341
Edgar, R. 418
Edge, M. D. 461–462
Edvinsson, S. 415
Edwards, L. 121, 136
Egerton, J. 442
Eilam, D. 123, 124, *124*
Eisenberger, N. I. 405
Ekstrom, T. J. 413
Elder, G. H., Jr. 7, 44–46, 151, 386, 444, 518
Ellefson, M. R. 140
Ellegren, H. 224
Elliott, L. 5
Ellis, B. J. 16, 47, 151, 154, 158, 160–163, 341, 427, 513, 515
Ellis, G. F. R. 250
Elman, J. L. 79, 135, 140–142, 141, 170–171
Emberly, E. 403
Emig, C. 490
Emmerich, W. 47, 519

Engel, A. K. 230
Enserink, M. 313
Epel, D. 217, 218
Ephrussi, B. 286
Epstein, R. 494
Epstein, W. 225
Erikson, E. H. 7, 38, 109–110, 327, 364
Essex, M. J. 403
Estrada, P. 487
Evans, D. M., 451
Evans, G. 434
Eysenck, H. J. 454

Falconer, D. S. 321
Fall, T. 452
Fancher, R. E. 301
Farber, S. L. 302, 303, 312, 444–445
Farrell, M. S. 434
Farrington, C. A. 483, 484, 486, 498
Fava, M. 413
Featherman, D. L. 152
Feiring, C. 36
Feldman, M. W. 17, 48, 108, 193, 201, 225, 258, 260, 267, 291, 295, 344, 372, 427, 438, 464
Felitti, V. J. 233
Felner, R. D. 480
Feng, Z. 403
Ferguson, A. A. 499
Fernald, R. D. 399
Fernandez, K. 141
Fine, M. A. 341
Firestine, A. M. 404
Fischbach, M. 438
Fischer, C. S. 433, 437, 462
Fischer, D. 413
Fischer, K. W. 44–46, 160, 183, 269, 496, 501, 505
Fiser, J. 140–141
Fisher, C. 48, 50, 182, 218, 426
Fisher, H. E. 154–155
Fletcher, R. 441
Flohr, H. 323, 361
Flynn, J. R. 310–311, 444, 448, 455–456, 460–462
Fodor, J. A. 225
Fok, A. K. 401
Fontdevila, A. 201
Forber, P. 225
Ford, D. H. 43–44, 64, 183, 260, 336–337
Foret, S. 412
Forgays, D. G. 177

Forgays, J. W. 177
Fortuna, L. R. 494–495
Fosse, R. 294, 451
Foster, R. 245, 247
Foucault, M. 498
Fowler, J. H. 405
Fox, C. W. 196
Fox Keller, E. 404, 405
Fraga, M. F. 403, 412
Fragaszy, D. M. 197
Francis, D. 413, 418
Francis, K. 500
Franklin, T. B. 414
Fraser, H. B. 403
Fraser, S. 433
Fredrickson, B. L. 404
Freeberg, T. M. 183
Freedman, D. G. 16, 322, 324, 326–327, 329–330, 360, 362–364, 366
Freiberg, H. J. 481
Freud, S. 7, 38, 44, 327, 364
Freund, G. G. 405
Friedlaender, D. 479, 481
Friedman, D. P. 411, 413
Friedman, J. N. 494
Friedman, N. 401
Friso, S. 413
Fromkin, V. 113
Fronius, T. 485
Frudakis, T. N. *291*
Fry, D. 234
Fuller, J. L. 27
Funk, A. J. 413
Furman, J. 496
Futuyma, D. J. 181, 284, 287–288

Galea, S. 494
Galef, B. G. 286
Galjaard, R. J. 411
Galland, N. 141
Gandhi, A. 490
Gangestad, S. W 232
Gapenne, O. 64
Gardner, D. S. 414
Gardner, H. 458
Garet, D. 401, 404
Garibaldi, M. L. 499
Gariépy, J.-L. 183
Garstang, W. 171–172, 337

Gastmans, C. 520–521
Gaulin, S. J. 226
Gay, G. 482
Gehring, W. J. 182
Gelman, S. A. 135
Georgoudi, M. 39
Gergen, K. J. 42
Gerken, L. 140
Gerstein, M. B. 28
Gibson, G. 399, 400, 405
Gibson, J. J. 229
Giddens, A. 498
Gilad, Y. 413
Gilbert, S. F. 133–134, 171, 181, 215, 217, 218
Gillborn, D. 434
Gillis, S. 141
Ginsburg-Block, M. D. 488
Gissis, S. B. 159, 242, 385–386
Gitonga, V. W. 296
Glauberman, N. 433
Gleason, T. 72
Glickman, S. E. 179
Gluckman, P. 125, 196, 215, 242, 384, 414
Godfrey-Smith, P. 225
Goffman, E. 498–499
Goldberger, A. S. 260, 272, 454
Golding, J. 415
Goldman, R. D. 249
Goldman, S. 487
Goldschmidt, R. 171–172, 337
Goldstein, M. H. 137, 141
Gollin, E. S. 44
Goltser-Dubner, T. 401
Gómez, R. L. 140
Gomez-Cabrero, D. 413
Gomez-Robles, A. 72, 233
Gomperts, N. 498
González, T. 494
Goodwin, B. 84, 181, 287, 514
Gore, A. C. 415
Gorey, K. M. 342–343
Gottesman, I. I. 451
Gottesman, I. L. 267, 295
Gottfredson, L. S. 436, 440
Gottfried, N. 80
Gottlieb, G. 23, 25, 28–29, 44–45, 48, 50, 66–67, 72, 77–80, 83, 84, 86, 106–107, 110, 121–122, 124, 135–137, 142–143, 151, 157, 164–165, 171–172, 181–183, 215, 216, 218, 222, 224, 231, 233, 258, 260,
267, 268, 271–272, 279–281, 285, 287, 336–338, 345, 376, 377, 384, 386, 415, 419, 438
Gould, S. J. 5, 46, 48, 152–156, 159, 224–225, 260, 329, 332–338, 344–345, 352, 357, 367, 373–376, 386, 436, 441, 464, 513, 518
Graff, J. 414
Graham, S. 48
Gramsci, A. 498
Granic, I. 85
Gray, R. D. 72, 79, 121, 136, 197, 204–205, 222, 226, 295
Greenberg, G. 5, 8–9, 21–24, 26–28, 48, 50, 76, 78, 80–82, 85, 108, 110–111, 152, 260, 513–514
Greendale, G. 401, 404
Greenough, W.T. 279
Grewen, K. M. 404
Gribnau, J. 411
Griesemer, J. 205
Griffiths, P. E. 72, 79, 121, 136, 204–205, 222, 226, 233, 295
Gross, J. J. 483
Grouse, L. D. 280, 285
Gruber, J. 260
Grzimek, B. 329, 366
Guckenburg, S. 485
Gugushvili, A. 444
Guhl, A. 22
Guillemin, C. 411, 413
Guo, S. W. 296
Gupta, P. 140
Gurdon, J. B. 283
Gurven, M. D. 222
Gusev, A. 434, 437–439, 458
Gutierrez, K. 482
Gutzwiller, E. 71
Gwaltney, J. M. 391

Hacker, P. M. S. 61, 69
Haeckel, E. 152
Haig, D. 201–202
Hailman, J. P. 120–121, 138, 162
Haith, M. M. 136
Haldane, J. B. S. 202
Haley, A. 295
Hall, B. K. 217, 222, 224, 287
Hall, G. S. 152
Hall, W. G. 125
Hallett, M. 403, 413
Halpern, C. T. 67, 78
Hamedani, M. G. 481, 484

Han, L. Y. 403
Hansen, K. D. 399, 401, 403, 413
Hanson, M. 196, 414
Haraway, M. H. 48
Haraway, M. M. 24, 26, 80, 108, 111
Harden, K. P. 5, 17, 19, 47, 108, 267, 269, 272, 512
Harding, S. 296
Hardt, O. 139
Hardy, A. C. 174
Hare, H. 354
Harley, D. 154
Harlow, H. F. 113, 129, 177
Harlow, M. K. 129, 177
Harpalani, V. 501
Harpending, H. 158–159
Harper, L. 427
Harris, D. B. 83
Harris, J. R. 261
Harris, Y. T. 296
Harvey, G. 235
Harvey, P. W. 413
Hattfield, G. 513
Hawkley, L. C. 399, 400, 405
Hawkley, L. G. 400, 401
Hayes, S. C. 39
Head, M. L. 306, 448
Hearnshaw, L. 260, 272
Hearnshaw, L. S. 441
Hebb, D. O. 111, 263–264, 369–370, 515
Heckman, J. J. 396, 399, 401
Heijmans, B. T. 413, 414
Heine, S. J. 71
Heine-Suner, D. 412
Helson, H. 22–23
Hemphill, S. A. 485
Hendren, N. 494, 496
Henley, T. 231–232
Henrich, J. 71, 233
Henry, D. A. 500
Hermer, L. 139
Hermer-Vazquez, L. 139
Hermes, G. L. 403
Herrenkohl, T. I. 485
Herrnstein, R. J. 16, 48, 266, 268, 272, 299, 429–430, 433, 438–439, 441–443, 445–446, 449–453, 456, 461–464
Hertzman, C. 403, 413
Hess, E. H. 111
Heth, C. D. 138

Hettema, J. M. 294
Heumann, J. 500
Hewitt, J. K. 262
Hewlett, B. S. 72, 232
Heyes, C. 222
Heying, S. 164, 260
Hickey, T. 40
Hiernaux, J. 340
Hill, K. R. 232
Hill, N. E. 494
Hillenmeyer, M. E. 245
Hinde, R. A. 111, 129
Hirsch, H. V. 214
Hirsch, J. 5, 48, 260, 264–267, 331, 371–372, 437–438, 442, 443, 454, 514, 516
Ho, E. 413
Ho, M. W. 24–25, 48, 158–159, 181, 251, 258–260, 284, 383, 386, 437–438, 514
Hobbes, T. 234
Hod, E. 411
Hodgson, G. M. 24
Hoehler, F. 413
Hofer, M. A. 286
Hoffman, H. S. 112
Hoffman, L. W 271
Hoffman, R. F. 151
Hogan, R. 151, 154, 223
Holden, C. 446
Holden, C. J. 226
Holdheide, L. 505
Hollenstein, T. 85, 513, 516
Holliday, R. 171
Holsboer, F. 413
Holt-Lunstad, J. 400
Honey, R. C. 121, 137, 178
Honeycutt, H. 29, 48, 73, 157, 159–160, 223, 295, 386, 513
Hood, K. E. 23, 28, 78
Hopkins, W. D. 72
Horgan, J. 302–303, 445
Horn, G. 137
Horobin, K. 138
Horowitz, F. D. 48, 257, 260, 266–267, 270–271, 280, 344, 427
Horvath, J. 141
Hossenfelder, S. 24–25
Howe, L. J. 439
Hrdy, S. 232, 234
Huang, M. H. 496

Hubbard, R. 76
Hubel, D. H. 113
Hultsch, D. F. 40
Hunt, G. R. 197
Huntley, J. F. 414
Hupbach, A. 139
Hur, Y. M. 445
Hurd, P. L. 139
Hurley, N. 485
Hutchins, E. 68
Hutto, D. D. 228–230
Huxley, J. 178, 216, 223, 352
Huxley J. 216
Huzinec, A. C. 481
Hymovitch, B. 176–177

Idaghdour, Y. 399, 405
Ihsen, E. 138
Immordino-Yang, M. H. 430, 487, 497, 517
Inagaki, T. K. 405
Ingber, D. E. 250
Ingersoll-Dayton, B. 520
Ingold, T. 72, 224, 231–234
Irvine, J. J. 482
Irwin, M. R. 399, 400, 401, 403, 404, 405
Isaac, G. L. 154
Isles, A. R. 418
Itsukaichi, T. 172
Iwamoto, K. 420

Jablonka, E. 7, 137, 156–157, 159, 164–165, 187, 192, 194, 197–198, 202–203, 217, 224, 227, 233, 242, 338, 345, 383, 385–387, 414, 518
Jackson, D. D. 451
Jacob, F. 247
Jacobs, D. T. 71, 235
Jacoby, R. 433
Jagers, R. J. 500, 505
Jaisson, P. 183
James, K. 444
Jenkins, J. J. 41–42
Jennings, H. S. 78–79, 82–83
Jennings, J. A. 505
Jennings, P. A. 505
Jensen, A. R. 16, 260, 263, 265–266, 302, 309, 368, 369, 440–443, 446, 458–460, 463
Jerison, H. J. 178, *179*
Jeynes, W. H. 482
Jiang, L. 411

Jiang, Y 495
Jirtle, R. L. 412
Johannsen, W. 189, 192, 244
Johanson, D. C. 154–155, 341
John, L. K. 307, 435, 448
Johnen, A. G. 89
Johnson, C. G. 16
Johnson, D. W. 484
Johnson, M. H. 121, 137, 141
Johnson, R. 405, 484
Johnson, S. 137, 141
Johnson, W. 295, **304**, 308, 447
Johnson, Z. P. 399, 401, 403, 413
Johnston, T. D. 121, 125, 135–136, 142, 181–183, 182, 279–280
Jones, F. S. 182
Jones, M. C. 434
Jones, M. R. 494, 496
Jones, S. M. 484
Jones, S. S. 136, 142
Jonkers, I. 411
Jorgensen, R. A. 201
Joseph, J. 260, 268, 272, 294, 300, 302–303, 306, 310–313, 434, 438–445, 447, 449–451, 454–456, 455
Joseph, M. G. 404
Joyce, R. 495
Joynson, R. B. 441
Juel-Nielsen, N. 301–302, 306, 441, 443
Julian, M. M. 519
Junien, C. 414
Juraska, J. M. 279

Kaati, G. 415
Kagan, J. 38
Kaliman, P. 404
Kamat, A. A. 403
Kamin, L. J. 302–303, 307, 312, 315, 440, 442–444, 454, 457–458, 460, 462, 464
Kaminski, N. 401
Kantor, J. R. 77–78, 80
Kaplan, G. 29
Karouzakis, E. 420
Kasai, K. 420
Kashyap, S. 403
Katsnelson, A. 139
Kauffman, M. B. 37, 40–43, 183, 280, 375
Kaufman, S. B. 436, 455
Kawakami, K. 450
Kearsey, M. J. 203

Keller, E. F. 28, 68, 78, 83, 192, *297*, 384, 386, 437–438
Keller, L. 296
Kelley, K. W. 405
Kelly, P. 505
Kemeny, M. E. 396
Kempe, V. 141
Kendler, K. S. 294, 451, 454
Kendler, T. S. 40
Kendziora, K. 498–500
Kenney, M. 113
Kerkel, K. 411
Kermoian, R. 138
Keuck, G. 152
Khanna, K. 434
Kidner, D. W. 71–72, 234
Kim, J. 405
Kim, S. Y. 403
Kincheloe, J. L. 433
King, A. P. 125, 137, 144, 183, 226–227, 231, 295
King, P. E. 517–520
Kinsey, A. C. 326, 363
Kinzler, K. D. 67, 135, 137, 139, 142
Kipling, R. 332
Kirschbaum, C. 401
Kitcher, P. 321, 359
Kiverstein, J. 228, 230
Kleckner, J. H. 334
Klironomos, F. D. 242
Klopfer, P. H. 136
Kluckhohn, C. 47, 519
Knudsen, E. I. 232
Ko, E. P. 84, 514
Koball, H. 495
Kobor, M. S. 401
Koch, S. 86
Kocherginsky, M. 403
Kohl, P. *245*, 252
Kohlberg, L. 7
Kollar, E. J. 182, 218
Konishi, M. 287
Konner, M. 231–232, 321–324, 327, 333, 336, 359–361, 364, 374, 375
Koposov, R. 403
Kornblihtt, A. R. 420
Kosek, J. 411
Kraft, P. 463
Krashen, S. D. 111–112
Krause, J. 141
Krausz, T. 403

Kreitzman, L. 245, 247
Kreppner, K. 50
Krimsky, S. 260
Kringlen, E. 451
Krone, C. R. 497
Kucharski, R. 412
Kuczynski, L. 64
Kuhl, P. 141, 487
Kuhn, T. S. 3, 39, 71, 73, 502
Kunkel, J. G. 174
Kuo, Z. Y. 23, 25, 50, 78, 80, 84, 174, 260, 279, 338
Kupiec, J.-J. 246, 252
Kyburz, D. 420

Labonte, B. 411, 413
Lafaite, M. 138
Laird, P. W. 418
Lakatos, I. 61, 63
Lakhotin, S. C. 284
Laland, K. N. 48, 197, 223–227, 295
Lam, L. L. 403
Lamarck, J. B. 174, 182
Lamb, M. E. 6, 35–37, 45, 72
Lamb, M. J. 7, 137, 156–157, 164–165, 187, 194, 217, 224, 227, 232–233, 260, 338, 345, 385–387, 518
Lambdin, C. 76, 80
Lamey, A. V. 85
Landau, B. 67, 74, 142
Langer, E. J. 40
Langstrom, N. 452
Lanius, R. A. 233
Larson, A. 174, 180–181
Lash, A. E 418
Latour, B. 73
Laudan, L. 60–61
Laven, J. S. E. 411
Layton, J. B. 400
Layzer, D. 85, 265
Lazar, J. W. 23
Learmonth, A. E. 139
Lee, C. 172, 478, 486, 488, 497–498
Lee, M. 403
Lee, R. B. 232, 234
Lee, S. 139, 399, 521
Lee-Painter, S. 36
Lehrman, D. S. 16, 38, 77, 106–109, 135–137, 163, 260, 279, 370
Lenneberg, E. 113, 130
Lenoci, M. 401

Leonovicová, V. 174
LePage, P. 481
Lepecq, J. C. 138
Lerner, J. V. 6, 151
Lerner, R. M. 3–9, 15, 23, 26, 28–29, 37–38, 40, 42–48, 50, 61, 64, 67, 77–78, 80, 82–83, 86, 151–152, 162, 183, 260, 263, 265–266, 271, 280, 321, 336–337, 344–345, 359, 371, 375–377, 383–384, 386, 425–427, 436, 438, 457, 498, 512–513, 516, 518–521
Lester, B. M. 161, 268
Le Texier, T. 308, 435
Levenson, J. M. 194, 418
Levin, R. 226
Levine, S. 176
Levins, R. 155–156, 336
Levy, A. A. 193, 201
Lewin, K. 50, 428
Lewis, A. C. F. 458
Lewis, C. 517
Lewis, D. M. G. 222
Lewis, M. 35–37, 40, 85, 160, 260, 513, 517
Lewis, R. 484
Lewontin, R. C. 48, 77–80, 155–156, 225–226, 233, 260, 267, 293, *293*, 315, 321, 336–337, 344, 359, 372, 373, 375–376, 437–438, 444, 453–454, 458, 461–462, 464, 515
Li, J. 519
Li, K. 411
Li, M. 80, 84
Li, P. W. 84
Lickliter, R. 6, 8, 16, 25, 29, 48, 64, 68, 73, 107, 151, 157, 159–160, 164–165, 223–227, 258, 269, 295, 386
Lieberman, M. D. 401, 404
Lifton, R. J. 5
Lilienfeld, S. O. 312
Lim, A. 401
Lin, S. P. 414
Linder, N. 414
Lindholm, M. E. 413
Lipinski, J. 140
Litman, L. C. 342
Liu, D. 145, 413, 418
Liu, L. 403
Lockard, J. S. 354
Locke, J. L. 171
Loehlin, J. C. 451, 453, 454
Logan, C. A. 183
Looft, W. R. 38, 40
Lopizzo, N. 419

Lorde, A. 500
Lorenz, K. Z. 16, 45, 106–108, 112–113, 119–121, 129, 131, 136–137, 162–163, 322–323, 327, 353, 360–361, 365, 426
Lorenz, S. E. 131
Losen, D. J. 485
Lotan, R. A 488
Lovejoy, C. O. 154–155
Loveland, D. H. 433
Løvtrup, S. 181, 284
Lubin, F. D. 413
Luca, F. 403
Luciano, M. 452
Luco, R. F. 420
Lumey, L. H. 413, 414
Lumsden, C. J. 321–322, 359, 360
Lunenburg, F. C. 490
Lupien, S. J. 232
Luppescu, S. 481
Lurie, N. 494
Lush, J. L. 291
Lutgendorf, S. 401, 404
Lutgendorf, S. K. 393–395, 401, 404
Lutz, A. 404
Lykken, D. T. 299, 446, 449
Lynch, M. 225
Lynn, R. 322, 442, 446, 460
Lyon, M. F. 411
Lyon, P. 231

Ma, R. 401
Mabry, P. L. 85
MacDonald, K. 111, 321–322, 327, 339, 359, 360, 364
MacDonald, M. 140
Mace, R. 226
Machado, A. 60
Macphail, E. M. 178
Magnuson, K. 494
Magnusson, D. 44–47, 260, 280, 336–337
Maher, B. 439
Maier, A. 490
Maier, N. R. F. 269
Major, B. 483
Malcolm, N. 71
Maleszka, J. 412
Maleszka, R. 412
Maloney, C. A. 414
Manduca, R. 494
Mangan, S. 517–520

Author Index

Manolio, T. A. 439
Mansuy, I. M. 414
Marabita, F. 413
Marcus, G. F. 136, 140–141
Margulis, L. 233
Marin, T. 401
Markman, E. 67, 142
Marks, J. 458
Marler, P. 129
Marlowe, F. 232
Marr, D. 69
Marshall, P. J. 69–70, 383, 515
Marsit, C. 268
Martin, D. I. 412, *412*, 414
Martin, N. G. 451
Martines, P. 214
Martinez, K. J. 505
Masataka, N. 171
Mascolo, M. F. 6, 8, 45–46, 160, 269, 496, 501
Mashal, N. M. 405
Mason, W. A. 113, 132, 177
Massart, R. 411, 413
Massey, D. 499
Masters, R. D. 151, 154
Matarazzo, J. D. 458
Matsuzawa, T. 81
Mattick, J. S. 413
Maye, J. 140
Mayer, M. J. 500
Maynard Smith, J. 251
Mayr, E. 25, 174, 181, 214, 216, 223–224, 244
Mazzocchi, F. 29, 85
McAdams, D. P. 500
McAdoo, H. 341
McArdle, W. 403, 413
McBruney, D. H. 226
McCabe, B. J. 137
McCarthy, M. M. 453
McClearn, G. E. 328, 366
McClelland, M. M. 386
McClintock, B. 77, 247
McCrae, R. R. 16
McCubbin, H. I. 341
McDonald, P. 183
McEwen, B. S. 258
McGivney, E. 494, 495
McGowan, P. O. 411, 413
McGraw, M. B. 111
McGue, M. 272, 299, 301, 311, 441, 454, 456

McGuire, T. R. 27, 371–372, 438, 454
McKenzie, B. E. 138
McKie, R. 435
McKim, D. B. 401
McLaughlin, B. 113
McLoyd, V. C. 48
McMorris, B. J. 485
McMurray, B. 140, 142
Meaney, M. 159, 161, 258, 268, 384–386, 410, 411, 412, 413, 415, 418–419
Mehler, J. 141
Melnick, M. 294
Menary, R. 69
Mensh, E. 458
Mensh, H. 458
Merchant, R. M. 494
Mergendoller, J. 486
Merrill, M. A. 459
Meyer, M. N. 434
Meza-Zepeda, L. A. 404
Micale, V. 413
Micceri, T. 437
Michalon, A. 414
Michel, G. F. 25, 48, 81, 114, 127, 133, 183
Michelini, K. 413
Michopoulos, V. 399, 401, 403, 413
Miklos, G. L. G. 172
Miller, G. E. 393, 395, 399, 401, 403
Miller, P. 500
Millstein, R. L. 225
Minelli, A. 215
Minnici, A. 505
Mintz, T. H. 140
Mischel, W. 42
Mischoulon, D. 413
Misteli, T. 384–385, 420
Mistry, J. 386
Mitchell, D. E. 113
Mitchell, K. J. 463
Mivart, S. G. 172, 215
Mix, K. S. 136
Molenaar, P. C. M. 7–8, 46–47, 84–86, 85–86, 266, 386, 499, 513, 518–519
Molinaro, A. 418
Moloney, D. P. 307, 449
Moltz, H. 21–23, 25, 84, 105, 109, 111–113
Monaghan, P. 140–141
Mongin, D. 513
Montagu, A. 80, 231

Moon, R. 462
Moore, C. A. 294
Moore, C. L. 25, 48, 127, 133, 135, 183
Moore, D. S. 5, 29, 48, 51, 107, 159, 161, 224–227, 233, 258, 260, 263–266, 268, 271–272, 295, 296, 338, 343–344, 384, 386–387, 409, 410, 414, *414*, 415, 438, 454, 455
Moore, K. 222
Moore, K. A. 490
Morgan, H. D. 412, 414
Morgan, T. H. 286
Moroney, D. 500
Morris, P. A. 7, 43–46, 160, 260, 271, 497
Mosack, V. 80, 152
Moss, L. 80
Mousseau, T. A. 196
Müller, G. 242, *243*, 248–249, 252
Müller, U. 45, 77, 81, 83, 85–86, 385
Müller-Hill, B. 5, 48, 258
Munsinger, H. 453
Mural, R. J. 84
Murgatroyd, C. 413
Murphy, M. C. 483
Murphy, M. L. M. 401
Murray, C. 16, 48, 272, 299, 429–430, 433, 438–439, 449, 452, 456–458, 461–463
Murray, H. 47, 519
Murry, V. M. 494
Murty, V. V. 411
Muscatell, K. A. 405
Mussen, P. H. 37, 43
Myers, E. W. 84
Myers, K. P. 125
Myers, M. M. 286
Myers, M. P 172
Myin, E. 228–230
Myrianthopoulos, N. C. 294

Nadeau, J. H. 242
Nadel, L. 139
Najm, W. I. 413
Nance, W. E. 203
Narvaez, D. 71–72, 71–73, 164–165, 231–233, 235, 345
Nash, J. 111
Nasir, N. S. 478, 487, 498
Nath, A. P. 400, 405
Nathanielsz, P. 414
Naumova, O. Y. 403
Neale, M. C. 294

Neisser, U. 437, 461
Nelson, V. R. 242
Nespor, M. 141
Nesselroade, J. R. 7, 38, 46–47, 85–86, 499
Neumann, S. M. 403
Newcombe, N. 139
Newen, A. 223, 229
Newman, H. H. 301–302, 306, 441, 443, 446
Newman, S. A. 248–249
Newport, E. L. 67, 135–136, 138, 140, 170
Ng, A. E. 521
Nichols, R. C. 451
Nieuwkoop, P. D. 80
Niggeschmidt, K. F. 438
Nisbett, R. E. 434–436, 454–456, 460
Nishioka, M. 420
Noble, D. 9, 26, 28, 45, 157–158, 160, 165, 224, 227, 233, 242–243, 245–246, 248–252, 322, 345
Noë, A. 69
Noguera, P. 481, 487, 488
Non, A. L. 434
Norenzayan, A. 71
Norgate, S. 294, 436, 450, 454, 456
Northstone, K. 415
Norton, H. L. 458
Nosek, B. A. 441
Novak, M. A. 113
Novák, V. J. A. 174
Novikoff, A. B. 50, 269
Nucci, L. 520
Nugent, B. M. 453

Oakes, J. 490
Oberle, E. 484
O'Connor, J. C. 405
O'Connor, M. F. 401
Odell, G. 287
Odling-Smee, F. J. 197, 227–228, 231, 295
O'Donovan, A. 401
Offenbacher, S. 419
Ogbu, J. U. 462
O'Grady, C. 308, 435
Okbay, A. 439
Okonofua, J. A. 499
Olafsrud, S. M. 404
Olmstead, R. 401, 404
Olson, W. 217
Oransky, I. 435
Orgel, L. E. 335

O'Riley, C. 455
Ortman, L. L. 22
Orzack, S. H. 225
Oser, F. K. 44
Osgood, N. D. 85
Osher, D. 430, 477, 479, 482, 485, 490, 496–501, 498, 500, 505–506, 518, 520
Osher, T. W. 482, 496
Ospelt, C. 420
Oster, G. 287
Ott, S. R. 218
Otu, H. H. 404
Ourth, L. 520
Overton, W. F. 3–4, 6–8, 26, 37–39, 43–50, 60–61, 67, 70, 73, 77, 80–83, 85–86, 107, 160, 165, 258, 260, 263, 269, 383–386, 425, 428, 501, 516–518
Oyama, S. 72, 79, 121, 136–137, 142, 181, 204, 224, 227, 233

Paas, F. 487
Page, E. B. 459
Pakstis, A. 498
Pam, A. 451
Pang, D. 403
Panksepp, J. 72, 228
Panofsky, A. 5, 260, 271, 434
Papakostas, G. I. 413
Parikh, V. 172
Parsons, P. A. 181
Partanen, J. 370
Partridge, T. 22, 24, 28, 76, 80, 82, 85, 513–514
Patchev, A. V. 413
Patel, F. N. 403
Patel, J. A. 495
Payne, C. M. 499
Paz, M. F. 403, 412
Pea, R. 478, 497–498
Pearson, H. 192
Pedersen, N. L. 304, 312, 444, 445, 447, 456
Peereman, R. 141
Pelizzola, M. 418
Pellegrini, A. D. 222, 229
Pellegrino, J. 487
Pembrey, M. 195, 413, 415
Peña, M. 141
Penfield, W. 113
Penkoff, C. 505
Pepper, S. C. 3, 40, 43
Pereira, A. 141

Perkins, P. 196
Perruchet, P. 141
Perry, S. 197
Persson, H. 485
Pervin, L. A. 36
Peters, M. 340
Peterson, A. C. 415
Petrinovich, L. 42, 131
Petrosino, A. 485
Pew, T. 403
Philipp, E. E. 372
Phillips, D. A. 227, 232
Piaget, J. 7, 37–38, 40, 44, 174, 322, 330, 367, 376
Pianka, E. R. 368
Pierre, P. J. 411, 413
Piferrer, P. 214
Pigliucci, M. 223–224, 227, 242, *243*, 385
Pimentel, M. A. 403
Pinker, S. 135
Pisula, W. 22
Pittman, K. 498
Ploeger, A. 222
Plomin, R. 4–5, 17–19, 38, 45, 47–48, 106–108, 257–262, 260, 267–270, 272, 310, 426, 434, 436–437, 439, 445, 450–455, 454, 512, 515
Plotkin, H. C. 174
Plumert, J. M. 139
Poincaré, H. 252
Polderman, T. J. C. 419, 452
Poldrack, R. A. 227
Popper, K. 60
Porche, M. V. 494
Porges, S. W. 232
Porter, S. R. 494, 496
Powell, N. D. 399, 401, 403
Power, C. 413
Pradeu, T. 215
Prager, E. M. 174
Prechtl, H. F. R. 23
Preminger, M. 405
Priceman, S. J. 403
Prinsloo, S. 401
Pritchard, D. J. 280, 284
Proctor, R. N. 5, 40, 47
Proctor, R. W 40
Pronko, N. H. 22, 23, 27, 81–82, 86
Provencal, N. 411, 413
Provine, W. 214
Przeworski, M. 434

Psych-Agitator. 433
Puro, P. 486
Putter, H. 414

Qu, S. 404
Quine, W. V. 141
Quinn, B. 481

Rabin, B. S. 391
Rabuy, B. 5
Racine, A. 403, 413
Raeff, C. 3, 8, 151, 160, 164, 269
Raines, S. 108
Ram, N 513, 516
Rapp, R. A. 196, 201
Rathunde, K. 44–46
Ratliff, K. R. 139
Ratner, A. M. 112
Rausch, M. K. 485
Rauscher, F. 222–223, 231, 234
Rawson, R. 445, 446
Raz, G. 414
Razran, G. 178
Reader, S. M. 197
Rechavi, O. 242
Redman, B. 435
Reed, T. 294
Rees, W. D. 414
Reese, H. W. 3, 37–39
Regolin, L. 138
Reid, R. G. B. 174
Reilly, S. 178
Reinsch, S. 413
Remez, R. 140
Rempel, H. 401
Rendell, L. 197
Renkl, A. 487
Rensch, B. 178
Ressler, K. J. 242
Ribas, L. 214
Richards, E. J. 411, 415
Richards, R. 106, 108, 163
Richardson, K. 260, 262, 268, 272, 294, 434, 436, 450, 454–456, 459
Richardson, M. K. 152
Richardson, R. C. 157–158, 225
Richerson, P. J. 199, 226
Rider, E. A. 138
Ridley, M. 292

Riegel, K. F. 37–42
Riesen, A. 129
Rieser, J. J. 138
Riskin, J. 108, 258, 260, 427, 438
Ritchie, S. 435, 450
Rivas-Drake, D. 505
Robert, J. S. 23, 77
Roberts, L. 113
Robette, N. 438
Robinson, G. E. 399
Robles-Ramamurthy, B. 494
Roeser, R. W. 479, 488
Rogers, L. J. 29
Rogers, S. M. 218
Rogoff, B. 482
Romanes, G. J. 251
Romm, C. 308
Ronca, A. E. 122
Ronnlund, M. 456
Ropcro, S. 403
Ropero, S. 412
Rosch, E. 68
Rose, H. 225, 386
Rose, R. J. 294, 444
Rose, R. M. 400
Rose, S. 225, 386, 458, 464
Rose, T. 7, 46–47, 477, 516, 518–519
Roseman, C. C. 462
Rosen, D. E. 181
Rosen, J. 434
Rosenberg, K. M. 176
Rosenblatt, J. R. 77
Rosenblatt, J. S. 109–111, 131–132
Rosenblum, L. A. 36, 40
Rosenkranz, M. A. 404
Rosnow, R. L. 39
Ross, K. G. 296
Rossant, J. 415
Rost, G. 140
Roth, T. L. 385, 412, 413
Rothenfluh-Hilfiker, A. 172
Rovine, M. J. 454
Rowe, D. C. 38, 45, 47–48, 259, 261, 267, 269
Rowlands, M. 69
Ruck, M. D. 482
Rudwick, M. J. S. 357
Ruggerio, A. M. 399, 401
Ruggiero, A. 411, 413
Rundqvist, H. 413

Rushton, J. P. 4, 16, 45, 47, 48, 257, 266–268, 308, 321, 327, 329, 339–343, 359, 460, 463
Russig, H. 414
Rust, R. T. 496
Rutherford, A. 433, 458
Rutter, M. 444, 454
Ryan, W. 499

Saatcioglu, F. 404
Sackett, G. P. 177
Saffran, J. 140
Sagan, C. 24
Sahlins, M. 72, 151
Salzen, E. 111–113
Sameroff, A. 37–38, 40, 44, 260, 280
Sampson, R. J. 494
Samuelson, L. K. 142–143
Sanjuan, M. A. F. 26, 513
Sapienza, C. 335, 415
Sapolsky, R. M. 232, 394, 396
Sarbin, T. R. 42
Sarich, V. 180
Sarikey, C. 498
Sasaki, A. 411, 413
Saunders, P. T. 24–25, 158–159, 181, 251, 514
Savage, J. E. 439
Savenkova, M. I. 415
Scarr, S. 260, 451, 462, 464
Scarr-Salapatek, S. 371
Schaie, K. W. 444
Schapiro, S.J. 72
Schellinger, K. B. 484
Scher, S. J. 222–223, 231, 234
Schiff, M., 453–454
Schimmack, U. 436
Schleicher, A. 495
Schleidt, W. 108
Schlomer, G. L. 341
Schlosberg, H. 21
Schmader, T. 483
Schneider, B. 481
Schneirla, T. C. 16, 21–23, 25–28, 40, 50, 77–78, 80, 83–84, 105, 107–111, 114, 131–132, 151, 154, 260, 269, 279, 330, 337, 344, 367, 512, 515, 517
Schnitker, S. A. 519
Scholl, B. J. 141
Schonbrodt, F. D. 306, 448
Schor, I. E. 420
Schore, A. 72, 227, 233

Schraiber, J. G. 461–462
Schultze-Florey, C. R. 401
Schupf, N. 411
Schweber, S. S. 353
Schwekendiek, D. 295
Scott, J. P. 109–112, 129, 131–132
Scoville R. P. 182
Scriver, C. R. 270
Scticn, F. 403
Seamans, R. 496
Sear, R. 222, 460
Seay, B. 80
Sebastian, S. 414
Sebring, P. B. 481
Seckl, J. R. 413
Sedikides, C. 463
Seeman, T. E. 391, 395, 400
Segal, N. 299, 303–304, **304**, 306, 309–310, 445–447, 449, 454
Segman, R. H. 401
Seidenberg, M. 141
Seitsinger, A. M. 480
Sela, M. 242
Selzam, S. 260, 269
Sesardic, N. 463
Setien, F. 412
Sewertzoff, A. N. 174
Shackelford, T. K. 154
Shakespeare, W. 6
Shallcross, W. 139
Sham, P. C. 454
Shanahan, M. J. 44–46, 183, 386, 444
Shane, D. 452
Shapiro, D. Y. 284
Shapiro, J. A. 192
Shapiro, L. 22, 69, 225
Sharma, S. 413
Shatz, C. J. 287
Shea, D.T. 401
Shefer, G. 123, 124, *124*
Shefi, N. 401
Sheldon, B. C. 224
Shenk, D. 258, 260, 264–266, 272, 438
Sheridan, J. F. 77
Sherrington, C. 516
Sherwood, C. C. 73
Shianna, K.V. 399
Shields, J. 267, 301–302, 306, 441, 443, 446
Shindeldecker, R. D. 286

Shonkoff, J. P. 227, 232
Shusterman, A. 139
Shweder, R. A. 44, 46, 48
Shyu, I. 413
Sidana, A. 505
Silva, F.J. 60
Silver, J. 131
Silverberg, J. 321, 359
Simmons, J. P. 307
Simons, R. L. 437–438, 451, 454
Simpson, G. G. 352
Simpson, S. J. 218
Sims, A. 499
Sinclair, K. D. 414
Singh, A. K. 284
Singh, R. 414
Singh, R. S. 181
Sinibaldi, E. 521
Sjostrom, M. 415
Skiba, R. J. 485
Skinner, B. F. 4, 38, 77, 80, 162–163, 426, 515, 520
Skinner, M. K. 415, 463
Skoner, D. P. 391
Skor, M. N. 403
Slagboom, P. E. 413, 414
Slavich, G. M. 260, 268, 384–387, 403, 404, 520
Slemmer, J. 141
Slife, B.D. 60
Sloan, E. K. 393, 395, 399, 401, 403, 404
Sloane, E. H. 28
Small, D. L. 231
Smedslund, J. 60
Smith, C. L. 342
Smith, L. B. 44–45, 85, 140–141, 142, 231, 259–260, 336–337
Smith, T. B. 400
Smolin, L. 73
Soja, N. N. 142
Sokol, D. K. 294
Solé, R. 77–78, 84, 514
Sood, A. K. 401
Sorenson, E. R. 232, 234
Southard, C. 403
Spadola, A. 411
Spelke, E. 67, 135–140, 139, 142, 170
Spemann, H. 128
Spence, M. J. 171
Spencer, H. 234
Spencer, J. P. 114, 139

Spencer, M. B. 44, 46–48, 497–498, 500–501, 505
Spencer, S. J. 483
Spengler, D. 413
Spetch, M. L. 139
Spier, E. 499
Spitz, R. 520
Sprague, J. 485
Squire, J. M. 286
Sroges, R. W. 179
Stagl, J. 401, 404
Stanley, C. A. 500
Stattin, H. 44–45, 47, 260
St Cyr, N. M. 401, 404
Steele, C. M. 483
Steele, D. M. 479
Stefan, A. M. 306, 448
Stein, A. D. 414
Steinberg, L. 341
Stemlieb, B. 401, 404
Sterelny, K. 197
Stern, W. 50
Sternberg, R. J. 313, 436
Stettner, L. J. 111–113
Stevens, M. 411
Stewart, D. L. 113
Stewart, J. 64, 69
Steyer, L. 477
Stoltenberg, S. F. 438
Stoolmiller, M. 454
Stotz, K. 222, 224, 227–228, 231, 233
Strawson, P. F. 60
Strohman, R. C. 29, 77, 79, 183, 258–259
Stryker, M. P. 287
Sturm, R. A. *291*
Suderman, M. J. 403, 411, 413
Sullivan, P. F. 434
Sun, B. 401
Sun, Y. H. 248
Sun, Y. V. 418
Sunday, S. R. 330, 367
Sundberg, C. J. 413
Sung, C. 401
Sung, C. Y. 400, 401
Suomi, S. J. 113, 399, 401
Susser, E. S. 414
Sussman, E. 141
Suter, C. M. *412*
Sutherland, H. G. 412, 414
Swanson, D. P. 501

Swartz, D. 154
Sweatt, J. D. 194, 413, 418
Sweller, J. 487
Szalai, J. 28
Sze, J. 401
Szyf, M. 403, 411, 413, 415

Tabery, J. 68, 263, 271, 343, 344, 437
Takahashi, R. 401, 403, 404
Takaki, R. T. 500, 506
Tallis, R. 230
Tandon, P. N. 77
Tanghe, K. B. 224
Tanner, J. M. 340, 342
Tarr, A. J. 401
Tarsha, M. S. 232
Taylor, H. F. 302–303, 312, 315, 438, 443, 444, 454, 464
Taylor, L. 489
Taylor, R. D. 484
Team, A. S. 415
Tegner, J. 413
Tellegen, A. 299, 307, 449
Templeton, S. M. 481
ten Cate, C. 131, 136
Tenenbaum, H. R. 482
Terman, L. M. 459–460
Thaker, P. H. 403
Tharp, R. G. 487
Thelen, E. 44–45, 85, 142, 231, 259–260, 336–337
Thiessen, E. D. 140
Thomas, A. 42, 151
Thomas, C. A. 181
Thomas, D. 454, 462
Thompson, E. 68, 230
Thompson, R. A. 36–37
Thompson, S. P. 140
Thompson, W. R. 27
Thornhill, R. 232
Thorpe, W. 111
Thurston, A. 414
Tilley, E. H. 341
Timney, B. 113
Tishkoff, S. A. 227
Tobach, E. 8, 22–23, 48, 77, 81, 110, 151, 154, 260, 330, 333, 337, 344, 367, 374, 376, 515
Tobi, E. W. 413, 414
Tobin, W. 502
Tobis, J. S. 413
Tolou-Shams, M. 494

Tomaka, J. 520
Tomblin, J. B. 140
Tompkins, L. 214
Tooby, J. 222, 224, 228–229
Topoff, H. 183
Torgersen, S. 451
Toth, S. 71
Toufexis, D. 413
Toumbourou, J. W. 485
Tretiakova, M. 403
Trevathan, W. R. 72, 231
Trivers, R. 354–355, 357
Trooster, W. J. 137
Tucker, M. S. 495
Tucker, W. H. 301, 440–442, 457–458
Tung, J. 399, 401, 403, 413
Turecki, G. 411, 413
Turkewitz, G. 171
Turkheimer, E. 260, 262, 295, 434
Turner, J. S. 197
Twyman, A. D. 139
Tyack, D. 479, 502
Tyler, A. N. 114
Tyler, M. 141

Unterman, R. 480
Uphouse, L. L. 280, 285
Urabe, I. 84, 514
Urban, J. B. 85
Urioste, M. 412
Uttal, W. R. 76
Uzumcu, M. 415

Vallortigara, G. 138, 139
Valsiner, J. 44–46, 280
van den Berg, I. M. 411
van den Berghe, P. L. 324–325, 362–363
van der Kolk, B. 232
van der Weele, C. 183
van Doorninck, J. H. 411
van Geert, P. 160
van Gelder, T. 228–229
van Marthens, E. 280, 284–285
van Sluyters, R. C. 113
Varela, F. J. 68, 229
Velez, G. 500
Veller C. 434
Venter, J. 48, 84, 258–259
Vijayan, S. 141

Vilain, E. 411
Vinas, A. 214
Vishton, P. M. 141
Visscher, P. M. 399
Vizi, S. 414
Volden, P. A. 403
von Bertalanffy, L. 6–7, 50, 280, 332, 345, 352, 373
von Eye, A. 265–266, 321, 345, 386
von Hippel, W. 5, 17–18, 157, 515
von Noorden, G. K. 113
von Saint Paul, U. 138
von Stumm, S. 434, 439
von Uexküll, J. 81
Votruba-Drzal, E. 494, 500
Vrba, E. 333–337, 374–376, 518
Vygotsky, L. 502

Wachtel, P.L. 60
Waddington, C. H. 202, 224, 250–252, 410
Wade, N. 17, 292, 344
Wager-Smith, K. 172
Wahlsten, D. 131, 260, 268, 387, 419, 438, 464
Wakefield, J. C. 61
Wald, E. 76
Waldron, M. 295
Walker, H. A. 401
Waller, N. G. 307
Wallman, J. 171, 286
Walls, T. 8, 336–337
Walsten, D. 371–372
Walton, G. M. 499
Walton, K. D. 296
Wandersman, A. 498
Wang, D. 411, 413
Wang, J. 403
Wang, W. 141
Wanscher, J. H. 244
Warne, R. T. 460, 463
Washburn, S. L. 151, 154
Waterland, R. A. 412
Watkins, J. W. N. 60
Watson, J. B. 50
Weaver, I. C. 135, 413, 415
Wechsler, D. 436, 459–460
Wehby, G. L. 452
Weinberg, R. A. 462
Weinrich, J. D. 354
Weiss, D. J. 140
Weiss, E. 22

Weiss, I. C. 414
Weiss, P. 79, 280, 283
Weissberg, R. P. 484
Weist, M. D. 505
Wells, K. 498
Wendel, J. F. 196, 201
Werker, J. F. 140
Werner, H. 384
Wesseldijk, L. W. 452
Wessells, N. K. 284
West, M. J. 125, 137, 144, 183, 226, 231, 295
West-Eberhard, M. J. 72, 136, 202, 215, 217, 227, 385
Wheeler, M. 69, 228, 230
White, S. H. 43
Whitehead, A. N. 7
Whitehead, H. 197
Whitelaw, E. 414
Whiten, A. 182, 197
Whitten, C. 455
Wiens, A. N. 458
Wiesel, T. N. 113
Wilkerson, I. 498, 501
Wilkinson, L. S. 418
Williams, C. J. 294
Williams, C. M. 439
Williams, G. C. 333, 374
Williams, J. B. 403
Williams, R. L. 458
Williams, R. N. 60
Williamson, S. K. 498, 500
Willis, P. 499
Willis, P. E. 499
Wilson, A. C. 174
Wilson, D. S. 5, 227
Wilson, E. O. 5, 224, 260, 321–322, 321–323, 327–331, 344, 354, 356, 359–361, 364–368, 399, 405, 515
Wilson, M. E. 413
Winegard, B. 460
Winkler, I. 141
Winston, A. S. 344, 464
Winthrop, R. 494, 495
Witarama, T. 401, 404
Witherington, D. C. 6, 8, 16, 25, 50, 64, 68, 151, 159–160, 164, 226–228, 232–233, 260, 272, 384–385, 515–516
Witherspoon, D. 494
Witonsky, D. 403
Witters, D. 521
Wittgenstein, L. 73

Author Index

Wohleb, E. S. 401
Wohlhueter, A. L. 404
Wolpert, L. 287
Wonder, E. L. 403
Wong, R. 141
Woodbury, R. 232
Woodruff, D. 499
Woodworth, R. 21
Wotjak, C. T. 413
Wray, G. A. 224
Wright, L. 307
Wright, S. 280, 283
Wright, W. 445–446
Wu, W. 403
Wu, Y. 413
Wyles, J. S. 174, 180
Wynne, C. D. L. 225

Xu, X. 495

Yamaguchi, L. A. 487
Yang, C. 140
Yang, X. F. 517

Yoder, N. 505
Yomo, T. 84, 514
Yong, E. 434
Yoshida, H. 142
Young, A. I. 439
Young, M. W. 172
Yu, C. 140–141
Yu, S. 403
Yuan, E. 411

Zablotsky, B. 521
Zamenhof, S. 280, 284–285
Zanforlin, M. 138
Zerbini, L. F. 404
Zhang, B. 415
Zhang, C. 403
Zhang, T. Y. 393, 395, 413
Zheng, T. 403
Zheng, X. 481
Zigler, E. 50
Zou, M. 403
Zuk, O. 454

SUBJECT INDEX

Note: Locators in *italics* represent figures and **bold** indicate tables in the text.

ACES *see* adverse childhood experiences
adaptationist program 334–336, 375, 376
Adaptation Level Theory (AL) 22
adaptations 373–377
adverse childhood experiences (ACES) 233, 489
AFQT *see* Armed Forces Qualification Test
AI *see* artificial intelligence
AL *see Adaptation Level Theory*
American Psychological Association (APA) 18, 27, 42, 436
anagenesis and integrative levels 25–26
anatomy, physiology, and behavior development 281, *281*
Animal Breeding Plans (J. L. Lush) 291
anomalous individuals and developmental plasticity 122–123
APA *see* American Psychological Association
approach/withdrawal (A/W) hypothesis 22–23, 27, 108–109
APS *see* Association for Psychological Science
Aristotle 65, 66, 409
Armed Forces Qualification Test (AFQT) 437
artificial intelligence (AI) 495
Association for Psychological Science (APS) 308
autopoiesis 383, 515–516

behavioral epigenetics 387, 409; epigenetic effects of experience 412–413; epigenetic mechanisms 410–411; epigenetic regulation 411; epigenetics, defining 409–410; father's epigenetics 414–415; genes 410
behavioral neophenogenesis 177; modes of 178, **178**; pathway of evolutionary change 175, **175**
behavioral neophenotypes 176, 177, 180, 181, 337–338; evolutionary pathway initiated by **175**; induction of 174–175
behavioral plasticity, determinants of 175–178
behavior genetics 27–28, 257, 260; criticisms of 268–269; fatal flaws in 258–260; heritability, calculating 263–265; heritability analysis, shortcomings of 262–263; heritability analysis, statistical problems associated with 265–266; heritability and inherited 266–268; history, rewriting 269–270; replicate, failures to 261–262
Behavior Genetics (Jerry Hirsch) 27
Behavior Genetics Association (BGA) 434, 464
The Bell Curve 433; dispute, psychometric approach in 436–437; in favor of genetic ethnic and class differences in IQ 457–463; in favor of substantial within-group IQ heritability 443–457; gene discoveries in 434; heritability fallacy 437–438; Herrnstein's 1970s writings on IQ 441–442; IQ and intelligence 436; Kamin, Leon 442–443; main themes of 434–435; molecular genetic researchers 439–440; science and psychology, replication crisis in 435–436; substantial genetic influences 438–439; twins reared apart (TRA) studies 440–441; twin studies 440
BGA *see* Behavior Genetics Association
Big Bang 24, 83
biogenetic analysis in psychology 76; cultural set 81; development as probabilistic 82; experiential set 80–81; individual set 81; methodological issues 84–86; ontogenetic set 80; ontological structure of psychology 80–81; phylogenetic set 80; relational,

Subject Index

developmental systems view of psychology 82–83; role of development 82; role of epigenesis in development 83–84
biological evolution 24
biological flaw 294–295
BIT *see* blood-informative transcripts
blood-informative transcripts (BIT) 405
Blueprint (Robert Plomin) 17
"blueprint" 248
Bolder Speculation 182
"book of life" 248–249
Bouchard's response on MISTRA analysis 300–301; built-in genetic bias 308; data collection stop point 307; design of the study 301; failed attempt to invalidate earlier critics of TRA research 302–303; failure at the "important first step" 304–305; MISTRA DZA control group data, suppression of 303–304; MISTRA MZA pairs 301–302; P-hacking in MISTRA IQ study 306–307; replicating/overturning previous TRA studies 306; replication crisis and academic psychology 307–308; "small sample" and "space limitations" 305–306
brain:body ratios of birds and mammals vs. fish, amphibians, reptiles *179*
brain development 155; as malleable 477; and sociocultural functioning 154
brain size in relation to rate of anatomical evolution 180, **180**
broad heritability (H^2) 368, 371
Brown, Jerry 16
Brownian movement 86
Brown v. Board of Education 499
Burt, Cyril 440

calico cat 411, *411*
California Department of Corrections and Rehabilitation 16
cAMP response element-binding protein (CREB) 401
candidate gene association (CGA) studies 261
Cartesian-Split-Mechanistic research paradigm 62, 63, **63**, 64, 65, 69
Cartesian-split-mechanistic scientific paradigm 428
CASEL *see* Collaborative for Academic, Social, and Emotional Learning
caste-like minority groups 462–463
cellular epigenetic inheritance 194–196
CGA studies *see* candidate gene association studies
chicken embryos, normal and experimentally modified beaks of *219*

children actively constructing knowledge 478
classroom communities, creating 481
coaction for individual development 282–283
"code" 246–247
cohort effect concept 309, 444
Collaborative for Academic, Social, and Emotional Learning (CASEL) 483
Committee on Publication Ethics (COPE) 314
complex traits, "inheritability" of 296
computationalism 228; Narrow Evolutionary Psychology (NEP) and conceptual agenda of 228–229; rise of 4E cognition and its challenge to 229–230
The Concept of Development (Dale Harris) 26, 83
Concepts and Theories of Human Development (Richard M. Lerner) 26
conceptual analysis 60–61
conserved transcriptional response to adversity (CTRA) 401
contexts determining development 477
contextualism 26, 40–42, 43, 377
COPE *see* Committee on Publication Ethics
correlational IQ adoption studies 453
COVID-19 pandemic 494–495, 505
CREB *see* cAMP response element-binding protein
critical-periods hypothesis 109, 127–134; Schneirla's critique of 110; weak and strong versions of 111–113
CTRA *see* conserved transcriptional response to adversity
culturally competent classrooms, enabling 482
culture grounding learning 478
C-value paradox 173, *173*

Darwinian evolution and epigenesis 24–25
Darwinian theory 187, 191, 205, 353, 356
Darwinism: Modern Synthesis and Integrated Synthesis 242, *243*; transformations of 188–191; *see also* neo-Darwinism
Darwin's theory of evolution 215
dead-reckoning 138–139
DEPTH *see* developmental evolutionary psychology theory
desert locust (*Schislocerca gregaria*) 218, *218*
The developing genome. An introduction to behavioral epigenetics (David S. Moore) 268
development: contexts determining 477; epigenesis role in 83–84; from genes to development to evolution 191–193; language 139–142; as probabilistic 82; relationship to evolution 172–174; role of 82

developmental and evolutionary biology 171–172
developmental and evolutionary synthesis 170; behavioral neophenotypes, induction of 174–175; behavioral plasticity, determinants of 175–178; bolder speculation 182; change in morphology without change in genes 180–181; changes arising in one generation persisting across generations 182–183; evolutionary change, exploratory behavior and rate of 178–180; population-genetic model, integration of individual development into 181
developmental causality 280–282, 284, 285, 286, 287
developmental cognitive psychology 170–171, 172
developmental complexity, two approaches to 160–163
Developmental Contextualism 26
developmental endowments and ecological legacies 196–197
developmental evolution 214; evolutionary change, moving on from narrow views of 216–217; links between development and evolution 217–219; roles of development in evolution 215–216
developmental evolutionary psychology theory (DEPTH) 165, 221. 231, 233, 234, 235
developmentally-nurturant relationships 515, 519–520
developmental plasticity 227; anomalous individuals and 122–123; integrating with epigenetics 233–234
developmental regulation in human development 516
developmental resources, inheriting 295–296
developmental science 3, 60, 425–429; embodiment 68–70; epigenesis 65–68; human nature and development, baselines for 71–72; metatheoretical divide in 61–65
developmental science, implications of RDS metatheory for 45; from deficit to diversity 46–47; reductionist models, vestiges of 47–49
dialectical philosophy 39
dialectic development, Riegel's model of 40
dialecticism 40–41
differentiation, unresolved problem of 286–287
dispute, psychometric approach in 436–437
dizygotic twins (DZAs) 303–309, 447, 448
DNA methylation 195, 403, 410, *411*, 412, 413, 414, 419
domain-general learning 140–141
downward causation 79, 250
dynamic, relational developmental systems 520–521
dynamic developmental systems 517; developmentally-nurturant relationships and reciprocating self 519–520; embodiment 517–518; holism 518; individual ⇔ context relations 518–519
dynamic heredity 418–420

dynamic systems 6
dynamic systems, complexity of 512–513; autopoiesis 515–516; developmental regulation in human development 516; mutual beneficence 516–517; relationism 516
DZAs *see* dizygotic twins

Eck, Johnny 122–123
EDN *see* evolved developmental niche
EDP *see* evolutionary developmental psychology
educational practice, science of learning and development to transform 477; brain development as malleable 477; children actively constructing knowledge 478; contexts determining development 477; culture grounding learning 478; human relationships catalyzing healthy development and learning 477–478; learning as social, emotional, and academic 478; variability in human development 478; *see also* schools, implications for educative and restorative approaches to discipline 484–485
EEA *see* equal environment assumption
EES *see* Extended Evolutionary Synthesis
EISs *see* epigenetic inheritance systems
embodied, embedded, extended and enactive (4E) cognition 229–230
embodiment 68–70, 517–518
emergence, concept of 24
empirical, theoretical, and metatheoretical scientific discourse 62, *62*
environmental flaw 294
EPBM *see* evolved probabilistic behavioral mechanism
EPCMs *see* evolved probabilistic cognitive mechanisms
epigenesis 65–68; Darwinian evolution and 24–25; role in development 83–84
epigenetic development 279, **280**
epigenetic inheritance 193; cellular epigenetic inheritance 194–196; developmental endowments and ecological legacies 196–197
epigenetic inheritance systems (EISs) 194, 201
epigenetics 383–386, 413–414; defining 409–410; effects of experience 412–413; father's epigenetics 414–415; mechanisms 410–411; regulation 411
epistemology 3
EP *see* evolutionary psychology
equal environment assumption (EEA) 294, 450–452
equipotentiality 282, 287
ethnic differences and "race" 457–458
ethnic-differences context, Flynn effect in 460–461

Subject Index

ethological approach to instinct 119–120
eugenics, genesis of 93–94
evolution 151; developmental complexity, two approaches to 160–163; development and evolution, links between 217–219; evolutionary developmental psychology 158–160; from genes to development to evolution 191–193; Gould's views of ontogeny and phylogeny 152–154; paleoanthropological perspectives 154–156; probabilistic epigenesis and human evolution 151–152; relationship of development to 172–174; role of the organism in its own evolution 336–337; roles of development in 215–216
evolutionary biology, evolution of 225; developmental plasticity 227; Extended Evolutionary Synthesis (EES) 227–228; integrative developmental evolutionary psychology theory 230–234; niche construction 226–227; proximate and ultimate causes of phenotypes 225–226
evolutionary change: behavioral neophenogenetic pathway of 175, *175*; exploratory behavior and rate of 178–180; moving on from narrow views of 216–217
evolutionary developmental biology 217
evolutionary developmental psychology (EDP) 151, 158–161
evolutionary psychology (EP) 157
Evolution in Four Dimensions 187; Darwinism, transformations of 188–191; epigenetic inheritance 193–197; from genes to development to evolution 191–193; information-transmission systems, evolution of 203–204; interactions between genetic, epigenetic, behavioral, and symbolic variations 200–203; socially mediated learning, transmission through 197–199; symbol-based information transmission 199–200
evolved developmental niche (EDN) 231
evolved probabilistic behavioral mechanism (EPBM) 163
evolved probabilistic cognitive mechanisms (EPCMs) 16, 160, 161, 162
exaptation, concept of 333–335
Extended Evolutionary Synthesis (EES) 227–228

FAP *see* fixed action pattern
father's epigenetics 414–415
fixed action pattern (FAP) 162, 163
Flynn effect 310–311, 455–456; in ethnic-differences context 460–461
formal operations 40

Galton, Francis 14, 92–93
Galton's brainchild 94–95
Galton's eugenics 91; genesis of eugenics 93–94; heritability 96–101; seeing double 95–96
Galton's legacy 101–102
gametic potential 323; sex differences in 324, 361–362; and social and sexual development 325–326, 362–364
GATA 401
"gene" 244–246
gene$_J$ 244–245
gene$_M$ 244–246, 249
gene expression: in human immune cells *393*; social regulation of 391–394
gene regulation 410
genes, environment and phenotype characters *245*
genesis of eugenics 93–94
genetic, epigenetic, behavioral, and symbolic variations 200–203
genetically identical mice 412, *412*
genetic determinism 321–322; and human development 327, 364–365
genetic determinism as sociobiology's key to interdisciplinary integration 360; gametic potential and social and sexual development 362–364; genetic determinism and human development 364–365; sex differences in gametic potential 361–362; sociobiology and human aggression 360–361
The genetic lottery: Why DNA matters for social equality (Kathryn Paige Harden) 17
genetic reductionism 17, 48; rejecting 512–513
genetics: vs. genomics 391; new genetics 396–397
Genome Wide Association Studies (GWAS) 264, 434
genomics: genetics vs. 391; of human thriving 403–404
genotype copies 324
genotype–environment correlation 265–266
Gestalt psychology 82
Girden, Edward 21
glucocorticoid receptor (GR) 394
GM-CSF *see* granulocyte-macrophage colony-stimulating factor
Gottleib, Gilbert 27
Gould's views of ontogeny and phylogeny 152–154
grammar 140–141
granulocyte-macrophage colony-stimulating factor (GM-CSF) 401
gravity as an inheritance 121–122
group vs. individual flaw 292–294
GR *see* glucocorticoid receptor
GWAS *see* Genome Wide Association Studies

Haeckel's theory of recapitulation 152
Handbook of Child Psychology (Damon) 37, 46
Harris, Dale 26
head-scratching in birds 120, *120*
heat-shock genes 284
heritability 96; calculating 263–265; concept of 368–373; exquisite specificity 99; and inherited 266–268; not about the relative importance of genes 97–99; unrelated to "inheritability" 99–101
heritability analyses: shortcomings of 262–263; sociobiology and 331–332; statistical problems associated with 265–266
heritability fallacy 290, 437–438; appropriation of "heritable" 290–291; biological flaw 294–295; complex traits, "inheritability" of 296; developmental resources, inheriting 295–296; environmental flaw 294; group vs. individual flaw 292–294; twin studies 291–292
heritable epigenetic modification 195
Herman, David 22
Herrnstein, Richard 438–439, 449–450, 461–462; 1970s IQ writings 440, 441–442
Hertwig, Oscar 66
hierarchical systems view 283–285
Hirsch, Jerry 27
histone acetylation 403, 410, 413, 418
holism 7, 50, 64, 518
Holmes, Oliver Wendell 15–16
Homo erectus 182
homology, concept of 328–330, 366–368
Homo sapiens 182
HPA axis *see* hypothalamic-pituitary-adrenal axis
human aggression, sociobiology and 322–323, 360–361
human development 82; developmental regulation in 516; genetic determinism and 327; genetic determinism and 364–365; mechanistic ideas in the study of 4–5; sociobiological view of 338–339; variability in 478
human evolution, probabilistic epigenesis and 151–152
human eye colors, natural variability in 291, *291*
human gene expression, social regulation of 391, *400*; genetics vs. genomics 391; new genetics 396–397; remodeling the body 395–396; social regulation of gene expression 391–394; social signal transduction 394–395
human immune cells, gene expression in *393*
human nature and development, baselines for 71–72
human relationships catalyzing healthy development and learning 477–478

humans and nonhumans, comparisons of 366–368
human social genomics 399–401; conserved transcriptional response to adversity 401–403; genomics of human thriving 403–404; network genomics and social recursion 404–405; prescriptive genomics 405–406
human sociobiology 354; adaptation in humans need 355; failure of the research program for 355–357; limited evidence and political clout 354–355
human thriving, genomics of 403–404
hypothalamic-pituitary-adrenal (HPA) axis 394

identical twins 294
immunoglobulin G_1 (IgG_1) antibodies 401
imprinting 136–137
individual ⇔ context relations 518–519
Individual development and evolution: The genesis of novel behavior (Gilbert Gottlieb) 271
individual development integration into population-genetic model 181
infant development, theory and research about 35–37
information-transmission systems, evolution of 203–204
"inheritability" of complex traits 296
inheritance, gravity as 121–122
instinct 119; anomalous individuals and developmental plasticity 122–123; ethological approach to 119–120; gravity as an inheritance 121–122; Lorenz, response to 120–121
instinct and critical periods, concepts of 105; critical-periods hypothesis 109–113
integrated student services 490
integrated student support (ISS) programs 490
integrated systems creation of support to address student needs 489–490
integrative developmental evolutionary psychology theory 230; developmental plasticity and epigenetics, integrating 233–234; niche provision, examining 231–232; species-typical baselines, establishing 232–233
IQ 99, 101; evidence in support of within-group genetic influences on 456–457; family environments irrelevant for 454–455; hereditarianism 440; Herrnstein's 1970s writings on 441–442; and intelligence 436
IQ, *The Bell Curve* in favor of genetic ethnic and class differences in 457; caste-like minority groups 462–463; ethnic differences and "race" 457–458; Flynn effect in ethnic-differences context 460–461;

Subject Index

Herrnstein and Murray 461–462; IQ, gender, and ethnicity 458–459; Terman, Lewis 459–460; transracial adoption studies 462; Wechsler, David 459–460
ISS programs *see* integrated student support programs

Jensen, Arthur 440

Kamin, Leon 440, 442–443
Kantor, J. R. 21
knowledge, children actively constructing 478

Lamb, Michael 35, 36
language development 139; phonology, grammar, and domain-general learning 140–141; word learning 141–142
language learning 140
large- versus small-brained species, exploratory behavior and rate of evolutionary change in 178–180
learning: culture grounding 478; as social, emotional, and academic 478
learning and development, science of 494; current research, limits of 500–501; racism and caste, rooting out the tentacles of 498–500; whole-child development, learning, and thriving 501–503
Lerner, Richard M. 26
Lewis, Michael 35–36
Lewontin, Richard 97
Lewontin's thought experiment 293, *293*
Lorenz, Konrad 16, 120–121, 129
lumbering robots 322, 324, 360, 361, 520

mainstream developmental science 14, 61
mastery-oriented assessment 488
McClintock, Barbara 247
meaning making 500, 505, 518
mechanism elucidation 263
mechanistic paradigm 3–4
memetics 199
mental defectives 15
metaorganisms 399
metatheoretical divide in developmental science 61–65
metatheory 4, 6–8; developmental science, role within 37–43; relational developmental systems (RDS) metatheory implications for developmental science 45–49
metatheory and primacy of conceptual analysis in developmental science 60; embodiment 68–70; epigenesis 65–68; human nature and development, baselines for 71–72; metatheoretical divide 61–65

methylation, life cycle of 414, *414*
methyl groups 385, 410, 412, 413
Mid-Range Metatheories or Metamodels 62
militant enthusiasm 16, 323, 360
mindsets, motivation, and learning 486–487
Minnesota Study of Twins Reared Apart (MISTRA) 272, 299, 445–448, 449; age and sex effects 311; authority, appeals to 313; Bouchard's response 300–308; cohort influences 309; converging evidence argument 311–313; critical analysis on 300–301; environmental (cohort) influences, conflicting statements on 309–310; Flynn effect 310–311; MISTRA DZA control group data, suppression of 303–304; MISTRA MZA pairs 301–302; near-full-sample MISTRA IQ correlations **304**; overlooked natural experiments 310; raw data, access to 308–309; reared-apart MZ twins 309–311; reported and nonreported correlations **303**; science and 313–315
MISTRA *see* Minnesota Study of Twins Reared Apart
molecular genetic researchers 439–440
Moltz, Howard 21
monozygotic twins (MZT) 294, *294*, 440, 448, 450
morphology, change in, without change in genes 180–181
MTSS *see* multi-tiered systems of support
multipotentiality 282
multi-tiered systems of support (MTSS) 489
Murray, Charles 438–439, 449–450, 452, 461–462
mutual beneficence 516–517, 521
MZT *see* monozygotic twins

Narrow Evolutionary Psychology (NEP) 222–231; and conceptual agenda of computationalism 228–229; and conceptual agenda of neo-Darwinism 224–225
narrow heritability (h^2) 371
National Longitudinal Survey of Youth (NLSY) 437
nativist-empiricist debate 135; dead-reckoning 138–139; domain-general learning 140–141; grammar 140–141; imprinting 136–137; language development 139–142; phonology 140–141; spatial cognition 138–139; spatial reorientation 139; word learning 141–142
natural selection as storytelling 351–352
nature-nurture coaction 35; infant development, theory and research about 35–37; metatheory's role within developmental science 37–43; relational developmental systems (RDS), emergence of 43–45; relational developmental systems (RDS) metatheory implications for developmental science 45–49

Nazi Germany 27
near-full-sample MISTRA IQ correlations **304**
neo-Darwinism 223, 242, 244; "blueprint" 248; "book of life" 248–249; "code" 246–247; developmental plasticity 227; evolution of evolutionary biology and its challenge to 225–228; Extended Evolutionary Synthesis (EES) 227–228; "gene" 244–246; language of 243–244, 249; Narrow Evolutionary Psychology (NEP) and conceptual agenda of 224–225; niche construction 226–227; "program" 247–248; proximate and ultimate causes of phenotypes 225–226; representation, alternative form of 250–251; "selfish" 246
NEP *see* Narrow Evolutionary Psychology
Nerve Growth Factor (*NGF*) gene 393, 403
network genomics analyses 405
network genomics and social recursion 404–405
NF-kB *see* nuclear factor kappa-light-chain-enhancer of activated B cells
NGF gene *see* Nerve Growth Factor gene
niche construction 197, 226–227
niche provision, examining 231–232
NLSY *see* National Longitudinal Survey of Youth
nonlinear causality 285–286
nuclear factor kappa-light-chain-enhancer of activated B cells (NF-kB) 401
number of genes and neurons in different lineages **173**

Occam's razor 4
On Human Nature (Wilson) 323, 361
ontogeny, Gould's views of 152–154
ontological structure of psychology 80; cultural set 81; experiential set 80–81; individual set 81; ontogenetic set 80; phylogenetic set 80
ontology 3
organism, from gene to 279; developmental causality 280–282; differentiation, unresolved problem of 286–287; hierarchical systems view 283–285; nonlinear causality 285–286; significance of coaction for individual development 282–283
organismic-type theories 38

paleoanthropological perspectives of individual 154–156
paradigms 3, 6–7
"parent-of-origin" effects 415
PGI *see* polygenic index
PGS *see* polygenic score studies
P-hacking in MISTRA IQ study 306–307, 308, 448

phonology 140–141
phylogeny, Gould's views of 152–154
PISA test *see* Programme for International Student Assessment test
plasticity 7; anomalous individuals and developmental plasticity 122–123; behavioral plasticity, determinants of 175–178; developmental plasticity 227–234; relative plasticity 384
Plomin, Robert 96
pluripotency 282
polygenic index (PGI) 439
polygenic score studies (PGS) 434, 463
polymorphism 261
population-genetic model, integration of individual development into 181
positive classroom and school environments 479; classroom communities, creating 481; creating structures that support strong, continuous relationships 479–481; culturally competent classrooms, enabling 482; relational trust and family engagement, building 481–482; stereotype threat, eliminating 482–483
prenatal nonlinear causality 286
prescriptive genomics 405–406
probabilistic epigenesis 84; framework 66; and human evolution 151–152
probabilistic functionalism 42
process-relational paradigm 6, 45, 50, 63, **63**, 64, 67, 70, 72
Programme for International Student Assessment (PISA) test 495
"program" 247–248
Pronko, N. H. 22
psychology 76; cultural set 81; development, role of 82; as a developmental science 26–27; development as probabilistic 82; epigenesis' role in development 83–84; experiential set 80–81; individual set 81; methodological issues 84–86; ontogenetic set 80; ontological structure of 80–81; phylogenetic set 80; relational, developmental systems view of 82–83
psychology, principles of 23; anagenesis and integrative levels 25–26; approach/withdrawal (A/W) 27; Darwinian evolution and epigenesis 24–25; as a developmental science 26–27; emergence 24
psychometric approach in dispute 436–437

QRPs *see* questionable research practices
questionable research practices (QRPs) 307, 314, 435, 448

Subject Index

Raab, David 21
race differences, quality of Rushton's hereditarian views of 343
racism and caste, rooting out the tentacles of 498–500
random mutation 193
RB *see* reactional biography
RDS *see* relational developmental systems
reactional biography (RB) 80
reared-apart MZ twins 309; cohort influences 309; conflicting statements on environmental influences 309–310; Flynn effect 310–311; overlooked natural experiments 310
reared-apart relatives sharing common environmental influences 444–445
recapitulation, Haeckel's theory of 152
reciprocal altruism 355
reciprocating self 517, 519–520
recursive network genomics, social signal transduction and *402*
reductionist models, vestiges of 47–49
relational developmental systems (RDS) 6, 26, 64–65, 107, 151, 160, 257, 269, 337–338, 383, 425, 428; for developmental science 45–49; emergence of 43–45; metamodel 63; metatheoretical orientation 35; metatheory 6, 7, 8
relational process paradigm 7
relational trust and family engagement, building 481–482
relationism 516; embracing 512–513
relative plasticity 7, 64, 165, 270, 383, 384, 385, 425, 520
repetitive DNA 334
Riegel's model of dialectic development 40
RNA transcription 395–396, *396*, 420
Rohles, Fred 22
root metaphors 3
Rushton, J. Philippe 339; evaluations of evidence of 342–343; hereditarian views of race differences 343; ideas about different reproductive strategies across race groups 340–342; tripartite theory of race, evolution, and behavior 339–340

Saint-Hilaire, Etienne Geoffroy 215
SATSA *see* Swedish Adoption/Twin Study on Aging
SBHG *see* small-band hunter-gatherer communities
Schislocerca gregaria 218, *218*
Schneirla's critique of critical periods hypothesis 110
scholarly brinksmanship 458
School Development Program (SDP) 490

schools, implications for 478; building relational trust and family engagement 481–482; classroom communities, creating 481; creating structures that support strong, continuous relationships 479–481; culturally competent classrooms, enabling 482; educative and restorative approaches to discipline 484–485; integrated student services 490; mastery-oriented assessment 488; mindsets, motivation, and learning 486–487; social, emotional, and cognitive habits, skills, and mindsets 483–485; stereotype threat, eliminating 482–483; student motivation and learning, providing supports for 485–488; teaching for understanding 487–488; *see also* educational practice, science of learning and development to transform
science and society, heredity hoax in 13–19
science of learning and development (SoLD) *480*, 498
SDP *see* School Development Program
selfish genes 205, 243, 323, 327, 360, 361, 364
"selfish" 246
self-organization 50, 65, 67, 81, 85
self-regulation 65, 481, 484, 488
sequenced human genome 391, 394
serotonin transporter gene (5HTT) 396
SES scale *see* socioeconomic scale
sex differences in gametic potential 324, 361–362, 363
shape bias 142
SHR rat strain *see* spontaneously hypertensive rat strain
single-nucleotide polymorphisms (SNPs) 434
small-band hunter-gatherer communities (SBHG) 72
SNPs *see* single-nucleotide polymorphisms
SNS *see* sympathetic nervous system
social, emotional, and cognitive habits, skills, and mindsets 483–485
social and sexual development, gametic potential and 362–364
socially mediated learning, transmission through 197–199
social recursion, network genomics and 404–405
social regulation of human gene expression 391–394, *393*, *400*; genetics vs. genomics 391; new genetics 396–397; remodeling the body 395–396; social signal transduction 394–395
social signal transduction *392*, 394–395; and recursive network genomics *402*
sociobiological claims, evaluating 327, 365; adaptations 332–333, 373–377; behavioral neophenotypes 337–338; exaptation, concept of 333–335; heritability, concept of 368–373; human development, sociobiological view of 338–339; humans and nonhumans, comparisons of 328–330,

366–368; role of the organism in its own evolution 336–337; Rushton, J. Philippe 339; Rushton's evidence, evaluations of 342–343; Rushton's hereditarian views of race differences 343; Rushton's ideas about different reproductive strategies across race groups 340–342; Rushton's tripartite theory of race, evolution, and behavior 339–340; sociobiology and heritability analyses 331–332
sociobiological stories 352–354
sociobiology 321, 354; adaptation in humans need 355; failure of the research program for 355–357; gametic potential, sex differences in 324; gametic potential and social and sexual development 325–326; genetic determinism 321–322; genetic determinism and human development 327; and human aggression 322–323, 360–361; limited evidence and political clout 354–355; scientific goals of 321–327; sociobiology and human aggression 322–323
sociobiology and human development 359; gametic potential and social and sexual development 362–364; genetic determinism and human development 364–365; sex differences in gametic potential 361–362
sociobiology theory 365
socioeconomic (SES) scale 453
SoLD *see* science of learning and development
sources of action in human development 8–9
spatial cognition 138; dead-reckoning 138–139; spatial reorientation 139
species-typical baselines, establishing 232–233
spontaneously hypertensive (SHR) rat strain 286
S-R connections *see* stimulus-response connections
Stanford-Binet IQ test 459
statistical learning 140–141
stereotype threat 482–483
stimulus-response (S-R) connections 38, 39
student motivation and learning, providing supports for 485; mastery-oriented assessment 488; mindsets, motivation, and learning 486–487; teaching for understanding 487–488
student services, integrated 490
substantial genetic influences 438–439
Swedish Adoption/Twin Study on Aging (SATSA) 302, 449
symbol-based information transmission 199–200
sympathetic nervous system (SNS) 394, 401

T. C. Schneirla Conference Series 23
T. C. Schneirla Research Fund 23

teaching for understanding 487–488
Terman, Lewis 459–460
Tobach, Ethel 22–23
totipotency 282
transgenerational stability 182
transracial adoption studies 453, 462
TRA studies *see* twins reared apart studies
Twain, Mark 263, 369
twins reared apart (TRA) studies 300, 440–441, 456
Type I interferon genes 401, 403

U.S. National Longitudinal Survey of Youth 437

variability in human development 478
visual cortex RNAs, sequence complexities of **285**
von Baer, Karl Ernst 66

Waddington's epigenetic landscape network 250, *250*
WAIS *see* Wechsler Adult Intelligence Scale
Wechsler, David 459–460
Wechsler Adult Intelligence Scale (WAIS) 299, 445
Weinstock, Solomon 21–22
WEIRD countries *see* Western, European, Industrialized, Rich, and Democratic countries
Weismann, August 66
Western, European, Industrialized, Rich, and Democratic (WEIRD) countries 233–234, 428
Whack-A-Mole 17, 344
whole-child development, learning, and thriving 501–503
within-group IQ heritability, *The Bell Curve* in favor of 443; adoption studies 453–454; classical twin method 450–452; equal environment assumption (EEA) 450–452; evidence in support of 456–457; family correlations 454–455; Flynn Effect 455–456; Herrnstein and Murray in support of 449–450; invalid findings 448–449; "Minnesota Study of Twins Reared Apart" (MISTRA) 445–448; Murray reverses course 452; reared-apart relatives sharing common environmental influences 444–445; twins reared apart (TRA) studies 456
Wolff, Caspar Friedrich 65
word learning 141–142

X chromosomes 100, 195, 411
XX chromosome pair 264, 369
XY chromosome pair 264, 369

zip codes 494, 499

Made in the USA
Monee, IL
03 May 2026